T0221084

# OLIVER
## (Includes some Cockshutt and Minneapolis-Moline Models)

## SHOP MANUAL O-202

**Oliver & Cockshutt Models** ■ 1550 ■ 1555
■ 1600 ■ 1650 ■ 1655
**Minneapolis-Moline Models** ■ G-550 ■ G-750

**Oliver & Cockshutt Models** ■ 1750 ■ 1800A ■ 1800B ■ 1800C
■ 1850 ■ 1900A ■ 1900B ■ 1900C ■ 1950 ■ 1950-T

**Oliver & Cockshutt Models** ■ 1755 ■ 1855 ■ 1955
**Minneapolis-Moline Models** ■ G-850 ■ G940

**Oliver Model** ■ 2255
**Minneapolis-Moline Models** ■ G955 ■ G1355

# I&T
## SHOP MANUALS

# Information and Instructions

This shop manual contains several sections each covering a specific group of wheel type tractors. The Tab Index on the preceding page can be used to locate the section pertaining to each group of tractors. Each section contains the necessary specifications and the brief but terse procedural data needed by a mechanic when repairing a tractor on which he has had no previous actual experience.

Within each section, the material is arranged in a systematic order beginning with an index which is followed immediately by a Table of Condensed Service Specifications. These specifications include dimensions, fits, clearances and timing instructions. Next in order of arrangement is the procedures paragraphs.

In the procedures paragraphs, the order of presentation starts with the front axle system and steering and proceeding toward the rear axle. The last paragraphs are devoted to the power take-off and power lift systems. Interspersed where needed are additional tabular specifications pertaining to wear limits, torquing, etc.

## HOW TO USE THE INDEX

Suppose you want to know the procedure for R&R (remove and reinstall) of the engine camshaft. Your first step is to look in the index under the main heading of ENGINE until you find the entry "Camshaft." Now read to the right where under the column covering the tractor you are repairing, you will find a number which indicates the beginning paragraph pertaining to the camshaft. To locate this wanted paragraph in the manual, turn the pages until the running index appearing on the top outside corner of each page contains the number you are seeking. In this paragraph you will find the information concerning the removal of the camshaft.

## More information available at Clymer.com
### Phone: 805-498-6703

**Haynes Publishing Group**
Sparkford Nr Yeovil
Somerset BA22 7JJ England

**Haynes North America, Inc**
859 Lawrence Drive
Newbury Park
California 91320 USA

**ISBN 10: 0-87288-372-8**
**ISBN-13: 978-0-87288-372-7**

© **Haynes North America, Inc. 1990**
With permission from J.H. Haynes & Co. Ltd.
**Clymer is a registered trademark of Haynes North America, Inc.**
**Printed in Malaysia**

Cover art by Sean Keenan

All rights reserved. No part of this book may be reproduced or transmitted in any form or by any means, electronic or mechanical, including photocopying, recording or by any information storage or retrieval system, without permission in writing from the copyright holder.

While every attempt is made to ensure that the information in this manual is correct, no liability can be accepted by the authors or publishers for loss, damage or injury caused by any errors in, or omissions from, the information given.

# OLIVER
## (Includes some Cockshutt and Minneapolis-Moline Models)

Oliver & Cockshutt Models ■ 1550 ■ 1555
■ 1600 ■ 1650 ■ 1655
Minneapolis-Moline Models ■ G-550 ■ G-750

Previously contained in I&T Shop Manual No. O-23

# SHOP MANUAL

# OLIVER & COCKSHUTT

**SERIES 1550, 1555, 1600,**
**1650, 1655**

## Also Covers MINNEAPOLIS-MOLINE

**SERIES G-550, G-750**

NOTE: Servicing the Minneapolis-Moline Series G-550 can be accomplished by using the information contained in this manual for Series 1555. To service Minneapolis-Moline Series G-750, use information for Series 1655.

### SERIAL NUMBER LOCATION

Tractor serial number plate is located on the rear side of the instrument panel support. Engine serial number is stamped on right front flange of engine directly below the generator or alternator.

### BUILT IN THESE VERSIONS

Rowcrop, Industrial, Wheatland, High Clearance and Utility. Rowcrop tractors are available in single front wheel, dual wheel tricycle and adjustable axle versions, while Industrial and Wheatland models are available with non-adjustable front axles only.

# INDEX (By Starting Paragraph)

# CONDENSED SERVICE DATA

| GENERAL | Model 1550 1555 G-550 | Model 1600 | Model 1650 1655 G-750 |
|---|---|---|---|
| Engine Make | Own | Own | Own |
| Cylinders, Number of | 6 | 6 | 6 |
| Cylinder Bore, Non-Diesel-Inches | 3⅝ | 3½(1), 3⅝ | 3¾ |
| Cylinder Bore, Diesel-Inches | 3⅝ | 3¾ | 3⅞ |
| Stroke-Inches | 3¾ | 4 | 4 |
| Displacement, Non-Diesel-Cu. In. | 232 | 231(1), 248 | 265 |
| Displacement, Diesel-Cu. In. | 232 | 265 | 283 |
| Compression Ratio, Gasoline | 8:1 | 8.5:1 | 8.5:1 |
| Compression Ratio, LP-Gas | 8.75:1 | 9:1 | 9:1 |
| Compression Ratio, Diesel | 16.5:1 | 16:1 | 16:1 |
| Pistons Removed From | Above | Above | Above |
| Main Bearings, Number of | 4 | 4 | 4(2) |
| Cylinder Sleeves Type | Wet | Wet | Wet |
| Forward Speeds, Number of | 6 | 6 | 6 |
| Battery Terminal Grounded | Neg. | Pos. | Neg. |

(1) Prior to Tractor Serial No. 137613.
(2) Series 1650 and 1655 diesel engine has 7 main bearings.

## TUNE-UP

| | | | |
|---|---|---|---|
| Firing Order | 1-5-3-6-2-4 | 1-5-3-6-2-4 | 1-5-3-6-2-4 |
| Valve Tappet Gap, Non-Diesel, Int. | 0.009-0.011 | 0.013-0.015 | 0.019-0.021 |
| Valve Tappet Gap, Diesel, Int. | 0.009-0.011 | 0.009-0.011 | See Par. 63 |
| Valve Tappet Gap, Non-Diesel, Exh. | 0.015-0.017 | 0.022-0.024 | 0.029-0.031 |
| Valve Tappet Gap, Diesel, Exh. | 0.019-0.021 | 0.019-0.021 | 0.029-0.031 |
| Valve Face Angle, Degrees | 44½ | 44½ | 44½(5) |
| Valve Seat Angle, Degrees | 45 | 45 | 45(5) |
| Ignition Distributor Make | Holley | D-R | Holley |
| Ignition Distributor Model | D-2563AA | 1112632 | D-2563AA |
| Generator & Regulator Make | D-R | D-R | D-R |
| Generator Model | See Par. 148 | 1100419 | See Par. 148 |
| Regulator Model | See Par. 148 | 1118997 | See Par. 148 |
| Starting Motor Make | D-R | D-R | D-R |
| Starting Motor Model, Non-Diesel | ——————See Par. 149—————— | | |
| Starting Motor Model, Diesel | ——————See Par. 149—————— | | |
| Ignition Distributor Contact Gap | 0.016, 0.025(4) | 0.016 | 0.025 |
| Ignition Distributor Timing, Gasoline | ——————See Par. 145—————— | | |
| Ignition Distributor Timing, LP-Gas | TDC | 5° BTDC | TDC |
| Injection Pump Make | ——————Roosa-Master—————— | | |
| Injection Pump Model | ——————See Par. 122—————— | | |
| Injection Pump Timing | 2° ATDC | 7° BTDC | 2° BTDC(3) |
| Injector Make | CAV, Bosch | CAV | CAV(3) |
| Carburetor Make, Gasoline | ——————Marvel-Schebler—————— | | |
| Carburetor Make, LP-Gas | Bosch-Ensign | Zenith or Bosch | Bosch-Ensign |
| Carburetor Model, Gasoline | ————See Par. 89———— | | TSX807 |
| Carburetor Model, LP-Gas | CBX | PC2J10 or CBX | CBX |
| Engine Low Idle RPM, Gasoline | 400, 800(4) | 300 | 450, 800(6) |
| Engine Low Idle, RPM, LP-Gas | 550, 800(4) | 525 | 525, 800(6) |
| Engine Low Idle RPM, Diesel | 675, 800(4) | 675 | See Par. 127 |
| Engine High Idle RPM, Non-Diesel | 2420 | 2100 | 2420 |
| Engine High Idle RPM, Diesel | 2420 | 2100 | 2420 |
| Engine Rated RPM | 2200 | 1900 | 2200 |
| Belt Pulley RPM, Rated engine— | | | |
| Single Speed PTO | 1018 | 1188 | .... |
| Dual Speed PTO | 1035 | .... | 1035 |
| PTO RPM, Rated Engine— | | | |
| Single Speed (540) | 541 | 581 | .... |
| Single Speed (1000) | 1015 | 1008 | .... |
| Dual Speed (540) | 550 | .... | 550 |
| Dual Speed (1000) | 994 | .... | 994 |

(3) Series 1650 after tractor serial number 187 586-000 and all 1655 diesel are equipped with Roosa-Master "pencil" injectors, and injection pump timing is 4° BTDC. (4) Model 1555. (5) Series 1650 S/N 187 586-000 and up, and Series 1655, valve face angle is 29½ degrees and valve seat angle is 30 degrees (6) Model 1655.

## CONDENSED SERVICE DATA (CONT.)

| SIZES-CAPACITIES-CLEARANCES | Model 1550 1555 G-550 | Model 1600 | Model 1650 1655 G-750 |
|---|---|---|---|
| Crankshaft Main Journal Diameter ................ | 2.249-2.250 | 2.624-2.625 | 2.624-2.625 |
| Crankpin Diameter, Non-Diesel .................. | 2.249-2.250 | 2.249-2.250 | 2.249-2.250 |
| Crankpin Diameter, Diesel .................... | 2.249-2.250 | ————2.4365-2.4375———— | |
| Rod Length, Center to Center .................. | 6.745-6.750 | ————6.749-6.750———— → | |
| Camshaft Journal Diameter, Front .............. | ————1.749-1.750———— | | |
| Camshaft Journal Diameter, Others ............. | ———— 1.7485-1.7495 ———— | See Par. 74 | |
| Piston Pin Diameter ......................... | ———— 1.2494-1.2497 ———— | | |
| Valve Stem Diameter, Inlet .................... | ————0.372-0.373———— | | |
| Valve Stem Diameter, Exhaust ................. | ————0.371-0.372———— | | |
| Compression Ring Width ...................... | ————See Par. 77———— | | |
| Oil Ring Width ............................. | ————See Par. 77———— → ‣ | | |
| Main Bearings Clearance, Non-Diesel ........... | 0.0002-0.0032 | ———→ 0.0005-0.0035———— | |
| Main Bearings Clearance, Diesel ............... | 0.0002-0.0032 | 0.0005-0.0035 | 0.0015-0.0045 |
| Rod Bearings Clearance, Non-Diesel ............ | ————0.0005-0.0015———— | | |
| Rod Bearings Clearance, Diesel ............... | ————0.0005-0.0015———— | | |
| Piston Skirt Clearance ....................... | ←————See Par. 76———— | | |
| Camshaft End Play .......................... | ————Spring Loaded———— | | |
| Crankshaft End Play ........................ | ————See Par. 80———— | | |
| Camshaft Bearings Clearance ................. | ————0.0015-0.003———— | | |
| Cooling System, Gallons ..................... | 5 | 5 | 5 |
| Engine Crankcase, Without Filter, Quarts ........ | 5, 6(7) | 8 | 8 |
| Crankcase Filter, Quarts ..................... | ¾ | 1½ | 1½ |
| Fuel Tank, Gasoline, Gallons .................. | 27.5 | 27.5 | 27.5 |
| Fuel Tank, LP-Gas (80% Fill), Gallons ........... | 34 | 34 | 34 |
| Fuel Tanks, Diesel, Gallons ................... | 27.5 | 27.5 | 27.5 |
| Transmission & Final Drive, Quarts ............. | 40 | 32 | 32 |
| Hydra-Power Drive, Quarts .................... | 6, 7(7) | 6 | 6, 7(8) |
| Creeper Drive, Quarts ....................... | 5 | 5 | 5 |
| Reverse-O-Torc, Quarts ...................... | 10 | 10 | 10 |
| Hydraulic Lift (With 3-Point Hitch), Quarts ...... | 20 | 20 | 20 |
| Hydraulic Lift (Without 3-Point Hitch), Quarts ... | 20 | 24 | 24 |
| Hydraulic Lift (Depth Stop Cylinder System), Quarts | 14 | .... | 14 |
| Belt Pulley, Pints ........................... | 6½ | 6½ | 6½ |
| **TIGHTENING TORQUES—FT.-LBS. (Oiled)** | | | |
| Cylinder Head, Non-Diesel ................... | 112-117 | 112-117 | 112-117 |
| Cylinder Head, Diesel ....................... | 112-117 | 112-117 | See Par. 62 |
| Cylinder Head Oil Screw, Non-Diesel ........... | 96-100 | 96-100 | 96-100 |
| Cylinder Head Oil Screw, Diesel ............... | 96-100 | 96-100 | See Par. 62 |
| Main Bearings, Non-Diesel ................... | 87-92 | 108-112 | 108-112 |
| Main Bearings, Diesel ....................... | 87-92 | 108-112 | 129-133 |
| Connecting Rods, Non-Diesel ................. | 46-50 | 46-50 | 46-50 |
| Connecting Rods, Diesel ..................... | 46-50 | 56-60 | See Par. 79 |
| Flywheel .................................. | 66-69 | 66-69 | 66-69 |
| Rocker Arm Brackets ........................ | 25-27 | 25-27 | 25-27 |
| Manifold Nuts ............................. | 25-27 | 25-27 | 25-27 |
| Injector Flange Nuts ........................ | 13-17 | 13-17 | 13-17 |
| Energy Cell Caps ........................... | 96-100 | 96-100 | 96-100 |

(7) Series 1555. (8) Series 1655.

# FRONT SYSTEM (AXLE TYPE)

## AXLE MAIN MEMBER
## AND PIVOT PIN

### Rowcrop Models

1. Rowcrop tractors may be equipped with an adjustable front axle as shown in Fig. O1. The main (center) member (16 or 16A) is carried on a pin which is retained in the front support (bolster) by a pin.

To remove the front axle, stay rod (26) and wheels as an assembly, sup-port front of tractor, then unbolt stay rod support (32) from front main frame. Disconnect inner ends of the tie rods from steering arm. Slide front axle from pivot pin on manual steer-ing models, or, on power steering equipped models, remove pivot pin and lift front of tractor and remove the axle assembly. Remove nut (27) from rear end of stay rod and with-draw stay rod support (32). Bushings (31) and (17 or 17A) should be reamed, if necessary, to provide 0.001-0.002 diametral clearance for the pins.

NOTE: High clearance models are similar except that stay rod has an integral bracket on rear end and two adjustable brace rods are attached be-tween stay rod bracket and axle ex-tensions.

### Utility—Short Wheel Base Models

2. On Series 1600, 1650 and 1655 Utility models, the pivot pin is an in-tegral part of the axle main member (16—Fig. O2). Bushing (24) in rear carrier (20) and bushing (not shown) in front carrier for pivot pin are pre-

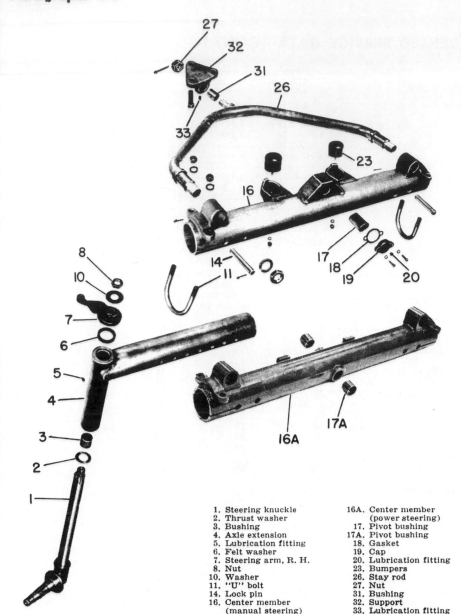

1. Steering knuckle
2. Thrust washer
3. Bushing
4. Axle extension
5. Lubrication fitting
6. Felt washer
7. Steering arm, R. H.
8. Nut
10. Washer
11. "U" bolt
14. Lock pin
16. Center member
    (manual steering)
16A. Center member
    (power steering)
17. Pivot bushing
17A. Pivot bushing
18. Gasket
19. Cap
20. Lubrication fitting
23. Bumpers
26. Stay rod
27. Nut
31. Bushing
32. Support
33. Lubrication fitting

Fig. O1 — Exploded view of Rowcrop adjustable front axle assembly. Models equipped with manual steering use center (main) member (16); those equipped with power steering use center member (16A). Narrow tread power steering models use center member similar to (16A), but shorter than standard part.

To remove axle, stay rod (21) and wheels as an assembly, support front end of tractor, remove stay rod ball socket cap from support (28) and disconnect tie rods from center steering arm. Remove pin which retains axle pivot pin in front bolster, drive out pivot pin and remove front axle assembly. Bushing (20) should be reamed, if necessary, to provide 0.001-0.002 clearance for pin.

### TIE RODS AND TOE-IN

#### All Axle Type

4. Tie rod ends are of the non-adjustable automotive type and are serviced by either renewing the end assembly or the complete tie rod assembly.

Toe-in for all axle type models is $\frac{9}{16}$-inch. Obtain correct toe-in by varying length of each tie rod an equal amount.

### STEERING KNUCKLES AND BUSHINGS

#### Rowcrop and Utility Models

5. To remove the steering knuckles (1—Fig. O1) and renew bushings (3), support front end of tractor and remove tire and wheel assembly. Straighten tab of washer (10) and remove nut (8). Remove steering arm (7) from knuckle and withdraw knuckle from axle extension.

sized and should not require reaming if carefully installed.

On Series 1550, 1555, 1650 and 1655 short wheel base models, pivot pin (22) is separate from the axle main member and is retained in rear pivot pin support (21) by a cap screw. No bushing is provided for the pivot pin. Bushing (not shown) for front pivot pin is in front carrier (axle support). Bushing is pre-sized and should not require reaming if carefully installed.

To remove front axle as an assembly, support front end of tractor, disconnect tie rods from center steering arm and remove thrust washer (17) from front end of pivot pin. Support axle, remove pivot pin rear carrier

(20 or 21) and slide axle back out of front carrier. Remove bushings from front and rear carriers and install new bushings with suitable driver. Be sure that grease hole in each bushing is aligned with lubrication fitting.

### Wheatland and Industrial Models

3. The Wheatland and Industrial model tractors are fitted with a non-adjustable, live spindle axle as shown in Fig. O3. Note: Industrial axle is shown, but Wheatland axle differs only in that the axle is arched.

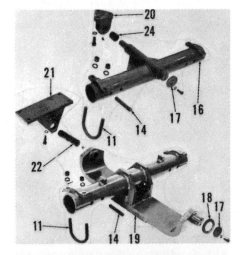

Fig. O2 — Axle center member used on short wheel base and utility model tractors.

11. "U" bolt
14. Lock pin
16. Center member
    (1600, 1650, 1655
    utility)
17. Thrust washer
18. Washer(shim)
19. Center member
    (1550, 1555, 1650,
    1655 short wheel
    base)
20. Support (rear)
21. Support (rear)
22. Pivot pin
24. Bushing

Drive the bushings (3) from axle extension and install new bushings flush with bore. Bushings are pre-sized and reaming should not be necessary if bushings are installed with suitable driver.

A thrust washer (2—Fig. O1) is used between knuckle and axle extension on Rowcrop models; Utility models are fitted with a thrust bearing. Install thrust bearing on steering knuckle with washer side down. Install thrust washer on Rowcrop knuckle with either side up. Insert knuckle into axle extension, install new felt (6) and install steering arm (7) extending to rear and at approximately 90° to wheel spindle

## LIVE SPINDLES AND SPINDLE CARRIERS

### Wheatland and Industrial Models

6. **SPINDLES.** Front wheels on Wheatland and Industrial models are attached to live (rotating) spindles (1—Fig. O3) which are carried in tapered roller bearings mounted in the spindle carriers (5). Spindles can be removed by removing the cap (14) from inner side of spindle carrier and removing the cotter pin (13) and slotted nut (12) from inner end of spindle. Refer to paragraph 8 for bearing and seal service information.

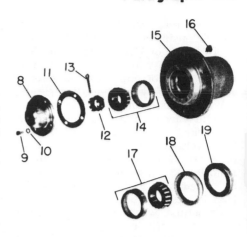

**Fig. O4 — Exploded view of front wheel hub, bearing and seal assembly used on Rowcrop and Utility models.**

| | |
|---|---|
| 8. Hub cap | 15. Hub |
| 9. Capscrews | 16. Plug |
| 11. Gasket | 17. Inner bearing |
| 12. Nut | 18. Seal |
| 13. Cotter pin | 19. Wear cup |
| 14. Outer bearing | |

7. **SPINDLE CARRIERS.** To renew the spindle carrier pivot pin (15—Fig. O3) and/or pivot pin needle bearing (7), first remove wheel and disconnect tie rod outer end from carrier. Remove nut (17) from bottom end of pivot pin and drive the pin up out of axle (18) and carrier. Drive bearings (7) from carrier and install new bearings by driving or pressing on lettered side of cage only. To renew carrier, remove spindle as outlined in paragraph 6. When reassembling, renew seal (8) and, if damaged, renew thrust bearing (10) and washer (9). Refer to paragraph 8 for wheel bearing and seal information.

## WHEEL BEARINGS AND SEAL

### All Models

8. Refer to Fig. O4 for exploded view of Rowcrop and Utility model front wheel bearing and seal installation and to Fig. O3 (items 1 through 14) for Wheatland and Industrial models.

To install new seal (3—Fig. O3, or 18—Fig. O4), first apply thin coat of gasket sealer to outer metal rim of seal. On Rowcrop or Utility model, install seal in wheel hub (15—Fig. O4) with Oliver ST-97 or ST-145 Seal Driver and ST-125 Mandrel or equivalent tool. On Wheatland or Industrial model, install seal in spindle carrier (5—Fig. O3) with Oliver ST-123 or ST-146 Seal Driver and ST-125 Mandrel or equivalent tool. Apply thin coat of gasket sealer to inside of seal wearing cup (19—Fig. O4, or 2—

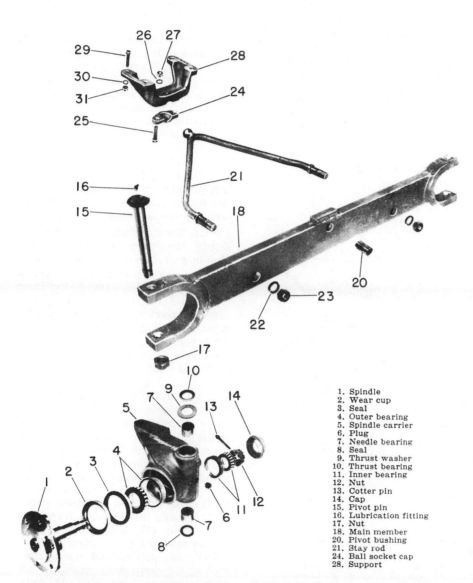

1. Spindle
2. Wear cup
3. Seal
4. Outer bearing
5. Spindle carrier
6. Plug
7. Needle bearing
8. Seal
9. Thrust washer
10. Thrust bearing
11. Inner bearing
12. Nut
13. Cotter pin
14. Cap
15. Pivot pin
16. Lubrication fitting
17. Nut
18. Main member
20. Pivot bushing
21. Stay rod
24. Ball socket cap
28. Support

**Fig. O3 — Exploded view of Industrial non-adjustable, live (rotating) spindle front axle. Wheatland axle assembly is similar except main member (18) is arched.**

Fig. O3) and install wearing cup on spindle using Oliver ST-98 Driver on Rowcrop or Utility model, or ST-124 Driver on Wheatland or Industrial model. Face of cup must be smooth and flat after installation.

When reassembling, wheel hub or spindle carrier and bearings should be packed with ½-pound of lithium base multi-purpose grease. Threads of plug (16—Fig. O4) in wheel hub or (6—Fig. O3) in spindle carrier should be coated with gasket sealer or white lead if plug has been removed.

To adjust wheel bearings, tighten nut until definite drag is felt when rotating wheel, then back nut off to remove preload. Continue to tighten and loosen nut until a point is located where torque required to tighten nut definitely increases, then tighten nut from this point to where cotter pin can be installed. Bearing adjustment is correct when torque of 20 inch-pounds is required to rotate wheel. Renew gasket (11—Fig. O4) when installing cap on Rowcrop or Utility model hub.

## MANUAL STEERING GEAR

### Models So Equipped

**9. R&R STEERING GEAR UNIT.** Disconnect tie rods from steering gear arm. Loosen steering shaft support bearing at clutch housing and disconnect universal joint from steering gear wormshaft. Support front end of tractor and unbolt stay rod support on Rowcrop models, remove stay rod ball socket cap on Wheatland or Industrial models, or on Utility models, unbolt rear pivot pin bracket. Remove capscrews that retain front axle bolster (support) or carrier to tractor main frame, raise front of tractor and roll complete front axle assembly away from tractor. Remove steering arm from steering gear sector shaft and remove steering gear unit from front bolster or carrier.

NOTE: When gear unit is removed from front bolster or carrier, do not turn wormshaft hard against stop in either direction as damage to ball guides could result.

**10. WORMSHAFT ADJUSTMENT.** Remove steering gear unit as outlined in paragraph 9. Loosen lash adjusting screw (6—Fig. O5) several turns to relieve any load that may be imposed by the meshing of sector and worm gears. Loosen lock nut (27) and turn adjuster (26) inward until 8-12 in.-lbs. torque is required to rotate the worm shaft. Tighten lock nut and recheck. Adjust sector mesh as outlined in paragraph 11.

**11. SECTOR MESH (BACKLASH) ADJUSTMENT.** This adjustment is controlled by lash adjuster screw (6—Fig. O5); however, before making any adjustment, make sure that the wormshaft adjustment is correct as specified in paragraph 10.

Disconnect tie-rods from sector shaft arm. Loosen lock nut (7). Locate mid-position of steering gear sector by rotating wormshaft from full right to full left, counting the total number of turns; then, rotate wormshaft back exactly half way to the center or mid-position. With steering gear sector in its mid-position of travel, rotate lash adjuster (6) in a clockwise direction until 16 to 20 inch-pounds of torque is required to turn wormshaft through mid-position of gear travel. Tighten lash adjuster locknut.

Note: Backlash adjusting screw (6) should have from zero to 0.002 end play in gear. If end play exceeds 0.002 it will prevent correct adjustment of backlash; in which case, sector cover (1) should be removed and a shim (5) of correct thickness be installed at head of adjuster screw to remove the excess backlash. Shims are available in a kit of four shims, 0.063, 0.065, 0.067 and 0.069 thick.

**12. OVERHAUL GEAR ASSEMBLY.** To disassemble steering gear unit, remove assembly as outlined in paragraph 9; then, proceed as follows: Remove pitman arm. Remove cap screws from sector cover (1—Fig. O5) and withdraw sector and cover as a unit from housing. Remove worm (screw) shaft cover (23) and withdraw wormshaft through this opening. Wormshaft bearing cup (17) and/or oil seal (11) can be renewed at this time.

Ball nut (20) should move along grooves in wormshaft smoothly and with minimum end play. If worm shows signs of wear, scoring or other derangement, it is advisable to renew

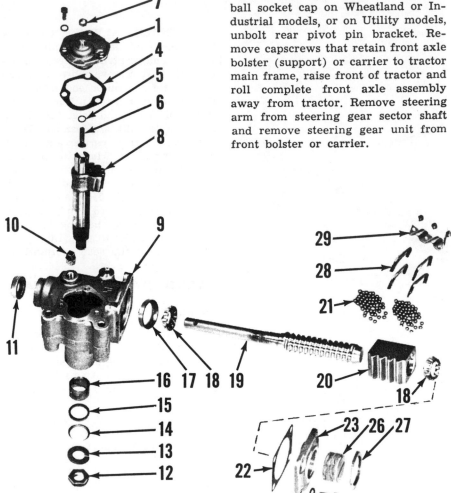

**Fig. O5 — Exploded view of Saginaw steering gear assembly used on manual steering models.**

| | | |
|---|---|---|
| 1. Cover | 9. Housing | 15. Packing | 22. Gasket |
| 4. Gasket | 10. Plug | 16. Bushing | 23. End cover |
| 5. Shim | 11. Seal | 17. Bearing cup | 26. Bearing adjuster |
| 6. Lash adjuster | 12. Nut | 18. Bearing cone | 27. Lock nut |
| 7. Jam nut | 13. Lock washer | 19. Steering shaft | 28. Ball guides |
| 8. Pitman shaft | 14. Packing retainer | 20. Ball nut | 29. Guide retainer |
|    (sector gear) | | 21. Steel balls (106) | |

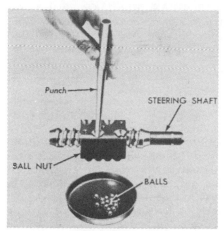

Fig. O6 — Using punch to align ball nut with wormgear and to fill ball circuit.

the worm and nut as a unit. To disassemble nut (20) from worm (screw) shaft (19), remove ball retainer clamp (29), ball return guides (28), balls and worm nut.

To reassemble ball nut, place nut over middle section of worm as shown in Fig. O6. Drop bearing balls into one retainer hole in nut and rotate wormshaft slowly to carry balls away from hole. Continue inserting balls in each circuit until circuit is full to bottom of both holes. If end of worm is reached while inserting balls and rotating worm in nut, hold the balls in position with a clean blunt rod as shown in Fig. O6 while shaft is rotated in an opposite direction for a few turns. Remove the rod and drop the remaining balls in the circuit. Make certain that no balls are outside regular ball circuits. If balls remain in groove between two circuits or at ends, they cannot circulate and will cause gear failure. Next, lay one-half of each split guide (28—Fig. O5) on the bench and place 13 balls in each. Place the other halves of each retainer over the balls. While holding the halves together, plug their ends with heavy grease to prevent the balls from dropping out; then, insert complete retainer units in worm nut and install guide clamp.

Sector shaft large bushing (16—Fig. O5) has an inside diameter of 1.375. Other sector shaft bushing has an inside diameter of 1.0625. I&T suggested clearance of shaft in bushings is 0.0015-0.003.

Note: Sector cover (1) and its bushing are not available separately.

Select and insert a shim (5) on sector mesh adjusting screw (6) to provide zero to 0.002 end play, before reinstalling sector and shaft in gear

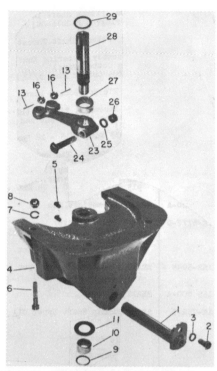

Fig. O7—Wheatland and Industrial front bolster used on models equipped with power steering. While shape of bolster differs for other models, operating parts remain basically the same for all axle models with power steering.

| | |
|---|---|
| 1. Pivot pin | 11. Washer |
| 4. Bolster | 23. Steering arm |
| 9. Seal | 27. Bearing |
| 10. Bearing | 28. Shaft |
| | 29. Seal |

housing. Adjust worm (screw) shaft bearings and sector mesh as outlined in paragraphs 10 and 11. Fill unit with No. 1 grade multi-purpose lithium or calcium base grease.

## FRONT BOLSTER (SUPPORT) OR CARRIER

### All Models

13. **R&R AND OVERHAUL.** On models equipped with manual steering, remove steering gear assembly as outlined in paragraph 9, then remove bolster or carrier from axle assembly. Front bushing for pivot pin is renewable in carrier on Utility models. Bushing is pre-sized and should not require reaming if carefully installed.

On models equipped with power steering, remove the power steering cylinder as outlined in paragraph 57 and the front axle assembly as outlined in paragraph 1, 2 or 3. Then, unbolt and remove bolster or carrier from tractor main frame. Remove clamp bolt from steering arm and

remove steering arm (23—Fig. O7), shaft (28) and thrust washer (11) from bolster or carrier. Oil seals (9 and 29) and needle bearings (10 and 27) can now be removed. When installing new needle bearings, drive or press on lettered end of bearing cage only. Oil hole in each bearing cage must align with lubrication fitting when bearing is installed. Press small bearing (10) in from top until flush with boss. Press large bearing (27) in from top until top of bearing is $\frac{3}{16}$-inch below flush with boss. Drive lower seal (9) in from bottom with lip out (towards bottom of bolster or carrier). Place thrust washer (11) against lower bearing boss and insert steering arm, with offset downward, between thrust washer and upper bearing boss. Insert shaft through bearings, arm and washer and install clamp bolt. Then, install upper seal (29) over shaft with lip outward (up) and drive seal into housing. Thoroughly lubricate shaft bearings with pressure type grease gun.

# FRONT SYSTEM (Tricycle Type)

## ADJUSTMENTS

### Models With Manual Steering

14. **AXLE POST OR WHEEL FORK BEARINGS.** On dual wheel tricycle models, adjust axle post (28—Fig. O8) to 0.006 or less end play by adding or removing shims (20) located between sector gear (8) hub and eccentric bushing (10). Shims are available in thicknesses of 0.007 and 0.0312. Bend washer (19) against flat of nut (9) when adjustment is correct. Access to shims is by removing front support as outlined in paragraph 20, or by removing radiator as outlined in paragraph 138, and then removing gear cover (13), nut (9) and sector gear.

On single front wheel models, adjust wheel fork (3—Fig. O9) to 0.006 or less end play by tightening or loosening nut (22). Stake nut to slot in fork after correct adjustment is obtained. Access to nut is by removing front support as outlined in paragraph 20, or by removing the radiator as outlined in paragraph 138, and then removing cover (6), from gear housing (26).

15. **WORMSHAFT.** Wormshaft bearings should be adjusted with the sector gear (8—Fig. O8, or 24—Fig. O9) removed.

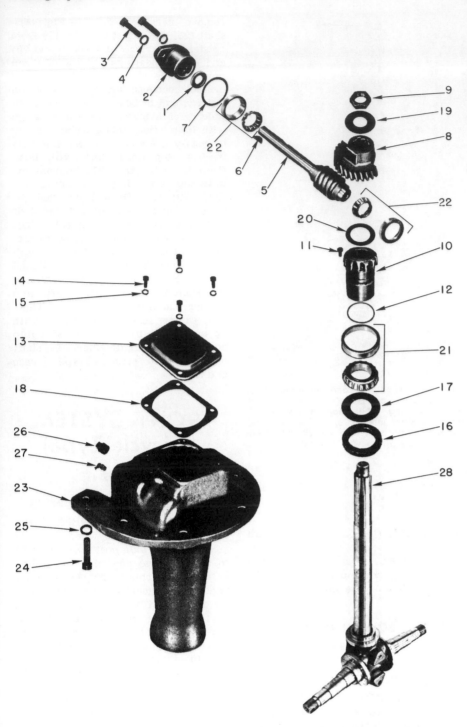

16. **GEAR MESH.** On dual wheel tricycle models, remove cover (13—Fig. O8) and turn wormshaft so that sector gear (8) is in mid-position. Remove lock screw (11) from eccentric bushing (10) and rotate bushing in bore until all backlash between sector gear and wormshaft gear is removed. Temporarily lock bushing in this position and turn wormshaft in each direction to each sector stop position; if no binding condition is noted, tighten lock screw securely. If binding occurs, rotate bushing slightly to just remove binding condition and lock bushing in position.

On single front wheel models, rotate bearing cage (31—Fig. O9) to a bolting position providing minimum backlash between sector gear (24) and wormshaft without causing any binding condition when wormshaft is rotated throughout range of travel.

## Models With Power Steering

17. **AXLE POST OR WHEEL FORK BEARINGS.** To adjust the dual front wheel axle post or single front wheel fork bearings, the power steering cylinder must first be removed as outlined in paragraph 57. If unit has been disassembled, be sure that bearing cups and cone and roller assemblies are fully seated, then adjust nut (5—Fig. O12 or 14—Fig. O13) so that axle post or wheel fork bearings have a preload of 70-80 in.-lbs. (including seal drag). After bearings are correctly adjusted, stake nut to slot in axle post or wheel fork shaft.

## All Dual Wheel Tricycle Models

18. **WHEEL BEARING ADJUSTMENT.** Refer to Fig. O10 for exploded view of wheel hub and related parts. Remove cap (8) and cotter pin (13), then tighten nut until a definite drag is felt when rotating wheel. Back nut off to remove preload, then continue to tighten and loosen nut until a definite point is located where torque required to tighten nut definitely increases, then tighten nut to where cotter pin can be installed. Bearing adjustment is correct when approximately 20 inch-pounds torque is required to rotate front wheel. Reinstall cap with new gasket (11).

## All Single Front Wheel Models

19. **WHEEL BEARING ADJUSTMENT.** To adjust the single front wheel bearings, the wheel assembly must be mounted in the wheel fork and the axle spindle retaining caps

**Fig. O8 — Exploded view of dual wheel tricycle front support and related parts as used on manual steering models. Refer to Fig. O12 for front support for power steering models.**

| | | | |
|---|---|---|---|
| 1. Oil seal | 9. Nut | 16. Felt | 22. Worm shaft bearings |
| 2. Bearing cover | 10. Eccentric bushing | 17. Felt retainer | 23. Steering column |
| 5. Steering worm | 11. Cap screw | 18. Gasket | (front support) |
| 6. Woodruff key | 12. "O" ring | 19. Washer | 26. Plug |
| 7. Shims | 13. Gear housing cover | 20. Shims | 28. Axle and post |
| 8. Steering sector | | 21. Tapered roller bearing | |

To adjust wormshaft bearings, add or remove shims (7—Fig. O8 or 21—Fig. O9) located between bearing retainer (2—Fig. O8 or 16—Fig. O9) and the column or gear housing to eliminate all wormshaft end play but permitting wormshaft to rotate without binding. When wormshaft bearings are properly adjusted, 30 inch-pounds torque should be required to rotate the wormshaft (including seal drag). Shims are available in thicknesses of 0.002, 0.004 and 0.0156.

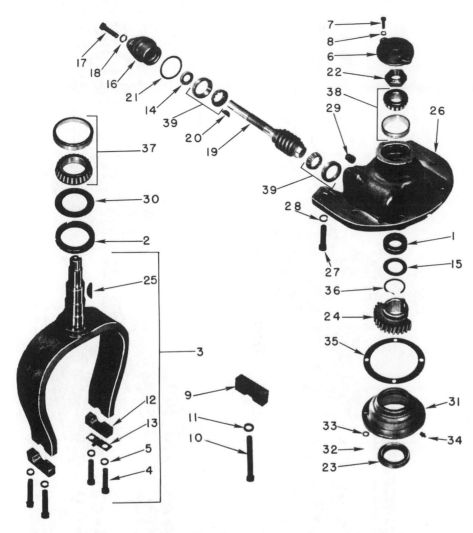

**Fig. O9 — Exploded view of single wheel front support (gear housing) and related parts used on manual steering models. Refer to Fig. O13 for power steering models.**

| | | | |
|---|---|---|---|
| 1. Felt washer | 14. Seal | 23. Oil seal | 31. Bearing cage |
| 2. Felt washer | 15. Felt retainer | 24. Sector gear | 35. Gasket |
| 3. Fork assembly | 16. Bearing cover | 25. Woodruff key | 36. Snap ring |
| 6. Bearing cap | 19. Worm shaft | 26. Gear housing | 37. Lower bearing |
| 9. Adapter (7.50-16 tire) | 20. Woodruff key | 29. Plug | 38. Upper bearing |
| 12. Bearing cap | 21. Shims | 30. Felt retainer | 39. Wormshaft bearings |
| | 22. Nut | | |

tightened, but with the nut lock (13—Fig. O9 and 8—Fig. O13) removed. Then, tighten the bearing nut (31—Fig. O11) until a definite drag is felt when rotating wheel. Back nut off to remove preload, then continue to tighten and loosen nut until a definite point is located where torque required to tighten nut definitely increases, then tighten nut to where finger on nut lock will enter a slot in nut when the nut lock is installed on spindle cap retaining screws.

## OVERHAUL FRONT ASSEMBLY

### All Models

**20. R&R FRONT ASSEMBLY AS A UNIT.** On power steering equipped models, first remove the power steering cylinder as outlined in paragraph 57. On manual steering models, remove the dirt guard bolted to rear of column, housing or support. Loosen steering shaft support bearing at clutch housing and disconnect universal joint from steering gear wormshaft. Then, on all models, support front of tractor and remove capscrews retaining column, gear housing or support to tractor main frame. Raise front end of tractor and remove the assembly.

### Dual Wheel Tricycle With Manual Steering

**21. OVERHAUL FRONT ASSEMBLY.** Remove assembly from tractor as outlined in paragraph 20 and re-

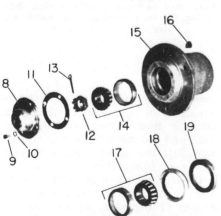

**Fig. O10 — Exploded view of front wheel hub, bearing and seal assembly used on dual wheel tricycle models.**

| | |
|---|---|
| 8. Hub cap | 15. Hub |
| 9. Cap screws | 16. Plug |
| 11. Gasket | 17. Inner bearing |
| 12. Nut | 18. Seal |
| 13. Cotter pin | 19. Wear cup |
| 14. Outer bearing | |

**Fig. O11 — Exploded view of wheel, bearing, seal and axle shaft assembly used on single wheel tricycle models. Wheel disc (24) is for optional 7.50-16 tire and rim, and requires use of adapters (9—Fig. O9).**

| | | | |
|---|---|---|---|
| 18. Seals | 20. Axle spindle | 23. Plug | 26. Rim clamps |
| 19. Wear cups | 21. Bearing assemblies | 24. Wheel disc | 31. Nut |
| | 22. Wheel | (7.50-16 rim) | 32. Flange |

move wheels from hubs. Refer to Fig. O8 for exploded view of unit. Remove plug (26) and drain lubricating oil from gear unit. Remove cover (13), straighten tab of washer (19) and remove nut (9). Bump or pull axle post (28) out of column (23). Remove lower bearing cone, felt retainer (17) and felt washer (16) from axle post. Withdraw sector gear (8), remove shims (20), eccentric bushing (10) and "O" ring (12) from cover (13) opening. Remove lower bearing cup from bottom of column. Remove wormshaft bearing retainer (2), shims (7) and wormshaft (5) with bearings (22) and seal (1) from rear of column.

Carefully inspect all parts and renew any that are excessively worn, scored or damaged. Reassemble unit using new seal (1), "O" ring (12), gasket (18) and felt (16) as follows:

Install seal (1) in retainer (2) with lip of seal inward (to front). Install wormshaft (5), bearings (22) and retainer (2) in column with correct thickness of shims (7) for bearing adjustment as outlined in paragraph 15.

Place new "O" ring (12) on eccentric bushing (10), lubricate "O" ring and install bushing in bore of column. Pack tapered roller bearing cone with No. 1 multi-purpose grease and install felt (16), retainer (17) and bearing cone on axle post being sure cone is firmly seated. Install bearing cup in bottom of column being sure cup is firmly seated and pack cavity between bearing cup and eccentric bushing with ½-lb. of No. 1 multi-purpose grease. Install axle post in column taking care not to dislodge eccentric bushing. Install sector gear (8), washer (19) and nut (9) with correct number of shims (20) for proper axle post adjustment as outlined in paragraph 14. Adjust gear mesh as outlined in paragraph 16. Fill gear unit to level of plug (26) with SAE 80 multi-purpose gear lubricant conforming to military specification MIL-L-2105. Install cover (13) with new gasket (18).

## Single Wheel Tricycle With Manual Steering

**22. OVERHAUL FRONT ASSEMBLY.** Remove assembly as outlined in paragraph 20. Refer to Fig. O9 for exploded view of unit. Remove plug (29) and drain lubricant from gear housing (26). Remove cap (6) and unstake and remove nut (22) from wheel fork shaft (3). Push fork shaft out of bearing (38) and gear housing.

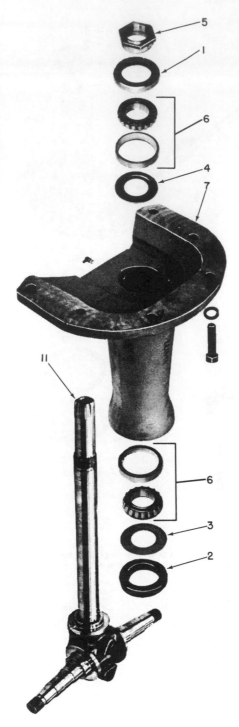

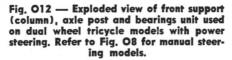

Fig. O12 — Exploded view of front support (column), axle post and bearings unit used on dual wheel tricycle models with power steering. Refer to Fig. O8 for manual steering models.

| | |
|---|---|
| 1. Seal | 5. Nut |
| 2. Felt | 6. Bearing assemblies |
| 3. Felt retainer | 7. Pedestal (column) |
| 4. Grease retainer | 11. Axle post |

Remove snap ring (36), sector gear (24), Woodruff key (25), and bearing cage (31) from fork shaft. Remove seal (23) and bearing cup from cage. Remove bearing cone, felt retainer (30) and felt (2) from fork shaft. Re-

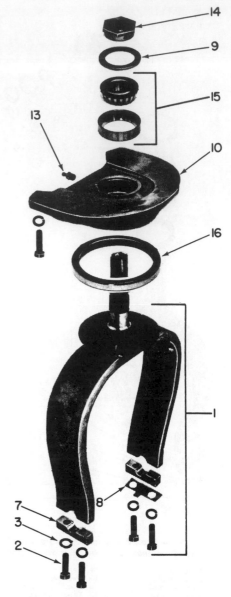

Fig. O13 — Exploded view of single front wheel front support (carrier) and related parts used on power steering models. Refer to Fig. O9 for manual steering models, and for 7.50-16 tire adapter.

| | |
|---|---|
| 1. Fork assembly | 14. Nut |
| 7. Cap | 15. Upper bearing |
| 9. Seal | 16. Lower bearing |
| 10. Carrier | |

move upper bearing (38), felt retainer (15) and felt (1) from gear housing. Remove retainer (16), shims (21) and wormshaft and bearings (19 and 39) from gear housing. Remove seal (14) from retainer.

Carefully inspect all parts and renew any that are excessively worn, scored or damaged. Reassemble unit using new seals (14 and 23), felts (1 and 2) and gasket (35) as follows:

Install seal (14) in retainer (16) with lip of seal inward (to front). Install wormshaft (19), bearings (39) and retainer (16) in gear housing with correct thickness of shims (21) for proper bearing adjustment as outlined in paragraph 15.

Drive cup of upper bearing (38) into gear housing (26) until firmly seated and install new felt (1) and felt retainer (15). Drive cup of lower bearing (37) into bearing cage (31) until firmly seated and install new felt (2) and felt retainer (30) on fork shaft. Pack cone and roller assemblies of bearings (37 and 38) with No. 1 multi-purpose grease and drive cone of lower bearing (37) onto fork shaft until firmly seated. Drive seal (23) into bearing cage (31) with lip up (away from bearing cup). Place bearing cage on fork shaft and gasket (35) on cage. Install Woodruff key (25), sector gear (24) and snap ring (36) on fork shaft. Place gear housing over fork shaft and install cone and roller of upper bearing (38) and nut (22) and adjust nut as outlined in paragraph 14. Adjust gear mesh as outlined in paragraph 16 and install bearing cage retaining capscrews. Install cover (6). Fill gear housing to level of plug (29) with SAE 80 multi-purpose gear lubricant conforming to military specification MIL-L-2105.

## Dual Wheel Tricycle With Power Steering

23. **OVERHAUL FRONT ASSEMBLY.** Remove assembly from tractor as outlined in paragraph 20. Refer to Fig. O12 for exploded view of unit. Unstake and remove nut (5), then bump or push axle post (11) from upper bearing (6) and column (7). Remove cone and roller assembly of lower bearing (6), felt retainer (3) and felt (2) from axle post. Remove seal (1), bearing assembly (6) and grease retainer (4) from upper bore of column and drive cup of lower bearing (6) from bottom of bore.

Carefully inspect all parts and renew any that are excessively worn, scored or damaged. Reassemble unit using new seal (1), felt (2) and grease retainer (4) as follows:

Install grease retainer (4) with ridge upward and drive upper bearing cup in firmly against retainer. Drive lower bearing cup in firmly against shoulder in bottom of bore. Pack cone and roller assemblies of both bearings with No. 1 multi-purpose grease and pack ½-pound of same type grease in cavity above lower bearing cup in column. Install new felt (2) and retainer (3) on axle post, then drive cone and roller of lower bearing firmly against shoulder of axle post.

Insert axle post in column and install upper bearing cone and roller assembly. Install seal (1) with lip up (away from bearing), lubricate seal contact surface of nut (5) and install nut for proper bearing adjustment as outlined in paragraph 17.

## Single Wheel Tricycle With Power Steering

24. **OVERHAUL FRONT ASSEMBLY.** Remove assembly from tractor as outlined in paragraph 20. Refer to Fig. O13 for exploded view of unit. Unstake and remove nut (14); then, press or bump fork shaft from carrier (10). Remove seal (9) and upper bearing assembly (15) from carrier and thrust bearing (16) from fork.

Carefully inspect all parts and renew any that are excessively worn, scored or damaged. Reassemble unit using new seal (9) as follows:

Drive cup of upper bearing into carrier until firmly seated. Pack upper bearing cone and roller assembly and lower thrust bearing assembly with No. 1 multi-purpose grease. Thrust bearing should be a hand push fit on shoulder of wheel fork; dress shoulder with file or stone to remove burrs if bearing fits tightly. Place carrier (10) over fork shaft and thrust bearing and install cone and roller assembly of upper bearing. Install seal (9) with lip upward (away from bearing), lubricate seal contact surface of nut (14) and install nut for proper bearing adjustment as outlined in paragraph 17.

## WHEEL BEARINGS AND SEAL UNITS

### All Dual Wheel Tricycle Models

25. **RENEW WHEEL BEARINGS AND/OR SEALS.** Refer to Fig. O10 for exploded view of hub, bearing and seal unit used on dual wheel tricycle models.

To install new seal (18), first apply a thin coat of gasket sealer to outer metal rim of seal, then install seal in hub with Oliver ST-97 or ST-145 Seal Driver and ST-125 Mandrel or equivalent tools.

To install new seal wear cup on axle post spindle, first apply a thin coat of gasket sealer to inner surface of wear cup, then install cup on spindle (not in hub) with Oliver ST-98 Wearing Cup Driver or equivalent tool. Face of cup must be smooth and flat after installation.

When reassembling, bearing cone and roller assemblies and each hub should be packed with ½-lb. of No. 1 multi-purpose grease. Threads of plug (16) should be sealed with white lead or gasket sealer. Adjust wheel bearings as outlined in paragraph 18.

### All Single Wheel Tricycle

26. **RENEW WHEEL BEARINGS AND/OR SEALS.** Refer to Fig. O11 for exploded view of wheel and bearings unit.

To install new seals (18), first apply a thin coat of gasket sealer to outside metal rim of seals, then install seal in wheel using Oliver ST-97 or ST-145 Driver and ST-125 Mandrel or equivalent tools.

Before installing new seal wearing cups (19) on axle spindle (20) or nut (31), apply a thin coat of gasket sealer to inner surface of the cups. Install wearing cup on nut (31) with Oliver ST-98 Wearing Cup Driver or equivalent tool. Install wearing cup on spindle by carefully driving cup in place with soft wood block. Face of cup must be smooth and flat after installation.

When reassembling, bearing cone and roller assemblies and hub of wheel should be packed with one pound of No. 1 multi-purpose grease. Adjust wheel bearings as outlined in paragraph 19 when reinstalling wheel assembly in wheel fork.

# FRONT SYSTEM
## (Front-Wheel Drive Axle)

Series 1600, 1650 and 1655 tractors are available with a front drive axle which is driven from the transmission bevel pinion shaft via a transfer case and a drive shaft fitted with two universal joints. A shifting mechanism in the transfer case allows connecting or disconnecting power to the front drive axle.

All four-wheel drive tractors are equipped with power steering. All models are equipped with a Saginaw Hydramotor steering unit. Refer to POWER STEERING section for information on the Saginaw Hydramotor and the two steering cylinders.

## FRONT AXLE AND CARRIER

### All Models So Equipped

**27. R&R AXLE ASSEMBLY.** The complete front axle assembly can be removed from tractor as follows: Remove drive shaft shield and shield front support. Disconnect drive shaft from companion flange of differential pinion shaft. Disconnect both power steering cylinders from axle and spindle supports and lay cylinders on top of axle carrier. Remove bolts retaining axle to axle carrier, then raise tractor and roll the complete axle and wheels unit forward and away from tractor.

Note: A rolling floor jack can be placed under differential pinion shaft to keep axle from rotating as tractor is lifted from axle.

If necessary, wheels and tie-rod can now be removed and procedure for doing so is obvious.

Reinstall axle by reversing the removal procedure and be sure piston rod ends of steering cylinders are attached to steering spindle supports. Tighten the cylinder attaching bolt lock nuts until they just contact the mounting flanges. Further tightening may distort mounting flanges and cause cylinder to bind.

**28. R&R AXLE CARRIER.** To remove axle carrier, first remove the front axle as outlined in paragraph 27, then, secure steering cylinders to tractor frame. Place a rolling floor jack under axle carrier and take weight of carrier. Remove pivot pin retaining capscrews, slide pivot pins from pivot supports and lower the axle carrier from tractor. If necessary, pivot supports can be removed from tractor frame.

Bushings in axle carrier can now be renewed. Bushings are pre-sized and should not require reaming if carefully installed.

**29. OVERHAUL FRONT AXLE.** Overhaul of the front drive axle assembly will be discussed as four operations; the planet spider assembly, the hub assembly, the spindle support assembly and the differential and carrier assembly. All operations except the differential and carrier overhaul can be accomplished without removing the front drive axle from tractor. Both outer ends of axle are identical, hence, only one outer end will be discussed.

**30. PLANET SPIDER.** To overhaul the planet spider assembly, support outer end of axle and remove the tire and rim. Remove relief valve from center of planet spider, remove plug from wheel hub and drain oil

**Fig. O14 — Remove pinion shaft pins by driving them inward.**

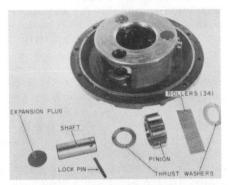

**Fig. O15 — View showing one pinion assembly removed from spider.**

**Fig. O16 — Sun gear can be removed from outer end of axle shaft after snap ring is removed. Note method of safety wiring the internal gear capscrews (early units).**

**Fig. O17 — Removing wheel hub assembly from spindle support.**

from planet spider. Remove capscrews that retain planet spider to wheel hub and the two puller hole capscrews. Use two of the removed retaining capscrews in the puller holes to remove planet spider assembly from wheel hub.

With unit removed, remove the three pinion shaft lock pins by driving them toward center of unit as shown in Fig. O14. Remove pinion shafts and expansion plugs by driving pinion shafts toward outside of planet spider. Remove planet pinions, rollers (34 in each pinion) and thrust washers. Discard the expansion plugs. Refer to Fig. O15.

Clean and inspect all parts and renew as necessary. Pay particular attention to the pinion rollers and thrust washers.

**31.** When reassembling, use heavy grease to hold rollers in inner bore of pinions. Be sure tangs of thrust washers are in the slots provided for them and that holes in pinion shaft and mounting boss are aligned before pinion shafts are final positioned. Coat mating surfaces of planet spider and wheel hub with No. 2 Permatex or equivalent sealer and install planet spider in wheel hub. Tighten retaining capscrews to a torque of 52-57 Ft.-Lbs.

**32. HUB ASSEMBLY.** To overhaul the wheel hub assembly, first remove planet spider assembly as outlined in paragraph 30. With planet spider removed, remove snap ring and sun gear from outer end of axle shaft. See Fig. O16. If necessary, the internal gear (separate on early units) can be removed at this time by clipping the lock wires and removing retaining capscrews. Straighten tabs of spindle nut lockwasher, then use OTC tool JD-4 or equivalent and remove spindle outer nut and lockwasher. Now loosen but **do not** remove the spindle inner nut. Unbolt spindle from spindle support and remove wheel hub assembly as shown in Fig. O17.

Place hub assembly on bench with spindle nut on top side and block up assembly so spindle will be free to drop several inches. Remove the spindle inner nut, then place a wood block over end of spindle and bump spindle from internal gear hub. Lift internal gear hub and bearing from wheel hub and be careful not to allow bearing to drop from hub of internal gear hub. Complete removal of spindle from wheel hub. All bearings and seals, thrust washers and dirt shield can now be removed and renewed if necessary. Bushing and oil seal in inner bore of spindle (items 33 and 35—Fig. O18) can also be renewed at this time.

33. Use Fig. O18 as a reference and reassemble wheel hub unit as follows: Install bearing cups (7 and 37) in hub with smallest diameters toward inside of hub. Place inner bearing in inner bearing cup, then install oil seal (34) with lip facing bearing. Bump seal into bore until it bottoms. Place dirt shield (8) on hub so flat side is toward flange of spindle, then using caution not to damage seal, install spindle in wheel hub. Hold spindle in that position and turn unit over so threaded end of spindle shaft is on top. Place outer bearing over end of spindle and push bearing down into cup. Start bearing hub of internal gear hub (1) into outer bearing cone and, if necessary, tap gear lightly with a soft faced hammer to position. Install thrust washer (38) and spindle inner nut (39) and tighten nut finger tight. Coat mating surfaces of spindle and spindle support with No. 2 Permatex or equivalent sealer and install dirt shield and spindle on spindle support. Tighten retaining capscrews to a torque of 80-88 Ft.-Lbs.

Adjust inner nut as required until a pull of 33-38 pounds on a spring scale attached to a wheel stud is required to keep hub in motion. See Fig. O19. Install lockwasher (40—Fig. O18) and outer nut (41). Tighten outer nut and recheck hub rolling torque. When adjustment is correct, bend tabs on lockwasher to secure both nuts. Install sun gear (46) and snap ring (44) on outer end of axle shaft. If internal gear (early models) (52) was removed, install gear so smallest outside diameter is in counterbore of internal gear hub. After retaining capscrews are tightened, secure in pairs with safety wire as shown in Fig. O16. Coat mating surfaces of wheel hub and planet spider with No. 2 Permatex or equivalent sealer and install planet spider. Tighten retaining capscrews to a torque of 52-57 Ft.-Lbs. Install the puller hole capscrews and the tire and rim.

34. SPINDLE SUPPORT. The spindle support can be serviced after planet spider and wheel hub assembly are removed as outlined in paragraphs 30 and 32. However, if service is required only on the spindle support, the planet spider, wheel hub and axle shaft can be removed as a unit as follows:

Raise outer end of axle and remove tire and rim. Attach hoist to wheel stud, then unbolt spindle from spin-

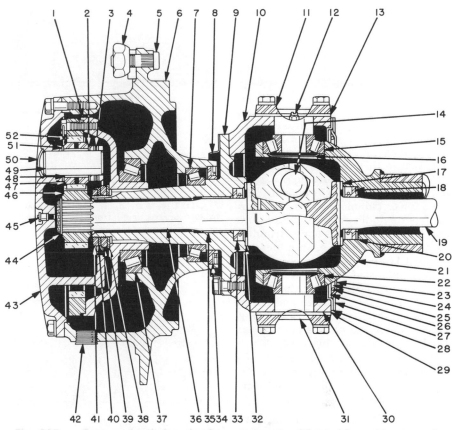

Fig. O18 — Cross-sectional view showing components of front drive axle outer end.

| | | | |
|---|---|---|---|
| 1. Internal gear hub | 15. Upper trunnion bearing | 26. Felt seal | 40. Lock washer |
| 2. Thrust washer | 16. Grease retainer | 27. Seal retainer | 41. Outer nut |
| 3. Pinion shaft lock pin | 17. Thrust washer | 28. Seal retainer | 42. Filler hole plug |
| 6. Wheel hub | 18. Oil seal | 29. Gasket | 43. Planet spider |
| 7. Hub inner bearing | 19. Axle shaft (inner) | 30. Shims | 44. Snap ring |
| 8. Dirt shield | 20. Washer | 31. Lower trunnion | 45. Relief valve |
| 9. Spindle | 21. Axle housing | 32. Thrust washer | 46. Sun gear |
| 10. Spindle support | 22. Lower trunnion | 33. Oil seal | 47. Pinion |
| 11. Upper trunnion | bearing | 34. Oil seal | 48. Pinion rollers |
| 12. Grease fitting | 23. Dust seal retainer | 35. Bushing | 50. Expansion plug |
| 13. Shims | 24. Dust seal | 36. Axle shaft (outer) | 51. Thrust washer |
| 14. Universal joint | 25. Spring | 37. Hub outer bearing | 52. Internal gear |
| | | 38. Thrust washer | |
| | | 39. Inner nut | |

Fig. O19 — Use method shown to check wheel hub bearing preload.

Fig. O20 — Planet spider, wheel hub and axle shaft assembly can be removed as shown.

dle support and pull complete hub assembly and axle shaft from outer end of axle. Refer to Fig. O20. **Do not allow weight of assembly to be supported by axle shaft or damage to oil seal in axle housing outer end will result.**

With the complete hub assembly and axle shaft removed, disconnect tie-rod and power steering cylinder from spindle support. Remove the capscrews from the two-piece retainer ring on inner side of spindle support and separate the retainers, seals and gasket from spindle support as shown in Fig. O21.

Note: At this time, it is desirable to remove the grease from cavity formed by spindle support and outer end of axle housing.

Remove upper trunnion, pull top of spindle support outward and remove spindle support from outer end of axle housing as shown in Fig. O22. Keep shims present under top trunnion tied to the trunnion for use during reassembly. Remove upper trunnion bearing from axle housing. Remove lower trunnion, shims and bearing from spindle support. Both trunnion bearing cups and upper trunnion bearing grease retainer can now be removed if necessary. If necessary to remove axle shaft thrust washer, oil seal and oil seal washer from axle outer end, a slide hammer and puller attachment can be used. Seals (Fig. O21) on outer end of axle housing can also be removed at this time.

Clean and inspect all parts and renew as necessary. It is recommended that new seals be used during reassembly.

35. To reassemble spindle support, proceed as follows: Install axle shaft seal washer and oil seal with lip toward inside and be sure oil seal is bottomed. Install axle shaft thrust washer. Install seal components over outer end of axle housing in the following order: Inner seal retainer with step toward inside of tractor; dust seal spring, rubber dust seal and felt grease seal and be sure bevel in inside diameter of both seals is toward bell of axle outer end; outer seal retainer with step toward outside of tractor and gasket. Install grease retainer (cup side up) and upper bearing cup (smallest I.D. down) in the upper trunnion bearing bore. Bolt lower trunnion to spindle support using original shims and tighten cap screws to a torque of 80-88 Ft.-Lbs. Place lower trunnion bearing over lower trunnion. Install lower trunnion bearing cup in outer end of axle housing with smallest I.D. of cup up. Place upper trunnion bearing in the upper trunnion bearing cup, then while tipping upper side of spindle support slightly outward, position spindle support over outer end of axle housing and install upper trunnion with original shim pack. Tighten trunnion retaining capscrews to a torque of 80-88 Ft.-Lbs.

Before attaching tie-rod, power steering cylinder or seal assembly to spindle support, check adjustment of trunnion bearings as follows: Con-

**Fig. O21 — Axle outer end seals and retainers are removed from spindle support as shown.**

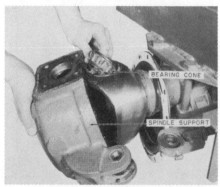

**Fig. O22 — Remove spindle support from outer end of axle housing as shown.**

**Fig. O23 — Use a spring scale in tie-rod stud hole to check trunnion bearing adjustment.**

nect a spring pull scale to tie-rod hole of spindle support and check pull required to rotate the spindle support. Refer to Fig. O23. Adjustment is correct when 12 to 18 pounds pull is required. To adjust bearings, vary number of shims located under the trunnions as required to obtain proper adjustment keeping the total thickness of shims under the top trunnion and lower trunnion as equal as possible. Shims are available in thicknesses of 0.003, 0.005 and 0.010.

Use grease to hold seal retainer gasket in place, then install seal assembly on spindle support, being sure

that the split ends of outer seal assembly do not align. Attach tie-rod and tighten tie-rod stud nut to a torque of 200 Ft.-Lbs. Attach power steering cylinder and tighten attaching bolt lock nut until it just contacts mounting flange.

Place approximately four pounds of grease in the cavity of spindle support and pack universal joint of axle shaft. Coat mating surfaces of spindle and spindle support with No. 2 Permatex or equivalent sealer and install the planet spider, wheel hub and axle assembly on the spindle support. Tighten attaching capscrews to a torque of 80-88 Ft.-Lbs. Then, install tire and rim.

## DIFFERENTIAL AND CARRIER

The Four Wheel Drive front axle can be equipped with either a conventional differential assembly or a "No-Spin" differential assembly. Removal procedure will be the same for both types. For service information, refer to paragraphs 37 and 40A.

36. **REMOVE AND REINSTALL.** To remove the differential and carrier assembly, raise front of tractor, block axle carrier to prevent front axle assembly from rocking and remove tires and rims. Drain differential housing. Disconnect power steering cylinders and lay them on top side of axle carrier. Remove drive shaft shield and shield front support. Disconnect drive shaft from companion flange of differential pinion shaft.

Attach hoist to one of the wheel studs and take up slack of hoist. Unbolt spindle from spindle support and pull complete assembly from outer end of axle housing. Refer to Fig. O20. Remove opposite assembly in like manner.

Note: Do not allow weight of hub assembly to be supported by axle shaft or damage to oil seal in axle housing outer end will result.

Place a rolling floor jack under front axle, unbolt axle from axle carrier and lower the axle from tractor. Position axle on supports with differential pinion shaft up and secure assembly in this position with blocks. Disconnect one end of tie-rod and swing it out of way. Remove capscrews retaining differential carrier to axle housing and remove the assembly from housing.

Reinstall by reversing removal procedure and coat carrier retaining capscrews and mating surfaces of carrier and axle housing with No. 2 Permatex or equivalent sealer. Tighten capscrews to a torque of 37-41 Ft.-Lbs. Tighten tie-rod stud nut to a

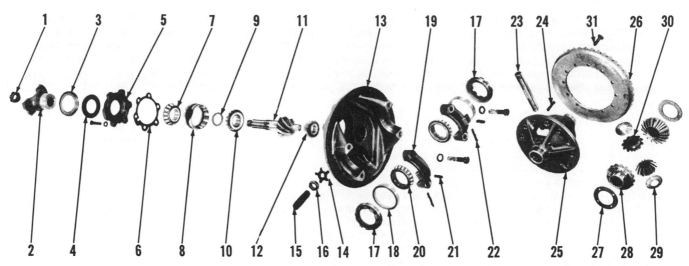

**Fig. O24 — Exploded view of the differential and carrier assembly used in front drive axle of all four-wheel drive models.**

| | | |
|---|---|---|
| 1. Nut | 6. Gasket | 11. Pinion shaft | 16. Lock nut | 21. Lock pin | 26. Bevel ring gear |
| 2. Companion flange | 7. Bearing cone | 12. Pilot bearing | 17. Adjusting nut | 22. Cotter pin | 27. Thrust washer |
| 3. Dirt shield | 8. Bearing cup | 13. Carrier | 18. Bearing cup | 23. Pinion pin (shaft) | 28. Side gear |
| 4. Oil seal | 9. Spacer and shim kit | 14. Locking washer | 19. Bearing cap | 24. Lock pin | 29. Thrust washer |
| 5. Bearing retainer | 10. Bearing cone | 15. Thrust screw | 20. Bearing cone | 25. Differential case | 30. Pinion gear |

torque of 200 Ft.-Lbs. When joining spindle to spindle support, coat mating surfaces with No. 2 Permatex or equivalent sealer and tighten retaining capscrews to a torque of 80-88 Ft.-Lbs. Piston rod end of power steering cylinders are attached to spindle supports. Tighten cylinder attaching bolt lock nuts until they just contact mounting flanges.

**37. OVERHAUL. (CONVENTIONAL)** With differential and carrier removed as outlined in paragraph 36, disassemble unit as follows: Straighten tab of locking washer (14—Fig. O24) and remove lock nut, locking washer and thrust screw (15), if used. Punch mark carrier bearing caps (19) so they can be reinstalled in original position, then remove cotter pins and the adjusting nut lock pins (21). Cut lock wires and remove the carrier bearing caps. Lift differential from carrier and keep bearing cups (18) identified with their bearing cones (20). Bearing cones can now be removed from differential case if necessary. Unbolt and remove bevel ring gear from differential case if necessary to renew gear. Drive pinion pin (shaft) lock pin (24) out of differential case and remove pinion pin (23), pinions (30), side gears (28) and thrust washers (27 and 29) from differential case.

Remove cotter pin and nut (1) from pinion shaft (11), then using a puller, remove the companion flange (2) and dust shield. Remove pinion shaft bearing retainer (5) and press pinion shaft and bearing from carrier. Use a split bearing puller to support pinion bearing cup (8) on edge nearest

pinion shaft gear and press pinion shaft from rear bearing and bearing cup. Remove spacer (9) and any shims which may be present from pinion shaft. Remove front bearing and inner (pilot) bearing (12) in a similar manner.

Clean and inspect all parts. Pay particular attention to bearings, bearing cups and thrust washers. If any of the differential side gears or pinions are damaged or excessively worn, renew all gears and thrust washers. Pinion shaft and bevel ring gear are available in a matched set only.

**38.** The differential and carrier unit is assembled as follows: Place inner (pilot) bearing on inner end of pinion shaft and stake in four places. Use a piece of pipe the size of inner pinion shaft bearing race to press forward bearing cone (10) onto shaft with taper facing threaded end of pinion shaft. Place bearing spacer and any shims which were present during disassembly over pinion shaft, then position the bearing cup over forward bearing. Press the rear pinion shaft bearing (7) on shaft with taper away from threaded end of pinion shaft. Check and, if necessary, renew the dust shield (3) on companion flange. Position companion flange so it will not obstruct cotter pin hole in end of pinion shaft, slide bearing retainer oil seal on its land on companion flange, then press companion flange on pinion shaft. Install retaining nut, clamp companion flange in a vise and tighten nut to a torque of 300 Ft.-Lbs.

Note: Pressure of oil seal will generally hold bearing retainer away from bearing. If it does not do so, tie retainer to companion flange.

**39.** With pinion shaft assembled as outlined above, clamp the bearing cup in a soft jawed vise just tight enough to prevent rotation, then using an inch-pound torque wrench on companion flange retaining nut, check torque required to rotate pinion shaft. Pinion shaft bearing adjustment is correct if 13 to 23 inch-pounds is required to rotate shaft. If rolling torque is not as specified, disassemble the pinion shaft assembly and vary thickness of spacer and/or shims as required to obtain proper rolling torque. A spacer and shim kit is available under Oliver part number 155 342-A.

With pinion shaft assembled and correct rolling torque (bearing adjustment) obtained, press pinion shaft assembly into carrier, install bearing retainer and tighten capscrews to a torque of 25 Ft.-Lbs. Install cotter pin to lock the nut in place .

**40.** Reassemble differential case assembly as follows: Place side gears, pinions and thrust washers in differential case and install pinion pin (shaft). Secure pinion pin with lock pin and, if lock pin is straight type, stake pin in position. It is not necessary to stake the spring type lock pin. If bevel ring gear was removed, reinstall with bolt heads on ring gear side of assembly and tighten the nuts to a torque of 78-86 Ft.-Lbs. Press bearings on differential case with tapers facing away from case. Place bearing cups over differential bearings and place differential assembly in carrier. Position bearing adjusting nuts in carrier and install the carrier bearing caps. Tighten the bearing cap screws until caps are snug but **be sure** threads of caps and adjusting nuts are in register. Maintain some clearance between gear teeth and tighten adjusting nuts until bearing cups are seated and all end play of differential is eliminated. Mount a dial indicator and shift differential assembly as required to obtain a backlash of 0.008-0.011 between bevel pinion shaft and bevel ring gear. Differential is shifted by loosening one adjusting nut and tightening the opposite nut an equal amount. Note: Mesh position of the bevel pinion shaft is not adjustable.

With gear backlash adjusted, tighten the bearing cap retaining capscrews to a torque of 65 Ft.-Lbs. and secure with lock wire. Install adjusting nut lock pins and cotter pins.

Note: If lock pins will not enter slots of adjusting nuts after backlash adjustment has been made, tighten rather than loosen the adjusting nut, or nuts. Recheck gear backlash.

If used, install thrust screw and turn screw in until it contacts back side of bevel ring gear, then back screw out ¼ to ½ turn. Apply sealer to threads of thrust screw at surface of carrier, then while holding screw from turning, install locking washer and nut. Secure nut by bending one tang of locking washer over nut and another tang over boss of carrier.

**40A. OVERHAUL (NO-SPIN).** With differential and carrier removed as outlined in paragraph 36, unit is disassembled as follows: Straighten tab of lock washer (14—Fig. O24) and remove lock nut, lock washer and thrust screw (15). Punch mark the carrier bearing caps (19) so they can be reinstalled in their original positions, then remove cotter pins and the adjusting nut lock pins (21). Cut lock wires and remove the carrier bearing

caps. Lift differential from carrier and keep bearing cups (18) identified with their bearing cones (20). Bearing cones can now be removed from differential case, if necessary.

Remove cotter pin and nut (1) from pinion shaft (11), then using a puller, remove the companion flange (2) and dust shield. Remove pinion shaft bearing retainer (5) and press shaft and bearings from carrier. Use a split bearing puller to support pinion bearing cup (8) on edge nearest pinion shaft gear and press pinion shaft from rear bearing (7) and bearing cup. Remove spacer (9) and any shims which may be present from pinion shaft. Remove front bearing (10) and inner (pilot) bearing (12) in a similar manner.

Unbolt and remove bevel ring gear from differential, if necessary. Remove differential case bolts and hold case together as last bolt is removed to keep assembly from flying apart due to the internal spring pressure. See Fig. O24A. Hold out rings (6) can be removed with snap ring spreaders.

**40B.** Clean and inspect all parts. Check splines on side gears and clutch members and remove any burrs or chipped edges with a stone or burr grinder. Renew any parts which have sections of splines broken away. Inspect springs (3) for fractures, or other damage, and renew springs which do not have a free height of 2¼-2½ inches. Center cam in central driver (7) must be free to rotate within the limits of keys in central driver. Check the weld between driven clutch (5) and cam ring on clutch by tapping lightly on cams of cam ring. If cam ring rotates in driven clutch, weld is defective (failed). Inspect teeth on central driver and driven clutches. Small defects can be dressed with a stone. If central driver or driven clutch is renewed, also renew the part it mates with. A smooth wear pattern up to 50 percent of face width is acceptable for the cams on center cam and driven clutch.

40C. The differential and carrier unit is assembled as follows: Place the inner (pilot) bearing on inner end of pinion shaft and stake in four places. Use a piece of pipe the size of inner race of pinion shaft bearing and press the forward pinion shaft bearing (10—Fig. O24) on shaft with taper facing threaded end of pinion shaft. Place bearing spacer, and any shims which were present during disassembly over pinion shaft, then position the bearing cup over forward bearing. Press the rear pinion shaft bearing (7) on shaft with taper away from threaded end of pinion shaft. Check, and renew if necessary, the oil seal (4) in pinion shaft bearing retainer (5). Seal is installed with lip toward inside. Place gasket and bearing retainer over pinion shaft. Check, and renew if necessary, the dust shield (3) on companion flange. Position companion flange so it will not obstruct cotter pin hole in end of pinion shaft, slide bearing retainer oil seal on its land on companion flange, then press companion flange on pinion shaft. Install retaining nut, clamp companion flange in a vise and tighten nut to a torque of 300 ft.-lbs.

NOTE: Pressure of oil seal will generally hold bearing retainer away from bearing. If it does not, tie retainer to companion flange.

40D. With pinion shaft assembled as outlined above, clamp the bearing cup in a soft jawed vise only tight enough to prevent rotation, then using an inch-pound torque wrench attached to companion flange retaining nut, check the rolling torque (bearing preload) of the pinion shaft. This rolling torque should be 13-23 in.-lbs.

If rolling torque is not as specified, it will be necessary to disassemble the shaft assembly and vary the spacer and/or shims as required. A spacer and shim kit is available under Oliver part number 155 342-A.

With pinion shaft assembled and the rolling torque of shaft determined, press pinion shaft assembly into carrier, install bearing retainer and tighten cap screws to a torque of 25 ft.-lbs. Install cotter pin to lock nut (1) in place.

40E. Lubricate all parts and reassemble no-spin differential by reversing disassembly procedure. Be sure to position spring retainers so that spring seats inside cupped section of retainer. Be sure key in central driver is aligned with slot in hold out ring. Tighten the case bolts to 36 ft.-lbs. torque and use a single wire through holes in bolt heads to lock bolts. Place

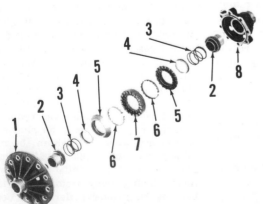

**Fig. O24A — Exploded view of the "No-Spin" differential assembly available for the Four Wheel Drive front axle.**
1. Case half
2. Side gear
3. Spring
4. Spring retainer
5. Driven clutch
6. Hold out ring
7. Central driver and center cam
8. Case half

an axle shaft into each side of differential and turn axle back and forth. It should be possible to feel backlash between clutch teeth. Backlash should be about $\frac{5}{32}$-inch. If no backlash can be felt and side gear is locked solid, check differential for incorrect assembly.

If bevel ring gear was removed, reinstall with bolt heads next to bevel ring gear and tighten nuts to a torque of 78-86 ft.-lbs. Press bearings on differential case with tapers facing away from case. Place bearing cups over differential bearings and place differential assembly in carrier. Position bearing adjusting nuts in carrier and install the carrier bearing caps. Tighten the bearing cap retaining cap screws until caps are snug but BE SURE threads of caps and adjusting nuts are in register. Maintain some clearance between gear teeth and tighten adjusting nuts until bearing cups are seated and all end play of differential is eliminated. Mount a dial indicator and shift differential assembly as required to obtain a backlash of 0.008-0.011 between bevel pinion shaft gear and bevel ring gear. Differential is shifted by loosening one adjusting nut and tightening the opposite an equal amount.

Note: Fore and aft position of the bevel pinion shaft is not adjustable.

With gear backlash adjusted, tighten the bearing cap retaining cap screws to a torque of 65 ft.-lbs. and secure with lock wire. Install adjusting nut lock pins and cotter pins.

NOTE: If lock pins will not enter slots of adjusting nuts after backlash adjustment has been made, tighten rather than loosen the adjusting nut, or nuts. Recheck gear backlash.

If used, install thrust screw and turn screw in until it contacts bevel ring gear, then back it out ¼-½ turn. Apply sealant to threads of thrust screw at surface of carrier, then install lock washer and lock nut. Secure lock nut by bending one tang of lock washer over nut and one tang over boss of carrier.

## TRANSFER DRIVE
### All Models So Equipped

41. **R&R AND OVERHAUL.** To remove the transfer drive assembly, it will be necessary to split the tractor front main frame from the rear main frame. If tractor is equipped with Hydra-Power, Creeper or Reverse-O-Torc Drive, it will also be necessary to remove the engine and drive unit before tractor can be split.

42. To split tractor equipped with Hydra-Power, Creeper or Reverse-O-Torc Drive, first remove engine and drive unit as outlined in paragraph 60, then proceed as follows: Disconnect speedometer drive cable from adapter. Disconnect shifting rod from transfer drive shifter arm and pull it from front main frame. Disconnect drive shaft from companion flange of transfer drive output shaft. Unclip and disconnect power steering oil lines. Disconnect safety starting switch wires. Disconnect rear light wires and remove clips. On non-diesel models, unhook governor control bellcrank spring. Support tractor front frame in manner which will prevent tipping, then support rear frame with rolling floor jack. Remove front frame to rear frame retaining cap screws and separate tractor.

43. To split direct drive models, proceed as follows: Remove PTO drive shaft as outlined in paragraph 221 or 226, or the hydraulic pump drive shaft as outlined in paragraph 248, if so equipped. Remove coupling or coupling chain from clutch shaft to transmission shaft. Disconnect shifting rod from transfer drive and pull it from hole in front main frame. Disconnect drive shaft from companion flange of transfer drive output shaft. Unclip and disconnect power steering lines. Disconnect rear light wires and remove clips. On non-diesel models, unhook governor control bellcrank spring. Support tractor front frame in a manner which will prevent tipping, then support rear frame with a rolling floor jack. Remove front frame to rear frame retaining capscrews and separate tractor.

44. With tractor split as outlined in paragraph 42 or 43, drain transmission and transfer housings, refer to Fig. O25 and proceed as follows: Place transmission and transfer drive in gear to unscrew nut (56), then remove companion flange and spacer from transfer drive output shaft. Disconnect oil line from transfer drive case, then unbolt transfer drive cover (43) from case and remove cover along with idler gear (32), output gear (26) and shaft (24). Remove shifter coupling (25) from fork if necessary. Remove snap ring (37) and gear (36) from forward end of transmission bevel pinion shaft; discard snap ring. Unbolt and remove transfer case (12) from tractor rear frame. Shifter fork (63) and actuator (67) can be removed after shifter arm (71) and Woodruff key (69) are re-

moved. Any further disassembly required will be evident. Save shims (22 and 58) located under bearing retainers (18 and 60) for reassembly.

45. Clean and inspect all parts and renew as necessary. New seals, "O" rings and gaskets should be used during assembly. The unit should be partially reassembled and the end play of idler and output shafts checked and adjusted before unit is installed on tractor rear main frame.

Install idler shaft rear bearing cup in transfer drive case and install output shaft rear bearing cup in shaft rear bearing retainer. Be sure both bearing cups are firmly seated and install output shaft rear bearing retainer (18) along with original shim pack (22). Install output shaft front bearing retainer (49) and idler shaft front bearing retainer (60) with original shim pack. Be sure both bearing cups are against the bearing retainers. Install dowels (30) in transfer case if removed. Place output shaft and idler shaft assemblies in the transfer drive case cover, then using a new gasket (42), secure cover to case. Use dial indicator to check end play of both shafts; end play should be 0.001-0.003 and is adjusted by varying number of shims under the bearing retainers. If shims are added, be sure bearing cups are seated against retainers before rechecking shaft end play.

With shaft bearings (end play) adjusted, remove cover, shafts and bearing retainers. Renew the "O" rings (21 and 59) on bearing retainers and install new oil seals as follows: Bevel pinion shaft oil seal (35) in case with lip rearward. Bevel pinion shaft seal (38) in cover with lip forward. Shift fork actuator seal (70) with lip inward. Install wide seal (48) at rear with lip facing rearward and the narrow oil seal (53) at front with lip facing forward in the output shaft front bearing retainer.

Coat all seals and "O" rings with grease prior to installation and fill cavity between the two seals in output shaft front bearing retainer with grease.

Reassemble unit by reversing the disassembly procedure. Renew the "O" ring on bevel pinion shaft front bearing retainer and install case with new gasket (17). Use a new snap ring (37) and "O" ring (34) when in-

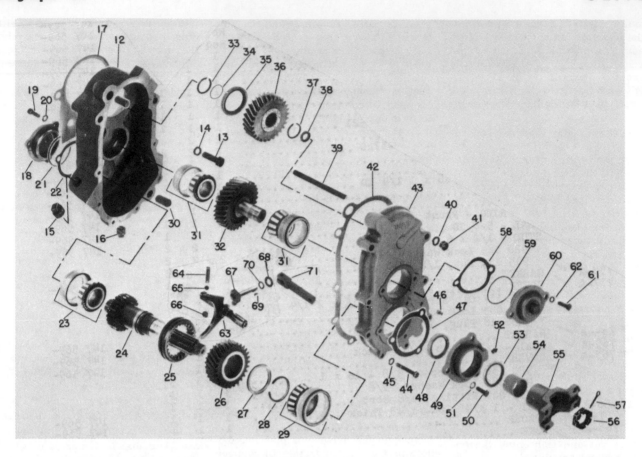

**Fig. O25 — Exploded view of the four-wheel drive transfer case and shift linkage. Breather filter element (not shown) should be renewed yearly or after each 1000 hours of operation. Maintain oil level at plug (15) opening; drain plug is (16).**

| | | | | |
|---|---|---|---|---|
| 12. Case | 23. Bearing | 31. Bearing | 39. Stud bolt | 55. Companion flange | 64. Detent spring |
| 15. Filler plug | 24. Shaft | 32. Idler gear | 42. Gasket | 56. Castellated nut | 65. Detent ball |
| 16. Drain plug | 25. Coupling | 33. Snap ring | 43. Cover | 58. Shims (0.004, 0.0075, | 66. Plug |
| 17. Gasket | 26. Gear (27 teeth) | 34. "O" ring | 47. Gasket | 0.015) | 67. Actuator |
| 18. Bearing retainer | 27. Washer | 35. Oil seal | 48. Seal | 59. "O" ring | 69. Woodruff key |
| 21. "O" ring | 28. Snap ring | 36. Gear | 49. Retainer | 60. Retainer | 70. Seal |
| 22. Shim (0.004, 0.0075, | 29. Bearing | 37. Snap ring | 53. Seal | 63. Shift fork & rod | 71. Arm |
| 0.015) | 30. Dowels | 38. Seal | 54. Spacer | | |

stalling drive gear (36) on transmission bevel pinion shaft. Use heavy grease on sliding coupling (25) to hold it in position in shifter fork. When installing companion flange (55), be sure not to obstruct cotter pin hole.

## DRIVE SHAFT
### All Models So Equipped

46. The front axle drive shaft is of conventional design. Removal and overhaul procedure is evident after removal of shield and an examination of unit. Spider and bearing assemblies are available as units only. All other parts are available separately.

When installing drive shaft, yoke end is installed to rear.

# FRONT SYSTEM
## (Hydraulic Power Front Wheel Drive)

Series 1650 and 1655 tractors are available with a hydraulic powered front wheel drive which uses pressurized oil to furnish power for the front wheels. The power unit is mounted in front of the tractor radiator and produces two hydraulic circuits which control and drive the tractor front wheels.

The delivery (primary) circuit furnishes the high volume of pressurized

oil used to power the front wheels and is produced by a variable displacement piston type pump driven by a shaft attached to the engine crankshaft pulley. This oil is routed through hoses to fixed displacement wheel motors which drive the front wheels through reduction gears.

The control (secondary) circuit furnishes control pressure to engage and

disengage the front wheel clutches as well as signal (control) pressure to regulate the delivery circuit operation. The signal pressure controls the delivery circuit by actuating (varying) a stroke control valve which is mechanically connected to a trunnion (swash plate) which regulates the displacement (stroke) of the piston pump.

The system also has an electrical

system consisting of interlock switches on the transmission shift rails and the clutch release lever as well as a solenoid lock on the drive control quadrant. The interlock switches permit front wheel drive operation only when transmission is in a forward position and the clutch pedal released (clutch fully engaged). The quadrant solenoid lock prevents "on-the-go" shifting from two wheel to four wheel operation. Whenever transmission is in a forward gear position and ignition switch is "ON", the quadrant lever is locked in "OFF" position. Transmission gear shift lever must be moved to neutral position to break electrical circuit to quadrant solenoid lock before front wheel drive can be engaged.

## TROUBLE SHOOTING

When testing the hydraulic front wheel drive assembly, checks must be made in the following sequence: Control circuit, electrical circuit and delivery circuit. Any other sequence will not give valid results.

**46A. CONTROL CIRCUIT.** The control circuit should maintain 90-130 psi with front wheel drive engaged and engine running. If pressure is less than 90 psi, check the following:

a. Check level of tractor hydraulic lift reservoir and if low, add fluid as required.

b. Check the manually controlled valve cable for breaks or maladjustment. Renew cable and/or adjust cable as outlined in paragraph 46I.

c. If doubt exists as to condition of quadrant pressure gage, attach a gage of known condition to gage line connector at front of valve housing and compare gage readings. Renew quadrant pressure gage if necessary.

**46B. ELECTRICAL CIRCUIT.** A system of interlock switches on the transmission shifter rods and clutch release lever is used to provide automatic start and stop control of the hydraulic front wheel drive. Delivery circuit of the hydraulic unit will not go into operation unless an electrical signal is present at the solenoid valve.

To check the interlock electrical system, connect a test light to the solenoid valve terminal, turn ignition switch to "ON" and move gear shift lever to a forward position. With clutch pedal released, test light should be on. Repeat test with gear shift lever on other side of shift pattern.

If test light does not react as stated, check for a blown fuse and if a blown fuse is found, locate cause before proceeding further. See interlock switch wiring diagram shown in Fig. O25A. Also check clutch interlock switch adjustment as follows: Attach test light to **blue** wire terminal of clutch interlock switch, turn ignition switch to "ON" position and move gear shift lever to a forward position. Test light should now be on and should remain on until clutch pedal is depressed 1½ inches from its rest position; at which time, light should go off and remain off through remainder of pedal downward travel. If adjustment of clutch interlock switch is required, loosen housing retaining screws, turn housing until correct adjustment is obtained and tighten retaining screws.

If necessary, the shifter rail interlock switches can be isolated from the clutch interlock switch for testing as follows: Connect test light to **red** wire connection of clutch interlock switch and turn ignition switch on. Place shift lever in a forward position on both sides of the shift pattern and note test light which should be on in

both cases. If test light is off in either case, renew the interlock switch on the side affected.

Install a new clutch interlock switch if an electrical signal is present at **red** wire connection of switch but switching action cannot be obtained. Adjust new switch as previously outlined.

**46C. DELIVERY CIRCUIT.** To test the delivery circuit, first remove test cap located on lower right side of gear pump housing and attach a 5000 psi gage. Lock tractor brakes and to assure complete safety, attach tractor drawbar to a substantial stationary object. Start engine and place front drive quadrant lever in No. 6 position, then adjust engine speed to 1800 rpm. Check quadrant gage to be sure that control circuit pressure is at least 90 psi, then connect a jumper wire from any battery source to terminal of solenoid.

NOTE: Jumper wire must not be connected before engine is started and 90 psi pressure is showing on quadrant pressure gage.

Note reading on test gage which should be approximately 3000 psi.

If delivery circuit pressure is not as stated, adjust pressure control cable housing as outlined in paragraph 46E. If pressure cannot be adjusted by shifting cable, refer to paragraph 46G for maximum pressure adjustment.

**46D. DRIVE WHEEL.** The condition of drive wheel hydraulic clutch "O" rings and wheel motor oil seals can generally be determined as follows:

a. If oil flows from the primary housing breather with tractor stationary and front wheel drive engaged, the hydraulic clutch piston "O" rings are defective and should be renewed.

b. If oil flows from the primary gear housing breather with tractor moving and front wheel drive engaged, the wheel motor oil seal is defective and should be renewed.

## ADJUSTMENTS

Three adjustments can be made to control the delivery circuit pressure. They are: Pressure control cable adjustment, Minimum Stroke adjustment and Maximum Pressure adjustment. All adjustments must be made with hydraulic fluid at operating temperature and are made as follows.

**46E. PRESSURE CONTROL CABLE ADJUSTMENT.** Remove the test opening cap located at right bottom side of gear pump housing and attach a pressure gage of approximately 5000

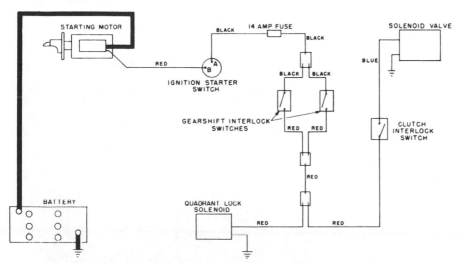

**Fig. O25A—Wiring diagram of the hydraulic powered front wheel drive system.**

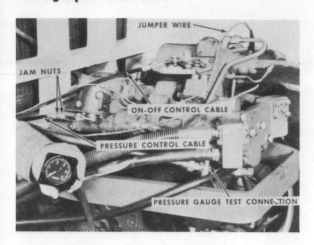

Fig. O26C—View showing method of making minumum stroke adjustment. Refer to text for procedure.

Fig. O26B—When making pressure control cable adjustment, install test gage as shown and refer to text.

psi capacity. See Fig. O26B. Lock the tractor brakes and to assure complete safety, secure drawbar to a substantial stational object. Start engine and place quadrant control lever in No. 6 position. Adjust engine speed to 1800 rpm and note quadrant gage which must show at least 90 psi before proceeding with test.

Connect a jumper wire between a battery voltage source such as the alternator "BAT" terminal and the unit solenoid valve terminal. Note the reading on test gage which should be 3000 psi. If pressure is not as stated, loosen the jam nuts at forward end of control cable and shift cable forward and rearward as required and tighten jam nuts to hold adjustment. If the recommended pressure adjustment cannot be obtained by changing the control cable length, check and adjust, if necessary, the maximum pressure as outlined in paragraph 46G.

**46F. MINIMUM STROKE ADJUST-MENT.** With unit shroud off, raise either of the front wheels. Start engine, run at 1800 rpm and place quadrant lever in the No. 1 position. Check the raised front wheel which should turn approximately 2 rpm at the 1800 engine rpm indicating a correct minimum stroke (pump) adjustment.

If minimum stroke adjustment is required, stop engine, remove valve housing to stroke control oil line and the fitting and plug outlet on top of valve housing. Remove snap ring and stroke control valve end cap, then loosen jam nut and turn adjusting screw about half way into end cap. See Fig. O26C. Reinstall end cap and secure with snap ring. Install snap ring with chamfered O.D. next to end cap. Start engine and run at 1800 rpm then using a $\frac{3}{16}$-inch Allen wrench, turn adjusting screw clockwise until the specified 2 rpm of raised front wheel is obtained. Stop engine, again remove end cap and tighten jam nut to hold the adjustment. Reinstall end cap and BE SURE chamfered O.D. of snap ring is next to end cap. Recheck adjustment and if satisfactory, reinstall plug, fitting and valve housing to stroke control oil line.

**46G. MAXIMUM PRESSURE AD-JUSTMENT.** With shroud removed, remove front jam nut from pressure control cable, or remove the cable if desired. Connect a 5000 psi gage to opening in gear pump housing (Fig. O26B). Lock tractor brakes, and to insure complete safety, chain drawbar to a substantial stationary object.

Start engine, place quadrant control lever in No. 1 position and run engine at 1800 rpm. BE SURE quadrant gage registers at least 90 psi, then connect a jumper wire between a battery voltage source such as the alternator "BAT" terminal and terminal on unit solenoid valve. Rotate cam control arm rearward until cam is fully bottomed and note reading of test gage. Test gage should register 3300-3500 psi.

If pressure is not as stated, remove cam housing, cam follower, shims and spring from stroke control valve housing and vary shims as required. Each shim will change pressure about 140 psi. Reassemble and recheck pressure. Adjust pressure control cable as outlined in paragraph 46E.

### FLUSH SYSTEM

**46H.** To flush (filter) the hydraulic front wheel drive system, obtain two flushing filters which are available from Oliver Corporation under parts number STS-165 and proceed as follows: Raise front of tractor until both front wheels are free, then support under front axle to maintain this position. Disconnect the high pressure hoses from both wheel motors, then install a 0.097 x 0.755 I.D. "O" ring (Oliver No. 162 573-A) on inlet connectors of wheel motors. Install filters with "out" end on wheel motor inlet fittings, then connect the pressure hoses to upper ends of filters. See Fig. O26D. Start engine, place quadrant control lever in No. 2 position and check quadrant pressure gage to see that it registers at least 90 psi, then connect a jumper wire between any battery source and solenoid valve.

NOTE: Be sure engine is running and quadrant pressure gage registers at least 90 psi before connecting jumper wire. To do otherwise will very likely damage system.

With jumper wire connected, run engine for 10 minutes at 1000 rpm,

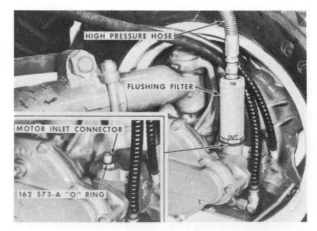

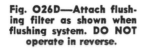

Fig. O26D—Attach flushing filter as shown when flushing system. DO NOT operate in reverse.

10 minutes at 1800 rpm and 10 minutes at high idle. With flushing cycle completed, remove jumper wire and shut off engine. Remove filters and "O" rings and reinstall high pressure hoses to wheel motors. Remove blocking from under front axle.

NOTE: Always have front wheels raised when flushing. Driving the tractor with filters installed may cause the operating pressure to exceed the maximum pressure rating of the filters. In addition, should the tractor be inadvertently operated in reverse, the filters will be back flushed and any contaminents present will be forced back into hydraulic system.

### CONTROL QUADRANT

Control for the hydraulic front wheel drive unit is provided by a dual cable control quadrant assembly mounted on right side of instrument panel. Included in the quadrant assembly is an electrically actuated solenoid lock and a pressure gage which registers the control circuit pressure of front wheel drive unit.

**46I. R&R AND OVERHAUL.** To remove the quadrant and cables, first remove shroud from the front wheel drive unit. Place quadrant lever in No. 6 position, then loosen clamping nipples from forward ends of cables and push cam lever and "OFF-ON" valve bellcrank from cable ends. Disconnect pressure gage line and solenoid wire from quadrant. Disconnect cable support and clip from engine and radiator grille, then unbolt and remove quadrant, mounting bracket and control cables.

To disassemble the quadrant assembly, remove cable clamp (29—Fig. O26E), loosen pressure gage bracket nuts, then remove the cap screws holding the two housings together and remove mounting bracket (37), left housing (35) and pressure gage (17). Remove control hubs (19 and 32), and pivot pin (18) from right housing (2), then remove control lever (33). Disconnect cables (27 and 28) from control hubs, then remove hubs and leaf spring (26) from pivot. Remove control hub latch (20) and spring (21). Remove solenoid (9) and bracket (7) from right housing. If solenoid or lock (15) requires renewal, disassemble solenoid and bracket by removing roll pins (13 and 14).

Inspect all parts for cracks, burrs, excessive wear or other damage and renew as necessary.

Assemble quadrant as follows: Assemble and install solenoid, solenoid bracket and lock in right housing and secure solenoid ground wire to hous-

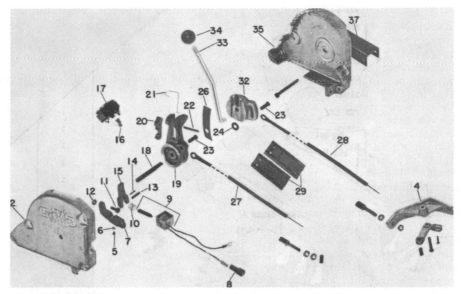

Fig. O26E—Exploded view of the control quadrant used on hydraulic powered front wheel drive tractors.

| | | |
|---|---|---|
| 2. Housing, R.H. | 15. Lock | 27. OFF-ON control |
| 4. Bracket | 16. Flared connector | cable |
| 7. Solenoid bracket | 17. Pressure gage | 28. Pressure control |
| 8. Connector | 18. Pivot | cable |
| 9. Solenoid and | 19. OFF-ON control | 29. Clamp |
| plunger | hub | 32. Pressure control |
| 10. Spring | 20. Latch | hub |
| 11. Adjusting screw | 21. Spring | 33. Control lever |
| 12. Lock nut | 22. Roll pin | 34. Knob |
| 13. Roll pin | 23. Headed pin | 35. Housing, L.H. |
| 14. Roll pin | 24. Washer | 37. Mounting bracket |
| | 26. Spring | |

ing. Place latch (20) and spring (21) in position in control hub (19) and install roll pin (22). Place control hubs and leaf spring over pivot pin. Hubs should be centered and slot in pin should face lock notch in the "OFF-ON" control hub (19). Install control lever (33) in pressure control hub (32). Attach cables to control hubs with longest cable attached to "OFF-ON" control hub (19), then install the assembly in right housing with "OFF-ON" hub against housing. Rotate hub rearward until latch drops down behind housing ramp, then adjust screw in solenoid bracket to obtain $\frac{1}{16}$-inch clearance between hub and solenoid lock. Locate pressure gage and bracket in right housing, then with control lever in No. 1 position, install left housing and mounting bracket. Secure pressure gage bracket and install cable clamp.

Mount quadrant and bracket to instrument panel and connect pressure line and solenoid wire. Place quadrant level in "OFF" position, secure the longer cable housing to bracket (4—Fig. O26E) and attach control cable to "OFF-ON" valve bellcrank while valve spool is "at rest" position. Place quadrant lever in No. 1 position, rotate compensating control cam arm rearward until cam contacts follower, then while holding cam arm in this position, attach shortest cable housing to the support bracket and the control

cable to cam arm. Reposition cable support bracket, if necessary, to allow full travel of cam arm without control cable binding.

Secure cables and pressure gage line to engine and grille. Check pressure control cable adjustment as outlined in paragraph 46E, then install front wheel drive unit shroud.

### REMOVE AND REINSTALL (HYDRAULIC UNIT)

**46J.** To remove the hydraulic unit remove shroud, disconnect "off-on" cable and pressure control cable, then disconnect quadrant pressure gage line from hydraulic unit. Disconnect wire from solenoid valve. Disconnect oil cooler outlet and inlet lines from hydraulic unit. Drive roll pins from forward ends of both wheel motor high pressure hoses and pull hoses from hydraulic unit manifold. Do not lose the two special retaining washers. Disconnect the wheel motor return hoses and leak-off hoses at hydraulic unit end. Remove mounting guard. Attach a hoist to hydraulic unit, unbolt unit from mounting bracket, then slide unit forward until drive shaft clears piston pump drive shaft and lift unit away from tractor. Mounting bracket can now be removed, if necessary.

To reinstall hydraulic unit, proceed as follows: Install mounting bracket, if necessary. Position hydraulic unit and start drive shaft on piston pump drive shaft, then mate unit with mounting bracket and install retaining cap screws. Reconnect all hoses, quadrant pressure gage line and solenoid valve wire. Place quadrant control lever in "OFF" position, secure "OFF-ON" (longest) cable housing to its support bracket and the cable to the manually controlled valve bellcrank with valve in its at rest position. Move quadrant control lever to No. 1 position, then rotate pressure compensating cam control arm rearward until cam contacts cam follower and temporarily attach pressure control cable (shortest) to cam control arm and cable housing to support bracket. Check full travel of cable and control arm and reposition cable support bracket to prevent binding, if necessary.

If service has been performed on hydraulic unit, flush system as outlined in paragraph 46H, then check and adjust if necessary the piston pump minimum stroke as outlined in paragraph 46F and the delivery circuit maximum pressure as outlined in 46G.

Reinstall shroud after adjustments are complete.

## OVERHAUL

If desired, all sub-assemblies of the hydraulic power front wheel drive unit, except piston pump and drive shaft, can be removed, serviced, and reinstalled without removing unit from tractor. However, because of the inter-relation of assemblies, should any damage occur within the unit, the possibility exists that contaminants could be circulated throughout the entire front wheel drive system. Therefore, it is advisable to disassemble the complete unit for cleaning and inspection should trouble be experienced.

The following procedure is based on the unit being removed from tractor as outlined in paragraph 46J, and each sub-assembly will be removed, disassembled and reinstalled.

**46K. MANIFOLD.** Remove the manifold-to-piston pump housing oil line, then unbolt and remove manifold from gear pump housing. Remove and discard the two "O" rings from recesses in rear of manifold. Remove the pressure hose port end caps, "O" rings and back-up (Teflon) rings. Discard "O" rings and back-up rings.

Clean manifold and inspect for cracks, warping and damaged threads. Pay particular attention to the pressure hose ports.

To reassemble and reinstall manifold, install pressure port end caps,

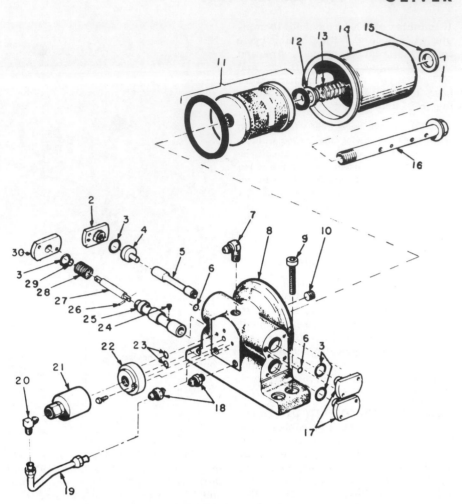

**Fig. O26F—Exploded view of filter and valves assembly of the hydraulic powered front wheel drive unit.**

|  |  |  |
|---|---|---|
| 2. End cap | 11. Filter element and gaskets | 21. Solenoid |
| 3. "O" ring | 12. Spring seat | 22. Solenoid base |
| 4. Piston | 13. Spring | 23. "O" rings |
| 5. Valve spool, hyd. controlled | 14. Shell | 24. Screw plug |
| 6. "O" ring | 15. Gasket | 25. Valve spool, man. controlled |
| 7. Elbow | 16. Center stud | 26. Roll pin |
| 8. Housing | 17. End caps | 27. Valve actuator |
| 9. Socket head screw | 18. Connector | 28. Spring |
| 10. Pipe plug | 19. Oil pipe | 29. "O" ring |
|  | 20. Elbow | 30. End cap |

"O" rings and back-up (Teflon) rings. The back-up rings are installed on outer sides of the smaller "O" rings. Place "O" rings in recesses in rear of manifold and install manifold on gear pump housing. Install manifold-to-piston pump housing oil line.

**46L. FILTER AND VALVES ASSEMBLY.** To remove the filter and valves assembly, first remove valve housing-to-stroke control oil line. Remove the four socket head cap screws and lift filter and valves assembly from gear pump housing. If valve housing-to-gear pump housing oil line remains with valve housing, it can now be removed. Discard all "O" rings as filter and valve assembly is removed and disassembled.

To disassemble filter and valve assembly, remove center stud (16—Fig. O26F), remove filter assembly and

discard filter element (11). Remove plug (10) for cleaning purposes. Remove solenoid valve oil line (19), then remove solenoid (21), solenoid base (22) and "O" rings (23). Remove manual control valve bellcrank and bellcrank support, then remove end caps (2, 17 and 30), spring (28) and "O" rings (3). Identify end caps for reinstallation. Remove piston (4) and the hydraulically controlled valve spool (5) from upper bore. Remove valve actuator (27) and manually controlled valve spool (25).

Clean and inspect all parts. Pay particular attention to valve spools and their bores. Spools should be a snug fit in bores yet slide freely when lubricated. Light scoring can be cleaned up with crocus cloth, however spools and/or housing should be renewed if heavy scoring is present. No

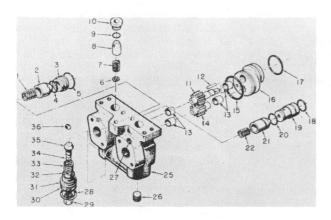

**Fig. O26G — Exploded view of the hydraulic powered front wheel drive gear pump assembly.**

13. Bushings
14. Pump driven gear
15. "O" ring
16. Housing cover
17. "O" ring
18. "O" ring
19. Valve seat
20. "O" ring
21. Monitoring circuit relief valve
22. Spring
25. Pump housing
26. Pipe plug
27. Pipe plug
28. Snap ring
29. Cap
30. "O" ring
31. Valve guide
32. "O" ring
33. Spring
34. "O" ring
35. Reverse flow relief valve
36. Pipe plug

1. Spring
2. Reverse flow check valve
3. Valve seat
4. "O" ring
5. "O" ring
6. Spring seat
7. Spring
8. Overspeed check valve
9. "O" ring
10. Valve seat
11. Pump drive gear
12. Dowel

test specifications are available for spring (28), however spring should be free of fractures or other defects and should be capable of returning valve spool to "OFF" position with comparative ease.

Reassemble by reversing the disassembly procedure, however keep the following points in mind. End cap (2) is thicker than end cap (17) and is installed at piston end of valve housing.

Index the oil holes in solenoid base (22) with similar oil holes in valve housing. Install plug (10) in right side opening of filter base as viewed from filter side. Do not block left side opening.

When reinstalling filter and valves assembly, be sure "O" rings are installed and position valve housing-to-gear pump housing oil line as assembly is installed. Tighten the valve

housing retaining socket head cap screws to 70-77 Ft.-Lbs. torque.

**46M. GEAR PUMP.** To remove the gear pump assembly, first remove the manifold as in paragraph 46K and the filter and valves assembly as in paragraph 46L. Gear pump assembly can now be removed from piston pump housing plate.

With gear pump assembly removed, disassemble as follows: Remove overspeed check valve seat (10—Fig. O-26G), "O" ring (9), overspeed check valve (8), spring (7) and spring seat (6). Remove valve seat (3), reverse flow check valve (2) and spring (1). Remove pressure test port cap (29), then snap ring (28), guide (31), spring (33) and reverse flow relief valve poppet (35). Remove cover (16), alignment dowel (12) and pump gears and shafts (11 and 14). Thread a ¼-20 inch cap screw into tapped hole in seat (19) and remove seat, monitoring circuit relief valve plunger (21) and spring (22). Removal of pipe plugs will be dictated by the need for cleaning.

Clean and inspect all parts. Bushings (13) in housing (25) and cover (16) can be renewed if necessary. None of the valves in the gear pump assembly are adjustable, however all springs, as well as valves and valve seats, should be closely examined. Renew any parts which are cracked, worn or otherwise damaged.

Assemble gear pump assembly as follows: Install spring (22), monitoring circuit relief valve plunger (21) with "O" ring (20) and seat (19) with "O" ring (18). Lubricate pump gears liberally and place gears in housing with drive gear on top side, then install alignment dowel (12) and housing cover (16) with "O" rings (15 and 17). Install reverse flow relief valve poppet (35) and "O" ring (34), spring (33), guide (31) with "O" rings (30 and 32) and secure with snap ring (28). Install spring (1), reverse flow check valve (2) and valve seat (3) with "O" rings (4 and 5). Install spring seat (6) with flat side down, spring (7), overspeed check valve (8) and valve seat (10) with "O" ring (9).

After gear pump unit is assembled, secure unit to piston pump housing plate and tighten retaining cap screws to 42-48 Ft.-Lbs. Install filter and valves assembly as outlined in paragraph 46L and manifold as outlined in paragraph 46K.

**46N. STROKE CONTROL.** With control cable and valve housing-to-stroke control line removed, the stroke control unit can be disassembled as follows:

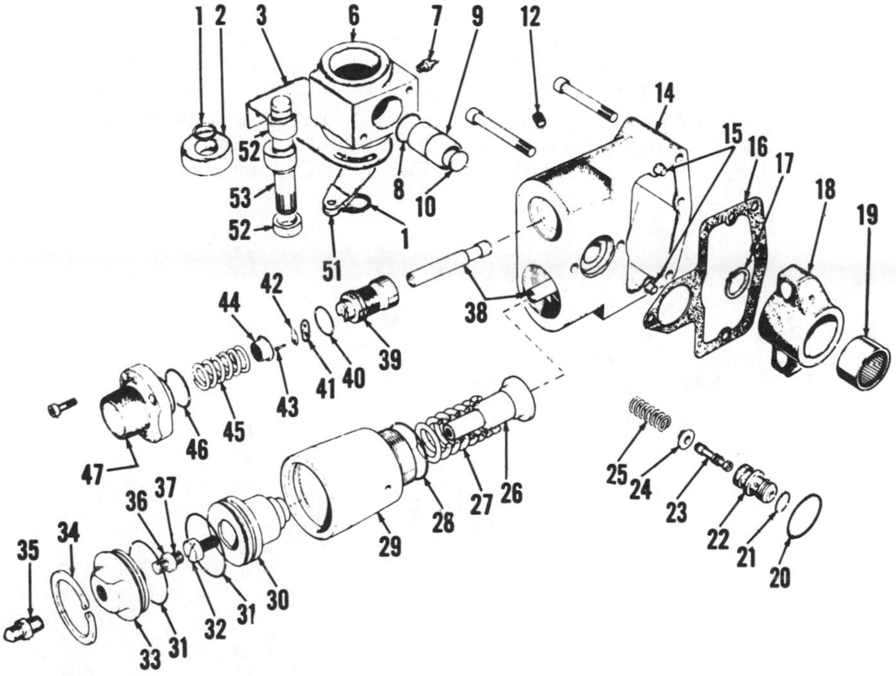

**Fig. O26H—Exploded view of stroke control used on hydraulic powered front wheel drive unit.**

1. Snap ring
2. Bearing support
3. Cable bracket
6. Cam housing
7. Grease fitting
8. "O" ring
9. Cam follower
10. Shim
12. Pipe plug
14. Housing
15. Dowels
16. Gasket

17. Snap ring
18. Rocker arm
19. Needle bearing
20. "O" ring
21. "O" ring
22. Valve sleeve
23. Valve pilot
24. Spring seat
25. Spring
26. Spring guide
27. Spring
28. "O" ring

29. Control cylinder
30. Control piston
31. "O" ring
32. Screw
33. End cap
34. Snap ring
35. Connector
36. Lock nut
37. Min. stroke adj. screw
38. Piston rods
39. Compensating piston

40. "O" ring
41. Reed valve
42. Back-up plate
43. Screw
44. Screen cap
45. Spring
46. "O" ring
47. End cap
51. Control arm
52. Needle bearing
53. Camshaft

Fig. O26J—Rocker arm is parallel with pump mounting surface when correctly installed.

Turn cylinder (29—Fig. O26H) counter-clockwise and remove cylinder assembly (items 26 through 37) from housing (14). Pull piston rod (38) from housing (14). Remove snap ring (34) and end cap (33) from cylinder. Remove retaining screw (32), then remove piston (30), spring (27) and spring guide (26) from cylinder. Discard "O" rings (28 and 31). Remove compensating cylinder end cap (47), then remove spring (45), screen cap (44), piston (39) and piston rod (38) from housing (14). Remove reed stop (42) and reed valve (41) from piston (39), if necessary. Discard "O" rings (40 and 46). Unbolt cam housing (6) from stroke control housing (14). Remove snap rings (1), place match marks on control arm (51) and camshaft (53), then press camshaft and bearing support (2) from cam housing. Remove bearing support (2) from camshaft. Remove cable support bracket (3). Remove camshaft bearings (52) if renewal is required. Remove cam follower (9), shims (10), spring (25) and spring seat (24) from stroke control housing, then remove stroke control housing from piston pump housing. Discard "O" rings (8 and 20) and gasket (16). Remove valve pilot (23) and valve sleeve (22) from bore in piston pump housing. Discard "O" ring (21). Rocker arm (18) and bearing (19) can be removed from control shaft if necessary by removing snap ring (17).

Clean and inspect all parts. Pay particular attention to springs (25, 27 and 45) and renew if any doubt exists as to their condition. Camshaft bearings (52) and rocker arm bearing (19) can be renewed, if necessary.

To reassemble stroke control, proceed as follows: Install rocker arm and bearing on control shaft so that rocker arm is parallel with mounting surface of piston pump housing as shown in Fig. O26J. Secure rocker arm with snap ring (17—Fig. O26H). Place new "O" ring (20) in counter-

bore, and new gasket (16) on dowels of housing (14), then install housing on piston pump housing and tighten retaining cap screws to 20-27 Ft.-Lbs. torque. Install spring seat (24), spring (25), shims (10) and cam follower (9) with new "O" ring (8). Install lower bearing (52) in cam housing (6) and upper bearing (52) in bearing support (2), then install bearing and bearing support on upper end of camshaft and secure with snap ring (1). Place camshaft and support and bearing assembly in cam housing, install control cable support bracket and control arm and align the previously affixed match marks on control arm and camshaft.

NOTE: If match marks were not affixed to control arm and control shaft prior to disassembly, place control arm on control shaft so that centerline of control arm is 37½ degrees clockwise from flat on end of control shaft.

Install retaining snap ring (1) to secure control arm. Tighten cam housing retaining cap screws to 20-27 Ft.-Lbs. torque. Be sure orifice in reed valve (41) is clear, then install reed valve and back-up plate (42) on compensating piston (39). Place connecting rod (38) in upper bore of housing (14) with end having annular groove next to rocker arm (18), then install compensating piston assembly screen cap (44), spring (45) and end cap (47) with "O" ring (46). Tighten end cap retaining cap screws to 10-12 Ft.-Lbs. torque. Place "O" ring (31) on piston (30) and install piston in cylinder

(29). Assemble spring (27) and spring guide (26); insert assembly in rear of cylinder (29) and secure to piston (30) with the retaining screw (32). Install connecting rod (38) in lower bore of housing (14), with end having annular groove next to rocker arm (18), then using new "O" ring (28), install cylinder assembly on housing (14). If not already done, loosen the minimum stroke adjusting screw lock nut (36), install connector (35) in end cap, then install end cap in cylinder. Secure end cap with snap ring (34) and be sure chamfered side of snap ring faces end cap.

NOTE: A minimum stroke adjustment must be made after hydraulic front wheel drive unit is completely assembled and installed. Refer to paragraph 46F.

460. **PISTON PUMP.** To disassemble and service the piston pump assembly, the hydraulic front wheel drive unit must be removed from tractor and the manifold, filter and valves assembly, gear pump assembly and stroke control valve assembly removed.

To disassemble the piston pump, proceed as follows: Remove snap ring and rocker arm from control shaft if not already done. Remove oil seal retainer (33—Fig. O26K) and gasket (32). Oil seal (1) can now be renewed. Support pump housing, drive end down, and remove all but two diagonally opposed port plate (17) retaining cap screws, then slowly loosen the remaining two cap screws to relieve pumping element spring pressure. Carefully remove port plate (17), cyl-

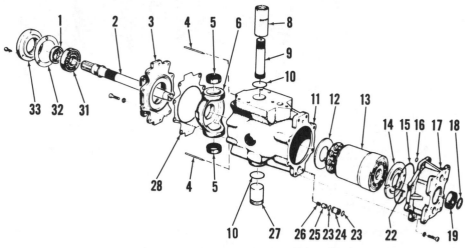

Fig. O26K—Exploded view of the hydraulic powered front wheel drive piston pump assembly showing component parts and their arrangement.

| | | | |
|---|---|---|---|
| 1. Oil seal | 10. "O" ring | 16. "O" ring | 25. Control circuit relief valve |
| 2. Drive shaft | 11. Housing | 17. Port plate | 26. Spring |
| 3. Housing cover | 12. Cam plate | 18. Snap ring | 27. Trunnion pin |
| 4. Lock pins | 13. Cylinder block assy. | 19. Bearing | 28. Gasket |
| 5. Bearing | 14. Valve plate | 22. Dowel | 31. Bearing |
| 6. Trunnion | 15. "O" ring | 23. "O" ring | 32. Gasket |
| 8. Trunnion pin | | 24. Valve guide | 33. Oil seal retainer |
| 9. Control shaft | | | |

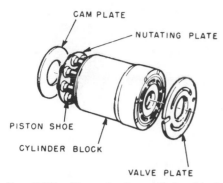

Fig. O26L—View showing cylinder block (piston pump) assembly. None of the piston pump parts are available separately.

the complete cylinder block assembly. Axial play of nutating plate (Fig. O26L) in piston shoes, or axial play of piston shoes in ball sockets must not exceed 0.005. If clearance exceeds that stated, renew complete cylinder block assembly. Examine surfaces of piston shoes that slide on cam plate and if any wear (more than 0.0001) exists, renew complete cylinder block assembly.

**46Q. CAM PLATE.** Carefully ex- amine cam plate. Fine scratches in running surface are permissible but if any wear or scoring (more than 0.0001) is present, renew the cam plate. **DO NOT** attempt to refinish cam plate.

**46R. VALVE PLATE.** Examine sur- face of valve plate that mates with cylinder block. Renew valve plate if running surface is 0.0005 or more lower than original surface. Also, carefully check the areas between the kidney-shaped slots. Any wear or scoring in these areas will result in poor efficiency even though other sur- faces are satisfactory. **DO NOT** at- tempt to refinish valve plate.

**46S. CONTROL SHAFT.** Check con- dition of control shaft splines. Splines should have an interference fit in trunnion and not more than 0.005 backlash with rocker arm.

**46T. ASSEMBLY.** If necessary, in- stall new needle bearings (5—Fig. O26K) in trunnion (6). Bearings are installed flush, or slightly below, outer edges of bearing bores. Install trun- nion pin "O" rings (10) in bores of pump housing, locate trunnion (6) in housing, then install lower trunnion pin (27) with tapped hole toward out- side and install lock pin (4). Place upper trunnion pin (8) on control shaft (9) with lock pin groove toward smallest end of control shaft. Position control shaft so machined step on trunnion end will provide clearance for cam plate when control shaft is installed in trunnion as shown in Fig.

Fig. O26N—When assembling piston pump install valve plate as shown. Refer to text.

O26M, then install rocker arm so it is parallel with the machined step and has stroke control connecting rod sockets toward cam plate side of trun- nion. Secure rocker arm with snap ring. Align lock pin groove of upper trunnion pin with lock pin hole, sup- port lower trunnion pin, then drive control shaft and upper trunnion pin into trunnion and housing and install lock pin. Install front bearing (19— Fig. O26K) in port plate (17). Install dowels (22), then place valve plate (14) over dowels and be sure the ar- row on side of valve plate is on the same side of port plate as the small "O" ring (16). See Fig. O26N. Lubri- cate valve plate liberally, set cylinder block assembly, port end down, on valve plate and align the drive shaft bores. Insert drive shaft (2—Fig. O26K) through cylinder block assembly, port plate and front bearing and secure shaft with snap ring (18). Place cam plate (12) in trunnion (6) with tap- ered surface of inside diameter facing rearward. Place "O" rings (23) on valve guide (24), then install spring (26), control circuit relief valve (25) and guide assembly in bore of pump housing. Install "O" rings (15 and 16) to port plate, then carefully install port plate and cylinder block assem- bly to pump housing. Tighten the re- taining cap screws to 100-115 Ft.-Lbs. torque. Install gasket (28) and cover (3) so that lock pin holes are covered and tighten retaining cap screws to 42-48 Ft.-Lbs. torque. Install rear bearing (31) on drive shaft. Install new oil seal (1) in seal retainer (33) with lip facing forward. Lubricate seal, then install gasket (32) and re- tainer on cover (3) and tighten re- taining cap screws to 10-12 Ft.-Lbs. torque.

Reinstall stroke control as in para- graph 46N, gear pump as in paragraph 46M, filter and valves as in paragraph 46L and manifold as in paragraph 46K. Install the assembled unit on tractor as outlined in paragraph 46J.

inder block (13) and drive shaft (2) as an assembly. Remove snap ring (18) and pull port plate assembly from drive shaft. Valve plate (14), "O" rings (15 and 16), dowels (22) and bearing can now be removed from port plate. Pull cylinder block (13) from drive shaft and stand it on a clean soft surface. Pull control cir- cuit relief valve assembly (items 23 through 26) from bore in housing. Re- move cam plate (12) from trunnion (6). Remove housing cover (3) and gasket (28) from pump housing and remove bearing (31) from cover. Re- move both lock pins (4). Thread a ¼-inch cap screw in tapped hole in lower trunnion pin and pull trunnion pin from housing. Insert a drift through lower trunnion hole and bump control shaft (9) and upper trunnion pin (8) from housing. Re- move "O" rings (10) from pump hous- ing. Needle bearings (5) can now be renewed if necessary.

With piston pump assembly disas- sembled, thoroughly clean all parts and inspect as follows:

**46P. CYLINDER BLOCK.** Use a straight edge to check the front face of cylinder block. There should be no signs of wear or scoring of the seal- ing lands around the kidney-shaped slots. Any difference in height be- tween outer land and the two inner sealing lands will require renewal of

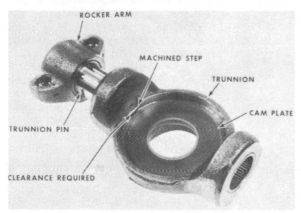

Fig. O26M—View show- ing relationship of rocker arm, control shaft and trunnion pin. Note po- sition of machined step on end of control shaft.

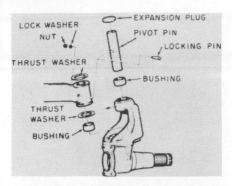

**Fig. O26P—Steering kunckle and pivot pin component parts. Refer to text for removal procedure.**

**Fig. O26S—When assembling front wheel drive motor, install dowels and valve plate as shown.**

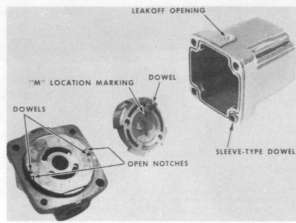

## SPINDLES AND DRIVE WHEELS

**46U. PIVOT PINS.** Front drive wheels spindles are fitted with renewable pivot pins and bushings which can be renewed as follows: Raise front of tractor and support front axle in raised position. Disconnect hoses from wheel motor and cap all openings. Remove wheel and axle outer member from axle center member.

Refer to Fig. O26P and remove expansion plug and pivot pin locking pin. Thread a ¾-inch puller in tapped hole in upper end of pivot pin and remove pivot pin. Remove axle outer member and thrust washers. Remove grease fittings and bushings from spindle.

When installing bushings, index holes in bushings with lube holes in spindle and install grease fittings. Place a thrust washer on each side of pivot pin hole of axle outer member, position axle member in spindle and install pivot pin with tapped hole toward top and locking pin groove next to locking pin hole. Install locking pin and a new expansion plug. Lubricate bushings with gun grease. Reinstall wheel and outer axle assembly, reconnect hoses to wheel motor and remove blocking.

**46V. DRIVE WHEEL MOTOR.** To remove drive wheel motor, first remove rock guard, then disconnect hoses from wheel motor and remove motor-to-clutch line. Drain lubricant from primary housing then remove housing cover (26—Fig. O26T). Turn

wheel hub until a rim attaching lug is in line with wheel motor, then pry primary drive gear away from wheel motor until bearing (48) contacts driven gear (29). Disconnect and pull motor from primary drive gear and housing. Remove spacer (46).

With wheel motor removed, refer to Fig. O26R and proceed as follows: Remove line fittings from motor, then clamp port plate (8) in a vise. Remove housing (27) and ramp (22) assembly and use caution not to allow thrust bearing (17) and thrust plates (16 and 18) to drop. Mark cylinder block (12) and drive shaft (5) so they can be reinstalled in the same position and remove cylinder block from drive shaft. Identify pistons (15) and their bores and remove pistons, spring seats (14) and springs (13). Mark valve plate (11) and port plate (8) so they can be reinstalled in the same position and remove valve plate and dowels (10) from port plate. Remove mounting flange (2) with oil seal (3) and "O" ring (4). Pull drive shaft (5) from port plate, then remove snap rings (6) and bearing (7). Lift piston plate from ramp (22). Remove seal ring (25) from housing. Remove screw (19) and lift ramp assembly from housing. Needle bearing (20) can be removed from ramp after removing snap ring (24) and screw seat (23).

Clean and inspect all parts. Bearing (9) can be removed from port plate (8) if necessary. Pay particular attention to pistons (15) and cylinder block (12). Pistons and/or cylinder block should not have any heavy scoring or wear. Renew valve plate if it shows signs of wear or scoring. Thrust bearing (17), plates (16 and 18) and ramp (22) should be in good condition with no wear, scoring or other damage.

Reassemble wheel motor as follows: Install ramp locating dowel in hole adjacent to "M" location marking as shown in Fig. O26S. If removed, press needle bearing (20—Fig. O26R) into

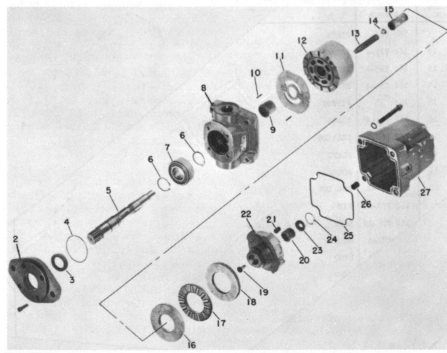

**Fig. O26R—Exploded view of front wheel drive motor. Refer to text when installing valve plate (11) and ramp (22).**

| | | |
|---|---|---|
| 2. Mounting flange | 9. Bearing | 15. Piston | 22. Ramp |
| 3. Oil seal | 10. Dowels | 16. Piston plate | 23. Screw seat |
| 4. "O" ring | 11. Valve plate | 17. Thrust bearing | 24. Snap ring |
| 5. Drive shaft | 12. Cylinder block | 18. Thrust plate | 25. Seal ring |
| 6. Snap ring | 13. Spring | 19. Retaining screw | 26. Dowel (sleeve) |
| 7. Bearing | 14. Spring seat | 20. Dowel | 27. Housing |
| 8. Port plate | | 21. Dowel | |

ramp (22) until end of bearing is flush with bearing bore, then install screw seat (23) with tapered side next to bearing and install snap ring (24). Place ramp in housing (27) so thickest side is toward the leakoff opening, insert drive shaft in ramp bearing and center drive shaft and ramp in housing, then install ramp retaining screw (19). Place thrust plate (18) on ramp with largest chamfer on inside diameter away from thrust bearing, then install thrust bearing (17) and piston plate (16). If necessary, install new bearing (9) in port plate so rear of bearing is flush with rear of bearing bore. Install snap rings (6) and bearing (7) on drive shaft (5) and install assembly in port plate. Install new seal (3) in mounting flange so lip of seal is facing bearing (7). Lubricate lip of seal and tape shaft splines, then install flange on port plate so that mounting hole lugs are 180 degrees from port plate inlet and outlet ports. Tighten mounting flange retaining cap screws to 12 Ft.-Lbs. torque. Install valve plate dowels (10) in port plate, then align previously affixed match marks and install valve plate (11) on dowels and be sure the open notches in valve plate are counter-clockwise from the dowels as shown in Fig. O26S. Install piston springs (13—Fig. O26R) in piston bores of cylinder block (12), tapered end first and hook springs to pins at bottom of bores. Place spring seats (14) on springs and install pistons (15). Align the previously affixed cylinder block and drive shaft marks and install cylinder block on drive shaft. If removed, install sleeve dowel (26) in housing at location shown in Fig. O26S. Install seal ring (25—Fig. O26R) in seal ring groove of housing. Clamp port plate assembly in a vise and install housing and ramp assembly so dowel (26) enters dowel counterbore of port plate. Tighten housing retaining cap screws to 37 Ft.-Lbs. torque. Reinstall all line fittings.

NOTE: When correctly assembled, wheel motor will have pressure port at top and leak-off port forward when unit is installed on tractor.

To install wheel motor on tractor, install spacer and loosely secure motor to primary gear housing. Install motor to clutch line. Bump primary reduction gear into housing until inner bearing bottoms, then reinstall primary gear housing cover and rock guard. Tighten housing cover cap screws to 30-35 Ft.-Lbs. torque. Tighten wheel motor cap screw and stud nut to 50-60 Ft.-Lbs. torque. Fill primary gear housing with lubricant,

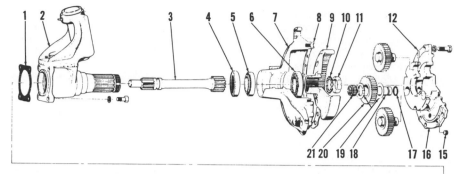

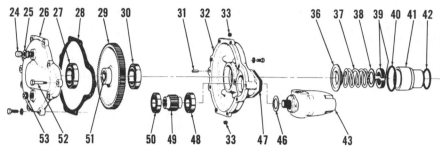

**Fig. O26T—Exploded view of front wheel reduction drive assembly. Refer to Fig. O26R for exploded view of drive motor (43).**

| | |
|---|---|
| 1. Gasket | 15. Pipe plug |
| 2. Steering knuckle | 16. Planet carrier |
| 3. Drive shaft | 17. "O" ring |
| 4. Oil seal | 18. Pinion pin |
| 5. Bearing assy. | 19. Thrust washer |
| 6. Bearing assy. | 20. Pinion |
| 7. Wheel hub | 21. Bearing |
| 8. Gasket | 24. Breather |
| 9. Internal gear | 25. Bushing |
| 10. Nut lock | 26. Housing cover |
| 11. Adjusting nut | 27. Bearing |
| 12. Roll pin | 28. Gasket |

| | |
|---|---|
| 29. Primary reduction gear | 40. "O" ring |
| 30. Bearing | 41. Clutch piston |
| 31. Dowel | 42. "O" ring |
| 32. Primary gear housing | 43. Drive motor |
| 33. Pipe plug | 46. Spacer |
| 36. Spring guide | 47. Gasket |
| 37. Spring | 48. Bearing |
| 38. Washer | 49. Primary drive gear |
| 39. Thrust washer halves | 50. Bearing |
| | 51. Bushing |
| | 52. Dowel bolt |
| | 53. Pipe plug |

connect all hoses and flush systems as outlined in paragraph 46H.

## DRIVE WHEEL

46W. To disassemble drive wheel assembly, proceed as follows: Remove wheel weights if so equipped. Raise front of tractor and support axle in raised position, then drain primary and planetary gear housings. Remove tire and rims. Remove planet carrier (16—Fig. O26T) from wheel hub (7). Straighten tabs of lock (10), then remove adjusting nut (11) and lock. Pull internal gear (9) from spindle and take care not to drop outer bearing (6) which generally is a loose fit on gear hub. If bearing is stuck, it can be forced from internal gear hub by using two $\frac{7}{16}$-inch cap screws in the tapped holes provided in gear. Remove wheel hub (7) from spindle and remove oil seal (4) and inner bearing (5). Bearing cups can be bumped from hub if necessary. Disassemble planet carrier by driving roll pins (12) into pinion shafts (18), driving pinion inward until roll pin holes are exposed and driving roll pins from pinion shafts. Pinion shafts can now be driven out toward outside of carrier and the pinions (20), pinion bearings (21) and thrust washers (19)

removed from carrier. Discard pinion shaft "O" rings (17).

Remove rock guard and disconnect inlet, outlet and leak-off lines from wheel motor. Remove wheel motor to clutch line and cap all openings. Remove primary gear housing cover (26), then remove primary gear housing (32) and gears (29 and 49) from steering knuckle. Remove gears from housing. Remove spring guide (36), spring (37) and washer (38), then pull on end of shaft (3) to release the thrust washer halves (39). Remove the thrust washer halves and pull shaft (3) from outer end of spindle. Clutch piston (41) and "O" rings (40 and 42) can now be removed from spindle. Remove bearings (48 and 50) from primary drive gear if necessary. Bearings (27 and 30) and bushing (51) can also be removed from primary reduction gear (29).

Clean and inspect all parts. Discard all "O" rings and gaskets. Renew any parts which show undue wear or scoring, or any other damage. If service is required on wheel motor, refer to paragraph 46V.

Reassemble drive wheel as follows: Install "O" rings (40 and 42) in knuckle, lubricate "O" rings freely and install clutch cylinder in knuckle (2). Insert drive shaft (3), small end first, through spindle and install thrust washer halves (39) in shaft groove so chamfered sides are toward clutch cylinder, then push shaft outward to seat thrust washer halves in clutch cylinder. Install flat washer (38), spring (37), spring guide (36) and gear housing (32). Be sure gear housing is installed so motor mounting is straight forward. If new bushing (51) is required, install bushing so outer end is flush with the fully splined end of gear hub. Install primary reduction gear inner bearing (30) so shielded side is toward clutch cylinder and on side opposite bushing. Install bearing (27) on opposite hub of primary reduction gear and bearings (48 and 50) on primary drive gear (49). Install both gear and bearing assemblies in housing and be sure driven gear (29) is installed with bushing (fully splined gear hub end) away from clutch housing. Be sure breather (24) is clean, then loosely attach cover (26) to housing (32). Install dowel (31) and dowel bolt (52). Install spacer (46) and motor (43) and be sure pressure port is on top and leak-off port is toward front. Install motor to clutch oil line, then tighten the wheel motor retaining cap screw and dowel bolt nut to 37 Ft.-Lbs. torque. Install rock guard, then tighten cover retaining cap screws to 30-35 Ft.-Lbs.

If new bearing cups are being installed in hub (7), install them with smaller I.D.'s to inside and be sure they are bottomed. Place bearing cone (5) in hub and install oil seal (4). Install bearing cone (6) on hub of internal gear (9). Drive hub assembly on spindle, place internal gear assembly on spindle then install lock (10) and adjusting nut (11) and adjust bearings as follows: Tighten adjusting nut until snug, then back-off until all pressure is released from bearings. Use a spring scale attached to rim lug hole and check effort required to keep hub rotating; then tighten adjusting nut until an additional 3½-6 pounds pull is required to keep hub rotating. Bend down tabs of lock (10) to hold nut in this position.

Place "O" rings (17) in grooves of pinion shafts (18). Place bearings (21) in bores of pinions (20), place a thrust washer (19) on each side of pinion gear, then insert pinion shaft in bore with roll pin hole aligned with roll pin hole in carrier web. Drive pinion shafts into position and install roll pins so outer ends are flush. Install planet carrier on hub and tighten

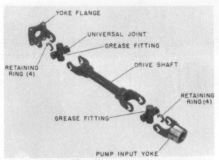

**Fig. O26U—Drive shaft for the hydraulic powered front wheel drive unit.**

retaining cap screws to 70-80 Ft.-Lbs. torque.

Connect hoses to wheel motor, then fill primary gear housing and planetary gear housing to recommended level with correct lubricant. Install tire and rim and tighten attaching bolts to 160-170 Ft.-Lbs. torque.

Flush system as outlined in paragraph 46H.

### DRIVE SHAFT

46X. Drive shaft for the hydraulic powered front wheel drive unit can be removed after disconnecting yoke flange (Fig. O26U) from crankshaft pulley adapter. Remove adapter from crankshaft after drive shaft is removed, if necessary.

Drive shaft universal joints can be disassembled after removing retaining rings, and are serviced with kits which include spider, bearing cups and retaining rings.

When reinstalling drive shaft, DO NOT secure the drive shaft to the front wheel drive piston pump shaft. To do so may limit engine crankshaft end play and could result in crankshaft bearing failure.

# POWER STEERING

### All Models So Equipped

47. All models having power steering utilize a Saginaw Steering Gear Hydramotor unit which, in direct relationship to turning the steering wheel, directs pressurized fluid in metered amounts to actuate the front wheel steering cylinder. (Two steering cylinders are used on four-wheel drive models.)

On models not equipped with a hydraulic lift system, pressure for the power steering system is supplied by a belt driven gear type pump mounted at front end of engine, and a separate power steering fluid reservoir is mounted on top of engine.

On models equipped with a hydraulic lift system, hydraulic fluid is utilized as the power steering fluid and pressure for the power steering system is supplied by the hydraulic pump via a flow divider valve mounted on the hydraulic lift housing.

### LUBRICATION (FLUID) AND BLEEDING

### All Models

48. On models not equipped with a hydraulic system, power steering fluid reservoir is located on a bracket attached to engine rocker arm cover

retaining bolts and reservoir filler pipe extends through opening in engine hood. Maintain fluid level at "FULL" mark on dipstick attached to reservoir filler cap. Recommended power steering fluid is SAE 10W Supplement 1 motor oil or "Type A" automatic transmission fluid.

On models with hydraulic lift system, refer to paragraph 233.

Steering system is self-bleeding and any air trapped in lines, cylinder or Hydramotor should be eliminated by turning steering wheel to move steering cylinder, or cylinders, through full range of travel several times in each direction.

### SYSTEM PRESSURE TEST

### All Models

49. On models with hydraulic lift system, refer to paragraph 243. On models without a hydraulic lift system, proceed as follows:

Install a pressure gage (2000 psi) and shut-off valve in the Hydramotor pressure line. With power steering fluid at operating temperature and the test gage valve open, turn steering wheel in either direction until the steering cylinder, or cylinders, are at extreme end of travel and hold the steering wheel in this position only long enough to observe pressure gage reading.

Maximum allowable pump relief pressure is either 1200 psi or 1600 psi, depending on whether tractor is equipped with early type single vane Hydramotor as shown in Fig. O29, or late type dual vane Hydramotor as shown in Fig. O37. The late type dual vane Hydramotor is factory installed on all 1550 and 1650 models and 1600 models after approximate serial No. 141 288-000, and may be installed on 1600 models prior to serial No. 141 288-000 as a field installation. Rotary valve Hydramotor is used on all 1555 and 1655 models.

Factory setting for power steering pump relief valve pressure was 1000-1100 psi prior to tractor serial No. 147 695-000 and 1500-1600 psi after this serial number on 1600 model tractors; factory setting for all 1550, 1555, 1650 and 1655 model tractors is 1500-1600 psi. To improve steering performance, it is recommended that the pump relief pressure be increased to 1200 psi for tractors equipped with early type Hydramotor, and to 1500-1600 for all tractors equipped with late type Hydramotor. CAUTION: Do not adjust pump relief pressure to above 1200 psi if tractor is equipped with Hydramotor having separate steel pivot bracket as shown in Fig. O29, or to above 1600 psi on tractors equipped with late type Hydramotor as shown in Fig. O37.

If pressure is less than specified, check pump by closing the shut-off valve only long enough to obtain a reading. If pressure is now approximately that specified the pump can be considered satisfactory, and the trouble is located in the power steering unit. If pump relief pressure remains the same as when checked with shut-off valve open, readjust pump relief pressure as follows:

Remove plug (1—Fig. O27) from pump and add or remove thin shims (3) to effect minor change in pump relief pressure; adding or removing one shim should change relief pressure approximately 75 psi. To change pump relief pressure from the 1000-1100 psi range to the higher 1500-1600 psi setting, a conversion package is available. The package includes six adjusting shims (3), a heavier relief valve spring (4), a different spacer shim (5), a heavy duty pump drive belt and a plug sealing "O" ring (2).

If pump relief pressure cannot be satisfactorily increased by adding shims, remove and overhaul the pump assembly as outlined in paragraph 50.

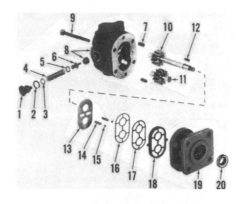

**Fig. O27 — Exploded view of separate power steering pump used on all models except late 1650 diesel and all 1655 diesel models.**

| | |
|---|---|
| 1. Plug | 11. Idler gear & shaft assy. |
| 2. "O" ring | 12. Key |
| 3. Shim, thin | 13. Diaphragm |
| 4. Spring | 14. Springs |
| 5. Shim, thick | 15. Check valve balls |
| 6. Valve poppet | 16. Backup gasket |
| 7. Pins | 17. Protector gasket |
| 8. Rear plate & valve seat assy. | 18. Diaphragm seal |
| 9. Capscrews | 19. Front plate |
| 10. Drive shaft & gear assy. | 20. Shaft seal |

## PUMP

### Models Without Hydraulic Lift

All models without hydraulic lift, except Series 1650 diesel models serial number 213 044-000 and up, and Series 1655 diesel models, use the separate power steering pump shown in Fig. O27. Series 1650 diesel models serial number 213 044-000 and up, and Series 1655 diesel models, use the double sheave, increased capacity pump shown in Fig. O27A.

**Fig. O27A — Exploded view of separate power steering pump used on late 1650 diesel and all 1655 diesel model tractors.**

| |
|---|
| 1. Plug |
| 2. "O" ring |
| 7. Pins |
| 8. Rear plate & valve seat assy. |
| 9. Capscrews |
| 10. Drive shaft & gear assy. |
| 11. Idler shaft & gear assy. |
| 12. Key |
| 13. Diaphragm |
| 14. Springs |
| 15. Check valve balls |
| 17. Backup gasket |
| 18. Diaphragm seal |
| 19. Front plate |
| 20. Shaft seal |
| 21. Spring, flow control |
| 22. Spring sleeve |
| 23. Body |

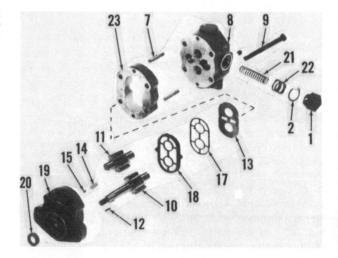

50. **R&R AND OVERHAUL.** To remove and overhaul the pump shown in Fig. O27, proceed as follows: Remove right hood panel (1600 and 1650 models), or left hood panel (1550 and 1555 models), disconnect pump to reservoir hose to drain reservoir and disconnect Hydramotor pressure tube at pump. Unbolt pump and mounting bracket from support and remove assembly from tractor.

Prior to disassembly, thoroughly clean exterior of pump and scribe assembly marks across mounting bracket, front plate (19—Fig. O27) and back plate (8). Remove pulley retaining nut, pulley and key (12) from pump drive shaft and remove mounting bracket from front plate. Remove back plate retaining capscrews (9) and bump end of drive shaft against wood block to separate the back plate from the front plate. Remove gear and shaft assemblies (10 and 11), diaphragm (13), gaskets (16 and 17), diaphragm seal (18), check springs (14) and check balls (15) from front plate. Using suitable tool, remove drive shaft seal (20). Remove plug (1) and relief valve components (2, 3, 4, 5 & 6) from back plate.

Thoroughly clean, dry and inspect all parts. Drive shaft and gear assembly (10) or idler gear and shaft assembly (11) should be renewed if shaft bearing diameter measures less than 0.4360, or gear width measures less than 0.566. (Gears are not renewable separately from shafts, although one gear and shaft assembly may be renewed without renewing the other.) Remove minor burrs from gear teeth with fine emery cloth.

**Fig. O28 — Install diaphragm with bronze side up and grooves towards low pressure side of pump as shown.**

Renew back plate if gear pocket diameter exceeds 1.1695, or wear at bottom of gear pocket is more than 0.001. Renew the back plate and/or front plate if bushing inside diameter is more than 0.4375. Back plate and relief valve seat (8) are available as an assembly only; renew back plate if seat is excessively worn or damaged. Relief valve components (items 1 through 6) are available as a repair kit which includes pins (7), or are available separately.

Reassemble as follows: Using a dull pointed tool, tuck diaphragm seal (18) into grooves in front plate with open part of "V" sections down. Drop the steel balls (15) into their bores and insert springs (14) on top of balls. Install protector gasket (17) and backup gasket (16) into diaphragm seal and install diaphragm over backup gasket with bronze face up and grooves toward low pressure side of pump as shown in Fig. O28. Install gear and shaft assemblies (10 and 11 —Fig. O27) in bores of front plate. Apply thin coat of heavy grease to milled faces of rear plate and front plate and install rear plate to front plate with previously affixed scribe marks aligned. Install retaining capscrews and tighten to torque of 7-10 Ft.-Lbs. Lubricate lip of seal (20), work seal over the drive shaft with lip inward and then drive the seal into front plate with suitable tools. Install relief valve poppet (6), thick shim (5), spring (4) and plug (1) with new "O" ring (2) and same number of thin (adjusting) shims as were removed. Install mounting bracket to front plate with previously affixed scribe marks aligned, then in-

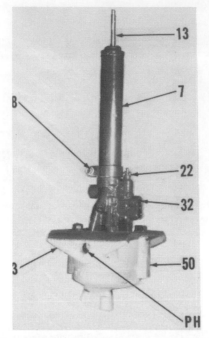

**Fig. O29 — View of early production Hydramotor steering unit removed from tractor. Pivot holes (PH) for unit are in separate steel bracket (3), while pivot holes for late production units are in the cover (50) casting; see (PH—Fig. O37).**

| | |
|---|---|
| PH. Pivot holes | 13. Adjuster bolt |
| 3. Pivot bracket | 22. Blocking valve |
| 7. Column jacket |     lockout |
| 8. Jacket clamp | 32. Housing |
| | 50. Cover |

stall key, pulley and pulley retaining nut.

Reinstall pump on tractor with test gage and shut-off valve installed as shown in Fig. O26, fill reservoir with proper fluid and, if necessary, adjust pump relief pressure as outlined in paragraph 49.

**50A. R & R AND OVERHAUL.** To remove and overhaul the pump shown in Fig. O27A, proceed as follows: Remove left front side panel, disconnect pump to reservoir hose to drain reservoir and disconnect Hydramotor pressure line. Unbolt pump bracket

and remove pump and bracket as an assembly.

Remove pulley nut, pulley and pulley key from pump shaft and remove bracket from pump. Remove plug (1— Fig. O27A) with "O" ring (2), sleeve spring (22) and flow control spring (21). DO NOT attempt to remove relief valve assembly. Place a scribe line across front plate (19), body (23) and rear plate (8), remove cap screws (9), then bump end of drive shaft against a wood block to separate pump. Remove gear and shaft assemblies (10 and 11), diaphragm (13), gasket (17), diaphragm seal (18), springs (14) and check balls (15). Remove shaft seal (20).

Check gear and shaft assemblies at bearing and seal surface. Renew shafts and gears as assemblies if wear is excessive or if damaged. Inspect bearing in both end plates and if bearings are excessively worn or damaged, renew end plates as bearings are not available separately. Renew pump body if it shows signs of erosion or other damage. Inspect relief valve, flow control spring and sleeve spring for wear, breakage or any other damage. Renew complete back plate if relief valve is worn or damaged.

To reassemble pump, install diaphragm seal (18) in front plate with open part of "V" sections down, using a dull pointed tool if necessary. Drop steel balls (15) in their bores and place springs (14) on top of balls. Install gasket (17) and diaphragm (13) into diaphragm seal and make sure diaphragm is positioned with bronze side up and grooves are on low pressure side of pump. See Fig. O28. Install dowel pins (7—Fig. O27A) and body (23) on front plate with previously affixed scribe line in register, then install gear and shaft assemblies. Apply a thin coat of heavy grease to face of rear plate, align scribe line and install back plate on pump body.

**Fig. O30 — Removing early production type Hydramotor unit out through instrument panel opening.**

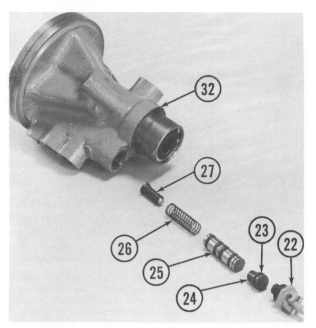

Fig. O31—Exploded view of Hydramotor housing and blocking valve components. Blocking valve can be removed without disassembling Hydramotor steering unit.

22. Lockout
23. "O" ring
24. Plug
25. Valve spool
26. Spring
27. Spring & guide assy.
32. Housing

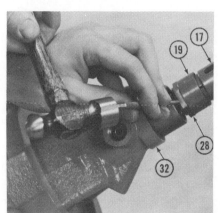

Fig. O32 — After installing steering shaft on Hydramotor stub shaft, stake hex nut to slot in steering shaft (17) with center punch.

17. Outer shaft
19. Tapered collar
28. Hex nut
32. Housing

Secure with retaining cap screws. Install springs (21 and 22) and cap (1) with new "O" ring (2). Coat lip of seal (20) with grease, carefully work seal over pump drive shaft and seat seal with a suitable driver. Install mounting bracket on pump, then install pulley key, pulley and pulley nut.

## HYDRAMOTOR STEERING UNIT

Hydramotor steering units used in tractors prior to tractor serial number 217 900-000 are referred to by Oliver Corporation as directional valve Hydramotors and are shown in composite exploded view in Fig. O33. Early production directional valve Hydramotors were fitted with a separate pivot (mounting) plate and have a maximum working pressure of 1200 psi. Late production units have pivot holes in unit cover and have a maximum working pressure of 1600 psi. Service on both early and late production units is similar; any differences in service will be noted in the following text.

Beginning with tractor serial number 217 900-000, a rotary valve Hydramotor is used and this unit is shown exploded in Fig. O47A. Refer to paragraph 56A for service information on this unit.

### All Models

51. **R&R HYDRAMOTOR STEERING UNIT.** First, disconnect the battery ground strap, then proceed as follows: Remove emblem from steering wheel adjuster lock knob, unscrew jam nut and the lock knob from adjuster bolt, remove flat washer from bolt and then pull steering wheel and inner shaft from steering column. Remove control panel hood and the instrument panel retaining screws; then, disconnect tachourmeter cable, snap-out the instrument panel lights, disconnect wiring from gages and remove the instrument panel. Loosen the outer nuts on steering column clamp and remove instrument panel slot "door" from column. Tag the four Hydramotor hoses for ease in reconnecting same, then disconnect hoses from unit and immediately cap all openings.

On late production units, loosen pivot pin lock nuts and unscrew the pivot pins until clear of the Hydramotor cover casting. Then, lift the unit out from opening in instrument panel support.

On early production units, drive the spring (roll) pin from governor control shaft arm, remove control shaft nut from steering unit R. H. support and pull control shaft from arm and the L. H. steering unit support. Be careful not to lose bushing from L. H. support as shaft is withdrawn. Loosen lock nuts on pivot pins and unscrew the pins until they are clear of the Hydramotor pivot bracket. Lift Hydramotor, turn it ¼-turn counter-clockwise, slip pivot bracket over front of the instrument panel opening and lift unit up out of opening as shown in Fig. O30.

Reverse removal procedures to reinstall either the early or late production unit. On models without hydraulic system, refill power steering reservoir as outlined in paragraph 48.

### Directional Valve Hydramotor

52. **R&R BLOCKING SPOOL VALVE.** The blocking spool valve and related parts can be removed and reinstalled after the Hydramotor steering unit has been removed as outlined in paragraph 51. Refer to Fig. O31 and proceed as follows:

Remove the lockout adjuster nut (22). Plug (24) and spool valve (25) may now be removed by pushing the plug into bore against spring pressure with screwdriver or other tool, then quickly releasing the plug to allow spring to pop it out of the bore. Remove plug and, if spool sticks in bore, invert the unit and tap housing (32) with soft faced mallet to jar spool out. Invert unit and allow spring (26) and the spring and guide assembly (27) to drop from bore.

If spool (25) is excessively worn or badly nicked, it should be renewed. A brightly polished wear pattern is normal. Spool should slide and rotate freely in bore.

To reassemble, install parts in bore of housing (32) as shown in exploded view, renewing the "O" ring (23) on plug (24) and tightening adjuster nut to a torque of 10-15 Ft.-Lbs.

### Directional Valve Hydramotor

53. **R&R STEERING COLUMN JACKET AND SHAFT ASSEMBLIES.** With Hydramotor unit removed as outlined in paragraph 51, proceed as follows:

Loosen clamp (8—Fig. O33) and pull column jacket assembly (7) from control valve housing (32). Hold steering shaft with large screwdriver inserted through slot in outer shaft (pull adjuster bolt up out of way) and unscrew the hex nut (28) until it nearly contacts control valve housing. Drive the tapered collar (19) towards nut until collar is loose, then turn collar until hole in collar is over locking ball hole in outer shaft (17), then shake the ball (20) out of hole. The outer shaft, tapered collar, hex nut and adjusting bolt (13) with nut (14) can then be removed from the stub shaft (37).

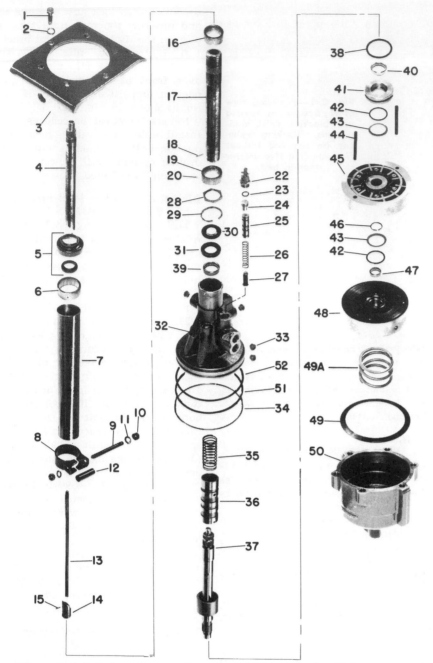

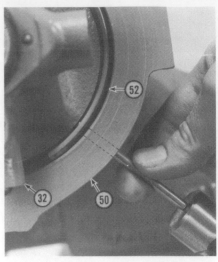

**Fig. O34—To remove cover (50) retaining snap ring (52) on late production units, drive pin or punch through hole in cover (dotted lines) to disengage snap ring in groove. Housing is (32).**

**Fig. O35 — In event coil pressure plate spring does not push cover from housing, tap around edge of cover with soft faced mallet.**

**Fig. O33 — Exploded view of composite early and late production Hydramotor steering unit. On early production units, cover (50) is retained to housing (32) by pivot bracket (3) and capscrews (1); "O" ring (51) is located in groove of housing and a Belleville washer (49) pressure plate spring is used. On late production models, cover is retained to housing by snap ring (52); "O" ring (51) and backup ring (34) are located in groove in I.D. of cover and a coil type pressure plate spring (49A) is used.**

| | | | |
|---|---|---|---|
| 3. Pivot bracket | 17. Outer shaft | 30. Dust seal | 43. Teflon seals |
| 4. Inner shaft | 18. Pin | 31. Oil seal | 44. Dowel pins |
| 5. Dust cap & scraper | 19. Tapered collar | 32. Housing | 45. Pump ring & |
| 6. Needle bearing | 20. Locking ball | 33. Connector seats | rotor assy. |
| (not serviced) | 22. Lockout assembly | 34. Backup ring | 46. Snap ring |
| 7. Column jacket | 23. "O" ring | 35. Spring | 47. Needle bearing |
| 8. Jacket clamp | 24. Plug | 36. Valve spool | 48. Pressure plate |
| 12. Spacer | 25. Valve spool | 37. Stub shaft assembly | 49. Spring (early |
| 13. Adjuster bolt | 26. Spring | 38. "O" ring | production) |
| 14. Adjuster nut | 27. Spring & guide assy. | 39. Needle bearing | 49A. Spring (late |
| 15. Pin | 28. Hex nut | 40. Needle bearing | production) |
| 16. Bearing race | 29. Snap ring | 41. Bearing support | 50. Cover (early |
| (not serviced) | | 42. "O" rings | production shown) |

If necessary to disassemble the steering shaft or column jacket assemblies, procedure for doing so is obvious. When installing new needle bearing (6), drive or press on lettered side of bearing cage only. Bearing race (16) is used on early production units only. Reassemble the unit before reinstalling on stub shaft as follows:

Engage splines of outer shaft on splines of stub shaft being sure that locking ball hole in outer shaft is away from flat on stub shaft. Align hole in tapered collar, hole in stub shaft and groove around the stub shaft, then drop locking ball in hole and groove and turn tapered collar ¼-turn. Tighten the hex nut to a torque of 20-30 Ft.-Lbs. and stake nut into slot in outer shaft as shown in Fig. O32.

**Early Production Directional Valve Hydramotor**

54. **R&R PIVOT (MOUNTING) BRACKET.** The pivot bracket (3—Fig. O33) retains cover (50) to control valve housing on early production Hydramotor units. With unit re-

Fig. O36 — To reinstall cover retaining snap ring (52), place unit in arbor press and push housing (32) into cover (50) with sleeve as shown.

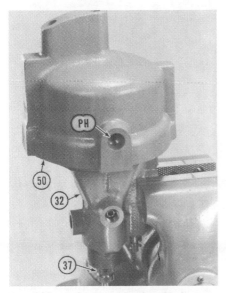

Fig. O37 — Late production Hydramotor mounted in vise for overhaul. Note pivot hole (PH) in cover (50).

PH. Pivot holes     37. Stub shaft
32. Housing          50. Cover

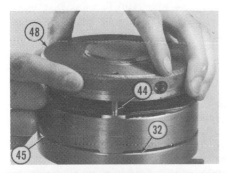

Fig. O38 — Removing pressure plate (48) from dowel pins (44).

32. Housing      45. Pump ring & rotor assy.
44. Dowel pins   48. Pressure plate

moved from tractor as outlined in paragraph 51, the pivot bracket can be removed as follows: Loosen capscrews (1) retaining pivot plate to cover and remove all the capscrews except the one where interference with control valve housing is encountered. Bump pivot plate loose and rotate the control valve housing within the cover so that remaining capscrew can be removed. Then remove the pivot plate. For further service, clamp the Hydramotor unit in vise as shown in Fig. O37 and refer to paragraph 56.

Reinstall pivot plate by reversing removal procedure. Tighten the plate retaining capscrews to a torque of 45-60 Ft.-Lbs.

## Late Production Directional Valve Hydramotor

55. **R&R COVER RETAINING SNAP RING.** On late production units, a snap ring (52—Fig. O34) is used to retain cover (50) to control valve housing (32). To remove the snap ring, first remove unit from tractor as outlined in paragraph 51, then proceed as follows: If end gap of snap ring is not near hole in cover as shown, bump the snap ring into this position with hammer and punch. Insert a pin or punch into hole and drive pin or punch inward to dislodge snap ring end from groove. With the pin or punch under the snap ring, use screwdrivers to pry ring from cover. After removing the snap ring, place Hydramotor unit in vise as shown in Fig. O37 and refer to paragraph 56 for further service procedure.

NOTE: Usually, spring (49A—Fig. O33) will push control valve housing from cover. However, if burrs or binding condition exists, it may be necessary to bump cover loose by tapping around edge of cover with mallet as shown in Fig. O35.

To reinstall the cover retaining snap ring, control valve housing must be held in cover against pressure from spring (49A—Fig. O33). It is recommended that the unit be placed in an arbor press and the housing be pushed into cover by a sleeve as shown in Fig. O36. CAUTION: **Do not** push against end of stub shaft (37—Fig. O37). Place snap ring over housing before placing unit in press. Carefully apply pressure on housing with sleeve until flange on housing is below snap ring groove in cover. Note that lug on housing must enter slot in cover. If housing binds in cover, **do not** apply heavy pressure; remove unit from press and bump cover loose with mallet as shown in Fig. O35. When hous-

Fig. O39 — Removing dowel pins (44) from pump ring and rotor assembly (45) and housing (32). Discard snap ring (46) after it is removed from stub shaft (37) and install new snap ring on reassembly.

Fig. O40 — Lifting the pump ring and rotor assembly (45) from stub shaft (37) and housing (32). Bearing support is (41).

Fig. O41 — Tap on end of stub shaft (37) with soft faced mallet to bump bearing support (41) out of housing.

ing has been pushed sufficiently far into cover, install snap ring in groove with end gap near hole in cover as shown in Fig. O34.

Fig. O42 — Lifting the Teflon seal (43) and "O" ring (42) from bearing support (41). Needle bearing (40) is serviced separately. Install new "O" ring in groove (G) before installing bearing support. Identical seals (42 and 43) are used in pressure plate.

Fig. O44 — To remove connector seats (33), thread inside diameter with tap as shown, then use puller bolts to remove seats.

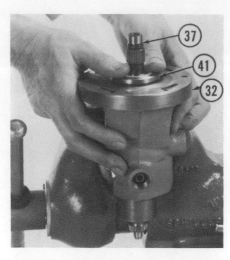

Fig. O46 — When pushing bearing support (41) for stub shaft (37) into housing (32), take care not to damage sealing "O" ring.

Fig. O43 — Removing the actuator assembly (stub shaft and control valve spool) from housing. Be careful not to cock spool in bore of housing.

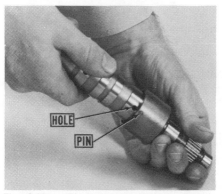

Fig. O45 — Pin in actuator sleeve must be engaged in hole in end of spool valve before actuator assembly is installed. If spool cannot be pulled out of sleeve, pin is engaged.

Fig. O47 — Be sure all vane springs are engaged behind the rotor vanes. Springs can be pried into place with screwdriver as shown.

## Directional Valve Hydramotor

56. **OVERHAUL HYDRAMOTOR STEERING UNIT.** With steering shaft and column unit removed as in paragraph 53 and the pivot bracket or cover retaining snap ring removed as outlined in paragraph 54 or 55, proceed as follows:

Remove cover (50—Fig. O37) by pulling upward with twisting motion. Remove the pressure plate spring, then lift off pressure plate as in Fig. O38. Note: Pressure plate on late production units is retained by two cap screws. Remove dowel pins as in Fig. O39, then remove snap ring (46) from stub shaft (37) with suitable snap ring pliers and screwdriver; discard snap ring. Pull pump ring and rotor assembly (45) up off of stub shaft as shown in Fig. O40. Tap end of stub shaft with soft faced mallet as shown in Fig. O41 until bearing support (41) can be removed, then carefully remove actuator assembly from housing as shown in Fig. O43. Note: It is recommended that the actuator assembly not be disassembled as it is a factory

balanced unit and serviced as an assembly only.

Housing (32—Fig. O33), blocking valve spool (25) and actuator assembly, which includes spring (35), spool (36) and stub shaft assembly (37) are serviced as a matched set only. If these parts are otherwise serviceable, needle bearing (39), seals (30 and 31), connector seats (33) and blocking valve components except the valve spool (25) may be renewed as necessary. Install new needle bearing (39) by pressing on lettered side of bearing cage only until bearing cage is flush with counterbore. Connectors (33) may be removed by threading inside diameter with tap as shown in Fig. O44, then removing connector with puller bolt. Refer to paragraph 52 for information on blocking valve unit.

Needle bearings (40 and 47—Fig. O33) in bearing support (41) and pressure plate (48) may be renewed if support and/or plate are otherwise serviceable. Install new bearings by pressing on lettered side of bearing cage only.

Rotor, ring, vanes and vane springs are serviced as a complete assembly

(45) only; however, the unit may be disassembled for cleaning and inspection. Reassemble by placing rotor in ring on flat surface. Insert vanes (rounded side out) in rotor slots aligned with large diameter of ring, turn rotor ¼-turn and insert remaining vanes. Hook the vane springs behind each vane with screwdriver as shown in Fig. O47; be sure that vane springs are in proper place on both sides of rotor.

To reassemble Hydramotor unit, place housing, with needle bearing, seals and snap ring installed, in a vise with flat (bottom) side up. Check to be sure that pin in actuator is engaged in valve spool; if spool can be pulled away from actuator as shown in Fig. O45, push spool back into actuator and be sure that the pin is engaged into hole in spool. Then, carefully insert actuator assembly into bore of housing. Place bearing support assembly on stub shaft and carefully push the assembly in flush with housing as shown in Fig. O46. Place the pump ring and rotor assembly on stub shaft and housing; on late production units, chamfered outer edge of pump ring must be away from housing (up). Install a **new** rotor retaining ring and insert the dowel pins through ring into housing. Stick the "O" ring and Teflon seals (See Fig. O42) into pressure plate with heavy grease, then install pressure plate over stub shaft, pump ring and rotor assembly and the dowel pins. NOTE: Beginning with approximate tractor serial number 194 491-000, two self locking bolts are used to retain pressure plate. If steering irregularities due to pressure plate displacement are encountered, install a new pressure plate (part no. 165 999-AS), rotor and ring set (part no. 166 001-A) and two retaining bolts (part no. 166 022-A).

When installing the two self locking bolts, tighten them to 96-120 In.-Lbs. torque, then back them off ¼-½ turn and be sure bolt heads do not protrude beyond outer circumference of pressure plate. With pressure plate and rotor and ring final positioned, and bolts backed off, about 10 In.-Lbs. torque should remain on the bolts. Place pressure plate spring on pressure plate. On early production units a Belleville washer type spring is used; cup side of spring must be away from pressure plate. A coil spring is used on all late production units. On early production units, install new sealing "O" ring in groove on outside diameter of housing. On late production units, install new "O" ring and backup ring in groove in cover (refer to Fig. O48). Then, install cover over the assembled steering unit and install pivot bracket on early production units as in paragraph 54, or cover retaining snap ring on late production units as in paragraph 55.

### Rotary Valve Hydramotor

56A. **OVERHAUL HYDRAMOTOR STEERING UNIT.** With unit removed as in paragraph 51, proceed as follows: Loosen lock nut (3—Fig. O47A) and remove jacket (2) and adapter (43) from Hydramotor. Remove lock nut and press dust cover and bearing (1) from jacket. Pull upward on outer steering shaft to remove rotary valve and steering shaft assembly from housing (32). Secure outer steering shaft assembly in a vise so hole in tapered ring (10) is upward, then loosen lock nut (11) and drive tapered ring toward lock nut. Remove shaft assembly from vise and rotate tapered

ring until hole in ring is over ball (12), then tap ball from shaft assembly. Separate shaft assembly from valve assembly, then remove nut (11), tapered ring (10), retaining spring (8), inner shaft (4), dowel pin (5), adjuster bolt (7) and nut (6) from outer shaft (9). Pull bearing plug assembly (19) from rotary valve assembly (23), then remove snap ring (14) and press seals (15 and 16) and bearing (17) from plug. Push valve spool and shaft out of valve body, unhook spool from shaft and slide spool from shaft. Remove "O" rings and backup rings from valve body. Remove "O" ring dampener from valve spool.

Position housing (32) with end plug (41) on top, then if necessary, position end of retaining ring (42) to within 15 degrees of hole in bottom side of housing. Use a ⅛-inch punch in hole in housing to unseat and remove retaining ring. Remove end plug (41) and spring (40). Remove backup ring (39) and "O" ring (38) from housing. Remove pressure plate (37), rotor assembly (36), thrust plate (35), drive shaft (24) and dowels (33) from housing. If necessary, push rotor through rotor ring to disassemble rotor vanes and springs. Remove snap rings (27), plugs (28), springs (30) and washers (31), then remove spool (32) from housing. Identify end position of spool (32) for reassembly.

56B. Discard all "O" rings, backup rings, seal rings and oil seals. Renew bearing (1 or 17) if rollers are broken, pitted or rusted. Renew shaft (4) if bearing surface is burred or pitted. Examine valve for signs of leakage and if leakage exists, renew complete valve assembly. Also inspect valve and valve shaft pins and grooves for undue wear or other damage. Small nicks or burrs can be removed from valve assembly by using a fine hone or crocus cloth, however do not remove any metal from valve spool or body. After installing teflon ring (26) on drive shaft (24), delay installation of shaft about 30 minutes to allow the teflon ring to shrink to correct size. Renew both transfer valve (32) and housing if valve or valve bore are deeply scored, but small nicks and burrs can be removed with crocus cloth. Use an air hose to check operation of pressure plate (37) valve. Air entering angled holes will cause valve to shift so that air comes out small center hole. If valve malfunctions, or if large burrs or deep scoring exists, renew pressure plate. Be sure that vanes slide freely in slots of rotor and bear in mind that vanes may appear to stick at extreme top and bottom of ro-

tor slots. If nicks or burrs cannot be removed from thrust plate (35) or end plug (41) without undue honing, renew parts.

NOTE: In some early models of the rotary valve Hydramotors, a problem developed in that displacement of the transfer valve plugs (28) and/or the plug in pressure plate (37) occurred. Beginning in June of 1970, this problem was corrected at the factory by redesigning the transfer plugs to make the retaining snap rings captive and the use of dowel pins in pressure plate to prevent pressure plate plugs from blowing out. These factory corrected units can be identified by a dab of white paint directly below hose ports on round surface of Hydramotor housing.

Should these problems be encountered in the field, an improvement package is available from the Oliver Corporation.

56C. To reassemble Hydramotor, proceed as follows: Install transfer spool, spring washers (31), springs (30) and plugs (28) with "O" rings (29) installed, then secure plugs with snap rings (27). If disassembled, install rotor, vanes and springs inside of rotor ring. Vanes are installed with rounded edges next to ring and springs must be installed in the same sequence on both sides of vanes. Install "O" ring (25) on drive shaft (24), then install teflon ring (26) over "O" ring and allow time for teflon ring to shrink to size. Place drive shaft in housing, use a 1 x 2 x 2½ inch long wood block to support shaft and turn housing bottom side up on bench. Install dowel pins (33) in housing. Install the two small "O" rings (34) in thrust plate (35), then with "O" rings toward control valve, align notches in plate with dowel pins and install plate in housing. Align dowel pin notches of rotor assembly (36) and pressure plate (37) and install in housing using care not to damage seals on drive shaft. Place "O" ring (38) and backup ring (39) in groove of housing with "O" ring next to pressure plate. Place large end of spring (40) on pressure plate, then install end plug (41) and secure with retaining ring (42).

NOTE: Face of end plug (41) has a cast recess and recess inside diameter is not concentric with outside diameter of plug. If interference with dowel pin is encountered during assembly, rotate plug until interference is eliminated.

Place "O" rings (22) on valve body and install seal rings (12) over "O" rings. Place dampener "O" ring (20) on valve spool, position spool so shaft engagement hole is downward and push spool in top of valve body until bottom of spool is flush with bottom of valve body. Insert spool shaft into bottom of spool and engage pin on shaft with engagement hole of spool. Align outer notch in shaft base with pin inside body and slide shaft in position.

Install bearing (17) in top of plug so that top of bearing is flush with lower edge of tapered flange. Install seals (15 and 16) with lips downward and largest seal on bottom. Install

snap ring (14) and "O" ring (18), then slide the bearing plug assembly over top of valve assembly shaft.

Install adjuster nut (6) on adjuster bolt (7), insert bolt, nut, dowel pin (5) and inner steering shaft (4) into outer steering shaft (9) and install spring retainer (8). Place outer shaft in a vise with ball hole upward and loosely install tapered ring (10) and nut (11). Align hole in tapered ring with hole in outer shaft, then insert valve shaft into outer steering shaft

and when ball groove in valve shaft aligns with holes in tapered ring and outer steering shaft, drop ball (12) into hole. Turn the tapered ring 180 degrees and tighten nut (11) to 40-50 Ft.-Lbs. torque. Install valve and steering shaft into housing and be sure notch in valve body aligns with pin on drive shaft.

If installing bearing (1), press bearing into jacket until top side is flush with end of jacket and install dust cover. Install lock nut (3), screw jacket into housing until it butts against plug

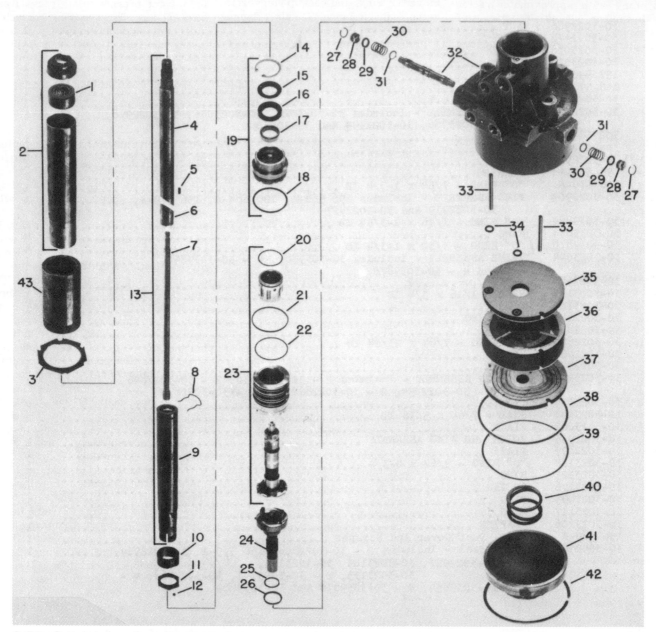

**Fig. O47A—Exploded view of rotary valve Hydramotor unit used on Series 1555 and 1655 tractors. Refer to Fig. O33 for an exploded view of the directional spool valve Hydramotor used on earlier tractors.**

| | | | | |
|---|---|---|---|---|
| 1. Bearing | 9. Outer steering shaft | 16. Oil seal (large) | 25. "O" ring | 34. "O" rings |
| 2. Jacket assy. | 10. Tapered ring | 17. Bearing | 26. Seal ring | 35. Thrust plate |
| 3. Nut | 11. Nut | 18. "O" ring | 27. Snap ring | 36. Rotor assy. |
| 4. Inner steering shaft | 12. Steel ball | 19. Plug assy. | 28. Plug | 37. Pressure plate |
| 5. Dowel | 13. Steering shaft assy. | 20. Dampener "O" ring | 29. "O" ring | 38. "O" ring |
| 6. Nut | 14. Snap ring | 21. Seal ring | 30. Spring | 39. Backup ring |
| 7. Adjuster bolt | 15. Oil seal (small) | 22. "O" ring | 31. Spring washer | 40. Spring |
| 8. Retainer spring | | 23. Rotary valve assy. | 32. Transfer valve & housing | 41. End plug |
| | | 24. Drive shaft | 33. Dowel pins | 42. Retainer ring |
| | | | | 43. Adapter |

assembly (19) then tighten jacket to 10 Ft.-Lbs. torque. Back-off jacket ⅛-¼ turn and tighten locknut to 50-100 Ft.-Lbs. torque. Reinstall tilt bracket on housing.

## POWER STEERING CYLINDER

### All Models Except Four Wheel Drive

**57. R&R POWER STEERING CYL-INDER.** On models equipped with oil cooler, disconnect oil cooler clamps from their supports and remove grille from main frame. Then, on all models, remove radiator as outlined in paragraph 138. Loosen power steering line clamps at side of engine, disconnect lines from power steering cylinder and immediately cap all openings. Refer to Fig. O52; remove plug (4) from cap (1) and remove cap from power steering cylinder. Pry plug (7) from pinion (11). Pull pinion from splines of front assembly shaft by turning puller bolt (9) counter-clockwise.

On Row Crop and Utility models, lift cylinder upward to disengage stud in cylinder from torque link, turn cylinder counter-clockwise on front assembly shaft and push cylinder downward to free pinion from shaft. Withdraw pinion from cylinder, then remove cylinder from front end of main frame.

On Wheatland and Industrial models, remove nut and washer from tapered stud at end of torque link and drive the stud upward out of main frame and the link. Then remove the power steering cylinder, with pinion installed, from front end of main frame.

Reverse removal procedures to reinstall cylinder. Punch mark on tooth of pinion (11) must be aligned with center tooth groove in piston as shown

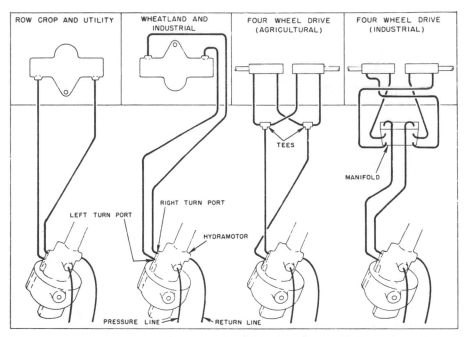

Fig. O49 — Schematic views showing Hydramotor and steering cylinder lines for the different front axle types.

Fig. O50 — View showing power steering cylinder on Row Crop and Utility models with cylinder cap and pinion plug removed. On Wheatland and Industrial models, cylinder is mounted to front side of front assembly shaft instead of to rear as shown.

H. Hole for removing snap ring
9. Puller bolt
11. Pinion
14. Cylinder

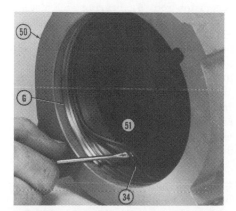

Fig. O48 — On late production units "O" ring (51) and backup ring (34) are installed in cover (50); be sure backup ring is to outside (open side of cover). Groove (G) is for cover retaining snap ring.

in Fig. O51. Tighten the puller bolt to a torque of 200 Ft.-Lbs., and tighten the nuts on torque link and tapered stud securely. Install plug (7—Fig. O52) with new "O" ring (6). Fill cavity around pinion with proper power steering fluid (see paragraph 48) before reinstalling cap (1) with new "O" ring (5).

**58. OVERHAUL POWER STEER-ING CYLINDER.** After removing the cylinder assembly as outlined in paragraph 57, proceed as follows: On Wheatland and Industrial models, remove pinion (11—Fig. O52). Remove snap ring (8), puller bolt (9) and washer (10). To remove end plug retaining snap ring (24), insert pin or punch in hole (H—Fig. O50) and push snap ring from groove in cylinder.

Grasp one of the lugs on end plug with vise-grip pliers and, with a twisting motion, pull end plug from cylinder. Remove plug sealing "O" ring (22—Fig. O52) from end of cylinder and remove the piston (21). Remove the Teflon seals (16) and "O" rings (17) from groove at each end of piston and remove pinion sealing quad ring (12) from bottom of cylinder. If necessary to remove the check valve assembly (18, 19 and 20) from either end of cylinder, insert punch in hole from inner side of piston head and drive valve ball, spring and spring pin from piston. If the hydraulic line connector seats (13) are damaged, they may be removed in a manner similar to that shown in Fig. O44. Carefully clean and inspect all parts and renew any that are excessively

**Fig. O51 — View of pinion (11) and piston (21) removed from cylinder to show timing marks (TM). Puller bolt (9) is retained in pinon with snap ring. White ring is Teflon seal ring (16).**

worn, deeply scored or damaged beyond further use.

Reassemble cylinder, using all new sealing rings as follows: Insert check ball (20) into bore in end of piston (21), insert spring (19) with small end next to ball and drive the retaining spring pin (18) in flush with end of piston. Install "O" ring (17) in bottom of groove in end of piston, then install Teflon ring (16) on top of the "O" ring. Lubricate piston and install in bore of cylinder. Install "O" ring (22) in groove in open end of cylinder, lubricate and install end plug (23) and install plug retaining snap ring (24). Install quad ring (12) in bottom of pinion bore in cylinder casting and lubricate ring and bore. Install pinion (11), and washer (10), puller bolt (9) and retaining snap ring (8) in top of pinion. On Wheatland and Industrial models, insert pinion in cylinder with timing marks on pinion and piston aligned as shown in Fig. O51. Reinstall cylinder as outlined in paragraph 57.

## POWER STEERING CYLINDERS

### Four Wheel Drive Models

**59. R&R AND OVERHAUL.** To remove either power steering cylinder, disconnect the hoses, cap all openings, then remove the bolts from cylinder end assemblies and remove the cylinder from tractor. Loosen the lock nut on rod end of assembly and the clamp on cylinder end, then remove the end assemblies. Bushings in either end assembly may be renewed if worn or damaged.

To disassemble cylinder, refer to Fig. O53 and proceed as follows: Remove end plate (19), push bearing

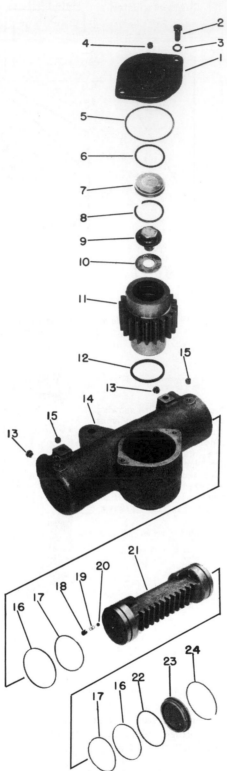

**Fig. O52 — Exploded view of power steering cylinder assembly used on all models except those equipped with four wheel drive.**

| | |
|---|---|
| 1. Cap | 14. Cylinder |
| 4. Plug | 15. Plugs |
| 5. "O" ring | 16. Teflon rings |
| 6. "O" ring | 17. "O" rings |
| 7. Plug | 18. Spring pins |
| 8. Snap ring | 19. Springs |
| 9. Puller bolt | 20. Check valves |
| 10. Washer | 21. Piston |
| 11. Pinion | 22. "O" ring |
| 12. Quad ring | 23. End plug |
| 13. Connectors | 24. Snap ring |

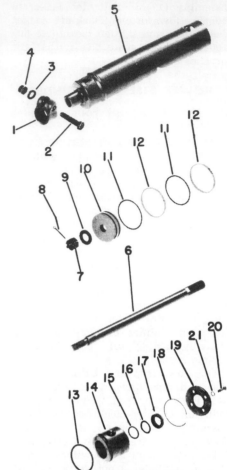

**Fig. O53 — Exploded view of power steering cylinder used on four wheel drive models. Refer to Fig. O49 for schematic view of cylinder line connections.**

| | |
|---|---|
| 1. Clamp | 13. "O" rings |
| 5. Cylinder barrel | 14. Bearing |
| 6. Piston rod | 15. "O" ring |
| 7. Slotted nut | 16. Backup ring |
| 9. Washer | 17. Wiper |
| 10. Piston | 18. Snap ring |
| 11. "O" rings | 19. Retainer plate |
| 12. Teflon rings | |

(14) into cylinder to expose snap ring (18) and remove the snap ring. Pull the piston rod (6), bearing and piston (10) from cylinder as an assembly. Hold rod from turning at flat near outer end, then remove cotter pin and piston retaining nut (7). Slide piston and bearing from inner end of rod. Remove the sealing rings and wiper from piston and bearing. Clean and carefully inspect all parts and renew any that are excessively worn, scored or damaged.

Reassemble, using all new seals, as follows: Install "O" ring (15) and backup ring (16) in bore of bearing with backup ring at outer side of "O" ring. Install wiper (17) with lip outward, lubricate bore of bearing and install bearing on piston rod with "O" ring groove on outer diameter of bearing towards small (inner) end of rod. Install new "O" ring (13) on bearing. Install new "O" rings (11) in bottom of grooves on piston and install the Teflon rings (12) on top of the "O" rings. Install piston, washer (9) and nut (10) on piston rod and tighten nut securely. Note: If nut threads onto piston rod too far and slots in nut will not engage cotter pin, remove the nut and add washers

(9) as necessary. Install cotter pin (8), lubricate piston and cylinder bore and carefully insert piston and bearing into bore but avoid damaging sealing rings on snap ring groove. Using a dull tool, compress the sealing rings as they pass hole for hydraulic fitting in cylinder tube. Push bearing far enough into bore to install snap ring (18), bump bearing out against snap ring with piston and install plate (19).

Before reinstalling cylinder, install the end assemblies and, with cylinder rod fully retracted, adjust position of end assemblies so that center-to-center measurement of the mounting holes is 18½ inches. Then, tighten the lock nut and clamp to secure end assemblies in this position and install cylinder on tractor. Refer to Fig. O49 for line connection diagram.

# ENGINE AND COMPONENTS

Oliver 1550, 1555, 1650 and 1655 Series tractors are fitted with six cylinder engine. Series 1550 and 1555 non-diesel and diesel engines have a bore and stroke of 3.625 x 3.750 and a displacement of 232 cubic inches. Early 1600 non-diesel engines have a bore and stroke of 3.500 x 4.000 inches and displacement of 231 cubic inches. Late 1600 non-diesel engines have a bore and stroke of 3.625 x 4.000 inches and displacement of 248 cubic inches. Series 1650 and 1655 non-diesel engines have a bore and stroke of 3.750 x 4.000 inches and a displacement of 265 cubic inches. Series 1600 diesel engines have a bore and stroke of 3.750 x 4.000 inches and displacement of 265 cubic inches. Series 1650 and 1655 diesel engines have a bore and stroke of 3.875 x 4.000 inches and a displacement of 283 cubic inches. Series 1550, 1555, 1600, 1650 and 1655 non-diesel engines and 1550, 1555 and 1600 diesel engines have a four-main bearing crankshaft; Series 1650, and 1655 diesel engines have a seven-main bearing crankshaft.

## R&R ENGINE ASSEMBLY

On direct drive models, the engine clutch can be removed without removing engine from tractor, or the engine and clutch assembly may be removed as a unit. On models equipped with Hydra-Power, Hydraul-Shift, Creeper or Reverse-O-Torc Drive, the engine and drive unit must be removed as an assembly, then the drive unit removed from engine for service of either the engine or drive unit and/or clutch.

60. To remove engine and clutch, Hydra-Power Drive, Hydraul-Shift, Creeper Drive or Reverse-O-Torc Drive as a unit, first disconnect battery ground strap, then proceed as follows:

Drain cooling system, and if engine is to be disassembled, drain oil pan. On models equipped with Reverse-O-Torc Drive, drain the drive unit. On models equipped with Hydra-Power, Hydraul-Shift or Creeper Drive, drain drive unit if unit is to be disassembled.

If so equipped, remove the single speed PTO drive shaft and clutch unit as outlined in paragraph 221. On models so equipped, remove dual speed PTO drive shaft as outlined in paragraph 226, or remove hydraulic pump drive shaft as outlined in paragraph 248.

Remove both hood side panels and hood. Remove fan assembly from water pump and set fan in radiator shroud. Disconnect the radiator hoses and if tractor is equipped with Hydra-Power, Hydraul-Shift, Creeper or Reverse-O-Torc Drive, remove the radiator and fan from tractor. Note: On those models with cooler in front of radiator, disconnect oil lines and remove cooler. Shut-off the fuel supply valve, disconnect fuel line and, on diesel models, disconnect fuel return line. Unhook fuel gage wire and remove fuel tank. Disconnect wiring from engine accessories. Disconnect or remove the diesel fuel injection pump control rod and stop wire or the governor (throttle) rod and choke wire.

On manual steering models, disconnect steering shaft support bearing from bracket at clutch housing.

On models equipped with auxiliary drive units which use radiator mounted oil coolers, disconnect the oil cooler lines at drive unit.

On power steering models not equipped with hydraulic lift, disconnect Hydramotor lines from power steering pump and reservoir.

On direct drive models, except 1550 and 1555 remove drive shaft shield, disconnect coupling chain, disengage snap ring from groove in shaft at front of coupling, slide snap ring and front coupling half forward, then remove rear coupling half from between transmission and rear end of drive shaft. Front half of coupling can then be removed from drive shaft. On 1550 and 1555 models remove drive shaft shield, then disconnect clutch shaft

coupling from transmission input shaft.

On Hydra-Power, Hydraul-shift, Creeper or Reverse-O-Torc Drive, remove chain coupling from between drive unit and transmission.

On Series 1550, 1555, 1650 and 1655 models, remove the dry type air cleaner from top of engine. Remove other engine accessories as indicated by service to be performed on engine or drive unit.

Attach lifting hooks to the lift bolts on engine head and attach hoist to lift hooks. Remove the four bolts retaining engine to main frame and lift engine and clutch or drive unit from tractor. NOTE: The left front and right rear engine retaining bolts are dowel bolts and must be reinstalled in same position from which they were removed.

To reinstall engine, reverse removal procedure. Bleed the diesel fuel system as in paragraph 113 or 113A.

## R&R CYLINDER HEAD

### All Models

61. To remove cylinder head, first remove both hood side panels and engine hood, drain the cooling system, then proceed as follows:

On power steering models not equipped with hydraulic lift, drain the power steering reservoir, remove Hydramotor to pump and reservoir hoses and remove the pump and reservoir assemblies from engine.

Remove upper radiator hose, air cleaner and air cleaner hose. On models with Hydra-Power Drive, (early type) disconnect Hydra-Power coolant hose at front of cylinder head and from bracket on rocker arm cover. Disconnect or remove breather tube from rocker arm cover. On diesel models, disconnect wiring from manifold heater solenoid. On non-diesel models, remove ignition coil, disconnect spark plug wires and remove the

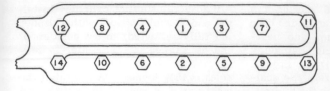

Fig. O55 — Cylinder head tightening sequence for Series 1650 indirect injection diesel engines only. Tighten all capscrews to a torque of 129-133 Ft.-Lbs. Oil screw is No. 11.

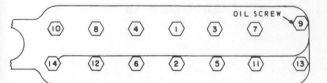

Fig. O55A — Cylinder head tightening sequence for Series 1650 and 1655 direct injection diesel engines only. Tighten cap screws 1 through 6 to 150 Ft. Lbs. torque and cap screws 7 through 14 to 133 Ft.-Lbs. torque.

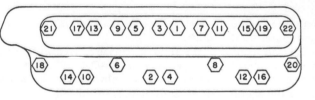

Fig. O56 — Cylinder head capscrew tightening sequence for Series 1600 and Series 1650 and 1655 non-diesel engines. Tighten the oil cap screw (22) to a torque of 96-100 Ft.-Lbs. and tighten all other capscrews to a torque of 112-117 Ft.-Lbs.

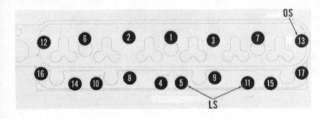

Fig. O56A — Cylinder head capscrew tightening sequence for Series 1550 and 1555 non-diesel engines. Tighten the oil cap screw (13) to a torque of 96-100 Ft.-Lbs. and tighten all other cap-screws to a torque of 112-117 Ft.-Lbs.

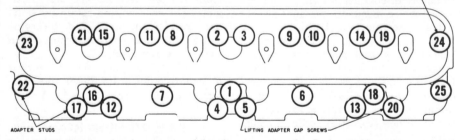

Fig. O57—Cylinder head capscrew tightening sequence for Series 1550, 1555 and 1600 diesel engines. Tighten the oil capscrew (24) to a torque of 96-100 Ft.-Lbs. and all other capscrews to a torque of 112-117 Ft.-Lbs.

carburetor, intake manifold and exhaust manifold as an assembly. On diesel models, disconnect injector pressure lines and the fuel return lines; immediately cap all openings.

Remove rocker arm cover and disconnect rocker arm shaft oil line. Remove rocker arm shaft assembly from cylinder head, then remove cylinder head from engine.

62. When reinstalling cylinder head, refer to Figs. O55 through O57 for location of oil screw and tightening sequence of cylinder head cap screws.

On all models except 1650 and 1655 diesel, tighten oil screw to 96-100 Ft.-Lbs. torque and all other cap screws to a torque of 112-117 Ft.-Lbs.

On Series 1650 indirect injection diesel engine models, tighten all cyl-

inder head cap screws to 129-133 Ft.-Lbs. torque. On Series 1650 and 1655 direct injection diesel engine models, tighten cap screws 1 through 6 (Fig. O55A) to 147-150 Ft.-Lbs. torque and cap screws 7 through 14 to 129-133 Ft.-Lbs. torque.

Rocker arm shaft bracket nuts and manifold nuts on all models should be tightened to a torque of 25-27 Ft.-Lbs.

## VALVES, GUIDES AND SEATS

### All Models

63. Intake and exhaust valves on all models except Series 1650 and 1655 direct injection engine models, have a face angle of 44½° and a seat angle of 45° Series 1650 and 1655 direct injec-

tion engine models use an intake valve with a face angle of 29½° with a seat angle of 30°.

Desired seat width for Series 1550 and 1555 engines is 0.056-0.066 for all intake valves, 0.080-0.090 for non-diesel exhaust valves, or 0.066-0.076 for diesel exhaust valves.

Desired seat width for Series 1650 and 1655 direct injection engines is 0.087-0.097 for intake valves, or 0.062-0.072 for exhaust valves.

Desired seat width for all other models is 0.080-0.090.

Exhaust valve seat inserts are standard equipment on all non-diesel engines and intake valve seat inserts are standard equipment on Series 1550, 1555 and 1600 diesel engines. Series 1650 diesel engines prior to tractor serial number 187 586-000 have both intake and exhaust valve seat inserts. Series 1650 tractor, serial number 187 586-000 and up, and series 1655 tractors are not equipped with valve seat inserts. Inserts are installed with a 0.002-0.0035 interference fit.

Early production non-diesel intake valves were provided with neoprene oil guards to prevent oil from passing into the combustion chamber via the valve stems. Late production non-diesel valve guides are fitted with Perfect Circle valve stem seals. Install new oil guards or valve stem seals each time the valves are reseated. Refer to Fig. O58 for Perfect Circle seal installation procedure. Early production valve guides can be fitted with Perfect Circle seals after machining the guide with proper tools. Refer to paragraph 64 for proper valve guide height.

Adjust valve tappet gap cold to the following values:

Intake, 1550-1555
  Non-Diesel .......... 0.009-0.011
Intake, 1600
  Non-Diesel .......... 0.013-0.015
Intake, 1650-1655
  Non-Diesel ..........0.019-0.021
Intake, 1550-1555, 1600
  Diesel ............. 0.009-0.011
Intake, 1650 Diesel (Indirect
  Injection) ...........0.019-0.021
Intake, 1650 Diesel (Direct
  Injection) -1655 ......0.029-0.031
Exhaust, 1550-1555
  Non-Diesel .......... 0.015-0.017
Exhaust, 1600
  Non-Diesel .......... 0.022-0.024
Exhaust, 1650-1655
  Non-Diesel ..........0.029-0.031
Exhaust, 1550-1555, 1600
  Diesel .............0.019-0.021
Exhaust, 1650 Diesel (Indirect
  Injection) ...........0.029-0.031
Exhaust, 1650 Diesel (Direct
  Injection) -1655 ......0.029-0.031

64. Intake and exhaust valve guides are not interchangeable. Intake guides for Series 1550 and 1555 non-diesel engines are $2\frac{13}{32}$ inches long. Intake guides for all other non-diesel engines are $2\frac{25}{32}$ inches long ($2\frac{5}{8}$ inches when equipped with Perfect Circle seals). Intake guides for Series 1550, 1555 and 1600 diesel are $3\frac{3}{16}$ inches long. Intake guides for Series 1650 diesel, with indirect injection, are $2\frac{1}{4}$ inches long, whereas intake guides for Series 1650 and 1655 diesel, with direct injection, are $3\frac{1}{2}$ inches long.

Exhaust guides for Series 1550 and 1555 non-diesel are $2\frac{1}{2}$ inches long. Exhaust guides for Series 1600, 1650 and 1655 non-diesel are $2\frac{5}{8}$ inches long. Exhaust guides for Series 1550, 1555 and 1600 diesel are $2\frac{5}{16}$ inches long. Exhaust guides for Series 1650 diesel, with indirect injection, are $2\frac{5}{8}$ inches long, whereas exhaust guides for Series 1650, with direct injection and 1655 diesel are $3\frac{1}{2}$ inches long.

When properly installed, valve guide height above machined surface of cylinder head, or valve spring counterbore, is as follows: Non-diesel intake (without Perfect Circle seal), $\frac{15}{16}$-inch. Non-diesel intake with Perfect Circle seal (not including seal), $\frac{27}{32}$-inch. Non-diesel exhaust and diesel intake (except Series 1650 with direct injection), $\frac{15}{16}$-inch. Series 1600 diesel exhaust (with indirect injection), $\frac{15}{16}$-inch. Series 1650 diesel intake and exhaust (with direct injection), and 1655, $\frac{7}{8}$-inch.

New guides for all series except 1650 (with direct injection) and 1655, have a nominal inside diameter of 0.374-0.375 with a maximum allowable inside diameter (wear limit) of 0.377 for intake valve guides and 0.378 for exhaust valve guides. Guides for series 1650 (with direct injection) and 1655 with spiral groove end have an inside diameter of 0.373-0.374 and those with plain end, 0.3745-0.3757.

## VALVE ROTATORS

### All Models

65. Exhaust valves of gasoline engines and both intake and exhaust valves of Series 1600 and 1650 diesel engines are equipped with positive type valve rotators. Only exhaust valve of Series 1550 and 1555 diesel engines are equipped with rotators. These rotators cannot be serviced but should be observed while engine is running to be sure each valve equipped with rotator turns slightly each time the valve is opened. Renew the rotator of any valve that does not turn.

Fig. O58 — Views A through E illustrate method of installing Perfect Circle valve stem seals on models so equipped. Photos courtesy of Perfect Circle Corporation.

## VALVE SPRINGS

### All Models

66. Valve springs of three different free lengths are used. Usage and test specifications are as follows:

**106 573-A**—Used as an intake valve on 1550, 1555, 1650, 1600 and 1655 gasoline and 1655 diesel engines; exhaust valve on 1550, 1555, 1600, 1650 direct injection and 1655 diesel en-

gines. Also used on all valves of LP-Gas engines.

Free length...................2.562
Lbs. pressure @ 1.906 in.....55-63
Lbs. pressure @ 1.506 in.....91-99

**107 625-A** — Used as intake valve spring on Series 1600 diesel engines and as exhaust valve spring on Series 1600 gasoline, and as intake and exhaust valve springs on Series 1650 indirect injection diesel engines.

Free length...................2.667
Lbs. pressure @ 1.750 in.....53-61
Lbs. pressure @ 1.350 in.....86-94

**159 026-A** — Used as exhaust valve spring on Series 1550, 1555, 1650 and 1655 gasoline engine and as an intake valve spring on Series 1550, 1555 and 1650 direct injection diesel engines.

Free length...................2.440
Lbs. pressure @ 1.750 in.....53-61
Lbs. pressure @ 1.350 in.....86-94

## VALVE LIFTERS (CAM FOLLOWERS)

### All Models

67. The mushroom type valve lifters operate directly in machined bores of cylinder block. Valve lifters are supplied only in standard size of 0.6240-0.6245 and should have an operating clearance of 0.0005-0.002 in their bores. Renew valve lifters if their diameter is less than 0.619. If valve lifter bore exceeds 0.631, renew cylinder block.

Valve lifters can be removed from cylinder block after removing camshaft as outlined in paragraph 74.

## VALVE ROCKER ARMS AND SHAFTS

### All Models

68. The rocker arm shaft assembly can be removed as follows: Remove hood side panels and hood. On Series 1550 and 1650 models, remove the dry type air cleaner. On power steering models not equipped with hydraulic lift, remove the power steering reservoir. Disconnect or remove the rocker arm breather and, on non-diesel models, remove the ignition coil. Remove rocker arm cover and disconnect the oil line. Then, unbolt and remove rocker arm shaft assembly.

Disassembly of rocker arm shaft assembly is obvious. Note that rocker arms are right and left hand assemblies and that offset valve contact ends are towards shaft support brackets.

Rocker arm shaft diameter, new, is 0.742-0.743; renew shaft if worn to 0.740 or less. Inside diameter of rocker arm bushing is 0.7445-0.7455 new. Desired operating clearance is 0.0015-0.0035; if clearance exceeds 0.005, renew rocker arm assembly as bushings are not available separately from rocker arms. Rocker arm contact radius can be refaced providing the original radius is maintained and face is kept parallel with rocker arm shaft.

The hollow rocker arm shaft is fitted with a sealing plug in each end and a restrictor plug near mid-length of shaft bore. The plugs are serviced separately from shaft and should be renewed if removed to clean the shaft bore. Also, location of the restrictor plug should be checked if the rear rocker arms are receiving more lubricating oil than the front arms. Install new restrictor plug to a depth of 13½ inches from rear end of shaft for all except Series 1550 and 1555 which should be 12¾ inches. Install new sealing plugs cup side out with cup edge flush with end of shaft.

Rocker arm shaft spring free length should be 2½ inches and spring should exert 10 pounds pressure when compressed to a length of ⅝-inch.

When reinstalling rocker arm shaft assembly, tighten bracket retaining nuts to a torque of 25-27 Ft.-Lbs.

### VALVE TIMING

#### All Models

69. Valves are correctly timed when the mark "C" on camshaft gear is aligned with identical mark on crankshaft gear as shown in Fig. O59.

### TIMING GEAR COVER

#### All Models

70. To remove timing gear cover, proceed as follows: Remove hood side panels and engine hood. Drain cooling system and remove radiator hose. On models so equipped, remove oil cooler (or coolers) located in front of radiator. Remove fan blades, set blades in radiator shroud and remove the radiator with the fan blade assembly. Remove water pump, generator or alternator and fan belt. On Row Crop or Utility models with power steering, remove the grille and the power steering cylinder. On non-diesel engines, disconnect governor linkage and remove the governor. Remove capscrew from crankshaft pulley and using suitable pullers attached to capscrews threaded into the pulley, remove pulley from crankshaft. Drain and remove the oil pan. Remove timing gear cover retaining capscrews and remove cover from engine. Front oil seal can now be renewed as outlined in paragraph 81.

Fig. O59 — Valves are properly timed on all models when "C" marked tooth on crankshaft gear is meshed with "C" marked tooth groove on camshaft gear as shown.

### TIMING GEARS

#### All Models

71. Desired camshaft gear to crankshaft gear backlash is 0.004-0.006; camshaft and crankshaft gears should be renewed if backlash exceeds 0.007.

To renew the camshaft gear, follow procedure outlined in paragraph 74 for renewing camshaft. Crankshaft gear can be renewed without removing crankshaft as follows: Pull gear from shaft by using suitable puller attached to capscrews threaded into the gear; a suitable centering adapter should be used to prevent damage to front end of shaft. Press new gear onto shaft with a length of 2-inch pipe, suitable spacers and a ¾-inch, 16 thread puller bolt threaded into front end of crankshaft. Be sure that gear is installed with "C" timing mark to front and that the timing gears are properly meshed as outlined in paragraph 69. On diesel models, retime fuel injection pump drive gear as outlined in paragraph 122A. If the crankshaft is removed and crankshaft gear is being installed in a press, support crankshaft at front throw to avoid springing the shaft.

Note: A timing gear lubrication tube is installed in front face of cylinder block. If necessary to renew the lubrication tube, be sure hole in side of tube is toward the camshaft gear.

### DIESEL INJECTION PUMP DRIVE GEAR

#### All Diesel Models

72. The diesel injection pump drive gear and shaft unit can be withdrawn from the fuel injection pump after removing the timing gear cover as outlined in paragraph 70, or after removing the cover plate from front of timing gear cover; refer to paragraph 124.

Before removing drive gear and shaft unit, shut off the fuel supply

valve and remove timing window from injection pump. Turn engine in normal direction of rotation until timing marks of the injection pump cam and governor weight retainer are both visible in timing window and the 7° BTDC timing mark (Series 1600 models) or 2° BTDC timing mark Series 1550, 1555 and 1650 (with indirect injection), or 4° BTDC on Series 1650 (with direct injection) and 1655, on engine flywheel is aligned with timing pointer. Refer to paragraph 122A. Then, remove gear and shaft unit taking care not to lose the thrust button or spring from front end of shaft.

Cut the locking wire and remove the two cap screws retaining drive gear to gear hub and remove gear from hub. Clamp the tang end of drive shaft in a soft jawed vise and remove the hub retaining nut, hub and drive key. Remove the two cup type shaft seals.

Install two new seals on drive shaft with lip of each seal pointing away from other seal. Use of Roosa-Master No. 13369 seal installation tool is recommended. Clamp tang of drive shaft in soft jawed vise and install drive key, hub and retaining nut. Tighten the nut to a torque of 35-40 Ft.-Lbs. Lubricate seals and bore of pump with Lubriplate or equivalent grease. Be sure that engine is still positioned on 7° BTDC (Series 1600) or 2° BTDC Series 1550, 1555 and 1650 (with indirect injection), or 4° BTDC on Series 1650 and 1655 (with direct injection). Insert drive shaft into pump bore with offset dimple in end of tang aligned with similar offset mark in drive slot of pump rotor. Carefully work lip of rear seal into bore of pump body with a small dull pointed tool. Note: If lip of seal is rolled back when shaft is inserted, remove the shaft and renew the seal. With shaft tang engaged in pump rotor, align pump timing marks and insert small screwdriver tip in notch below timing marks to keep rotor from turning. Place drive gear on hub with teeth meshed into idler gear teeth and move gear counter-clockwise as far as possible to eliminate backlash. Insert gear retaining screws in the two holes of gear and hub that are most nearly aligned, tighten the screws securely and install locking wire. Insert spring and thrust button in front end of shaft and install timing gear cover or gear cover plate to timing gear cover.

Check pump timing by turning engine two turns to where proper degree mark on flywheel is aligned with pointer and No. 1 cylinder is on top dead center. If pump timing marks

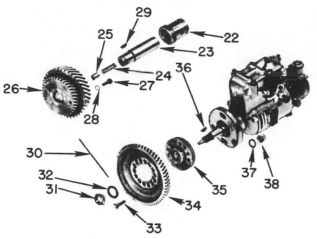

**Fig. O60—Exploded view of diesel fuel injection pump idler and drive gears. Matched set of idler gears (26) are fastened together by two capscrews (27).**

22. Idler bushing
23. Idler shaft
24. Plunger spring
25. Thrust plunger
26. Idler gears
27. Capscrews (2)
28. Lockwashers (2)
29. Woodruff key
30. Locking wire
31. Nut
32. Lockwasher
33. Capscrews (2)
34. Drive gear
35. Gear hub
36. Woodruff key
37. Washers
38. Nuts

are not exactly aligned, loosen the pump mounting bolts and turn pump body as required to align the marks. Tighten the pump mounting bolts securely while holding pump in proper position.

## DIESEL INJECTION PUMP IDLER GEARS

### All Diesel Models

73. After removing timing gear cover as outlined in paragraph 70 and the injection pump drive gear and shaft unit as in paragraph 72, the idler gear and shaft assembly can be withdrawn from bushing and sleeve assembly in front face of cylinder block.

Desired idler gear to crankshaft gear backlash is 0.004-0.006; if backlash exceeds 0.007, the idler gear assembly, crankshaft gear and camshaft timing gears should be renewed.

Desired idler gear shaft to bushing clearance is 0.0015-0.003; if clearance exceeds 0.005, the shaft and/or sleeve and bushings assembly should be renewed. Shaft diameter, new, is 0.999-1.000. A thrust button and spring in front end of idler shaft controls idler gear end play.

The idler gear assembly is keyed and press fitted to the idler shaft. The sleeve and bushing assembly is pressed into front of cylinder block and can be removed with OTC 927 Puller and OTC 933 Extractor or similar tools. Install new sleeve and bushing assembly with suitable driver so that oil holes in sleeve and crankcase bore are aligned and flange on sleeve is tight against crankcase.

Thrust button spring should exert 6½-8½ pounds pressure when compressed to a length of ³¹/₃₂-inch. Renew thrust button if scored or excessively worn.

## CAMSHAFT

### All Models

74. To remove camshaft, first remove hood side panels and hood, grille and radiator and remove timing gear cover as outlined in paragraph 70. Remove ignition distributor and fuel pump on non-diesel engines so equipped. On all engines remove rocker arm cover, rocker arm assembly, push rods, oil pan and oil pump. If engine has been removed from tractor, turn the engine upside down so that cam lifters will fall away from camshaft. With engine in tractor, push cam lifters up away from camshaft. and if they do not stick in their bores, support them in some way. Withdraw camshaft and gear from front of cylinder block. Camshaft gear can be removed in press or by using suitable pullers attached to ⅜-16 capscrews threaded into front of gear.

On Series 1550, 1555 and 1600 diesel and all non-diesel engines, front camshaft journal is supported in a renewable bushing and the three rear journals rotate in unbushed bores in cylinder block. On Series 1650 and 1655 diesel, camshaft is supported in four renewable bushings. Maximum allowable bushing inside diameter (wear limit) is 1.7560. Install new bushings with closely piloted driver such as OTC T-812 Driver and mandrel. Bushings are pre-sized and reaming should not be necessary if they are carefully installed. Note: On non-diesels and Series 1550, 1555 and 1600 diesel models, cylinder block bores can be line bored to 1.8745-1.8755 and bushings be installed for the three rear camshaft journals if necessary.

Camshaft journal diameter, new, is 1.749-1.750 for the front journal and 1.7485-1.7495 for the three rear journals. Desired journal to bushing or bore clearance is 0.0015-0.003; maxi-

mum allowable clearance is 0.005.

Camshaft end play is controlled by a spring loaded thrust button. Thrust spring free length should be 1³/₁₆ inches and spring should exert 15½ to 18½ pounds pressure when compressed to a length of ²⁵/₃₂-inch.

Before installing camshaft, press gear onto shaft making sure that side of gear with "C" timing mark is to front. Thrust spring and button can be inserted in camshaft after shaft is installed in cylinder block.

## ROD AND PISTON UNITS

### All Models

75. Connecting rod and piston units are removed from above after removing cylinder head and oil pan. Before removing connecting rod caps, note if cylinder number is stamped on camshaft side of rod and cap; if not, stamp the cylinder number into each rod and cap prior to removing rod and piston assembly.

Tighten connecting rod bolt nuts to a torque of 46-50 Ft.-Lbs. on all Series 1550, 1555 and all other non-diesel engines; to a torque of 56-60 Ft.-Lbs. on Series 1600 diesel engines and to a torque of 56-58 Ft.-Lbs. on Series 1650 diesel engines prior to tractor serial number 147 586-000, or 46-50 Ft.-Lbs. torque on Series 1650 diesel tractors, serial number 147 586-000 and up and all Series 1655.

## PISTONS, RINGS AND SLEEVES

### All Models

76. **PISTONS AND SLEEVES.** All engines are fitted with wet type sleeves which are sealed at bottom end with two packing rings and at top end by the cylinder head gasket. Sleeves can be pulled from cylinder block using an OTC 938 puller and proper size sleeve attachment. Sleeve bore diameter (new) is as follows:

Series 1550-1555,
  all .................3.6250-3.6265
Series 1600 non-diesel
  (early) ............3.500-3.5015
  Late production ....3.6250-3.6265
Series 1600 diesel......3.750-3.7515
Series 1650-1655
  non-diesel .........3.750-3.7515
Series 1650-1655 diesel.3.875-3.8765

Sleeves and pistons are available in a package containing a sleeve, piston, piston pin, pin retaining rings, piston ring set and sleeve packing rings for one cylinder only. Sleeves and pistons are not available separately. Install new sleeve and piston packages if sleeve taper exceeds 0.004, sleeve out-of-round condition exceeds 0.002 or if ring end gap, measured at top of ring travel with new compression ring, is excessive. Also, install new sleeve and piston package if side clearance of new compression ring in top ring groove exceeds 0.006, or if piston and/or sleeve is scored, cracked or otherwise damaged.

Prior to installing sleeves, check sleeve standout or protrusion above machined surface of cylinder block as follows: Thoroughly clean the sleeve and sleeve bore in cylinder block, then insert sleeve in bore **without** sealing rings. Clamp sleeve in bore with short cap screws and flat washers. Check sleeve protrusion with straight edge and feeler gage. Protrusion for all models except 1650 S/N 187 586-000 and up, and 1655 is 0.001-0.004. Protrusion for models 1650 S/N 187 586-000 and up, and 1655 is 0.003-0.006. If protrusion is less than specified, a shim must be installed between flange of sleeve and bottom of counterbore in cylinder block. If the protrusion is more than specified, clean the sleeve flange and cylinder block counterbore more thoroughly or try a different sleeve. See Fig. O61.

Note: On Series 1650 and 1655 diesel models, sleeve protrusion does not include sleeve "fire wall"; measure sleeve protrusion outside of firewall on these models.

To install sleeve, proceed as follows: Coat new packing rings with Lubriplate or multi-purpose grease, install the rings in **clean** grooves at bottom of sleeve and run a small round rod or punch around the sleeve **under** the rings several times to eliminate any twist in the rings and to seat rings evenly in the grooves. Be sure bores in cylinder block, especially the taper at upper edge of lower bore, are clean and smooth. Then, insert sleeve into bore and push into place by hand. Do not force the sleeve into place; if excessive pressure is required, it indicates the sealing rings have been distorted. Remove sleeve, check and renew the rings if necessary before proceding with sleeve installation.

When installing new pistons in new sleeves, piston skirt to cylinder wall clearance should be checked as follows: After installing sleeves in cylinder block, insert piston (without rings) in sleeve with a ½-inch wide feeler gage of specified thickness inserted between cylinder wall and piston skirt at right angle to piston pin. Piston skirt clearance is correct if 3 to 6 pounds pull (3-5 pounds pull

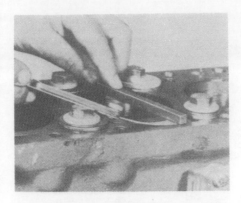

**Fig. O61 — Checking sleeve protrusion above machined surface of cylinder block. Note that sleeve is clamped in place by short capscrews and spacer washers.**

on Series 1650 diesel indirect injection) on a pull scale is required to withdraw the feeler gage. Specified feeler gage thickness is as follows:

| | |
|---|---|
| Series 1550-1555, all | 0.002 |
| Series 1600 non-diesel | 0.0025 |
| Series 1650-1655 non-diesel | 0.003 |
| Series 1600 diesel | 0.003 |
| Series 1650 diesel (indirect injection) | 0.004 |
| Series 1650-1655 diesel (direct injection) | 0.002 |

77. **PISTON RINGS.** Series 1600 and 1650 non-diesel, and Series 1650 diesel (direct injection) and 1655 diesel, have pistons fitted with two compression rings and one oil control ring. Series 1600 diesel, all Series 1550 and 1555 and Series 1650 (indirect injection) have pistons fitted with three compression rings and one oil control ring. Ring installation information is included with ring sets.

Refer to the following specifications when checking piston rings:

**Compression Ring Width:**

| | |
|---|---|
| All 1600-1650 non-diesel | 0.0925-0.0935 |
| Series 1600 diesel | 0.1235-0.1240 |
| Series 1650-1655 diesel | 0.0930-0.0935 |
| All Series 1550-1555 | 0.0930-0.0935 |

**Oil Ring Width:**

| | |
|---|---|
| Series 1600 diesel | 0.1860-0.1870 |
| Series 1650-1655 diesel | 0.1860-0.1865 |

**Ring End Gap:**

| | |
|---|---|
| Compression rings (all models) | 0.010-0.020 |
| Oil ring (diesel) | 0.010-0.023 |
| Maximum allowable (all) | 0.045 |

**Steel Oil Ring Rail End Gap:**

| | |
|---|---|
| All models so equipped | 0.015-0.055 |

**Ring Side Clearance:**

Top ring,
| | |
|---|---|
| all 1600-1650-1655 non-diesel | 0.0025-0.0045 |
| Series 1600 diesel | 0.0020-0.0035 |
| Series 1650-1655 diesel | 0.0025-0.0040 |
| Series 1550-1555, all | 0.0025-0.0040 |

2nd ring,
| | |
|---|---|
| all 1600-1650-1655 non-diesel | 0.0020-0.0040 |
| Series 1600 diesel | 0.0015-0.0030 |
| Series 1650-1655 diesel | 0.0020-0.0035 |
| Series 1550-1555, all | 0.0015-0.0030 |

3rd ring,
| | |
|---|---|
| Series 1600 diesel | 0.0015-0.0030 |
| Series 1650 diesel | 0.0020-0.0035 |
| Series 1550-1555, all | 0.0015-0.0030 |

Oil ring,
| | |
|---|---|
| Series 1600 diesel | 0.0010-0.0030 |
| Series 1650-1655 diesel | 0.0015-0.0030 |
| Series 1550-1555 diesel | 0.010-0.023 |

Maximum allowable side clearance, all rings (except flexible oil rings), all models .......... 0.006

## PISTON PINS AND BUSHINGS

### All Models

78. The full floating type piston pins are retained in piston bosses by snap rings. Piston pins are available for service in oversizes of 0.005 and 0.010 as well as the standard size 1.2494-1.2497.

Connecting rods are fitted with two split type steel backed bronze bushings in each rod. Bushings are available for service and when being installed, the notch in inner side of each bushing must be aligned with oil hole in rod and outer side of each bushing flush with rod. After bushings are installed, hone the bushings to provide a thumb press fit of pin in bushings at room temperature with surfaces dry. With pin and bushings lubricated, pin should just fall through bushings from its own weight. Desired pin to bushing clearance is 0.0004-0.0009, with maximum allowable clearance of 0.0019.

When installing oversize piston pins, the piston bosses must be honed to provide a 0.0002-0.0004 clearance for pin. Maximum allowable pin to piston bore clearance is 0.0008.

## CONNECTING RODS AND BEARINGS

### All Models

79. Connecting rod bearings are of the non-adjustable precision insert type and can be renewed after removing oil pan and connecting rod caps. When installing new bearing inserts, make certain that tangs on inserts fit into slots in connecting rod and cap and that the cap is installed with slots

in rod and cap on the same side of the assembly. Note: Factory installed connecting rods and caps are not numbered; cylinder numbers should be stamped on camshaft side of rod and cap prior to removing cap, if necessary.

All Series 1550, 1555 and 1600 diesel connecting rods are offset 0.120-0.130; connecting rods on all other models have no offset.

Bearing inserts are available in undersizes of 0.003 and 0.020 as well as standard size.

Check the crankpin and bearing inserts against the following values:
Crankpin diameter,
Series 1600-1650-1655
  Non-diesel .............2.249-2.250
Series 1550-1555, all .....2.249-2.250
Series 1600-1650-1655
  Diesel .............2.4365-2.4375
Rod bearing running clearance,
  all models ..........0.0005-0.0015
Maximum running clearance,
  all models .................0.0025
Connecting rod side play,
  all models ..........0.0075-0.0135
Connecting rod bolt torque, Ft.-Lbs.:
Series 1550-1555, All .........46-50
Series 1600-1650-1655
  (Non-Diesel) ..............46-50
Series 1600 diesel ...........56-60
Series 1650 diesel (indirect
  injection) .................56-58
Series 1650-1655 diesel
  (direct injection) ..........46-50

## CRANKSHAFT AND BEARINGS

### All Models

80. Crankshaft in the Series 1650 and 1655 diesel engine is supported in seven main bearings. In all other models, crankshaft is supported in four main bearings. Bearings are of the non-adjustable precision insert type. Crankshaft end play is controlled by the flanged number five main bearing inserts on Series 1650 and 1655 diesel crankshaft and by the flanged number three main bearing inserts on other models. Excessive crankshaft end play can be corrected by renewing the crankshaft and/or main bearings, or by regrinding the crankshaft main bearing thrust journal diameter on Series 1600, 1650 and 1655 to 0.020 undersize and 0.017 overwidth, then installing the 0.020 undersize, 0.017 overwidth bearing inserts available.

Note: The 0.020 undersize main bearing thrust journal inserts for Series 1600 engines were of standard width. However, the 0.020 undersize, 0.017 overwidth thrust journal inserts for Series 1650 non-diesel models can be installed in either the Series 1600

non-diesel or Series 1600 diesel engine.

To remove the crankshaft, first remove engine as outlined in paragraph 60. Remove auxiliary unit (if so equipped), clutch housing and clutch. Then, remove flywheel, oil pan and oil pump, rear oil seal retainer and timing gear cover. On diesel models, remove the fuel injection pump idler gear, injection pump drive gear and shaft, then unbolt fuel injection pump from engine front plate and remove the plate from front of cylinder block. Remove the connecting rod and main bearing caps and lift crankshaft from engine. Remove crankshaft timing gear as outlined in paragraph 71, if necessary.

Refer to the following specifications for checking the crankshaft and main bearings:

Desired crankshaft end play:
  Series 1650-1655
    diesel ...............0.0045-0.0095
  All other models ......0.0045-0.0085
Maximum allowable crankshaft end play:
  Series 1650-1655
    diesel ...................0.011
  All other models ............0.010
Crankshaft thrust journal standard width:
  Series 1650-1655
    diesel .............1.5025-1.5045
  Series 1650-1655
    non-diesel .........1.7525-1.7545
  Series 1600
    (all models) ........1.7525-1.7545
  Series 1550-1555
    (all models) .......1.6275-1.6295
Thrust bearing width (std. and 0.003 undersize):
  Series 1650-1655
    diesel .............1.4950-1.4980
  Series 1650-1655
    non-diesel .........1.7460-1.7480
  Series 1600
    (all models) ........1.7460-1.7480
  Series 1550-1555
    (all models) .......1.6210-1.6230
Thrust bearing width (0.020 undersize):
  Series 1650-1655
    diesel .............1.5120-1.5150
  Series 1650-1655
    non-diesel .........1.7630-1.7650
  Series 1600
    (all models) ........1.7630-1.7650
  Series 1550-1555
    (all models) .......1.6210-1.6230
Crankshaft main journal diameter, standard:
  Series 1600-1650-1655..2.6240-2.6250
  Series 1550-1555 ......2.2490-2.2500

Crankshaft crankpin diameter, standard:
  Series 1600-1650-1655
    non-diesel .........2.2490-2.2500
  Series 1600-1650-1655
    diesel .............2.4365-2.4375
  Series 1550-1555
    (all models) .......2.2490-2.2500
Main bearing running clearance:
  Series 1650-1655
    diesel .............0.0015-0.0045
  Max. allowable ....0.0055
  Series 1650-1655
    non-diesel ........0.0005-0.0035
  Max. allowable .....0.0045
  Series 1600
    (all models) .......0.0005-0.0035
  Max. allowable .....0.0045
  Series 1550-1555
    (all models) .......0.0002-0.0032
  Max. allowable .....0.0045
Journal out-of-round or taper:
  Series 1600-1650-1655 .......0.0003
  Series 1550-1555 ...........0.0005
Flywheel mounting flange runout, max.:
  All models .................0.001
Main bearing bolt torque, Ft.-Lbs.
  Series 1650-1655
    diesel .................129-133
  Series 1650-1655
    non-diesel .............108-112
  Series 1600
    (all models) ...........108-112
  Series 1550-1555
    (all models) .............87-92

## CRANKSHAFT OIL SEALS

### All Models

Original crankshaft front oil seal was a cork face type which mounted on crankshaft and sealed against inner surface of timing gear cover. Conversion kits were made available at gasoline engine serial number 131509, and diesel engine serial number 131056, to convert from the cork face seal to a lip type seal. Kits included the lip seal, new timing gear cover, crankshaft pulley and the necessary gaskets.

The lip type seal was subsequently modified to include spiral flutes in the seal lip and the fluted seal should be used for all repairs.

NOTE: Timing gear covers with a $2\frac{13}{16}$ and 2¾ inch oil seal bores have been used. Be sure to identify timing gear cover to insure correct cover and seal installation.

81. **FRONT OIL SEAL.** After removing the timing gear cover as outlined in paragraph 70, the crankshaft front oil seal (early models) can be removed from the crankshaft. Install a new sealing gasket on the crankshaft, place new seal on shaft with sealing face out and place a second sealing gasket on shaft in front of the oil seal. Thoroughly lubricate cork surface of seal. Check condition of the seal con-

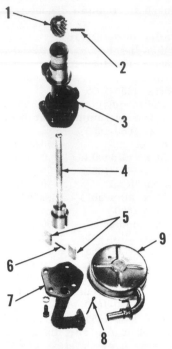

**Fig. O62 — Exploded view of vane type oil pump used in all models except Series 1650 and 1655 diesel.**

| 1. Drive gear | 6. Spring |
| 2. Pin | 7. Pump cover |
| 3. Pump body | 8. Cotter pin |
| 4. Pump shaft | 9. Floating screen |
| 5. Vanes | |

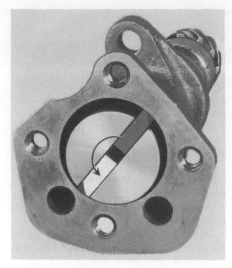

**Fig. O63 — When installing pump vanes, be sure beveled edges are placed as shown.**

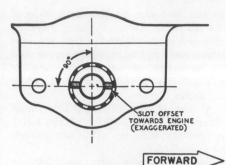

**Fig. O64 — When installing oil pump in non-diesel models, slot in drive gear must be timed. Refer to text.**

tact surface on timing gear cover before reinstalling the cover. Note: On early production, a gasket was fitted between the oil seal and crankshaft timing gear only. However, install a new gasket at both front and rear of the oil seal when servicing these engines.

Later lip type seals are installed in cover with lip toward rear. Seals are installed $\frac{3}{16}$-inch (non-diesel), or $\frac{3}{8}$-inch (diesel) below outer edge of seal bore. If pulley surface is grooved from old seal, install new seal an additional $\frac{1}{32}$-inch deeper.

**82. REAR OIL SEAL.** On direct drive models, the crankshaft rear oil seal can be renewed without removing the engine from tractor.

To renew the crankshaft rear oil seal, remove the flywheel as outlined in paragraph 83 and drain and remove the oil pan. Remove the oil seal retainer capscrews and carefully pry retainer loose from the locating dowels and gasket seal. Carefully press new seal into retainer with lip forward. Reinstall retainer with new gasket, tighten the retaining capscrews and then install oil pan with new gasket. If crankshaft flange is grooved, add another gasket behind retainer so seal will run on a new surface.

Note: If the oil seal retainer is warped or otherwise damaged, the seal and retainer should be renewed as a unit. Late production retainers are ribbed to prevent warpage and if an early production retainer without reinforcing ribs is encountered, it is recommended that the late type ribbed retainer and seal assembly be installed.

## FLYWHEEL

### All Models

83. On models equipped with Reverse-O-Torc drive, remove engine from tractor as outlined in paragraph 60, remove converter housing and torque converter as in paragraph 182. The flywheel can then be unbolted and removed from the crankshaft flange and dowel pins.

On models equipped with auxiliary drive unit, remove engine from tractor as outlined in paragraph 60 and remove the drive unit as outlined in paragraph 155 or 172. Then remove clutch unit and flywheel following same general procedure as outlined for direct drive models.

On direct drive models, remove the engine clutch as outlined in paragraph 151, then proceed as follows: Remove cap screws retaining PTO drive hub and flywheel to engine crankshaft. Using two of the capscrews as puller bolts, remove drive hub from the flywheel and two dowel pins. Then remove flywheel from the dowel pins and crankshaft flange.

To install a new flywheel ring gear, heat gear evenly to a temperature of 400°F. to 450°F. and install gear with beveled end of teeth toward front of flywheel. If necessary to renew the timing tape, thoroughly clean flywheel, peel backing from tape and

apply tape to flywheel with edge of tape even with edge of flywheel and the TDC marks on flywheel and tape aligned.

When installing flywheel, and PTO drive hub if so equipped, coat flywheel capscrews with Loctite sealant C or CV and tighten the retaining cap screws to a torque of 66-69 Ft.-Lbs. After flywheel is installed, check flywheel runout which should not exceed 0.005.

NOTE: Late flywheels will not have dowel holes, however dowels should not be removed from crankshafts which have them installed.

When installing late flywheel on dowelled crankshaft, cut off dowels as close to crankshaft as possible without damage to crankshaft, then center punch protruding portions and drill them flush with flange face using a ½-inch drill.

## OIL PUMP

### All Models Except Series 1650-1655 Diesel

84. All models except the Series 1650 and 1655 diesel are fitted with a vane type oil pump as shown in the exploded view in Fig. O62. To remove the pump, drain and remove oil pan. After removing the one retaining cap screw, pump can be withdrawn from bottom of engine crankcase.

Pump vanes (5) and spring (6) can be removed after removing cover (7). To remove shaft (4), place assembly marks on gear (1) and shaft, then remove the pin (2) and press or pull gear from shaft. Clean all parts and inspect for wear. Although all parts are available separately, a new pump assembly should be installed if both the pump body and shaft are worn excessively.

If new shaft or gear is being installed, press gear on shaft to a position so that shaft end play is $\frac{1}{32}$-inch, drill a hole through gear and shaft with a No. 13 drill and install pin. If

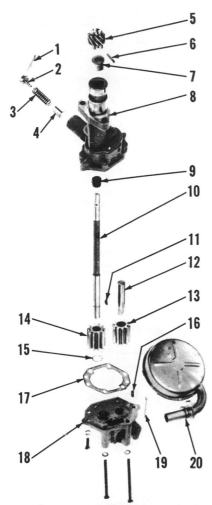

Fig. O65—Exploded view of gear type oil pump used on Series 1650 and 1655 diesel engines. Engine oil pressure is controlled by main gallery relief valve shown in Fig. O66. Pump relief valve (items 1 through 4) is set at 42 psi by adjusting screw (2).

| | |
|---|---|
| 1. Cotter pin | 11. Woodruff key |
| 2. Adjusting screw | 12. Idler shaft |
| 3. Spring | 13. Idler gear |
| 4. Plunger | 14. Driven gear |
| 5. Drive gear | 15. Snap ring |
| 6. Pin | 16. Dowel pins (2) |
| 7. Upper bushing | 17. Gasket |
| 8. Pump body | 18. Pump cover |
| 9. Lower bushing | 19. Cotter pin |
| 10. Drive shaft | 20. Floating screen |

reinstalling original gear and shaft, press gear onto shaft with previously affixed assembly marks aligned and so that pin can be installed in the hole through gear and shaft. Also see Fig. O63.

85. When installing pump assembly on diesel engines, remove the tachourmeter drive unit if so equipped. After installing the oil pump, insert tachourmeter drive shaft into slot in pump drive gear. Then, reinstall tachourmeter drive housing and retaining clamps.

86. On non-diesel models, remove the ignition distributor and turn the engine to where the No. 1 piston is at TDC on compression stroke. Then,

install oil pump with slot in drive gear positioned as shown in Fig. O64. Reinstall and time ignition distributor as outlined in paragraph 145.

### Series 1650-1655 Diesel

87. A gear type oil pump as shown in the exploded view in Fig. O65 is used on the Series 1650 diesel engine. To remove pump, drain oil pan, remove the pan and the two pump retaining capscrews and withdraw pump from bottom of engine crankcase.

Place alignment marks on adjusting screw (2) and pump body (8) prior to removing screw, then count the number of turns required to remove the screw and record for proper reassembly. Also, place assembly marks on the drive gear (5) and shaft (10) in the event that the original gear and shaft may be reinstalled.

Refer to the following specifications when overhauling pump:
Relief valve setting............42 psi.
Relief valve spring
  free length ...............$2\frac{11}{16}$ inch
Relief valve plunger
  diameter ...............0.745-0.747
Idler gear to shaft
  clearance, desired ....0.0035-0.0045
  Maximum allowable ........0.006
Drive shaft to bushing clearance,
  Upper bushing, desired .0.0005-0.003
  Maximum allowable ........0.045
  Lower bushing,
    desired ............0.0015-0.004
  Maximum allowable ........0.006
Drive gear to body
  clearance .............0.004-0.008
If installing new upper drive shaft bushing (7), be sure oil hole in bushing is aligned with hole in pump body (8). If installing new drive gear (5) and shaft (10), press gear onto shaft so that drive gear to body clearance is 0.004-0.008, drill a $\frac{3}{16}$-inch hole through gear and shaft and install pin (6).

When installing relief valve, thread adjusting screw in same number of turns required to remove screw and to a position where the previously affixed marks are aligned, then install cotter pin (1). Note: If new screw is being installed, or original position of screw is not known, reinstall screw flush with pump body, then turn screw in to first position that cotter pin can be installed.

Install the oil pump as outlined in paragraph 85 or 86.

### MAIN OIL GALLERY PRESSURE RELIEF VALVE

### All Models

88. Oil pressure on all models is controlled by a spring loaded poppet

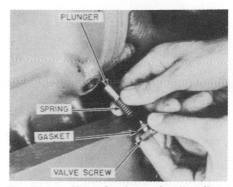

Fig. O66 — View showing main oil gallery relief valve.

type relief valve in the main oil gallery as shown in Fig. O66. To check engine oil pressure, remove the oil pressure warning sender switch from engine and install a master oil pressure gage. With engine at normal operating temperature, oil pressure should be as follows:

**Series 1600, All Models:**
  Pressure @ 400
    engine RPM ...........5-10 psi
  Pressure @ 2000
    engine RPM ...........25-30 psi

**Series 1650 (Indirect Injection) Diesel Models:**
  Pressure at slow
    idle speed ...............10 psi
  Pressure @ 2200
    engine RPM ...........25-42 psi

**Series 1650 (Direct Injection)— 1655 Diesel Models:**
  Pressure at slow
    idle speed ...............20 psi
  Pressure @ 2200
    engine RPM ...........30-45 psi

**All Other Models**
  Pressure at slow
    idle speed ...............5 psi
  Pressure @ 2200
    engine RPM ...........20-40 psi

If pressure is not as specified, the gallery oil pressure relief valve should be removed, cleaned and inspected. The 0.497-0.498 diameter plunger should slide freely in bore of cylinder block. Check the spring against the following specifications:

**Series 1650-1655 Diesel Models:**
  Spring free length......1⅜-inches
  Pressure @ 1 in. length ......10 lbs.

**All Models Except Series 1650-1655 Diesel:**
  Spring free length........2 inches
  Pressure @ 1 in. length 5½-6½ lbs.
On Series 1650 and 1655 diesel models, a 42 psi safety relief valve is incorporated in the gear type oil pump. If, after reinstalling the main gallery relief valve on these models, pressure is still not as specified, a faulty or sticking pump relief valve should be suspected.

# CARBURETOR (Except L-P Gas)

## All Gasoline Models

89. Series 1600, 1650 and 1655 gasoline engines are equipped with Marvel-Schebler Model TSX carburetors as follows: Early production Series 1600 (231 cu. in. displacement), TSX-868; late production Series 1600 (248 cu. in. displacement), TSX-880; and Series 1650 and 1655 TSX-807.

Series 1550 and 1555 gasoline engines are equipped with Marvel-Schebler Model TSX-903 carburetors.

With carburetor throttle body inverted, float level should be ¼-inch above gasket ($\frac{9}{32}$-inch above machined surface) on all models. Repair data follows:

**Model TSX-807:**

| | |
|---|---|
| Repair kit | 286-1318 |
| Gasket set | 16-594 |
| Inlet needle and seat | 233-581 |
| Idle jet | 49-203 |
| Nozzle | 47-416 |
| Power jet | 49-479 |
| Economizer jet | 49-145 |

**Model TSX-868:**

| | |
|---|---|
| Repair kit | 286-1418 |
| Gasket set | 16-594 |
| Inlet needle and seat | 233-543 or 233-608 |
| Idle jet | 49-203 |
| Nozzle | 47-796 |
| Power jet | 49-216 |
| Economizer jet | 49-264 |

**Model TSX-880:**

| | |
|---|---|
| Repair kit | 286-1437 |
| Gasket set | 16-594 |
| Inlet needle and seat | 233-543 or 233-608 |
| Idle jet | 49-203 |
| Nozzle | 47-A96 |
| Power jet | 49-369 |
| Economizer jet | 49-264 |

**Model TSX-903**

| | |
|---|---|
| Repair kit | 286-1480 |
| Gasket set | 16-654 |
| Inlet needle and seat | 233-543 |
| Idle jet | 49-101L |
| Nozzle | 47-220 |
| Power jet | 49-289 |
| Economizer jet | 49-265 |

# L-P GAS SYSTEM (Zenith)

Early production Series 1600 LP-Gas models were equipped with a Zenith vaporizer and primary regulator unit and a Zenith pressure regulating carburetor. For American Bosch-Ensign LP-Gas equipment used on Series 1550 and 1555, late production Series 1600, and Series 1650 and 1655 LP-Gas models, refer to paragraph 98.

## ADJUSTMENTS

### Models With Zenith System

90. Initial adjustments on the carburetor are 2 to 3 turns open for the idle mixture screw and 1 to 2 turns open for the main fuel adjusting screw.

Start engine and bring to operating temperature. Adjust throttle stop screw to obtain an engine low idle speed of 475-525 rpm. Turn the idle mixture screw in or out as required until engine runs smoothly. Correct adjustment should be obtained with idle mixture screw open approximately 2⅜ turns. Recheck engine low idle rpm.

Place load on engine and run at high idle (2200 rpm). Turn the main adjusting needle inward until engine starts to lose power, then turn needle out until full power is restored. Repeat this operation until a definite setting is obtained. Correct setting should be obtained when the needle

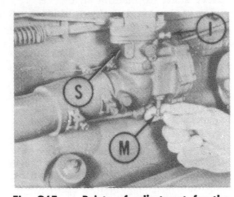

**Fig. O67 — Points of adjustment for the Zenith LP-Gas pressure regulating carburetor.**

I. Idle mixture screw
M. Main fuel needle
S. Throttle stop screw

is 1¾ turns open for tractors using commercial propane; 1½ turns open for tractors using 50 percent propane and 50 percent butane and 1¼ turns open for tractors using 20 percent propane and 80 percent butane.

If stationary loading of engine is not possible, make adjustments then operate tractor under drawbar load. Refer to Fig. O67.

### FILTER

### Models With Zenith System

91. The LP-Gas filter, located at left side of battery tray, is fitted with a renewable element and should be

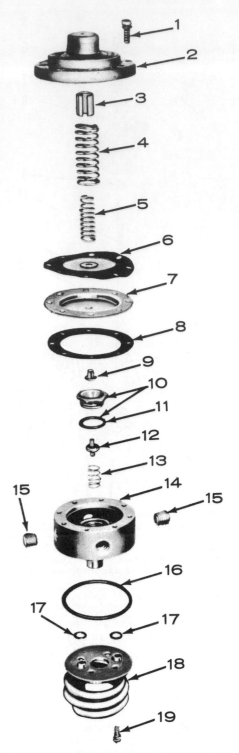

**Fig. O68 — Exploded view of Zenith LP-Gas vaporizer assembly.**

| | | | |
|---|---|---|---|
| 2. Diaphragm cover | | 11. "O" ring |
| 3. Vibration dampener | | 12. Fuel valve |
| 4. Outer spring | | 13. Valve spring |
| 5. Inner spring | | 14. Body |
| 6. Diaphragm | | 15. Allen plug |
| 7. Baffle plate | | 16. "O" ring |
| 8. Gasket | | 17. "O" ring |
| 9. Fuel valve cap | | 18. Coil & mounting |
| 10. Valve seat | | plate |

cleaned monthly as follows: Turn off the vapor and liquid withdrawal valves, remove plug from bottom of filter bowl and allow any accumulations to drain. Slowly open the vapor

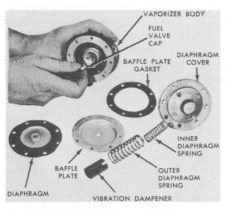

Fig. O71 — To test vaporizer assembly, install pressure gage and line as shown. Refer to text for procedure.

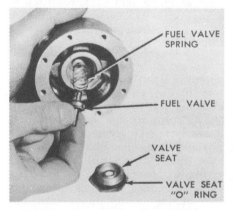

Fig. O69 — Partially disassembled view of vaporizer assembly. Note fuel valve cap.

Fig. O70 — View showing fuel valve being removed from body.

valve and blow out any accumulations which remain.

Every year, unscrew the filter bowl and renew the element and gasket.

## VAPORIZER

### Models With Zenith System

92. The vaporizer used on LP-Gas models of the early Series 1600 tractors is a Zenith model number A962A-1.

93. R&R AND OVERHAUL. Before disconnecting any lines, be sure that all fuel is out of the lines, vaporizer, regulator and carburetor by closing the tank withdrawal valves and allowing the engine to run until it stops. Turn off ignition switch, then drain cooling system and remove hood and side panels. Disconnect and remove air cleaner and bracket assembly. Disconnect fuel lines from vaporizer, then remove the retaining cap screw at bottom of vaporizer housing and withdraw vaporizer assembly.

Refer to Fig. O68 and proceed as follows: Remove the four cap screws and separate vaporizer coil and mounting plate assembly (18) from vaporizer body (14). Discard all three "O" rings. Remove any alternate four of the diaphragm cover screws and

install in their place four aligning studs (Zenith No. C161-195, or equivalent). Maintain pressure on the diaphragm, remove the remaining four cap screws, then carefully release the pressure on diaphragm cover and remove cover (2), diaphragm springs (4 and 5) and vibration dampener (3). Remove studs; then remove diaphragm (6), baffle plate (7), baffle plate gasket (8) and fuel valve cap (9) as shown in Fig. O69. Use a 1-inch socket wrench and remove valve seat (10—Fig. O68), fuel valve (12) and spring (13) as shown in Fig. O70.

Clean all parts in a grease solvent, however, DO NOT use carburetor cleaning solvents for cleaning any of the parts as carburetor cleaners will destroy the impregnation used in the casting and the coating on the coil. Inspect all parts for undue wear or other damage and renew as necessary.

94. To reassemble, proceed as follow: Place fuel valve spring (13—Fig. O68) over boss in center of vaporizer housing (14), then place fuel valve (12) on spring with short stem toward housing. Be sure spring is resting on machined shoulder of fuel valve, then with new "O" ring on valve seat (10), install valve seat in vaporizer body and tighten. Place fuel valve cap (9) over the protruding fuel valve stem. Reinstall the aligning studs in four alternate holes in top of vaporizer body, then install new baffle plate gasket (8), baffle plate (7), with recessed side toward body, and diaphragm (6), with flanged disc away from body. Be sure fuel valve cap enters center hole of baffle plate. Place vibration dampener (3) inside the outer diaphragm spring (4). Place inner diaphragm spring (5) over center of diaphragm plate, then install outer spring over the inner spring. Place diaphragm cover (2) over aligning studs and depress cover until cap screws can be installed in the remaining four holes. Continue to depress the diaphragm cover and seat the cap screws lightly. Remove the aligning studs and install the remaining four cap screws. Tighten all the cap screws evenly and to a moderate tightness. Install a new "O" ring (16) on vaporizer body and two new "O"

rings (17) on ends of vaporizer coil. Align mounting holes and install coil and mounting plate assembly (18) to vaporizer body.

95. PRESSURE AND LEAKAGE TESTS. To conduct pressure and leakage tests, proceed as follows: Install a pressure gage capable of registering at least 30 psi in the vaporizer outlet as shown in Fig. O71. Connect the vaporizer inlet to a source of compressed air, loosen the previously installed gage until leakage occurs, then retighten gage. Gage should register a pressure of 9 to 11 psi and the reading should remain steady. If gage reading creeps up, it indicates a leaking fuel valve or fuel valve seat. If a leak is indicated, correct it by cleaning and/or renewing valve parts as necessary.

To check for leakage, use a bubble solution and proceed as follows:

Cover vent hole in diaphragm cover. If bubble forms, the diaphragm is leaking. Renew diaphragm and recheck.

Check around the diaphragm cover. If bubbles form, renew baffle plate gasket and recheck.

Check pipe plugs and if bubbles form, remove plugs; then reinstall them using a pipe plug compound and recheck.

Check the vaporizer coil mounting plate and if bubbles form, remove the vaporizer coil and mounting plate and renew "O" rings; then recheck.

## PRESSURE REGULATING CARBURETOR

### Models With Zenith System

96. OPERATION. The Zenith PC2 J 10 pressure regulating carburetor serves both as a secondary regulator and carburetor. See Fig. O72 for a cross-sectional view of the carburetor.

The fuel valve seat (2) is adjustable so that the relationship of the diaphragm lever (9) and the diaphragm flange can be varied to meet the specifications for the particular engine operation. The fuel valve seat (2) is locked in position in the carburetor body by means of the lock screw and plug.

Fuel enters the carburetor at the fuel inlet (1) from the primary regu-

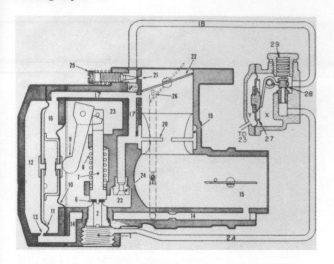

Fig. O72 — Cross-sectional schematic view of Zenith LP-Gas pressure regulating carburetor.

1. Fuel inlet
2. Fuel valve seat
7. Fuel valve
8. Fuel valve spring
9. Diaphragm lever
10. Inner diaphragm
11. Outer diaphragm
12. Outer diaphragm chamber
13. Air passage orifice
14. Air passage
15. Air intake
16. Inner diaphragm chamber
17. Idle fuel passage
19. Annulus
20. Venturi
21. Idle needle seat
22. Throttle disc
23. Pressure chamber
24. Main jet
25. Idle needle
26. Economizer orifice
27. Idle diaphragm
28. Idle mixture valve
29. Spring

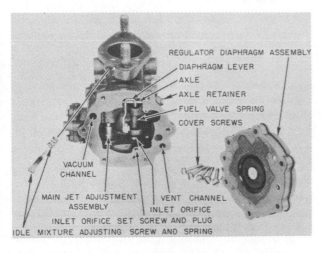

Fig. O73 — Zenith LP-Gas pressure regulating carburetor shown with diaphragm assembly removed.

REGULATOR DIAPHRAGM ASSEMBLY
DIAPHRAGM LEVER
AXLE
AXLE RETAINER
FUEL VALVE SPRING
COVER SCREWS
VACUUM CHANNEL
MAIN JET ADJUSTMENT ASSEMBLY
VENT CHANNEL
INLET ORIFICE
INLET ORIFICE SET SCREW AND PLUG
IDLE MIXTURE ADJUSTING SCREW AND SPRING

Fig. O74 — When adjusting diaphragm lever, use step number two of Zenith gage number C161-194 as shown.

ZENITH C161-194 GAUGE
STEP NO. 2 0.045-0.055"
LEVER
FUEL INLET ORIFICE

(7). In engine idle operation, throttle valve (22) is closed, causing a high vacuum above the throttle valve, and fuel flows from the pressure chamber through orifice (18) past idle needle (25) to provide fuel for engine idle. At full throttle operation, the restriction of the venturi (20) causes a pressure drop at this point, drawing fuel into the carburetor and mixing it with air at the venturi. At part throttle operation, the action of the throttle valve (22) causes a decrease of pressure at orifice (26) which is transmitted through a drilled passage to passage (14) causing a slight lowering of pressure in chamber (12) back of diaphragm (11), resulting in a progressively leaner fuel mixture at part throttle operation.

97. R&R AND OVERHAUL. To remove the carburetor, first close both withdrawal valves and exhaust the fuel in the regulator and lines by allowing the engine to run until it stops. Turn off the ignition switch, disconnect the choke and throttle linkage, air cleaner hose and the fuel inlet line. Unbolt and remove the carburetor assembly. Remove the six screws, securing the diaphragms to the carburetor and remove the diaphragms as shown in Fig. O73. Remove the fuel inlet fitting from the bottom of the carburetor and the inlet valve seat set screw and plug; then remove the inlet valve seat while holding the valve off its seat with the diaphragm lever.

Remove the diaphragm lever shaft plug in the side of the carburetor body and remove the shaft, lever and fuel valve assembly. Remove the main and idle needle valves and the main valve body using Zenith tool No. C161-193, or equivalent. Remove the throttle and choke flys and shafts, and the venturi by removing the venturi locking screw in the side of the carburetor body.

Clean all parts, with the exception of the diaphragms and fuel valve seal, in a suitable solvent and examine for wear or damage. Examine the diaphragm for cracks, pin holes or deterioration and renew if necessary. Reassemble the carburetor by reversing the disassembly procedure using new gaskets and a new fuel valve seal.

While holding the fuel valve open, install the fuel valve seat and adjust same until fuel valve lever just touches the land of step number three of Zenith gage C161-194 as shown in Fig. O74. Retain fuel valve in this position by installing plug and set screw.

lator at a pressure of 9-11 psi; but is prevented entering the carburetor by the neoprene seal (6) on the regulating (fuel) valve (7). When the engine is not running, pressures are equal in all parts of the carburetor and no fuel can flow. When the engine choke is closed and the engine turned over with the starter, a partial vacuum is built up in the carburetor air horn beyond the choke fly and transmitted to the chamber (16) between the two diaphragms by means of passage (17).

The greater pressure in chamber (12) causes the large diaphragm (11) to move inward, acting through the spacer on diaphragm (10), moving the lever arms (9) to overcome the pressure of spring (8) and raise the fuel valve (7) off its seat admitting fuel to the pressure chamber (23). Fuel continues to enter the pressure chamber until pressure in chamber (23), combined with the pressure of spring (8), overcomes the pressure of diaphragm (11) and closes the fuel valve

Note: If Zenith gage is not available, the distance between machined surface of body and tip of lever is $\frac{3}{32}$-inch.

When reinstalling the diaphragm, diaphragm cover and gasket, make sure that the parts are arranged so that the diaphragm passages are aligned with the passages in the carburetor body.

# LP-GAS SYSTEM (Bosch-Ensign)

Series 1550 and 1555, late production Series 1600 and Series 1650 and 1655 tractors are available with an LP-Gas system manufactured by the Ensign Products Section, American Bosch Arma Corporation. Like other LP-Gas systems, this system is designed to operate with the fuel tank not more than 80% filled.

The Bosch-Ensign Series CBX carburetors and Series RDG Regulator have three points of mixture adjustment, plus an idle stop screw. Application and repair data follows:

Bosch-Ensign Carburetor Part No.:
Series 1550-1555 ........CBX125A5561A
Series 1600.............CBX125A5525A
Series 1650-1655 ......CBX150A5530A-2
Carburetor Repair Kit Part No.:
All Models ....................KT46115
Bosch-Ensign Regulator Part No.:
All Models ..................RDG100A1
Regulator Repair Kit Part No.:
All Models ....................KT46150

## ADJUSTMENTS
### Models With Bosch-Ensign System

98. **INITIAL ADJUSTMENTS.** After overhauling or installing new carburetor or regulator, make the following initial adjustments: Open idle screw on regulator 1½ turns. Open starting adjustment screw on carburetor 1 turn. Open load adjustment screw on carburetor 6½ turns open for Series 1550 and 1555 or 7½ to 8 turns for Series 1600 or 3½ turns for Series 1650 and 1655. Close choke and open throttle halfway to start engine.

99. **STARTING SCREW ADJUSTMENT.** After engine has been started and is at normal operating temperature, fully open throttle while leaving choke closed. Adjust starting screw to give highest engine speed, then turn screw out slightly until engine speed just starts to drop and tighten lock nut. Return throttle to slow idle position and open choke.

100. **IDLE STOP SCREW.** Adjust idle stop screw on carburetor throttle to obtain an engine slow idle speed of 525 RPM for Series 1600 and 1650; 550 RPM for Series 1550; or 800 RPM for Series 1555 and 1655.

101. **IDLE MIXTURE SCREW.** Adjust idle mixture screw on regulator to obtain best engine idle performance when engine is at operating temperature. Readjust idle stop screw if necessary.

102. **LOAD SCREW ADJUSTMENT (WITH ANALYZER).** Be sure to follow the gas analyzer operating instructions and set load screw to give a reading of 12.8 on gasoline scale or 14.3 on LP-Gas scale with engine at normal operating temperature and running at high idle speed. Recheck idle mixture and stop screw adjustment.

103. **LOAD SCREW ADJUSTMENT (WITHOUT ANALYZER).** With engine running at full throttle and at normal operating temperature, apply load until governor opens carburetor throttle wide open. Find adjustment point where engine speed begins to drop from mixture being too rich, then too lean. Set adjustment midway between these two points and tighten jam nut. Recheck idle mixture and stop screw adjustment .

104. **LOAD SCREW ADJUSTMENT. (WITHOUT ANALYZER OR LOAD).** Make idle adjustment carefully as outlined in paragraph 101. With engine running at high idle speed and at normal operating temperature, adjust load screw to give maximum engine RPM; then, slowly turn load screw in until engine speed begins to fall. Set load screw midway between these two positions.

## FILTER
### Models With Bosch-Ensign System

105. The Bosch-Ensign filter (See Fig. O75) is equipped with a felt filtering element and a magnetic ring. When servicing the LP-system or on major engine overhauls, it is advisable to remove the filter housing and clean or renew the filtering element. CAUTION: Shut off both liquid and vapor valves at fuel tank and run engine until fuel is exhausted before attempting to remove the filter housing.

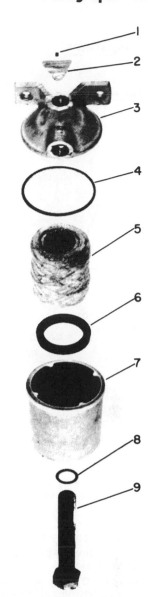

Fig. O75 — Exploded view of the Bosch-Ensign FLP25B1 fuel filter used on late Series 1600, all Series 1550 and 1555, and Series 1650 and 1655, LP-Gas equipped models.

1. Screw
2. Nameplate
3. Cover
4. Gasket
5. Element
6. Magnet
7. Housing
8. Gasket
9. Bolt

## REGULATOR
### Models With Bosch-Ensign System

The Bosch-Ensign Series RDG regulator combines a heat exchanger to vaporize liquid LP-Gas with a two stage regulator to reduce fuel pressure to slightly below atmospheric. The primary regulator reduces the fuel from tank pressure to approximately 4 psi as it enters the final regulator. Heat for vaporizing the liquid fuel is obtained from coolant from the tractor cooling system. Refer to Fig. O76 for cross-sectional views of regulator and principles of operation.

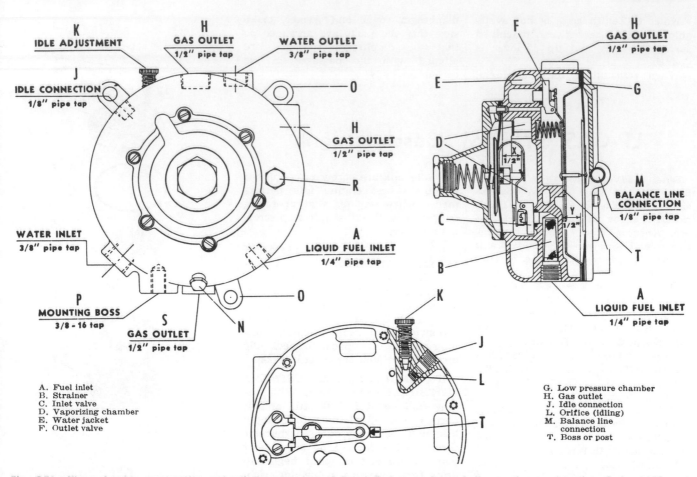

A. Fuel inlet
B. Strainer
C. Inlet valve
D. Vaporizing chamber
E. Water jacket
F. Outlet valve

G. Low pressure chamber
H. Gas outlet
J. Idle connection
L. Orifice (idling)
M. Balance line
    connection
T. Boss or post

Fig. O76—Views showing construction and adjustment points of Bosch-Ensign regulator similar to that used on late Series 1600, and all Series 1650 and 1655 and Series 1550 and 1555 tractors. For exploded view of regulator, refer to Fig. O77.

**106. R&R REGULATOR.** Shut off both the liquid and vapor withdrawal valves at fuel tank and run engine until fuel in system is exhausted. Drain engine cooling system. Remove water connections to engine cooling system. Disconnect fuel and balance lines and remove regulator from tractor. Reverse removal procedure to reinstall regulator.

**107. OVERHAUL REGULATOR.** Refer to exploded view of regulator in Fig. O77. Disassemble the regulator and clean parts with solvent and dry with air hose. Inspect and renew valve seats, valves, diaphragms and springs as necessary. Parts are available separately or in a repair kit.

After inlet valve (primarily regulator valve) assembly is installed, open and close valve several times; then, check distance from face of housing to bottom of groove in inlet valve lever (See Fig. O78 and dimension "X" in Fig. O76). Bend inlet valve lever as necessary so that this measurement is exactly ½-inch.

When installing outlet valve (final regulator valve) assembly, align center of valve lever with arrow on boss

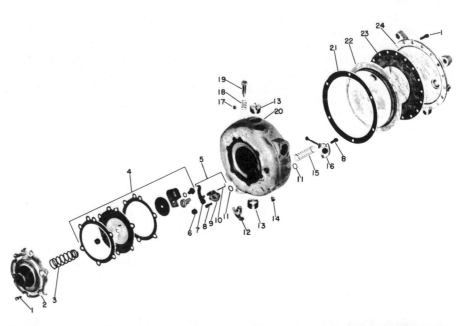

Fig. O77—Exploded view of the Bosch-Ensign regulator used on Series 1550, 1555, 1650, 1655 and late Series 1600 tractors. Refer to Fig. O76 for views showing construction and adjustment points.

| | | | |
|---|---|---|---|
| 1. Screws | 7. Lever | 13. Plug | 19. Idle adj. screw |
| 2. Cover | 8. Screw & washer | 14. Strainer | 20. Regulator body |
| 3. Spring | 9. Valve body | 15. Spring | 21. Gasket |
| 4. Diaphragm assy. | 10. Pin | 16. Valve assy. | 22. Plate |
| 5. Valve assembly | 11. "O" ring | 17. Screw | 23. Diaphragm |
| 6. Plug | 12. Drain cock | 18. Spring | 24. Cover plate |

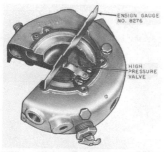

**Fig. O78 — Using Ensign gage 8276 to set the fuel inlet valve lever to dimension "X" as indicated in Fig. O76.**

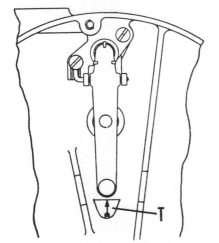

**Fig. O79 — Location of post or boss with stamped arrow for the purpose of setting the fuel inlet valve lever.**

(See Fig. O79) and, after operating valve several times, bend valve lever as necessary so that lever is flush with top of boss.

**108. TROUBLE SHOOTING REGULATOR PROBLEMS.** Generally, difficulties encountered with regulator are due to leakage of gas past valves. Trouble will generally show up as excessive fuel consumption, decrease in power, inability to properly adjust fuel mixtures and/or loss of gas through carburetor when engine is not running. To test regulator, remove plug (R—Fig. O76) and install a 0-10 psi gage in opening. If gage pressure gradually builds up after engine is stopped, inlet valve is leaking and same should be cleaned or renewed. Remove fuel hose from regulator to carburetor. Soap bubble should hold over fuel opening at regulator. If not, fuel outlet valve or low pressure diaphragm is leaking. Clean or renew valve and check diaphragm.

## CARBURETOR

### Models With Bosch-Ensign System

The Bosch-Ensign Series CBX carburetor is equipped with starting and load adjust-

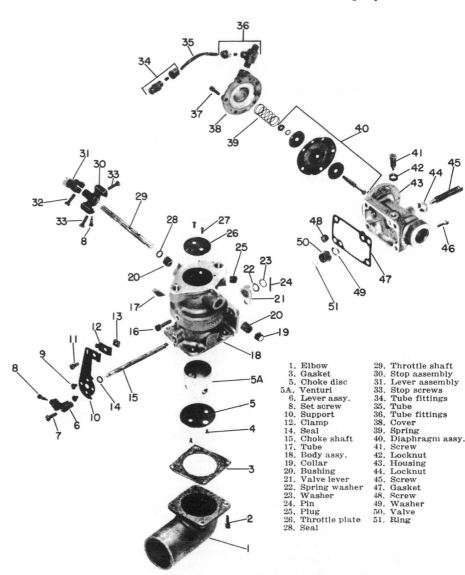

| | |
|---|---|
| 1. Elbow | 29. Throttle shaft |
| 3. Gasket | 30. Stop assembly |
| 5. Choke disc | 31. Lever assembly |
| 5A. Venturi | 33. Stop screws |
| 6. Lever assy. | 34. Tube fittings |
| 8. Set screw | 35. Tube |
| 10. Support | 36. Tube fittings |
| 12. Clamp | 38. Cover |
| 14. Seal | 39. Spring |
| 15. Choke shaft | 40. Diaphragm assy. |
| 17. Tube | 41. Screw |
| 18. Body assy. | 42. Locknut |
| 19. Collar | 43. Housing |
| 20. Bushing | 44. Locknut |
| 21. Valve lever | 45. Screw |
| 22. Spring washer | 47. Gasket |
| 23. Washer | 48. Screw |
| 24. Pin | 49. Washer |
| 25. Plug | 50. Valve |
| 26. Throttle plate | 51. Ring |
| 28. Seal | |

**Fig. O80—Exploded view of the Bosch-Ensign LP-Gas carburetor used on Series 1550, 1555, 1650, 1655 and late Series 1600 tractors.**

**Fig. O80A — To assemble fuel adjusting valve, insert valve through fuel inlet opening, screw the adjusting screw inward through valve and secure valve on screw with snap ring. Cup of spring washer is towards valve.**

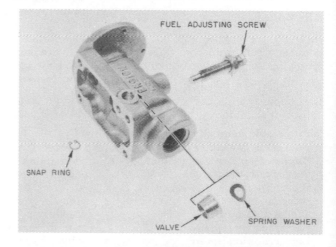

ment screws and with an economizer unit that richens the fuel mixture under load conditions. See Fig. O80 for exploded view of carburetor. Idle fuel mixture adjustment is provided on the regulator.

**109. OVERHAUL CARBURETOR.** Refer to Fig. O80. Repair parts are available separately, also a repair kit is available for the carburetor. Other than renewing throttle shaft and throttle shaft bushings, carburetor servicing generally concerns renewal of economizer diaphragm (40) and making sure that starting valve (21) is working properly. Check to see that economizer diaphragm will hold vacuum. Inspect starting valve for proper position when choke is closed. Valve should fit tightly against carburetor body and completely cover main fuel opening in carburetor body when choke is closed. Refer to Fig. O80A for assembly of fuel adjusting screw.

Fig. O81 — Remove plug (1) to purge air from primary fuel filter.

# DIESEL FUEL SYSTEM

When servicing any unit associated with the diesel fuel system, the maintenance of absolute cleanliness is of the utmost importance.

Probably the most important precaution that servicing personnel can impart to owners of diesel powered tractors is to urge them to use an approved fuel that is absolutely clean and free from foreign material. Extra precautions should be taken to make certain that no water enters the fuel storage tanks. This last precaution is based on the fact that all diesel fuels contain some sulphur. When water is mixed with sulphur an acid is formed which will quickly erode the closely fitting parts of the injection pump and fuel injection nozzles.

Engines for all series 1550, 1555, 1600 and series 1650 prior to tractor serial number 187 586-000, are the indirect injection type and use energy cells in conjunction with the injector nozzle shown in Fig. O84. Series 1650 engines, tractor serial number 187 585-000 and up, and all Series 1655, are the direct injection type which do not use an energy cell and are fitted with the Roosa-Master "pencil" type injector shown in Fig. O89A.

## TROUBLE SHOOTING

### All Models

110. If the engine will not start or does not run properly, refer to the following for possible cause of trouble:

111. **ENGINE WILL NOT START.** If the engine will not start, possible causes of trouble are as follows:

    Fuel tank empty
    Fuel supply valve closed
    Engine stop control applied or improperly adjusted
    Water in fuel
    Inferior fuel
    Clogged fuel filter
    Air traps in system
    Low cranking speed
    Pump installed out of time
    Pump shaft broken (pump seized)
    Faulty or worn fuel injection pump
    Faulty injector nozzles
    Low engine compression

112. **ENGINE DOES NOT RUN PROPERLY.** If the engine will start but does not run properly (smokes excessively, mis-fires, etc.), possible causes of trouble are as follows:

    Water in fuel
    Inferior fuel
    Fuel filter partially clogged
    Air cleaner element clogged
    Air traps in fuel system
    Pump out of time
    Faulty fuel injectors
    Low compression on one or more cylinders
    Faulty fuel injection pump

### BLEEDING THE SYSTEM

#### Indirect Injection

113. To bleed the fuel system on indirect injection models, proceed as follows: Open fuel tank shut-off valve, then remove the plug from primary fuel filter (1—Fig. O81) and allow fuel to flow until all bubbles disappear. Reinstall plug. Loosen bleed screw (2—Fig. O82) on the final fuel filter and acuate hand primer pump (3), if so equipped, or on later models (except series 1550), the primer lever on full supply pump, close bleed screw when bubble free fuel appears. Continue to operate the hand primer pump or primer lever on fuel supply pump an additional fifteen or twenty strokes to force any remaining air in the low pressure lines through the lines and into the tank.

NOTE: Series 1550 and 1555 tractors are not fitted with a fuel pump, however same procedure is followed using gravity force.

Attempt to start engine and if engine fails to start, or runs unevenly, loosen lines at injectors and bleed air from them either by placing fuel stop in run position and operating starting motor, or allowing engine to run at low idle. Tighten connectors when bleeding is completed.

#### Direct Injection

113A. To bleed fuel system on direct injection models, remove bleed

Fig. O82 — View showing method of bleeding final fuel filter. Refer to text.

Fig. O82A—Fuel supply pump and filters used on direct injection models of Series 1650.

plug on primary filter (Fig. O82A) and actuate primary lever on fuel supply pump until bubble free fuel flows from opening, then reinstall plug. If primary lever of supply pump fails to operate pump, turn engine until camshaft lobe releases pump rocker arm. Loosen bleed screw on final fuel filter, actuate primer lever on fuel supply pump until bubble free fuel flows from bleed screw opening, then tighten bleed screw. Continue to operate primer lever on fuel supply pump an additional fifteen or twenty strokes to force any remaining air in the low pressure lines through the lines and into the tank.

Attempt to start engine and if engine fails to start, or runs unevenly, loosen lines at injectors and bleed air from them either by placing fuel stop in run position and operating starting motor, or allowing engine to run at low idle. Tighten connections when bleeding is completed.

## INJECTORS

Series 1550, 1555 and 1600, as well as series 1650, prior to tractor serial number 187 586-000, are equipped with throttling pintle type fuel injection nozzles. In operation, some fuel is atomized to start the combustion process, but much of the fuel is emitted from the nozzle as a solid "core" which crosses the combustion chamber and enters the energy cell. As the power stroke continues, the fuel-air mixture is ejected from the energy cell into the combustion chamber where burning of the fuel is completed.

Beginning with tractor serial number 187 586-000, the series 1650 engines, as well as series 1655 engines, are fitted with Roosa-Master "pencil" type injection nozzles which inject directly into the combustion chamber and energy cells are no longer fitted in the cylinder head.

CAUTION: Fuel leaves the injector nozzle with sufficient force to penetrate the skin. Keep your person clear of nozzle spray when testing nozzles.

### All Models

**114. LOCATING A FAULTY NOZZLE.** If rough or uneven engine operation, or misfiring, indicates a faulty injector, the defective unit can usually be located as follows:

With the engine operating at low idle speed, loosen the high pressure connection at each injector in turn. As in checking spark plugs, the faulty unit is the one which, when its line is loosened, least affects the running of the engine.

If a faulty nozzle is found and considerable time has elapsed since the injectors have been serviced, it is recommended that all injectors be removed and new or reconditioned units be installed, or the nozzles be serviced as outlined in the following paragraphs. Also, energy cells should be removed and cleaned as in paragraphs 128 and 129 as a faulty injector can cause an energy cell to become carbon-fouled.

### Indirect Injection Models

**115. REMOVE AND REINSTALL.** Before loosening any fuel line connec-

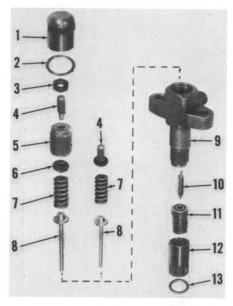

Fig. O84—Composite view showing early and late type injector used in Series 1550, 1555, 1600 and early 1650 diesel engines. Note separate adjusting screw (4) and spring seat (6) used with early type injector.

| | |
|---|---|
| 1. Cap nut | 8. Seat & spindle |
| 2. Copper gasket | 9. Holder |
| 3. Lock nut | 10. Nozzle valve |
| 4. Adjusting screw | 11. Nozzle body |
| 5. Retainer nut | 12. Cap nut |
| 6. Spring seat | 13. Copper washer |
| 7. Spring | |

tions, thoroughly clean the head surface, lines and injectors by washing with diesel fuel or a suitable solvent. After disconnecting the pressure and leak-off lines, cap all connections to prevent entry of dirt or other foreign material into fuel system. Loosen pressure line connections at injection pump to prevent bending the lines. Remove the retaining screws and carefully withdraw the injector assembly from cylinder head being careful not to strike the tip end of nozzle against any hard surface.

Thoroughly clean the nozzle recess in cylinder head before reinserting the injector assembly. No hard or sharp tools should be used for cleaning. A piece of wood dowel or brass stock properly shaped, or an approved nozzle bore cleaner should be used. Install injector with a new copper gasket and loosely install retaining nuts until all fuel lines are connected; then, tighten injector retaining nuts to a torque of 13-17 Ft.-Lbs. Bleed the injectors and lines as outlined in paragraph 113. Tighten the fuel injector inlet lines after bleeding.

**116. NOZZLE TEST.** A complete job of testing and adjusting the injector requires the use of a special tester.

The injector should be tested for spray pattern, seat leakage, back

leakage and opening pressure as follows:

**117. SPRAY PATTERN.** Operate tester handle until oil flows from injector connection, then attach the injector assembly. Close the valve to tester gage and operate tester handle a few quick strokes to purge air from injector and tester pump, and to make sure injector is not plugged or inoperative.

If a straight, solid core of oil ejects from nozzle tip without undue pressure on tester handle, open valve to tester gage and remove cap nut (1—Fig. O84). Slowly depress the tester handle and observe the pressure at which the nozzle valve opens. If opening pressure is not approximately 1925 psi (new spring) or 1750 psi (used spring), loosen locknut (3) and turn adjusting screw (4) until opening pressure is 1925 psi when testing a new injector or an injector assembly in which a new spring has been installed, or 1750 psi when testing an injector with used spring.

When opening pressure has been adjusted, again close valve to tester gage and operate tester handle at approximately 100 strokes per minute while observing spray core. Fuel should emerge from nozzle opening in one solid core in a straight line with injector body, and with no branches, splits or dribbling.

NOTE: The tester pump cannot duplicate the injection velocity necessary to obtain the operating pattern of the throttling pintle type nozzles. Also absent will be the familiar popping sound associated with the nozzle opening of spray type nozzles. Under operating velocities, the observed solid core will cross the combustion chamber and enter the energy cell. In addition, a fine conical mist surrounding the core will ignite in the combustion chamber area above the piston. The solid core cannot vary more than 7½ degrees in any direction and still enter the energy cell. While the core is the only spray characteristic that can be observed on the tester, absence of any core deviation is of utmost importance.

**118. SEAT LEAKAGE.** The nozzle valve should not leak at pressures of 150 psi less than opening pressure. To check for seat leakage, open the valve to tester gage and actuate tester handle slowly until gage pressure is 150 psi less than opening pressure of the nozzle. Maintain this pressure for at least 10 seconds, then observe the flat

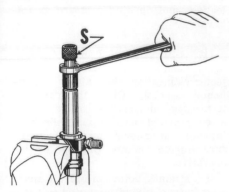

Fig. O85 — Clamp squared portion of injector holder in soft jawed vise when disasembling injector. Use centering sleeve (S) when reassembling nozzle and holder.

surface of nozzle body and the pintle tip for drops or undue wetness. If drops or wetness appear, the injector must be disassembled and overhauled as outlined in paragraph 120A.

119. BACK LEAKAGE. A back leak test will indicate the condition of the internal sealing surfaces of the nozzle assembly. Before checking the back leakage, first check for seat leakage as outlined in paragraph 118; then proceed as follows:

Turn the adjusting screw (4—Fig. O84) inward until nozzle opening pressure is set at 2300 psi. Operate the tester handle to bring gage pressure to just under the 2300 psi opening pressure, then release tester handle and observe the length of time required for gage needle to drop from 2200 psi to 1500 psi. The time should not be less than 6 seconds. A faster drop would indicate wear or scoring between piston surface of needle (10) and nozzle (11), or leakage due to improper sealing of pressure face surfaces of nozzle and holder body (9). NOTE: Leakage at tester connections or at tester check valve will show up as fast leak back in this test. If all injectors tested fail to pass the back leakage test, the tester, rather than the injectors, should be suspected.

120. OPENING PRESSURE. To assure peak engine performance, it is recommended that the six injectors installed in any engine be adjusted as nearly as possible to equal opening pressure. The recommended opening pressure is 1925 psi for a new nozzle (or used nozzle with new spring); or 1750 psi for used nozzle with a used spring. When a complete new injector and holder assembly, or a used assembly in which a new spring has been installed, is installed in an engine, the injector opening pressure

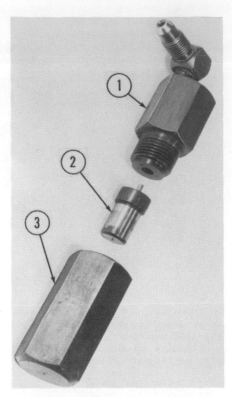

Fig. O86 — Hydraulic nozzle valve extractor for use with injector tester to remove stuck valves.

will drop as the spring becomes seated during constant operation.

After the opening pressure has been adjusted, tighten lock nut (3) and install cap nut (1). Then, recheck opening pressure to be sure adjusting screw did not move while tightening the lock nut and cap nut.

120A. MINOR OVERHAUL. The maintenance of absolute cleanliness in the overhaul of injector nozzle assemblies is of utmost importance. Of equal importance is the avoidance of nicks or scratches on any of the lapped surfaces. To avoid damage to any of the highly machined parts, only recommended cleaning kits and oil base carbon solvents should be used in the injector repair sections of the shop. The nozzle valve and body are individually fit and hand lapped, and these two parts should always be kept together as mated parts.

Before disassembling a set of injectors, immerse the units in a clean carbon solvent and thoroughly clean the outer surfaces with a **brass** wire brush. Be extremely careful not to damage the pintle end of the nozzle valve extending out of nozzle body. Rinse the injectors in clean diesel fuel.

Disassemble only one injector at a time or provide a way to keep parts of one injector assembly together. To

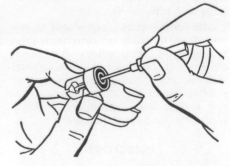

Fig. O87 — Using special scraper to clean carbon from fuel gallery in nozzle body as shown in cut-away view above.

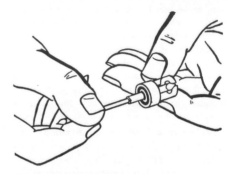

Fig. O88 — Use brass seat tool to remove carbon from valve seat.

disassemble injector, proceed as follows: Clamp the squared upper part of injector holder body in a soft jawed vise, tightening vise only enough to keep injector from slipping, or use a holder fixture. Remove cap nut (1—Fig. O84), loosen lock nut (3) and back out the adjusting screw (4) until all tension is removed from spring. Then, remove the nozzle holder nut (12). Withdraw the nozzle valve (10) from nozzle body (11) with the stem end, or if valve is stuck, use the special extractor as shown in Fig. O86. NEVER loosen valve by tapping on exposed pintle end of valve. Remove spring retainer cap nut (5), spring seat (6) (early type) spring (7) and the seat and spindle (8). Renew spring if rusted, cracked or distorted in any way. Carefully examine the spring seats and spindle and renew if chipped, cracked or damaged in any way.

Examine the lapped pressure faces of nozzle body (11) and holder (9) for nicks or scratches, and examine piston portion (large diameter) of nozzle valve (10) for scratches or scoring. Clean the fuel gallery with the special hooked scraper as shown in Fig. O87 by applying side pressure as the nozzle body is rotated. Clean the valve seat with the brass seat tool as shown in Fig. O88. Polish the seat with the pointed wooden

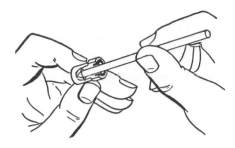

**Fig. O89 — Use wooden stick and small amount of tallow to polish valve seat.**

polishing stick and a small amount of tallow as shown in Fig. O89. Clean the pintle orifice from the inside using the proper size probe. Polish the nozzle valve seat and pintle with a piece of felt and some tallow, loosening any particles of hardened carbon with a pointed piece of brass stock. Never use a hard or sharp object such as a knife blade as any scratches will cause distortion of the injection core.

As the parts are cleaned, immerse them in clean diesel fuel in a compartmented pan. Insert the nozzle body while holding both parts below the fuel level in pan and assemble to nozzle holder while wet. Use a centering sleeve when reassembling as shown in Fig. O85 and tighten the holder nut to a torque of approximately 45-50 Ft.-Lbs. Install spindle and seat (8—Fig. O84), spring (7), seat (6) and retainer nut (5) and tighten retainer nut to a torque of 50-60 Ft.-Lbs. On early type, install adjusting screw (4) and lock nut (3). On later type, install adjusting nut (4) in retainer nut (5) prior to installing retainer nut. Adjust screw (4) to obtain proper nozzle opening pressure. Test the overhauled injector assembly as outlined in paragraphs 116 thru 120.

## Direct Injection Models

**121. REMOVE AND REINSTALL.** To remove an injector, remove side panels and hood and clean injectors, lines and surrounding area with clean diesel fuel. Remove exhaust manifold. Use hose clamp pliers to expand clamp and pull leak-off boot from injector. Disconnect high pressure line and cap all openings. Remove cap screws from nozzle clamp and remove clamp and spacer. Pull injector from cylinder head.

NOTE: Unless carbon stop (dam) seal has failed causing injector to stick, the injectors can easily be removed by hand. If injectors cannot be removed by hand, use OTC puller JDE-28, or equivalent, and be sure to pull injector straight out of bore. DO NOT attempt to pry injector from

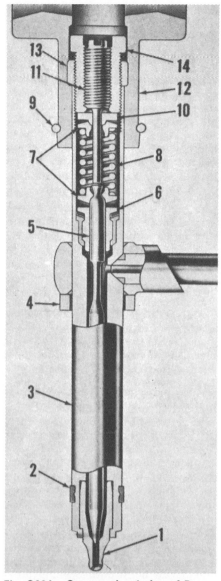

**Fig. O89A—Cross sectional view of Roosa-Master "pencil" injector. Nozzle tip (1) and valve guide (6) are parts of finished body and are not serviced separately.**

1. Nozzle tip
2. Carbon seal (dam)
3. Body
4. Seal washer
5. Nozzle valve
6. Valve guide
7. Pressure spring
8. Pressure spring
9. Boot clamp
10. Ball washer
11. Lift adjusting screw
12. Boot
13. Lock nut
14. Pressure adjusting screw

cylinder head or damage to injector could result.

When installing injector, be sure injector bore is free of carbon. If necessary, use OTC reamer JDE-39, or equivalent, and be sure to clean bore thoroughly after reaming. Use new compression seal and carbon stop (dam) seal on injector body and insert injector into its bore using a slight twisting motion. Install hold-down clamp and spacer and tighten retaining cap screw to 20 Ft.-Lbs. torque. Install exhaust manifold, then bleed fuel system as outlined in paragraph 113A.

Reinstall hood and side panels.

**121A. NOZZLE TEST.** A complete job of nozzle testing and adjusting requires the use of an approved nozzle tester. Only clean, approved testing oil should be used in the tester tank. The nozzle should be tested for spray pattern, opening pressure, seat leakage and back leakage (leak-off). Injector should produce a distinct audible chatter when being tested and cut off quickly at end of injection with a minimum of seat leakage.

NOTE: When checking spray pattern, turn nozzle about 30 degrees from vertical position. Spray is emitted from nozzle tip at an angle to the centerline of nozzle body and unless injector is angled, the spray may not be completely contained by the beaker. Keep your person clear of the nozzle spray.

**121B. SPRAY PATTERN.** Attach injector to tester and operate tester at approximately 60 strokes per minute and observe the spray pattern. A finely atomized spray should emerge at each of the four nozzle holes and a distinct chatter should be heard as tester is operated. If spray is not symmetrical and is streaky, or if injector does not chatter, overhaul injector as outlined in paragraph 121F.

**121C. OPENING PRESSURE.** The correct opening pressure is 2550-2650 psi for used nozzles, or 2750-2850 psi for new nozzles or when new spring has been installed. There should not be more than 100 psi difference between the injectors used in any engine except when intermixing new and used nozzle assemblies.

To check nozzle opening pressure, attach injector to tester and pump tester several times to actuate injector and a low valve to seat normally. Now observe tester gage and quickly pump pressure until gage falls off rapidly. This indicates the nozzle opening pressure and should be 2550-2650 psi for used nozzles, or 2750-2850 psi for new nozzles or when new spring has been installed.

If opening pressure is not correct but nozzle will pass all other tests, adjust opening pressure as follows: Loosen the pressure adjusting screw lock nut (13—Fig. O89A), then hold the pressure adjusting screw (14) and back out the valve lift adjusting screw (11) at least one full turn. Actuate tester and adjust nozzle pressure by turning adjusting screw as required. With the correct nozzle opening pressure set, gently turn the valve lift adjusting screw in until it bottoms, then back it out ½-turn which will provide the correct valve lift of 0.009 Hold pressure adjusting screw and tighten lock nut to a torque of 70-75 In.-Lbs.

NOTE: A positive check can be made to see that the lift adjusting screw is bottomed by actuating tester until a pressure of 250 psi above nozzle opening pressure is obtained. Nozzle valve should not open.

121D. SEAT LEAKAGE. To check nozzle seat leakage, proceed as follows: Attach injector on tester in a horizontal position. Raise pressure to approximately 2400 psi, hold for 10 seconds and observe nozzle tip. A slight dampness is permissible but should a drop form in the 10 seconds, renew the injector or overhaul as outlined in paragraph 121F.

121E. BACK LEAKAGE. Attach injector to tester with tip slightly above horizontal. Raise and maintain pressure at approximately 1500 psi and observe leakage from return (top) end of injector. After first drop falls, the back leakage should be 5 to 8 drops every 30 seconds. If back leakage is excessive, renew injector or overhaul as outlined in paragraph 121F.

121F. OVERHAUL. First wash the unit in clean diesel fuel and blow off with clean, dry compressed air. Remove carbon stop seal and sealing washer. Clean carbon from spray tip using a brass wire brush. Also, clean carbon or other deposits from carbon seal groove in injector body. DO NOT use wire brush or other abrasive on the Teflon coating on outside of nozzle body between the seals. Teflon coating can be cleaned with a soft cloth and solvent. Coating may discolor from use, but discoloration is not harmful.

Clamp the nozzle in a soft jawed vise, loosen lock nut (13—Fig. O89A) and remove pressure adjusting screw (14), ball washer (10), upper spring washer (7), spring (8) and lower spring seat (7).

If nozzle valve (5) will not slide from body when body is inverted, use a suitable valve extractor; or reinstall on nozzle tester with spring and lift adjusting screw removed, and use hydraulic pressure to remove the valve.

Nozzle valve and body are a matched set and should never be intermixed. Keep parts for one injector separate and immerse in clean diesel fuel in a compartment pan, as unit is disassembled.

Clean all parts thoroughly in clean diesel fuel using a brass wire brush and lint-free wiping towels. Hard carbon or varnish can be loosened with a suitable, non-corrosive solvent.

Clean the spray tip orifices first with an 0.008 cleaning needle held in a pin vise; then with a 0.011 needle.

Clean the valve seat using a Valve Tip Scraper and light pressure while rotating scraper. Use a Sac Hole Drill to remove carbon from sac hole.

Piston area of valve and guide can be lightly polished by hand, if necessary, using Roosa Master No. 16489 lapping compound. Use the valve retractor to turn valve. Move valve in and out slightly while rotating, but do not apply down pressure while valve tip and seat are in contact.

If back leakage was not correct and polishing of valve does not bring unit within specifications, a new valve can be installed.

Valve and seat are ground to a slight interference angle. Seating areas may be cleaned up, if necessary, using a small amount of 16489 lapping compound, very light pressure and no more than 3-5 turns of valve on seat. Thoroughly flush all compound from valve body after polishing.

NOTE: Never lap a new valve to an old injector body seat. If lapping with the old valve did not restore seat, lapping with a new valve will only destroy angle of the new valve.

When assembling the nozzle, back lift adjusting screw (11) several turns out of pressure adjusting screw (14), and reverse disassembly procedure using Fig. O89A as a guide.

121G. Connect nozzle to tester and turn pressure adjusting screw (14) into nozzle body until opening pressure is 2550-2650 psi with a used spring, or 2750-2850 psi with a new spring.

After spring pressure has been adjusted and before tightening lock nut (13), valve lift must be adjusted as follows: Hold pressure adjusting screw from turning and using a small screwdriver, thread lift adjusting screw (11) into nozzle unit until it just bottoms on valve (5). To be sure screw is bottomed, actuate tester handle until a pressure 250 psi above opening pressure is obtained. Nozzle valve should not open.

With lift adjusting screw bottomed, turn screw counter-clockwise ½-turn to establish the recommended 0.009 valve lift. Tighten the lock nut to a torque of 70-75 In.-Lbs. while holding pressure adjusting screw from turning. Recheck the opening pressure; them recheck spray pattern, seat leakage and back leakage. Install the unit as outlined in paragraph 121.

NOTE: When adjusting opening pressure and injector has not been disassembled, back out the lift adjusting screw at least one full turn before moving pressure adjusting screw. This will prevent accidental bottoming of lift screw while attempting to set the pressure.

## FUEL INJECTION PUMP

### All Diesel Models

122. All diesel model tractors are equipped with Roosa-Master fuel injection pumps. Following is the tractor and injection pump application.

Series 1600, tractor serial number 124 420-000—131 791-000, DBGFC 629-7BH. Series 1600, tractor serial number 131 782-000—147 568-000, DBGFC 629-1ED. Series 1650, tractor serial number 147 569-000—187 585-000, DBGFC 629-1DH. Series 1650 and 1655 tractor serial number 187 586-000—Up, DBGFC 633-12DH. Series 1550 and 1555 tractor serial number 160 000-000—Up, DBGFC 627-2DH.

The series 1600 pumps (7BH and 1ED) are not fitted with an automatic advance and can be used interchangeably on all series 1600 models. All other pumps (1DH, 12DH and 2DH) are equipped with integral hydraulic automatic advance mechanism.

Because of the special equipment needed, and the skill required of the servicing personnel, service of injection pumps is generally beyond the scope of that which should be attempted in the average shop. Therefore, this section will include only timing of pump to engine, removal and installation and the linkage adjustments which control engine speeds.

If additional service is required, the pump should be turned over to an Oliver facility that is equipped for diesel injection pump service, or to an authorized Roosa-Master diesel service station. Inexperienced personnel should never attempt to service diesel fuel injection pumps.

122A. PUMP TIMING. To check injection pump timing, shut off the fuel supply valve and remove timing window cover from fuel injection pump. Turn engine until No. 1 piston is coming up on compression stroke, then continue to turn engine slowly until the following degree timing marks on engine flywheel align with timing pointer.

Series 1600 (All) ...........7°BTDC
Series 1650 (147 569-187 585) 2°BTDC
Series 1650 (187 586-)
-1655 ..................4°BTDC
Series 1550-1555 ..........2°ATDC

With engine flywheel positioned as indicated above, the timing lines in injection pump timing window should be exactly aligned as shown in Fig. O90. If pump timing lines are not in register as shown, loosen the pump mounting stud nuts and turn pump as required to bring lines into register, then tighten mounting stud nuts securely. Recheck setting by turning engine two complete revolutions forward; or turn engine about one-half turn backward, then forward again, until the correct flywheel timing mark is again aligned with timing pointer and recheck the timing marks in injection pump timing window. Readjust timing if necessary.

CAUTION: Do not loosen the pump mounting stud nuts while engine is running if timing is being set by dynamometer method.

NOTE: When aligning flywheel timing mark with pointer, engine must be turned in normal direction of rotation. If mark is turned past pointer, continue to turn the engine through two revolutions to again align mark and pointer, or back engine up at least ¼-turn and then turn forward to timing mark. Turning the engine backwards to just the point where timing mark is aligned with pointer will result in incorrect timing due to gear backlash.

**123. R&R PUMP.** Thoroughly clean outside of pump, lines and connections. Shut off the fuel supply valve, remove timing window cover from fuel injection pump and turn engine so that both timing marks are visible in timing window and the flywheel timing mark given in paragraph 122A is aligned with pointer. Then reinstall injection pump timing window cover

Disconnect battery ground strap to prevent engine from being turned accidentally while pump is removed; then, proceed as follows: Disconnect throttle rod, stop control wire and cable, fuel supply line and excess fuel return line from fuel injection pump. Remove the injector pressure line clamps and disconnect the lines at connections near pump end of lines. Loosen the injector pressure line connections at injectors so lines can be spread to remove pump and immediately cap all open lines and connectors. Protective caps and plugs are available and are listed in Oliver parts catalog. Remove the three pump mounting stud nuts and remove pump from mounting studs and drive shaft, spreading the injector lines to allow pump to be moved rearward. After removing pump, remove seals from pump drive shaft and carefully in-

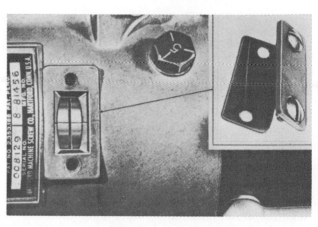

Fig. O90 — When No. 1 cylinder is on compression stroke and proper flywheel timing mark is aligned with pointer, injection pump timing marks should be aligned as shown.

spect the shaft. If shaft is broken at the safety groove or is otherwise damaged, renew shaft as outlined in paragraph 124.

NOTE: Aligning flywheel and pump timing marks prior to removing pump is not required procedure, but is an aid in reinstalling pump. If pump drive shaft is removed or renewed in conjunction with pump removal, either reinstall pump without regard to timing procedure and reinstall drive shaft as outlined in paragraph 125, or install shaft in pump and reinstall pump and shaft as assembly as outlined in paragraph 126.

Before reinstalling pump, check to be sure engine was not turned while pump was removed. If engine was turned, or pump was removed without aligning timing marks as outlined in removal procedure, proceed as follows: Either remove the rocker arm cover and turn engine until intake valve of No. 1 cylinder just closes or loosen the No. 1 cylinder injector mounting stud nuts and turn engine until compression leak occurs around the loosened injector. Then, continue to turn engine slowly until the flywheel timing mark given in paragraph 122A is aligned with pointer and reinstall the rocker arm cover, or remove the No. 1 cylinder injector and reinstall same with new sealing washer.

Remove the timing window cover from outer side (throttle control rod side) of fuel injection pump and with a screwdriver inserted in rotor drive slot, turn pump rotor so that pump timing marks are aligned. Install two new pump drive shaft seals on shaft with lips of seals facing away from each other. Lubricate the seals and shaft area between seals with Lubriplate or equivalent grease and carefully install the pump over drive shaft and seals. Take care not to roll lip of rear seal back as pump is being installed; if this happens, remove pump and renew the seal. (Note: Use

of Roosa-Master seal installation tool 13369 is recommended to install seals on shaft and use of Roosa-Master seal compressor tool 13371 is recommended when installing pump over the drive shaft seals). When pump is installed over drive shaft far enough to engage drive shaft tang with slot in pump rotor, it may be necessary to rock the rotor slightly with pencil eraser tip inserted through timing window or to rock the pump assembly slightly to engage the drive shaft tang and rotor slot. When pump mounting slots are engaged with the mounting studs and pump body contacts mounting plate, rotate the pump assembly in the mounting slots as required to bring the pump timing marks into exact alignment, then securely tighten the pump mounting nuts. Turn the engine through two revolutions, or back engine up at least ¼-turn, then slowly turn forward until flywheel timing mark is again aligned with pointer and recheck pump timing marks. Loosen the pump mounting stud nuts and adjust timing if necessary.

Bleed the fuel system as outlined in paragraph 113 or 113A.

**124. RENEW PUMP DRIVE SHAFT.** To remove the pump drive shaft, gear and hub assembly, remove the gear cover from front side of timing gear cover. If injection pump is installed, shut off the fuel supply valve and remove the timing window cover from outer side of injection pump to drain fuel; otherwise, diesel fuel will run into engine when shaft is removed. Then, withdraw the drive gear, hub and shaft assembly from opening in front of timing gear cover. Cut the locking wire and remove the drilled head cap screws retaining gear to hub. Clamp tang of drive shaft in soft jawed vise and remove the hub retaining nut. Pull or press hub from shaft and Woodruff key. If fuel injection pump was not removed, install new drive shaft as outlined in paragraph 125. If pump is removed, new shaft can be installed in pump, then install pump and shaft as outlined in paragraph 126; or if desired, pump can be reinstalled without shaft and the shaft can then be installed as in paragraph 125.

125. To install new drive shaft with pump installed on tractor, proceed as follows: Check to see that injection pump is mounted so that mounting studs are centered in slots; if not, loosen the stud nuts, turn pump so that studs are centered in slots and tighten the nuts. Turn engine until No. 1 piston starts up on compression stroke, then continue to turn engine slowly until flywheel timing mark given in paragraph 122A is aligned with pointer. Clamp tang of drive shaft in soft jawed vise, install Woodruff key in slot and install drive hub with retaining nut and washer. Tighten the nut to a torque of 35-40 Ft.-Lbs. Remove shaft from vise and install new drive shaft seals with seal lips facing away from each other. Use of Roosa-Master seal installation tool 13369 is recommended. Lubricate the seals and shaft area between seals with Lubriplate or similar grease. Note the offset dimples in end of drive tang on shaft and in slot of pump rotor. Carefully insert shaft into pump with offset dimples on shaft tang and rotor slot aligned and work the rear seal lip into bore of pump body with a dull pointed tool. When drive tang is engaged in rotor slot, remove timing window cover from fuel injection pump and turn the drive shaft and hub so that pump timing marks are aligned as shown in Fig. O90. While holding pump rotor from turning with screwdriver inserted in slot below the timing marks in timing window, install drive gear on hub and turn gear counter-clockwise as far as possible to eliminate gear backlash. Install the two drilled head cap screws in set of two holes that are aligned in gear and hub, tighten the capscrews securely and install locking wire. Insert thrust spring and thrust button in front end of drive shaft and install drive gear cover to timing gear cover with new gasket. Recheck pump timing marks and adjust if necessary as outlined in paragraph 123. Bleed diesel fuel system as in paragraph 113 or 113A.

126. To install new drive shaft when fuel injection pump is removed, either reinstall pump and then install shaft as outlined in paragraph 125, or proceed as follows:

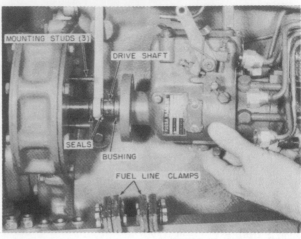

Fig. O91 — Remove fuel line clamps and spread fuel lines as shown to remove pump. Use Roosa-Master seal compressor tool 13371 as shown to aid in reinstalling pump over drive shaft seals.

Remove timing window cover from outer side of fuel injection pump. Install new seals on drive shaft with lips of seals facing away from each other. Use of Roosa-Master seal installation tool 13369 is recommended. Lubricate seals and shaft area between seals with Lubriplate or similar grease, then carefully insert shaft into pump so that offset dimple in end of shaft tang is aligned with offset dimple in slot of pump rotor. Using a dull pointed tool, carefully work lip of rear seal into bore of pump housing or compress seal with Roosa-Master seal compressor tool 13371 when installing shaft. With drive tang of shaft engaged in slot of pump rotor, install the pump on tractor with mounting hole slots centered on the mounting studs. Insert Woodruff key in slot in pump shaft. Be sure that drive gear is a free fit on hub, then install hub, washer and retaining nut on shaft. Temporarily install drive gear on hub in any position and with the drilled head cap screws finger tight. Then, tighten the hub retaining nut to a torque of 35-40 Ft.-Lbs. and remove the drive gear. Turn engine so that No. 1 piston is coming up on compression stroke, then continue to turn engine slowly until the flywheel timing mark given in paragraph 122A is aligned with pointer. Turn injection pump drive shaft so that the pump timing marks are aligned as shown in Fig. O90. While holding pump from turning with screwdriver inserted in slot below pump timing marks, install pump drive gear on hub and turn gear as far counter-clockwise as possible to eliminate all gear backlash. Install the two drilled head cap screws into a set of holes that are aligned in drive gear and hub, tighten the capscrews and secure with wire. Insert the thrust button spring and thrust button in front end of pump drive shaft and install drive gear cover to

timing gear cover using a new gasket. Recheck pump timing as outlined in paragraph 122A and readjust if necessary. Bleed the diesel fuel system as outlined in paragraph 113 or 113A.

127. **GOVERNOR ADJUSTMENT.** Both the high idle speed and low idle speed adjustments are preset at the factory and normally should not be disturbed. However, should it become necessary, adjust the governed speeds as follows:

Start engine and bring to normal operating temperature. Break the seals on the high idle (rear) adjusting screw and/or low idle adjusting screw as necessary (refer to Fig. O92); then, proceed as follows:

To adjust high idle speed, place throttle lever in high idle position, loosen lock nut and turn adjusting screw in or out as required to obtain a high idle speed of 2100 RPM on Series 1600 models or 2420 RPM on Series 1550, 1555, 1650 and 1655. Tighten the lock nut and install new wire seal.

To adjust low idle speed, position

Fig. O92 — Adjusting the high speed screw. Low idle speed is adjusted by front screw.

throttle lever at low idle position, loosen lock nut and turn adjusting screw (front screw) in or out to obtain a low idle speed of 675 RPM on Series 1550 and 1600 models, 650-675 RPM on Series 1650 models with indirect injection, or 800 RPM on Series 1650 with direct injection, 1555 and 1655. Tighten lock nut and install new wire seal.

NOTE: The throttle control lever on fuel injection pump is spring loaded and should slightly over-travel the high and low speed stops on the injection pump governor control shaft. To properly adjust the throttle linkage, proceed as follows: With engine not running, disconnect throttle control rod at fuel injection pump and place hand throttle lever in slow idle position. Loosen the lock nut on front end of control rod. While holding the injection pump throttle shaft against the low idle speed stop, adjust length of control rod so that injection pump throttle lever must be moved $\frac{1}{16}$-$\frac{1}{8}$ inch against spring pressure to reinstall the linkage pin. Then tighten lock nut on control rod and move the hand throttle lever to high idle speed position. Loosen the lock nut on stop screw in throttle bellcrank at instrument support panel, adjust the stop screw until the injection pump lever over-travels high idle speed stop $\frac{1}{16}$ to $\frac{1}{8}$-inch and then tighten the lock nut.

If throttle hand lever tends to creep at high idle speed, tighten the friction adjusting nut located at lever end of control lever shaft in instrument panel support.

## ENERGY CELLS

### Indirect Injection Nozzles

These assemblies (Fig. O93) are mounted directly opposite from the fuel injectors in the cylinder head.

In almost every instance where a carbon-fouled or burned energy cell is encountered, the cause is traceable either to a malfunctioning injector, incorrect fuel or incorrect installation of the energy cell. Manifestations of a fouled or burned unit are misfiring, exhaust smoke, loss of power or pronounced detonation (knock).

128. REMOVE AND REINSTALL. Any energy cell can be removed without removing any engine component. To remove an energy cell, first remove the threaded cell holder plug (1—Fig. O93) and take out the cell holder spacer (2). With a pair of thin nosed pliers, remove cell cap (3). To remove the cell body (4), screw a

Fig. O93 — Energy cells should be removed and cleaned when servicing injectors.

1. Holder plug
2. Spacer
3. Energy cell cap
4. Energy cell body

suitable threaded puller into end of the cell body. If puller is not available, remove the fuel injector; and use a brass rod to drift the cell body out of the cylinder head. Clean unit as outlined in paragraph 12 9and reinstall by reversing the removal procedure.

129. CLEANING. Clean all carbon from front and rear crater of cell body using a brass scraper or a shaped piece of hard wood. Be sure body orifice is clean. Clean the exterior of the energy cell with a brass wire brush, and soak the parts in carbon solvent. Reject any part which shows signs of leakage or burning. If parts are not burned, re-lap sealing surfaces of cap (3—Fig. O93) and body (4) by using a figure 8 motion on a lappling plate coated with fine compound.

## PRIMER PUMP

### Early Series 1600

130. R&R AND OVERHAUL. Shut off fuel, disconnect inlet and outlet lines from primer pump and cap off open ends of fuel lines. Remove the two mounting screws and remove pump from tractor.

Refer to Fig. O94. Loosen bail clamp from plunger, then disengage ends of bail wire from primer pump body. Remove plunger, plunger guide and

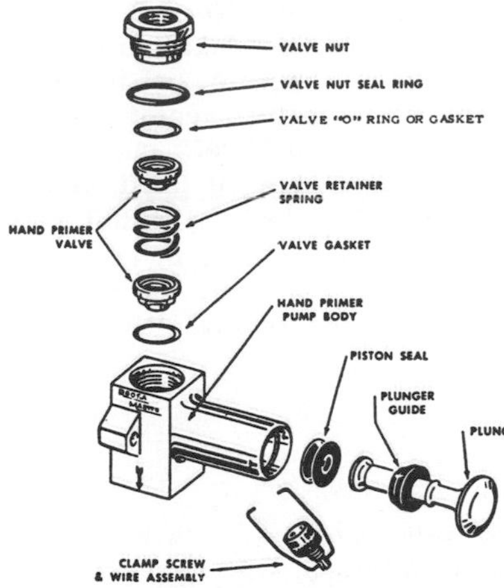

Fig. O94 — Exploded view of diesel fuel system primer pump used on early Series 1600; late Series 1600, 1650 and 1655 tractors are equipped with a fuel pump which incorporates a hand primer pump.

VALVE NUT
VALVE NUT SEAL RING
VALVE "O" RING OR GASKET
VALVE RETAINER SPRING
HAND PRIMER VALVE
VALVE GASKET
HAND PRIMER PUMP BODY
PISTON SEAL
PLUNGER GUIDE
PLUNGER
CLAMP SCREW & WIRE ASSEMBLY

piston seal from body. Place pump body in vise, then remove valve nut. Seals, valves and spring can now be removed.

Examine all parts for tears, undue wear, or other damage and renew as necessary. If plunger guide is to be renewed, old guide can be cut. Lubricate new guide and slide over plunger with chamfer toward inside of pump body. Install valves so that they open in the same direction as the arrow embossed on the primer pump body.

Bleed system as outlined in paragraph 113.

## FUEL SUPPLY PUMP

### Series 1650-1655—Late Series 1600

131. Late Series 1600 and Series 1650 and 1655 diesel models are equipped with a diaphragm type fuel supply pump which incorporated a

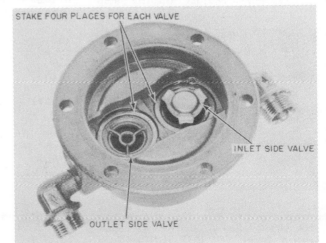

STAKE FOUR PLACES FOR EACH VALVE

INLET SIDE VALVE

OUTLET SIDE VALVE

Fig. O95—Inlet and outlet valves of fuel lift pump are retained by staking as shown.

fuel system primer. Note: Engine may stop in a position that hand primer lever will not operate fuel pump diaphragm; if so, turn engine so that camshaft is in different position.

Overhaul of the fuel lift pump is conventional. Inlet and outlet valves are retained by staking metal of cover at four points around the valve. To renew valves, remove the staking with a small sharp chisel and stake new valve in place as shown in Fig. O95.

## MANIFOLD PREHEATER

### All Models So Equipped

132. The inlet manifold is fitted with a solenoid controlled preheater unit which is installed in lower side of manifold.

Operation of the preheater can be checked by depressing the control switch located on the instrument panel for about 30 seconds and then placing your hand on the manifold to see if it is warm.

Service on the preheater and solennoid is accomplished by renewing the units.

# NON-DIESEL GOVERNOR

All Series non-diesel engines are fitted with a flyweight type governor that is mounted on the front face of engine timing gear cover and is driven by the camshaft gear. While changes have been made on the governor, governors remain basically similar.

## ADJUSTMENT

### All Non-Diesel Models

133. **GOVERNOR TO CARBURETOR LINKAGE.** Refer to Fig. O96 and check adjustment of carburetor to governor linkage as follows: With engine not running, disconnect link from governor to carburetor at governor lever end and place throttle lever in high idle speed position. With carburetor throttle shaft in wide open position, forward end of link should be $\frac{1}{16}$-inch forward of the top hole in governor lever. If not, loosen lock nut at carburetor end of link and turn link in or out until length is correct. Reinstall link in governor lever and tighten lock nut.

134. **LOW IDLE SPEED.** After adjusting carburetor to throttle linkage as outlined in paragraph 133, start engine and bring to normal operating temperature. Place throttle lever in slow idle position and check slow idle speed. Readjust throttle stop

**Fig. O97 — Adjusting throttle linkage on Series 1600. Refer to Fig. O98 for Series 1550, 1555, 1650 and 1655.**

screw on carburetor if necessary to obtain a slow idle speed of 400 RPM on Series 1550 gasoline models, 550 RPM on Series 1550 LP-Gas models, 300 RPM on Series 1600 gasoline models, 525 RPM on Series 1600 and 1650 LP-Gas models, 450 RPM on Series 1650 gasoline models or 800 RPM on all Series 1555 and 1655.

135. **HIGH IDLE SPEED.** Prior to checking high idle speed, check carburetor to governor linkage adjustment as outlined in paragraph 133 and slow idle speed as in paragraph 134. Then, with engine running and at normal operating temperature, place throttle lever in high idle position

and check engine RPM. High idle speed should be 2420 RPM on Series 1550, 1555, 1650 and 1655 and 2100 RPM on Series 1600. If high idle speed is not as specified, stop engine, place throttle lever in slow idle position, loosen bumper screw lock nut (see Fig. O99) and back-out bumper screw until governor operating lever is released. Then, proceed as follows:

SERIES 1600 MODELS. Refer to Fig. O97 and loosen adjusting nuts on governor control rod at rear of carburetor. Start engine and with throttle lever in high idle speed position, adjust length of control rod by turning adjusting nuts in or out to obtain high idle speed of 2100 RPM. If length of threads on rod does not permit proper high idle speed adjustment, stop engine and readjust position of stop screw on bellcrank in instrument panel support at rear end of control rod. When proper high idle speed is obtained with adjusting nuts on control rod and jam nut on bellcrank stop screw tight, readjust bumper screw as outlined in paragraph 136.

SERIES 1550, 1555, 1650 and 1655 MODELS. Refer to Fig. O98 and with engine not running, place throttle lever in high idle speed position. If spring lever (see inset) is not parallel

**Fig. O96 — View showing adjustment of governor arm to carburetor throttle shaft link. Although Series 1600 is shown, adjustment of Series 1550, 1555, 1650 and 1655 is similar.**

**Fig. O98 — Adjustment points for Series 1550, 1555, 1650 and 1655 throttle linkage. Refer to text for adjustment procedures. Series 1600 throttle linkage adjustment is shown in Fig. O97.**

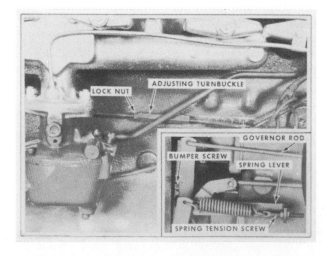

**Fig. O99 — Adjust bumper screw to minimize surging at high idle no load speeds. Refer to text for proper adjustment procedure.**

**Fig. O100 — Exploded view of Series 1550, 1555, 1650 and 1655 governor assembly. Series 1600 governor is similar except that a single control arm as shown in Fig. O96 is used.**

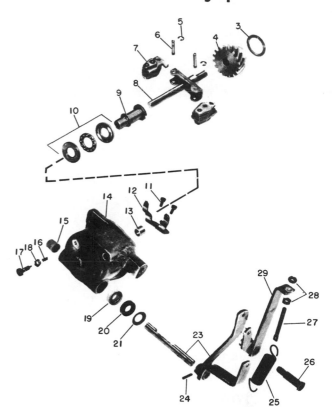

3. Thrust washer
4. Drive gear
5. Snap rings
6. Pins
7. Governor weights
8. Shaft & spider
9. Sleeve
10. Bearing
11. Screw & lock washer
12. Fork
13. Bushing
14. Housing
15. Bushing
16. Spring
17. Bumper screw
18. Nut
19. Bearing
20. Seal
21. Retainer
23. Shaft & lever assy.
24. Pin
25. Spring
26. Pivot bolt
27. Adjusting screw
28. Lock nuts
29. Lever

with engine crankshaft, loosen lock nut and turn adjusting turnbuckle on governor control rod to bring spring lever to this position. If length of threads on governor control rod do not permit this adjustment, loosen jam nut on stop screw in bellcrank at rear end of governor control rod in instrument panel support and adjust stop screw so that spring lever is parallel with engine crankshaft. Tighten lock nut and/or jam nut. Start engine, place throttle lever in high idle position and adjust spring tension screw (see inset) so that engine high idle speed is 2420 RPM with spring tension screw lock nuts tight. Then, readjust bumper screw as outlined in paragraph 136.

**136. BUMPER SCREW.** Engine surging at high idle speed can be eliminated by adjustment of the bumper screw (Fig. O99). Stop engine, loosen bumper screw lock nut and turn bumper screw inward slightly. Restart engine and check for surging at high idle speed. Continue to make slight adjustments to bumper screw until surging is eliminated. DO NOT turn bumper screw while engine is running or turn screw in farther than necessary to eliminate surging. Tighten bumper screw lock nut and recheck low idle RPM as in paragraph 134.

### R&R AND OVERHAUL

NOTE: Information in the following paragraph applies only to those governors which have laminated flyweights. Beginning with engine serial number 123687 (gasoline) or 123385 (LP-Gas), die-cast flyweights were introduced and subsequently at engine serial number 126968, the die-cast flyweights were replaced by forged steel flyweights. Service for governors with die-cast, or forged steel flyweights, requires complete shaft, weights and gear assembly; housing and lever assembly; or complete governor assembly.

In addition, lever shaft bushing (15-0100)

was replaced by a bearing and spacers at engine serial number 126968.

**137. Governor can be removed after** disconnecting linkage, loosening fan belt and removing the governor to timing gear cover retaining cap screws. Carefully withdraw the unit from timing gear cover. DO NOT lose the thrust washer which is located between governor gear and crankcase.

Refer to Fig. O100 for exploded view of the governor used on all Series except Series 1600. Series 1600 governor is similar except that a single control arm as shown in Fig. O96 is used. Governor gear (4—Fig. O100) can be removed by using a suitable puller. Disassembly of weight unit and thrust bearing assemblies is evident.

Governor gear hub rotates in a sleeve and bushing assembly which is pressed into front face of crankcase. The sleeve and bushing assembly can be renewed by removing timing gear cover and using a suitable puller. Inside diameter of governor gear bushing is 1.0015-1.002. Outside diameter of governor gear is 0.999-0.9995 and should have 0.002-0.003 operating

clearance in bushing. Renew bushing and/or gear when operating clearance exceeds 0.005.

NOTE: Approximately March 1, 1966, a one-piece governor bushing replaced the two-piece sleeve and bushing assembly on all non-diesel production engines. The thrust washer (3—Fig. O100) is NOT used with the one-piece bushing.

When installing a new one-piece governor bushing, oil groove is toward rear and front of bushing must protrude exactly 3/16-inch from front of crankcase. Use caution during this operation as depth of bushing is critical.

Thrust washer (3) will still be included in governor shaft and weight repair packages, however, if the one-piece bushing is installed, the thrust washer MUST NOT be used.

Any further disassembly and/or overhaul is obvious after an examination of the unit and reference to Fig. O100. Use caution when renewing parts as both $\frac{3}{8}$ and $\frac{7}{16}$-inch shafts (8) are available for series 1555, 1650 and 1655.

After reinstallation, readjust governor as outlined in paragraphs 133 through 136.

# COOLING SYSTEM

### RADIATOR

**138. To remove the radiator, first** drain cooling system, then remove hood side panels and hood. On models

so equipped, disconnect oil cooler from radiator. Unbolt fan blades from hub and let fan rest in fan shroud. Disconnect upper and lower hoses, unbolt radiator from tractor front frame

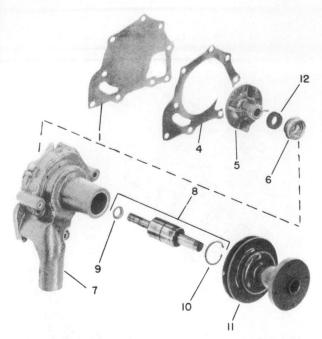

**Fig. O102 — Exploded view of water pump assembly typical of that used on all models.**

1. Cover
4. Gasket
5. Impeller
6. Seal
7. Body
8. Shaft & bearing assy.
9. Slinger
10. Snap ring
11. Pulley
12. Seal seat

of shaft, then using a press, push the shaft and bearing assembly out of impeller and pump body. Seal, seal seat and slinger can now be renewed. Shaft and bearing assembly are available as a unit only.

When reinstalling impeller, press same on shaft until rear side is flush with rear end of pump shaft. Pulley is correctly installed when distance between front face of pulley and rear face of housing is $6\frac{21}{32}$ inches for Series 1600, 1650 and 1655, or $8\frac{29}{32}$ inches for Series 1550 and 1555.

NOTE: A spacer (not shown) is used between pulley and fan on Series 1650 and 1655.

## THERMOSTAT

140. To remove thermostat, remove hood side panels and hood. Drain coolant and disconnect upper radiator hose. Remove cap screws from air cleaner bracket and remove air cleaner, thermostat housing and thermostat.

Thermostat may be either a bellows or a pellet type. Bellows type should start to open at 167-172 degrees F. and be fully open at 200 degrees F. The pellet type should start to open at 175 degrees F. and be fully open at 190 degrees F.

and lift same from tractor. Grille need not be disturbed.

### WATER PUMP

139. **R&R AND OVERHAUL.** To remove water pump, drain cooling system and remove left side panel. Unbolt fan blades from hub and let fan blades rest in fan shroud. Remove fan belt, disconnect lower radiator hose from pump, then unbolt and remove pump from left side of tractor.

To disassemble the water pump, refer to Fig. O102 and proceed as follows: Remove pump pulley and pump cover. Remove snap ring from front

# IGNITION AND ELECTRICAL SYSTEM

### SPARK PLUGS

#### All Non-Diesel Models

141. Recommended normal duty spark plugs for all Series 1550 and 1555 non-diesel engines are: AC, C-85 Com.; Autolite, BT-4; Champion, D-16; Prestolite, 18-4. Recommended heavy duty spark plugs are: AC, 83-S Com.; Autolite, BT-4; Champion, D-14; Prestolite, 18-4.

Recommended normal and heavy duty spark plugs for all Series 1600, 1650 and 1655 gasoline engines are: AC, 44XLS; Autolite, AG42; Champion, N-11-Y; Prestolite, 14GT42. Recommended spark plugs (all duty) for LP-Gas engines are: AC, C44XL; Autolite, AG4; Champion, N6; Prestolite, 14G3.

Electrode gap is 0.025 and installation torque value is 30-34 Ft.-Lbs.

### DISTRIBUTOR

#### Series 1600 Non-Diesel Models

142. Series 1600 non-diesel models are equipped with a Delco-Remy 1112632 ignition distributor. Refer to Fig. O103. Specifications are as follows:

**D-R 1112632 Distributor**

Breaker contact gap............0.016
Cam angle, degrees...........31-34
Breaker arm spring tension, oz..17-21
Start advance..........0.7-3.5 @ 300
Intermediate advance ...2.5-4.5 @ 370
7.5-9.5 @ 850
Maximum advance.....11-13 @ 1200

Advance data is in distributor degrees and distributor RPM. Breaker arm spring tension is taken at center of breaker contact point and is 19-23 ounces if taken at rear of contact point.

### Series 1550-1555-1650-1655 Non-Diesel Models

143. Series 1550, 1555, 1650 and 1655 non-diesel models are equipped with a Holley D-2563AA ignition distributor. Refer to exploded view in Fig. O104. Specifications are as follows:

**Holley D-2563AA**

Breaker contact gap............0.025
Cam angle, degrees @
idle speed..................35-38
Centrifugal advance curve:

| Distributor RPM | Degrees Advance |
|---|---|
| 225 | 0-1½ |
| 275 | ½-2¾ |
| 325 | 2-4 |
| 900 | 8-10 |
| 1200 | 11-13 |

Vacuum advance @ 1000 distributor RPM:

| Inches Hg. | Degrees Advance |
|---|---|
| 7 | 0 |
| 10 | 0-3 |
| 13 | 3-6 |
| 16 | 6-8 |
| 20 | 6-8 |

### IGNITION TIMING

#### All Non-Diesel Models

144. Ignition distributor can be installed and timing can be adjusted by static timing as outlined in paragraph 145. However, to obtain more accurate timing adjustment, ignition timing should be adjusted by using timing light as outlined in paragraph 146.

145. **DISTRIBUTOR INSTALLATION AND STATIC TIMING.** Remove flywheel timing hole cover and turn

engine until the No. 1 piston is starting up on compression stroke. Then, continue turning engine slowly until the proper flywheel timing mark is aligned with pointer. Be sure ignition breaker point gap is properly adjusted as specified in paragraph 142 or paragraph 143. If engine has been overhauled or oil pump has been removed, check to see that oil pump drive gear is properly installed in time as shown in Fig. O64. Then, install distributor so that rotor is aligned with No. 1 cylinder terminal of distributor cap and breaker points are just starting to open. Tighten the clamp cap screws to hold distributor in this position.

Static ignition timing specifications (in flywheel degrees) are as follows: Series 1600 gasoline models,

231 cu. in. displacement..0° (TDC)
248 cu. in. displacement..2° BTDC

All Series 1600 LP-Gas
models ................5° BTDC

All Series 1550-1555-1650-1655
non-diesel ..............0° (TDC)

**146. ADJUST TIMING WITH TIMING LIGHT.** To check timing with timing light, remove the flywheel timing hole cover and connect timing light according to manufacturer's instruction. On Series 1550, 1555, 1650 and 1655 models, disconnect the vacuum advance line from distributor and plug open end of line. Start engine and run at recommended RPM and check to see that correct flywheel timing mark aligns with pointer by using timing light. If correct timing mark is not aligned with pointer, loosen distributor clamp cap screws and rotate distributor body as required to bring mark into alignment with pointer. Recheck timing after tightening the distributor clamp cap screws.

Specifications in flywheel degrees and engine RPM for adjusting timing by timing light are as follows:

Series 1600 gasoline models,

231 cu. in.
displ.....19° BTDC @ 1900 RPM
248 cu. in.
displ.....21° BTDC @ 1900 RPM

All Series 1600
LP-Gas...24° BTDC @ 1900 RPM

All Series 1550-1555-1650-1655
Non-Diesel..22° BTDC @ 2200 RPM

## GENERATOR AND REGULATOR

## All Series 1600 Models

**147.** All Series 1600 models are equipped with a Delco-Remy 1100419 generator and a Delco-Remy 1118997

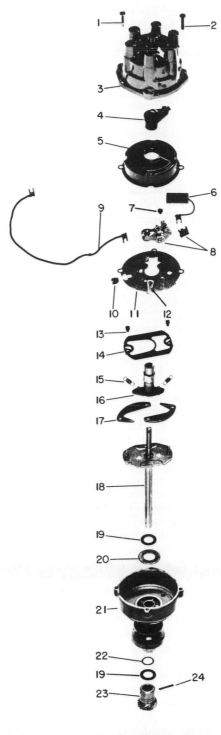

Fig. O103 — Exploded view of Delco-Remy distributor used on Series 1600 non-diesel models. Refer to Fig. O104 for exploded view of ignition distributor used on Series 1550, 1555, 1650 and 1655.

| | |
|---|---|
| 1. Screw, No. 8-32 | 13. Screw & lock washer |
| 2. Screw, No. 10-32 | 14. Plate |
| 3. Cap | 15. Springs |
| 4. Rotor | 16. Cam |
| 5. Dust cover | 17. Weights |
| 6. Condensor | 18. Shaft & weight plate |
| 7. Screw | 19. Washers |
| 8. Contact set | 20. Seal |
| 9. Low tension lead | 21. Housing |
| 10. Grommet | 22. Shim(s), 0.005 & 0.010 |
| 11. Breaker plate | 23. Gear & coupling |
| 12. Screw | 24. Pin |

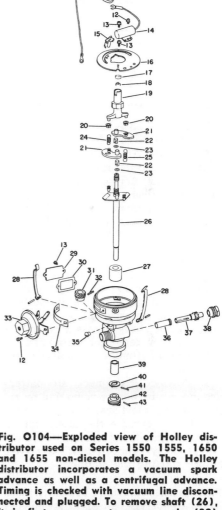

Fig. O104—Exploded view of Holley distributor used on Series 1550 1555, 1650 and 1655 non-diesel models. The Holley distributor incorporates a vacuum spark advance as well as a centrifugal advance. Timing is checked with vacuum line disconnected and plugged. To remove shaft (26), it is first necessary to remove pin (32) from Tachometer drive gear (31) and pins (41 and 43) from lower end of shaft. Pin (32) is accessible after removing plate (29).

| | |
|---|---|
| 1. Cap | 26. Shaft & weight plate |
| 2. Rotor | 27. Bushing |
| 3. Dust cover | 28. Clamps |
| 7. Low tension lead | 29. Cover plate |
| 8. Ground wire | 30. Gasket |
| 9. Contact set | 31. Tachourmeter drive |
| 11. Breaker plate | gear |
| 14. Condensor | 32. Pin |
| 15. Spring | 33. Diaphragm assy. |
| 16. Base plate | 34. Gasket |
| 17. Wick | 35. Plug |
| 18. Retainer | 36. Bushing |
| 19. Cam | 37. Gear & sleeve |
| 20. Bushings | 38. Retainer |
| 21. Weights | 39. Bushing |
| 22. Bearing strip | 40. Washer |
| 23. Washer | 41. Pin |
| 24. Primary spring | 42. Coupling |
| 25. Secondary spring | 43. Pin |

voltage regulator. Specification data follows:

**1100419 Generator:**
Brush spring tension...........28 oz.
Field draw @ 80° F.
  Volts .........................12
  Amperes .................1.5-1.62
Output (cold)
  Amperes (max.) .............25
  Volts .........................14.7
  RPM .........................2710

**1118997 Regulator:**
Cut-out relay
  Air gap .....................0.020
  Point gap ...................0.020
  Closing voltage range.....11.8-14.0
  Adjust to ...................12.8
Voltage regulator
  Air gap .....................0.075
  Voltage range ...........13.6-14.5
  Adjust to ...................14.0
Ground polarity ...........positive

## ALTERNATOR AND REGULATOR

### All Series 1550-1555-1650-1655 Models

CAUTION: An alternator (A.C. generator) is used to supply charging current for the Series 1550, 1555, 1650 and 1655 electrical system. Because certain components of the alternator can be seriously damaged by procedures that would not affect a D.C. generator, the following precautions must be observed:

1. Always be sure that when installing batteries or connecting a booster battery, the negative posts of all batteries are grounded.

2. Never short across any of the alternator or regulator terminals.

3. Never attempt to polarize the alternator.

4. Always disconnect all battery ground straps before removing or replacing any electrical unit.

5. Never operate the alternator on an open circuit; be sure that all leads are properly connected and tightened before starting the engine.

148. All Series 1550, 1555, 1650 and 1655 models are equipped with a Delco-Remy (Delcotron) alternator (A. C. generator) and a Delco-Remy voltage regulator. Alternator models 1100725, 1100731, 1100720, 1100771 and 1100808 and regulator models 1119511, 1119515, 1119516 and 1119517 have been used.

Specification data follows:

**Alternator—1100725-1100720-1100731-1100771**
Ground ....................negative
Field current @ 12V.
  (at 80° F.)...........2.2-2.6 Amps.
Field resistance @ 80°F. 4.6-5.5 ohms
Cold output:
  Specified voltage ............14

Amps. @ 2000 RPM............21
Amps. @ 5000 RPM............30
Rated Hot Output.........32 Amps.

**Alternator—1100808**
Ground ....................negative
Field current at 12 V.
  (at 80° F.) .........2.2-2.6 Amps
Cold output:
  Specified voltage ............14
  Amps at 2000 RPM .........32
  Amps at 5000 RPM .........50
Rated Hot Output .........55 Amps

**Regulator—1119516-1119517**
Relay unit:
  Air gap .....................0.015
  Point opening ..............0.030
  Closing voltage ...........1.5-3.2
Regulator unit:
  Air gap .....................0.067[1]
  Point opening (1119516) .....0.014
  Point opening (1119517) .....0.015
  Voltage setting:

| Temp., °F.[2] | Volts[3] |
|---|---|
| 65 | 13.9-15.0 |
| 85 | 13.8-14.8 |
| 105 | 13.7-14.6 |
| 125 | 13.5-14.4 |
| 145 | 13.4-14.2 |
| 165 | 13.2-14.0 |
| 185 | 13.1-13.9 |

[1] Air gap setting of 0.067 is for starting adjustment after bench repairs only; correct air gap is obtained by adjusting unit for proper voltage regulation.

[2] Ambient temperature measured ¼-inch away from voltage regulator cover; adjustment should be made only when at normal operating temperature.

[3] Regulated voltage when regulator is working on upper set of points; when unit is working on lower set of points, voltage should be 0.05-0.4 volts for model 1119516, or 0.1-0.4 volts for model 1119517, less than given in table.

**Regulator—1119515**
Relay Unit:
  Air gap .....................0.015
  Point opening ..............0.030
  Closing voltage ...........1.5-3.2
Regulator unit:
  Air gap .....................0.067(1)
  Point opening ..............0.014
  Voltage setting at degrees F.
    13.9-15.0 at 65
    13.8-14.8 at 85
    13.7-14.6 at 105
    13.5-14.4 at 125
    13.4-14.2 at 145
    13.2-14.0 at 165
    13.1-13.9 at 185

**Regulator 1119511**
Regulator Unit:
  Air gap ....................0.67 (1)
  Point opening ..............0.015

Voltage setting at degrees F.
  13.9-15.0 at 65
  13.8-14.8 at 85
  13.7-14.6 at 105
  13.5-14.4 at 125
  13.4-14.2 at 145
  13.2-14.0 at 165
  13.1-13.9 at 185

(1) Starting point for point adjustment after bench repair. Subsequent adjustment is made if necessary so that operation on lower contacts is 0.1-0.4 volts lower than on upper contacts.

## STARTING MOTOR

149. Series 1600 non-diesel engines are equipped with a Delco-Remy 1107682 starter and Series 1550, 1555, 1650 and 1655 non-diesel engines are equipped with a Delco-Remy 1107358 or 1108431 starter. All diesel engines are equipped with a Delco-Remy 1113098 or 1113139 starter. Specification data follows:

**Starter—1107682, 1107358**
Brush spring tension (min.), 35 oz.

No-load test:
  Volts ........................11.6
  Max. amps....................94
  Min. RPM ...................3240

Resistance test:
  Volts ........................3.5
  Min. amps....................325
  Max. amps....................390

**Starter—1113098**
Brush spring tension (min.), 48 oz.

No-load test:
  Volts ........................11.5
  Min. amps. (includes solenoid).57.0
  Max. amps. (includes solenoid).70.0
  Min. RPM ...................5000
  Max. RPM ...................7400

Lock test:
  Volts ........................3.4
  Amperes .....................500
  Torque (Ft.-Lbs.) .............22

**Starter—1108431**
Brush spring tension (min.) ...35 oz.
No load test:
  Volts ........................9.0
  Max. amps ...................80
  Min. RPM ...................3500

**Starter—1113139**
Brush spring tension (min.) ...80 oz.
No-load test:
  Volts ........................11.5
  Min. amps (includes solenoid)..57.0
  Max. amps (includes solenoid).70.0
  Min. RPM ...................5000
  Max. RPM ...................7400

Lock test:
  Volts ........................3.4
  Amps ........................500
  Torque (Ft.-Lbs.) ............22

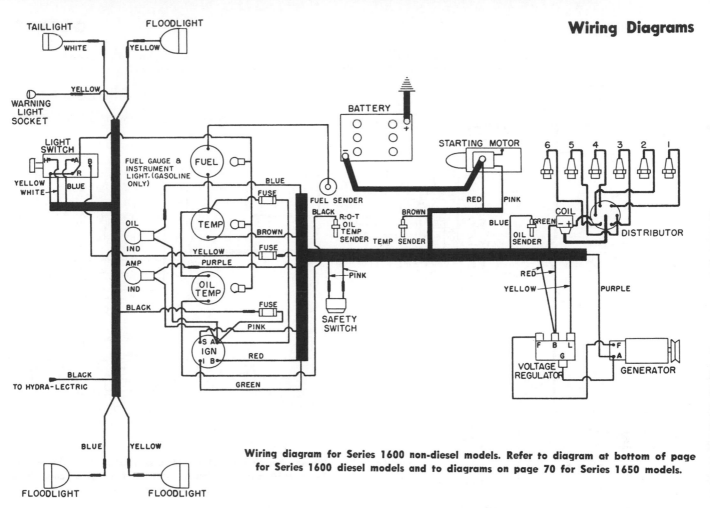

Wiring diagram for Series 1600 non-diesel models. Refer to diagram at bottom of page for Series 1600 diesel models and to diagrams on page 70 for Series 1650 models.

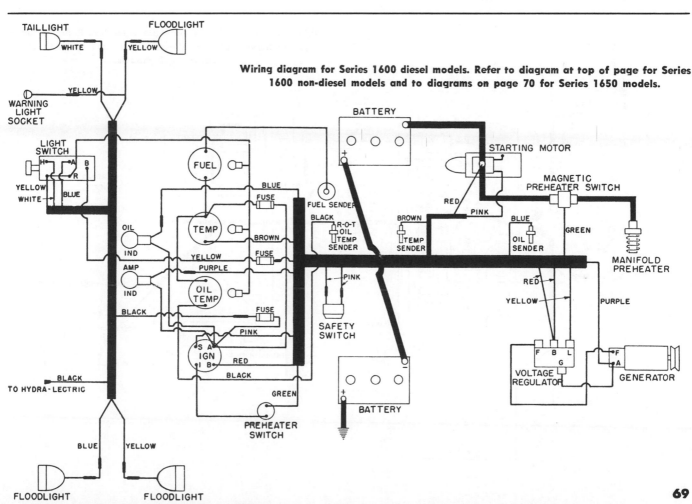

Wiring diagram for Series 1600 diesel models. Refer to diagram at top of page for Series 1600 non-diesel models and to diagrams on page 70 for Series 1650 models.

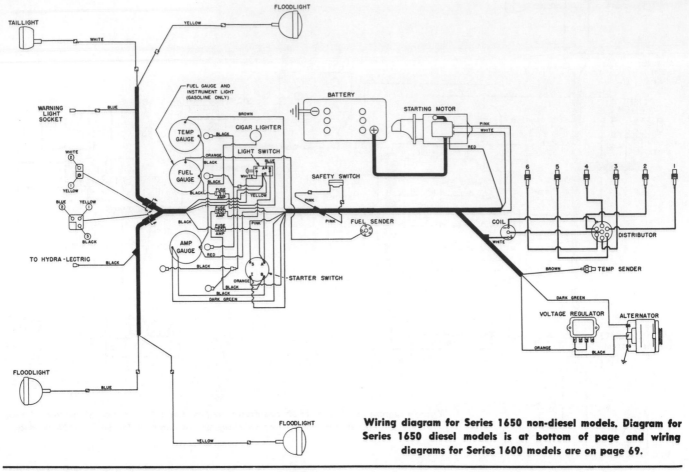

**Wiring diagram for Series 1650 non-diesel models. Diagram for Series 1650 diesel models is at bottom of page and wiring diagrams for Series 1600 models are on page 69.**

**Wiring diagram for Series 1650 diesel models. Refer to top of page for Series 1650 non-diesel models and to page 69 for Series 1600 models.**

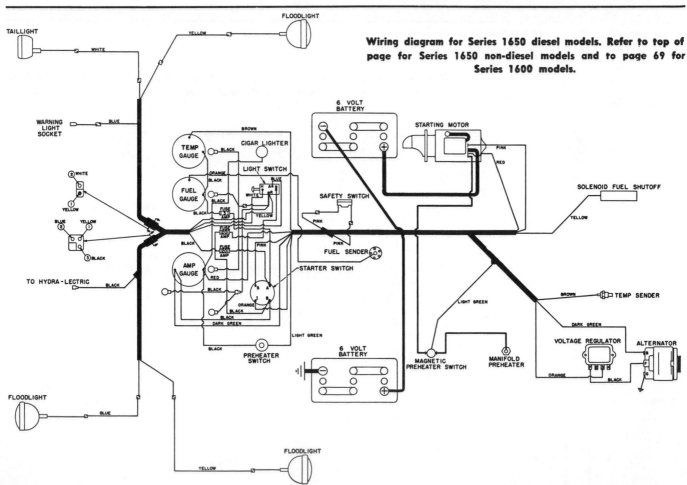

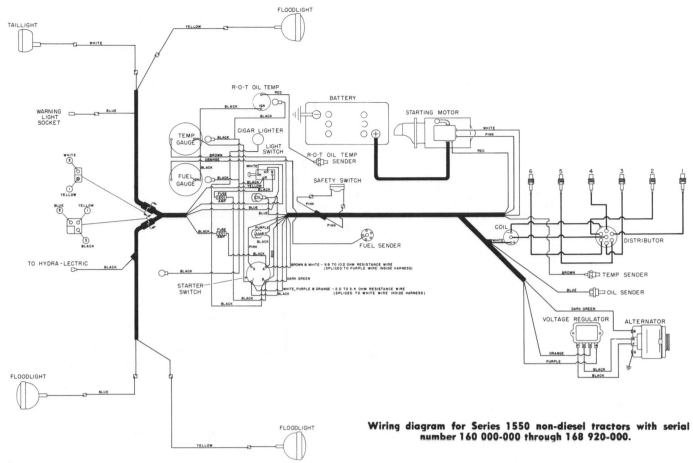

Wiring diagram for Series 1550 non-diesel tractors with serial number 160 000-000 through 168 920-000.

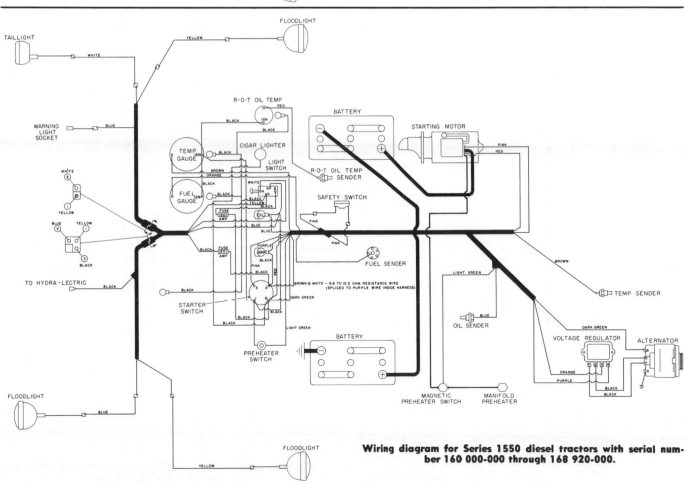

Wiring diagram for Series 1550 diesel tractors with serial number 160 000-000 through 168 920-000.

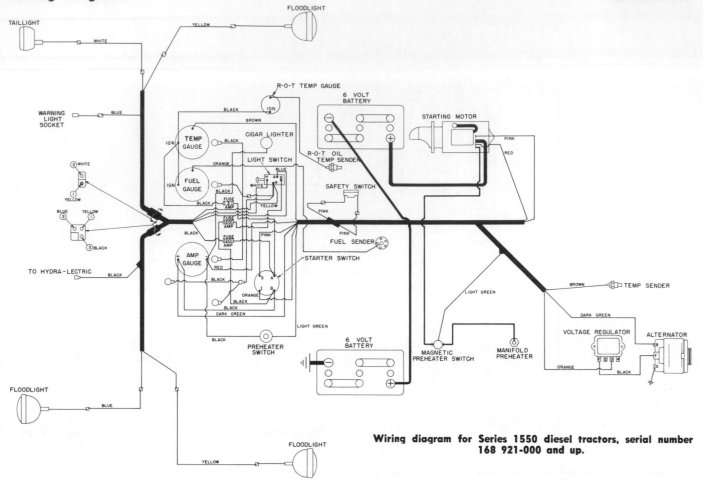

**Wiring diagram for Series 1550 diesel tractors, serial number 168 921-000 and up.**

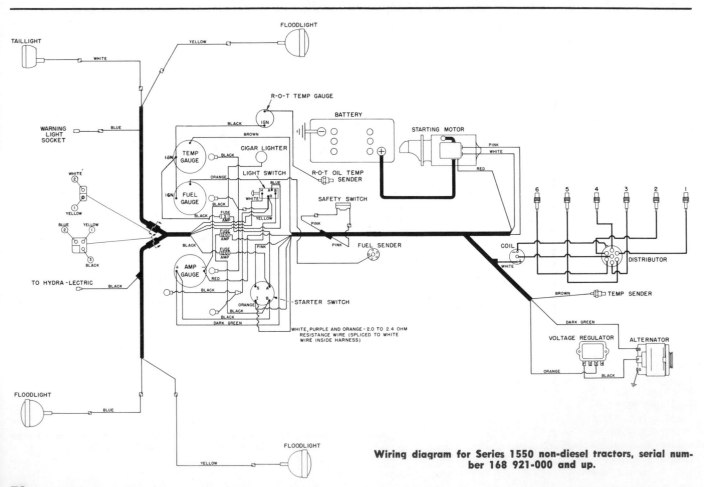

**Wiring diagram for Series 1550 non-diesel tractors, serial number 168 921-000 and up.**

# ENGINE CLUTCH

The engine clutch for all models is a dry type single plate spring loaded clutch. Series 1550, 1555 and 1600 models are equipped with a 11 inch diameter clutch; Series 1650 and 1655 models have a 12 inch clutch. Service procedure is similar for both the 11 and 12 inch clutch assemblies.

## ADJUSTMENT

### All Models

150. Clutch pedal free travel is measured at clutch pedal pad and should be ¾-inch for all models except those having the button type clutch. Button type clutches should have 1½ inches pedal free travel. To adjust pedal free travel, loosen jam nut at clevis end of clutch rod, disconnect clevis from clutch release shaft lever and turn clevis as required to obtain free travel. When adjustment is correct, install new clevis pin retaining cotter pin and tighten the jam nut.

## R&R ENGINE CLUTCH

### All Direct Drive Models Except With 4-Wheel Drive

151. If tractor is equipped with a power take-off, remove the pto shaft and clutch on Series 1600 as outlined in paragraph 221 or Series 1550, 1555, 1650 and 1655 pto input shaft as in paragraph 226. If equipped with a hydraulic system, but not equipped with power take-off, remove the hydraulic pump drive shaft as outlined in paragraph 248. Disconnect the battery ground strap, then proceed as follows:

Remove left hood side panel and remove governor or fuel injection pump control rod. Disconnect clutch control rod from release shaft lever. Remove drive shaft shield. On all except Series 1550 and 1555, disconnect coupling sprocket chain at master link and remove chain from sprockets. Disengage snap ring from groove in clutch shaft at front side of front sprocket and slide the snap ring and front sprocket forward on clutch shaft. Slide the rear sprocket forward off of transmission input shaft and remove sprocket from between clutch shaft and input shaft. Slide front sprocket back off of clutch shaft and remove from between clutch shaft and input shaft. On Series 1550 and 1555, remove cap screws from coupling and slide coupling forward on clutch shaft.

On all models, remove starting motor and remove dust shield from lower front side of clutch housing. On models with manual steering, loosen the steering shaft from bearing supports at right side of clutch housing and disconnect the shafts at "U-joint" at rear of shaft bearing support. Remove the clutch shaft felt seal and seal retainer from rear of clutch housing. Unbolt clutch housing from engine rear plate and slide housing rearward until front side of housing can be tipped upward. Remove clutch shaft by pushing it out rear opening of housing and down through opening in main frame. If so desired, the clutch housing can be removed at this time as follows: Shut off the fuel supply valve and remove the fuel line and sediment bowl. With front (open) side of clutch housing up and top side of housing to rear, withdraw the housing from between fuel tank and main frame. If not necessary to remove housing, return it to upright position and move it rearward out of way. Loosen the clutch cover retaining cap screws evenly to avoid distorting the clutch cover and remove the cover assembly from flywheel. Remove the clutch driven (friction) disc assembly.

Reinstall clutch assembly by reversing the removal procedure. Position driven disc so that wide hub flange is to rear and use clutch shaft as a pilot to align disc. Apply a light coat of molybedenum disulphide grease to splines of clutch shaft and transmission input shaft when assembling coupling sprockets.

### Models With 4-Wheel, Hydra-Power, Hydraul Shift or Creeper Drive

152. To remove clutch from tractors equipped with four-wheel drive, Hydra-Power Drive, Hydraul Shift or Creeper Drive, the engine must first be removed from tractor with clutch housing or the auxiliary drive unit attached as outlined in paragraph 60. On four wheel drive models with direct drive, remove the clutch housing as follows: Remove starting motor and dust shield from front side of engine front plate and clutch housing. Unbolt and remove the clutch housing and clutch shaft as a unit. On Hydra-Power, Hydraul-Shift or Creeper Drive, remove the drive unit as outlined in paragraph 155, 171C or 172.

Fig. 0104A—When installing button type clutch conversion package, chamfer edge of bore as shown, if necessary.

Loosen the clutch cover retaining cap screws evenly to avoid distorting the cover and remove the cover assembly and friction disc from engine flywheel. Clutch cover assembly can be overhauled as outlined in paragraph 153.

When reinstalling clutch on four wheel drive models with direct drive, the clutch shaft can be removed from rear of clutch housing and used as a pilot for the friction disc. On Hydra-Power, Hydraul-Shift or Creeper Drive Models, a spare clutch shaft or suitable pilot must be obtained for use in reinstalling friction disc. Tighten the cover retaining cap screws evenly so avoid distorting the cover.

NOTE: On 1650 tractors, a button type clutch conversion package is available which can be used to replace the original disc clutch on models prior to tractor serial number 187 586-000. Conversion package consists of a new clutch cover, six spacers and a button type clutch disc. A button type clutch is being factory installed on all 1655 models and 1650 models with tractor serial number 187 586-000 and up and at tractor serial number 215 506-000 the button type clutch was modified to include higher rate springs, heavier eyebolts and relieved release levers.

To install the button clutch conversion package in Series 1650 tractors having original disc clutch, proceed as follows: Remove clutch as in paragraph 151 or 152; or remove Auxiliary Drive unit as in paragraph 155, then remove clutch assembly from flywheel. After clutch is removed, remove pto drive hub and flywheel from crankshaft.

If inside diameter of flywheel face is not chamfered, chuck flywheel in a lathe and machine a ¼-inch by 45 degree chamfer on inside counterbore edge as shown in Fig. O104A. Chuck pto drive hub in a lathe and remove 1/16-inch of material from flywheel side of hub as shown in Fig. O104B; or, grind 1/16-inch from heads of pto hub cap screws.

Place three pieces of 0.380 keystock between pressure plate and flywheel, then install spacers so cover retaining cap screws pass through spacers, then secure cover to flywheel. Adjust clutch levers (fingers) so that bearing contacting surface of levers is 2.125 inches from flywheel face. Remove cover, spacers and keystock, then install flywheel and pto hub to crankshaft. Install button clutch disc with short hub toward flywheel, install clutch pilot tool, then install cover and spacers to flywheel.

Complete assembly of tractor and adjust clutch pedal free travel to 1½ inches.

## OVERHAUL CLUTCH COVER ASSEMBLY

### All Models

153. Remove the clutch assembly as outlined in paragraph 151 or 152 and proceed as follows: Place cover assembly in press, pressure plate side down, on a block that will permit downward movement of the cover. Place a bar across the cover and compress the unit slightly. Remove the nuts from the three eye bolts and slowly release pressure from the clutch cover. Lift cover from pressure plate and remove springs, release levers, eye bolts and pins and the struts from pressure plate. Remove the anti-rattle springs from cover.

Carefully inspect all parts and renew any that are excessively worn or otherwise damaged. Springs of two different pressure ranges are used in each clutch assembly. Renew springs if rusted, cracked or distorted, or if they fail to meet the following specifications:

**Oliver Part No. L-559-A** (9 used in 11 and 12 inch disc clutch; 6 used in early button clutch; 3 used in late button clutch)

    Color .................Unpainted
    Spring
    pressure.....189-201 lbs. @ 1½ in.

**Oliver Part No. M-559** (3 used in 11 inch disc clutch)

    Color ......................Tan
    Spring
    pressure.....155-165 lbs. @ 1½ in.

**Oliver Part No. H-559** (3 used in 12-inch disc clutch)

    Color ...................Orange
    Spring
    pressure.....165-175 lbs. @ 1½ in.

**Fig. O104B—When installing button type clutch conversion package, remove 1/16-inch of material from pto drive hub as shown, or remove 1/16-inch material from hub retaining cap screw heads.**

**Oliver Part No. L-559-3** (6 used in late button clutch)

    Color .....................Black
    Spring
    pressure ....223-237 lbs. @ 1½ in.

Reverse disassembly procedure to reassemble the clutch unit. When placing the twelve springs on pressure plate position them as follows: On 11 inch disc clutch, place a tan spring at the second spring position in a counter-clockwise direction from each lever assembly and place the nine unpainted springs in the remaining spring positions.

On 12 inch disc clutch, and early button type clutch, use same spring positions as used for the 11 inch clutch except use 3 orange springs in place of the 3 tan springs.

On late button type clutch, place a black spring on each side of clutch lever assemblies and an unpainted spring in the second spring position in a clockwise direction from each lever assembly. After the cover and pressure plate unit is reassembled, adjust release lever position as follows:

With flywheel removed from crankshaft and pilot bearing retainer removed from flywheel, place flywheel on bench with clutch friction surface up. Place three pieces of 0.330 (disc

clutch), or 0.380 (button clutch) thick key stock equidistantly apart on clutch friction surface of flywheel, then attach clutch cover to flywheel tightening the retaining cap screws evenly to avoid distortion of cover. Adjust the three eye bolt nuts so that distance from release bearing contact surface of each release finger to face of flywheel is 2.000 inches for disc clutches; 2.125 for early button clutches; or 2.225 for late button type clutches. Stake the nuts to the eye bolts when adjustment is correct. Remove the cover from flywheel and reinstall the flywheel and clutch assembly on engine crankshaft.

NOTE: Between tractor serial numbers 159 549-000 and 187 586-000 an increased capacity clutch was installed in Series 1650 tractors. This same clutch was available for installation in both Series 1600 and 1650 tractors prior to tractor serial number 159 549-000. The increase in clutch capacity was obtained by using twelve L-559-A clutch springs instead of the original nine L-559-A and three M-559 or H-559 springs. In addition, the driven plate incorporates stronger damper springs. Refer to beginning of this section for information on the button type clutch now being used in the Series 1650 and 1655 tractors. The 12 inch increased capacity clutch cover (part no. 160 970-AS) and driven plate (part no. 160 971-AS) replaces the original equipment 11 inch clutch used in Series 1600 tractors. The original equipment clutch for Series 1650 tractors can be rebuilt by using twelve L-559-A springs and the new driven plate. Service parts for the original equipment 11 and 12 inch clutches are still available.

Service procedure for reworking the Series 1650 clutch is the same as already outlined, however, use 0.315 thick keystock when adjusting clutch fingers and adjust fingers to 1.97-2.03 distance between finger and flywheel face.

# HYDRA-POWER DRIVE UNIT

Series 1550, 1555, 1600, 1650 and 1655 tractors have available as optional equipment a Hydra-Power drive unit which is located between the engine clutch and tractor transmission. This auxiliary drive unit provides both direct drive and underdrive speed ratios for the transmission input shaft. It is a self-contained unit and includes a hydraulically operated multiple disc clutch, a hydraulic pump, filter and heat exchanger. The gear type pump is mounted on inner side of drive unit housing cover and furnishes pressurized oil to operate the multiple disc clutch as well as providing lubrication

for the drive unit. In direct drive position, the multiple disc clutch is engaged and a "straight through" power flow is obtained. In the under-drive position, the multiple disc clutch is disengaged and power flow is directed to the countershaft which reduces the speed of the drive unit output shaft, and the transmission input shaft, approximately 26 per cent. The Hydra-Power drive unit can be shifted while tractor is in motion.

Some modifications have been made in the Hydra-Power drive unit and where these changes affect service procedures, it will be noted in the following paragraphs.

## TROUBLE SHOOTING

154. The following symptoms and their causes will be helpful in trouble shooting the Hydra-Power unit.

1. **OPERATING TEMPERATURE TOO HIGH.**
   a. Coolant hoses, heat exchanger or oil lines obstructed.
   b. Multiple disc clutch slipping.
   c. Engine cooling system overheated.
   d. Low pump pressure.
   e. Defective or improperly adjusted bearings.

2. **INOPERATIVE OR HESITATES IN HYDRA-POWER DRIVE.**
   a. Defective over-running clutch.
   b. Defective clutch.
   c. Weak retractor spring.

3. **INOPERATIVE OR HESITATES IN DIRECT DRIVE.**
   a. Multiple disc clutch slippage or defective.
   b. Clutch operating pressure low.

4. **MULTIPLE DISC CLUTCH SLIPPING.**
   a. Clutch operating pressure low.
   b. Oil passages obstructed.
   c. Defective clutch.
   d. Oil collector sealing rings worn or broken.
   e. Defective oil collector "O" ring.
   f. Foreign material in clutch.

5. **ABNORMAL LUBRICATION CIRCUIT PRESSURES (TOO HIGH OR TOO LOW).**
   a. Defective pump.
   b. Pump inlet clogged.
   c. Defective by-pass spring.
   d. By-pass valve plunger not seating properly.
   e. Low oil level.

6. **LOW CLUTCH OPERATING PRESSURE.**
   a. Low oil level.
   b. Defective pump.
   c. Pump inlet clogged.
   d. Oil collector sealing rings worn or broken.
   e. Defective clutch "O" ring.
   f. Defective clutch piston ring.
   g. Defective regulator spring.
   h. Regulator valve spool stuck open.
   i. Clogged oil passage.

Fig. O105 — When checking operating pressure of the Hydrapower drive unit, install pressure gage as shown.

7. **HIGH CLUTCH OPERATING PRESSURE.**
   a. Defective regulator spring.
   b. Regulator valve spool stuck closed.
   c. Clogged oil passage.

8. **ERRATIC CLUTCH OPERATING PRESSURE.**
   a. Regulator valve spool installed backward.

## OPERATING PRESSURE

154A. Hydra-Power drive unit has two oil pressure circuits; a clutch operating circuit and a lubrication circuit. Both circuits can be checked at the same point using a single gage.

To check the Hydra-Power drive operating pressures, proceed as follows: Remove the pipe plug, which is located in side cover directly below the filter to heat exchanger tube, and install, as shown in Fig. O105, a pressure gage capable of registering at least 300 psi. Be sure oil level in Hydra-Power drive is at the proper level, then start engine and run until oil in Hydra-Power drive is at operating temperature. With oil warmed, operate engine at approximately 2000 rpm, place Hydra-Power Drive unit in direct drive and observe pressure gage. Gage should read 140-190 psi. for tractors prior to serial number 201 091-000, or 140-160 psi for later tractors. If pressure is not as specified, and no malfunctions of the unit are evident, add shims (181—Fig. O106) behind the regulator valve spool spring to raise pressure. Shims are 0.0359 thick and each shim will raise the pressure approximately 5 psi.

Place Hydra-Power unit in the under-drive position and with engine running at 2000 rpm, observe the gage pressure which should be 20-60 psi. If gage reading is not as specified, refer to the Trouble Shooting section or the section pertaining to overhaul.

## REMOVE AND REINSTALL

155. To remove the Hydra-Power drive unit, it is necessary to remove engine and Hydra-Power drive as a unit as outlined in paragraph 60. With engine and drive unit removed, support engine on blocks so that oil pan will not be damaged, then remove clutch housing lower dust shield and starter. Disconnect coolant lines. Un-

9. Snap ring
23. Control lever
25. Pin
26. Bushing
27. Snap ring
28. Drive gear
29. Woodruff key
30. Plug
31. Crush washer
32. Regulator spring
33. Regulator spool
34. By-pass spring
35. Plug
36. By-pass valve plunger
37. Gasket
38. Cover
41. Plug
42. Plug
43. Gasket
44. Oil pump
47. Elbow
48. "O" ring
49. Control valve spool
50. Detent ball
51. Detent spring
55. Adapter
56. Oil filter
57. Oil line
58. Heat exchanger
62. Bracket
65. Oil line
67. Hose clamp
68. Inlet hose
69. Outlet hose
180. "O" ring
181. Shim

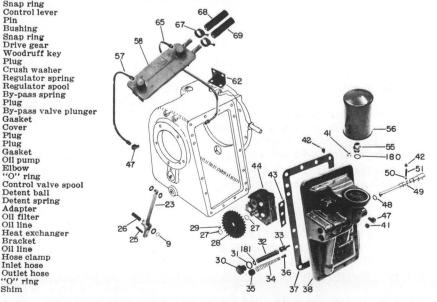

Fig. O106 — Exploded view of the early Hydra-Power drive cover showing component parts. Later cover assemblies are similar.

**Fig. O107—Composite view of Hydra-Power unit showing parts used in various units. Identify unit by serial number before attempting to service Hydra-Power units. The one-piece spider (139) replaced spider (137) and retainer (149) at unit serial number 48968. The one-piece input shaft, gear and clutch hub assembly (162) replaced items (152, 153 and 155) at unit serial number 48528. The one-piece support (166) replaced the earlier two-piece retainer and support at unit serial number 44200. Very early units prior to unit serial number 16235 did not use spacer (120). Refer to text for difference in service procedure on early and late output shaft assemblies.**

| | | | | |
|---|---|---|---|---|
| 94. Washer | 107. Master link | 122. "O" ring | 137. Clutch spider | 150. Snap ring | 164. Oil seal |
| 95. Cap | 108. Cotter pin | 123. Retainer | 138. Cap screw | 151. Snap ring | 165. "O" ring |
| 96. Element | 109. Sprocket, rear | 126. Support shaft | 139. Clutch spider | 152. Clutch hub | 166. Support |
| 97. Cup | 110. "O" ring | 127. "O" ring | 140. Lock plate | 153. Input gear | 169. Tube |
| 98. Gasket | 111. Oil seal | 128. Housing | 141. Piston | 154. Oil seal | 170. Thrust washer |
| 99. Cap | 112. Bearing assy. | 129. Drain plug | 142. Piston ring | 155. Input shaft | 171. Gear |
| 100. Dipstick | 113. Retainer | 130. Dowel | 143. Seal ring | 156. Ball bearing | 172. Clutch gear |
| 101. Nut | 114. Stud | 131. Lever shaft | 144. Release spring | 157. Snap ring | bearing |
| 102. "O" ring | 117. "O" ring | 132. Output shaft | 145. Spring retainer | 158. Driven plate | 173. Sprag clutch |
| 103. Filler tube | 118. "O" ring | 133. Output gear | 146. Snap ring | 159. Drive plate | 174. Clutch gear |
| 104. Sprocket, front | 119. Shim (.004, | 134. Oil collector | 147. Drive plate | 160. Plate retainer | 175. Snap ring |
| 105. Nut | .007, 015) | 135. Sealing rings | 148. Driven plate | 161. Snap ring | 176. Bearing |
| 106. Coupling chain | 120. Spacer | 136. "O" ring | 149. Plate retainer | 162. Input shaft | 177. Countershaft |
| | 121. Bearing assy. | | | 163. Snap ring | |

bolt and remove clutch housing and Hydra-Power unit from engine. Remove clutch release bearing and lubrication hose. Remove cap screw and key from release bearing fork and pull cross shaft from fork and clutch housing. Unbolt and remove the clutch housing from Hydra-Power unit.

Reinstall the Hydra-Power drive unit by reversing the removal procedure.

### OVERHAUL

**156. DISASSEMBLY PROCEDURE.** With unit removed as outlined in paragraph 155, drain unit and on early models so equipped, remove front control rod. Remove valve control lever and, if necessary, lever bushing can be renewed at this time. On early

units, remove filter to heat exchange line. On late units, remove filter to countershaft oil line, then disconnect oil hose from oil collector stem and remove fitting. Remove filter, if desired, then unbolt and remove housing cover and attached pump.

On early units with two-piece input shaft support and retainer, remove clutch bearing tube support and retainer, then remove snap ring from outer race of input shaft bearing and bump off input shaft bearing support.

Remove support shaft retainer (123 —Fig. O107), pull support shaft (126) and allow countershaft to rest on bottom of housing. Remove and discard "O" rings from support shaft and shaft front bore in housing.

On models prior to unit serial num-

ber 48528, pull input shaft (155), bearing (156), input gear (153) and clutch hub (152) as an assembly from housing. On models with unit serial number 48528 and up, pull input shaft (162) and support (166) as an assembly from housing.

Remove snap ring (150) from front end of output shaft (132), then unbolt output shaft bearing retainer (113). Pull output shaft rearward and remove the clutch and oil collector assembly and the output gear (133) from housing as the shaft is withdrawn. Countershaft assembly can now be removed from housing.

If necessary, the heat exchanger (early units), filler neck and valve spool control lever shaft can also be removed.

**Fig. O112 — Exploded view of the Hydra-Power drive oil pump.**

1. Rear plate
5. Dowel
6. Screen
9. Center plate
10. Drive shaft and gear
11. Idler shaft and gear
12. Front plate

At this time all of the components of the Hydra-Power drive unit are accessible for inspection and/or overhaul. Refer to the appropriate following paragraphs:

**157. CONTROL VALVE SPOOL, REGULATOR VALVE AND BY-PASS VALVE.** The control valving can be removed from housing cover as follows: Remove the Allen plug (42—Fig. O106) on top side of control valve bore and remove detent spring (51) and ball (50). Pull control valve spool (49) forward out of its bore, then remove and discard "O" ring. Remove plugs, then remove regulator valve (33) and by-pass valve (36) assemblies. Save any shims (181) which may be present behind the regulator valve spring.

NOTE: Removal of the regulator and by-pass valves can be accomplished with Hydra-Power drive unit on tractor.

Check springs against the values which follow:

Detent spring
Free length................$1\frac{9}{32}$ in.
Test lbs. @ inches.15.3-18.7 @ $\frac{7}{8}$

Regulator spring
Free length..............$3\frac{5}{8}$ in.
Test lbs. @ inches...55-67 @ $2\frac{3}{4}$

By-pass spring
Free length..............$3\frac{1}{2}$ in.
Test lbs. @ inches.$5\frac{1}{2}$-$6\frac{1}{2}$ @ $2\frac{7}{16}$

Outside diameter of regulator valve spool and land of control valve spool is 0.6865-0.6869 and their bore inside diameters are 0.6875-0.6883.

When reassembling, use new "O" ring on control valve spool and install regulator valve with pin inward. If any shims were present behind regulator spring during disassembly, be sure to reinstall them. Prior to installing cover on housing, renew "O" ring on stem of oil collector and in oil collector stem bore of housing cover (early models not converted).

**159. PUMP OVERHAUL.** After removing cover from Hydra-Power drive unit as outlined in paragraph 156, unbolt pump from cover and discard gasket. Remove drive gear outer retainer snap ring, remove gear and key from pump drive shaft, then remove inner gear retainer snap ring. Remove the through bolts and separate pump. Dowels can be driven out and the screen removed if necessary. See Fig. O112.

NOTE: If the pump gears and shafts are to be reinstalled, mark idler gear so that it will be reinstalled in its original position.

Pump dimensional data are as follows:

Gear width..........0.9975-0.9980
Pump body width.....0.9995-1.0000
Gear shaft diameter...0.6247-0.6250
Shaft bore diameter...0.6265-0.6275
Shaft operating
clearance .........0.0015-0.0028

Reassemble by reversing disassembly procedure and use new gasket when mounting pump on housing cover. Heads of through bolts should be on drive gear side of pump and dowels should be installed prior to tightening through bolts.

**160. INPUT SHAFT ASSEMBLY.** Remove input shaft assembly as in paragraph 156. To disassemble input shaft assembly in units prior to serial number 48528, remove snap ring (151 —Fig. O107) and pull clutch hub (152) and input gear (153) from input shaft. Remove snap ring (157) and press shaft from bearing. Oil seal (154) can be removed from rear end of input shaft at any time.

To disassemble input shaft assembly in units with serial number 48528 and up, disengage snap ring (163) from support (166) and pull input shaft and bearing from support. Remove snap ring (157), then remove bearing (156) and snap ring (163) from input shaft. Oil seal (154) can be removed from rear end of input shaft at any time.

Inspect input shaft, input gear and clutch hub for damaged or worn splines or teeth. Check condition of input shaft bearing. Renew parts as necessary.

Reassemble input shaft as follows: Install oil seal in rear end of input shaft with lip facing toward front of shaft. On units prior to serial number 48528, press bearing on shaft with snap ring groove in bearing outer race toward front of shaft, install retaining snap ring, then press bearing toward snap ring to insure that bearing inner race is tight against snap ring. Install input gear on shaft with hub next to bearing, then install the clutch drive hub on shaft with its hub toward input gear and install the retaining snap ring.

On units serial number 48528 and up, place snap ring (163) over front end of input shaft, then install bearing (156) and snap ring (157). Install input shaft assembly in support (166) and secure with snap ring (163).

**161. CLUTCH ASSEMBLY. (PRIOR TO UNIT S/N 48967).** Remove clutch assembly as outlined in paragraph 156. To disassemble the clutch assembly, first pull oil collector (134—Fig. O107) from clutch spider hub, then remove the two sealing rings (135). Straighten tabs of lock plates (140), remove clutch plate retaining cap screws and lift clutch spider (137) from plate retainer (149). Clutch plates can now be removed from plate retainer. Place clutch spider in a press, depress the spring retainer (145) slightly and remove retainer snap ring (146). Release press and remove retainer and retractor spring (144), then remove piston (141) from clutch spider either by bumping spider hub on a wood block, or by carefully applying air pressure to the oil hole located in land between oil collector sealing rings.

Clean and inspect all parts. Use the following specification data as a guide for renewing parts.

Clutch retractor spring
test.........155-185 lbs. @ $1\frac{3}{16}$ in.
Driven plate thickness.....0.112-0.120
Driving plate thickness....0.064-0.067
Max. allowable cone.........0.010

Clutch piston ring:
Width .................0.123-0.124
End gap (desired)......0.005-0.015
Max. allowable ...........0.025
Side clearance (desired).0.003-0.006
Max. allowable ...........0.025
Oil collector hub O.D.....3.544-3.546
Oil collector I.D..........3.550-3.552

Oil collector seal ring:

Width .............0.0930-0.0935

End gap (desired)......0.003-0.008

Max. allowable ...........0.020

Reassemble clutch as follows: Install new seal ring in I. D. of clutch piston. Install piston ring on O. D. of clutch piston. Start piston into spider with flat side toward oil cavity and install by compressing piston ring with fingers. Place clutch spider in a press, position retractor spring and retainer and compress retainer until snap ring can be installed. Start with an external lug clutch plate (driven) and alternate with an internal spline plate (drive) and install the six driven plates and five drive plates. Install the clutch spider on plate retainer, install lock plates and cap screws and tighten cap screws to 14-15 Ft.-Lbs. torque. Secure cap screws with lock plates. Install sealing rings on oil collector hub and compress rings with soft wire. Start oil collector over hub with word "Front" toward clutch spider and remove wire as oil collector goes over sealing rings. Lubricate the clutch plates.

**161A. CLUTCH ASSEMBLY (UNIT S/N 48968 UP).** Remove clutch assembly as outlined in paragraph 156. To disassemble the clutch assembly, first pull oil collector (134—Fig. O107) from clutch spider (139) hub and remove the two sealing rings (135). Remove snap ring (161), retainer (160) and clutch plates (158 and 159). Place clutch spider in a press, depress spring retainer (145) slightly and remove retaining snap ring (146). Release press and remove retainer and retractor spring (144). Use ⅛-inch brazing rod and fabricate four U-pins to the dimensions shown in Fig. O112A. Place the U-pins equidistantly around edge of clutch spider so inner leg will prevent the piston outer seal ring from expanding. Hold pins in position, turn spider over and bump spider gently on bench until piston comes out of bore.

Clean and inspect all parts. Use the following specifications data as a guide for renewing parts.

Clutch retractor spring
  test .........155-185 lbs. @ 1¾ in.
Driven plate thickness ....0.066-0.072
Driving plate thickness ...0.066-0.072
Clutch piston ring:
  Width .................0.123-0.124
  End gap (desired) ......0.005-0.015
    Max. allowable .............0.025

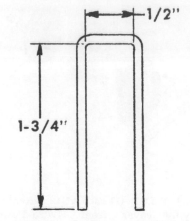

Fig. O112A — Service on one-piece clutch spider piston can be facilitated by using U-pins made from ⅛-inch brazing rod formed to dimensions shown. Refer to text.

Side clearance (desired) 0.003-0.006
  Max. allowable ............0.025
Oil collector
  hub O.D. ...............3.544-3.546
Oil collector
  I.D. ...................3.550-3.552
Oil collector seal ring:
  Width ...............0.0930-0.0935
  End gap (desired) ......0.003-0.008
  Max allowable .............0.020

Reassemble clutch by reversing the disassembly procedure. Clutch piston is installed with flat side inward. When installing clutch plates, start with a driven plate (external teeth) and alternately install the five drive plates and five driven plates, ending with a drive plate (internal teeth). Compress oil collector seal rings with soft wire, start oil collector over hub with word "Front" toward clutch spider and remove wire as oil collector goes over sealing rings. Lubricate clutch plates.

**162. OUTPUT SHAFT ASSEMBLY.** Early Hydra-Power units, prior serial number 16235, were equipped with an output shaft which was not shouldered. At unit, serial number 16235, a shouldered output shaft was installed, along with a spacer and shims to provide for bearing adjustment. Subsequently, at unit serial number 44200, an output gear with a larger inside diameter and full length internal splines, an output shaft with a larger outside diameter and a clutch spider with a larger inside diameter to accommodate the larger shaft were installed. At the same time, two "O" rings replaced the output shaft bearing retainer gasket.

Although the late individual parts are not interchangeable with earlier parts, the three parts can be installed in prior units if operating conditions require it.

163. To disassemble the output shaft assembly, unstake lock nut (105—Fig. O107) and using a spanner wrench (Oliver No. ST-149 or equivalent), remove nut. Remove sprocket (104) and "O" ring (110) from shaft, then press shaft from rear bearing and retainer. On early units with shoulderless shaft, note location of front bearing on shaft, then press bearing from shaft. On later units having shouldered shaft, remove shims and spacer, then press front bearing from shaft. Remove oil seal and rear shaft bearing from retainer and, if necessary, pull bearing cups from retainer.

Clean and inspect all parts and renew as necessary. Always use new "O" ring and oil seal when reassembling unit.

164. Assemble early output shaft as follows: Start front bearing on shaft with smaller diameter towards threaded end (rear) of shaft and press bearing onto shaft until inner race is even with rear edge of ground (polished) surface on shaft. Note: This is to insure sufficient room for installation of spider retaining ring.

If necessary, install bearing cups in bearing retainer with smaller diameters toward center, then insert shaft and bearing in retainer. Use coupling sprocket as a driver and press rear bearing on shaft. Remove coupling sprocket, install new "O" ring and oil seal (lip inward), then reinstall coupling sprocket. Install new lock nut and tighten only finger tight at this time. Refer to paragraph 168 in assembly procedure for information concerning bearing adjustment.

165. Assemble later output shaft assembly as follows: Place front bearing on rear of output shaft with smaller diameter toward rear and press bearing on shaft until it bottoms against shoulder. If necessary, install bearing cups in retainer with smaller diameters towards inside. Place spacer (120) on output shaft with the counterbored end next to front bearing cone, then install shims (119) that were previously removed or, if service package is being installed, install all shims that are included in package. Place shaft in bearing retainer (113), install rear bearing sprocket (104) and the lock nut (105); tighten lock nut to a torque of 200 Ft.-Lbs. Check shaft bearing adjustment which should be from a maximum of 0.001 shaft end play to a maximum shaft rolling torque of 5 inch pounds. If bearing adjustment is not within these limits,

remove lock nut, sprocket and rear bearing and vary the number of shims (119) as necessary. Shims are available in thicknesses of 0.004, 0.007 and 0.015. When bearing adjustment is within the desired limits, remove the lock nut and sprocket and install new "O" ring (110) and oil seal (111). Seal is installed with lip forward. Install sprocket and lock nut, tightening the lock nut to a torque of 200 Ft.-Lbs. Stake nut in position.

### 166. COUNTERSHAFT ASSEMBLY.

Lift countershaft from housing, if necessary, and retrieve both thrust washers. Pull gear (171—Fig. O107) from rear of shaft. Note: On series 1600 a double gear is used as shown in Fig. O113. Remove retaining ring (Series 1600) and front bearing from front end of shaft and pull clutch gear from the sprag clutch. Note position of sprag clutch, then remove clutch from shaft. Remove snap ring (Series 1600) and rear bearing from shaft. Remove snap rings and needle bearings from I. D. of countershaft.

Clean and inspect all parts and renew as necessary. Use the following data as a guide for renewing parts.

Front thrust washer
  thickness ..............0.060-0.062
Rear thrust washer
  thickness ..............0.060-0.062
Countershaft support
  shaft O. D............1.3745-1.3750
Countershaft O. D.
  clutch end ...........2.8429-2.8434
  Max. allowable taper........0.0002
Clutch gear I. D........3.4988-3.4998
  Max. allowable I. D. taper...0.0003

Front and rear clutch gear bearing:
  Bearing surface O. D...3.4968-3.4978

To reassemble countershaft, proceed as follows: Install needle bearings in I. D. of countershaft only far enough to allow installation of snap rings. Place rear clutch gear bearing on shaft with bearing surface toward front and install rear snap ring. Note: On Series 1600, rear clutch gear bearing is notched; on all other models, front and rear clutch bearings are identical and no snap rings are used. Install the sprag clutch on countershaft so that the drag strips point toward rear bearing. Install clutch gear over the sprag clutch and rotate clutch gear clockwise around the sprag clutch to assist in installing gear. After installation, clutch should overrun (turn freely) when gear is turned clockwise but lock up when counterclockwise gear rotation is attempted. Install the

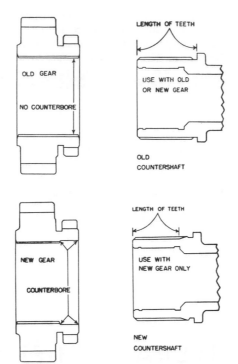

**Fig. O113 — When servicing Hydra-Power drive on series 1600, be sure to use correct countershafts and double gears as shown.**

front gear bearing and on Series 1600, install retaining ring and the output double gear with smaller gear next to shaft shoulder. On all other models, install rear gear next to pump drive teeth on countershaft.

NOTE: On Series 1600, some differences exist in countershafts and the double countershaft gears as shown in Fig. O113. Note that the later double gear has a counterbore in the front I. D., whereas the early gear is not counterbored. Early countershaft will work with either gear but late countershaft must be used only with the late double gear. Part numbers remain the same, so identification of parts must be visual. Therefore, when ordering new parts, be sure the correct combination of double gear and countershaft is obtained. On all other models, a single gear is used at rear of countershaft and pump is driven from a gear which is integral with countershaft.

### 167. REASSEMBLY PROCEDURE.

Install oil filler tube, heat exchanger and bracket (early units) control valve spool lever shaft or input shaft support studs if removed.

Use heavy grease on the countershaft thrust washers and position washers in housing with tangs in the slots provided. On early models, the thrust washer with the largest out-

side diameter is the rear washer; on late models, front and rear washer are identical. Place assembled countershaft in case with double output gear to rear and allow the assembly to rest in bottom of case. Take care not to dislodge the thrust washers. Delay installation of countershaft support shaft until input shaft assembly has been installed.

Use a new bearing retainer gasket, or "O" rings, and start output shaft into rear of housing. As shaft is moved forward, install output gear with hub to rear and clutch assembly with oil collector to rear. Align notches of gasket (when used) and retainer with oil drain hole in housing. On late units, be sure small "O" ring is in place in counterbore in front face of bearing retainer and aligned with drain hole in housing. Install lock washers and nuts on bearing retaining studs and tighten the nuts securely. Install clutch spider retaining ring on front of output shaft.

168. At this time, adjust output shaft bearings on units not having a shouldered shaft. Bump front end of output shaft to insure that snap ring is seated against clutch spider. Install "O" ring and coupling sprocket on output shaft, if not already installed, then install and tighten lock nut until only a slight amount of shaft end play exists. Now, use an adapter such as Oliver tool No. ST-152 with an inch-pound torque wrench and check torque required to rotate shaft; record this torque which is the amount of oil seal and bearing drag. Continue to tighten lock nut until torque required to rotate shaft is 2 to 5 inch-pounds greater than torque due to seal and bearing drag. Note: If inch-pound torque wrench and adapter is not available, tighten nut so that shaft end play is 0.001 or slightly less than 0.001. Bump output shaft both ways when adjusting bearings to be sure bearings are seated. When proper adjustment point is reached, bump forward end of output shaft to be sure snap ring is seated against clutch spider and recheck shaft rolling torque or end play. Readjust bearings if necessary. When certain that bearing adjustment is correct, stake lock nut to slot in shaft and at 180 degrees away from slot to maintain bearing adjustment.

169. On units prior to serial number 44200, align splines of clutch drive (internal splined) plates and insert input shaft assembly. All five of the driven discs will rotate if drive hub

is properly positioned. Use new gasket and install bearing support so notch in support aligns with notch in housing. Then install snap ring on outer race of input shaft bearing. Install new seal in bearing retainer with lip of seal toward rear, then install the bearing retainer and the release bearing tube and support. Align notches in gasket and bearing retainer with oil drain hole in bearing support.

On units, serial number 44200 and up, place new "O" ring on support and install support and input shaft as an assembly. Seal is installed in support with lip toward rear.

NOTE: A new release bearing tube can be installed in tube support if necessary. Install tube so that it protrudes 4 inches from front flange of tube support on units with two-piece support, or 3⅝ inches on units with one-piece support, and be sure tube is not cocked during assembly. Any misalignment of release bearing tube will result in rapid wear of the clutch release bearing.

170. Install the countershaft support shaft as follows: Install new "O" ring in shaft bore in front of housing. Install new "O" ring on rear of shaft and, if necessary, install oil line elbow in rear of shaft. Lubricate both "O" rings, make certain that both thrust washers are in position, then lift countershaft assembly to align bores and install countershaft support shaft from rear. Install shaft retainer and support shaft to heat exchanger oil line.

171. To install housing cover, install "O" ring on stem of oil collector (early units only) and install "O" ring in oil collector stem bore in housing cover (all units). Place new cover gasket on housing, lift stem of oil collector and start stem into bore in cover, then carefully install cover to housing by rotating cover slightly as stem enters cover bore. Tighten cover retaining cap screws securely.

Fill unit with 5 quarts of Type A Automatic Transmission fluid.

# HYDRAUL SHIFT

The Hydraul Shift is a three-speed auxiliary unit interposed between engine clutch and transmission and provides on-the-go shifting in over, under, or direct drive. This auxiliary unit is a self-contained unit and includes its own reservoir, lubrication and pressure systems and control valving. Oil operating temperature is controlled by a cooler located in front of the radiator. Gears are in constant mesh and are helical cut. Drive selection is accomplished by engaging or disengaging multiple disc clutches and the operation of a planetary gear set and an overruning (Sprag) clutch. The pressurized oil used to operate clutches, as well as oil for lubrication, is furnished by a gear type pump mounted on inner side of housing cover.

In overdrive position, oil pressure engages overdrive (rear) clutch and power flow goes through countershaft which drives the planet gear carrier. Overdrive clutch holds sun gear shaft stationary and the speed increase is obtained by the planet gears driving output shaft through the ring gear.

In underdrive position, both direct and overdrive clutches are disengaged and power flow is through the counter shaft which drives the planet gear carrier. The overruning clutch locks sun gear shaft and output shaft together causing them to rotate as a unit. This makes the planet drive inactive and results in a speed reduction.

In direct drive, the direct (front) clutch is engaged. This locks input shaft and output shaft together and power flow is straight through the unit. Overruning clutch over-runs and permits free rotation of sun gear shaft planet gears and countershaft.

## TROUBLE SHOOTING

### All Models So Equipped

171A. The following symptoms and their causes will be helpful in trouble shooting the Hydraul Shift unit.

1. OPERATING TEMPERATURE TOO HIGH.
    a. Coolant hoses, oil cooler or oil lines obstructed.
    b. Low pump pressure or pump intake clogged.
    c. One or both clutches slipping.
    d. Engine cooling system overheated.
    e. Defective or improperly adjusted bearings.
    f. Reservoir oil level low.

2. INOPERATIVE OR HESITATES IN OVERDRIVE.
    a. Ovedrive clutch slipping or defective.
    b. Low pump pressure or pump intake clogged.
    c. Oil passage obstructed.

3. INOPERATIVE OR HESITATES IN UNDERDRIVE.
    a. Overunning clutch worn or defective.
    b. Damaged or sticking underdrive and/or overdrive clutch.

4. INOPERATIVE OR HESITATES IN DIRECT DRIVE.
    a. Low pump pressure.
    b. Direct drive clutch faulty or slipping.
    c. Oil passage obstructed.

5. LUBRICATION CIRCUIT PRESSURE ABNORMAL (TOO HIGH OR TOO LOW).
    a. Defective by-pass valve spring.
    b. By-pass valve plunger sticking or not seating.
    c. Pump defective or pump intake clogged.
    d. Oil passage obstructed.

6. LOW CLUTCH OPERATING PRESSURE.
    a. Pump defective or pump intake clogged.
    b. Reservoir oil level low.
    c. Defective regulator valve spring.
    d. Regulator valve spool stuck open.

7. HIGH CLUTCH OPERATING PRESSURE.
    a. Defective regulator valve spring.
    b. Regulator valve spool stuck closed or sticking in bore.
    c. Oil too heavy or too cold.

8. ERRATIC CLUTCH OPERATING PRESSURE.
    a. Regulator valve spool incorrectly installed.
    b. Sticky regulator valve spool.

## SYSTEM PRESSURE CHECKS

### All Models So Equipped

171B. The Hydraul Shift unit has two oil pressure circuits; a lubrication circuit and a clutch operating pressure circuit. When making any system pressure check, be sure reservoir oil is at correct level, oil is at operating temperature and engine is operating at rated rpm.

To check lubrication circuit, proceed as follows: Install a pressure gage either in place of the cover-to-countershaft oil line, or in place of heat exchanger (oil cooler) line, at the housing cover. Location will depend on space available as either position will suffice.

NOTE: If desired a hole can be cut in the top of an oil filter and a gage welded in place. This assembly can be installed in place of the regular oil filter and the lubrication circuit pressure checked.

Start engine, run at rated rpm and observe gage. Gage should register 20-60 psi with control valve in any drive position. If lubrication pressure is not as stated, refer to item 5 in Trouble Shooting section. Also, be sure hose on pressure gage has at least ½-inch I.D. as a smaller hose might restrict oil flow and give a false reading.

To check clutch operating pressure circuit, remove pipe plug (T—Fig. O113D) from bore below oil filter assembly and install gage. NOTE: At approximately tractor serial number 207 500-000, two passageways were drilled in housing cover behind oil filter base to aid in checking clutch pressure. Start engine, run at rated rpm and observe gage. Gage should register 140-190 psi for tractors prior to serial number 201 091-000, or 140-160 psi for tractors serial number 201 091-000 and up, with control valve in either overdrive or direct drive (clutch pressurized). If clutch pressure is not as stated, refer to items 6 and 7 in Trouble Shooting section. Shims (46—Fig. O113A) may be used behind the pressure regulator valve spring (45), if necessary. Shims are 0.0359 thick and each shim will vary pressure approximately 5 psi.

### REMOVE AND REINSTALL

### All Models So Equipped

171C. To remove the Hydraul Shift drive unit it is necessary to remove the engine and Hydraul Shift unit as outlined in paragraph 60. With engine removed, support it on wood blocks so oil pan will not be dam-

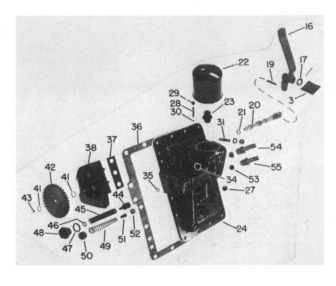

**Fig. O113A — Exploded view of Hydraul Shift unit cover, valves and filter unit. Refer also to Fig. O113D.**

24. Cover
27. Plug (ctsk)
28. Spring
29. Plug
30. Detent ball
31. Stud
34. Insert
35. "O" ring
36. Gasket
37. Gasket
38. Pump
41. Snap ring
42. Pump gear
43. Key
44. Regulator valve spool
45. Spring
46. Shim (.035)
47. Washer
48. Spring
50. Plug (ctsk)
51. By-pass valve plunger
52. Seat
53. Insert
54. Oil cooler line, right
55. Oil cooler line, left

aged, then remove lower dust shield and starter. Unbolt and remove clutch housing and Hydraul Shift unit from engine. Remove clutch release bearing and lubrication hose on models so equipped. Remove cap screw and key from release bearing fork and pull clutch shaft from fork and clutch housing. Unbolt and remove the clutch housing from Hydraul Shift unit.

Reinstall by reversing removal procedure.

### OVERHAUL

### All Models So Equipped

171D. **DISASSEMBLE HYDRAUL SHIFT UNIT.** With Hydraul Shift unit removed as outlined in paragraph 171C, drain unit and remove dipstick and filler cap assembly. Remove valve spool control lever. Straighten tabs of lock washer (72—Fig. O113B), then use a spanner wrench (Oliver No. ST-171, or equivalent) and remove nut (71). Remove sprocket (73) and "O" ring (74) from output shaft. Unscrew filter element (22—Fig. O113A) from adapter (23), remove the filter to countershaft line, then remove cover (24) and pump (38) assembly from unit housing (111—Fig. O113B). Pull input shaft support (153) and input shaft (164) from housing and direct drive clutch. See Fig. O113E. Remove cap screws which retain overdrive clutch housing (78—Fig. O113B), then rotate housing slightly to allow removal of countershaft support (110). See Fig. O113F. Support the countershaft (169—Fig. O113B) to prevent binding the support, bump on front end of support until sealing plug (108) is removed, then remove support (110) from front of housing. Allow countershaft to rest on bottom of housing. Remove output shaft (139), direct

drive clutch and internal ring gear (152) assembly from housing. Remove planet gear carrier (133), drive gear (127) and bearing (126) if not removed with output shaft, clutch and internal ring gear assembly. Remove thrust bushing (125) from housing. Remove overdrive clutch housing (78), clutch and sun gear shaft (103) from housing by bumping lightly on front end of sun gear shaft. Remove support (105) from housing if it was not removed with the overdrive clutch and housing assembly. Remove countershaft and countershaft thrust washers (167) from housing. Remove filler neck (115), studs, dowels or control lever shaft (113) as necessary.

With Hydraul Shift unit disassembled, discard all "O" rings, seals and gaskets and refer to the appropriate following paragraphs to service sub-assemblies. Refer to paragraph 171L for reassembly procedure.

171E. **COVER AND VALVES.** NOTE: If only pressure regulator valve, by-pass valve or oil filter are to be serviced, it can be accomplished without removing cover and with Hydraul Shift unit in tractor.

Remove plug (48—Fig. O113A or O113D), washer (47), spring (45) and pressure regulator spool (44) from bore in cover. DO NOT lose any shims (46) which may be used. A $\frac{5}{16}$-inch cap screw can be used to remove spool if it is stuck. Remove countersunk plug (50), spring (49) and by-pass plunger (51). Seat (52) is renewable, if necessary. Remove plug (29), spring (28) and detent ball (30), then pull control spool (20) from bore and discard "O" ring (21). Also See Fig. O113D. DO NOT attempt to remove spool (20—Fig. O113B) until detent assembly is removed or damage to

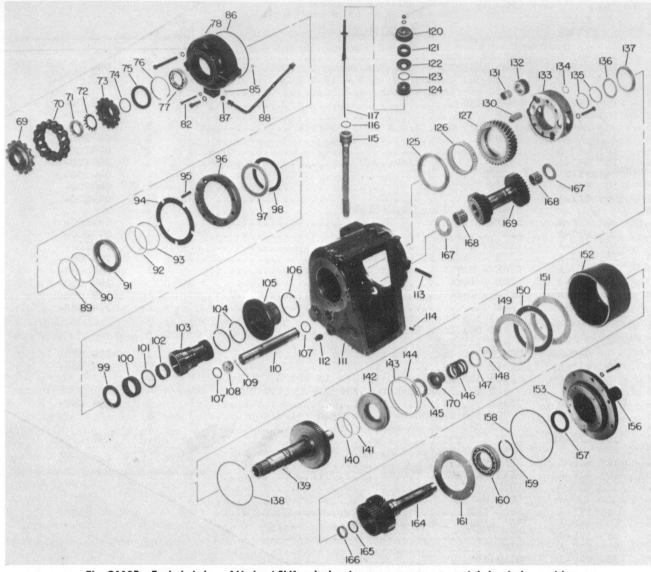

**Fig. O113B—Exploded view of Hydraul Shift unit showing component parts and their relative positions.**

| | | | | | |
|---|---|---|---|---|---|
| 69. Sprocket, rear | 89. Seal ring | 101. Spacer | 117. Dipstick | 137. Bushing | 152. Internal ring |
| 70. Coupling chain | 90. "O" ring | 102. Bearing | 120. Cap | 138. Retaining ring | gear |
| 71. Spanner nut | 91. Piston | 103. Sun gear shaft | 121. Filter element | 139. Output shaft | 153. Support |
| 72. Lock washer | 92. "O" ring | 104. Seal rings | 122. Cup | 140. Seal ring | 156. Tube |
| 73. Sprocket, front | 93. Seal ring | 105. Support | 123. Gasket | 141. "O" ring | 157. Oil seal |
| 74. "O" ring | 94. Spring plate | 106. Seal ring | 124. Cap | 142. Piston | 158. "O" ring |
| 75. Oil seal | 95. Return springs | 107. "O" ring | 125. Bushing | 143. "O" ring | 159. Snap ring |
| 76. Snap ring | (5 used) | 108. Plug | 126. Bearing | 144. Seal ring | 160. Bearing |
| 77. Bearing | 96. Retainer | 109. "O" ring | 127. Drive gear | 145. Thrust washer | 161. Bearing retainer |
| 78. Clutch housing | 97. Separator plates | 110. Support | 130. Planet gear pin | 146. Return spring | 164. Input shaft |
| 82. Cap screw | (6 used) | 111. Housing | 131. Bearing | 147. Retainer | 165. Oil seal |
| (¼ x 1½) | 98. Clutch plates | 112. Plug, magnetic | 132. Planet gear | 148. Snap ring | 166. Bearing |
| 85. "O" ring | (5 used) | 113. Lever shaft | 133. Planet carrier | 149. Pressure plate | 167. Thrust washer |
| 86. "O" ring | 99. Thrust washer | 114. Dowel | 134. Snap ring | 150. Clutch plate | 168. Bearing |
| 87. Insert | 100. Overrunning | 115. Filler tube | 135. Seal rings | (6 used) | 169. Countershaft |
| 88. Oil line | (Sprag) clutch | 116. "O" ring | 136. Thrust washer | 151. Separator plate | 170. Spring guide |
| | | | | (5 used) | |

spool could result. Remove oil filter adapter (23) so oil passages can be inspected and cleaned.

Clean and inspect all parts. Control spool (20) and pressure regulator spool (44) should be free of nicks or scoring and should slide freely in their bores when lubricated. By-pass plungers (51) and seat (52) should be free of nicks, burrs or grooves. Springs (28, 45 and 49) should be free of rust, bends or fractures.

Refer to the following data as a guide for parts renewal.

Pressure Regulator Valve Spring
    Free length—in. ...............3⅝
    Spring test lbs. @ in. ..55-57 @ 2¾
By-Pass Valve Spring
    Free length—in. ...............3½
    Spring test lbs.
      @ in. ...........5½-6½ @ 2⁷⁄₁₆
Detent Ball Spring
    Free length—in. ...............1⁹⁄₃₂
    Spring test lbs. @ in. 15.3-18.7 @ ⅞
Pressure Regulator Spool
    Diameter ...........0.6865-0.6869

Control Spool Land—
    Diameter ...........0.6865-0.6869
Pressure Regulator Spool
    Bore—I.D. ..........0.6875-0.6883
Counter Spool Bore
    I.D. .................0.6875-0.6883

Reassemble by reversing the disassembly procedure .

171F. PUMP. Unbolt pump from housing cover and discard gasket. Remove drive gear outer retainer snap ring, remove gear and key, then re-

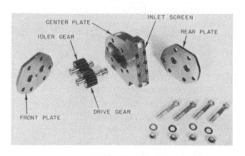

**Fig. O113C—Hydraul Shift oil pump disassembled. Refer to text for dimensional data.**

**Fig. O113E — Hydraul Shift unit with input shaft assembly removed. Refer to Fig. O113B for legend.**

move gear inner retainer snap ring.

Remove through bolts and separate pump. Dowels can be driven out and screen removed, if necessary. See Fig. O113C.

NOTE: If same pump gears and shafts are to be reinstalled, mark idler gear so it will be reinstalled in original position. Do not remove gears from shafts as they cannot be serviced separately.

Pump dimensional data are as follows:

Gear width .......... 0.9975-0.9980
Pump body width .... 0.9995-1.0000
Gear shaft diameter .. 0.6247-0.6250
Shaft bore diameter .. 0.6265-0.6275
Shaft operating
  clearance .......... 0.0015-0.0028

Reassemble by reversing the disassembly procedure and use new gasket when mounting pump on housing cover. Heads of bolts should be on drive gear side of pump and dowels should be installed prior to tightening through bolts.

**171G. INPUT SHAFT AND SUPPORT.** Remove "O" ring (158—Fig. O113B) from support (153), then remove bearing (166) and seal (165) from bore in rear of input shaft. Unbolt bearing retainer (161) and press input shaft (164) and bearing (160) from support. Oil seal (157) can now be removed from support, and if renewal is required, clutch release bearing tube (156) can also be removed at this time. Do not remove bearing (160) from input shaft unless renewal of bearing or bearing retainer (161) is indicated. If either part is to be renewed, remove snap ring (159) and press input shaft from bearing.

To reassemble input shaft and support, place bearing retainer on input shaft, then press bearing on shaft and install snap ring. Press bearing tube, (if renewed) into support until tube protrudes 4 inches from front flange of support and be sure tube is parallel with bore. Install oil seal (157) in support with lip of seal toward rear. Press input shaft and bearing into

**Fig. O113F—Countershaft support and the plug (108) must have cap screw hole positioned as shown to allow installation of support retaining cap screw.**

support and secure with bearing retainer. Install oil seal (165) in rear bore of input shaft with lip facing toward rear, then install bearing (166). Install "O" ring (158) in groove of support.

Refer to paragraph O171L for installation procedure.

**171H. OUTPUT SHAFT, INTERNAL RING GEAR AND DIRECT DRIVE CLUTCH.** To disassemble the output shaft, internal ring gear and direct drive clutch, proceed as follows: If planet gear carrier remained with output shaft assembly during disassembly, separate it from output shaft, then remove snap ring (138—Fig. O113B) and output shaft (139) from internal ring gear (152). With a press, compress spring (146), remove snap ring (148), retainer (147), return spring (146), guide (170) and thrust washer (145) from output shaft. Remove clutch piston (142) from output shaft (gear), either by bumping end of output shaft on a block of wood, or by directing compressed air into oil pas-

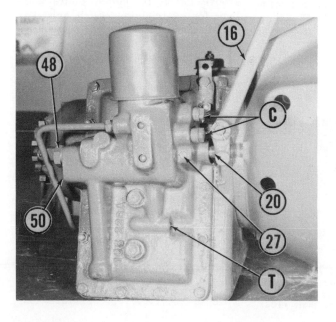

**Fig. O113D — Hydraul Shift unit cover installed. Refer to Fig. O113A for legend. (T) is test port for checking clutch pressure. Shipping plugs (C) are installed in oil cooler line ports.**

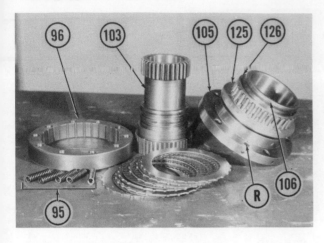

Fig. O113G—Component parts of overdrive clutch, sun gear shaft and support except clutch housing and spring plate. Note small "O" ring (R) in front side of support (105).

sage of output shaft, then remove seal rings (140 and 144) and "O" rings (141 and 143) from I.D. and O.D. of piston. Remove seal rings (135) and thrust washer (136) from rear of output shaft. NOTE: Bushing (137) is catalogued as a separate part, however, if renewal is not indicated, do not attempt to remove it from output shaft. Remove pressure plate (149), clutch plates (150) and separator plates (151) from internal ring gear.

Clean and inspect all parts. Sprag (overunning) clutch surface of output shaft must be smooth and show no signs of wear. Clutch plates must have no broken lugs, splines or warpage. Refer to the following data as a guide for parts renewal.

Return Spring
  Free length .................2.270
  Spring test lbs. @ in. 230 @ 1.750
Clutch Plate Thickness ....0.076-0.084
Separator Plate
  Thickness ..............0.087-0.093
Pressure Plate
  Thickness ..............0.248-0.252
Piston Ring Groove—Outside
  Width .................0.156-0.162
  Depth .................0.193-0.194
Piston Ring Groove—Inside
  Width .................0.128-0.132
  Depth .................0.127-0.128

Reassemble output shaft, internal ring gear and direct drive clutch assembly as follows: Start with a clutch plate (150) and alternately install the six clutch plates and five separator plates (151) in internal ring gear (152), then install pressure plate (149). Install "O" rings (141 and 143) and seals (140 and 144) in O.D. and I.D. grooves of piston (142), grease seals and press piston in front side of output shaft gear. Use caution when installing piston not to damage seals. Install washer (145), guide (170), return spring (146) and retainer (147) on front end of output shaft, compress

spring and install snap ring (148). Install output shaft assembly in internal ring gear with clutch piston next to clutch plates and install snap ring (138). Install thrust washer (136) over splined end of output shaft, then install the two seal rings (135) in grooves of output shaft.

**171J. PLANET GEAR CARRIER.** To disassemble the planet gear carrier assembly, first remove drive gear (127—Fig. O113B) from carrier (133) by removing the four retaining cap screws. Remove the planet gear pin retaining rings (134) and drive planet gear pins (130) fom carrier. Remove planet gears (132) from carrier and bearings (131) from planet gears.

Inspect planet gears, bearings and pins for wear, scoring or other damage and renew as necessary. Inspect drive gear teeth and inside diameter. Inside diameter of gear should not be unduly worn or scored. Check condition of bearing (126) and with bearing in gear, check fit of gear and bearing on support (105). Gear and bearing should have a free fit on support without excessive clearance. At this time, also inspect bushing (125) for wear or scoring.

Fig. O113H—Install bearing (102), spacer (101) and Sprag clutch (100) in sequence shown. Drag strips (bronze) of Sprag clutch are toward front of sun gear shaft (103).

To reassemble planet carrier, place planet gear pin retainer in groove at front of pin bore. Place bearing in planet gear, position gear in carrier, then insert planet gear pin from rear of carrier until pin is against retainer and step on rear of shaft is flush with lower surface of carrier. Repeat operation for the three remaining planet gears. Place drive gear on rear of carrier, align cap screw holes, then install retaining cap screws and tighten them to 48-52 Ft.-Lbs. torque. Bushing (125) and bearing (126) are placed on support (105) when unit is being assembled.

**171K. OVERDRIVE CLUTCH AND SUN GEAR.** If removed with overdrive clutch assembly, remove sun gear shaft (103—Fig. 113B) from overdrive clutch. Remove overrunning clutch (100), spacer (101), bearing (102) and seal rings (104) from sun gear shaft. Bump support (105) from housing, if necessary, and remove seal ring (106). Remove clutch plates (98), separator plates (97), retainer (96), return springs (95) and return spring plate (94) from housing (78). See Fig. O113G. Remove clutch piston (91—Fig. O113B) from housing by bumping housing on a wood block. Remove seal rings (89 and 93) and "O" rings (90 and 92) from O.D. and I.D. of piston. Remove oil seal (75), snap ring (76) and bearing (77) from bore of housing.

With overdrive clutch disassembled, discard oil seal (75), seals (89 and 93) and "O" rings (90 and 92), then clean and inspect all parts. Clutch plates and separator plates must not be warped, scored or unduly worn. Clutch splines of sun gear shaft must not have deep grooves worn by clutch plates and bore for overrunning clutch must be smooth and show no wear. On units prior to serial number 60885, inspect retainer (96) and if chamfer on splines exceeds $\frac{1}{32}$-inch, replace retainer with one of later design. Refer to the following data as a guide for parts renewal.

Return springs
  Free length .................1.320
  Spring test lbs.
    @ in. .................10 @ 1.159
Clutch Plate Thickness ....0.073-0.078
Separator Plate Thickness 0.077-0.082
Piston Ring Groove—Outside
  Width .................0.128-0.132
  Depth .................0.195-0.197
Piston Ring Groove—Inside
  Width .................0.128-0.132
  Depth .................0.151-0.152

Partially assemble overdrive assembly as follows: Install bearing (77) in housing (78) and secure with snap ring (76). Install oil seal (75) with lip toward bearing. Install "O" rings in outer and inner grooves of piston (91), install seals over both "O" rings, then grease seals and install piston in housing (78). Install seals (104) in grooves of sun gear shaft (103). Place bearing (102) and spacer (101) in bore of sun gear shaft, then install the overrunning clutch so the drag strips (bronze) are toward bearing (102). See Fig. O113H. Delay any further assembly of the overdrive clutch until the Hydraul Shift unit is assembled.

**171L. REASSEMBLE HYDRAUL SHIFT UNIT.** When reassembling the Hydraul Shift unit, proceed as follows: Install seal ring (106—Fig. O113G) on hub of support and the small "O" ring (R) in oil hole counterbore on front side of support flange, then install support in rear bore of housing so oil hole in support flange (with "O" ring) is aligned with the oil hole in housing which is on right side of lower cap screw hole. Temporarily install one of the short cover retaining cap screws finger tight to keep support from shifting when planet carrier and output shaft assembly are installed.

Coat countershaft thrust washers (167—Fig. O113B) with grease and stick them to inner sides of countershaft support bores and be sure tangs of thrust washers are in slots. With bearings (168) installed, carefully place countershaft in housing, with double gear rearward, so thrust washers are not dislodged, then pull countershaft toward cover opening until gear teeth clear front bore of housing. Place bushing (125) on support with cupped side rearward, then install bearing (126) on support and use fingers to compress seal ring, if necessary. Install planet carrier and drive gear (gear rearward) in housing with gear (127) over bearing (126). Be sure bushing (137), thrust washer (136) and seal rings (135) are installed on splined (long) end of output shaft, start output shaft through planet carrier and support and install internal ring gear over the planet gears.

Place new "O" ring (107) on front end of countershaft support (110). Move countershaft into position, align thrust washers, lift countershaft so support will not bind, then slide support rearward until support is ready to enter rear bore. At this time, posi-

tion tapped hole in rear end of support at the nine o'clock position and push shaft rearward until front of support is flush with housing.

NOTE: When overdrive clutch housing (78) is finally installed, holes in housing, plug (108) and support (110) must be aligned to allow installation of the 1/4 x 1 1/2-inch support retaining cap screw, as well as to place the oil hole in support on top side. Support is slotted at forward end if subsequent positioning is required.

With countershaft support positioned, place new "O" ring (107) on O.D. of plug (108), then use grease to stick the small "O" ring (109) in counterbore of cap screw hole in front of plug. Insert cap screw (82) through plug and small "O" ring (109) and thread a few turns into support, then bump plug into housing bore. Remove cap screw after plug is installed. See Fig. O113F.

Obtain a small 1/2-inch thick wood block about 3x3 inches square, place block against forward end of output shaft, then tip assembly forward so output shaft is supported by the wood block. With small end forward, place sun gear shaft assembly over output shaft and push sun gear shaft on output shaft until it bottoms.

NOTE: Installation of overrunning clutch can be checked at this time. Sun gear shaft will turn (clutch overrun) clockwise but will lock up when attempt is made to turn shaft counterclockwise.

Remove cap screw that was installed earlier in flange of support (105—Fig. O113B). To ease assembly of overdrive clutch, install two guide studs in cap screw holes of housing (111). Install retainer (96) over guide studs with chamfer in I.D. of retainer on top side. Start with a separator plate (97) and alternately install the six separator plates and the five clutch plates (98). Place thrust washer (99) over end of output shaft with flat side toward sun gear shaft and be sure it does not hang up on shoulder of out-

put shaft. Place return springs (95) in blind holes of retainer (96), then place plate (94) over return springs. Place new "O" ring (86) over flange of support (105) and small "O" rings (85) in counterbores of front face of housing (78), then carefully install housing (78) over guide studs. Remove guide studs and secure clutch housing with cap screws. Install countershaft support retaining cap screw (82). Place new "O" ring (74), sprocket (73), lock washer (72) and nut (71) on end of output shaft and using a spanner wrench (Oliver No. ST-171 or equivalent), tighten nut (71). Secure nut by bending a tab of lock washer into slot in nut.

Set unit in an upright position, align splines and center direct drive clutch plates as nearly as possible. Install new "O" ring (158) on pilot of front support (153), then insert input shaft hub into direct drive clutch and rotate input shaft as necessary to engage the clutch plates. NOTE: Two cap screws can be temporarily installed to help in overcoming resistance of "O" ring; however, BE SURE input shaft can be rotated while drawing support into housing to prevent damage to inner clutch plates. Cut-out (C—Fig. O113E) in O.D. of support flange must mate with a similar cut-out (C) at right side of housing.

Place the wood block used earlier under front of housing to tip it slightly rearward, then remove the two temporarily installed cap screws from support. Place engine clutch housing on support and secure clutch housing and support to Hydraul Shift unit housing. Install clutch throw-out bearing and lubrication hose (if used), throw-out bearing fork and clutch shaft. Use new gasket and install cover and pump assembly. Install cover-to-countershaft oil line and the control spool lever. When installing new oil filter, tighten filter not more than 1/2-turn after gasket contacts base. Install drain plug if necessary and fill unit with 3 1/2 quarts of Automatic Transmission Fluid, Type "A".

# CREEPER DRIVE UNIT

Some models of the Series 1550, 1555, 1600, 1650 and 1655 may be equipped with a Creeper Drive unit which provides direct drive and creeper drive speeds. The Creeper Drive operation can be used only in conjunction with 1st, 2nd, 4th and first reverse

speed positions of the tractor main transmission as a lock-out system prevents creeper drive operation in the other speeds. The Creeper Drive unit must not be shifted while tractor is in motion.

## REMOVE AND REINSTALL

### Models So Equipped

172. To remove the Creeper Drive unit, first remove engine with Creeper Drive unit attached, as outlined in paragraph 60. Support engine on wood blocks so oil pan will not be damaged. Remove starting motor and dust shield from lower front of clutch housing. Disconnect throw-out bearing grease tube (when used), then remove throwout bearing, clutch cross shaft and release fork. Creeper Drive unit can now be removed from clutch housing.

Reinstall by reversing the removal procedure.

### OVERHAUL

### Models So Equipped

173. With unit removed as outlined in paragraph 172, drain unit, if necessary, then refer to the following appropriate paragraphs for disassembly and reassembly of the cover assembly, input shaft assembly, output shaft assembly and the countershaft assembly.

174. **COVER ASSEMBLY.** To disassemble the cover assembly, remove the countershaft support shaft oil line, then unbolt and remove cover assembly from housing. Drive out roll pin (68—Fig. O114), pull shift lever (67) from shifter arm (80), then pull arm from cover. Seal (66) can now be renewed. Remove baffle plate (74) and gasket (73). Remove one plug (62), lift snap rings (79) from their grooves, then bump out shifter rail (72) and catch detent ball (78) and spring (77) as shift rail clears shifter hub (81). Note: Remaining plug (62) will be removed by shift rail as it is bumped out. Remove shifter fork (82) if necessary, then drive out roll pin (70) and remove interlock (lockout) rod (69) and hub (81) from cover. "O" ring (71) can now be renewed.

Clean and inspect all parts and renew as necessary. Use new "O" ring and seal and reassemble by reversing disassembly procedure.

175. **INPUT SHAFT ASSEMBLY.** To remove the input shaft assembly, first remove cover as outlined in paragraph 174. Remove shaft retainer (97), pull countershaft support shaft (100) and let countershaft assembly rest in bottom of housing. On early units, remove throwout bearing tube and support (43) and bearing retainer (48), then pull input shaft assembly from bearing support (38) and housing.

**Fig. O114—Composite view of Creeper Drive unit showing parts used in both early and late type. Late type unit was installed at tractor serial number 185 191-000 and uses items 20, 21A, 21B, 102, 103, 104 and 105.**

| | | |
|---|---|---|
| 1. Washer | 22. Stud | 49. Gasket | 80. Shifter arm |
| 2. Cap | 25. Shim (.004, | 50. Snap ring | 81. Hub |
| 3. Element | .007. 015) | 51. Knob | 82. Shifter fork |
| 4. Cup | 26. Spacer | 52. Clamp | 85. Dowel |
| 5. Gasket | 27. Front bearing | 54. Bushing | 86. Thrust washer |
| 6. Cap | 28. Housing | 55. Control rod | 87. Snap ring |
| 7. Dipstick | 29. Drain plug | 58. Gasket | 88. Countershaft |
| 8. Nut | 32. Output shaft | 59. Cover | gear |
| 9. "O" ring | 33. Output gear | 62. Plug | 89. Snap ring |
| 10. Filler tube | 34. Snap ring | 63. Stud | 90. Bearing |
| 11. Oil seal | 35. Coupling | 66. Oil seal | 91. Countershaft |
| 12. "O" ring | 36. Input gear | 67. Shifter lever | 92. Thrust washer |
| 13. Sprocket, front | 37. Gasket | 68. Groove pin | 93. Oil line |
| 14. Nut | 38. Bearing support | 69. Lock-out rod | 94. Connector |
| 15. Coupling chain | 39. Oil seal | 70. Roll pin | 95. Elbow |
| 16. Master link | 40. Input shaft | 71. "O" ring | 96. "O" ring |
| 18. Sprocket, rear | 41. Ball bearing | 72. Shifter rail | 97. Retainer |
| 19. Rear bearing | 42. Tube | 73. Gasket | 100. Support shaft |
| 20. Retainer | 43. Tube support | 74. Plate | 101. "O" ring |
| 21. Gasket | 44. Stud | 77. Detent spring | 102. Support |
| 21A. "O" ring | 47. Oil seal | 78. Detent ball | 103. "O" ring |
| 21B. "O" ring | 48. Retainer | 79. Snap ring | 104. Ball bearing |
| | | | 105. Snap ring |

Bump off bearing support (38). Remove snap ring (50) and press input shaft (40) from bearing (41). Remove snap ring from outer race of bearing (41). Push gear (36) and coupling (35) forward on input shaft to ease removal of snap ring (34), then remove snap ring, coupling and gear from input shaft. Seal (39) can now be removed from rear of input shaft and seal (47) can be removed from retainer (48).

On late units, remove front support (102) and input shaft (40) assembly from housing, then remove snap ring (105) and pull input shaft and bearing from front support. Remove snap ring (50) from front of bearing (104), then press input shaft from bearing. Oil seals (39 and 47) can now be removed.

Clean and inspect all parts and renew as necessary. Use new gaskets and seals during assembly. Seal (39) is installed with lip facing toward front of input shaft. Seal (47) is installed with lip toward rear.

Reassemble the input shaft by reversing the disassembly procedure.

176. **OUTPUT SHAFT ASSEMBLY.** To remove the output shaft assembly, remove the cover as outlined in paragraph 174, then unbolt bearing retainer (20—Fig. O114), pull shaft assembly rearward and remove gear (33) from side of housing.

With shaft assembly removed, unstake lock nut (14), then using a spanner wrench (Oliver No. ST-149, or equivalent), remove nut, coupling sprocket (13) and "O" ring (12). Press output shaft from rear bearing and retainer, remove shims (25) and spacer (26), then press output shaft from front bearing. Remove oil seal (11) and rear bearing from retainer. If necessary, bearing cups can now be removed from retainer.

Clean and inspect all parts and pay particular attention to the bearings and bearing cups. Use new "O" ring, seal and gasket (or "O" rings) during assembly. Assemble output shaft components and adjust bearings as outlined in paragraph 176A prior to installing shaft assembly in housing.

**176A.** To assemble shaft and adjust bearings, proceed as follows: Install bearing cups in retainer (20) with smaller diameters toward inside. Press front bearing on shaft with smaller diameter toward rear. Place spacer (26) and the original shims (25) on shaft, insert shaft in retainer and press on rear bearing with smaller diameter toward front. Sprocket coupling can be used as a driver. Install a new lock nut (14) on shaft and tighten nut to a torque of 200 Ft.-Lbs. Check the shaft bearing adjustment which should be between 0.001 shaft end play and 5 inch pounds rolling torque. If adjustment is not as stated, remove nut, coupling sprocket and rear bearing and vary shims (25) as required. Shims are available in thicknesses of 0.004, 0.007 and 0.015.

With shaft adjustment made, remove nut and coupling sprocket and install new "O" ring (12) and oil seal (11) with lip toward front. Reinstall coupling sprocket and lock nut, tighten nut to a torque of 200 Ft.-Lbs. and stake nut in position.

**177. COUNTERSHAFT ASSEMBLY.** With cover assembly off, remove retainer (97—Fig. O114) and pull countershaft support shaft (100). Lift rear snap ring (89) from its groove and slide it about ¼-inch rearward. The countershaft assembly (91) and thrust washers (86 and 92) can now be removed from housing. Removal of gear (88) and bearings (90) is obvious.

Clean and inspect all parts and renew as necessary.

When reassembling, install gear (88) on shaft with hub rearward. Install needle bearings (90) into countershaft only far enough to install the retaining snap rings.

### ASSEMBLY

**178.** To reassemble Creeper Drive unit, proceed as follows: Start output shaft into housing and place gear (33 —Fig. O114) on shaft with shifter fork groove toward front. On early units align notches in gasket (21) and retainer (20) with oil drainback hole in housing and secure retainer. Use new gasket (37) and place bearing support (38) on housing with cut-out in O.D. of support at the 9 o'clock position. Insert input shaft assembly into housing and over pilot of output shaft. Install snap ring on O. D. of input shaft bearing. Align notches of gasket (49) and bearing retainer (48) with oil drain-back hole in bearing support and install bearing retainer and throw-out bearing tube and support (43). On late units, use new "O" ring (103) and install front support (102) and input shaft as an assembly. Use grease to hold thrust washers (86 and 92) in place and install thrust washer with larger O. D. at rear. Place countershaft assembly in position, insert support shaft (100) and install retainer (97). Align shifter fork with fork groove of gear (33), then install cover and oil line (93). Disassemble breather assembly and clean or renew the breather filter element (3) and fill unit with 5 quarts of Automatic Transmission Fluid, Type A.

# REVERSE-O-TORC DRIVE UNIT
## (Series 1600-1650 Prior S/N 204 145-000)

The Reverse-O-Torc Drive for Series 1600 and 1650 prior to tractor serial number 204 145-000 is a forward and reverse auxiliary transmission mounted in front of the tractor transmission providing a 1:1 direct (forward) drive ratio and a 1.1:1 reverse drive ratio to the transmission input shaft. The unit consists of a planetary gear set, a multiple disc forward clutch and a multiple disc reverse clutch which are connected coaxially and are mounted in a cast iron housing. The clutches are controlled by hydraulic pistons that are actuated by a hydraulic pump and valving located within the unit. A hydraulic torque converter mounted on the engine flywheel drives the forward and reverse unit. Operating fluid and charging pressure for the converter is provided by the Reverse-O-Torc pump via a pressure regulating valve.

Reverse-O-Torc Drive is available on Industrial models only. Models so equipped do not have an engine clutch or reverse gears in the tractor transmission as the forward and reverse clutches within the auxiliary drive are utilized for starting and stopping the tractor as well as providing a reverse ratio for each of the transmission forward gears. Also, models equipped with Reverse-O-Torc Drive cannot be equipped with a power take-off or a hydraulic system.

### Industrial Models So Equipped

**179. OPERATION.** The Reverse-O-Torc unit is controlled by two foot pedals located at right side of operator's platform. Depressing the foot pedal marked "F" will start the tractor in forward motion when tractor transmission is shifted into gear. Depressing the pedal marked "R" will start the tractor in a reverse direction. Additional downward movement of either pedal will increase engine speed. In addition to the pedals marked "F" and "R", a foot accelerator pedal is located at the right side of operator's platform and depressing the accelerator pedal will increase engine speed without activating the Reverse-O-Torc unit.

Shift transmission only when Reverse-O-Torc pedals are in neutral. Engage the Reverse-O-Torc pedals only when engine is at slow idle speed. Stop tractor with brakes before changing direction of travel. If torque converter heat indicator is in "SHIFT" range on dial, slow engine to 1000 RPM until indicator drops to "Normal" range.

**180. TROUBLE SHOOTING.** The following will serve as a guide when trouble shooting problems encountered with the Reverse-O-Torc Drive unit:

1. NOISY IN NEUTRAL. Could be caused by:
   a. Worn bushings in pump assembly.
   b. Worn sprag or sprag races in converter assembly.
   c. Low oil level.

2. NOISY IN REVERSE. Could be caused by:
   a. Worn or rough planetary gears.

3. OVERHEATING. Could be caused by:
   a. Improper operation.
   b. Worn or damaged pump.
   c. Converter sprag clutch worn and slipping.

4. WILL NOT PULL. Could be caused by:
   a. Converter drive lugs sheared or not engaged in pump.
   b. Pump gears seized and converter drive lugs sheared.
   c. Low oil level.
   d. Worn or damaged bushings.

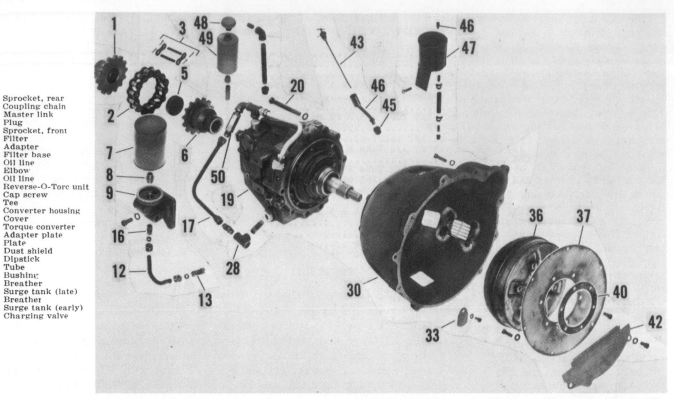

1. Sprocket, rear
2. Coupling chain
3. Master link
5. Plug
6. Sprocket, front
7. Filter
8. Adapter
9. Filter base
12. Oil line
13. Elbow
17. Oil line
19. Reverse-O-Torc unit
20. Cap screw
28. Tee
30. Converter housing
33. Cover
36. Torque converter
37. Adapter plate
40. Plate
42. Dust shield
43. Dipstick
44. Tube
45. Bushing
46. Breather
47. Surge tank (late)
48. Breather
49. Surge tank (early)
50. Charging valve

Fig. O115—Reverse-O-Torc Drive unit (19) is driven by torque converter (36) mounted on engine flywheel. A large surge tank (47) replaced surge tank (49) on late production units. The unit shown was replaced at tractor serial number 204 145-000 by the unit shown in Fig. O116A.

5. WILL NOT PULL IN FORWARD DIRECTION. Could be caused by:
   a. Worn or broken sealing rings in forward clutch.
   b. Damaged sealing rings in forward clutch cylinder.
   c. Clutch plates worn or broken in forward clutch assembly.

6. WILL NOT PULL IN REVERSE DIRECTION. Could be caused by:
   a. Worn or damaged sealing rings reverse clutch piston.
   b. Broken or worn reverse clutch plates.

7. HARD SHIFTING (PEDAL ACTION). Could be caused by:
   a. Rotary control valve burred or nicked.
   b. Linkage bent or seizing.

181. **R&R REVERSE - O - TORC DRIVE.** To remove the Reverse-O-Torc Drive unit, first remove the engine and drive unit as an assembly as outlined in paragraph 60. Then, unbolt and remove the drive unit (19 —Fig. O115) from torque converter housing (30).

When reinstalling the drive unit, lubricate rear hub of converter (36) and seal in pump housing on front face of drive unit. Carefully install drive unit to rear of torque converter housing and tighten the retaining cap

screws (20) to a torque of 40-45 Ft.-Lbs. Reinstall engine and drive unit in tractor as outlined in paragraph 60.

182. **R&R TORQUE CONVERTER.** To remove torque converter (36—Fig. O115), first remove Reverse-O-Torc Drive unit as outlined in paragraph 181. Then, proceed as follows:

Remove dust shield (42) from lower front side of torque converter housing and starter. Unbolt and remove torque converter housing from engine. Unbolt adapter plate (37) from flywheel and remove adapter plate and converter as a unit. Unbolt adapter plate from front face of converter.

The torque converter is serviced as a complete assembly only. Renew converter if damaged or if it is not operating properly. Reinstall by reversing removal procedure.

183. **OVERHAUL REVERSE - O - TORC DRIVE UNIT.** After removing the drive unit as outlined in paragraph 181, remove unit oil filter and external oil lines. Then, refer to Fig. O116 and proceed as follows:

Place the drive unit in a suitable stand or support with front end of unit up. Remove the cap screws (82) retaining pump assembly (80) to front adapter (85) and lift the pump assembly from unit. To disassemble pump, remove the two flat head

screws (81) (one on late units). Pry oil seal (83) from pump housing.

Remove the socket head cap screws (86) retaining front adapter (85) to housing (18) and lift the front adapter and piston (79) assembly from unit. Be careful not to drop pressure plate (77) if it sticks to piston. Piston can be removed from front adapter by applying air pressure to clutch oil passage in front adapter. Remove sealing rings (78 and 89) from piston and hub of front adapter.

Lift pressure plate (77) from unit if not removed with front adapter and piston. Remove the 12 pressure plate springs (76) and the three dowel pins (75) from outer perimeter of housing. Remove the three inner clutch plates (73) and the two outer plates (74). Depress converter regulator valve (44) to be sure valve is free in housing and then remove valve and valve spring (43).

Remove thrust washer (72) from thrust face of the forward clutch cylinder (67). Remove the snap ring (70) from input shaft at front side of bearing (69); however, DO NOT remove the snap ring (71) from clutch cylinder (67) at this time. Lift the input shaft and forward clutch assembly from unit. With a soft hammer, tap on front end of input shaft to drive the shaft and clutch hub (52) assembly rearward out of the forward

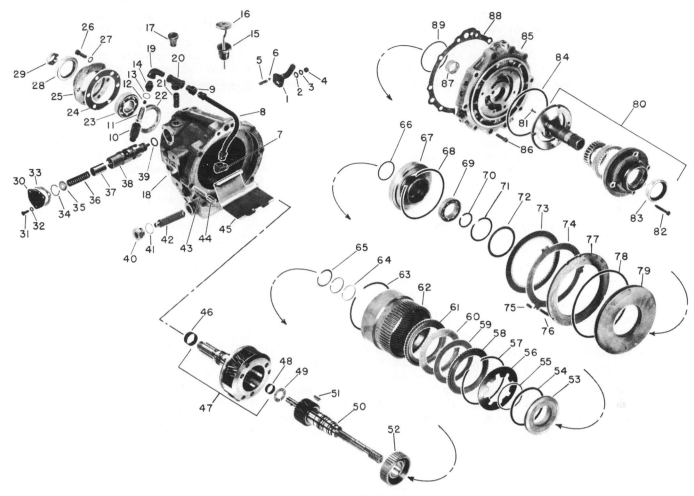

**Fig. O116 — Exploded view of the Reverse-O-Torc drive unit showing component parts and their relative positions.**

| | | | | |
|---|---|---|---|---|
| 1. Shift lever | 16. Dip stick | 34. Snap ring | 47. Output shaft | 60. Clutch plate (outer) | 75. Dowel |
| 2. Flat washer | 17. Breather | 35. Spring retainer | assembly | 61. Pressure plate (rear) | 76. Spring |
| 3. Lock washer | 18. Housing | 36. Spring | 48. Bushing | 62. Ring gear | 77. Pressure plate |
| 4. Nut | 19. Elbow | 37. Regulator valve | 49. Thrust washer | 63. Snap ring (selective) | 78. Seal ring |
| 5. Detent spring | 20. Tee | 38. Control valve | 50. Drive shaft | 64. Seal ring | 79. Piston |
| 6. Detent ball | 21. Nipple | 39. "O" ring | 51. Woodruff key | 65. Snap ring | 80. Oil pump assembly |
| 7. Elbow | 22. Sleeve | 40. Reducing bushing | 52. Clutch hub (forward) | 66. "O" ring | 81. Flat head screw |
| 8. Oil tube | 23. Bearing | 41. Copper gasket | 53. Piston, front clutch | 67. Clutch cylinder | 83. Oil seal |
| 9. Connector | 24. Gasket | 42. Strainer | 54. Piston ring | 68. Snap ring | 84. Gasket |
| 10. Check valve body | 25. Bearing retainer | 43. Spring | 55. Spring bearing ring | 69. Bearing | 85. Front adapter |
| 11. Spring | 28. Oil seal | 44. Converter regulator | 56. Clutch spring | 70. Snap ring | 86. Socket head screw |
| 12. Check valve ball | 29. Nut | valve | 57. Snap ring | 71. Snap ring | 87. Bushing |
| 13. Copper gasket | 30. Cover | 45. Baffle | 58. Pressure plate (front) | 72. Thrust washer (selective) | 88. Gasket |
| 14. Valve seat | 33. Gasket | 46. Bushing | 59. Clutch plate (inner) | 73. Clutch plate (inner) | 89. "O" ring |
| 15. Filler | | | | 74. Clutch plate (outer) | |

clutch assembly and ring gear (62). Remove the clutch sealing (cast iron) rings (64) from input shaft. If clutch hub (52) or input shaft (50) is to be renewed, remove snap ring (65) and press input shaft from clutch hub.

To disassemble the forward clutch unit, place the clutch and ring gear assembly in a press with splines on outer diameter of ring gear up. With a suitable arbor placed on hub of clutch cylinder (67), press the cylinder just far enough into ring gear to allow removal of snap ring (68). Slowly release pressure and remove the clutch cylinder (67) and piston (53) assembly. Piston can be removed from cylinder by applying air pres-

sure to holes in inner diameter of cylinder. Remove snap ring (71) and bump bearing (69) from clutch cylinder.

Remove bearing ring (55) if not removed with clutch piston. Lift out spring (56) and remove snap ring (57), then remove the clutch discs (59 and 60) and pressure plates (58 and 61). Remove snap ring (63) from ring gear (62).

To remove the planetary gear unit and output shaft assembly (47), remove the plug (5—Fig. O115) from sprocket (6). Then, hold sprocket with suitable tool and remove nut (29—Fig. O116) from rear end of output shaft. With a suitable soft drift,

bump rear end of output shaft until free of bearing (23) and remove the shaft and planetary assembly from housing. Unbolt retainer (25) from rear end of housing. Drive bearing (23) and seal (28) out of retainer. Sleeve (22) can be pressed from rear of housing (18) and is available for service; however, renewal of sleeve should not be necessary. Baffle (45) can be snapped out of housing for cleaning purposes; note position of baffle before removing same.

To remove the control valve (38), remove nut (4) and lever (1) being careful not to lose detent ball (6) and spring (5). Remove plate (30), then push the valve assembly out of bore in housing. To remove regulating plunger (37), support valve in arbor press so that valve will not be damaged and push retainer (35) in so that snap ring (34) can be removed. Slowly release pressure and remove retainer (35), spring (36) and plunger (37). Remove and clean the suction screen (42).

The two bushings (46) (early units only) in housing (18) for the output shaft, bushings (48) in front bore of output shaft (47) and bushing (87) in front adapter (85) can be renewed at this time if excessively worn or scored. Bushings are pre-sized and reaming should not be necessary if bushings are carefully installed. Be sure that regulator valve (44), control valve (38) and valve plunger (37) are not scored and fit snugly, yet slide freely in their bores. Renew any parts that are excessively worn, scored or otherwise damaged.

184. To reassemble the Reverse-O-Torc Drive unit, refer to Fig. O116 and proceed as follows: If removed, snap the baffle (45) into place inside housing; two holes in rear end of baffle should fit over two bosses cast into the housing. Reinstall filter screen (42).

Lubricate the pressure regulator valve (37) and insert it, hollow end out, in bore of control valve (38). Insert spring (36) in bore of regulator valve, place retainer (35) on top of spring and compress retainer and spring into control valve in an arbor press so that snap ring (34) can be installed. Note: Support valve carefully in arbor press to avoid damage to valve surfaces. Lubricate and install control valve in bore of housing and then install retaining plate (30) with new gasket (33). Tighten the cap screws (31) to a torque of 10-15 Ft.-Lbs. Insert detent spring (5) and ball (6) in bore at opposite side of housing and install control lever (1). Tighten the lever retaining nut to a torque of 12-14 Ft.-Lbs.

Drive the ball bearing (23) into retainer (25) until firmly seated. Install new seal (28) into rear side of retainer with lip of seal forward (to inside). Then, using new gasket (24), install retainer on rear of housing tightening the retaining cap screws to a torque of 40-45 Ft.-Lbs. Insert the pinion gear and output shaft assembly (47) through rear bore of housing and bump the assembly rearward until shoulder on shaft is firmly seated against ball bearing. Lubricate seal and seal contact surface of sprocket

(6—Fig. O115) and place on rear end of output shaft. Install a new self-locking nut (29—Fig. O116) and tighten the nut to a torque of 150 Ft.-Lbs. Install a new plug (5—Fig. O115) in rear end of sprocket (6).

To reassemble the forward clutch unit, place ring gear (62—Fig. O116) on bench with front end (end with splines on outside) up. Place rear pressure plate (61) in ring gear with flat machined side up. Lubricate the clutch plates (59 and 60) and place the plates alternately in ring gear with plate having internal teeth next to the pressure plate. Install front pressure plate (58) with flat machined side down and secure pressure plate with snap ring (57). Note: Be sure proper snap ring is used; snap ring thickness should be 0.090-0.093 and free diameter should be approximately $5\frac{19}{32}$ inches. Place clutch spring (56) on snap ring with concave side of spring towards snap ring. Lubricate and install "O" rings (54 and 66) in grooves of piston (53) and cylinder (67). Install piston in cylinder with flat side in. With heavy grease, stick spring bearing ring (55) in groove in face of piston. Install the assembled cylinder in ring gear being careful that ring (55) does not drop out. Place the unit in an arbor press and with a proper size arbor, compress cylinder against clutch spring and install snap ring (68). Release pressure and turn the assembly over with ring gear (rear) end up. Place a proper size arbor against clutch pressure plate (61) and compress the clutch components against snap ring (68). Measure the clearance between pressure plate and rear side of snap ring groove inside the ring gear. Then, install the proper thickness of snap ring (63) to provide 0.056-0.086 clearance between pressure plate and snap ring. Snap rings are available in three thicknesses of 0.050-0.054, 0.074-0.078 and 0.096-0.100 which are color coded respectively, green, orange and white.

NOTE: The front clutch information above applies to early production units. While no unit serial numbers are available, in mid-1965, the forward clutch hub (item 67—Fig. O166) was chamfered and clutch clearance was changed to 0.046-0.066. When a unit prior to this change is disassembled, a chamfered hub should be installed and the clutch clearance set to 0.046-0.066.

If clutch hub (52) was removed from input shaft (50), insert Woodruff key (51) in shaft and press hub onto shaft being sure that flat side of hub is towards gear and that slot

in hub is aligned with key. With hub bottomed against drive gear, install hub retaining snap ring (65). Lubricate and install the two sealing rings (64) in ring grooves of input shaft. Lock ends of rings together and be sure they turn freely in grooves.

Fabricate an assembly tool by drilling a hole that will accommodate rear (pilot) end of input shaft in a hard wood block; block should be larger than the ring gear.

Set the input shaft pilot end in the hole in wood block. Lower the assembled ring gear and forward clutch unit down over the input shaft. To align clutch discs with splines on clutch hub, partially support the clutch unit and turn unit back and forth. Clutch unit is properly installed when ring gear is resting on wood assembly block (rear end of ring gear is flush with rear face of drive gear).

CAUTION: **Do not** remove the unit from assembly block or lift by the input shaft at this time as any forward movement of the input shaft beyond proper location in clutch will disengage the clutch discs and the seal rings on input shaft. Move the unit on the assembly block to a press and install bearing (69) on input shaft. Then install the bearing retaining rings (70 and 71) in cylinder and on input shaft; the unit can then be removed from the assembly block.

Support the housing (18) with front end up and proceed as follows: Using light grease, stick the thrust washer (49) onto rear face of drive gear on input shaft (50). Install the assembled shaft and clutch unit in housing so that drive gear engages the pinion gears in the carrier on output shaft. Lay a straight edge across front face of housing and measure distance between straight edge and thrust face of the forward clutch cylinder. If distance is 0.405 or less, install a 0.061-0.063 thrust washer (72); if distance is more than 0.405, install a 0.085-0.087 thrust washer on the forward clutch cylinder thrust surface.

Insert converter regulator valve spring (43) and valve (44) in bore of housing with closed end of valve out. Insert the twelve clutch pressure plate springs (76) in the four sets of three holes in reverse clutch cavity of housing (18). Using grease, stick the three dowels (75) in grooves in side of reverse clutch cavity.

Insert one of the clutch discs (73) with internal teeth into the clutch cavity with teeth engaged in splines on outside of ring gear. Insert one of the discs (74) with external lugs with notches in the lugs engaged on dowel pins (75) placed in the housing.

NOTE: One of the three lugs is odd shaped; that is, the notch is not centered in the lug. This lug must engage the dowel nearest the oil screen retaining fitting (40) with the notch offset towards bottom (filter screen) side of housing. Repeat this procedure to complete the installation of the three internal notched plates and the two plates with external lugs. Install the reverse clutch pressure plate (77) with the cast notch in edge of plate aligned with large oil hole at top of housing (18) and with the three machined notches engaging the three dowel pins. The pressure plate should

drop into position approximately flush with front face of housing; if not, remove plate and check dowels and springs for misalignment.

NOTE: In later production units, thickness of plate (73) was increased from $\frac{1}{16}$ to $\frac{1}{8}$-inch. Late plates are not interchangeable with early plates.

Lubricate and install the "O" rings (78 and 89) on reverse clutch piston (79) and inner hub of front adapter (85). Insert piston in adapter with flat side in (forward); also note that chamfered side of hub must be inward. Install adapter and piston assembly on housing with new gasket (88). Tighten the adapter retaining cap screws to a torque of 28-30 Ft.-Lbs.

Install new oil seal (83) in pump cover with lip of seal to rear (inward) and reassemble pump using Fig. O116 as a guide. Tighten the flat head screws (81) (or screw) to a torque of 17-22 Ft.-Lbs. Pump gears should turn freely at this time. Lubricate the pump and install pump on front adapter plate with a new gasket (84). Tighten the retaining cap screws to a torque of 17-22 Ft.-Lbs. Note: Be sure that the long cap screws are installed at the thick pump bosses.

Reassemble the charge pressure and safety control valve parts (items 10 through 14) in order shown and reinstall oil filter and external oil lines.

# REVERSE-O-TORC DRIVE UNIT
# (Series 1650 S/N 204 145-000 and up - 1655 - 1550 - 1555)

The Reverse-O-Torc Drive for Series 1550, 1555, 1650 S/N 204 145-000 and up and 1655 is a forward and reverse auxiliary transmission mounted in front of the tractor transmission providing a 1:1 drive ratio to the transmission input shaft. The unit consists of two clutches, a countershaft, an idler gear and a charging pump mounted in a cast iron housing. The wet type multiple disc clutches are actuated by pistons and release springs and the clutch engaging oil pressure as well as the converter charging oil is furnished by the reversing gear oil pump. A hydraulic torque converter mounted on the engine flywheel drives the reversing gear.

Reverse-O-Torc Drive is available on Industrial model tractors only. Models so equipped do not have an engine clutch or reverse gears in the tractor transmission as the clutches in the reversing gear provide the neutral, forward and reverse functions. In addition, models equipped with the reversing gear cannot be fitted with power take-off or a hydraulic system.

## Industrial Models So Equipped

185. **OPERATION.** The Reverse-O-Torc unit is controlled by two foot pedals located at right side of operators platform. With the tractor transmission in gear, depressing the foot pedal marked "F" will start tractor in forward motion and depressing the foot pedal marked "R" will start tractor in reverse direction. Continued depressing of either pedal will bring pedal into contact with the foot ac-

celerator pedal yoke and increase the engine rpm. Depressing the foot accelerator pedal will increase engine rpm without actuating the Reverse-O-Torc unit.

Shift tractor transmission only when Reverse-O-Torc pedals are in neutral and engage Reverse-O-Torc pedals only when engine is at slow idle. Bring tractor to a stop before changing direction of travel. If torque converter indicator hand moves into the "SHIFT" range on indicator dial, stop tractor motion and run engine at 1000 rpm until indicator drops back onto "NORMAL" range.

186. **TROUBLE SHOOTING.** The following will serve as a guide when attempting to localize any malfunctions that might occur in the operation of the Reverse-O-Torc unit.

1. UNIT OVERHEATS. Could be caused by:
   a. Unit overloaded.
   b. Fluid level low.
   c. Oil lines to heat exchanger restricted.
   d. Water lines to heat exchanger restricted.
   e. Improper operation (rapid reversals).

2. NOISY OPERATION. Could be caused by:
   a. Fluid level low.
   b. Rough or worn gears, shafts or bearings.
   c. Loose mounting bolts.

3. UNIT STARTS WITH A JERK. Could be caused by:
   a. Engine low idle speed set too high.

4. SLUGGISH OPERATION. Could be caused by:
   a. Engine high idle speed set too low.
   b. Fluid level low.
   c. Control linkage binding.
   d. Contaminated or wrong fluid being used.

5. FLUID FOAMING OUT BREATHER. Could be caused by:
   a. Fluid level too high.
   b. Unit overheating
   c. Contaminated or wrong fluid being used.

187. **PERFORMANCE AND PRESSURE CHECKS.** A performance test can be made to determine the overall condition of Reverse-O-Torc, however before making test be sure the engine high idle speed is correctly set and that pedal linkage is free of binding and pedal travel is sufficient to obtain engine high idle speed. Start engine and run until Reverse-O-Torc reaches operating temperature. Lock brakes, place tractor transmission in sixth gear, then depress either Reverse-O-Torc pedal as far as possible and check engine speed which should be 1350-1600 rpm. Repeat test for the other pedal. If engine speed is more than 1350-1600 rpm with either Re-

verse-O-Torc pedal depressed, the unit may be faulty. If engine speed is less than 1350-1600 rpm, the engine may be at fault.

187A. The high (clutch) pressure and charging pressure regulator valves of the reverser unit can be checked by using a single gage. When making these pressure tests, check the converter charging regulator valve first and the high pressure regulator valve second.

To make these tests, proceed as follows: Remove plug from top front of reversing unit and install a gage of at least 200 psi capacity. Remove and clean the high pressure (upper) and converter charge (lower) regulator valves and do not intermix the valve parts. Reinstall the converter charge (lower) valve, spring and cap. Reinstall the high pressure (upper) valve and valve cap only. DO NOT install spring at this time. Start engine, and with both pedals released, run engine at 1800 rpm and note gage (converter charge) pressure which should be 80 psi. If converter pressure is not as stated, remove cap and vary shims between cap and spring as required. If pressure cannot be adjusted, overhaul reverser unit as outlined in paragraph 189.

With converter charge pressure checked and/or adjusted, remove cap and install the high pressure (upper) regulator valve spring. Start engine and run at 1800 rpm and observe gage which should read 160 psi. If pressure is not as stated, remove cap and vary shims between cap and spring as required. Place tractor transmission in sixth gear and "LOCK" brakes, then depress each shift pedal. Pressure will drop momentarily when either pedal is depressed but should return to within 5 psi of the neutral position pressure. If pressure cannot be adjusted as outlined above, overhaul reverser unit as outlined in paragraph 189.

**188. R&R REVERSE-O-TORC AND TORQUE CONVERTER.** To remove the Reverse-O-Torc drive unit, first remove engine and drive unit as outlined in paragraph 60, then on early units, disconnect hoses from heat exchanger, attach a hoist to drive unit and remove drive unit from converter housing.

188A. To remove torque converter, first remove Reverse-O-Torc, then remove starter motor and the dust shield from lower front of converter housing. Remove converter housing from engine. Unbolt converter drive plate

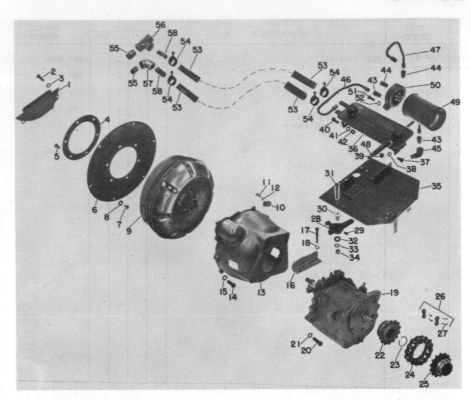

**Fig. O116A—General arrangement of the component parts which comprises the Reverse-O-Torc used in Series 1550, 1555, 1650 S/N 204 145-000 and up and 1655. Heat exchanger (36) is used on early units. Late units use a cooler mounted in front of radiator.**

| | | |
|---|---|---|
| 1. Dust shield | 13. Converter housing | 26. Master link |
| 4. Reinforcement plate | 16. Screen | 28. Bell crank |
| 6. Drive plate | 19. Reversing gear | 30. Bushing |
| 9. Torque converter | 22. Sprocket | 35. Mounting plate |
| 10. Timing hole cover | 23. Snap ring | 36. Heat exchanger |
| | 24. Coupling chain | 49. Filter element |
| | 25. Sprocket | 50. Cap |

from flywheel and remove converter. Remove reinforcement plate and drive plate from converter.

When reinstalling torque converter, use four cap screws and install drive plate and reinforcement plate to converter and tighten cap screws until they are just snug. Mount converter assembly on flywheel, then use a dial indicator and check converter hub run-out. Hub run-out must not exceed 0.015. Excessive hub run-out is corrected by shifting converter on drive plate and when acceptable converter hub run-out is obtained, remove converter and drive plate, install the remaining cap screws in reinforcement plate and tighten all cap screws to 28-32 Ft.-Lbs. Reinstall converter assembly on flywheel, then install converter housing on engine. NOTE: When installing the converter housing, install the cap screws in the reamed lower hole of starter flange and the hole nearly diametrically opposite before installing the remaining cap screws. With converter housing installed, mount a dial indicator on converter hub and check alignment of housing bore and converter hub. Hub

and housing bore must be concentric to within 0.012. If hub to housing bore is excessively misaligned, recheck housing installation.

Lubricate (grease) converter hub, install reverser unit on converter housing and be sure drive-tangs on converter hub align with charging oil pump drive gear. Complete balance of assembly and install engine and reversing gear assembly in tractor.

**189. OVERHAUL REVERSE-O-TORC.** With the reversing gear removed, drain oil from unit if necessary, then disconnect heat exchanger (early units) and filter lines from case and remove mounting plate, filter and heat exchanger. Remove dip stick and screen. Remove both regulator valves (73—Fig. O116B) and check valve (74). Keep the regulator valves identified so they will be reinstalled in their original bores. Remove control valve and if control valve is to be disassembled, mark closed end cap (55) and valve body (52) so closed end cap will be reinstalled on same end of valve body. Remove oil pump (81) and gasket (80), then remove the screws from pump rear plate and pull rear plate from pump body. Mark rear side of pump outer ring (gear) so it can be reinstalled with same edge rearward. Oil seal (85) can be removed at this time. Remove rear cover (9) and gasket (12). Oil seal (8) can be removed at this time. Remove the corks (63) from case at front of countershaft and idler shaft, then drive on FRONT of countershaft (35) and remove shaft from rear of case. Allow countershaft gear (37) to rest on bottom of case until clutches and shafts are removed.

NOTE: Countershaft and idler gear shaft are made with the forward ends 0.002 smaller than the rear ends. Therefore, it is mandatory that both shafts be driven rearward during removal to prevent damage to reverser case and/or shafts.

Remove clutches and shafts assembly out rear of case. Lift out countershaft gear and bearings. Drive on front of idler gear shaft (1), remove shaft from rear of case and remove idler gear (6) and spacer (2). Remove any pipe plugs or expansion plugs which show signs of leakage. If necessary, oil filler neck (61) can also be removed.

189A. Remove snap ring (18) from inside forward edge of front clutch cylinder (29) and remove input shaft (33). Do not remove the three sealing rings (34) from input shaft unless rings are damaged or have 0.015 or more clearance between ring and ring groove. Press bearing (13) from shaft if either bearing or shaft is renewed. Remove thrust washer (7), clutch plates (20 and 21) and clutch hub (32), then remove snap ring (31) and pull clutch cylinder assembly from output shaft (14). Place cylinder assembly in a press and slightly depress spring retainer (25), then remove snap ring (24). Remove press pressure slowly and remove snap ring, spring retainer and return spring (26). Plug both holes in inside of cylinder hub and remove piston (27) by carefully applying air pressure to hole in outer edge on back side of cylinder. Remove the small steel ball from counterbored hole in cylinder. Remove seals from piston and cylinder. Remove the large snap ring (18) and pull rear clutch shaft from output shaft. Disassemble rear clutch in the same manner as outlined for front clutch.

189B. INSPECTION. Use the following data to determine the need for parts renewal.

Regulator Valve Springs
  Free length .............. 2¼ in.
  Test .......... 1½ in. @ 15 lbs.

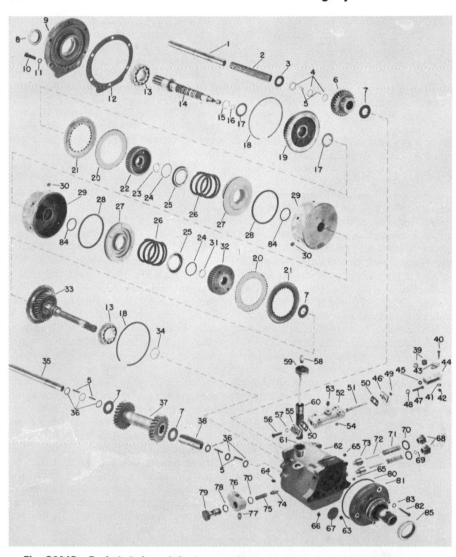

**Fig. O116B—Exploded view of the Reverse-O-Torc unit used in Series 1550 tractors.**

1. Idler shaft
2. Spacer
3. Washer
4. Spacer washers
5. Roller bearings
6. Idler gear
7. Thrust washer
8. Oil seal
9. Rear cover
11. Dyna-Seal (3 used)
12. Gasket
13. Ball bearing
14. Output shaft
15. Sealing ring (4 used)
16. Sealing ring (2 used)
17. Thrust washer
18. Snap ring
19. Output gear
20. Clutch plate (ext. spline)
21. Clutch plate (int. spline)
22. Rear clutch hub
23. Snap ring
24. Snap ring
25. Spring retainer
26. Clutch spring
27. Clutch piston
28. Piston outer seal
29. Clutch cylinder
30. Steel ball (¼")
31. Snap ring
32. Front clutch hub
33. Input shaft
34. Seal ring (3 used)
35. Countershaft
36. Spacer washers
37. Countershaft gear
38. Spacer
39. Rollers
43. Cage spring
44. Roller cage
46. Open valve cap
49. Oil seal
50. Gasket
51. Valve spool
52. Valve body
53. Plug
54. "O" ring (5 used)
55. Closed valve cap
58. Breather
59. Cap & dip stick
60. Strainer screen
61. Filler neck
62. Case
63. Cork
67. Expansion plug
68. Cap
69. Adjusting shim
70. "O" ring
71. Valve spring
72. Pin
73. Regulator valve
74. Check valve
75. Valve spring
76. Valve block
78. "O" ring
79. Cap
80. Gasket
81. Oil pump assy.
83. Dyna-Seal (4 used)
84. Piston inner "O" ring
85. Oil seal

Check Valve Spring
  Free length ................. 1¾ in.
  Test .......... 1¼ in. @ 2 lbs.
Drive plate cone ......... 0.015-0.018
Driven plate thickness .. 0.0605-0.0630
Countershaft journal
  diameter ............. 0.8070-0.8075
Idler shaft journal
  diameter ............. 0.7495-0.7500
Countershaft gear I.D. .. 1.1710-1.1717

Idler gear I.D. .......... 1.1135-1.1140
Output gear bushing I.D. 1.1885-1.1890
Clutch cylinder
  bushing I.D. .......... 1.0635-1.0640
Pilot bushing I.D. ...... 0.8765-0.8770
Roller bearing diameter 0.1814-0.1816
Output Shaft
  Pilot bearing journal
    O.D. .............. 0.8740-0.8475

Clutch cylinder journal
O.D. ...............0.0610-0.0615
Output gear journal
O.D. ...............1.1860-1.1865
End play ...............0.015-0.030

189C. To assemble Reverse-O-Torc unit, proceed as follows: Press output shaft bearing (13) on rear of output shaft (14) until it bottoms against shoulder. Place thrust washer (17) on front of shaft next to shoulder, then slide output gear on shaft with gear toward rear. Install second thrust washer (17) against output gear, then install clutch drive hub (22) with thrust face against thrust washer and secure with snap ring (23). If previously removed, install the four seal rings (15) in their grooves in output shaft. Start with an internal toothed clutch plate (21) and alternately install the internal and external toothed clutch plates and while installing clutch plates, be sure the external toothed plates all have the cones facing in the same direction and that the missing teeth are aligned. Place new seal (28) on piston (27) and new seal (84) on hub of cylinder (29). Install steel ball (30) into counterbored hole in inside end of cylinder, then install piston in cylinder with flat side inward and push piston into cylinder until it bottoms. Place return spring (26) and spring retainer (25) on top of piston, then using a press, depress spring and install snap ring (24). Install the cylinder assembly over clutch plates so that oil holes in cylinder outer wall are aligned with the missing teeth of the external toothed clutch plates and secure in position with the large snap ring (18).

Assemble front clutch cylinder and piston as already outlined, place assembly on output shaft and secure with snap ring (31). Place drive hub (32) on shaft with thrust face outward, then starting with an external toothed clutch plate, alternately install the clutch plates (20 and 21). Be sure to keep the cones of the external toothed clutch plates facing in the same direction and the missing external teeth aligned with oil holes in outer wall of clutch cylinder. If necessary, install the two small pilot seals (16) on forward end of output shaft. If input shaft bearing (13) is being renewed, press it on input shaft until it is tight against input shaft gear. Install input shaft over output shaft and secure in clutch cylinder with the large snap ring (18). Install the three sealing rings (34) on input shaft.

Stand reverser case on front end and place fiber thrust washer (7) in case over idler shaft front bore. Use grease in bore of idler gear, install the double row of rollers with a spacer washer between and at each end of the rollers and place idler gear in case with hub end forward (down) and on top of the fiber washer. Place second washer (3) on idler gear and spacer (2) on washer (3), then insert idler gear shaft (1), small end first, into case bore and the idler gear assembly. Bump shaft forward into position.

Place spacer (38) in center of countershaft (37), coat both ends of gear bore with grease and install a double row of rollers in each end of the countershaft gear, and be sure a spacer washer is between and at each end of the double roller bearing sets. Stick thrust washers (7) on ends of countershaft gear, place assembly in case, then align countershaft gear with shaft bores and insert a piece of heavy cord through shaft bores and countershaft gear assembly. Now shift countershaft gear so it will not interfere with installation of clutches and shafts and install the clutches and shafts assembly. Use caution not to damage the seal rings on input shaft. Stand assembly on front end, use the cord to align countershaft gear with shaft bores, then install countershaft, small end first, through countershaft gear and bump into place. Install rear cover (9), minus oil seal, and use enough gaskets (12) to introduce mea-

sureable end play in output shaft. Use a dial indicator and measure output shaft end play, then deduct enough gaskets to provide 0.015-0.030 end play for the output shaft. Gaskets are 0.015 thick. With output shaft end play adjusted, remove rear cover, install oil seal (8) with lip facing forward, then reinstall cover and use new Dyna-Seals on the three lower cap screws.

Install new oil seal (85) in oil pump body with lip facing rearward. Install the pump outer ring with previously identified edge rearward. Use new gasket (80) and Dyna-Seals (83) when installing pump. Bottom and left mounting cap screws are ¼-inch longer than the other two.

If control valve was disassembled, be sure to reinstall the closed end cap (55) on original end of valve body. Open end cap (46) may have "O" ring or lip type spool stem seals. If seals are the lip type, inner seal has lip toward valve and outer seal has lip away from valve. Use new "O" rings and install control valve on case with roller and cage extended to the left.

When installing converter check valve (74), insert valve, fluted end first, into valve bore. Install high pressure and converter charge regulator valves into their original bores. Complete assembly by attaching mounting plate, filter and heat exchanger (early units) to reverser case.

After Reverse-O-Torc is installed in tractor, check and adjust high pressure and converter charge pressure as outlined in paragraph 187.

# TRANSMISSION
## (Series 1600-1650-1655)

Series 1600, 1650 and 1655 are equipped with a constant mesh, helical gear type transmission having six forward and two reverse speeds. On Industrial models equipped with a Reverse-O-Torc auxiliary transmission, the transmission reverse gearing is omitted as the Reverse-O-Torc provides a forward and reverse drive for each of the six basic transmission speeds. When equipped with an auxiliary Hydra-Power Drive, or Hydraul-Shift Drive, two or three different speed ratios are available for each transmission gear. Optional transmission gears are available for some models which provide different transmission speed ratios. Also, a Creeper Drive is available which provides an underdrive in 1st, 2nd, 5th and R1 transmission speeds.

NOTE: On tractors with no auxiliary drive unit it is not absolutely necessary to split tractor for transmission overhaul, however due to space limitations and the shaft adjustments required, it is generally advisable to make the tractor split as outlined in paragraph 194. Also, overhaul of transmission will usually require removal of four-wheel drive transfer case (paragraph 41), bull gear cover (paragraph 193) or hydraulic lift system (paragraph 253), bull gears (paragraph 210), differential assembly (paragraph 204), power take-off shaft (paragraph 221 or 226) or hydraulic pump drive shaft (paragraph 248).

Fig. O117—Transmission lubrication circuit by-pass valve removed from bore in rear main frame. Valve plunger and spring are retained by an expansion plug (not shown).

## LUBRICATION

190. **LUBRICANT.** Transmission and differential are lubricated from a common oil supply. Oil level should be maintained at check plug opening located at rear side of right rear axle housing in the main frame. Filler cap is located in front end of bull gear cover or hydraulic lift system housing under the center platform plate. Two drain plugs are used; front plug is located under front center of main frame and drains transmission compartment and rear plug is located at rear end of main frame and drains the differential and final drive compartment. Also, pressure lubrication pump suction screen should be removed and cleaned whenever lubricating oil is drained. Screen is located on plug which is threaded into bottom of differential and final drive compartment just to rear of the transmission compartment drain plug.

Recommended transmission, differential and final drive lubricant is SAE 80 multi-purpose gear lubricant. SAE 90 multi-purpose gear lubricant may be used in temperatures above 32° F. All gear lubricant must conform to Military Specification MIL-L-2105. Sump capacity is 32 quarts. Lubricant should be changed yearly or after each 1000 hours of operation. Oil filter should be renewed after each 250 hours of operation.

191. **PRESSURE LUBRICATION SYSTEM.** The transmission and differential are lubricated by a pressure oiling system. Pressure is provided by a G-rotor type pump driven from rear end of transmission countershaft. Oil from pump is directed from pump to an externally mounted oil filter. From the oil filter, oil is directed to the differential shaft and to the trans-

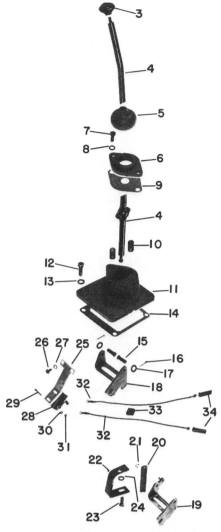

Fig. O118 — Exploded view of transmission shift and interlock cover. Interlock (18) is used on Row Crop and Wheatland models; interlock (19) is used on Industrial and Utility models and is optional on Row Crop and Wheatland models. Refer to Fig. O119 for safety starting switch adjustment.

| | |
|---|---|
| 4. Shift lever | 19. Interlock |
| 6. Cover | 20. Arm |
| 9. Gasket | 21. Snap ring |
| 10. Springs | 22. Guide |
| 11. Support | 25. Support |
| 14. Gasket | 28. Safety switch |
| 15. Pins | 32. Wires |
| 17. Washers | 33. Seal |
| 18. Interlock | 34. Connectors |

mission input and bevel pinion shafts. A by-pass (pressure relief) valve in the transmission circuit limits system pressure to 8-14 psi. Oil by-passed by this valve is returned to the transmission compartment via a hole in the tractor main frame. Refer to Fig. O117.

With the lubricating oil at operating temperature and engine running at 1900 RPM, system pressure should be 8-14 psi with the transmission shift lever in either neutral position. Pressure can be checked by teeing into the line to the transmission pinion

Fig. O119 — View showing adjustment of safety starting switch. Refer to text for procedure.

shaft at connection under the left foot platform section or by substituting an oil filter which has been modified by brazing a pipe nipple in top of filter and installing a pressure gage. If pressure is below 8 psi, drain transmission and differential compartments, renew the transmission oil filter and refill to proper level with specified lubricant. Refer to paragraph 190. Faulty oil pump, broken oil tube or faulty by-pass valve could also cause low oil pressure. Refer to paragraph 198 for pump servicing information. If pressure is higher than 14 psi, check the by-pass valve. (Fig. O117). If valve is OK, check for obstruction in oil tubes and passages.

## OVERHAUL TRANSMISSION

192. **TRANSMISSION SHIFT AND INTERLOCK COVER.** Remove the right hand and center platform sections and disconnect the wires that lead into the shift lever cover. Then, unbolt and remove the shift lever and interlock cover as an assembly.

Disassembly is evident after inspection of the unit and reference to Fig. O118. Before installing on transmission, check the safety switch (28) and if necessary, readjust position of switch so that it opens when roller is about half way up the ramp on interlock (18 or 19). Refer also to Fig. O119.

Use new gasket (14) and reinstall by reversing removal procedure.

193. **R&R BULL GEAR COVER.** On models equipped with hydraulic system, refer to paragraph 253. On models without hydraulic system, disconnect battery ground strap and remove the battery or batteries and case as a unit. On models so equipped, remove the platform enclosures and wheel guard extensions. Remove the side and center foot platforms. Re-

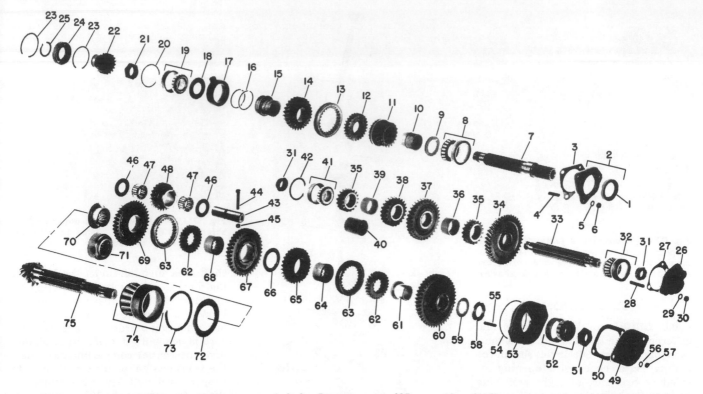

Fig. O120 — Exploded view of transmission gears and shafts. Reverse gears (35 rear, 48 and 69) are not used on models with Reverse-O-Torc Drive; spacers (40 and 71) instead of gears are used on output shaft and countershaft. Refer to Fig. O121 for parts used when four-wheel drive case is mounted on front end of main frame.

| | | | | |
|---|---|---|---|---|
| 1. Seal | 15. Sleeve | 25. Snap ring | 40. Spacer (R-O-T only) | 53. Bearing cage | 66. Washer |
| 2. Retainer | 16. Seal rings | 26. Cover | 41. Bearing assy. | 54. "O" ring (gasket | 67. Gear (38 or 41 teeth) |
| 3. Shims | 17. Oil collector | 27. Shims | 42. Snap ring | also used on 1650) | 68. Sleeve (1 9/32 in.) |
| 7. Input shaft | 18. Washer | 31. Nut | 43. Reverse idler shaft | 58. Nut | 69. Gear (38 teeth) |
| 8. Bearing assy. | 19. Bearing assy. | 32. Bearing assy. | 46. Washers | 59. Washer | 70. Sleeve |
| 9. Washer | 20. Snap ring | 33. Countershaft | 47. Bearings | 60. Gear (44 teeth) | 71. Spacer (R-O-T only) |
| 10. Sleeve | 21. Nut | 34. Gear (41 or 28 teeth) | 48. Reverse idler gear | 61. Sleeve | 72. Seal |
| 11. Gear (19 or 31 teeth) | 22. Hydraulic pump | 35. Gears (17 teeth) | 49. Cover | 62. Splined collar | 73. Snap ring |
| 12. Splined collar | drive gear | 36. Spacer (1¼ in.) | 50. Gasket | 63. Coupling | 74. Bearing assy. |
| 13. Coupling | 23. Snap rings | 37. Gear (37 teeth) | 51. Nut | 64. Sleeve (1⅜ in.) | 75. Bevel pinion shaft |
| 14. Gear (26 teeth) | 24. Bearing assy. | 38. Gear (22 or 19 teeth) | 52. Bearing assy. | 65. Gear (27 or 23 teeth) | gear (7, 8, 9 or 10 teeth) |

move the seat assembly from bull gear cover and remove shift lever and interlock cover as in paragraph 192. Unbolt and remove bull gear cover.

Reinstall bull gear cover by reversing removal procedure and using new gaskets.

194. SPLIT TRACTOR. To split tractor between the front and rear main frames, disconnect battery ground strap and proceed as follows: Disconnect battery cable to cranking motor and safety starting wires from connector. On models so equipped, remove the platform enclosure. On models with hydraulic system, disconnect the oil cooler lines from the flow divider valve. On models with four-wheel drive, remove the engine as outlined in paragraph 60. Disconnect the rear light wires. Remove the sprocket chain connector from between clutch shaft and transmission input shaft. Support tractor front frame in manner which will prevent it

from tipping and support front end of rear frame with rolling floor jack. Working through opening in bottom of rear frame (except on four-wheel drive models), remove the front frame to rear frame retaining cap screws and separate tractor. Note: These 12-point cap screws are torqued to 300 Ft.-Lbs. and will require a heavy duty socket for removal and re-torqueing when reassembling tractor.

195. INPUT SHAFT. Refer to the appropriate sections (see note preceding paragraph 190) and remove the pto assembly or hydraulic pump drive shaft, the hydraulic lift system or bull gear cover and either split the tractor as outlined in paragraph 194 or remove the clutch shaft as in paragraph 151 or the engine and drive unit as in paragraph 60. Then, proceed as follows:

Remove snap ring from bore in main frame at rear side of the hydraulic pump drive pinion bearing (24—

Fig. O120) and push the pinion (22) and bearing out to rear. Remove coupling sprocket and snap ring from front end of input shaft. Note: Snap ring is used on direct drive models only. Remove input shaft front bearing retainer (2) taking care not to lose or damage shims (3) and remove seal (1) from retainer. Remove oil tube that is connected to the oil collector (17). Unstake and remove nut (21) from rear of input shaft; discard nut as it should not be reused.

On Row Crop and Wheatland models, move shifter coupling (13) onto rear gear (14) and place a spacer between coupling and front side of main frame. The spacer can be made by cutting a 3½ inch I. D. pipe to a length of 2½ inches, then cutting the pipe in half lengthwise. Spacer is not required on Utility or Industrial models.

With Oliver ST-101 shaft driver or equivalent tool, drive input shaft forward until free of rear bearing cone;

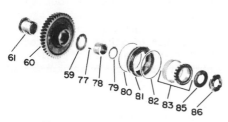

**Fig. O121 —** Above parts are used on four-wheel drive models to permit front wheel drive case to be attached to front of main frame.

| | |
|---|---|
| 59. Washer | 80. "O" ring |
| 60. Gear (44 teeth) | 81. Bearing cage |
| 61. Sleeve | 82. "O" ring |
| 77. Key | 83. Bearing assy. |
| 78. Sleeve | 85. Washer |
| 79. Shims | 86. Nut |

take care not to damage shaft threads. Remove shaft and components from main frame. Remove gears and collector ring from mounting sleeves and collector ring seals from rear sleeve. Press thrust washer and front bearing cone from shaft. Remove snap ring and rear bearing cup from main frame.

Carefully inspect all parts and renew as necessary. On Industrial models with Reverse-O-Torc, be sure that a sealing plug is installed in bore of input shaft; obtain and install plug if shaft is not so fitted.

196. Reinstall input shaft as follows: Install new seal (1) in retainer (2)

with lip of seal to rear. Install snap ring (20) in main frame and firmly seat rear bearing cup against snap ring. Press thrust washer (9) and front bearing cone (8) onto shaft. Install sealing rings (16) on rear mounting sleeve (15) and install collector (17) over the sealing rings with counterbore in collector over flange of sleeve. Place rear gear (14) onto rear sleeve with helical teeth toward collector ring. Move shifter fork to forward position. Insert front sleeve (10) in front gear (11), place shift coupling (13) on front gear and place the assembly in main frame with coupling engaged in shift fork. Insert input shaft through the front gear and sleeve and install drive collar (12) on shaft. Place rear bearing cone in cup, position rear sleeve with collector and gear installed in the main frame and push shaft into rear sleeve. Place the collector washer (18) between rear bearing cone and collector and push input shaft into bearing cone. With Oliver ST-101 shaft driver or equivalent tool, drive input shaft rearward. Install front bearing cup into main frame bore and install bearing retainer with shim pack originally removed. Install new nut (21) on rear end of input shaft and tighten nut to a torque of 150 Ft.-Lbs. Stake the nut in place on shaft and adjust input shaft bearings by varying the number of shims under front bearing retainer. Bearing adjustment is correct if shaft end play does not exceed 0.001, or if bearing preload is such that rolling torque is 5 inch pounds or less above oil seal drag. Shims are available in thicknesses of 0.004, 0.007 and 0.015. Reinstall oil tube, hydraulic pump drive pinion and pinion bearing retaining snap ring.

197. **SHIFT RAILS AND FORKS.** After removing input shaft assembly as in paragraph 195, proceed as follows:

Remove cap screws retaining shifter forks to shift rails. Remove the cap screws retaining shifter guide (60—Fig. O122) on Creeper Drive models, or the cover (56) on other models, and the poppet block (53) to front face of main frame and withdraw the shift rails and poppet block as a unit. Refer to exploded view in Fig. O122 as a guide for disassembly and reassembly of the shift rails and poppet block.

Reinstall the shift rails and forks by reversing removal procedure. Reassemble using new gaskets and new shakeproof washers under the shifter fork retaining cap screws.

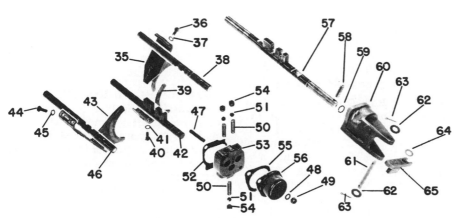

**Fig. O122 —** Exploded view of shift rails and detent block with related parts. On models equipped with Creeper Drive, special shift rail (57) and guide (60), with related parts (58 through 65), are used providing a lock out for transmission speeds 3, 4, 6 and R2 when Creeper Drive is engaged.

| | | | |
|---|---|---|---|
| 35. Shift fork | 46. Shift rail | 54. Plugs | 60. Guide |
| 38. Shift rail | 50. Detent springs | 55. Gasket | 61. Pin |
| 39. Shift fork | 51. Detent balls | 56. Cover | 62. Washer |
| 42. Shift rail | 52. Gasket | 57. Shift rail | 64. Washer |
| 43. Shift fork | 53. Detent block | 58. Pin | 65. Arm |
| | | 59. "O" ring | |

**Fig. O123 —** Exploded view of transmission lubrication pump. Cover (5) is used on all models except those equipped with Reverse-O-Torc Drive. On Reverse-O-Torc equipped models, valves (15) in pump cover (10) allow pump to pressurize the transmission lubrication circuit when rotating in either direction.

1. Shaft
2. Key
3. Body
4. Rotor set
5. Cover
8. Tubing seats
9. Tubing seats
10. Cover (R-O-T only)
13. Plug
14. Valve springs (4)
15. Valve balls (4)
16. Sealing washers (4)
17. Plugs (4)

**198. LUBRICATION PUMP.** The transmission pressure lubrication system pump can be removed after removing the hydraulic lift system or bull gear cover. To remove pump, proceed as follows:

Disconnect suction tube from pump and remove the pressure tube (pump to oil filter tube). Depress the pump housing lock with a small rod or punch inserted through hole in main frame casting at front of pump and withdraw pump assembly from bore in main frame.

Refer to exploded view of pump in Fig. O123 for disassembly and reassembly. Inner rotor should be a free fit on pump shaft (1) with drive key (2) installed. Polish shaft with crocus cloth or file key if binding occurs. Some shafts have two key slots; in some cases, binding condition can be corrected by moving key to other slot. On Industrial models with Reverse-O-Torc, pump cover (10) is fitted with four check balls (15), springs (14) and retainers (17). Install new sealing washers (16) when reinstalling the retainers (17). If damaged, tubing seats (8 and/or 9) can be removed by threading with tap and pulling the seats from pump cover.

To reinstall pump, insert lock spring and plunger in pump body (See Fig. O124), depress the lock and insert pump in bore of main frame. Tang on shaft (1—Fig. O123) must enter slot in end of transmission countershaft. Reconnect suction line and reinstall pressure line.

**199. TRANSMISSION COUNTERSHAFT.** The countershaft can be removed after removing input shaft as in paragraph 195, shift rails and forks as in paragraph 197 and the pressure lubrication pump as in paragraph 198. See note preceding paragraph 190 and refer to appropriate paragraphs for removing the bull gears and differential. Then, proceed as follows:

On models not equipped with four wheel drive, remove the transmission bevel pinion shaft oil line. On four-wheel drive models, remove the transfer case as outlined in paragraph 44. Remove the countershaft front bearing retainer (26—Fig. O120) taking care not to lose or damage shims (27). Remove and discard the nut (31) from rear end of countershaft. Rework a soft drift punch so that it will enter the slot in rear of countershaft and seat against the bottom of slot. Then, with the reworked drift, drive countershaft forward until front

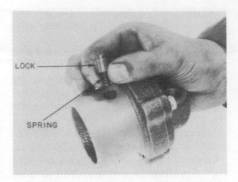

**Fig. O124 — Installing pump lock and spring.**

bearing cup is free of main frame and shaft is free of rear bearing cone. Remove the shaft and components from main frame. Press front bearing cone from shaft. To remove rear bearing cup from main frame, it will be necessary to first remove the reverse idler gear as in paragraph 201.

NOTE: On Row Crop and Wheatland models, front countershaft gear (34) has 41 teeth; gear (37) has 33 teeth. On Series 1650 Wheatland models only, gear (38) has 19 teeth; on all other models, gear (38) has 22 teeth. On Utility and Industrial models, and as optional usage on Row Crop and Wheatland models, front countershaft gear (34) has 28 teeth; gear (37) has 37 teeth. On Industrial models with Reverse-O-Torc, long spacer (40) is used instead of short spacer (39) and rear countershaft gear (35).

200. To install transmission countershaft, proceed as follows: If removed, reinstall the countershaft rear bearing cup and then install the reverse idler as in paragraph 201. Press front bearing cone onto countershaft, install **new** nut (31) and tighten nut to a torque of 150 Ft.-Lbs. Insert countershaft through bore in front face of main frame and install components in following order: Front countershaft gear (34) (41 or 28 teeth); gear (35) (17 teeth); spacer (36) (1¼ inches long); center countershaft gear (37) (33 or 37 teeth); gear (38) (22 or 19 teeth); spacer (39) (1 1/16 inches long) or long spacer (40) (2 5/32 inches long) and place rear bearing cone in cup and gear (35) (17 teeth) if used. Drive the countershaft rearward until nut (31) can be installed. Use **new** nut and tighten nut to a torque of 150 Ft.-Lbs. Install front bearing cup in bore of main frame, then install retainer (26) with shims (27) removed on disassembly. Be sure bearing cups are seated and check shaft end play with

dial indicator. Note: Do not check end play by placing indicator against gear as gears are loose on shaft. Vary the number of shims as required to obtain a shaft end play of 0.001-0.003. Shims are available in thicknesses of 0.004, 0.007 and 0.015.

**201. REVERSE IDLER GEAR.** The reverse idler gear and shaft (not used on models with Reverse-O-Torc) can be removed after removing the transmission input shaft as in paragraph 195 and the transmission countershaft as in paragraph 199. To remove idler shaft, first remove the bolt (44—Fig. O120) and self-locking nut (45). Then push shaft forward out of bore in main frame and remove the idler gear (48), thrust washers (46) and bearings (47). To reinstall, reverse the removal procedure and install gear with long hub forward. Secure the idler shaft with the bolt (44) and self-locking nut.

**202. BEVEL PINION SHAFT.** The transmission bevel pinion shaft can be removed after removing the input shaft as in paragraph 195 and the countershaft as in paragraph 199. To remove bevel pinion shaft, proceed as follows:

On four-wheel drive models, unstake and remove the nut (86—Fig. O121) from front end of bevel pinion shaft. Save nut for adjusting bearing preload on reassembly. Then, remove the washer (85) and shims (79) from shaft.

On all models except four wheel drive, remove cover (49—Fig. O120) from front of bearing cage (53), unscrew and discard the nut (51), then using two ½-13 cap screws threaded into the tapped holes in the bearing cage, pull bearing cage and bearing (52) from front of bevel pinion shaft and front face of main frame. Using a small chisel, unstake the spanner nut (58) and remove nut (left-hand threads) with Oliver ST-127 Spanner Wrench or equivalent tool. Discard the nut.

Then, on all models, slide the bevel pinion shaft rearward and remove components as shaft is withdrawn. Remove the oil seal (72), snap ring (73) and rear bearing cup from rear bore in main frame. Remove the front bearing cup from bearing cage (53), or (81—Fig. O121), and remove rear bearing cone from shaft.

NOTE: On models equipped with Reverse-O-Torc, spacer (71—Fig. O120) is used on bevel pinion shaft instead of sleeve (70) and gear (69). On Row Crop and Wheatland models, gear (65) has 27 teeth; on

Utility and Industrial models, and as option on Row Crop and Wheatland models, gear (65) has 23 teeth. On Series 1650 and 1655 Wheatland models only, gear (67) has 41 teeth; on all other models, gear (67) has 38 teeth. On early production Series 1600 models, thrust washer (66) had only one internal tooth to engage splines of bevel pinion shaft; if one of these washers is encountered, it should be renewed using late type washer with 14 internal teeth.

203. To install bevel pinion shaft, proceed as follows: Install snap ring (73—Fig. O120) in rear bearing bore in main frame. Install seal (72) against front side of snap ring with lip of seal forward. Firmly seat rear bearing cup against rear side of snap ring. Press rear bearing cone firmly against shoulder on bevel pinion shaft and install sleeve (70), or spacer (71) on models with Reverse-O-Torc, on shaft against the bearing cone. Insert shaft through rear bearing cup and install components in following order: Gear (69) (38 teeth-thin hub) (except on Reverse-O-Torc equipped models) with clutch teeth forward. Install rear drive collar (62). Place rear shifter coupling (63) over collar. Install splined sleeve (68) (1$\frac{9}{32}$ inches long). Place gear (67) (38 teeth-thick hub) over sleeve with clutch teeth rearward. Install washer (66) with 14 internal teeth. (Renew washer if it has only one internal tooth.) Install sleeve (64) (1⅜ inches long). Place gear (65) (27 or 23 teeth) on sleeve with clutch teeth forward. Install front drive collar (62). Place front shifter coupling (63) on drive collar. Install splined sleeve (61) with flange to rear. Install gear (60) (44 teeth) with clutch teeth rearward. Place washer (59) on shaft with bent tang engaged in slot of sleeve (61).

On models except those with four-wheel drive, install nut (58) and tighten to a torque of 150 Ft.-Lbs. Firmly seat the front bearing cup in the bearing cage (53) and install new "O" ring (54). (On Series 1650 and 1655, also use new gasket between bearing cage and front face of main frame.) Install the bearing cage and temporarily install and tighten retaining nuts (57). Buck up against rear end of bevel pinion and drive front bearing onto front end of shaft. Install new nut (51) and adjust bearing preload with nut as follows: Tighten nut so that there will be a small amount of end play in the pinion shaft bearings. It may be necessary to bump front end of shaft to seat front bearing cone against the nut. Wrap a cord around the front bevel pinion shaft gear (60)

and attach a pull scale to the cord and engage shift collar to gear. The pull required to rotate the shaft steadily will represent the seal drag on the shaft. Then, tighten nut so that a pull of 2½-4 pounds in excess of that determined as seal drag will be required to rotate the shaft steadily. Remove the bearing cage retaining nuts and install cover (49) with new gasket (50).

On four-wheel drive models, refer to Fig. O121 and proceed as follows: Insert key (77) in slot of bevel pinion shaft and install spacer (78). Install new "O" ring (80) on rear end of bearing cage (81) and firmly seat front bearing cup in cage. Install cage in front bore of main frame and place bearing cone in cup and over spacer sleeve (78). Place shims (79) removed on disassembly next to spacer and install washer (85) and old nut (86).

Tighten nut so that there is a slight amount of bearing end play and check shaft rolling torque by attaching pull scale to cord wrapped around gear (60). Engage shifter collar to gear. Pull required to rotate shaft will represent seal drag. Then, tighten nut to a torque of 150 Ft.-Lbs. and again check shaft rolling torque. Bearing adjustment is correct if pull required to rotate shaft with nut tightened is 2½-4 pounds greater than that noted when checked with end play in bearings. Add shims (79) between spacer (78) and washer (85) if shaft rolling torque (bearing preload) is excessive, or remove shims if tightening nut does not increase the shaft rolling torque 2½ to 4 pounds on pull scale. When bearing preload is correct, remove and discard the old nut (86), install and tighten a new nut to a torque of 150 Ft.-Lbs. and stake nut to slot in bevel pinion shaft.

# TRANSMISSION
# (Series 1550-1555)

Series 1550 and 1555 tractors are equipped with a sliding spur gear transmission having six forward and two reverse speeds. When equipped with an auxiliary drive, two different speed ratios are available for each transmission gear. On Industrial models equipped with Reverse-O-Torc drive, the reverse gearing of the tractor main transmission is omitted as the Reverse-O-Torc unit provides the reversing function for any of the six basic transmission forward speeds. Optional transmission gears are also available for some models which provide different transmission speed ratios.

NOTE: On tractors with no auxiliary units, it is not absolutely necessary to split tractor for transmission overhaul, however due to space limitations and the shaft adjustments required, it is generally advisable to make the tractor split as outlined in paragraph 203B. In addition, overhaul of transmission will require removal of bull gear cover or hydraulic lift housing, bull gears, differential assembly and power take-off shaft or hydraulic pump drive shaft.

## LUBRICATION

203A. LUBRICANT. Transmission and differential are lubricated from a common oil supply. Oil level should be maintained at check plug opening in main frame located at right rear side of right rear axle housing. Filler plug is located in front end of bull

gear cover or hydraulic lift housing under center platform. Two drain plugs are used; front plug is located under front center of main frame and drains transmission compartment and rear plug is located at lower right rear corner of main frame and drains the differential and final drive compartment. Recommended transmission and differential lubricant is SAE 80 multi-purpose gear lubricant, or SAE 90 multi-purpose gear lubricant if temperatures normally remain above 32°F. All lubricants must conform to Military Specifications MIL-L-2105. Lubricant should be changed yearly or after each 1000 hours. Sump capacity is 40 quarts and on tractors prior to serial number 172 931-000, raise the sump capacity from 32 quarts to 40 quarts by installing the following pipe fittings in the test plug opening and in the following order: ⅜ x 2 nipple, ⅜ x 90° elbow, ⅜ x 1 nipple and ⅜ plain cap.

## OVERHAUL TRANSMISSION

203B. SPLIT TRACTOR. To split tractor between front and rear main frames, proceed as follows: Remove the pto drive shaft (paragraph 221 or 226) or the hydraulic pump drive shaft (paragraph 248). Disconnect battery cables and remove battery, or

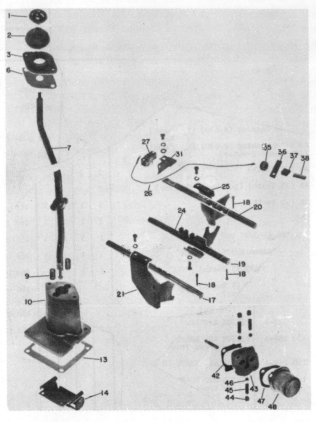

1. Knob
2. Cap (boot)
3. Cover
6. Gasket
7. Shift lever
9. Spring
10. Support
13. Gasket
14. Interlock
17. Shifter rod, right
18. Cotter pin
19. Shifter rod, center
20. Shifter rod, left
21. Shifter fork, right
24. Shifter fork, center
25. Shifter fork, left
26. Wire
27. Safety switch
31. Bracket
35. Grommet
36. Clip
37. Seal
38. Connector
42. Gasket
43. Poppet block
44. Plug
45. Spring
46. Ball
47. Gasket
48. Cover

Fig. O124B—Late type interlock assembly showing starting safety switch mounted. Refer to text for switch adjustment.

batteries, then disconnect battery cable from starting motor. On models so equipped, remove platform enclosure. On all models, remove left, right and center platform sections. Disconnect starter safety switch wires and if so equipped, the hydraulic lines from flow divider and lift housing. Disconnect the rear light wires. Disconnect clutch shaft coupler, or sprocket chain on models with Hydra-Power, Reverse-O-Torc or Creeper drive. Support tractor front frame so it will not tip and place a rolling floor jack under front of rear frame, then remove the front frame to rear frame retaining cap screws and separate tractor.

NOTE: These 12-point cap screws are tightened to 300 Ft.-Lbs. torque and will require a heavy duty socket for removal and installation.

**203C. R&R BULL GEAR COVER.** On models equipped with hydraulic lift system, refer to paragraph 253. On models with no hydraulic lift system, remove the transmission and bull gear cover as follows: Disconnect battery ground strap and remove battery or batteries. On models so equipped, remove the platform enclosures. Remove seat. Remove the left, right and center platform sections. Remove shift lever and shift lever support, disconnect starter safety switch wires, then unbolt and remove transmission and bull gear cover.

**203D. SHIFTING MECHANISM.** The gear shift lever (Fig. O124A) can be removed from transmission and bull gear cover, or hydraulic lift housing, after center platform section has been removed. Remove cover (3) and remove lever (7) and springs (9) from support (10).

If tractor is equipped with Hydra-Power, Reverse-O-Torc or Creeper drive, split tractor as outlined in paragraph 203B before attempting to remove the shifter rods. Remove shifter forks and rods as follows: Remove transmission and bull gear cover (paragraph 203C) or hydraulic lift housing (paragraph 253). Remove switch wires from interlock and remove interlock (14—Fig. O124A). Remove safety switch (27) and bracket (31) from left shifter rod (20). Remove cover (48) and gasket from poppet block (43), then pull poppet block and shifter rods from transmission as an assembly. Turn shifter rods 90 degrees, pull them from poppet block and be careful not to lose the poppet balls (46) as they are released by the shifter rods. Remove springs (45) and plugs (44) from poppet block.

NOTE: Beginning with tractor serial number 174 448-000 the starting safety switch was moved to a stationary position on a new interlock assembly. If a tractor is encountered that has the starting safety switch mounted on rear end of left hand shifter

rod it is recommended by Oliver Corporation that a new interlock, part number 164 090-A (or 164 092-A for low speed Row Crop) be installed and the safety switch installed and adjusted as follows: Remove old interlock and wire clip from rear frame, discard interlock but retain clip, cap screws and lock washers. Disconnect safety switch bracket from shifter rod and remove switch, bracket and wires. Discard wire grommet located in wall of transmission case. Remove switch from old bracket, discard bracket, then using old attaching hardware, loosely attach switch to new interlock. See Fig. O124B. Position switch so that it opens circuit when roller is about one-half up ramp and tighten screws to hold switch in this position. With wire clip installed on rear mounting screw of interlock, install interlock and switch and secure wires with clip.

To install shifter rods in poppet block, place a spring in spring hole, set a poppet ball on top of spring, then depress ball and spring and insert shifter rod. Shifter rods and poppet block are correctly assembled when left and right shifter rods have detent notches facing downward and shift lever notches inward, center shifter rod has shifter lug upward and oil drain holes in poppet block face toward rods (rearward). Be sure cotter pins (18—Fig. O124A) are in forward ends of shifter rods and use new gaskets (42 and 47) during reassembly.

If original starting safety switch assembly is being retained, adjust switch as follows: Position switch bracket until switch roller is centered in notch of right hand shifter rod and tighten bracket retaining cap screws. Loosen switch mounting screws and position switch so that switch opens when roller is about one-half up ramp of notch when transmission is shifted into gear. Tighten switch mounting screws to maintain this adjustment. An audible click can be heard as switch opens and closes. Pull wires rearward through grommet in transmission wall to provide slack enough for backward and forward movement of switch.

**203E. INPUT SHAFT.** To remove input shaft, drain transmission and split tractor as outlined in paragraph 203B, then remove transmission and bull gear cover (paragraph 203C) or hydraulic lift housing (paragraph 253). Remove snap ring (16—Fig. O124C) from bore in rear wall of rear main frame and bump hydraulic pump drive pinion (15) and bearing (17) rearward out of bore. Remove bearing retainer (3) from front of main frame and save shims (4). Oil seal (2) can be removed from retainer at this time, if necessary. Use driver, Oliver ST-101, or its equivalent and bump input shaft forward until front bearing cup comes out of bore in front wall of transmission housing and rear bearing cone comes off rear of input shaft. Complete removal of input shaft and lift input gear (12) out top of housing. Snap ring (14) and bearing cup can now be removed from bore and bearing (8), spacer (9) and snap ring (11) can be removed from input shaft.

NOTE: If input shaft assembly is being removed without removing shifter mechanism, place blocking between input gear and front wall of rear frame to prevent bending of shifter fork.

Clean and inspect all parts and renew as necessary. On Industrial models with Reverse-O-Torc, or models without hydraulics or pto, inspect plug (1) located in forward bore of input shaft and renew if necessary.

Reinstall input shaft assembly as follows: Install snap ring (11) and spacer washer (9) on forward end of shaft, then with taper facing toward front, press front bearing on input shaft until bearing is snug against spacer and snap ring. If removed, install snap ring (14) and rear bearing cup, with taper facing forward, and seat cup against snap ring. Start rear of input shaft in front of main frame, start input gear (12) on shaft with larger gear forward for all models except regular Row Crop which should have smaller gear forward, and if shift fork is installed, be sure it enters fork groove of gear. Place rear bearing cone in rear cup, then bump shaft into bearing cone until it bottoms. Install front bearing cup over shaft front bearing, bump into position, then install front bearing retainer and original shims. Bump shaft in both directions to insure that bearings and cups are seated and check the shaft

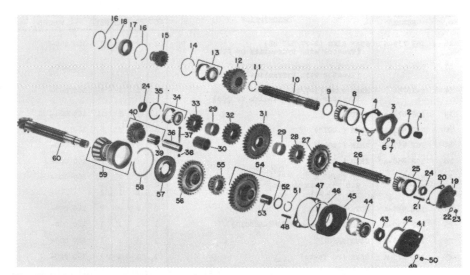

Fig. O124C — Exploded view of the transmission gears and shafts used in the Series 1550 and 1555 tractor. Reverse idler assembly (items 36 through 40) and gear (33) are not used in models equipped with Reverse-O-Torc. Spacer (30) replaces gear (33) and rear spacer (29) when Reverse-O-Torc is used. Input gear (12) is shown positioned for all models except regular Row Crop. On Row Crop models, gear should be reversed with smaller gear facing forward.

| | | | |
|---|---|---|---|
| 1. Plug (R-O-T only) | 18. Snap ring | 32. Countershaft gear (19 teeth) | 45. Bearing cage |
| 2. Oil seal | 19. Cover | 33. Countershaft gear (15 teeth) | 46. "O" ring |
| 3. Bearing retainer | 20. Shims | 34. Bearing | 47. Snap ring |
| 4. Shims | 24. Nut | 35. Snap ring | 51. Snap ring |
| 8. Bearing | 25. Bearing | 36. Reverse idler shaft | 52. Washer |
| 9. Spacer | 26. Countershaft | 37. Cap screw | 53. Bushing |
| 10. Input shaft | 27. Countershaft gear (28 teeth) | 38. Nut | 54. Drive gear (36 or 37 teeth) |
| 11. Snap ring | 28. Countershaft gear (15 or 16 teeth) | 39. Bushing | 55. Output gear (17 or 23 teeth) |
| 12. Input gear | 29. Spacer | 40. Reverse idler | 56. Output gear (33 teeth) |
| 13. Bearing | 30. Spacer (R-O-T only) | 41. Cover | 57. Oil shield |
| 14. Snap ring | 31. Countershaft gear (35 teeth) | 42. Gasket | 58. Snap ring |
| 15. Hyd. pump drive pinion | | 43. Nut | 59. Bearing |
| 16. Snap ring | | 44. Bearing | 60. Pinion shaft (7 or 8 teeth) |
| 17. Bearing | | | |

bearing adjustment. Bearing adjustment is correct when shaft end play does not exceed 0.001, or if bearing preload is such that rolling torque is 5 inch pounds or less above oil seal drag. Shims (4) are available in thicknesses of 0.004, 0.007 and 0.0149. After input shaft bearing adjustment is made, install the hydraulic pump drive gear and bearing assembly.

**203F. COUNTERSHAFT.** To remove countershaft, remove shifter forks and rods as outlined in paragraph 203D and the input shaft as outlined in paragraph 203E. Refer to Fig. O124C. Remove rear countershaft nut (24) and discard. Remove cover (19) and shims (20) and save shims for subsequent use. Drive countershaft forward until rear bearing cone comes off shaft and front bearing cup clears housing bore, then pull shaft forward and lift gears and spacers from top of housing as they come off countershaft. Remove front countershaft nut and discard, then press countershaft from front bearing. If it is necessary to remove countershaft rear bearing cup the reverse idler shaft and gear must first be removed and is accomplished by removing bolt (37) and pulling shaft (36) from its bore.

NOTE: If tractor is equipped with Reverse-O-Torc drive the reversing function is accomplished by the reversing gear unit and the reverse idler assembly (items 36 through 40) is not installed. In addition, a longer (2-13/64 in.) spacer (30) is installed on countershaft to replace shorter spacer (29) and gear (33).

Reinstall countershaft as follows: Place front countershaft bearing on shaft with taper facing forward, then install a new nut and tighten nut to minimum of 150 Ft.-Lbs. torque. If previously removed, install snap ring (35) and rear bearing cup with taper forward and seated against snap ring. Start countershaft into front bore of transmission housing and install gear (27) with rounded side of teeth rearward (gear has 35 teeth for regular Row Crop; 28 teeth for all others). Install gear (28) (gear has 15 teeth for regular Row Crop; 16 teeth for all others). NOTE: When the 15 tooth regular Row Crop gear (28) is being used, be sure not to confuse it with the 15 tooth reverse idler drive gear (33). The reverse idler drive gear (33) has an O.D. of 2$\frac{21}{32}$ inches whereas the 15 tooth countershaft gear (28) has an O.D. of 2$\frac{13}{16}$ inches. Install front spacer (29). Install 19 tooth gear (31) with rounded side of teeth toward rear, then install rear spacer (29). Place rear bearing in rear bearing cup, then install 15 tooth gear and push countershaft into rear bearing.

NOTE: If tractor is fitted with Reverse-O-Torc, install the long spacer (30) instead of spacer (29) and gear (33).

Drive countershaft into rear bearing until rear nut (24) can be started, then install a new nut and tighten to 150 Ft.-Lbs. torque minimum. Bump shaft both ways to seat bearings and cups, then using original shims (20), install cover (19). Use a dial indicator positioned on rear of shaft (not on gears) and check shaft end play which should be 0.001-0.003. Vary shims (20) as required to obtain correct shaft end play.

On models so equipped, reinstall reverse idler assembly.

203G. PINION SHAFT. To remove bevel pinion shaft, first remove shifter forks and rods as in paragraph 203D, input shaft as in paragraph 203E and countershaft as in paragraph 203F. Remove differential as outlined in paragraph 204. On models so equipped, remove retaining bolt (37—Fig. O124C), shaft (36) and idler gear (40). Remove cover (41) and gasket (42), then remove and discard nut (43). Use two ½-inch cap screws in the tapped holes of bearing cage (45) and pull front bearing and bearing cage. Remove bearing cup and "O" ring (46) from bearing cage. Remove snap ring (51)

and washer (52) from front of shaft, then pull pinion shaft rearward and lift gears from top of transmission housing as they come off shaft. Remove oil shield (57) (if used), bearing cup and snap ring (58) from bore in transmission compartment rear wall and press rear bearing from pinion shaft.

Clean and inspect all parts. Bushing (53) in bore of drive gear (54) is renewable and inside diameter of new bushing is 1.4375-1.4385. Outside diameter of pinion shaft at drive gear bushing journal is 1.4345-1.4350. Nominal operating clearance between bushing and shaft is 0.0025-0.0040 and if clearance appears excessive, renew bushing and/or shaft. Bushing must be sized after being installed in gear.

To install bevel pinion shaft, proceed as follows: Install snap ring (58), then from rear of transmission compartment rear wall, install bearing cup with taper facing rearward tight against snap ring. Install oil shield (seal) (57) (if used) tight against front side of snap ring with felt side next to snap ring and seal lip pointing rearward. Press rear bearing on pinion shaft with taper facing toward front end of shaft. NOTE: If shaft is being renewed, be sure shaft has same number of teeth as original. Shaft is available with 7 or 8 teeth. Insert pinion shaft through rear bearing cup and install the 33 tooth gear (56) with shifter fork groove toward front. Install gear (55) on shaft with shifter

fork groove toward rear (gear has 23 teeth for regular Row Crop; 17 teeth for all others). Install gear (54) on shaft with internal teeth toward rear (gear has 37 teeth for regular Row Crop; 36 teeth for all others). Install washer (52) and snap ring (51) on front of pinion shaft. Install front bearing cup in bearing cage with taper facing forward, then install "O" ring (46) and gasket (47). Secure bearing cage in place, block rear of pinion shaft and drive front bearing, taper facing rearward, on pinion shaft. Install a new nut on front end of pinion shaft and tighten nut to obtain the correct pinion shaft bearing preload which is 2¾-4½ lbs. pull on a spring scale attached to gear (56) with a cord. Use new gasket (42) and install cover (41).

203H. REVERSE IDLER GEAR. The reverse idler gear and shaft (not used on models with Reverse-O-Torc) can be removed after removing the transmission input shaft as in paragraph 203E and the transmission countershaft as in paragraph 203F. To remove idler shaft, first remove the bolt (37—Fig. O124C and nut (38). Push shaft (36) forward out of bore in main frame and remove the idler gear, thrust washers and bearings. To reinstall, reverse the removal procedure and install gear with long hub forward. Secure the idler shaft with the bolt (37) and self-locking nut.

# DIFFERENTIAL, BEVEL GEARS, FINAL DRIVE AND REAR AXLES

## DIFFERENTIAL

204. REMOVE AND REINSTALL. On models with a hydraulic lift system, first drain the hydraulic system fluid as outlined in paragraph 233. Then, drain the transmission and final drive lubricant and proceed as follows:

On models with single speed PTO, remove the PTO clutch and drive shaft as outlined in paragraph 221. On models with dual speed PTO, remove the PTO drive shaft as outlined in paragraph 226 and remove the PTO housing and gear assembly as in paragraph 229. On models not equipped with a PTO, but having a hydraulic lift system, remove the hydraulic pump drive shaft as outlined in paragraph 248. Remove the hydraulic lift

system as outlined in paragraph 253 or, on models not equipped with hydraulic system, remove the bull gear cover as outlined in paragraph 193 or 203C. Disconnect light wires, remove both rear fenders, support the tractor and remove both rear wheel and tire units.

If equipped with a three-point hitch, remove the draft control spring. Remove snap ring from inner end of each axle shaft and remove bearing retainers from outer end of each rear axle carrier (housing). Withdraw both axle shafts far enough to disengage the shafts from the bull gears and allow the shafts to rest in the carriers. It may be necessary to use a pry bar to move the axle shafts out of the bull gears. Lift the bull gears from rear main frame.

Remove brake adjusting nuts, both brake covers (and shims on models so equipped) and withdraw the brake discs from the brake housings and bull pinion shafts. Unbolt the left brake housing from rear main frame and withdraw the left bull pinion and brake housing taking care not to lose or damage the shims located between brake housing and rear main frame. Identify the shims so that they can be reinstalled in same position. Note that the left bull pinion (13—Fig. O126) of Series 1600, 1650 and 1655 has sealing rings installed in grooves (G). Also see Fig. O127.

Unbolt the right brake housing and remove the housing and right bull pinion from rear main frame. The right bull pinion of Series 1600, 1650 and 1655 also has grooves (G), but does not have sealing rings installed in the grooves.

Remove the thrust bearings (15—Fig. O125), thrust washers (16) and side gears (14) from the differential and identify them so that they can be reinstalled in their original positions. Turn the left end of differential towards left front corner of compartment and lift differential from tractor.

Reinstall differential by reversing removal procedure. Refer to paragraph 208 for information concerning the sealing rings on the Series 1600, 1650 and 1655 left bull pinion. Adjust the differential bearings and backlash of bevel ring gear to bevel pinion as outlined in paragraph 205.

**205. BEARINGS ADJUST.** Differential end play and gear backlash are controlled by the shim packs interposed between the brake housings and the rear main frame. When adjusting bearings, do not depend on a shim count or prior assembly. Use a micrometer to measure shim pack thickness. Shims are available in thicknesses of 0.004, 0.007 and 0.0149.

Mount a dial indicator and check the differential end play which should be 0.001-0.003. Add or subtract shims as necessary to obtain the specified end play, however, BE SURE to maintain some gear backlash during this operation.

With differential end play adjusted, use the dial indicator to check the backlash between bevel pinion and bevel drive gear. Transfer shims from one brake housing to the other until the dial indicator reading agrees with the backlash reading etched on bevel drive gear. Note: Only transfer shims. Do not add or remove shims or the previously adjusted end play will be changed.

Mesh position of bevel gears is fixed and non-adjustable.

**206. OVERHAUL.** Remove differential as outlined in paragraph 204. Remove pinion pin retaining snap rings, then with puller Oliver No. STS-100, or equivalent, threaded into tapped hole of pinion pin, pull pinion pins and remove pinions.

When reassembling, install pinions with taper facing inward and tapped hole of pinion pin facing outward. When mounting bevel drive gear to spider, install bolts with heads next to bevel drive gear. Tighten nuts alternately and evenly to a torque of 85-90 Ft.-Lbs.

Note: When installing new spider, it may be necessary to ream holes in

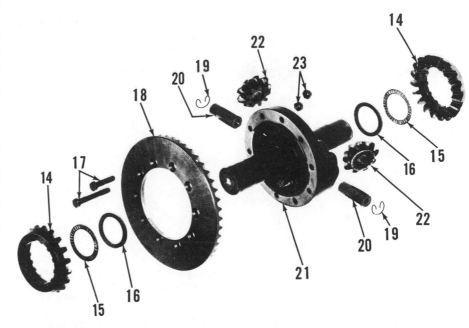

Fig. O125—Exploded view of differential assembly which is similar for all models. Differential side gears (14) are carried on inner ends of the final drive bull pinions (13—Fig. O126). Outer ends of differential pinion shafts (20) are drilled and tapped to aid in removing them from differential spider.

| | | |
|---|---|---|
| 14. Side gears | 16. Thrust washers | 18. Bevel ring gear | 22. Differential pinions |
| 15. Thrust bearings | 17. Ring gear retaining bolts | 19. Snap rings | 23. Nuts |
| | | 20. Pinion shafts | |

spider to 0.511-0.513 as follows: Place bevel drive gear on spider, ream one hole and install bolt. Now ream the hole which is directly opposite and install bolt. Tighten nuts to hold bevel drive gear in position and ream balance of holes. Install remainder of bolts and tighten nuts alternately and evenly to a torque of 85-90 Ft.-Lbs.

### BEVEL GEARS

**207.** The bevel pinion shaft extends through the transmission and carries the transmission driven gears. For information concerning removal and installation of the bevel pinion shaft, refer to paragraph 202 or 203G in TRANSMISSION section of this manual.

The bevel ring gear is bolted to the differential spider. Refer to paragraphs 204 and 206 for differential service information which includes procedure for renewing the bevel ring gear.

### FINAL DRIVE & REAR AXLE

**208. BULL PINION GEARS.** To renew a bull pinion gear, drain lubricant from final drive housing and proceed as follows:

Remove brake covers (and shims on models so equipped) from both sides of tractor. Remove brake adjusting nuts and remove the brake discs

and actuating assembly from side of tractor on which bull pinion shaft is to be removed. Mount a dial indicator against outer end of either bull pinion shaft and, with assistant, pry against ends of bull pinion shafts and note end play of differential unit on the dial indicator. Desired end play is 0.001-0.003; if end play is excessive, it is recommended that the hydraulic lift system (see paragraph 253) or bull gear cover (paragraph 193 or 203C) be removed so that bevel gear backlash can be checked after readjusting differential end play on reassembly. If differential end play is not excessive, bevel gear backlash can be considered satisfactory and it will not be necessary to remove hydraulic lift system or bull gear cover; record differential end play and proceed as follows:

Unbolt brake housing on side from which bull pinion is to be removed and, while holding bull pinion in place, remove the brake housing. Be careful not to lose or damage the shims located between brake housing and rear main frame. Carefully withdraw bull pinion from differential shaft so that differential side gear will be removed with the bull pinion. Remove and inspect the roller type thrust bearing (15—Fig. O125) and thrust washer (16). If any rollers fall from bearing, be sure to retrieve them from final drive compartment. Renew

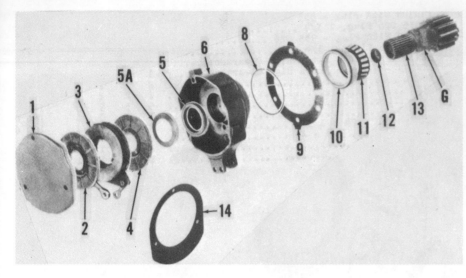

**Fig. O126 — Exploded view of brakes, brake housing and bull pinion unit. On left bull pinion only, of series 1600, 1650 and 1655, sealing rings are placed in grooves (G); refer to Fig. O127. Refer to Fig. O129 for exploded view of brake actuating assembly (3).**

| | | |
|---|---|---|
| 1. Brake cover | 5A. Cup | 11. Bearing cone & |
| 2. Brake disc | 6. Brake housing | roller assy. |
| 3. Actuating assy. | 8. "O" rings | 12. Expansion plug |
| 4. Brake disc | 9. Shims | 13. Bull pinion |
| 5. Oil seal | 10. Bearing cup | 14. Shims |

**Fig. O127—On left bull pinion, of Series 1600, 1650 and 1655, "O" rings are placed in the sealing grooves (G—Fig. O126) and Teflon rings are installed on top of the "O" rings. View shows installation of Teflon ring over the outer "O" ring.**

the thrust bearing and/or washer if worn or rough.

If the left bull pinion shaft has been removed, condition of the bevel pinion and bevel ring gear teeth can be checked through opening in housing. Note: If the hydraulic lift system or bull gear cover is not removed, it is recommended that only one brake housing and bull pinion be removed at a time.

To install a new bull pinion, be sure that expansion plug is installed in outer end and proceed as follows: Install thrust washer next to shoulder on differential shaft and place the thrust bearing next to washer.

The left bull pinion shaft of Series 1600, 1650 and 1655 is fitted with sealing rings and acts as a collector for oil under pressure from the transmission lubrication circuit for lubrication of the differential shaft bearing sur-

faces. If installing the left bull pinion on these models, install new "O" rings in bottom of grooves (G—Fig. O126) and then install new Teflon rings on top of the "O" rings. Refer to Fig. O127. After installing the Teflon rings wait for the plastic material to retract to normal size before proceeding with assembly. Note: Stretch the Teflon rings as little as possible when installing them.

Renew the oil seal in brake housing, installing seal with lip to inside. Install new "O" ring on shoulder of brake housing. Lubricate seal and, if installing left bull pinion on Series 1600, 1650 and 1655, lubricate the sealing rings on pinion shaft. Place differential side gear on inner end of bull pinion and carefully install bull pinion on differential shaft. Reinstall brake housing with same number of shims as were removed.

Check end play of the differential asembly as outlined in removal procedure. If end play is not within 0.001 to 0.003, refer to paragraph 205.

**209. BULL GEARS, WHEEL AXLE SHAFTS AND AXLE CARRIERS (HOUSINGS).** Removal of the bull gears, wheel axle shafts and axle carriers (housings) is inter-related; however, removal procedure may vary according to which elements are to be serviced.

If removing the bull gears to gain access for removing the differential assembly, or for renewal of a bull gear, proceed as outlined in paragraph 210.

To renew an axle shaft, axle carrier (housing) and/or axle shaft bearings, refer to procedure outlined in paragraph 211.

**210. R&R BULL GEARS.** To remove the bull gears, first drain and remove the hydraulic lift system, if so equipped, as outlined in paragraph 253. If not equipped with a lift system, remove the bull gear cover as outlined in paragraph 193 or 203C. Drain the final drive compartment, support rear of tractor and remove rear wheel and tire units. (If only one bull gear is to be removed, remove wheel and tire unit on that side only).

On models with single pto, remove the pto clutch and drive shaft assembly as outlined in paragraph 221. On models with dual pto, remove the pto drive shaft as outlined in paragraph 226 and remove the pto housing and gear unit as outlined in paragraph 229. On models not equipped with a power take-off, but having a hydraulic system, remove the hydraulic pump drive shaft as outlined in paragraph 248.

If equipped with a three point hitch, remove the draft control spring. Remove snap rings from inner ends of the axle shafts and remove bearing retainers from outer ends of the axle carriers. Take care not to lose or damage any of the shims located between retainer and outer end of carrier. Withdraw axle shafts far enough to disengage the shafts from bull gears and allow the shafts to rest in the carriers. It may be necessary to use a pry bar to force the shafts from the gears. Lift the bull gears from rear main frame.

Reinstall bull gears by reversing removal procedure and reinstall rear axle shaft bearing retainer to carrier with same number of shims as were removed.

**211. R&R AND OVERHAUL AXLE SHAFT, BEARINGS AND/OR AXLE CARRIER.** On models with dual pto, the bull gear retaining snap rings can be removed after removing the pto gear and housing unit as outlined in paragraph 229. Then, support rear of tractor and remove the rear wheel and tire units and rear fenders. Attach a hoist to the rear axle carrier, unbolt carrier from rear main frame and withdraw the axle and carrier as a unit from the rear main frame and bull gear.

On models with single speed pto, follow the general procedures as outlined for removal of bull gears in paragraph 210, except that the bull gears need not be removed from the rear main frame. To inspect or renew the axle shaft inner bearing cup, it is necessary to remove the carrier from the rear main frame and it is suggested that the axle and carrier be removed as a unit, instead of removing the bearing retainer and withdrawing the shaft.

After the axle and carrier unit is removed, proceed as follows: Unbolt and remove the bearing retainer from outer end of axle carrier and remove the axle shaft and bearing cone assemblies. Using a press or a bearing puller, remove the bearing cone assemblies from each end of axle shaft. Drive the inner bearing cup away from retaining snap ring, remove the snap ring, then pull bearing cup from inner end of axle carrier. Remove the outer bearing cup, seal assembly and felt dust seal from the bearing retainer.

Reassemble the axle shaft and carrier unit as follows: Drive inner bearing cup into inner end of axle carrier with tapered side of cup inward. Install the bearing cup retaining snap ring, then seat the cup against the snap ring. Heat the bearing cone assemblies in hot oil to approximately 200° F. and bump the assemblies onto axle shaft. (As optional procedure, install the bearing cone assemblies in a press or with a bearing puller attachment.) Place a new felt seal in bearing retainer, then drive new oil seal into retainer with lip to inside. Drive bearing cup into retainer with tapered side of cup to inside.

Install the axle shaft and bearings assembly into carrier, then install retainer with proper thickness of shims to provide a shaft rolling torque (bearing preload) of 84-108 inch-pounds (including seal drag). If bull gear is removed, the rolling torque can be checked by placing gear on inner end of axle shaft, wrapping a

**Fig. O128 — Exploded view of rear axle, carrier and bull gear assembly. Axle shaft bearing preload is adjusted by varying thickness of shims (7). Bull gears are retained on inner end of axle shafts by snap rings (17).**

| | |
|---|---|
| 1. Retainer assembly | 6. Grease seal |
| 4. Grease fitting | 7. Shims |
| 5. Felt dust seal | 8. Bearing assy. |
| | 9. Axle shaft |

| | |
|---|---|
| 10. Axle carrier (housing) | 15. Snap ring |
| 13. "O" ring | 16. Bull gear |
| 14. Bearing assembly | 17. Snap ring |

cord around the gear and checking pull required to rotate shaft with a spring pull scale. Bearing adjustment is correct if pull on cord required to rotate shaft and gear is 9 to 12 pounds. Shims are available in thicknesses of 0.004, 0.007 and 0.0149.

To reinstall axle, lubricate bore of rear main frame with grease, install a new "O" ring in groove at inside end of axle carrier, then reinstall the unit by reversing removal procedure.

### EXTRA HIGH CLEARANCE

#### Models So Equipped

**211A. R&R AND OVERHAUL.** An extra high clearance rear axle assembly, shown exploded in Fig. O128A, is available and can be removed and serviced as follows: Remove wheel guard (fender) and the wheel and tire assembly, then drain final drive housing (16). Remove bull gear cover or the hydraulic lift assembly. Remove bearing caps (21 and 34) and apply brake, then unstake and remove nuts (19). Attach a hoist to carrier (4), remove retainer assembly (9), then unbolt carrier from rear

frame and carefully remove rear axle assembly from bull gear and rear frame.

With unit removed, place on a suitable work surface, then remove cover (41) and seal retainer (24) with seal (13) and felt washer (28). Remove and discard seal (13). Attach a suitable pusher in tapped holes on inner side of housing (16) and push axle (31) out of bearings (32), spacer (17) and lower sprocket (38). Spacer (30) and bearing cone (29) can now be removed from axle. Remove a link from drive chain (39), then remove lower sprocket and chain from housing (16). Attach a suitable pusher to outer side of housing (16), push sprocket (upper) shaft from bearing cone (18), then separate and remove housing (16) from carrier (4). Remove upper sprocket (38) and spacer (17) from housing. Remove snap ring (14) from sprocket shaft. Remove sprocket shaft and seal (13) from carrier. NOTE: Always renew seal (13) whenever sprocket shaft is removed. Remove inner bearing cone (2) from sprocket shaft. Remove any bearing cups that require renewal.

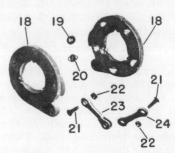

Fig. O129 — Exploded view of brake actuating assembly. Refer to Fig. O126 also.

| | |
|---|---|
| 18. Disc | 21. Studs |
| 19. Steel balls (7/8-inch) | 23. Yoke link |
| 20. Springs | 24. Plain link |

Fig. O128A—Exploded view of the extra high clearance rear axle assembly available for series 1600, 1650 and 1655. Note that spacer (30) is larger than spacers (17).

| | | | |
|---|---|---|---|
| 1. Bull gear | 12. Sprocket shaft | 19. Nut | 32. Bearing assy. |
| 2. Bearing assy. | 13. Oil seal | 20. Gasket | 33. Gasket |
| 3. "O" ring | 14. Snap ring | 21. Cap | 34. Cap |
| 4. Carrier | 15. Gasket | 24. Seal retainer | 35. Stud |
| 8. Dowels | 16. Housing, final | 27. Gasket | 38. Sprocket |
| 9. Retainer | drive | 28. Felt seal | 40. Gasket |
| 10. Cap screw | 17. Spacer | 29. Bearing assy. | 41. Cover |
| 11. Lock washer | 18. Bearing assy. | 30. Spacer | 44. Pipe plug |
| | | 31. Axle | |

Clean and inspect all parts. Be sure none of the spacers are damaged and that all bearing cones and their mating cups are in good condition.

Reassemble unit and adjust bearings as follows: Press inner bearing cone (2) on sprocket shaft (12) so taper will face carrier. Press bearing cup (2) in carrier with taper facing out until it bottoms. Install oil seal (13) (double lip seal, tractor serial number 193 154-000 and up) in outer end of carrier with lip of seal facing out. Carefully install sprocket shaft in carrier and use caution not to damage seal. Use a seal protector if available. Install snap ring (14). Install sprocket shaft outer bearing cup in final drive housing with taper facing out. Place a new gasket (15) over end of carrier and be sure cap screw holes align and dowels (8) are in position. Place upper sprocket (38) in final drive housing (16), then insert end of sprocket shaft into final drive housing and through sprocket. Install final drive housing retaining cap screws and tighten securely. Install small spacer (17) and bearing cone (18) with taper facing inward, on sprocket shaft, then install nut (19) but do not tighten at this time. If necessary, install axle shaft bearing cups with tapers facing outward. Place

chain (39) in final drive housing and pull it around upper sprocket. Place lower sprocket in final drive housing, place chain in position around sprocket and join chain by installing the previously removed link. Install outer axle bearing on axle with taper toward inner end of axle, then install large spacer (30) next to inner edge of bearing.

Insert axle into housing and through lower sprocket, then install small spacer (17), inner bearing (32) and nut (19). Use necessary adapters to install a torque wrench on nut (19) and tighten nut until it requires 84-108 In.-Lbs. torque to keep axle in motion. Recheck to make sure bearing preload is correct, then stake nut in three places. Install felt seal (28) and new oil seal (13) in retainer (24), then using new gasket (27), install assembly on final drive housing using extreme care not to damage seal. Use new gaskets and install cap (34) and cover (41). Place bull gear in rear main frame if necessary, then install the rear axle assembly and secure to main frame. Secure bull gear to sprocket shaft. Set tractor brake, then using a dial indicator, tighten nut (19) until sprocket shaft end play measures 0.001-0.003. When shaft end play is correct, stake nut in three places and

install new gasket (20) and bearing cap (21).

Complete reassembly of tractor and fill final drive housing with three quarts of recommended lubricant.

# BRAKES

Brakes are self-energizing disc type, mounted on outer ends of bull pinion shafts. Tractors after serial number 216 426-000 have shim adjusted brakes whereas brakes on prior models are adjusted by nuts on brake actuating rod.

### All Models

212. **ADJUSTMENT.** To adjust brakes on models with no shims, loosen jam nut on brake actuating rod and turn adjusting nut until the third or fourth notch of brake lock will engage the platform wear plate when the brake pedal is depressed moderately. Adjust both brakes the same.

On models with shim adjusted brakes, the brake pedal free travel should be 3½ inches. When pedal free travel exceeds 3½ inches, remove brake covers and subtract one shim at a time until correct free travel is established. To equalize pedal travel, loosen jam nut and turn actuating rod adjusting nut as required.

213. **R&R AND OVERHAUL.** Unhook brake return springs and remove step assembly from left side of tractor. Remove jam nut and adjusting nut from brake actuating rods. Remove brake housing covers and the brake assemblies.

Disassembly of the actuating assembly is accomplished by removing actuating rod from links and disengaging the three extension springs. If necessary, polish the steel balls and the ball ramps. A small amount of Lubriplate may be used on steel balls during reassembly.

Linings are bonded to brake discs

and renewal requires that a complete new disc be used.

Adjust brakes as outlined in paragraph 212.

# BELT PULLEY ASSEMBLY

A belt pulley assembly is available for use in conjunction with the single speed 540 RPM power take-off assembly or with the 540 RPM shaft of the dual speed power take-off assembly. Belt pulley is controlled by the pto clutch and is mounted on rear face of the pto housing.

## All Models So Equipped

**214. REMOVE AND REINSTALL.** Procedure for removal of belt pulley assembly is evident on inspection of unit. However, pulley must be to left side of unit when being installed.

If an original installation of belt pulley is being made, proceed as follows: Remove the pto safety shield. If tractor has three-point hitch, remove left lifting link and lower link. Mount belt pulley unit on pto housing so that pulley is on left side of tractor.

**215. OVERHAUL.** With belt pulley removed, drain oil and proceed as follows: Remove pulley from pulley shaft (5—Fig. O130); then, unbolt and remove drive shaft (30) and carrier

(25) assembly from housing (35). Remove "O" ring (39), shims (38) and oil seal (34) from carrier, then using Oliver ST-142 Spanner Wrench or equivalent tool, remove adjuster nut (2) (adjuster nut has left hand thread). Press the drive shaft (30) and gear (3) from carrier. Remove snap ring (15) and pull bevel drive gear from drive shaft. Press inner bearing from drive shaft and remove bearing

cups from carrier. Remove expansion plug (31) from drive shaft only if it shows signs of leakage.

Unbolt pulley shaft carrier (7) from housing and remove carrier and pulley shaft (5) assembly from housing. Remove shims (11) from carrier, unstake and remove adjuster nut (12) and use a puller to remove pinion gear (4) and bearing cone from pulley shaft. Press pinion gear from bearing. Press pulley shaft (5) from carrier, then remove outer bearing and oil seal (10). Remove bearing cups from carrier. Remove expansion plug (37A) from pulley housing only if it shows signs of leakage.

Remove cap screw (14) and disassemble breather assembly.

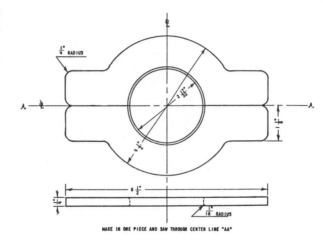

Fig. O131 — Seal installation tool can be made from piece of ¼-inch thick steel plate. Refer to dimensions shown on drawing.

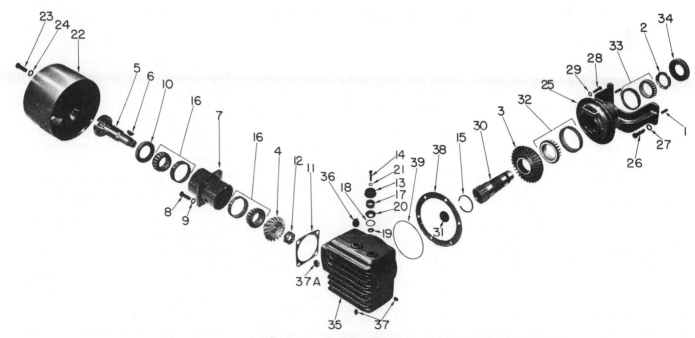

**Fig. O130 — Exploded view of belt pulley assembly. Unit can be installed on tractors equipped with either the single speed 540 RPM power take-off, or with the 540 RPM pto shaft installed in the dual speed power take-off.**

| | | |
|---|---|---|
| 1. Dowel pins | 12. Nut | 25. Carrier | 34. Seal |
| 2. Spanner nut | 13. Cap | 30. Shaft | 35. Housing |
| 3. Bevel drive gear | 15. Snap ring | 31. Expansion plug | 38. Shims |
| 4. Bevel pinion gear | 16. Bearing | 32. Bearing | 39. "O" ring |
| 5. Pulley shaft | 17. Breather element | 33. Bearing | |
| 6. Woodruff key | 18. Gasket | | |
| 7. Carrier | 19. Baffle | | |
| 10. Seal | 20. Cup | | |
| 11. Shims | 21. Seal | | |
| | 22. Pulley | | |

Fig. O132 — Measurement between points shown is used in determining thickness of shims to use for proper location of gear in housing. Refer to text for procedure.

Fig. O133—Assembly dimension is stamped on housing as shown.

216. Inspect all parts for excessive wear, scoring, chipping or other damage and renew parts as necessary. Reassemble unit as follows: Place new oil seal (10) on pulley shaft with lip of seal inward (away from pulley mounting flange), then install bearing cone with bevel towards housing. Install bearing cone on drive pinion with bevel away from gear. Install bearing cups in carrier, then with an oil seal installation tool similar to that shown in Fig. O131 positioned

between pulley flange of shaft and oil seal (10—Fig. O130), press oil seal and pulley shaft into carrier. Install the pinion and bearing on pulley shaft and screw the nut (12) onto shaft so that a slight end play of shaft in bearing remains. Use a spring scale and note effort required to rotate carrier on shaft. Tighten nut until the spring scale reads 5 to 6 pounds more than reading taken prior to tightening nut. This will give the desired 10-12 inch pounds bearing preload.

With the bearing preload set, the pulley shaft location in housing must be determined as follows: Measure distance between flat surface of pinion gear and mounting flange of carrier as shown in Fig. O132. To this measurement, subtract dimension stamped on housing as shown in Fig. O133. To the result, add 1.157. If a plus or minus value is etched on pinion gear, add or subtract this value accordingly and this will give thickness of shims (11—Fig. O130) required between carrier and housing. Shims are available in thicknesses of 0.005, 0.007 and 0.020.

Assemble drive shaft and install in drive shaft carrier. Tighten adjusting nut (2) until bearings have 10-12 inch pounds preload and stake nut (adjusting nut has left hand thread). Install oil seal (34) with lip facing inward. Note the backlash value stamped on bevel drive gear and vary shims (38) located between carrier and housing to obtain this backlash value. Shims are available in thicknesses of 0.004, 0.007 and 0.0149.

Clean and reinstall breather assembly with new element (17). Fill unit with 6½ pints of SAE 80 multipurpose lubricant conforming to military specification MIL-L-2105.

# SINGLE SPEED PTO

The single speed power take-off unit is mounted on the rear face of the rear main frame (transmission housing) and receives its drive directly from the engine via the drive shaft (8—Fig. O136 or O136A) which is splined into a drive hub mounted on the engine flywheel. Power to the external drive shaft is controlled by an over-center clutch to provide independent operation of the unit.

Optional gearing is available to convert the unit from a 540 RPM pto to a 1000 RPM pto.

Tractors having a Reverse-O-Torc Drive cannot be equipped with a power take-off. For information on the dual speed power take-off, refer to paragraph 224.

## All Series So Equipped

217. **ADJUST PTO CLUTCH.** The double plate over-center type clutch used in the single speed pto has a shim type adjustment for wear. Clutch should engage (go over-center) when a pull of 43-55 pounds is applied to control lever just below the lever knob or grip. A spring pull scale can

be hooked onto lever at this point to check for proper clutch adjustment.

If pull required to engage clutch is less than 43 pounds, remove the pto cover as outlined in paragraph 218 and proceed as follows: Loosen the six cap screws (10—Fig. O134) retaining the plate (9) to housing (25). Remove two of the cap screws holding a shim set (12) and remove two or three of the 0.007 brass shims from the set. Reinstall shims remaining with long lobed ends facing in a counter-clockwise direction. Repeat the procedure for the other two shim sets removing the same number of 0.007 brass shims as were removed from first set. Tighten the plate retaining cap screws, reinstall pto cover and recheck pull required to engage clutch. Remove more shims if pull is less than 43 pounds or add shims if pull required to engage clutch is more than 55 pounds.

218. To remove pto cover (43—Fig. O136), first move clutch lever to dis-

engaged position. Disconnect linkage from clutch lever to actuating shaft. Remove the two cap screws retaining safety shield to pto gear housing and the four cap screws securing pto cover to housing. Remove cover, safety shield and dust seal from pto gear housing.

219. **CLUTCH OVERHAUL.** The pto clutch can be disassembled and overhauled after removing pto cover as outlined in paragraph 218, or after removing the clutch and drive shaft assembly as outlined in paragraph 221.

To disassemble clutch, proceed as follows: Remove the six plate retaining cap screws (10—Fig. O134) and the adjusting shims (12), taking care not to damage or lose shims. Remove the back plate (9), over-center linkage and pressure plate (21) as an assembly. Remove the two discs (22) and center plate (23) taking care not to lose the three springs (24). The housing (25) can then be unbolted

1. Carrier
2. Snap ring
3. Bearing
4. Snap ring
5. Snap rings
6. Sleeve
7. Pins
8. Snap ring
9. Back plate
12. Shim sets
13. Belleville (spring) washer
14. Plate
15. Links
16. Rollers
17. Pins
18. Links
19. Pins
20. Plate
21. Plate & lever assy.
22. Lined discs
23. Plate
24. Springs
25. Housing

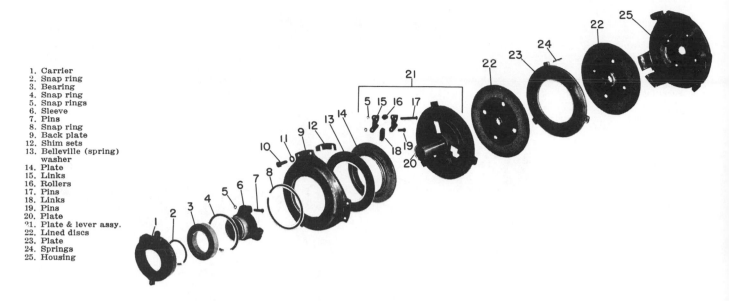

Fig. O134 — Exploded view of the dual disc pto clutch assembly used in single speed power take-off unit. Refer to Fig. O136 for exploded view of complete pto assembly.

and removed from the mounting sleeve (9 or 10—Fig. O136 or O136A) if necessary.

To disassemble the back plate, over-center linkage and pressure plate assembly, refer to Fig. O134 and proceed as follows: Disconnect overcenter linkage from pressure plate (20) and sleeve (6). Remove snap ring (4) from inner side of carrier (1) and remove sleeve and bearing from carrier. Remove snap ring (2) and remove bearing (3) from sleeve. Place back plate and plate (14) in a press to compress the spring washer (13) and remove snap ring (8).

220. Reverse the disassembly procedure to reassemble unit. If new lined discs (22) are being installed, reassemble using three new sets of shims (12) and check clutch adjustment as outlined in paragraph 217 after reinstalling unit. If reassembling clutch with used discs, but with new plates (20 and/or 23) or new housing (25) it may be necessary to add shims to those removed in disassembly. If desired, thickness of shim sets for initial adjustment of clutch (always recheck adjustment **after** reinstalling unit) can be determined as follows:

Remove the housing (25) from mounting sleeve and assemble the clutch unit in a press, but do not install the cap screws (10) or adjusting shims (12). With the over-center linkage in disengaged position, apply about 100 pounds force evenly to back plate (9) and check gap between back plate and housing at shim installa-

tion points with feeler gage. Average the three measurements and subtract 0.050; this should give the thickness of shim sets necessary for initial adjustment.

221. **R & R PTO DRIVE (INPUT) SHAFT AND CLUTCH ASSEMBLY.** To remove the pto drive shaft and clutch as an assembly on series 1600, remove pto cover as outlined in paragraph 218, remove cap screws (12—Fig. O136) securing retainer (11) to gear housing (24); then, withdraw the clutch and shaft assembly from rear of gear housing and tractor rear frame.

On series 1550 and 1555 follow the same procedure given above but notice the difference between the 540 rpm and 1000 rpm sleeve and pinion assemblies as shown in Fig. O136A. When reinstalling clutch and shaft assembly, renew gasket between adapter and housing, and if adapter and bearing retainer have been separated, renew the gasket between adapter and retainer. Check clutch adjustment as outlined in paragraph 217 after unit is installed.

222. **OVERHAUL PTO DRIVE (INPUT) SHAFT ASSEMBLY.** Remove the drive shaft and clutch assembly as outlined in paragraph 221 and remove clutch unit from mounting sleeve as outlined in paragraph 219. Refer to Figs. O136 and O136A and proceed as follows: Bump front (flywheel) end of shaft (8) to remove bearing (7) and seal (4) from rear end of mounting sleeve (9 or 10). Remove snap ring (5) from rear of spacer (6) and pull bearing and spacer from end of shaft.

Separate the adapter (19 or 25) from retainer (11, 13 or 16) and remove snap ring (17 or 23) from front of mounting sleeve (9). Use pullers and center plug to remove front bearing (16, 21 or 22) and slide pinion (15 or 19) from mounting sleeve. On series 1550, with 540 rpm pto, remove snap ring (20). Use pullers and center plug to pull retainer (11, 13 or 16) and rear bearing (14, 17 or 18) from mounting sleeve. Remove seal (10, 11 or 12) from retainer.

Renew all seals and other parts required and reassemble unit as follows: Install seal (10, 11 or 12) in retainer with lip toward front. Place retainer and seal assembly on mounting sleeve (9 or 10) and press rear bearing (14, 17 or 18) against shoulder of sleeve with snap ring on outer race of bearing toward rear. On series 1550 540 rpm units, install snap ring (20) and bearing (22) with open side toward rear. On all other models, install pinion (15 or 19) with long hub forward, bearing (16 or 21), with open side rearward, and snap ring (17 or 23). Install front snap ring (5) on shaft (8), press bearing (7) against snap ring, install spacer (6) against bearing with taper to rear and install rear snap ring (5). Bump mounting sleeve into position over bearing (7) from front of shaft and install oil seal (4) in sleeve with lip forward. Position adapter (19 or 25) over bearing (14, 17 or 18) with new gasket (18 or 24) between adapter and retainer (11, 13 or 16). Install the shaft assembly with new gasket (20 or 24); then, reinstall clutch assembly as in paragraph 221.

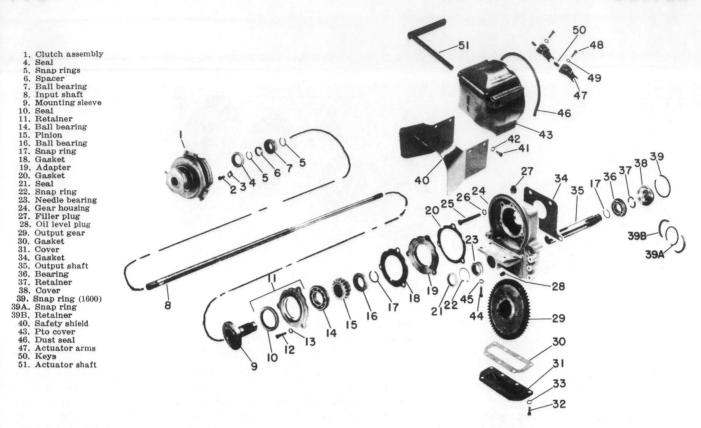

1. Clutch assembly
4. Seal
5. Snap rings
6. Spacer
7. Ball bearing
8. Input shaft
9. Mounting sleeve
10. Seal
11. Retainer
14. Ball bearing
15. Pinion
16. Ball bearing
17. Snap ring
18. Gasket
19. Adapter
20. Gasket
21. Seal
22. Snap ring
23. Needle bearing
24. Gear housing
27. Filler plug
28. Oil level plug
29. Output gear
30. Gasket
31. Cover
34. Gasket
35. Output shaft
36. Bearing
37. Retainer
38. Cover
39. Snap ring (1600)
39A. Snap ring
39B. Retainer
40. Safety shield
43. Pto cover
46. Dust seal
47. Actuator arms
50. Keys
51. Actuator shaft

Fig. O136—Exploded view of the single speed power take-off unit that is available for Series 1600 models. Refer to Fig. O134 for exploded view of clutch assembly (1).

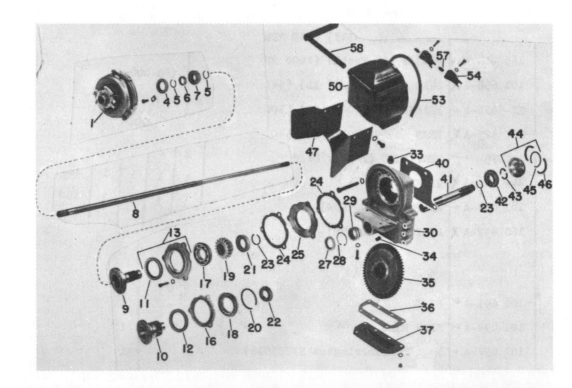

1. Clutch
4. Seal
5. Snap ring
6. Spacer
7. Bearing
8. Shaft
9. Sleeve (1000 rpm)
10. Sleeve (540 rpm)
11. Seal
12. Seal
13. Retainer
16. Retainer
17. Bearing
18. Bearing
19. Pinion
20. Snap ring
21. Bearing
22. Bearing
23. Snap ring
24. Gasket
25. Adapter
27. Seal
28. Snap ring
29. Bearing
30. Housing
33. Plug
34. Plug
35. Output gear
36. Gasket
37. Cover
40. Gasket
41. Output shaft
42. Bearing
43. Retainer
44. Cover package
45. Snap ring
46. Retainer
47. Safety shield
50. Cover
53. Seal strip
54. Actuator arm
57. Key
58. Actuator shaft

Fig. O136A—Exploded view of series 1550 and 1555 single speed pto unit. 1000 rpm units use mounting sleeve (9) and pinion (19). 540 rpm units have drive pinion integral with mounting sleeves (10). Refer to Fig. O134 for exploded view of clutch (1).

NOTE: The pto drive shaft is supported midway between the flywheel and pto unit by the hydraulic system pump drive pinion and bearing assembly. Refer to paragraph 249 for servicing the pump drive pinion and bearing. (Pump drive pinion and bearing are installed on tractors without hydraulic system to support the pto drive shaft and are accessible on these models after removing the bull gear cover).

**223. OVERHAUL PTO OUTPUT SHAFT AND GEAR.** To renew the output shaft, bearings, seal and/or gear, remove the pto clutch and input shaft as an assembly as outlined in paragraph 221, then use Fig. O136A as a guide and proceed as follows:

Remove the four cap screws retaining the gear housing (30—Fig. O136A) to the rear face of tractor rear frame and remove housing from tractor. On series 1600, remove snap ring (39—Fig. O136). On all other tractors, remove snap ring (45—Fig. O136A) and the two piece retainer (46), then

bump shaft (41) forward to remove bearing cover (44). Work through bottom opening in housing, disengage snap ring (23) from groove in shaft and move the snap ring about ¼-inch rearward. Support rear end of shaft and drive front bearing (42) to rear until bearing retainer (43) can be removed from front end of shaft. With a soft drift, bump shaft rearward about half-way through front bearing, move gear forward on shaft and move snap ring (23) to front side of snap ring groove. Then, drive shaft rearward out of bearing and gear and remove gear from housing. Drive the shaft and front bearing forward out of housing. Remove bearing from shaft. Remove oil seal (27), snap ring (28) and shaft rear bearing (29) from rear bore of housing.

To reassemble unit, proceed as follows: Press or drive only on lettered side of needle bearing cage (29) to install bearing, then install snap ring (28). Insert shaft (41) through needle

bearing and place snap ring on shaft about ¼-inch forward of groove in shaft. Place gear (35) in housing with long hub to rear, slide shaft through gear and move snap ring to rear of groove in shaft. Install front bearing (42) on shaft with snap ring side forward, install bearing retainers (43) and drive the shaft and bearing rearward into position. Engage snap ring (23) in groove of shaft at rear side of gear (35). Install bearing cover (44) and snap ring (series 1600) or retainer (46) and snap ring (45) (all except series 1600). Install oil seal (27) with lip toward front.

Reinstall unit on rear face of tractor main frame using new gasket (40). Install bottom cover (37) with new gasket (36) and fill housing to level of test plug (34) with SAE 80 multipurpose gear lubricant conforming to military specification MIL-L-2105. Reinstall pto clutch cover and check adjustment as outlined in paragraph 217.

# DUAL SPEED PTO

A 540 RPM, 1000 RPM or dual speed 540 and 1000 RPM pto is available for Series 1550, 1555, 1650 and 1655 models except those equipped with Reverse-O-Torc Drive. The pto unit is mounted on rear face of rear main frame and is utilized as a main frame cover. Power is taken directly from the engine flywheel and pto is controlled by a multiple disc over-center type clutch within the unit, providing independent control of the pto output shaft. Conversion from 540 RPM pto to 1000 RPM pto is made by changing the pto output shaft which automatically "shifts" the unit to the corresponding pto speed.

The pto control lever engages the clutch when in forward position. When lever is in rear position, the clutch is disengaged and further movement of the control lever rearward applies a pto holding brake. It should be noted that the pto holding brake is not to be used to stop motion of a pto driven implement.

**All Series So Equipped**

**224. CLUTCH ADJUSTMENT.** To adjust the pto clutch, stop engine and move control lever to point midway between engaged and disengaged positions and proceed as follows:

Remove the pipe plug (9—Fig. O138 or 16—Fig. O137A) from top of housing and rotate clutch assembly so that adjusting ring lock pin (12—Fig. O139) can be seen through plug open-

ing. Depress the lock pin and turn adjusting ring (11) with screwdriver. Turn ring to left to increase pressure, or to right to decrease pressure required to engage clutch. Clutch adjustment is correct when 54-62 pounds pull is required to engage clutch. Adjustment can be checked with a spring pull scale attached to control lever just below the lever grip. Move adjusting ring just one notch at a time and recheck clutch adjustment. When adjustment is correct, reinstall pipe plug.

**225. CONTROL LEVER ADJUSTMENT.** Refer to Fig. O137. The yoke (79) should be adjusted on rod (76) so that clutch control lever is locked firmly in disengaged position by latch (69). To adjust yoke, remove pin (80), loosen lock nut (82) and turn yoke in or out to obtain desired adjustment. Reinstall pin and tighten lock nut when adjustment is correct. Use same procedure on late models having similar right hand levers.

NOTE: On late Series 1650 and Series 1550 models equipped with depth stop cylinder hydraulic system, refer to Fig. O137A; adjustment of control lever latch differs on these models.

**226. R&R CLUTCH DRIVE (PTO INPUT) SHAFT.** Refer to Fig. O138 and proceed as follows. Unbolt and

remove cover (60) and "O" ring (59). Working through openings in hub of spider (58), disengage snap ring (57) from rear of bearing (56) in bore of clutch assembly (52). Then, withdraw spider, bearing and shaft (51) from rear of pto housing. Remove snap ring (54) and pull spider from shaft. Shaft is supported at mid length by the hydraulic system pump drive pinion and pinion bearings; refer to paragraph 249 for servicing the pinion and bearings.

Reverse removal procedures to reinstall the clutch drive shaft. Renew sealing "O" ring (59) when reinstalling cover (60).

**227. OVERHAUL PTO CLUTCH ASSEMBLY.** To remove the pto clutch assembly, first remove the drive shaft as outlined in paragraph 226. Then, remove snag ring (53—Fig. O138) from groove in rear end of clutch shaft (50), disconnect linkage between pto clutch lever and actuator lever (18) and withdraw clutch assembly from shaft and pto housing.

NOTE: In some isolated cases, a vibration dampner was installed on Series 1650 pto units. The dampner (116—Fig. O138) is mounted on clutch spider (58) by three cap screws. Dampner can be removed from spider, if necessary, either before or after clutch removal.

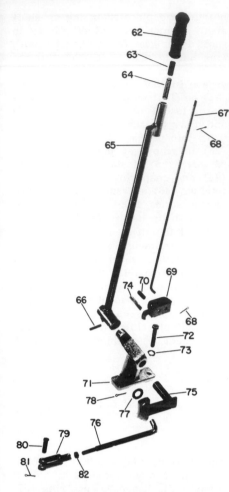

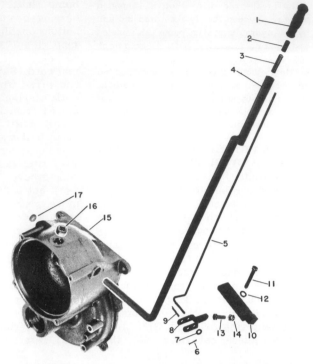

Fig. O137A — On some models equipped with depth stop cylinder hydraulic system, the PTO clutch lever is differently constructed than on other models. However, adjustment of PTO clutch remains the same. To adjust lever latch, move lever to disengaged position, loosen locknut (14) and adjust stop bolt so that it just contacts latch (8.

1. Grip
2. Plunger
3. Spring
4. Lever
5. Latch rod
8. Latch
9. Pin
10. Bracket
13. Latch stop bolt
14. Lock nut
15. Dual speed pto housing
16. Adjustment plug
17. Plug

**Fig. O137 — Pto clutch lever assembly and linkage for dual speed power take-off. Latch (69) locks lever in disengaged position. Late models may have a similar right hand lever.**

| | |
|---|---|
| 62. Grip | 71. Support |
| 63. Plunger | 74. Pin |
| 64. Spring | 75. Shaft |
| 65. Lever | 76. Link |
| 66. Roll pin | 79. Link end |
| 67. Rod | 82. Lock nut |
| 69. Latch | |

Should the occasion arise which requires installation of vibration dampner, order package Oliver part number 167 987-AS. When installing dampner, use dampner as a template and be sure holes are drilled 21/32-inch from spider tang end and at center of tang width. Use an $\frac{11}{32}$-inch drill and a $\frac{5}{16}$-18 thread cap.

For disassembled view of the clutch assembly, refer to Fig. O139. Drive out the roll pins (16), remove pins (15), sliding collar (5), levers and sleeve collar (1) from floating plate (14). Depress locking pin (12), unscrew adjusting ring (11) and remove adjusting ring, floating plate and clutch discs from clutch hub (19). Any further disassembly which may be required is evident from inspection of unit.

When reassembling, be sure to stake retaining screws (21) if drive

keys (20) were removed. Renew clutch discs (17 and/or 18) if coned more than 0.005. Install pins (6, 8 & 15) with heads in direction of clutch rotation. Lubricate clutch discs with transmission lubricant prior to assembly.

NOTE: Beginning with tractor serial number 187 586-000, the number of driven plates (17) was increased to 5 from 4 and the driver plates (18) from 5 to 6.

If desired, clutch can be adjusted prior to installation as follows: Place clutch assembly in a press and turn adjusting ring (11) as necessary so that a force of 195 pounds is required to engage clutch (snap lever arms over-center). If equipment to apply measured force of 195 pounds is not available, adjust clutch after installation as outlined in paragraph 224.

Reinstall clutch by reversing removal procedure. Clutch adjustment should be rechecked and adjusted, if necessary, as outlined in paragraph 224. Check and adjust lever linkage as outlined in paragraph 225.

**228. CHANGE PTO OUTPUT SHAFT.** The power take-off can be changed from 540 RPM to 1000 RPM output shaft speed, or vice versa, by changing to the appropriate output shaft (34—Fig. O138). To change output shaft, remove snap ring (38) from PTO housing and withdraw the shaft and bearing (37) assembly. Renew the "O" ring (33) in bore of housing if necessary; then, insert the alternate shaft and bearing assembly and secure with snap ring (38).

The alternate PTO output shaft furnished with tractor is fitted with bearing and bearing retaining snap rings. If necessary to renew either the bearing or the shaft, procedure for doing so is evident from inspection of unit and reference to exploded view in Fig. O138.

**229. R&R PTO HOUSING AND GEAR ASSEMBLY.** To remove the PTO assembly from rear of tractor main frame, first remove the drive (PTO input) shaft as outlined in paragraph 226 and remove the clutch unit as outlined in paragraph 227. Drain transmission lubricating oil, then proceed as follows:

On Series 1600, 1650 and 1655, disconnect rear oiling tube (45—Fig. O138) from front tube (48) at the elbow connector (47). Then, unbolt and remove PTO housing and gear assembly from rear of tractor main frame. Note that housing is located on tractor main frame by two dowels (11).

Reinstall PTO housing and gear assembly on tractor main frame using new gasket (12). Connect the oiling tube on Series 1600, 1650 and 1655, then reinstall clutch unit, drive shaft and spider.

**230. OVERHAUL PTO HOUSING AND GEAR ASSEMBLY.** Remove the housing and gear assembly as outlined in paragraph 229, then disassemble unit as follows:

Remove the cap screws retaining carrier (23—Fig. O138) to front of PTO housing and remove carrier from

**Fig. O138**—Exploded view of dual speed pto unit. A wet type multiple disc clutch (52) is used; over-center linkage can be actuated by either right or left hand lever (Fig. O137 or O137A). Refer to Fig. O139 for exploded view of clutch unit (52). Units used on Series 1550 and 1555 tractors do not use lubrication tube assembly (items 45 through 49). Cover (115) and snap ring (114) are used only on single speed 540 rpm pto.

| | | | |
|---|---|---|---|
| 1. 540 RPM gear | 13. 540 RPM pinion | 23. Carrier | 35. Snap ring | 46. Elbow | 56. Bearing |
| 2. Bearing assy. | 14. 1000 RPM pinion | 26. Dowel | 36. Snap rings | 47. Elbow | 57. Snap ring |
| 3. Snap ring | 15. Spacer | 27. 1000 RPM gear | 37. Bearing | 48. Oil tube | 58. Spider |
| 4. Bearing assy. | 16. Ball bearing | 28. Expansion plug | 38. Snap ring | 49. Elbow | 59. "O" ring |
| 5. "O" ring | 17. Nut | 29. Seal | 39. Safety shield | 50. Clutch shaft | 60. Cover |
| 6. Housing | 18. Arm & shaft | 30. Seal | 42. Disc | 51. Drive shaft | 114. Snap ring |
| 9. Pipe plug | 19. Seal | 31. Nut | 43. Oil collector ring | 52. Clutch assy. | 115. Cap |
| 10. Expansion plug | 20. Arms | 32. Snap ring | 44. Pin | 53. Snap ring | 116. Dampner |
| 11. Dowel | 21. Pins | 33. "O" ring | 45. Oil tube | 54. Snap rings | |
| 12. Gasket | 22. Nut | 34. Output shaft | | 55. Snap ring | |

the two locating dowel pins (26), front bearing (16) and housing. Unstake and remove nut (22) from hub of gear (27) and remove gear by bumping it rearward out of carrier and front bearing assembly (4). Remove rear bearing cone and roller assembly from gear and remove the bearing cups from carrier. Remove seal (28) from rear face of gear (26); it is not necessary to remove expansion plug (27) from gear if there is no evidence of oil leakage.

NOTE: Cover (115) and snap ring (114) are installed on forward side of gear (1) and are used only when pto is used as a single speed 540 RPM unit. If oil leakage through output shaft bearing on tractors prior to serial number 201 330-000 is encountered, install cover, snap ring and seal (29).

Unstake and remove nut (17) from

front end of clutch shaft (50) and bump the shaft rearward out of bearing (16), gears (13 and 14) and the rear bearing (37). Remove oil collector ring (43) and PTO brake washer (42) from shaft. Be careful not to lose oil collector ring pin (44). Remove snap ring (32) and drive the bearing (37) rearward out of housing.

Remove snap ring (38) and pull output shaft (34) and bearing assembly (37) from housing. Remove snap ring (32) and "O" ring (33) from bore and using Oliver Spanner Wrench ST-142 or equivalent tool, unscrew nut (31) from hub of gear (1). Remove "O" ring (5) from hub and bump gear forward out of bearing (4). Remove cone and roller assembly of front bearing from gear hub and the front bearing cup from housing. Bump rear bearing cup rearward away from snap ring (3), remove snap ring and drive the rear bearing assembly forward out

of housing. Drive the oil seal (30) rearward out of housing.

Clean and carefully inspect all parts. Renew parts as required and reassemble using all new "O" rings, seals and nuts (17, 22 and 31). To reassemble the unit, proceed as follows:

Drive new seal (30) into housing bore from rear until seated against shoulder and with lip of seal facing forward. Place rear bearing cone and roller assembly in bore and drive rear bearing cup into bore far enough to install snap ring (3) in groove. Then, bump bearing forward so that cup is firmly seated against snap ring. Drive front bearing cone and roller assembly onto gear (1) firmly against shoulder on hub and drive front bearing cup in firmly against shoulder in housing bore. Insert gear hub through housing bore and drive rear bearing cone and roller onto hub so that there is some end play of gear in bearings.

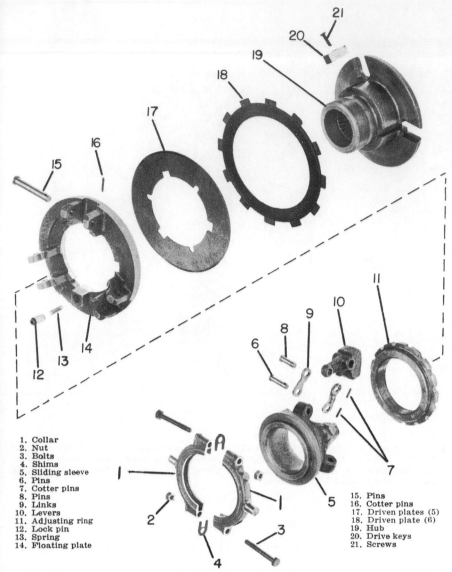

1. Collar
2. Nut
3. Bolts
4. Shims
5. Sliding sleeve
6. Pins
7. Cotter pins
8. Pins
9. Links
10. Levers
11. Adjusting ring
12. Lock pin
13. Spring
14. Floating plate

15. Pins
16. Cotter pins
17. Driven plates (5)
18. Driven plate (6)
19. Hub
20. Drive keys
21. Screws

**Fig. O139 — Exploded view of the multiple disc clutch assembly used in dual speed pto. Refer to Fig. O138 for complete pto assembly.**

Place "O" ring (5) in groove on hub, lubricate lip of seal and loosely install nut (31). Check the torque required to rotate gear in bearings (be sure that nut turns in seal and not on shaft); this value will represent seal drag. Then, tighten nut until an additional 5 to 10 inch-pounds torque is required to rotate gear in bearings and stake nut to slot in gear hub.

Drive bearing (37) into housing bore and install bearing retaining snap ring (32). If removed, drive oil collector locating pin (44) into hole in housing and make sure that oil collector will slide freely on pin. Place oil collector on shaft (50) with small hub against shoulder (PTO brake friction surface forward) and place PTO brake disc (42) on shaft next to oil collector. Insert shaft through bearing (37), align hole in oil collector

with pin (44) and bump shaft forward until PTO brake disc is held tightly between inner race of bearing and shoulder on shaft. Check to be sure oil collector will slide back and forth between shoulder on shaft and PTO brake disc. Place rear pinion (13) (18 teeth) on shaft with either side forward. Place front pinion (14) (28 teeth) on shaft with long hub forward (flat side against rear pinion). Install bearing (16) and nut (17) on shaft, tighten nut to a torque of 125-150 Ft. Lbs. and stake nut to shaft.

NOTE: In some cases on Series 1650 tractors, prior to tractor serial number 166 595-000, the pto output shaft continued to turn with the pto clutch disengaged. This is the result of an excessive flow of lubricant to the pto clutch causing the clutch plates to drag.

This condition can be corrected by drilling

four 3/16-inch holes equidistantly around shaft 3-1/16 inches from rear (unthreaded) end of shaft; or, installing a new shaft having the holes. Shaft having holes were installed in Series 1650 tractors at serial number 166 595-000. If shaft is drilled, be sure to remove any burrs or sharp edges.

If expansion plug (28) was removed from gear (27), install new plug. Drive rear bearing cone and roller assembly firmly against shoulder on hub of gear (26) and drive bearing cups into carrier (23) until firmly seated. Insert gear through bearing cups and carrier bore and drive front bearing cone and roller assembly onto gear hub so that there is some end play of gear in bearings. Install nut (22) and tighten nut to preload bearings so that a torque of 5 to 10 inch-pounds is required to rotate gear in bearings. Stake nut to slot in gear hub when bearing preload is properly adjusted. Drive new seal (29) into rear hub of gear with lip of seal to rear (out), and lubricate seal lip.

Carefully install carrier and gear assembly onto dowel pins (26), bearing (16) and hub of rear gear (1) taking care not to damage seal in gear (27). Tighten the carrier retaining cap screws securely.

NOTE: Foregoing information pertains to the dual speed pto assembly. The single speed 540 or 1000 RPM pto can be converted to the dual speed pto by changing gears, carrier and spacer (15—Fig. O138). Refer to paragraph 230A.

230A. **PTO CONVERSION.** To convert the single speed 540 RPM pto to dual speed pto, remove single speed carrier and replace pinion spacer (15—Fig. O138) with the 1000 RPM pinion (14). Install rear oil seal (30) and if used, remove cover (115) and snap ring (114) from front side of 540 RPM gear (1). Install front oil seal (29) and expansion plug (28) in the 1000 RPM gear (27), then install gear (27) and bearings (2 and 4) in dual speed carrier (23). Secure gear in carrier with nut (22). Install gear and carrier assembly on front of pto housing.

To convert the single speed 1000 RPM pto to dual speed pto, remove carrier (23) assembly and replace pinion spacer (15) with 540 RPM pinion (13). Install 540 RPM gear (1), bearings (2 and 4), snap ring (33), "O" ring (5) and rear seal (30) and secure assembly in pto housing with nut (31). Install oil seal (29) in front of 1000 RPM gear (27) and install gear and carrier assembly on pto housing.

# HYDRAULIC LIFT SYSTEM
# (INTERNAL VALVE)

The hydraulic system for all Series can be any of the combinations given below.

A. Hydra-Lectric Stop, or Depth Stop, internal valve with 3-point hitch, draft control and remote cylinders.

B. Hydra-Lectric Stop, or Depth Stop, internal valve with remote cylinders.

C. Depth Stop, external valve, with remote cylinders.

The service information in this section is based on the internal valve systems having a 3-point hitch with draft control and valving to operate two remote cylinders.

For information on the Depth Stop external valve system, refer to paragraph 265.

**231. TROUBLE SHOOTING.** The following are troubles and causes which may be encountered when operating the hydraulic system. Also, refer to schematic diagram of the hydraulic circuit in Fig. O140 and to the cross-sectional view of the Hydra-Lectric control unit in Fig. O141. When tests indicate that trouble is in the electrical portion of the hydraulic lift system, refer to paragraph 232.

**INTERNAL OIL LEAKS.** Could be caused by: Defective pump mounting gasket, pump shaft seal or housing gasket.

**UNIT OVERHEATS.** Could be caused by: Unit overloaded. Relief valve improperly adjusted. Low oil level, oil of wrong viscosity being used or oil contaminated.

**NOISY UNIT.** Could be caused by: Worn or damaged pump or pump drive gear. Internal high pressure leaks. Low oil level or oil too heavy. Oil filter plugged. By-pass valve plugged or defective manifold.

**OIL FOAMS OUT OF BREATHER.** Could be caused by: Oil level too high or too low. Oil of wrong viscosity.

**UNIT WON'T LIFT.** Could be caused by: Pressure relief valve improperly adjusted or defective. Pump or pump drive failure. Pump not primed. Dirty or faulty servo valve. Bent or broken internal linkage. Low oil level or excessive leakage. Broken housing, cylinder, piston or defective piston seal. By-pass valve leaking.

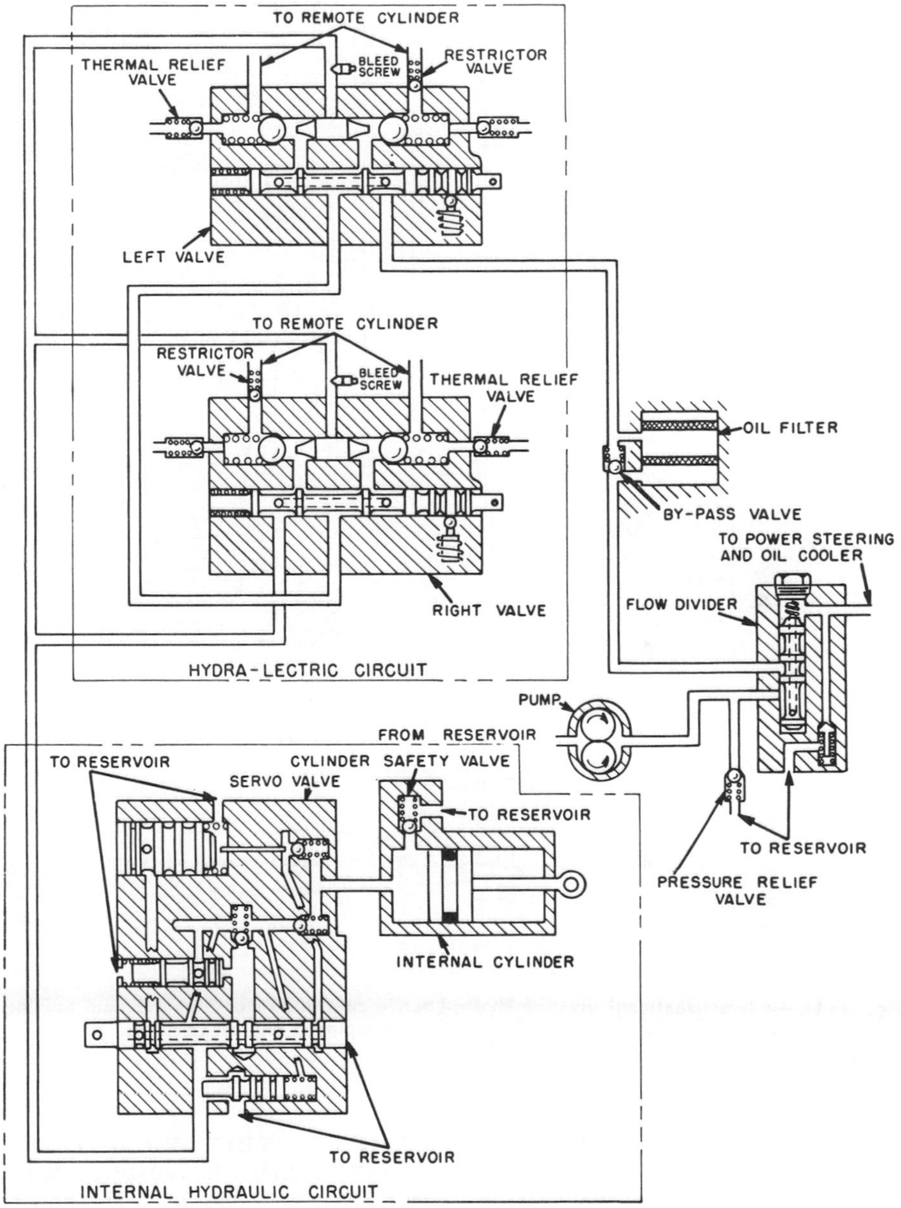

Fig. O140—Schematic diagram of Series 1600 hydraulic circuit which includes Hydra-Lectric valves for two remote cylinders and draft control for the 3-point hitch. Other series are similar except that the oil filter is located between the pump and flow divider valve and servo valve is slightly different.

Faulty remote cylinder. Mechanical failure of lifting mechanism parts.

**UNIT WILL NOT LOWER.** Could be caused by: Servo valve pilot spool sticking. Lowering valve worn or sticking. Lowering valve check valve not opening.

**RELIEF VALVE BLOWS AFTER LIFTING.** Could be caused by: Im-

proper adjustment of unit. Servo valve pilot spool not returning to neutral.

**UNIT GOES OUT OF ADJUST-MENT.** Could be caused by: Broken linkage spring inside pilot spool in servo valve. Loose servo valve mounting cap screws. Draft control linkage broken or worn. Lifting mechanism worn or loose.

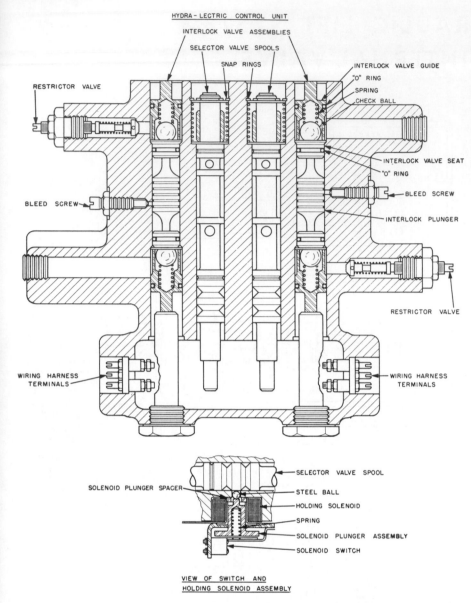

Fig. O141 — Cross-sectional view of Hydra-Lectric control unit and the solenoid holding switch assembly. Refer to Fig. O159 for exploded view of the unit.

**ERRATIC DRAFT CONTROL OPERATION.** Could be caused by: Servo valve sticking. Worn or binding linkage or lift mechanism. Unit out of adjustment. Broken or loose draft control spring.

**RELIEF VALVE BLOWS WHEN REMOTE CYLINDER IS EXTENDED.** Could be caused by: Control lever being held too long in up position. Selector valve sticking. Shorted "up" switch. Contact arm of "up" switch not opening. Hand control lever not returning to neutral due to defective or maladjusted solenoid or solenoid switch, valve centering springs or remote cylinder switches.

**RELIEF VALVE BLOWS WHEN**

**REMOTE CYLINDER IS CON-TRACTED.** Could be caused by: Control lever held too long in "down" position. Selector valve sticking. Defective or maladjusted "down" switch. Remote cylinder friction collar improperly adjusted. Hand control lever not returning to neutral due to defective or maladjusted solenoid or solenoid switch, valve centering springs or remote cylinder switches.

**RELIEF VALVE BLOWS DURING REMOTE CYLINDER EXTENDING STROKE.** Could be caused by: Excessive load. Restricted piston stroke. Binding lift linkage. Couplings improperly connected. Restrictor valve stuck or passageway plugged.

**RELIEF VALVE BLOWS DUR-ING REMOTE CYLINDER CON-TRACTING STROKE.** Could be caused by: Restrictor valve improperly adjusted, stuck or oil line blocked. Couplings improperly connected. Bleed screw not properly adjusted when single acting cylinder is used.

**UNIT WILL NOT HOLD LOAD.** Could be caused by: Fittings or couplings leaking. Thermal relief valve leaking. Cylinder scored or worn or piston seal leaking. Interlock seals or seats leaking. Oil passage connectors leaking.

**ERRATIC REMOTE CYLINDER OPERATION.** Could be caused by: Loose or corroded electrical connections. Defective or bent cylinder stop collar. Bent "up" switch contact arm. Defective solenoid assemblies. Weak or broken valve centering spring.

**REMOTE CYLINDER STOP COL-LAR FAILURE.** Could be caused by: Excessive air gap between collar and magnet. Cylinder solenoid not energized.

## ELECTRICAL TESTS

Electrical malfunctions fall into two classes; a dead circuit and a shorted circuit. When making electrical tests, determine first which kind of malfunction is involved.

It is necessary to have a test light assembly to perform the electrical tests and if one is not available, make one up as follows: Use two lengths of insulated wire and attach battery clips to one end of each wire. Place a 12-volt bulb and socket in series in one of the two wires. On remaining ends of the two wires, affix stiff, insulated probes.

Before initiating any electrical tests, be sure the trouble is not caused by a dead battery, blown fuses, loose or corroded connections or broken wiring. Use wiring diagram shown in Fig. O142 as a guide.

**232. TESTING ELECTRICAL CIR-CUITS.** Before starting test, remove hood and instrument panel cover. Disconnect all wire connectors except the connector for the fuse holder in the instrument panel support. Unplug wiring harness from Hydra-Lectric control unit and the cables from remote cylinders.

Attach test lamp assembly to battery, remove right hand fuse holder cap and remove the fuse from cap. Put fuse back into holder, place one probe on exposed end of fuse and the other on the Hydra-Lectric terminal of ignition switch. If bulb lights, circuit is satisfactory from ignition switch to fuse.

Replace fuse cap and fuse in holder, then move same probe that was used on fuse to the double connector located below the hand control levers. If the bulb lights the circuit between ignition switch and the double connector is satisfactory.

Check wiring and switches in each hand control lever by placing one probe on red wire and the other on black wire. Light should come on when switch is depressed. Now move the probe from the black wire to the white wire (other probe still on red wire). The light should come on but will go off when the switch is depressed. If both these conditions are met, the wiring and switches in hand control levers can be considered satisfactory.

Check the wiring, solenoids and switches in the Hydra-Lectric unit by placing a probe on each of the solenoid terminals on right side of tractor. Bulb should light when left hand control lever is pushed forward or rearward. Note: The solenoid switches are in off position when levers are in neutral position and bulb will not light. Repeat this operation for opposite side of tractor, using other control lever, if unit utilizes two remote cylinders. If bulb does not light as already indicated, remove the Hydra-Lectric control unit and check the solenoids and switches individually for complete circuits. Bulb should light when probes are placed on solenoid leads.

Test continuity of Hydra-Lectric wiring harness as follows: Place one probe on black pig tail and the other on the "B" terminal of the male plug. Bulb should light. Place one test probe on white pig tail and the other on "LW" terminal of 90 degree plug. Bulb should light. Test "G" and "W" circuits by placing a probe on corresponding letters of each plug. Bulb should light on both tests.

Test continuity of remote cylinder wiring harness by placing a test probe on same lettered terminals of each plug (W, B & G). Bulb should light in each case.

Check circuits of remote cylinder as follows: Remove cylinder head cover. Place one probe on "B" terminal of cable female connector and ground other probe to an unpainted surface of cylinder. Bulb should light indicating solenoid circuit is satisfactory. Move probe from "B" terminal to "G" terminal of cable. Bulb should

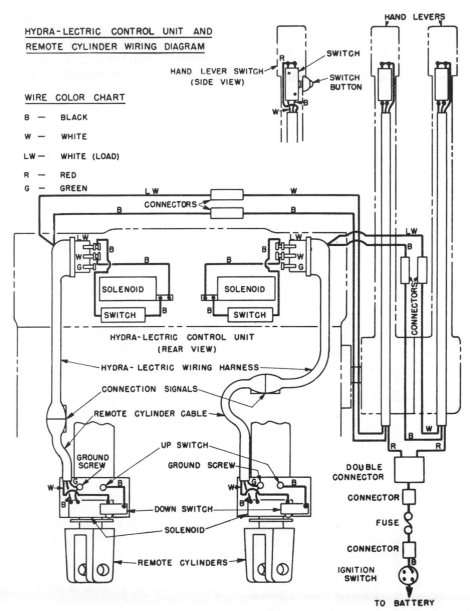

Fig. O142 — Wiring diagram for the Hydra-Lectric control unit and remote cylinders.

light indicating ground circuit is satisfactory. Move probe from "G" terminal to "W" terminal of cable. Bulb should light indicating circuit through "down" switch and "up" switch is satisfactory. Make a further check of down switch by actuating switch several times. Bulb should go off and on each time switch is operated, indicating satisfactory operation of "down" switch. Also move piston rod to the completely extended position at which time the bulb should go off. Now move piston rod in and out a short distance. Bulb should go off and on each time rod moves in and out, indicating that the "up" switch is operating satisfactorily.

If all the conditions outlined under electrical tests are met, the electrical

system can be considered satisfactory. However, if bulb does not react as indicated, the circuit is probably short circuited and a search should be made for loose or corroded connections, or broken, frayed or bare wires.

## LUBRICATION AND MAINTENANCE

NOTE: When performing service or maintenance of any kind on the hydraulic lift system, the practice of cleanliness is of the utmost importance. Pump, valves and cylinders are manufactured to close tolerances and dirt or foreign material is most detrimental and should be avoided at all times.

233. **DRAIN AND REFILL.** To drain the hydraulic system, place lift arms in down position and retract any remote cylinders; then, remove the drain plug as shown in Fig. O143 or Fig. O144, attach hose to opening, start engine and run engine at 1000 RPM until oil is pumped from reservoir. Stop engine as soon as oil stops flowing; DO NOT operate engine when hydraulic system is without oil. If lift system is equipped with remote cylinders, remove hose and move cylinder piston in and out to clear cylinders of oil.

Capacity of hydraulic unit is 20 quarts of oil if equipped with 3-point hitch, or 24 quarts of oil when equipped with Hydra-Lectric remote control only. Add ¾-quart for each 3 inch cylinder, 1½ quarts for each 4 inch cylinder and ½-pint for each 5 feet of hose. Recommended hydraulic fluid is White-Oliver Type 55 hydraulic oil. System should be drained and refilled with new oil each year or after every 1000 hours of operation.

Reconnect any hoses that were disconnected and reinstall drain plug. Fill reservoir to "FULL" mark on dipstick. Start engine and cycle all hydraulic cylinders several times. Recheck fluid level and add fluid, if necessary, to bring level to "FULL" mark on dipstick.

234. **BREATHER.** A micronic element type breather is located on top side of hydraulic housing near the right front corner or, on some models, on left front corner as a part of the oil dipstick assembly. Under normal conditions, the element should be renewed when the hydraulic fluid is being changed (see paragraph 233). In abnormally dusty conditions, more frequent renewal of the element is required.

Renewal of element is evident after removing cap screw or nut from top of the breather assembly. Use new gaskets and crush washers when reassembling breather.

235. **OIL FILTER.** The hydraulic lift system is equipped with a full flow renewable element type oil filter. Filter is located on left side of hydraulic housing on Series 1600 and on front end of housing on Series 1550, 1555, 1650 and 1655. A by-pass valve which opens at approximately 20 psi pressure differential is incorporated in the hydraulic circuit. The by-pass valve will open if the filter becomes plugged. Service of the by-pass valve on Series 1600 hydraulic system re-

Fig. O143 — View showing hydraulic drain plug location on Series 1600 models.

Fig. O144 — View showing hydraulic drain plug location on Series 1550, 1555, 1650 and 1655 models.

quires removal of hydraulic housing from tractor and removal of the lift cylinder from hydraulic housing. On Series 1550, 1555, 1650 and 1655 hydraulic system, bypass valve retaining plug is located just to rear of drain plug location shown in Fig. O144.

Renewal of filter element is evident after unscrewing filter body from hydraulic housing. When reinstalling, place a new "O" ring on filter body, lubricate the "O" ring and tighten the filter body to a torque of 150-200 Ft.-Lbs.

## SYSTEM ADJUSTMENTS

The adjustments in this section will include only those external adjustments which can be accomplished with no disassembly of the hydraulic system. For internal adjustments requiring disassembly, refer to appropriate overhaul paragraph.

236. **LIFT ARM FREE TRAVEL.** Refer to Fig. O145 and proceed as follows: With engine running, move the lift arm (3-point hitch) control lever to fully raised position and allow lift arms to raise to upper limit. If at this point, the lift arm free travel is not within the range of ½ to 1 inch as shown, loosen lock nut and turn linkage adjusting screw as necessary to obtain specified free travel of lift arms after lift arms reach the fully raised position. Shorten linkage to increase or lengthen linkage to decrease the lift arm free travel. Tighten lock nut and recheck.

237. **DRAFT SENSITIVITY.** Refer to Fig. O146. For position control only, loosen lock nut on cap screw in draft signal lever and turn the cap screw in against oil filter (or stop on Series 1550, 1555, 1650 and 1655). When this adjustment is completed, the pin in upper end of draft signal arm attached to lower link support should be against front end of slot in draft signal

link on Utility models, or against rear end of slot in draft signal link on all other models. Do not adjust the cap screw so that pin applies tension or pressure on the draft signal link.

NOTE: Beginning with tractor 160 940-000, the one-piece draft signal link was replaced with a two-piece link. Adjustment procedure remains the same and link is adjusted to position signal lever pin in link slot.

On Series 1550, 1555, 1650 and 1655 and late production Series 1600, lower link supports have three attaching points for the lower (draft) links. For position control only, attach the lower links in the bottom hole in supports on Utility models only, and in the upper hole in supports on all other models.

For draft control, loosen lock nut (Fig. O146) and back cap screw out until there is ¾-1 inch clearance between oil filter (Series 1600) or stop (Series 1550, 1555, 1650 and 1655) and end of cap screw; then tighten the lock nut. Also, on Series 1550, 1555, 1650 and 1655 and late production Series 1600, the lower links must be attached to the middle or upper set of holes in supports on Utility models, or in the middle or lower set of holes on all other models. Attaching the lower links in the upper set of holes on Utility models, or in the lower set of holes on all other models, provides maximum draft sensitivity. Attaching the lower links in the middle set of holes provides medium draft sensitivity on all models. Lower link supports on early Series 1600 have only one hole for attaching the lower links. Be sure by-pass valve is closed for automatic draft control.

238. **RESTRICTOR VALVE.** A restrictor valve is located in one of the remote outlets of each Hydra-Lectric control unit. Refer to cross-sectional view of Hydra-Lectric control unit

Fig. O145 — View showing lift link free travel adjustment.

Fig. O147 — Adjusting the Hydra-Lectric restrictor valve. Refer to text for procedure.

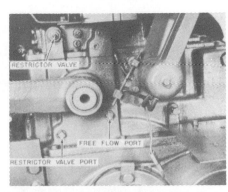

Fig. O149 — View showing location of restrictor valve port and free flow port for Hydra-Lectric control unit on right side of tractor; free flow port is at rear and restrictor port is at front on left side of unit.

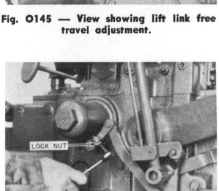

Fig. O146 — View showing draft sensitivity adjustment on Series 1600 models. Adjustment on Series 1550, 1555, 1650 and 1655 models is similar except that screw is adjusted against stop instead of the hydraulic oil filter body.

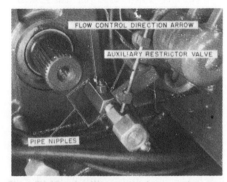

Fig. O148—Auxiliary restrictor valve on models so equipped. Mounting arrangement of valve may vary. Refer to text for procedure.

Fig. O150 — View showing location of by-pass valve on right side of hydraulic housing. Refer to text for adjustment information.

in Fig. O141. The hose attached to a single acting remote cylinder or to the base end of a double acting cylinder must be connected to the restricted outlet.

Location of the restrictor valve adjusting screw for the left Hydra-Lectric circuit is shown in Fig. O147. Adjusting screw for right circuit is located diagonally across the Hydra-Lectric control unit on right side of tractor. Note: Before making any adjustment to the restrictor valve, remove load from remote cylinder.

If bouncing or jerking is encountered or system relief valve blows when a remote cylinder is being contracted, refer to Fig. O147 and adjust the restrictor valve on side of control unit to which the remote cylinder hose is connected. To minimize bouncing or jerking, loosen the lock nut and turn the adjusting screw in (clockwise) to increase the restriction. If too much restriction results in the system relief valve blowing when the cylinder is being contracted, loosen the lock nut and turn the adjusting screw out (counter-

clockwise) as necessary to prevent blowing of the relief valve. Tighten the lock nut after making adjustment.

NOTE: On some Series 1550 and 1650 tractors leakage may occur around threads of restrictor valve adjusting screws. If this problem is encountered, install cap nuts (Oliver part no. 167 764-A) and copper sealing washers (B-234-A) on restrictor valve screws. Cap nuts were installed on production tractors at serial number 207 493-000.

239. AUXILIARY RESTRICTOR VALVE. An externally mounted valve is available for installation when it is desirable to restrict return flow of oil from the rod end of a remote cylinder. See Fig. O148. When installing an auxiliary restrictor valve, be sure that the arrow embossed on valve body points toward the valve port in hydraulic housing and that the valve is installed in the line connected to the free-flow port; see Fig. O149.

When equipped with an auxiliary restrictor valve, the control unit restrictor valve for the base end of the cylinder and the auxiliary restrictor valve for the rod end of the cylinder

should be adjusted for as nearly equal restriction as possible. If restriction is too great, system relief valve will blow when remote cylinder is being actuated.

To adjust the auxiliary restrictor valve, remove load from remote cylinder, refer to Fig. O148 and proceed as follows: Loosen the lock nut and turn the adjusting screw in to increase restriction or out to reduce restriction, then tighten the lock nut. When properly adjusted, the remote cylinder should extend or retract smoothly at engine high idle speed. If relief valve blows, back adjusting screw out to reduce restriction. Refer also to paragraph 238.

240. BY-PASS VALVE. When using a single-acting remote cylinder, the by-pass valve lock nut must be loosened and the adjusting screw backed out five to six turns. Refer to Fig. O150. When using a double-acting remote cylinder or the three-point hitch (draft control), the by-pass valve adjusting screw must be turned in tight.

The by-pass valve is a steel ball (16—Fig. O156) which is retained in

Fig. O151 — Adjusting the interlock bleed screw. Refer to text for adjustment procedure.

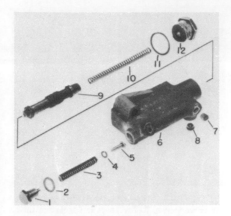

Fig. O152—Exploded view of priority type flow divider valve used on early model tractors. Pressure relief valve can be adjusted by use of shims (4). Shims are 0.0149 thick and will change pressure about 31 psi.

| | |
|---|---|
| 1. Plug | 7. Plug |
| 2. Washer | 8. Insert |
| 3. Spring | 9. Spool |
| 4. Shim | 10. Spring |
| 5. Relief valve | 11. Washer |
|    plunger | 12. Cap |
| 6. Spool body | |

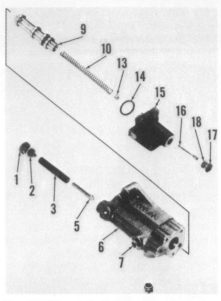

Fig. O152A — Exploded view of demand type flow divider valve used on late model tractors. Demand piston body (15) is sealed to spool body (6) with "O" ring (14) and two small "O" rings which are not shown. Shims (not shown) are available to adjust the relief valve.

| | |
|---|---|
| 1. Plug | 9. Spool |
| 2. Washer | 10. Spring |
| 3. Spring | 13. Retainer |
| 5. Relief valve | 14. "O" ring |
|    plunger | 15. Piston body |
| 6. Spool body | 16. Demand piston |
| 7. Plug | 17. Cap & washer |

the lift cylinder (14) by a snap ring (17). Turning the adjusting screw (18) in tight holds the by-pass valve ball seated.

**241. INTERLOCK BLEED SCREW.** When using a single-acting remote cylinder, the interlock bleed screw (see cross-sectional view of Hydra-Lectric control unit in Fig. O141) on the side of the unit to which the cylinder is connected must be backed out three full turns. When operating a double-acting remote cylinder, the interlock bleed screw, or screws, must be turned all the way in. Fig. O151 shows location of the interlock bleed screw on right side of tractor; interlock bleed screw on left side of tractor is located diagonally opposite on left side of unit.

NOTE: On some series 1550 and 1650 tractors leakage may occur around threads of interlock bleed screws. If this problem is encountered, install new interlock bleed screws which have heads relieved, cap nuts (Oliver part no. 167 764-A) and copper sealing washers (B-234-A). New interlock bleed screws, cap nuts and sealing washers were installed on production tractors at serial number 207 493-000.

**242. SYSTEM OPERATING PRESSURE.** The hydraulic system pressure relief valve should open at 1500-1600 psi for Series 1600, or 2000-2100 psi for Series 1550, 1555, 1650 and 1655. To check system relief pressure, remove the drain plug (refer to Fig. O143 for Series 1600, Fig. O144 for Series 1550, 1555, 1650 and 1655) and insert gage capable of registering sufficient pressure. Start engine and bring hydraulic fluid to normal operating pressure. If equipped with Hydra-Lectric control unit, overload the system by placing a remote lever in a power position.

If not equipped with Hydra-Lectric remote control unit, temporarily shorten the linkage for the 3-point hitch (refer to paragraph 236 and Fig. O145) so that there is no free travel of lift arms and overload the system by placing the control lever in fully raised position. With the engine running at high idle rpm, gage reading should be as stated.

If full flow relief pressure is not within the correct range, record the gage reading and readjust pressure as follows: Remove the relief valve plug (55—Fig. O154) located at right front corner of the hydraulic housing and withdraw the spring (53), shims (52) and plunger (51). Add or remove shims (52) as necessary to obtain correct relief pressure. Each 0.010 shim will change the relief pressure approximately 75 psi. Renew the sealing "O" ring (54) when reinstalling plug (55). Readjust linkage as outlined in paragraph 236, if necessary.

NOTE: At tractor serial number 221 295-000, a pilot relief valve (items 88 through 95 Fig.—O154) replaced previous relief valve. When full capacity of hydraulic system is required on tractors prior to serial number 221 295-000, a package (Oliver part number 30-3040097) can be installed. However, when installing the new pilot relief valve, the restrictor (32—Fig. O154) must be removed on all tractors except those with external valve remote only systems, serial number 212 449-000 and up, which have a relief valve return line.

The pilot relief valve is factory adjusted to 2000 psi and cleaning or

servicing is not recommended. Install a complete cartridge, screen and "O" ring if valve malfunctions.

**243. FLOW DIVIDER VALVE RELIEF PRESSURE.** A relief valve within the flow divider valve assembly (see Fig. O152 or O152A) controls pressure of hydraulic fluid flowing to the power steering system and/or hydraulic oil cooler.

On early Series 1600 models (prior to Serial No. 141 288) equipped with a single vane Hydramotor unit, flow divider valve relief pressure should be 1150 to 1200 psi. On late Series 1600 and all other Series equipped with dual vane Hydramotor, flow divider valve relief pressure should be 1450-1550 psi. On models not equipped with power steering, flow divider relief valve pressure is not of importance as the fluid is merely circulating through the hydraulic oil cooler.

To check flow divider valve relief pressure, remove the plug (7) adjacent to the pressure line to power steering connection and insert a 2000 psi capacity hydraulic gage. With the hydraulic fluid at normal operating temperature and engine running at high idle RPM, turn the front wheels in either direction against stop and observe gage pressure.

To adjust flow divider relief pressure, refer to Fig. O152 or O152A, remove the relief valve plug (1), spring (3) and plunger (5) and add or remove shims (4) as necessary to bring relief pressure within the desired limits. NOTE: Three different pressure relief valve springs (3) have been used in the priority type flow dividers. Early units were equipped with an unpainted spring and were adjusted to 1000 psi; later units were equipped with a blue painted spring and were adjusted to 1200 psi; latest units are equipped with a purple painted spring and are adjusted to 1500 psi, and the latest valve assembly is identified by "1500" stamped on the relief valve plug. If an early unit with unpainted relief valve spring is encountered, power steering performance can be improved by renewing the spring using Field Package (see paragraph 244) containing a blue or purple spring depending on whether the tractor is equipped with a single or dual vane power steering Hydramotor unit. Refer to paragraphs on Hydramotor steering unit in POWER STEERING section of this manual for means of identifying single or dual vane Hydramotor unit. CAUTION: Do not adjust flow divider relief valve pressure higher than 1200 psi when tractor is equipped with single vane Hydramotor unit.

### FLOW DIVIDER VALVE

Hydraulic power for the power steering system is supplied from the hydraulic lift system pump through the flow divider valve. The flow divider valve is bolted to front of lift housing on Series 1600 and to left side of lift housing on all other series. Tractors built prior to March 1968 were equipped with a priority type (one-piece housing) valve whereas all later models were equipped with a demand type (two-piece housing) valve.

The priority type valve diverts 5 gpm of the hydraulic lift pump output to steering and oil cooler circuit before diverting the balance of oil flow to the hydraulic lift unit. Steering system pressure is maintained at 1500 psi by the relief valve.

The demand type valve provides two controlled flow rates. When steering effort is at minimum and pilot flow is occurring, oil is metered through holes in flow control spool at about 3 gpm and pressure shifts spool so that oil in excess of 3 gpm is passed to the hydraulic lift system.

When steering effort is required, the flow divider valve goes into de-

mand position. The increased steering system pressure is communicated to the demand piston which is actuated to compress spool spring and shift spool. This restricts port to hydraulic lift system and increases the volume of oil to the steering system to 5 gpm. Steering system pressure is maintained at 1500 psi by the relief valve.

**244. R&R AND OVERHAUL. (PRIORITY TYPE).** Refer to Fig. O152 and proceed as follows: Disconnect the pressure tube from the flow divider valve, then unbolt and remove the valve assembly from front end of hydraulic housing on Series 1600 models, or from left side of hydraulic housing on Series 1550 and 1650 models.

Unscrew plug (12) and withdraw spring (10) and flow divider spool (9). Spool should slide smoothly in its bore and be free of any nicks or burrs. Free length of spring (10) is $4\frac{25}{32}$ inches and spring should exert 13 to 15 pounds force when compressed to a length of $3\frac{3}{8}$ inches. Renew spring if it does not meet specifications or any defects are noted. Lubricate the valve spool and reinstall spool and spring. Reinstall plug (12) with new crush washer (11).

Unscrew plug (1) and withdraw spring (3), shims (4) and plunger (5). If spring is unpainted, it should be renewed using a Field Package, Oliver Part No. 157 958-AS if tractor is equipped with a single vane Hydramotor steering unit or Oliver Part No. 158 315-AS if tractor is equipped with a dual vane Hydramotor steering unit. The Field Package includes a new hardened relief valve seat, a new blue coded spring (in Part No. 157 958-AS) or purple coded spring (in Part No. 158 315-AS) and adjusting shims. Check the relief valve spring against the following values:

**Part No. 157 875-A (Painted Blue):**

Free length.................$2\frac{27}{32}$ in.
Lbs. pressure @ $2\frac{1}{4}$ in..........59

**Part No. 157 953-A (Painted Purple):**

Free length.................$2\frac{3}{4}$ in.
Lbs. pressure @ $2\frac{1}{4}$ in..........74

The relief valve seat is serviced only in the Field Packages mentioned in previous paragraph or as an assembly with the valve body (6). When valve is reassembled and reinstalled on tractor, check relief pressure as outlined in paragraph 243 and adjust by adding or removing shims (4) as necessary. Shims are 0.0149 thick and each shim will change pressure approximately 31 psi.

**244A. R&R AND OVERHAUL (DEMAND TYPE).** To remove flow divider valve, disconnect oil line, then unbolt and remove flow divider valve and "O" rings from lift housing.

To disassemble, refer to Fig. O152A and remove plug (1) and crush washer (2), spring (3), valve plunger (5) and any shims which may be present. Unbolt demand body (15) from spool body (6) and remove spring (10) and spool (9). Remove cap (17), crush washer (18), piston (16) and retainer (13) from demand body. Oil line insert can be removed with a thread tap if necessary.

Spool (9) and piston (16) should be a snug fit and slide smoothly in their bores and should be free of nicks or burrs. Spring (10) is natural color, has a free length of $4\frac{25}{32}$ inches (approx.) and should test 13-15 lbs. when compressed to a length of $3\frac{3}{8}$ inches. Relief valve spring (3) is color coded purple, has a free length of $2\frac{3}{4}$ inches (approx.) and should test 69-79 lbs. when compressed to a length of $2\frac{1}{4}$ inches. Check seating surfaces of relief valve plunger (5) and its seat in body. Relief valve seat and body are not available separately. The addition, or subtraction of one shim (if used) will change the relief valve pressure approximately 31 psi.

Assemble flow divider valve as follows: Place piston (16) in demand body (15), then install cap (17) with new crush washer (18). Place retainer (13) in demand body with hole in retainer over end of piston. Install the small "O" rings in body (6), then install spool (9) and spring (10). On early flow dividers install "O" ring (14) in bore of body (6); on late flow dividers, install "O" ring (14) on pilot of demand body (15), then install demand body on spool body and be sure retainer (13) is correctly seated in spring (10). Install relief valve plunger (5), spring (3), any shims which are present, and cap (1) with new crush washer (2).

Use new "O" rings and attach flow divider to lift housing. Connect oil line and check flow divider as outlined in paragraph 243, if necessary.

### PUMP (GEAR TYPE)

**245. REMOVE AND REINSTALL.** On tractors prior to serial number 191 121-000, the hydraulic pump and the pump manifold are removed as a unit as follows: Remove the hydraulic lift unit as outlined in paragraph 253. Remove the cylinder and piston assembly, then remove the two cap screws which retain manifold to hydraulic

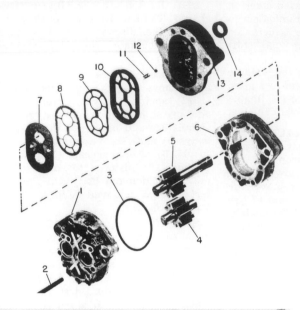

Fig. O153 — Exploded view of the hydraulic pump assembly. Check ball (12) and spring (11) are used on early Series 1600 pump assembly only and pump on models with Depth Stop hydraulic system.

1. Rear plate
2. Capscrews
3. "O" ring
4. Idler gear
5. Drive gear
6. Pump body
7. Diaphragm
8. Backup gasket
9. Protector gasket
10. Diaphragm seal
11. Spring
12. Check ball
13. Front plate
14. Shaft seal

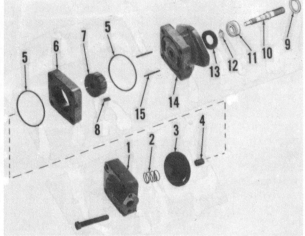

Fig. O153B — Exploded view of vane type pump used to replace the gear type pump used for tractor hydraulic lift system.

1. Cover
2. Spring
3. Plate
4. Bushing
5. "O" rings
6. Pump ring
7. Rotor
8. Vanes
9. Snap ring
10. Pump shaft
11. Bearing
12. Snap ring
13. Oil seal
14. Pump body
15. Dowels (pins)

housing and the two nuts from pump mounting studs. Lift pump and manifold from hydraulic housing.

**246. OVERHAUL.** With the pump and manifold removed as outlined in paragraph 245, remove cap screws and separate pump and manifold. Refer to Fig. O153 and proceed as follows: Place a scribe mark across front plate, body and back plate to insure proper assembly. Place pump in a vise and remove cap screws (2) which hold pump together. Remove pump from vise and while holding pump in hands, bump drive shaft against wood block and separate pump. If body (6) remains with either front plate (13) or back plate (1), remove by tapping with a soft faced hammer, or by installing drive shaft in bearing and bumping shaft. Removal of diaphragm (7), spring (11), ball (12), gaskets and oil seal (14) is obvious.

Clean all parts and inspect as follows: Check pump shafts for wear or scoring and if diameter of shafts at bearing area is less than 0.685, renew

shafts and gears as they are not available separately. Measure gear width and if less than 1.107, for Series 1600 and 1650, or 1.330 for Series 1550, renew shaft and gear assembly. Measure inside diameter of bearings and if greater than 0.691, renew complete front or back plate as bearings are not available separately. Bearing should be flush with pattern islands. If wear or scoring in face of back plate exceeds 0.0015, renew back plate. Measure gear pockets of body (16) and if diameter exceeds 1.719, renew body.

When reassembling, use of repair kit which contains new diaphragm, diaphragm seal, back-up gasket, protector gasket, shaft oil seal, back plate "O" ring, check ball and spring is recommended. Note: Check ball and spring are not used in Series 1550 and 1650 pump or in late Series 1600 pump. Also, install new manifold "O" rings which are not included in the repair kit.

Lubricate all parts with oil and reassemble pump as follows: Install

new shaft seal in front plate with lip of seal towards inside of pump. Using a dull tool, tuck diaphragm seal into grooves in front plate with "V" section of seal down. Press the protector gasket, and then the backup gasket into diaphragm seal. These gaskets must be pushed as far down into the grooves as possible and be completely inside the seal lip. On early pump only, drop steel ball into seat in suction side of front plate and place spring on top of ball; spring and ball are not used in late pump assemblies. Place diaphragm, with bronze side up, on top of backup gasket so that smaller hole in diaphragm is over the ball and spring, or on late pumps, towards suction side of front plate. Press diaphragm down into seal and be sure seal lip completely encircles diaphragm.

Dip the gear assemblies in oil and install through diaphragm and front plate; take care not to damage shaft seal when installing drive gear. Apply a thin coat of heavy grease to milled surfaces of body and front plate and install body over the gears with side having smallest port cavity openings toward the front plate. Place large "O" ring in the irregular groove in back plate, apply a thin coat of heavy grease to the "O" ring and milled surfaces of back plate and pump body, then install back plate. Tighten the retaining cap screws to a torque of 27-30 Ft.-Lbs. Install manifold with new "O" rings and tighten the manifold retaining cap screws to a torque of 23-26 Ft.-Lbs.

After pump is assembled, torque required to turn pump drive shaft should not exceed 3½ Ft.-Lbs. and after pump is installed in hydraulic housing, torque required to turn pump drive shaft should not exceed 4 Ft.-Lbs.

### PUMP (VANE TYPE)

NOTE: Beginning with 1650 tractor serial number 191 121-000, all models are equipped with a vane type hydraulic pump instead of the gear type pump previously used. At tractor serial number 208 960-000, the vane type pump was replaced with a new gear type pump. Conversion packages are available to convert from vane to gear type pumps.

246B. **R&R AND OVERHAUL.** To remove pump, first remove hydraulic lift unit as outlined in paragraph 253, then remove cylinder and piston assembly. Disconnect outlet line from pump. Remove pump mounting cap screws and remove pump from lift housing.

With pump removed, remove inlet pipe, place scribe mark across pump, then remove the four cover cap screws and separate pump cover (1—Fig. O153B) from pump ring (6). Lift spring (2), plate (3), "O" ring (5), ring (6) and dowels (15) from body (14). Remove vanes (8) from rotor (7), then remove snap ring (12) and lift rotor from shaft (10). Remove snap ring (9), then remove shaft (10), bearing (11) and seal (13) from pump body.

Clean and inspect all parts. Bushing (4) is available separately. Inspect "O" rings (5), ring (6), rotor (7) and vanes (8) and if any of these parts are faulty, they must all be renewed and are available only as a kit. NOTE: A seal kit is also available which includes seal (9) and two "O" rings (5). Inspect balance of parts and renew as necessary.

Reassemble by reversing disassembly procedure and be sure to lubricate parts and align the previously affixed scribe line. The arrow embossed on pump ring must point in the direction of pump rotation.

## PUMP (GEAR TYPE)

246C. R&R AND OVERHAUL. With pump removed as outlined in paragraph 246B, proceed as follows: Remove pump retaining cap screws and separate pump by bumping drive shaft on a block of wood. If body (4—Fig. 153C) remains with either front plate (12) or back plate (1), remove it by placing drive shaft in bearing and bumping end of shaft with a soft faced hammer. Remove diaphragm (7), gaskets (8 and 9) and seal (11) from front plate. Remove seal (13) if necessary. Thrust plate (2) can be removed after pump gears are out.

Clean all parts and inspect as follows: Check pump gear shafts for wear or scoring and if diameter of shafts at bearing area is less than 0.810, renew shafts and gears. Check gear width and if width is less than 0.953, renew shafts and gears. Gears and shafts are not available separately. Measure, inside diameter of bearings and if greater than 0.816, renew complete front or back plate as bearings are not available separately. Renew body if diameter of gear pockets exceeds 2.002.

When reassembling pump, use all new seals, gaskets and "O" rings. Reassemble pump as follows: Install seal (13) in front plate with lip of seal toward inside. Lubricate seal lip with a small amount of heavy grease. Work diaphragm seal (10) into place

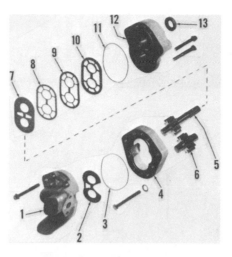

Fig. O153C—Exploded view of latest gear type hydraulic pump. Refer to text for installation of items 2 and 7.

| 1. Back plate | 8. Back-up gasket |
| 2. Thrust plate | 9. Protector gasket |
| 3. "O" ring | 10. Diaphragm seal |
| 4. Body | 11. "O" ring |
| 5. Drive shaft & gear | 12. Front plate |
| 6. Idler shaft & gear | 13. Oil seal |
| 7. Diaphragm | |

in front plate with open part of "V" section down and be sure inner lip of seal is not rolled. Install protector gasket (9), backup gasket (8) and diaphragm (7) in diaphragm seal and be sure they are under lip of diaphragm seal. If necessary, use a dull tool to work these parts into position.

NOTE: Diaphragm (7) is installed with bronze side toward pump gears and with the smallest of the two center cut-outs toward inlet (suction) side of pump. Surface of diaphragm should be flush with face of front plate when correctly installed.

Dip gear assemblies in oil and install in front plate but be careful not to damage shaft seal (13). Position body (4) so larger port openings will be toward back plate, install "O" ring (11) in groove of body, then apply a thin coat of grease to faces of front plate and body. Place body over gears and tap body with a soft faced hammer until it seats against front plate. Place thrust plate (2) over end of gear shafts with bronze side next to gears and the side with cut-out toward inlet (suction) side of pump. Thrust plate must fit in cut-out of body and be flush with rear side. Place cap screw with crush washer in mounting ear of body. Place "O" ring (3) in groove of body, then apply a thin coat of grease to faces of body and back plate (1). Place back plate over gear shafts and tap with a soft faced hammer until back plate seats against body. Install pump retaining cap screws and tighten to a torque

of 75-85 Ft.-Lbs. Check pump rotation which should require 3½ Ft.-Lbs. to turn drive shaft. If pump shows undue rotational resistance, recheck pump assembly.

## HYDRAULIC PUMP DRIVE SHAFT AND PINION

247. The hydraulic pump is driven from an idler gear which is in turn driven from a pinion gear mounted in the wall which separates the final drive and transmission compartments. This internally splined drive pinion (22—Fig. O120) is driven by the pto shaft on tractors so equipped, or by a separate shaft (67—Fig. O154) on those tractors having no power take-off. Refer to PTO section for information on the PTO shaft and to paragraph 249 for information on pinion gear.

248. R&R PUMP DRIVE SHAFT. Remove cap screws that retain bearing carrier (72—Fig. O154) to rear frame or rear frame cover and pull carrier and shaft assembly from tractor rear frame. Remove snap ring (68) and pull shaft and bearing from retainer (72) then remove snap ring (70) from rear end of shaft and remove bearing (69) from shaft.

When reinstalling shaft assembly in tractor, it may be necessary to engage starter momentarily to engage splines of shaft and flywheel.

249. R&R PUMP DRIVE PINION. Remove hydraulic housing as outlined in paragraph 253. Remove the PTO shaft (single speed) as outlined in paragraph 221, or on dual speed, as outlined in paragraph 226, or on models having no PTO, remove hydraulic pump drive shaft as outlined in paragraph 248. Remove snap ring from rear side of pinion bearing (24—Fig. O120) and bump pinion and bearing rearward out of housing wall. Remove snap ring (25) from rear of pinion hub and press bearing from pinion (22).

When reinstalling, lubricate bearing and install pinion gear with long hub on rear side.

## HYDRA-LECTRIC CONTROL UNIT

The Hydra-Lectric control unit serves as a top cover for the hydraulic unit and contained within it are the control valve spools, interlock assemblies, restrictor valves, interlock bleed screws and the Hydra-Lectric switches and solenoids. Refer to Fig. O159 for an exploded view of the unit and to Fig. O141 for cross-sectional view.

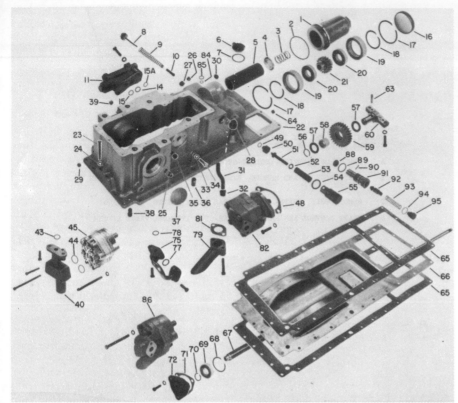

**Fig. O154—Exploded view of hydraulic housing used on 1650 models. Series 1550, 1555 and 1655 hydraulic housing is similar. Series 1600 hydraulic housing has oil filter located on left side of housing and flow divider is located on front side of housing. Refer to Fig. O159 for an exploded view of the Hydra-Lectric Stop control unit. On models not equipped with 3-point hitch, rockshaft holes are closed with plugs (37).**

| | | | |
|---|---|---|---|
| 1. Filter body | 22. Housing | 49. Washer | 71. Gasket |
| 2. "O" ring | 25. Plug | 50. Seat | 72. Retainer |
| 3. Spring | 26. Plug | 51. Relief valve | 75. Line |
| 4. Retainer | 27. Plug |     plunger | 77. "O" ring |
| 5. Filter element | 28. Plug | 52. Shim | 78. "O" ring |
| 6. Cap | 29. Plug | 53. Spring | 79. Inlet pipe |
| 7. Gasket | 30. Insert | 54. "O" ring | 81. Gasket |
| 8. Pin | 31. Nipple | 55. Retainer | 82. Pump (vane) |
| 9. Spring | 32. Restrictor | 56. Snap ring | 84. Ring |
| 10. By-pass plunger | 33. Crush washer | 57. Washer | 85. "O" ring |
| 11. Flow divider | 34. By-pass screw | 58. Bearing | 86. Pump (gear) |
|     valve | 35. Plug | 59. Gear | 88. Screen |
| 14. "O" ring | 36. Dowel | 60. Support | 89. "O" ring |
| 15. "O" ring | 37. Plug | 63. Dowel | 90. Dowel |
| 15A. "O" ring | 38. Dowel | 64. Plug | 91. Cartridge |
| 16. Plug | 39. Plug | 65. Gasket | 92. Valve |
| 17. Snap ring | 40. Manifold | 66. Reservoir | 93. Spring |
| 18. Snap ring | 43. "O" ring | 67. Drive shaft | 94. Washer |
| 19. Carrier | 44. "O" ring | 68. Snap ring | 95. Plug |
| 20. Bearing | 45. Pump (gear) | 69. Bearing | |
| 21. Pinion | 48. Gasket | 70. Snap ring | |

**250. R&R AND OVERHAUL.** Remove the seat and tool box, then unbolt and remove the Hydra-Lectric control unit from hydraulic lift housing.

Remove the interlock valve guide retainers (49—Fig. O159) and turn unit bottom side up. Identify and disconnect the wires from terminals in housing. Remove solenoid cap (71), then disconnect the wires from switches and remove switches from bracket and tube assembly (61). Remove the solenoid springs (70) and the solenoid plungers (67). Overtravel springs are riveted to plungers and should not be removed. If spring is defective, renew complete plunger assembly. Also, do not remove the small positioning roll pins from sole-

noid spring unless necessary. Remove the bracket and tube assembly (61), solenoids (60) and two small ¼-inch steel balls (68) from their recesses in the housing.

Drive roll pins (11) from inner ends of spools, remove end plate (1), then remove snap rings from spool bores and push spools out rear of housing. Centering spring (8), washers (7) and travel limiter (9) can be removed from spool, if necessary by removing snap ring (6). Grasp stem of rear interlock valve guides (12) with a pair of pliers and pull from housing. Remove spring (15) and check ball (13). Use a soft drift of not more than ½-inch diameter and drive front valve guide (12), spring (15) and check ball (13), valve seat (16)

and plunger (18) forward out of housing. Work from front of housing and bump valve seat (16) from housing.

Restrictor valves and interlock bleed screws can be removed at any time and the procedure for doing so is obvious.

The two (early models) or four (late models) thermal relief valves (56) located in bottom side of cover can be removed and inspected if necessary. Thermal relief valves are preset at the factory to relieve at 4400-5100 psi and are non-adjustable. A faulty valve is renewed as an assembly.

NOTE: Some loader operations cause extremely high pressure within the system and cause repeated opening and closing of the thermal relief valves. Leakage may occur as valves were not designed to withstand frequent cycling.

Heavy duty thermal relief valves are available for installation if this situation occurs. However, because of the larger size of the heavy duty valves, material will have to be ground off the cover to accommodate the left front valve. See Fig. O159C.

If heavy duty valves are disassembled for cleaning or repair, set valve opening pressure to 4375-4625 psi and stake adjusting plug in three places when adjustment is complete.

**251.** Clean and inspect housing bores, valve spools and all other parts for scoring, nicks or other damage. Spools and interlock plungers should be a snug fit in bores, yet slide freely. Check all springs for fractures or other damage. Check wiring, solenoids and terminals for damage. Check switch action and continuity. If deemed necessary, resistance of solenoid can be checked with an ohmmeter. Solenoid should show approximately 4 ohms resistance.

**252.** Reassemble by reversing the disassembly procedure. Coat all working parts with engine oil to provide initial lubrication. Coat all "O" rings and seals with heavy grease to prevent damage during installation and be sure to use new "O" rings, gaskets and seals.

After assembly, be sure to check operation of solenoid switches by actuating valve spools. A click should be heard when spools are moved either way. Be sure that all six passage connectors (55—Fig. O159) are in place before installing control unit to hydraulic housing.

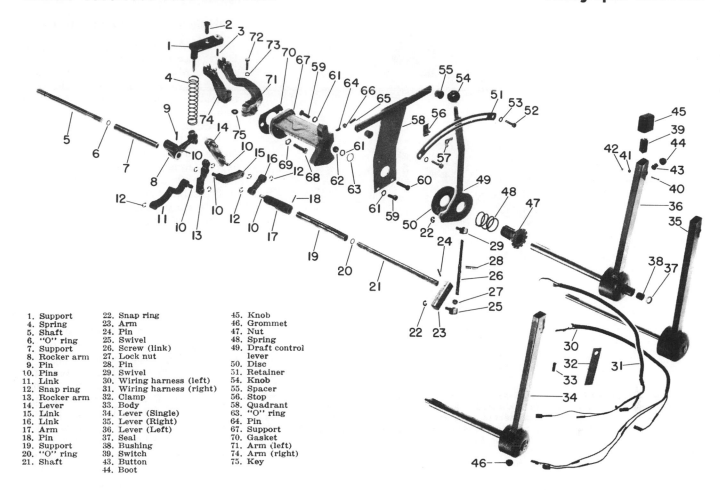

| | | |
|---|---|---|
| 1. Support | 22. Snap ring | 45. Knob |
| 4. Spring | 23. Arm | 46. Grommet |
| 5. Shaft | 24. Pin | 47. Nut |
| 6. "O" ring | 25. Swivel | 48. Spring |
| 7. Support | 26. Screw (link) | 49. Draft control |
| 8. Rocker arm | 27. Lock nut | lever |
| 9. Pin | 28. Pin | 50. Disc |
| 10. Pins | 29. Swivel | 51. Retainer |
| 11. Link | 30. Wiring harness (left) | 54. Knob |
| 12. Snap ring | 31. Wiring harness (right) | 55. Spacer |
| 13. Rocker arm | 32. Clamp | 56. Stop |
| 14. Lever | 33. Body | 58. Quadrant |
| 15. Link | 34. Lever (Single) | 63. "O" ring |
| 16. Link | 35. Lever (Right) | 64. Pin |
| 17. Arm | 36. Lever (Left) | 67. Support |
| 18. Pin | 37. Seal | 70. Gasket |
| 19. Support | 38. Bushing | 71. Arm (left) |
| 20. "O" ring | 39. Switch | 74. Arm (right) |
| 21. Shaft | 43. Button | 75. Key |
| | 44. Boot | |

Fig. O155 — Exploded view of hydraulic control linkage used on models equipped with draft control (3-point hitch) and two Hydra-Lectric control valves for remote cylinders. A dash pot piston (37—Fig. O156) is attached to arm (8). Arms (71 and 74) are connected to the Hydra-Lectric control valves and lever (14) is connected to the servo valve pilot spool.

## DEPTH STOP CONTROL UNIT

### (INTERNAL VALVE)

252A. The Depth-Stop internal valve control unit is shown in Fig. O159A and note that except for housing (19), piston assemblies (34, 35 and 36) and tube assembly (31) the unit is almost identical to the control unit used for Hydra-Lectric-Stop units shown in Fig. O159.

System adjustments for the Depth-Stop internal valve control units are the same as those given for the Hydra-Lectric-Stop units except that any reference to electrical units should be disregarded. Pistons (34) and springs (35) can be removed after cap (36) is off. See Fig. 0159B.

NOTE: In some cases, prior to tractor serial number 212 449-000, the right hand spool shifted through the neutral position and into "Down" position when cylinder reached end of the "Up" stroke. This was caused by back pressure on end of valve spool.

If this problem occurs, install a new spool (Oliver part no. 169 068-A) which has a plug welded in open (rear) end of spool.

## HYDRAULIC HOUSING

The hydraulic housing (22—Fig. O154) contains the hydraulic pump and manifold, servo valve and control lever linkage along with the rockshaft and its operating mechanism. The type of hydraulic system will determine what components are included in the hydraulic housing as well as whether a cover or the Hydra-Lectric control unit shown in Fig. O159 is used as a top cover.

Everything except the hydraulic pump, manifold, piston rod and the piston and cylinder assembly can be serviced without removing the hydraulic unit from tractor.

253. R&R HYDRAULIC HOUSING. Drain hydraulic unit as outlined in paragraph 233. Remove seat and tool box. Disconnect lift links and the upper link. Disconnect the electrical connections. Disconnect hoses and/or piping and plug the openings. Remove platforms and on tractors equipped with wheel guard extensions, also remove the platform splash panels. Disconnect lines from flow divider valve and plug the openings. Clean the entire unit thoroughly, then, on early models, remove the flow divider valve from front end of housing, and to provide a lifting point, install two ⅜ x 3 inch cap screws into the two tapped holes in housing. Place a chain under these two cap screws or, on late models, under the hydraulic filter body and the upper link support at rear, then attach a hoist. Unbolt hydraulic unit from rear frame and lift from tractor. With unit still supported, drive dowels from bottom left front and right rear of hydraulic housing and remove oil pan (66). Now set unit on a bench, or preferably, mount unit on an engine stand if complete overhaul is anticipated.

Reinstall by reversing the removal procedure.

## CONTROL LEVERS AND LINKAGE

**254. R&R AND OVERHAUL.** The Hydra-Lectric and draft control levers assembly can be removed as follows. Right fender must be in outermost position (or removed), then remove the Hydra-Lectric control unit as outlined in paragraph 250. Refer to Fig. O155. Remove the cap screws and keys from control arms (71 and 74) at inner ends of Hydra-Lectric lever shafts and remove control arms. Pull retaining rings (22) from swivels of adjusting link (screw) (26) and remove link from draft control lever (49) and external servo valve control arm (draft control arm) (23). Remove cap screws which retain the lever support (67) to hydraulic housing and pull support and levers assembly from hydraulic housing. Drive roll pin (9) from draft control shaft inner arm (8), pull shaft from support and remove arm. Balance of internal linkage can be removed after removing the actuating rocker arm spring support (1) and spring and servo valve. Draft control shaft support (19) and actuating rocker shaft support (7) can be removed from hydraulic housing if necessary.

When installing new draft control shaft support, drive support into housing until it protrudes about ⅜-inch inside housing. When installing new actuating rocker arm pivot, install with ring groove toward inside and drive pivot into housing until rocker arm (16) operates freely on pivot with the retaining ring installed.

Any further disassembly and/or overhaul of the levers assemblies or linkage is obvious.

### SERVO VALVE

**255. R&R AND OVERHAUL.** To remove and overhaul the servo valve, first drain hydraulic unit, then remove hydraulic housing cover or the Hydra-Lectric control unit as outlined in paragraph 250. Remove the actuating rocker support (1—Fig. O155) and spring, then unbolt and remove servo valve.

NOTE: Production changes have been made in the servo valve. Refer to Fig. O157 for exploded view of early valve and to Fig. O158 for exploded view of a later valve assembly. As valves are basically similar, the following procedures will refer to Fig. O157 only. On late valves, disregard reference to the two check valve units; only one is used as shown in Fig. O158: Be sure to identify the servo valve when ordering parts.

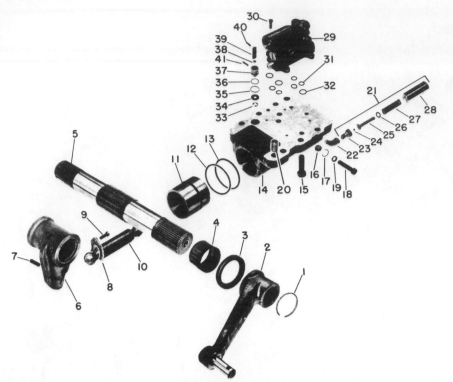

Fig. O156 — Exploded view of rockshaft and rockshaft lift cylinder with related parts. Exploded view of servo valve (29) is shown in Fig. O157 (early) or Fig. O158 (late production). Dash pot piston (37) is connected to rocker arm (8—Fig. O155) with link (39), and works in bore drilled in lift cylinder housing (14). Cylinder safety valve assembly is (21).

| | | | |
|---|---|---|---|
| 1. Snap ring | 11. Piston | 22. Elbow | 33. Snap ring |
| 2. Lift arm (right) | 12. Backup washer | 23. Seat | 34. Seal |
| 3. Seal | 13. "O" ring | 24. Valve ball | 35. Backup ring |
| 4. Bushing | 14. Cylinder | 25. Plunger | 36. "O" ring |
| 5. Rockshaft | 16. By-pass valve ball | 26. Shims | 37. Dash pot piston |
| 6. Arm | 17. Snap ring | 27. Spring | 38. Plug |
| 7. Pin | 18. Adjusting screw | 28. Body | 39. Link |
| 8. Retainer | 19. Lock nut | 29. Servo valve assy. | 40. Pin |
| 9. Piston rod | 20. Dowels | 31. "O" ring | 41. Pin |
| 10. Piston rod | 21. Safety valve assy. | 32. "O" ring | |

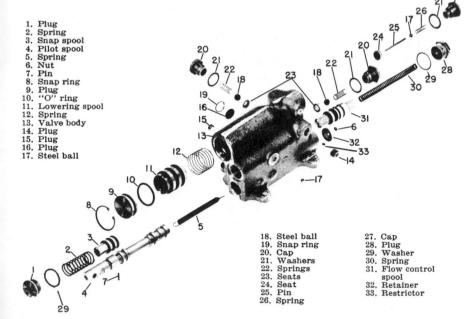

| | |
|---|---|
| 1. Plug | |
| 2. Spring | |
| 3. Snap spool | |
| 4. Pilot spool | |
| 5. Spring | |
| 6. Nut | |
| 7. Pin | |
| 8. Snap ring | |
| 9. Plug | |
| 10. "O" ring | |
| 11. Lowering spool | |
| 12. Spring | |
| 13. Valve body | |
| 14. Plug | |
| 15. Plug | |
| 16. Plug | |
| 17. Steel ball | |

| | | | |
|---|---|---|---|
| 18. Steel ball | | 27. Cap | |
| 19. Snap ring | | 28. Plug | |
| 20. Cap | | 29. Washer | |
| 21. Washers | | 30. Spring | |
| 22. Springs | | 31. Flow control spool | |
| 23. Seats | | 32. Retainer | |
| 24. Seat | | 33. Restrictor | |
| 25. Pin | | | |
| 26. Spring | | | |

Fig. O157—Exploded view of early servo valve (3-point hitch control valve). Later production valve (see Fig. O158) has redesigned snap spool (3) which eliminates need of extra check valve (18) shown at upper left side of valve body (13).

With servo valve removed it can be disassembled as follows: Refer to Fig. O157 and push pilot spool (4) in as far as possible; then tap lightly to bump out retainer (32). Remove pilot spool assembly from body (13), then remove nuts and retainer (32), drive out roll pin (7) and remove spring (5) from pilot spool. Remove plug (1), spring (2) and snap spool (3). Remove snap ring (8), install a cap screw in plug (9) and pull plug. Remove lowering spool (11) and spring (12). Remove the two check valve caps (20), springs (22) and balls (18). Remove cap (27), then remove spring (26), ball (17) and lowering valve pin (25) from body. Remove cap (28), spring (30) and flow control spool (31).

Any further removal of plugs, or seat (24), necessary for inspection of bores and passages is obvious.

Clean all parts and inspect for excessive wear or scoring. Spools should be a snug fit in bores yet move freely. Renew all parts showing excessive wear or damage.

All servo parts except check ball seats, pilot spool, snap spool and body are renewable separately. Renewal of any of these parts will require renewal of complete valve.

Flow control restrictor (33) is a drilled set screw which provides metered oil to rear of flow control spool. This screw is accessible from flow control spool bore but SHOULD NOT be removed. If set screw should come out or be lost, a new one can be obtained separately. Stake screw in position if new screw is installed.

Use new "O" rings and gaskets when reassembling. Lubricate all working parts with engine oil prior to assembly to provide initial lubrication. Use heavy grease on all steel balls and springs so they will stay in position during assembly.

NOTE: Several modifications have been made to the later servo valves to correct hitch lowering when engine is stopped, or excessive cycling of the hydraulic system when engine is running.

On tractors prior to serial number 168 536-000, install a field package, part number 165 686-AS, which includes a spring (27—Fig. O158), part number 160 215-A; a cap (28), part number 160 216-A and a spring guide, part number 165 637-A. Coils at ball end of the spring are close wound to prevent jamming of the ball and spring is color coded green. Cap (28) has a straight bore with a flat spring seat. On tractors between serial num-

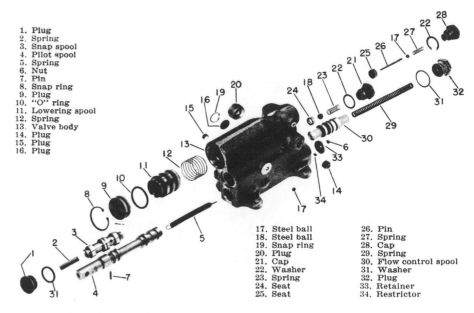

1. Plug
2. Spring
3. Snap spool
4. Pilot spool
5. Spring
6. Nut
7. Pin
8. Snap ring
9. Plug
10. "O" ring
11. Lowering spool
12. Spring
13. Valve body
14. Plug
15. Plug
16. Plug

17. Steel ball
18. Steel ball
19. Snap ring
20. Plug
21. Cap
22. Washer
23. Spring
24. Seat
25. Seat

26. Pin
27. Spring
28. Cap
29. Spring
30. Flow control spool
31. Washer
32. Plug
33. Retainer
34. Restrictor

**Fig. O158—Exploded view of later production servo valve assembly. Refer to Fig. O157 for early valve unit. Lowering spool (11) operates check valve ball (17) via the pin (26). Latest production servo valves have a plug installed instead of items (16) and (19).**

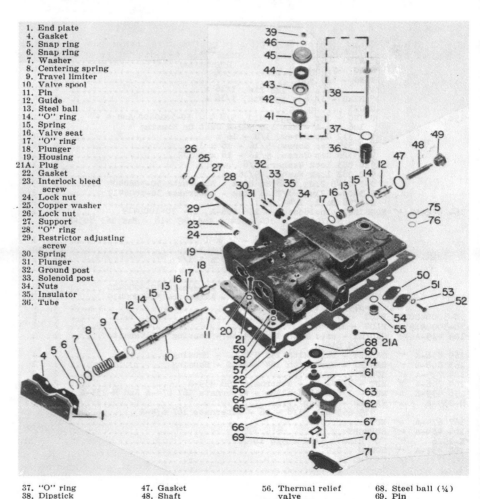

1. End plate
4. Gasket
5. Snap ring
6. Snap ring
7. Washer
8. Centering spring
9. Travel limiter
10. Valve spool
11. Pin
12. Guide
13. Steel ball
14. "O" ring
15. Spring
16. Valve seat
17. "O" ring
18. Plunger
19. Housing
21A. Plug
22. Gasket
23. Interlock bleed screw
24. Lock nut
25. Copper washer
26. Lock nut
27. Support
28. "O" ring
29. Restrictor adjusting screw
30. Spring
31. Plunger
32. Ground post
33. Solenoid post
34. Nuts
35. Insulator
36. Tube

37. "O" ring
38. Dipstick
39. Nut
41. Cap
42. Gasket
43. Cup
44. Element
46. Seal

47. Gasket
48. Shaft
49. Plug
50. Gasket
51. Cover
54. "O" ring
55. Connector

56. Thermal relief valve
60. Solenoid
61. Bracket and tube
62. Switch
66. Wire
67. Plunger assy.

68. Steel ball (¼)
69. Pin
70. Spring
71. Cap
74. Spacer
75. "O" ring
76. Ring

**Fig. O159—Exploded view of Hydra-Lectric-Stop control unit. Refer to Fig. O141 for cross-sectional view of the unit. Unit serves as a top cover for hydraulic lift housing.**

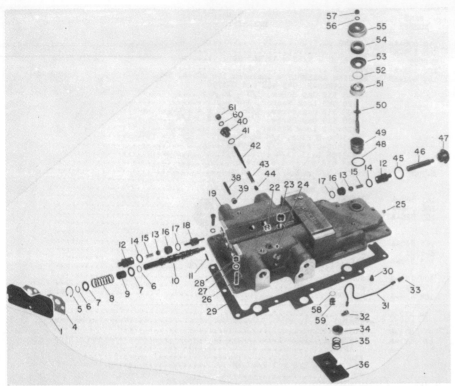

Fig. O159A—Exploded view of the Depth-Stop internal valve control unit. Note that this unit uses hydraulically operated detents rather than electrical as used in the Hydra-Lectric-Stop control unit.

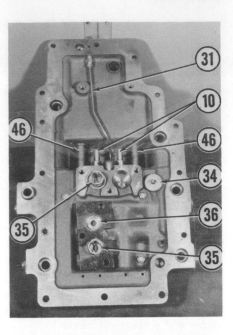

Fig. O159B—Bottom view of Depth-Stop control unit with cap, one spring and one piston removed. Note spring counterbores in cap (36) and detent pin on top side of piston (34).

| | |
|---|---|
| 10. Valve spools | 35. Spring |
| 31. Oil line | 36. Cap |
| 34. Piston | Shafts |

| | | | |
|---|---|---|---|
| 1. End plate | 17. "O" ring | 34. Piston | 47. Plug |
| 4. Gasket | 18. Plunger | 35. Spring | 48. "O" ring |
| 5. Snap ring | 19. Housing | 36. Cap | 49. Tube, filler |
| 6. Snap ring | 22. Plug | 37. Interlock bleed | 50. Dipstick |
| 7. Washer | 23. Plug | screw | 51. Cap |
| 8. Centering spring | 24. Sealing ball | 39. Jam nut | 52. Gasket |
| 9. Travel limiter | 25. Plug | 40. Support | 53. Cup |
| 10. Valve spool | 29. Gasket | 41. "O" ring | 54. Breather element |
| 11. Roll pin | 30. Thermal relief | 42. Restrictor adjusting | 55. Cap |
| 12. Guide | valve | screw | 56. Seal |
| 13. Steel ball | 31. Tube | 43. Spring | 57. Nut |
| 14. "O" ring | 32. Elbow | 44. Plunger | 58. "O" ring |
| 15. Spring | 33. Connector | 45. Gasket | 59. Connector |
| 16. Valve seat | | 46. Shaft | |

Fig. O159C—When installing heavy duty thermal relief valves, grind material from cover as shown. Depth-Stop units (as shown) must also have corner of cap removed.

bers 168 536-000 and 189 768-000, install a spring guide, part number 165 637-A.

Beginning with tractor serial number 189 536-000 and continuing to serial number 191 799-000 a stop pin, part number 165 421-A was added to cap (28) and part number of servo valve was changed from 160 665-AS to 163 389-AS. On tractors between these serial numbers, replace stop pin, part number 165 421-A with spring guide, part number 165 637-A.

When servicing any servo valve that is malfunctioning, the bore for pin (26) should be checked for alignment and fit. Check bore as follows: Use a test pin $\frac{1}{8}$-inch in diameter and at least six inches long. Insert pin in bore and check movement of pin at cap end of valve housing. If end of pin will vary $\frac{1}{16}$-inch, or more, renew the servo valve.

## ROCKSHAFT

256. R&R AND OVERHAUL. The rockshaft can be removed from hy-

draulic housing without removing housing from tractor as follows: Remove the Hydra-Lectric control unit, or cover, as outlined in paragraph 250. To provide removal clearance, remove either fender and the tire and wheel. Refer to Fig. O156. Disconnect lift links, then remove snap rings (1) from outer ends of rockshaft and remove lift arms (2). Use a soft faced hammer and bump rockshaft out of housing and rocker arm (6).

Note: This operation will force oil seal (3) and bushing (4) out on the side rockshaft is being removed from. The seal will be damaged and will require renewal, however, bushing can be salvaged if caution is used. Opposite oil seal and bushing can now be removed, if necessary. Rocker arm can be removed after disconnecting piston rod (10) and the rocker arm feed-back link.

When reinstalling, insert rockshaft from side of housing that has no bushing, align blind splines of rockshaft and rocker arm and slide rockshaft

into position. Use Oliver driver No. ST-144, or equivalent, and install bushing. Lubricate lip of oil seal and coat outer edge with sealant. Slide seal over rockshaft and drive into counterbore.

## ROCKSHAFT CYLINDER AND PISTON

257. R&R AND OVERHAUL. Refer to Fig. O156. The work cylinder and piston assembly can be removed after removing the hydraulic housing as outlined in paragraph 253.

With hydraulic housing and oil pan

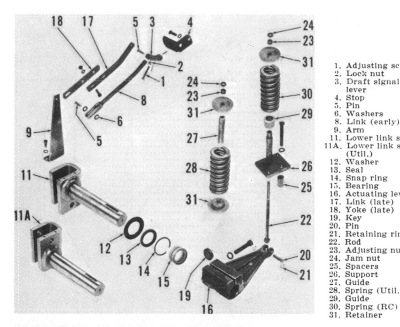

| | |
|---|---|
| 1. | Adjusting screw |
| 2. | Lock nut |
| 3. | Draft signal lever |
| 4. | Stop |
| 5. | Pin |
| 6. | Washers |
| 8. | Link (early) |
| 9. | Arm |
| 11. | Lower link support (RC) |
| 11A. | Lower link support (Util.) |
| 12. | Washer |
| 13. | Seal |
| 14. | Snap ring |
| 15. | Bearing |
| 16. | Actuating lever |
| 17. | Link (late) |
| 18. | Yoke (late) |
| 19. | Key |
| 20. | Pin |
| 21. | Retaining ring |
| 22. | Rod |
| 23. | Adjusting nut |
| 24. | Jam nut |
| 25. | Spacers |
| 26. | Support |
| 27. | Guide |
| 28. | Spring (Util.) |
| 29. | Guide |
| 30. | Spring (RC) |
| 31. | Retainer |

Fig. O160—Composite view of draft control actuating assembly used when tractors are equipped with draft control. Utility models do not use support (26) and have lower link supports installed opposite to those of other models.

removed, unscrew the by-pass valve cap screw (18—Fig. O156) from outside of housing until its clears cylinder; then unbolt and remove cylinder (14) and piston assembly. Piston (11) can be removed from cylinder assembly by bumping open end of cylinder on a wood block.

Note: At this time, the oil filter by-pass valve on Series 1600 hydraulic system can be serviced, if necessary. Valve is disassembled by removing the plug located in the left rear of the cylinder mounting flange. Piston rod can be removed after removing the piston rod retainer.

When reassembling, (tractors prior to serial number 201 135-000), be sure piston "O" ring (13) is on pressure side, that is, toward closed end of piston and leather backup ring (12) is towards open end of piston. Soak leather backup ring in hydraulic oil and lubricate "O" ring prior to installation into cylinder.

On tractors, serial number 201 135-000 and up, install "O" ring in groove then install teflon sealing ring over "O" ring. Allow teflon seal ring to shrink to normal and lubricate piston and seal rings prior to installation into cylinder.

## DRAFT CONTROL ACTUATING ASSEMBLY

258. This part of the hydraulic lift (three-point hitch) is located in the bottom rear of the tractor rear main

frame and includes the lower link supports, actuating lever assembly and the draft control spring.

To service these parts, refer to the appropriate sections and remove the pto clutch and drive shaft (single speed) or the complete pto assembly (dual speed). On models not equipped with pto, remove the rear frame cover. On all models, remove the hydraulic housing.

NOTE: The draft control actuating assembly on Utility models differs from that used on other models in that the lower links impart a counter-clockwise force (as viewed from right side of tractor) to the actuating lever when draft loads are encountered; whereas, on other models, the force is in a clockwise direction. Differences in construction can be viewed in Fig. O160, and will be noted in the following paragraphs where service procedure is affected.

259. **R&R AND OVERHAUL.** With units removed for access as noted in paragraph 258, refer to Fig. O160 and proceed as follows: Loosen jam nut (24), then remove the jam nut and adjusting nut (23). Remove retainer (31) and spring (30). On utility models, remove retainer (31) and spacer (27), then withdraw bolt (22) from bottom of main frame. On all other models, remove guide (29—Fig. O160), then unbolt and remove support (26) and spacers (25) from rear main frame. On all models, refer

to Fig. O160 and remove cap screws and keys (19) and pull lower link supports (11) and thrust washers (12) from rear main frame. Remove actuating arm (16) from rear main frame and, on all except utility models, remove pin (20) and rod (22) from actuating arm. Pry out oil seals (13), remove snap rings (14), then, using suitable driver, remove bearings (15) from housing. Note: Seals will be damaged beyond further use when removing from housing; do not remove seals and/or bearings unless necessary.

Reassemble by reversing removal procedure. Adjust draft control spring as outlined in paragraph 260.

260. **ADJUST DRAFT CONTROL SPRING.** On all except Series 1550 and 1555, turn nut (23—Fig. O160) down until all free play is removed from draft control spring, then continue to tighten the nut an additional 2½ turns, or until spring is compressed ⅛-inch from its free length. Then, tighten the jam nut (24—Fig. O160) to retain adjusting nut in this position.

On Series 1550 and 1555 models, adjust spring until overall length of spring and both retainers is $7\frac{25}{32}$ inches for Utility models; or $7\frac{7}{16}$ inches from top of support to top of upper retainer for all other models.

## HYDRA-LECTRIC (REMOTE) CYLINDERS

Hydra-Lectric cylinders are available in three and four inch sizes.

Both size cylinders are similar in construction, with the exception that the cylinder head is bolted to the barrel on three inch cylinders while the four inch cylinder head is retained to the barrel with a snap ring. Any differences in service procedure will be noted.

261. **R&R AND OVERHAUL.** Removal of cylinder for service work is obvious.

Clear fluid from cylinder by moving piston in and out, then thoroughly clean assembly. Remove the 45 degree elbow, pipe nipple and the cable clamp from cylinder. On three inch cylinders, unbolt cylinder head. On four inch cylinders, insert small punch through hole in cylinder barrel and push snap ring out of its groove. Remove snap ring using a small screw driver. Note: It may be necessary to bump cylinder head into barrel slightly to take pressure off snap ring. Use a series of sharp jerks on piston

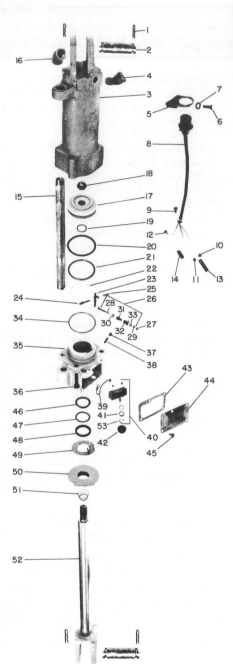

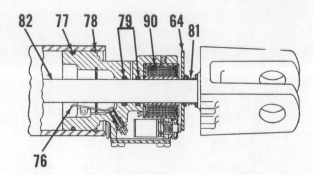

**Fig. O163 — Cross-section view of 4-inch Hydra-Lectric cylinder. Cylinder head is retained by snap ring (78); whereas 3-inch cylinder head is retained by capscrews as shown in Fig. O162.**

64. Stop collar
76. Contact arm
77. "O" ring
78. Snap ring
79. "O" ring
81. Spring
82. Piston rod
90. Solenoid

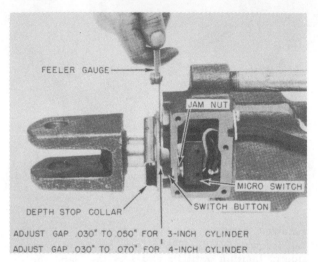

**Fig. O164—Position stop collar as shown when adjusting "Down" switch.**

**Fig. O165 — "Up" switch must be activated when piston is ⅛-inch from cylinder head as shown. Refer to text.**

**Fig. O162 — Exploded view of 3-inch Hydra-Lectric remote double acting cylinder.**

| | |
|---|---|
| 3. Barrel | 32. Sleeve |
| 5. Bracket | 33. Washer |
| 8. Cable | 34. "O" ring |
| 13. Tubing | 35. Head |
| 14. Tubing | 37. Set screw |
| 15. Pipe | 38. Pin |
| 17. Piston | 39. Wire |
| 18. Nut | 40. Switch assy. |
| 19. "O" ring | 42. Boot |
| 20. Washer, backup | 43. Gasket |
| 21. "O" ring | 44. Cover |
| 22. Spring | 46. Packing |
| 23. Arm | 47. Washer |
| 24. Pivot | 48. Seal |
| 25. Pivot | 49. Coil |
| 26. Terminal assy. | 50. Stop |
| 28. Terminal | 51. Spring |
| 30. "O" ring | 52. Rod |
| 31. Insulator | |

rod and bump cylinder head from barrel. Remove piston lock nut, then remove piston and cylinder head from piston rod. If desired, stop collar and spring can also be removed. "O" rings and back-up washers can now be removed from cylinder head and piston. Remove "Up" switch contact arm outer shaft from cylinder head by pushing on inner end. The four inch cylinder outer shaft has an "O" ring. Remove the "Up" switch contact arm, spring and inner shaft from inside cylinder head. Remove cover from cylinder head, identify and disconnect wire from "Up" switch terminal, then remove the set screw and nylon locking pin which retain "Up" switch terminal and remove terminal. Remove snap ring and rubber boot from "Down" switch, remove switch retaining nut, then pull switch from cylinder head. Identify and disconnect wires and remove switch. Identify and disconnect remaining wires, then loosen solenoid retaining set screw and remove solenoid. On three inch cylinders, the cylinder rod dust seal located directly behind the solenoid can also be removed, if necessary.

Any further disassembly required will be obvious. Clean and inspect all parts and renew any which show signs of undue wear, scoring, or other damage. Use all new "O" rings, gaskets and back-up washers when reassembling. Pay close attention to wiring and renew any that show signs of fraying or other deterioration. Adjust cylinder as outlined in paragraph 262.

**262. CYLINDER ADJUSTMENT.** Adjust the Hydra-Lectric "Down" switch as follows: Extend the piston rod a short distance and remove cover from cylinder head. Now push stop collar toward "Down" switch until a gap of 0.030-0.050 for three inch cylinders, or 0.030-0.070 for four inch cylinders, exists between stop collar and cylinder magnet as shown in Fig. O164. Loosen jam nut and adjust "Down" switch until it just breaks the electrical circuit with stop collar in the position stated above. This point can be determined by a click of the switch, or by the light going off if a test light is used. Tighten switch jam nut and recheck the adjustment.

**263.** To check the adjustment of the "Up" switch, a test light is required and is accomplished as follows: Connect one test light probe to the "Up" switch terminal, or the "W" terminal of plug connector, and ground the other probe to an unpainted surface of the cylinder. Light should come on as points of switch are normally closed. Now extend piston rod to its limit at which time light should go out. Measure the exposed portion of the piston rod, then push the piston rod inward ⅛-inch. Light should now come on. If light does not come on, loosen the "Up" switch terminal locking set screw and adjust terminal until the ⅛-inch piston rod movement described above will activate switch. Tighten set screw and recheck the adjustment.

# DEPTH STOP HYDRAULIC SYSTEM
## (External Valve)

Late production Series 1650 models and all Series 1655, 1550 and 1555 models are available with a "depth stop external valve hydraulic system" for use with one or two remote double acting, or single acting, "depth stop" cylinders. One or two additional control valve units may be added to the control valve assembly increasing the capacity of the system to operate either three or four cylinders.

### SYSTEM ADJUSTMENTS

**265. SYSTEM RELIEF PRESSURE.** System relief pressure is controlled by the plunger type relief valve (24—Fig. O171) and is adjusted by varying the number of shims (23) located between the valve and spring (22). To check relief pressure, connect a 3000 psi capacity hydraulic gage to one of the remote cylinder quick disconnect couplings and pressurize that port by holding the control valve lever in proper position. Note: Hold lever in pressurized position only long enough to observe pressure gage reading. With engine running at 2200 RPM and hydraulic fluid at normal operating temperature, system relief pressure should be 2050-2150 psi. If not within the range of 2050-2150 psi, stop engine, remove the valve plug (20), spring (22), shims (23) and plunger (24). Add or remove shims as necessary to obtain correct relief pressure and renew the sealing "O" ring (21) when reinstalling plug. Note: If adding shims does not increase the system relief pressure, a faulty hydraulic system pump should be suspected.

NOTE: At tractor serial number 221 295-000, a pilot relief valve (items 121 through 128—Fig. O171) replaced previous relief valve (items 20 through 26). When full capacity of hydraulic

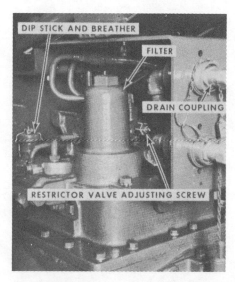

**Fig. O170 — View of left rear corner of depth stop cylinder hydraulic housing (with shroud removed) showing location of hydraulic fluid dipstick, filter, drain coupling and restrictor valve adjusting screw on left control valve unit.**

system is required on tractors prior to serial number 221 295-000, a package (Oliver part number 30-3040097) can be installed. However, when installing the new pilot relief valve, the restrictor must be removed from the nipple which returns oil to reservoir from relief valve. Tractors after serial number 212 449-000 with remote only, external valve systems require no further modification as they have a relief valve return line incorporated.

The pilot relief valve is factory adjusted to 2000 psi and cleaning or servicing is not recommended. Install a complete cartridge, screen and "O" ring if valve malfunctions.

**266. FLOW DIVIDER RELIEF PRESSURE.** A relief valve within the flow divider valve assembly should limit pressure within the power steering system to approximately 1500 psi. Refer to paragraph 243 for method of checking and adjusting the flow divider relief valve pressure.

**267. RESTRICTOR VALVE ADJUSTMENT.** A restrictor valve on the rear of each control valve unit (See Fig. O170) controls the flow of fluid returning to the hydraulic sump from the base end of the remote cylinder connected to that valve unit. An implement will have a tendency to bounce as it is lowered if flow is not sufficiently restricted. If bouncing is encountered, loosen lock nut (18—Fig. O173) on restrictor valve adjusting screw (17) and turn screw in (clockwise) to increase restriction. Note: Remove load on remote cylinder before attempting to adjust the restrictor valve. If adjusting screw is turned in too far resulting in too much restriction, the relief valve will blow causing a buzzing noise whenever remote cylinder is retracted. If this occurs, turn the screw out (counterclockwise) to correct the condition. Tighten lock nut to maintain restrictor valve adjustment.

**267A. THERMAL RELIEF VALVE.** The double acting control valve is fitted with two thermal relief valves (items 8, 9, 10 and 11—Fig. O173). Shims are available to adjust these valves if necessary, and valves can be checked and adjusted as follows: Use a hydraulic hand pump, with gage. Connect pump to front valve port, move valve spool "out" to pressurize port, then build pressure in valve until thermal relief valve opens. Thermal relief valve should open at 4200-4800 psi and must show no leakage at 2500-psi. If valve setting is not as stated, remove plug (8) and add or subtract

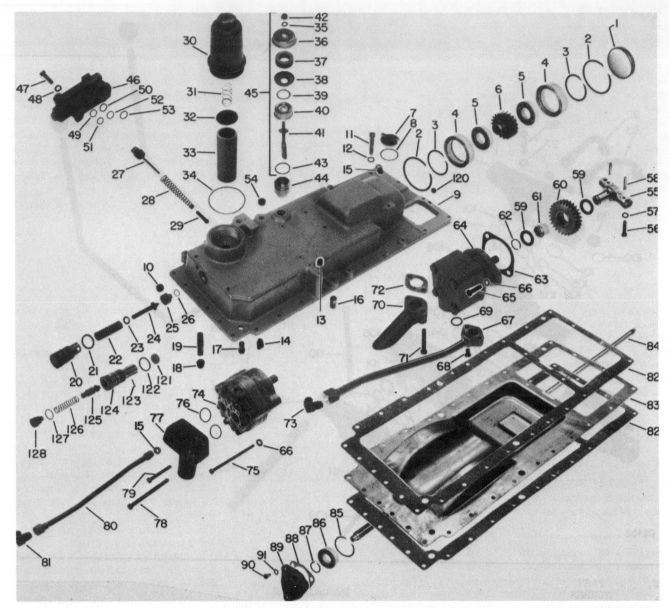

**Fig. O171—Exploded view of Depth-Stop external valve hydraulic housing and related parts. Refer to Figs. O172 and O173 for exploded views of control valve assembly and control valve unit. Pilot relief valve (items 121 through 128) replaced relief valve (items 20 through 26) at tractor serial number 221 295-000.**

| | | | | | |
|---|---|---|---|---|---|
| 1. Plug | 17. Rear dowel | 31. Spring | 45. Dipstick & | 62. Snap ring | 84. Pump drive shaft |
| 2. Snap ring | 18. Restrictor | 32. Retainer | breather assy. | 63. Gasket | 85. Snap ring |
| 3. Snap ring | 19. Nipple | 33. Element | 46. Flow divider | 64. Pump (vane) | 86. Bearing |
| 4. Carrier | 20. Retainer | 34. "O" ring | valve | 67. Pressure tube | 87. Snap ring |
| 5. Bearing | 21. "O" ring | 35. Seal | 49. "O" ring | 69. "O" ring | 88. Gasket |
| 6. Pinion | 22. Spring | 36. Cap | 50. "O" ring | 70. Inlet pipe | 89. Bearing housing |
| 7. Cap | 23. Shim (0.010) | 37. Element | 51. Backup ring | 72. Gasket | 120. Plug |
| 8. Gasket | 24. Plunger | 38. Cup | 52. Backup ring | 73. Elbow | 121. Screen |
| 9. Housing | 25. Seat | 39. Gasket | 53. "O" ring | 74. Pump (gear) | 122. "O" ring |
| 10. Plug | 26. Crush washer | 40. Cap | 54. Plug | 76. "O" ring | 123. Dowel |
| 13. Plastic plug | 27. Stop pin | 41. Dipstick | 55. Support | 77. Manifold (early) | 124. Cartridge |
| 14. Guide | 28. Spring | 42. Nut | 58. Dowels | 80. Pressure tube | 125. Valve |
| 15. Tube insert | 29. Plunger | 43. "O" ring | 59. Thrust washers | 81. Elbow | 126. Spring |
| 16. Center dowel | 30. Filter body | 44. Filler tube | 60. Idler gear | 82. Gasket | 127. Washer |
| | | | | 61. Bearing | 83. Reservoir | 128. Plug |

shims as necessary between spring (10) and plug. Shims are 0.010 thick.

Test the other thermal relief valve by connecting hand pump to rear valve port, move spool "in" and repeat procedure given above.

## FLUID AND SYSTEM FILTER

268. Recommended fluid for the system is White-Oliver Type 55 hydraulic oil. Reservoir capacity is 3½ gallons. System fluid should be changed yearly or after each 1000 hours of operation, and the filter should be renewed whenever the fluid is changed.

To drain the hydraulic system, connect an open end hose to the upper left quick disconnect coupling ("Drain Coupling"—Fig. O170), start engine and move inner (left) control lever forward to pump fluid from the unit. CAUTION: Stop engine as soon as flow has diminished. Remove hose from quick disconnect port. Disconnect hoses from cylinders and work piston rods in and out to expel all fluid. Renew the filter element, connect hoses and refill unit with new, clean recommended fluid. Start engine, extend and contract cylinders several times to bleed all air trapped in system and recheck fluid level. Add fluid as necessary to bring fluid to full level on dipstick.

To renew the filter, unscrew the cast iron filter body (See Fig. O170) and remove the old element. Clean the filter body, spring and retainer with clean solvent. Install new "O" ring on filter body, lubricate "O" ring and "O" ring seat in hydraulic housing with grease and assemble as follows: Hold element in upright position, place retainer on upper end of element and place spring on top of retainer. Position filter body over spring, retainer and element and carefully install the unit on hydraulic housing. Tighten the filter body to a minimum of 150 Ft.-Lbs. torque.

### OVERHAUL

**269. FLOW DIVIDER VALVE.** Except for tubing seats, flow divider valve used on "depth stop cylinder hydraulic system" is same as that used on models with Hydra-Lectric and/or three-point hitch hydraulic systems. Refer to paragraph 244 or 244A for information on servicing the flow divider valve.

**270. HYDRAULIC PUMP.** Tractors prior to serial number 191 121-000 were equipped with a gear type hydraulic pump. For information on this pump, refer to paragraph 246.

Beginning with tractor serial number 191 121-000 and running through tractor serial number 218 222-000, production tractors were equipped with a vane type hydraulic pump. Conversion packages were made available to convert from the early gear type pump to the vane type pump. For information on the vane type pump, refer to paragraph 246B.

At tractor serial number 218 223-000, the vane type pump was replaced by a new gear type pump. Conversion packages are available to convert from the vane type pump to the new gear type pump. For information on the latest gear type pump, refer to paragraph 246C.

**271. PUMP DRIVE SHAFT AND PINION.** After removing the hydraulic housing as outlined in paragraph 273, service procedure for the pump drive shaft and the pump drive pinion is the same as outlined in paragraphs 247 through 249.

**272. CONTROL VALVE UNITS AND ASSEMBLY.** Two, three or four control valve units are mounted on

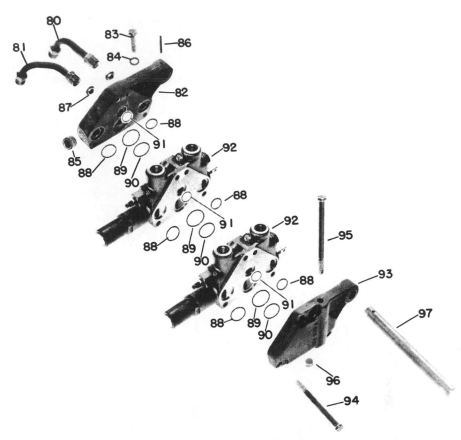

**Fig. O172 — Exploded view of control valve assembly. Refer to Fig. O173 for exploded view of each control valve unit (92). Two control valve units are standard; one or two more units may be added as an option.**

| | |
|---|---|
| 80. Return line | 85. Plug |
| 81. Pressure line | 86. Roll pin |
| 82. Manifold | 87. Tube inserts |
| | 88. "O" ring |

| | |
|---|---|
| 89. "O" ring | 92. Valve unit |
| 90. Backup ring | (See Fig. O173) |
| 91. "O" ring | 93. Plate |
| | 97. Lever shaft |

top of the hydraulic housing in an assembly as shown in Fig. O172. To remove the valve assembly, first remove the operator's seat and the shroud from the seat support bracket. Disconnect the remote cylinder tube and hose assemblies from the valve ports, then, unbolt and remove the seat support bracket, hose and quick disconnect coupling units from the tractor as an assembly. Drive the roll pin (86—Fig. O172) from valve assembly manifold (82) and lever pivot shaft (97), then remove the shaft and levers. Disconnect the pressure line (81) and return line (80); then, remove the cap screws (83 and 95) securing valve assembly to hydraulic housing and remove the assembly from tractor.

**272A.** To disassemble the double acting control valve assembly and individual units, use Figs. O172 and O173 as guides and proceed as follows: Remove the cap screws (94—Fig. O172) retaining the end plate (93) and valve units (92) to manifold (82)

and carefully separate the parts. "O" rings (88, 89 and 91) are used to seal passages between valve units, end plate and manifold. A backup ring (90) is used with "O" ring (89).

Thoroughly clean outside of control valve unit in suitable solvent, refer to Fig. O173 and proceed as follows: Unscrew detent cap (36) and remove complete detent and valve spool assembly from valve body. Clamp flat (lever) end of valve in vise and remove detent cap (36) from assembly by driving against hex nut portion with brass drift; this unseats detent balls from groove inside cap. Catch the four detent balls as cap is removed. Place small spanner wrench in detent ball holes of retainer (31) and unscrew retainer from valve spool. Remove washer (27), spring (28) and washer (29) from retainer. Remove snap ring (39) from detent retainer, then remove washer (38), spring (37), plunger (35) and piston (33), with "O" ring (34), from retainer. Slide bushing (26) and "O" ring (25) from spool. Run a small

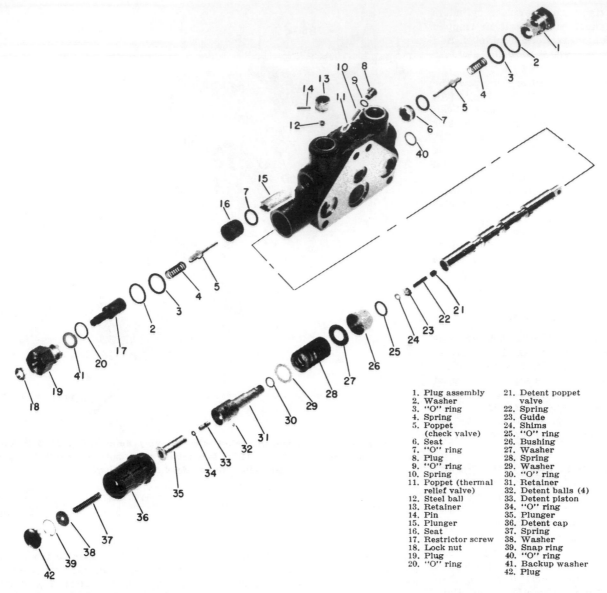

| 1. Plug assembly | 21. Detent poppet |
| --- | --- |
| 2. Washer |    valve |
| 3. "O" ring | 22. Spring |
| 4. Spring | 23. Guide |
| 5. Poppet | 24. Shims |
|    (check valve) | 25. "O" ring |
| 6. Seat | 26. Bushing |
| 7. "O" ring | 27. Washer |
| 8. Plug | 28. Spring |
| 9. "O" ring | 29. Washer |
| 10. Spring | 30. "O" ring |
| 11. Poppet (thermal | 31. Retainer |
|    relief valve) | 32. Detent balls (4) |
| 12. Steel ball | 33. Detent piston |
| 13. Retainer | 34. "O" ring |
| 14. Pin | 35. Plunger |
| 15. Plunger | 36. Detent cap |
| 16. Seat | 37. Spring |
| 17. Restrictor screw | 38. Washer |
| 18. Lock nut | 39. Snap ring |
| 19. Plug | 40. "O" ring |
| 20. "O" ring | 41. Backup washer |
| | 42. Plug |

Fig. O173—Exploded view of one double acting control valve unit; tractor may be equipped with two, three or four units. Refer to Fig. O172 for view showing assembly of two control units. The restrictor valve for each control unit must be individually adjusted; adjusting screw is (17).

(tag) wire into open end of valve spool through the shims (24), spring guide (23), spring (22) and poppet (21); then remove spool from vise and tip lever end up to let shims, etc., slide down the wire. Keep these parts on wire for cleaning. Unscrew plug (1) and remove spring (4), poppet (5) and seat (6). Remove lockout plunger (15). Unscrew plug assembly (19) and remove spring (4), poppet (5) and seat (16). Remove steel ball (12) and retainer (13); it is not necessary to remove pin (14) from retainer (13) unless pin is to be renewed. Unscrew lock nut (18) and remove restrictor adjusting screw (17). Remove both thermal-relief plugs (8) and shims (not shown), springs (10) and poppets (11). NOTE: Identify each thermal-relief plug, shim, spring and

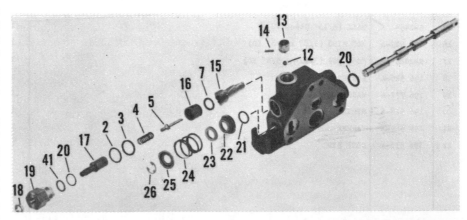

Fig. O173A—Exploded view of the single acting control valve used with depth stop systems.

| 2. Back-up washer | 12. Steel ball | 17. Adjusting screw | 22. Washer (deep) |
| --- | --- | --- | --- |
| 3. "O" ring | 13. Retainer | 18. Lock nut | 23. Spacer washer |
| 4. Spring | 14. Pin | 19. Plug | 24. Spring |
| 5. Poppet | 15. Plunger | 20. "O" ring | 25. Washer (shallow) |
| 7. "O" ring | 16. Seat | 21. "O" ring | 26. Retainer clip |

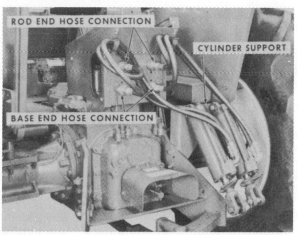

**Fig. O174 — Note that hose from base end of depth stop control cylinder must be coupled to lower connector and hose from rod end coupled to upper connector. That is, the hose from base end of cylinder must be connected to restrictor port of control valve unit.**

ROD END HOSE CONNECTION

CYLINDER SUPPORT

BASE END HOSE CONNECTION

poppet unit so that they may be re-installed in same location from which they were removed, using the same number of shims between plug and spring as were removed. Remove and discard all "O" rings and back-up washers.

Thoroughly clean all parts in solvent and dry thoroughly. Small nicks and burrs should be removed with fine emery cloth. The control valve spool and valve body are not serviceable items; if either of these parts are damaged, a new valve unit assembly must be installed. Inspect point of poppet (21) and mating seat in valve spool for excessive wear.

Check to be sure small orifice hole in valve spool is open and clean, and be sure small drilled holes in thermal-relief bores are clean. Inspect detent grooves inside cap (36) for excessive wear. Inspect large hole in side of lockout seat (16) for excessive wear from ball (12). Inspect all springs for excessive weakness, wear or breakage.

Lubricate all parts before reassembly and reassemble as follows: Install new "O" ring (34) on detent piston (33) and install piston in the detent retainer (31). Place detent release plunger (35) in detent retainer, place spring (37) inside plunger and washer

(38) on spring, then install snap ring (39) to retain washer, spring and plunger in detent retainer. Install new "O" ring (30) on outside of detent retainer, then place washer (29), centering spring (28) and washer (27) over retainer. Place shims (24), spring guide (23), spring (22) and poppet (21) on tag wire and slide parts into position inside valve spool. Be sure parts are seated correctly, then clamp valve spool flat end in vise with open end up. Install new "O" ring (25) and bushing (26) over spool, then thread detent sub-assembly into spool and, using small spanner wrench, tighten to a torque of 5-8 Ft.-Lbs. Stick detent balls in holes of detent retainer with heavy grease. Carefully lower detent cap (36) over detent retainer until cap comes in contact with the steel balls, then use plastic hammer to drive cap down over the balls until the balls engage the second detent groove in cap. Install new "O" ring (40) in spool bore of valve body, then install the detent and spool assembly tightening the detent cap snugly. NOTE: Be sure arrow on flat (lever) end of valve spool points toward ports of the valve unit. Install new back-up washer (41) in bore of plug (19) and place new "O" ring (20) in plug at inner side of back-

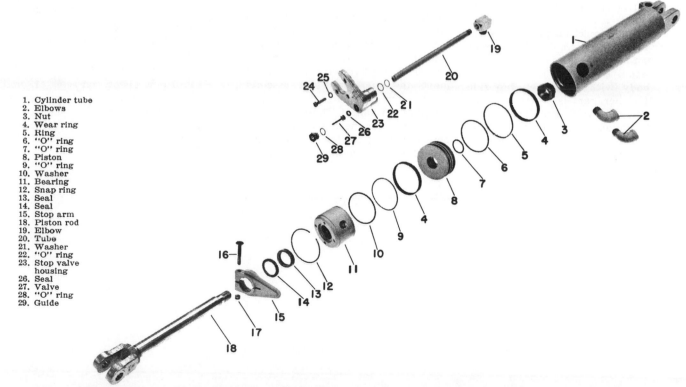

1. Cylinder tube
2. Elbows
3. Nut
4. Wear ring
5. Ring
6. "O" ring
7. "O" ring
8. Piston
9. "O" ring
10. Washer
11. Bearing
12. Snap ring
13. Seal
14. Seal
15. Stop arm
18. Piston rod
19. Elbow
20. Tube
21. Washer
22. "O" ring
23. Stop valve housing
26. Seal
27. Valve
28. "O" ring
29. Guide

**Fig. O175 — Exploded view of depth stop control cylinder. Stop arm (15) can be clamped in different positions on piston rod (18) to adjust length of cylinder stroke. When stop arm contacts valve (27), pressure build up in control valve opens poppet valve (21—Fig. O173) in valve spool releasing detent and allowing valve to return to neutral position.**

up washer. Thread restrictor adjusting screw (17) into plug (19), then install lock nut (18) on adjusting screw. Install back-up washer (2) and "O" ring (3) on outside of plugs (19 and 1). Install new "O" rings (7) on seats (16 and 6). Install seat (16) in lockout bore being sure large hole in seat is aligned with port to allow seating of ball (12). Install poppet (5), spring (4) and plug assembly (19). Install lockout plunger (15) in bore, then install remaining seat (6), poppet (5), spring (4) and plug (1). Install steel ball (12) and retainer (13) in port towards the restrictor adjusting screw (17). Install the thermal-relief poppets (11), springs (10), shims and plugs (8), with new "O" rings (9), in bores from which they were removed. Install button plug (42) in detent cap.

272B. To disassemble single acting control valve, remove retaining clip (26—Fig. O173A), slide spool out front of valve body and remove cup washer (25), spring (24) and washers (22 and 23). "O" rings (20 and 21) can now be removed from spool bore. Remove retainer (13) with pin (14) and check ball (12). Remove plug and adjusting screw assembly (items 19 and 17), then remove spring (4), poppet (5), seat (16) and lockout plunger (15). Remove jam nut (18) and unscrew adjusting screw (17) from plug (19). Remove "O" rings and back-up washers from plug and "O" ring from seat.

Clean and inspect all parts for scoring or excessive wear. Small nicks or burrs can be removed with fine emery cloth. Spool should be a snug fit in valve body with no perceptible side play. If either part is worn or damaged, renew both as spool and body are not available separately. See that small hole in seat (16) is not plugged and that large hole (ball seat) is not

worn or damaged. Check springs for fractures or other damage.

Assemble control valve as follows: Lubricate parts as they are installed. Install back-up washer (2) and "O" ring (20) in bore of plug (19) with "O" ring toward inside of valve. Thread adjusting screw (17) into plug and install jam nut (18). Install lockout plunger (15), small end first, into upper bore of valve body. Place "O" ring (7) in groove of seat (16), then install seat so large hole is on top side and aligned with cylinder port. Install poppet (5), long end first, and spring (4). Place back-up ring (2) and "O" ring (3) on plug (19) so "O" ring is toward valve body and install plug assembly. Install "O" ring (20), then install spool from rear of valve body until rear "O" ring groove is exposed and install "O" ring (21). Push spool back through "O" ring (21), then install washer (22) (deep), spacer washer (23), spring (24), washer (25) (shallow) and retaining clip (26).

NOTE: The single acting control valve has only a restrictor valve adjustment and is adjusted the same as the double acting control valve. Refer to paragraph 267.

273. R&R HYDRAULIC HOUSING. First, drain hydraulic system fluid as outlined in paragraph 268; then, proceed as follows:

If removing the housing for access to the hydraulic pump, bull gears, etc., it will not be necessary to remove the seat support, control valve assembly or remote hose and connectors from the hydraulic housing. Remove the operator's seat assembly, shroud from around seat support and the platform panels. Thoroughly clean the unit. Disconnect pressure line to power steering and/or oil cooler from flow divider valve and disconnect re-

turn line from hydraulic housing. Hook a chain sling to seat support, unbolt hydraulic housing from rear main frame and lift unit from tractor. While unit is still supported (if necessary to gain access to hydraulic pump), drive the dowels from hydraulic housing and remove the oil pan.

The seat support and control valve assembly can be removed from the hydraulic housing as outlined in paragraph 272 either before or after removing the housing from tractor. If removing hydraulic housing with seat bracket and hydraulic control valve assembly already removed, attach lift chain with the forward seat support bracket retaining cap screws.

Reverse removal procedure to reinstall unit on tractor and refill the system as in paragraph 268.

274. OVERHAUL CYLINDER. Refer to exploded view of cylinder in Fig. O175 and proceed as follows:

Loosen nut (17) and slide stop arm (15) to yoke end of piston rod (18). Remove cap screws (24) retaining stop valve housing (23) to bearing (11) and withdraw valve housing from end of tube (20). Unscrew guide (29) and remove stop valve (27).

Push bearing (11) into cylinder tube (1) far enough to allow removal of snap ring (12), then bump bearing from cylinder tube with piston (8) by working piston rod in and out.

Remove nut (3), piston (8) and bearing (11) from piston rod. Need and procedure for further disassembly is evident on inspection of unit.

Install new piston rod seals (13 and 14) in bearing (11) and reassemble unit using all new "O" rings, wear rings and backup rings. Reverse disassembly procedure to reassemble the cylinder.

# OLIVER
## (Includes some Cockshutt Models)

**Oliver & Cockshutt Models** ■ 1750 ■ 1800A ■ 1800B ■ 1800C
■ 1850 ■ 1900A ■ 1900B ■ 1900C ■ 1950 ■ 1950-T

**Previously contained in I&T Shop Manual No. O-21**

SHOP MANUALS

# SHOP MANUAL
# OLIVER

## SERIES

| | |
|---|---|
| 1750 | 1900A |
| 1800A | 1900B |
| 1800B | 1900C |
| 1800C | 1950 |
| 1850 | 1950-T |

## Also Covers COCKSHUTT
## SERIES

| | |
|---|---|
| 1750 | 1900B |
| 1800B | 1900C |
| 1800C | 1950 |
| 1850 | 1950-T |

## IDENTIFICATION

Series 1800 and 1900 tractors (series A) begin with serial number 90 525-000 and continue through serial number 124 395-000. 1800 and 1900 series B tractors begin with serial number 124 396-000 and continue through serial number 134 683-000. 1800 and 1900 series C tractors begin with serial number 134 684-000 and continue through serial number 150 420-000. 1850 and 1950 series tractors begin with serial number 150 421-000. 1750 series tractors begin with serial number 180 537-000. 1950-T series tractors begin with serial number 194 080-000.

Tractor serial number plate is located on rear side of instrument panel support. On series A 1800 tractors, engine serial number is stamped on right rear flange of engine. On 1800 series B and C, are 1750 and 1950-T and 1850 non-diesel tractors, engine serial number is stamped along outer edge of timing gear coer mounting flange directly below generator or alternator. On series 1850 diesel tractors, engine serial number is stamped on cylinder block directly below injection pump. On series 1900 and 1950 tractors, engine model and serial numbers are stamped on upper right rear of cylinder block.

## BUILT IN THESE VERSIONS

Rowcrop, Wheatland, Ricefield, Industrial and 4-Wheel Drive. Rowcrop tractors are available in either dual wheel tricycle or adjustable front axle versions, while Wheatland, Ricefield, Industrial and 4-Wheel Drive tractors are available with non-adjustable axles only.

## INDEX (By Starting Paragraph)

# INDEX (By Starting Paragraph)

# CONDENSED SERVICE DATA

## GENERAL

| | Series 1800 | Series 1800 B, 1800 C | Series 1850 | Series 1900, B, C, 1950 | Series 1750, 1950-T |
|---|---|---|---|---|---|
| Engine Make | Own | Own | Own | GM | Own |
| Engine Model | 1800 | 1800 | 1850 | 4-53 | 1750, 1950-T |
| Cylinders, No. of | 6 | 6 | 6 | 4 | 6 |
| **Cylinder Bore—Inches:** | | | | | |
| Non-Diesel | 3¾ | 3⅞ | 3⅞ | ...... | 3⅞ |
| Diesel | 3⅞ | 3⅞ | 3⅞ | 3⅞ | 3⅞ |
| **Stroke—Inches:** | | | | | |
| Non-Diesel | 4 | 4 | 4⅜ | ...... | 4 |
| Diesel | 4 | 4⅜ | 5 | 4½ | 4⅜ |
| **Displacement—Cubic Inches:** | | | | | |
| Non-Diesel | 265 | 283 | 310 | ...... | 283 |
| Diesel | 283 | 310 | 354 | 212.4 | 310 |
| **Compression Ratio:** | | | | | |
| Gasoline | 8.5 | 8.5 | 8.5 | ...... | 8.5 |
| LP-Gas | 9.0 | 9.0 | 9.0 | ...... | ...... |
| Diesel | 16.0 | 16.0 | 16.0 | 17.0 | 16.0 |
| Main Bearings, No. of | 7 | 7 | 7 | 5 | 7 |
| Cylinder Sleeves Type | Wet ...... | Wet ...... | ND - Wet D - Dry | Wet ...... | Wet ...... |
| Forward Speeds, No. of | 6 | 6 | 6 | 6 | 6 |
| Battery Terminal Grounded | Positive ...... | Positive ...... | Negative ...... | 1900 - Pos. 1950 - Neg. | Negative ...... |

## TUNE-UP

| | Series 1800 | Series 1800 B, 1800 C | Series 1850 | Series 1900, B, C, 1950 | Series 1750, 1950-T |
|---|---|---|---|---|---|
| Firing Order | 1-5-3-6-2-4 | 1-5-3-6-2-4 | 1-5-3-6-2-4 | 1-3-4-2 | 1-5-3-6-2-4 |
| **Valve Tappet Gap, Inlet:** | | | | | |
| Non-Diesel | See Par. 98 | See Par. 98 | See Par. 98 | ...... | 0.020 |
| Diesel | See Par. 98 | See Par. 98 | See Par. 98, 99 | None | 0.030 |
| **Valve Tappet Gap, Exhaust:** | | | | | |
| Non-Diesel | 0.022-0.024 | 0.022-0.024 | 0.023-0.025 | ...... | 0.030 |
| Diesel | See Par. 98 | See Par. 98 | 0.010 (hot) | See Par. 242 | 0.030 |
| Valve Face Angle, Degrees | 44½ | 44½ | ND - 44½, D - 45 | 30 | See Par. 98 |
| Valve Seat Angle, Degrees | 45 | 45 | 45 | 30 | See Par. 98 |
| Ignition Distributor Make | Delco-Remy | Delco-Remy | Holley | None | Holley |
| Ignition Distributor Model | 1112603 | 1112603, 1112632 | D - 2563AA | ...... | D-2563AA |
| Generator and Regulator Make | Delco-Remy | Delco-Remy | Delco-Remy | Delco-Remy | Delco-Remy |
| Generator Model | 1100400 | See Par. 233 | See Par. 234 | 1100419, 1100725 | See Par. 234 |
| Regulator Model | 1118997 | 1118997 | 1119517 | 118997, 1119517 | 1119517 |
| Starting Motor Make | Delco-Remy | Delco-Remy | Delco-Remy | Delco-Remy | Delco-Remy |
| **Starting Motor Model:** | | | | | |
| Non-Diesel | 1107682 | 1107682 | 1107358 | ...... | 1107358 |
| Diesel | See Par. 235 | See Par. 235 | 1113656 | 1113100, 1113136 | 1113139, 1113656 |
| Ignition Distributor Contact Gap | 0.022 | See Par. 229 | 0.025 | ...... | 0.025 |
| **Ignition Distributor Timing (Static):** | | | | | |
| Gasoline | 0° BTDC | 2° BTDC | 0° | ...... | 2° BTDC |
| LP-Gas | See Par. 230 | See Par. 230 | 2° BTDC | ...... | ...... |
| Injection Pump Make | Roosa-Master | Roosa-Master | CAV | ...... | Roosa-Master |
| Injection Pump Timing (Static) | 8° BTDC | 2° BTDC | 28° BTDC | ...... | 4° BTDC |
| Injector Timing | ...... | ...... | ...... | See Par. 243 | ...... |
| **Carburetor Make:** | | | | | |
| Gasoline | Marvel-Schebler | Marvel-Schebler | Marvel-Schebler | ...... | Marvel-Schebler |
| LP-Gas | Zenith | Zen.-Ensign | Ensign | ...... | ...... |
| **Carburetor Model:** | | | | | |
| Gasoline | See Par. 145 | See Par. 145 | USX-37 | ...... | USX-44 |
| LP-Gas | 12719 | 12858, CBX | CBX | ...... | ...... |
| **Engine Low Idle RPM:** | | | | | |
| Gasoline | 325 | 350 | 400 | ...... | 400 |
| LP-Gas | 500 | 500 | 400 | ...... | ...... |
| Diesel | 650 | 650 | 650 | 500 | 800 |
| **Engine High Idle RPM:** | | | | | |
| Gasoline | 2200 | 2425 | 2640 | ...... | 2640 |
| LP-Gas | 2200 | 2425 | 2640 | ...... | ...... |
| Diesel | 2200 | 2425 | 2565 | See Par. 246 | 2640, 2650 (1950-T) |
| Engine Rated RPM | 2000 ...... ...... | 2200 ...... ...... | 2400 ...... ...... | 2000(A), 2200(B), 2400(C). | 2400 ...... ...... |
| **Belt Pulley Rated RPM:** | | | | | |
| 540 rpm pto | 1011 | 1035 | 1033 | 1033 | 1033 |
| 1000 rpm pto | 1053 | 1056 | 1035 | 1035 | 1035 |
| **PTO RPM @ Engine Rated RPM:** | | | | | |
| 540 rpm pto | 537 | 550 | 549 | 549 | 549 |
| 1000 rpm pto | 1000 | 1003 | 984 | 984 | 984 |

4

## SIZES—CLEARANCES—CAPACITIES

| | Series 1800 | Series 1800 B, 1800 C | Series 1850 | Series 1900, B C, 1950 | Series 1750, 1950-T |
|---|---|---|---|---|---|
| Crankshaft Journal Diameter | 2.624-2.625 | 2.624-2.625 | See Par. 132 & 133 | ...... | 2.624-2.625 |
| Crankpin Diameter | 2.4365-2.4375 | 2.4365-2.4375 | See Par. 132 & 133 | ...... | 2.4365-2.4375 |
| Rod Length, Center to Center | 6.749-6.750 | 6.749-6.750 | (1) | ...... | 6.749-6.750 |
| Camshaft Journal Diameter: | | | | | |
| Front | 1.749-1.750 | 1.749-1.750 | See Par. 113 & 114 | ...... | 1.749-1.750 |
| Others | 1.7485-1.7495 | 1.7485-1.7495 | See Par. 113 & 114 | ...... | 1.7485-1.7495 |
| Piston Pin Diameter | 1.2494-1.2497 | 1.2494-1.2497 | See Par. 128 & 129 | ...... | 1.2494-1.2497 |
| Valve Stem Diameter: | | | | | |
| Inlet | 0.372-0.373 | 0.372-0.373 | See Par. 98 & 99 | ...... | 0.372-0.373 |
| Exhaust | 0.371-0.372 | 0.371-0.372 | See Par. 98 & 99 | ...... | 0.371-0.372 |
| Compression Ring Width | 0.093-0.094 | 0.093-0.094 | 0.0093-0.094 | ...... | 0.093-0.094 (5) |
| Oil Ring Width | 0.186-0.187 | 0.186-0.187 | (2) | ...... | 0.186-0.187 |
| Piston Ring Side Clearance | See Par. 125 | See Par. 125 | See Par. 125 & 127 | ...... | See Par. 125 |
| Main Bearing Clearance: | | | | | |
| Non-Diesel | 0.0015-0.0045 | 0.0015-0.0045 | 0.0015-0.0045 | ...... | 0.0015-0.0045 |
| Diesel | 0.0015-0.0045 | 0.0015-0.0045 | 0.0015-0.0045 | ...... | 0.0015-0.0045 |
| Rod Bearings Clearance: | | | | | |
| Non-Diesel | 0.0005-0.0015 | 0.0005-0.0015 | 0.005-0.0015 | ...... | 0.0005-0.0015 |
| Diesel | 0.0005-0.0015 | 0.0005-0.0015 | 0.0015-0.003 | ...... | 0.0005-0.0015 |
| Piston Skirt Clearance | See Paragraph 122 | | See Par. 122 & 126 | ...... | See Par. 122 |
| Camshaft End Play | Spring Loaded | | See Par. 113 & 114 | ...... | Spring Loaded |
| Crankshaft End Play | 0.0045-0.0095 | 0.0045-0.0095 | See Par. 132 & 133 | ...... | 0.0045-0.0095 |
| Camshaft Bearing Clearance | 0.0015-0.0035 | 0.0015-0.0035 | See Par. 113 & 114 | ...... | 0.002-0.0035 |
| Cooling System—Gallons | 5 | 5 | 5 | 5.25 | 5 |
| Engine Crankcase—Quarts | 6 | 8 | 8 - ND | 14.5 | 8 |
| Fuel Tank, Gasoline—Gallons | | | 12 - Dsl. | | |
| Fuel Tank, LPG (80%)—Gallons | 36.5 | 34.5 | 34.5 | ...... | 34½, 27½ (1950-T) |
| Fuel Tank, Diesel—Gallons | 42 | 42 | 42 | ...... | |
| Transmission—Quarts | 31.5 | 31.5 | 31.5 | 31.5 | 31½, 27½ (1950-T) |
| Final Drive—Quarts | 2.75 | 2.75 (3) | (3) | (3) | (3) |
| Transfer Case—Quarts | 36 | 36 (3) | (3) | (3) | (3) |
| Hydra-Power Drive—Quarts | ...... | 1 | 1 | 1 | 1 |
| Creeper Drive—Quarts | ...... | 6 | 6 | 6 | 6 |
| Reverse-O-Torc—Quarts | ...... | ...... | 5 | 5 | 5 |
| Hyd. Lift (Draft Control)—Quarts | ...... | 10 | 10 | 10 | 10 |
| Hyd. Lift (Hydra-Lectric)—Quarts | 20 | 20 | 20 | 20 | 20 |
| Belt Pulley—Pints | 24 | 24 | 24 | 24 | 24 |
| Diff. Housing (4WD)—Quarts | 6.5 | 6.5 | 6.5 | 6.5 | 6.5 |
| Planetary Drive (4WD) Quarts | ...... | 8 | 8 | 8 | 8 |
| | ...... | 2.5 | 2.5 | 2.5 | 2.5 |

## TIGHTENING TORQUES—Ft.-Lbs.

| | Series 1800 | Series 1800 B, 1800 C | Series 1850 | Series 1900, B C, 1950 | Series 1750, 1950-T |
|---|---|---|---|---|---|
| Cylinder Head: | | | | | |
| Non-Diesel | 129-133 | 129-133 | 129-133 | ...... | 129-133 |
| Diesel | 129-133 | 129-133 | 80-85 | ...... | See Par. 96 & 96A |
| Cylinder Head Oil Screw | 113-117 | 113-117 | 129-133 | ...... | See Par. 96 & 96A |
| Main Bearings | 129-133 | 129-133 | 140-150 (4) | ...... | 129-133 |
| Connecting Rods | 56-58 | 56-58 | 65-70 (4) | ...... | 46-50 |
| Flywheel | 66-69 | 66-69 | 74-80 (4) | ...... | 66-69 |

(1) 1850 diesel, 8.624-8.626; 1850 non-diesel, 6.749-6.750. (2) 1850 diesel, 0.249-0.250; 1850 non-diesel, 0.186-0.187. (3) Applies to 1800 and 1900 series A and B. All 1800 and 1900 series C and all 1850 and 1950 tractors have a common reservoir. 1800 series C Row Crop models require 43 quarts to fill transmission and final drive whereas all other models of the 1800 series C, 1750, 1850, 1950 and 1950-T tractors require 51 quarts to fill transmission and final drive. (4) Applies to 1850 diesel tractor; for 1850 non-diesel tractors, use the 1800 series C values. (5) All except top compression ring for series 1950-T which is 0.114-0.115 wide.

# FRONT SYSTEM (Axle Type)

## AXLE MAIN MEMBER AND PIVOT PIN

### All Rowcrop Models

1. Rowcrop tractors may be equipped with an adjustable front axle as shown in Fig. 010 or a short wheelbase front axle as shown in Fig. 010A.

To disconnect the axle shown in Fig. 010 from front main frame, support tractor and unbolt stay rod support (32) from front main frame. Disconnect inner ends of the tie rods from center steering arm. Slide front axle from pivot pin on models prior to 115 337-000, or lift front of tractor on models 115 337-000 and up, then roll axle and wheels assembly away from tractor. Remove nut (27) from rear end of stay rod and remove support (32). Bushings (31) and (17 or 17A) should be reamed if necessary, to provide 0.001-0.002 diametral clearance for pins.

To disconnect the axle shown in Fig. O10A from front main frame, support tractor, unbolt support (20) from front main frame and disconnect inner ends of tie rods from center steering arm. Remove front washer (retainer) (15), then move axle assembly rearward until pilot on front of axle center member clears bolster

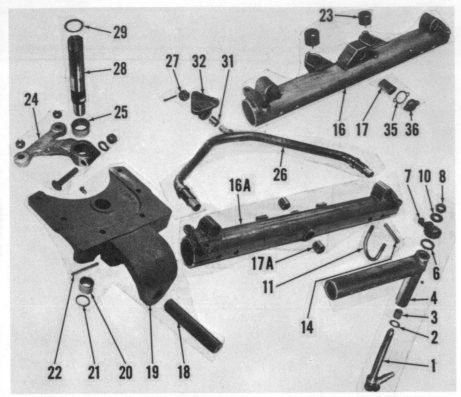

Fig. O10—Exploded view of the adjustable front axle and bolster used on Row Crop models. Item (16) is axle center main member on models prior to serial number 115 337-000. Item (16A) is axle center main member on models 115 337-000 and up. Bolster differs slightly on early models.

| | | |
|---|---|---|
| 1. Steering knuckle | 16. Center member (early) | 24. Center steering arm |
| 2. Thrust washer | 16A. Center member (late) | 25. Bearing |
| 3. Bushing | 17. Pivot bushing | 26. Stay rod |
| 4. Outer member | 17A. Pivot bushing | 27. Nut |
| 6. Felt washer | 18. Pivot pin | 28. Pitman shaft |
| 7. Steering arm | 19. Front bolster | 29. Seal |
| 8. Nut | 20. Bearing | 31. Support bushing |
| 10. Lock washer | 21. Seal | 32. Support |
| 11. "U" bolt | 22. Pin | 35. Gasket |
| 14. Lock pin | 23. Bumper | 36. Cap |

(24). Lift front of tractor and roll axle and wheels away from tractor. Remove retaining cap screw (5) and remove rear pivot pin (9) and support from axle main member. Bushing (12) in axle and bushing (19) in front bolster should be reamed if necessary, to provide 0.001-0.002 diametral clearance for rear pivot pin (9) and pilot on front of axle center main member.

2. After reinstalling the front axle and wheels assembly, check and adjust if necessary, the $\frac{3}{16}$-inch toe-in by varying the length of each tie-rod an equal amount.

### All Wheatland, Industrial and Ricefield models

3. The Wheatland, Industrial and Ricefield tractors are fitted with a non-adjustable, live-spindle axle such as that shown in Fig. O11. While the Wheatland and Ricefield axle is shown, the Industrial models differ only in that the axle is not arched.

4. To remove the axle, stay rod (20) and wheels as an assembly, support

front of tractor, disconnect stay rod support from front main frame and the tie rod inner ends from steering arm. Remove the pin which retains axle pivot pin in front axle bolster (carrier), then drive out pivot pin and remove front axle assembly. Remove nut (21) from aft end of stay rod if bushing (29) is to be renewed. Bushings (29 and 19) should be reamed if necessary, to provide 0.001-0.002 diametral clearance for the pin.

5. After reinstalling the front axle and wheels assembly, check and adjust, if necessary the ¼-inch toe-in by varying the length of the tie-rods. Adjust each tie-rod an equal amount.

### STAY-RODS

#### All Models So Equipped

6. The stay-rods (26—Fig. O10 or 20 —Fig. O11) are available as service items and the procedure for renewing same is obvious. Bushings (31 or 29) should be reamed, if necessary, to provide 0.001-0.002 diametral clearance for the pin.

### KNUCKLES & KNUCKLE BUSHINGS

#### All Rowcrop Models

9. To remove the steering knuckles (1—Fig. O10 or O10A) and bushings

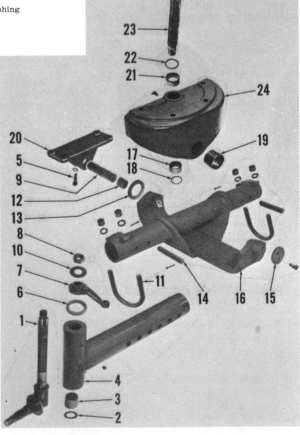

Fig. O10A — Exploded view of the short wheelbase front axle assembly and bolster assembly available for series 1750 rowcrop tractors.

1. Steering knuckle
2. Thrust washer
3. Bushing
4. Outer member
5. Lock screw
6. Felt washer
7. Steering arm
8. Nut
9. Pivot pin, rear
10. Lock washer
11. "U" bolt
12. Bushing
13. Rear washer
14. Pin
15. Front washer
16. Center member
17. Bearing
18. Seal, lower
19. Bushing
20. Support
21. Bearing
22. Seal, upper
23. Shaft
24. Bolster

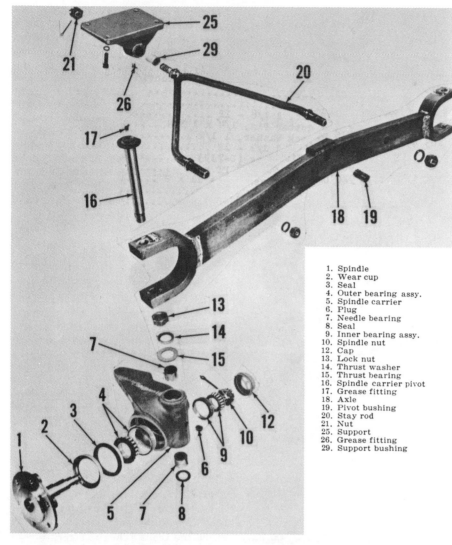

1. Spindle
2. Wear cup
3. Seal
4. Outer bearing assy.
5. Spindle carrier
6. Plug
7. Needle bearing
8. Seal
9. Inner bearing assy.
10. Spindle nut
12. Cap
13. Lock nut
14. Thrust washer
15. Thrust bearing
16. Spindle carrier pivot
17. Grease fitting
18. Axle
19. Pivot bushing
20. Stay rod
21. Nut
25. Support
26. Grease fitting
29. Support bushing

Fig. O11—Exploded view of the live spindle, non-adjustable axle typical of those used on Wheatland, Industrial and Ricefield models.

Seals, wear cups and bearing assemblies can be removed, or renewed, after removing the spindles from the spindle carriers. It is important that the wearing surface of wear cups (2) be smooth and square with seal (3). Use crocus cloth, or very fine sand paper to remove slight nicks or foreign matter from wear cup. Apply a thin coat of shellac or gasket cement to outside diameter of grease seal and to inside of wear cup prior to installation.

## MANUAL STEERING GEAR
### All Models So Equipped

12. **REMOVE AND REINSTALL.** To remove manual steering gear assembly, first disconnect tie-rods from steering gear arm. Loosen the steering shaft support bearing at clutch housing and the universal at the gear unit worm shaft. Pull universal from gear unit worm shaft. Unbolt stay-rod support from main frame. Raise and support front of tractor. Remove cap screws which retain front axle support (bolster) to main frame and roll complete assembly away from tractor. Gear unit can now be removed from bolster.

NOTE: When gear unit is removed from bolster do not turn worm shaft hard against stop in either direction as damage to ball guides could result.

13. **WORMSHAFT ADJUSTMENT.** Disconnect tie-rods from steering arm and steering shaft from worm gear shaft. Loosen lash adjusting screw (6—Fig. O12) several turns to relieve any load that may be imposed by the meshing of sector and worm gears. Loosen lock nut (27) and turn adjuster (26) inward until 8-12 in.-lbs. torque is required to rotate the worm shaft. Tighten lock nut and recheck. Adjust sector mesh as outlined in paragraph 14.

14. **SECTOR MESH ADJUSTMENT.** This adjustment is controlled by lash adjuster screw (6); however, before making any adjustment, make sure that the wormshaft adjustment is correct as specified in paragraph 13.

Disconnect tie-rods from sector shaft arm. Loosen lock nut (7). Locate mid-position of steering gear sector by rotating steering wheel from full right to full left, counting the total number of turns; then, rotate steering wheel back exactly half way to the center or mid-position. With steering gear sector in its mid-position of travel, rotate lash adjuster (6) in a clockwise direction until a slight drag is felt only when steering wheel passes

(3), support front of tractor and remove tire and wheel assembly. Straighten lock washer (10) and remove nut (8). Pull knuckle (1) from axle extension. Remove thrust washer (2) from knuckle and drive bushings (3) from axle extensions.

Install new bushings with outer ends flush with bore. Bushings should be reamed, if necessary, to provide a suggested diametral clearance of 0.002-0.005 for the steering knuckle.

## SPINDLES & SPINDLE CARRIERS
### All Wheatland, Industrial and Ricefield Models

10. **SPINDLES.** Front wheels are attached to rotating ("live") spindles (1—Fig. O11) which are carried in tapered roller bearings mounted in the spindle carriers (5). The spindle carriers are supported by pivot pins (16) anchored in the axle yokes and

fitted with needle roller bearings (7) and separate thrust bearings (15).

Spindle bearing adjustment is accomplished by removing cap (12) and the cotter pin from spindle unit, then adjusting the spindle nut to provide a slight rotational drag. Bearings and bearings cones can be renewed without removing the spindle carrier from axle.

11. **SPINDLE CARRIERS.** To remove or renew the pivot pin (16) and spindle carrier (5), first remove wheel; then, remove tie-rod end from spindle carrier. Remove the cone lock nut from bottom end of pivot pin and bump pivot pin up and out of axle yoke. The pivot pin needle bearings (7) can now be renewed. Press, or drive, on the end of the bearing which is stamped with the manufacturers name when installing new needle bearings.

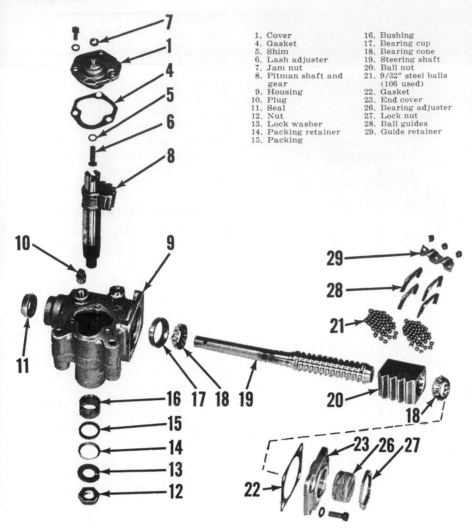

1. Cover
4. Gasket
5. Shim
6. Lash adjuster
7. Jam nut
8. Pitman shaft and gear
9. Housing
10. Plug
11. Seal
12. Nut
13. Lock washer
14. Packing retainer
15. Packing
16. Bushing
17. Bearing cup
18. Bearing cone
19. Steering shaft
20. Ball nut
21. 9/32" steel balls (106 used)
22. Gasket
23. End cover
26. Bearing adjuster
27. Lock nut
28. Ball guides
29. Guide retainer

Fig. O12—Exploded view of the manual steering gear unit used on axle type tractors.

through the mid-position. Wheel should revolve freely at all other points in its rotation. Tighten the lash adjuster lock nut.

Gear is correctly adjusted when 16 to 20 inch pounds of torque is required to pass through mid-position of gear travel. Measurement is taken at wormshaft.

Note: Backlash adjusting screw (6) should have from zero to 0.002 end play in gear. If end play exceeds 0.002 it will prevent correct adjustment of backlash; in which case, sector cover (1) should be removed and a shim (5) of correct thickness added at head of adjuster screw to remove the excess backlash. Shims are available in a kit of four shims, 0.063, 0.065, 0.067 and 0.069 thick.

15. OVERHAUL GEAR ASSEMBLY. To disassemble steering gear unit, remove assembly as outlined in paragraph 12; then, proceed as follows: Remove pitman arm. Remove cap screws from sector cover (1) and withdraw sector and cover as a unit

from housing. Remove worm (screw) shaft cover (23) and withdraw wormshaft through this opening. Wormshaft bearing cup (17) and/or oil seal (11) can be renewed at this time.

Ball nut (20) should move along grooves in wormshaft smoothly and with minimum end play. If worm shows signs of wear, scoring or other derangement, it is advisable to renew the worm and nut as a unit. To disassemble nut (20) from worm (screw) shaft (19), remove ball retainer clamp (29), ball return guides (28), balls and worm nut.

To reassemble ball nut, place nut over middle section of worm as shown in Fig. O13. Drop bearing balls into one retainer hole in nut and rotate wormshaft slowly to carry balls away from hole. Continue inserting balls in each circuit until circuit is full to bottom of both holes. If end of worm is reached while inserting balls and rotating worm in nut, hold the balls in position with a clean blunt rod as

shown in Fig. O13 while shaft is rotated in an opposite direction for a few turns. Remove the rod and drop the remaining balls in the circuit. Make certain that no balls are outside regular ball circuits. If balls remain in groove between two circuits or at ends, they cannot circulate and will cause gear failure. Next, lay one-half of each split guide (28—Fig. O12) on the bench and place 13 balls in each. Place the other halves of each retainer over the balls. While holding the halves together, plug their ends with heavy grease to prevent the balls from dropping out; then, insert complete retainer units in worm nut and install guide clamp.

Sector shaft large bushing (16—Fig. O12) has an inside diameter of 1.375 inches. Other sector shaft bushing has an inside diameter of 1.0625 inches. I&T suggested clearance of shaft in bushings is 0.0015-0.003.

Note: Sector cover and its bushing are not available separately.

Select and insert a shim (5) on sector mesh adjusting screw (6) to provide zero to 0.002 end play, before reinstalling sector and shaft in gear housing. Adjust worm (screw) shaft bearings and sector mesh as outlined in paragraphs 13 and 14.

## FRONT SUPPORT (BOLSTER)

### Models So Equipped

16. R & R AND OVERHAUL. On models having manual steering, remove the front axle and steering gear assembly as outlined in paragraph 12; then, unbolt and remove the manual steering gear assembly. Remove the pin which retains the axle pivot pin in the front bolster and drive pivot pin from bolster. Front support can now be removed from front axle.

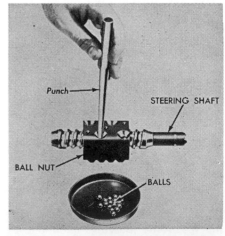

Fig. O13—Aligning ball nut on wormshaft while inserting balls in ball circuit. Insert one-half of the total number of balls (106) in each circuit and guide.

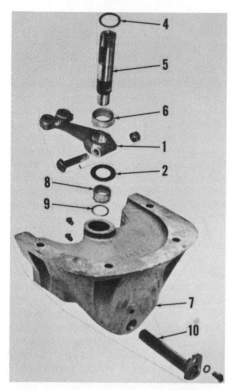

**Fig. O13A—View of front support from Wheatland model tractor with power steering. While shapes may differ, the operating parts of front supports remain basically the same.**

| | |
|---|---|
| 1. Steering arm | 6. Bearing |
| 2. Thrust washer | 7. Bolster |
| 4. Seal | 8. Bearing |
| 5. Pitman shaft | 9. Seal |
| | 10. Pivot pin |

On those models equipped with power steering, remove the power steering unit or power steering cylinder as outlined in paragraph 51 or 88 and the front axle assembly as outlined in paragraph 1. Unbolt bolster and remove from front main frame. Remove clamp bolt from pitman arm, mark the position of pitman arm and shaft, then remove pitman arm and thrust washer from front bolster. Oil seals and needle bearings can now be removed. See Fig. O13A.

When reinstalling needle bearings, drive on end which is stamped with number and align the oil holes in bearings with the oil holes in bolster. Install oil seals with lips toward outside (away from pitman arm).

# FRONT SYSTEM (Tricycle Type)

## FRONT SUPPORT ASSEMBLY

### All Rowcrop Models

**17. R & R VERTICAL SPINDLE.** To remove or renew the vertical spindle (12—Fig. O14), first remove the power

steering unit or cylinder as outlined in paragraph 51 or 88. Support front of tractor and remove wheels and hubs. Remove nut (1), raise front of tractor and either bump or pull vertical spindle from support (5). Lower bearing assembly (9), seal retainer (10) and felt seal (11) can be renewed at this time. Be sure to inspect grease retainer (4) closely and if damaged during removal of vertical spindle, renew same.

When reinstalling vertical spindle, adjust same with nut (1) until spindle has zero end play yet turns freely.

**18. R & R SUPPORT ASSEMBLY.** To remove the front support as an assembly, first remove the power steering unit or cylinder as outlined in paragraph 51 or 88. Raise front of tractor and unbolt support (5) from front main frame.

# FRONT SYSTEM (Four Wheel Drive)

Series 1750, 1800, 1850, 1900 and 1950 tractors are available with a front drive axle which is driven from the transmission bevel pinion shaft via a transfer case and a drive shaft fitted with two universal joints. A shifting mechanism in the transfer case allows connecting or disconnecting power to the front drive axle as desired.

All four wheel drive tractors are equipped with power steering. All tractors prior to serial number 134 684-000 were fitted with a Char Lynn control unit whereas all tractors after serial number 134 683-000 are fitted with a Saginaw control unit. These power steering units, along with the steering cylinders, will be discussed in the Power Steering section.

## FRONT AXLE AND CARRIER

### Models So Equipped

**20. R & R AXLE ASSEMBLY.** The complete front axle assembly can be removed from tractor as follows: Disconnect drive shaft from companion flange of differential pinion shaft. Disconnect both power steering cylinders from axle and spindle supports and lay cylinders on top of axle carrier. Remove bolts retaining axle to axle carrier, then raise tractor and roll complete axle and wheels assembly forward away from tractor.

Note: A rolling floor jack can be placed under differential pinion shaft

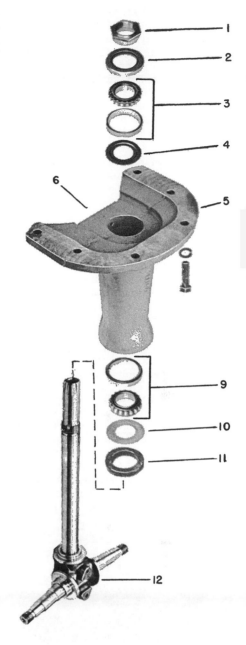

**Fig. O14—Exploded view of tricycle front support and associated parts.**

| | |
|---|---|
| 1. Nut | 6. Grease fitting |
| 2. Oil seal | 9. Lower bearing assy. |
| 3. Upper bearing assy. | 10. Felt retainer |
| 4. Grease retainer | 11. Felt washer |
| 5. Support | 12. Vertical spindle |

to keep axle from rotating as tractor is lifted from axle.

If necessary, wheels and tie-rod can now be removed and procedure for doing so is obvious.

Reinstall axle by reversing the removal procedure and be sure piston rod ends of steering cylinders are attached to steering spindle supports. Tighten the cylinder attaching bolt lock nuts until they just contact the mounting flanges. Further tightening may distort mounting flanges and cause cylinder to bind.

**21. R & R AXLE CARRIER.** To remove axle carrier, first remove front axle as outlined in paragraph 20, then secure steering cylinders to tractor frame. Place a rolling floor jack under axle carrier and take weight of carrier. Remove pivot pin retaining cap screws, slide pivot pins from pivot supports and lower axle carrier from tractor. If necessary, pivot supports can be removed from tractor frame.

Bushings in axle carrier can now be renewed. Bushings are pre-sized and should require no final sizing if carefully installed.

**22. OVERHAUL FRONT AXLE.** Overhaul of the front drive axle assembly will be discussed as four operations; the planet spider assembly, the hub assembly, the spindle support assembly and the differential and carrier assembly. All operations except the differential and carrier overhaul can be accomplished without removing the front drive axle from tractor. Both outer ends of axle are identical, hence, only one outer end will be discussed.

**23. PLANET SPIDER.** To overhaul the planet spider assembly, support outer end of axle and remove the tire and rim. Remove relief valve from center of planet spider, remove plug from wheel hub and drain planet spider. Remove cap screws which retain planet spider to wheel hub and

**Fig. O20 — Remove pinion shaft pins by driving them inward.**

the two puller hole cap screws. Use two of the removed retaining cap screws in the puller holes and pull planet spider assembly from wheel hub.

With unit removed, remove the three pinion shaft lock pins by driving them toward center of unit as shown in Fig. O20. Remove pinion shafts and expansion plugs by driving pinion shafts toward outside of planet spider. Remove planet pinions, rollers (34 each pinion) and thrust washers. Discard the expansion plugs. See Fig. O21.

Clean and inspect all parts and renew as necessary. Pay particular attention to the pinion rollers and thrust washers.

**24.** When reassembling, use heavy grease to hold rollers in inner bore of pinions. Be sure tangs of thrust washers are in the slots provided for them and that holes in pinion shaft and mounting boss are aligned before pinion shafts are final positioned. Coat mating surfaces of planet spider and wheel hub with Permatex No. 2 or equivalent sealer and install planet spider in wheel hub. Tighten retaining cap screws to a torque of 52-57 ft.-lbs.

**25. HUB ASSEMBLY.** To overhaul the wheel hub assembly, first remove planet spider assembly as outlined in paragraph 23. With planet spider removed, remove snap ring and sun gear from outer end of axle shaft. See Fig. O22. If necessary, the internal gear can be removed at this time by clipping the lock wires and removing the attaching cap screws. Straightened tabs of spindle nut lock washer, then use OTC tool JD-4, or its equivalent, and remove spindle outer nut and the lock washer. Now loosen, but DO NOT remove, the spindle inner nut. Unbolt spindle from spindle support and remove wheel hub assembly from spindle support as shown in Fig. O23.

Place hub assembly on bench with spindle nut on top side and block up assembly so spindle will be free to drop several inches. Remove the spindle inner nut, then place a wood block over end of spindle and bump spindle from the internal gear hub. Lift internal gear hub and bearing from wheel hub and be careful not to allow bearing to drop from hub of internal gear hub. Complete removal of spindle from wheel hub. All bearings and seals, thrust washers and dirt shield can now be removed and renewed, if necessary, and procedure for doing so is obvious. Bushing (35— Fig. O24) and oil seal (33) in inner bore of spindle can also be renewed at this time.

**26.** Use Fig. O24 as a reference and reassemble wheel hub assembly as follows: Install bearing cups (7 and 37) in hub with smallest diameters toward inside of hub. Place inner bearing in inner bearing cup, then install oil seal (34) with lip facing bearing. Bump seal into bore until it bottoms. Place dirt shield (8) on hub so flat side is toward flange of spindle, then using caution not to damage seal, install spindle in wheel hub. Hold spindle in position and turn unit over so threaded end of spindle shaft is on top side. Place outer bearing over end of spindle and push down into bearing cup. Start bearing hub of internal gear hub (1) into outer bearing cone, and if necessary, tap gear lightly with a soft faced hammer to position. Install thrust washer (38) and spindle inner nut (39) and tighten nut only finger tight. Coat mating surfaces of spindle and spindle support with Permatex No. 2 or equivalent sealer and install dirt shield and spindle on spindle support. Tighten retaining cap screws to a torque of 80-88 ft. lbs.

Adjust inner nut as required until a pull of 33-38 pounds on a spring

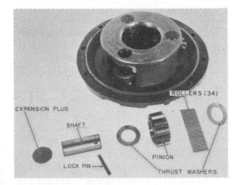

**Fig. O21—View showing one pinion assembly removed from spider.**

**Fig. O22—Sun gear can be removed from outer end of axle shaft after snap ring is removed. Note method of safety wiring the internal gear cap screws.**

**Fig. O23 — Removing wheel hub assembly from spindle support.**

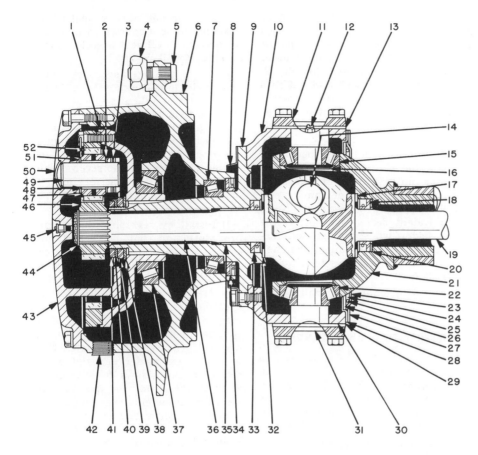

**Fig. O24 — Cross sectional view showing components of front drive axle outer end.**

| | | |
|---|---|---|
| 1. Internal gear hub | 19. Axle shaft (inner) | 36. Axle shaft (outer) |
| 2. Thrust washer | 20. Washer | 37. Hub outer bearing |
| 3. Pinion shaft lock pin | 21. Axle housing | assy. |
| 6. Wheel hub | 22. Lower trunnion bearing | 38. Thrust washer |
| 7. Hub inner bearing | assy. | 39. Inner nut |
| assy. | 23. Dust seal retainer | 40. Lock washer |
| 8. Dirt shield | 24. Dust seal | 41. Outer nut |
| 9. Spindle | 25. Spring | 42. Filler plug hole |
| 10. Spindle support | 26. Seal (felt) | 43. Planet spider |
| 11. Upper trunnion | 27. Retainer ring | 44. Snap ring |
| 12. Grease fitting | 28. Seal retainer | 45. Relief valve |
| 13. Shims | 29. Gasket | 46. Sun gear |
| 14. Universal joint | 30. Shims | 47. Pinion |
| 15. Upper trunnion bearing | 31. Lower trunnion | 48. Pinion rollers |
| assy. | 32. Thrust washer | 50. Expansion plug |
| 16. Grease retainer | 33. Oil seal | 51. Thrust washer |
| 17. Thrust washer | 34. Oil seal | 52. Internal gear |
| 18. Oil seal | 35. Bushing | |

scale attached to a wheel stud is required to keep hub in motion. See Fig. 025. Install lock washer (40—Fig. 024) and outer nut (41). Tighten outer nut and recheck the hub rolling torque. When adjustment is correct, bend tabs on lock washer to secure both nuts. Install sun gear (46) and snap ring (44) on outer end of axle shaft. If internal gear (52) was removed, install gear so smallest outside diameter is in counterbore of internal gear hub. After retaining cap screws are tightened, secure in pairs with safety wire as shown in Fig. 022. Coat mating surfaces of wheel hub and planet spider with Permatex No. 2 or its equivalent sealer and install planet spider. Tighten retaining cap screws to a torque of 52-57 ft.-lbs. Install the puller hole cap screws and the tire and rim.

**27. SPINDLE SUPPORT.** The spindle support can be serviced after planet spider and wheel hub assemblies are removed as outlined in paragraphs 23 and 25. However, if service

is required only on the spindle support, the planet spider, wheel hub and axle shaft can be removed as an assembly as follows:

Raise outer end of axle and remove tire and rim. Attach a hoist to wheel stud, then unbolt spindle from spindle support and pull complete hub assembly and axle shaft from outer end of axle. See Fig. 026. Do not allow weight of the hub assembly to be supported by axle shaft or damage to oil seal in axle housing outer end will result.

With the complete hub assembly and axle shaft removed, disconnect tie-rod and power steering cylinder from spindle support. Remove the cap screws from the two-piece retainer ring on inner side of spindle support and separate retainers, seals and gasket from spindle support as shown in Fig. 027.

NOTE: At this time it is desirable to remove the grease from the cavity formed by the spindle support and outer end of axle housing.

Remove upper trunnion, pull top of spindle support outward and remove spindle support from outer end of axle housing as shown in Fig. 028. Keep shims present under top trunnion tied to the trunnion for use during reassembly. Remove upper trunion bearing from axle housing. Remove lower trunnion, shims and bearing from spindle support. Both trunnion bearing cups and upper trunnion bearing grease retainer can now be removed, if necessary. If necessary to remove axle shaft thrust washer, oil seal and oil seal washer from axle outer end, a slide hammer and puller can be used. Seals (Fig. 027) on outer end of axle housing can also be removed at this time.

Clean and inspect all parts and renew as necessary. It is recommended that new seals be used during assembly.

**Fig. O25 — Use method shown to check wheel hub bearing preload.**

**Fig. O26 — Planet spider, wheel hub and axle shaft assembly can be removed as shown.**

28. To reassemble spindle support, proceed as follows: Install axle shaft seal washer and oil seal with lip toward inside and be sure oil seal is bottomed. Install axle shaft thrust washer. Install seal components over outer end of axle housing in the following order: Inner seal retainer with step toward inside of tractor. Dust seal spring, rubber dust seal and felt grease seal and be sure that bevel on inside diameter of both seals is toward the bell of axle outer end. Outer seal retainer with step toward outside of tractor and gasket. Install grease retainer, cup side down, and upper bearing cup, smallest I. D. down, in the upper trunnion bearing bore. Bolt the lower trunnion to spindle support using the original shims and tighten the cap screws to a torque of 80-88 ft.-lbs. Place the lower trunnion bearing over lower trunnion. Install lower trunnion bearing cup in outer end of axle with smallest I. D. up. Place upper trunnion bearing in the upper trunnion bearing cup, then while tipping upper side of spindle support slightly outward, position spindle support over outer end of axle and install the upper trunnion and original shim pack. Tighten trunnion retaining cap screws to a torque of 80-88 ft.-lbs.

Before attaching tie-rod, power steering cylinder or seal assembly to spindle support, check adjustment of trunnion bearings as follows: Connect a spring scale to tie-rod hole of spindle support and check the effort required to rotate the spindle support as shown in Fig. O29. This effort should be 12-18 pounds on the spring scale. If adjustment is required, vary the number of shims located under trunnions. Keep the number of shims equal on upper and lower trunnions. Shims are available in thicknesses of 0.003 (green), 0.005 (blue) and 0.010 (brown).

Use grease to hold seal retainer gasket in place, then install the seal assembly on spindle support and be sure none of the split ends of outer seal assembly are aligned. Attach tie-rod and tighten tie-rod stud nut to a torque of 200 ft.-lbs. Attach power steering cylinder and tighten attaching bolt lock nut until it just contacts mounting flange.

Place approximately four pounds of grease in the cavity of spindle support and pack universal joint of axle shaft. Coat mating surfaces of spindle and spindle support with Permatex No. 2 or equivalent sealer and install the planet spider, wheel hub and axle assembly on the spindle support. Tighten the attaching cap screws to a torque of 80-88 ft.-lbs.

Complete assembly by installing tire and rim.

## DIFFERENTIAL AND CARRIER

The Four Wheel Drive front axle can be equipped with either a conventional differential assembly or a "No-Spin" differential assembly. Removal procedure will be the same for both types. For service information, refer to paragraphs 31 and 34A.

30. **REMOVE AND REINSTALL.** To remove the differential and carrier assembly, raise front of tractor until tires and rims can be removed and support in this position. Block axle carrier to prevent axle assembly from rocking, then remove front tires and rims. Drain differential housing. Disconnect power steering cylinders and lay them on top side of axle carrier. Disconnect drive shaft from companion flange of differential pinion shaft.

Attach a hoist to one of the wheel studs and take up slack of hoist. Unbolt spindle from spindle support and pull complete assembly from outer end of axle. See Fig. O26. Repeat operation for opposite side.

NOTE: Do not allow weight of hub assembly to be supported by axle shaft or damage to oil seal in axle housing outer end will result.

Place a rolling floor jack under front axle, unbolt axle from axle carrier and lower axle away from tractor. Position axle on supports with differential pinion shaft up and secure in this position with blocks. Disconnect one end of tie-rod and swing it out of the way. Remove cap screws which retain differential carrier to axle housing and remove assembly from housing.

Reinstall by reversing the removal procedure and coat carrier retaining capscrews and mating surfaces of carrier and axle housing with Permatex No. 2 or equivalent sealer. Tighten cap screws to a torque of 37-41 ft.-lbs. Tighten the tie-rod stud nut to a torque of 200 ft.-lbs. When joining spindle to spindle support, coat the mating surfaces with Permatex No. 2 or equivalent sealer and tighten the retaining cap screws to a torque of 80-88 ft.-lbs. Piston rod ends of power steering cylinders are attached to spindle supports. Tighten cylinder attaching bolt lock nuts until they just contact the mounting flanges.

31. **OVERHAUL (EXCEPT NO-SPIN).** With differential and carrier removed as outlined in paragraph 30, unit is disassembled as follows: Straighten tab of lock washer (14—Fig. O30) and remove lock nut, lock washer and thrust screw (15). Punch mark the carrier bearing caps (19) so they can be reinstalled in their original positions, then remove cotter pins and the adjusting nut lock pins (21). Cut lock wires and remove the carrier bearing caps. Lift differential from carrier and keep bearing cups (18) identified with their bearing cones (20). Bearing cones can now be re-

Fig. O27 — Axle outer end seals and retainers are removed from spindle support as shown.

Fig. O28 — Remove spindle support from outer end of axle housing as shown.

Fig. O29 — Use a spring scale in tie-rod stud hole to check trunnion bearing adjustment.

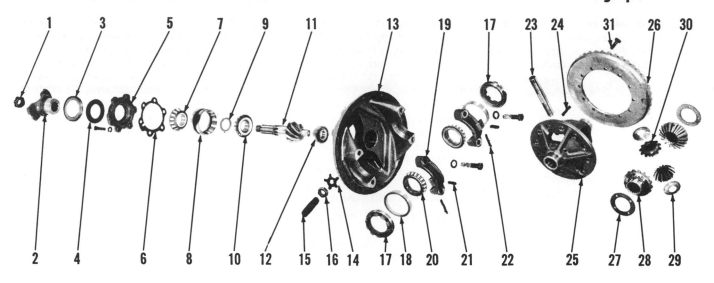

**Fig. O30—Exploded view of the differential and carrier assembly used in front drive axle of four-wheel drive tractors without "No-Spin "differential.**

| | | | |
|---|---|---|---|
| 1. Nut | 9. Spacer and shim kit | 16. Lock nut | 23. Pinion pin (shaft) |
| 2. Companion flange | 10. Bearing cone | 17. Adjusting nut | 24. Lock pin |
| 3. Dirt shield | 11. Pinion shaft | 18. Bearing cup | 25. Differential case |
| 4. Oil seal | 12. Pilot bearing | 19. Bearing cap | 26. Bevel ring gear |
| 5. Bearing retainer | 13. Carrier | 20. Bearing cone | 27. Thrust washer |
| 6. Gasket | 14. Lock washer | 21. Lock pin | 28. Side gear |
| 7. Bearing cone | 15. Thrust screw | 22. Cotter pin | 29. Thrust washer |
| 8. Bearing cup | | | 30. Pinion gear |

moved from differential case, if necessary. Unbolt and remove bevel ring gear from differential case if necessary. Drive pinion pin (shaft) lock pin (24) out of differential case and remove pinion pin (23), pinions (30), side gears (28) and thrust washers (27 and 29) from differential case.

Remove cotter pin and nut (1) from pinion shaft (11), then using a puller, remove the companion flange (2) and dust shield. Remove pinion shaft bearing retainer (5) and press shaft and bearings from carrier. Use a split bearing puller to support pinion bearing cup (8) on edge nearest pinion shaft gear and press pinion shaft from rear bearing (7) and bearing cup. Remove spacer (9) and any shims which may be present from pinion shaft. Remove front bearing (10) and inner (pilot) bearing (12) in a similar manner.

Clean and inspect all parts. Pay particular attention to bearings, bearing cups and thrust washers. If any of the differential side gears or pinions are damaged or excessively worn, renew all gears and thrust washers. Pinion shaft and bevel ring gear are available only as a matched set.

32. The differential and carrier unit is assembled as follows: Place the inner (pilot) bearing on inner end of pinion shaft and stake in four places. Use a piece of pipe the size of inner race of pinion shaft bearing and press the forward pinion shaft bearing (10) on shaft with taper facing threaded

end of pinion shaft. Place bearing spacer, and any shims which were present during disassembly over pinion shaft, then position the bearing cup over forward bearing. Press the rear pinion shaft bearing (7) on shaft with taper away from threaded end of pinion shaft. Check, and renew if necessary, the oil seal (4) in pinion shaft bearing retainer (5). Seal is installed with lip toward inside. Place gasket and bearing retainer over pinion shaft. Check, and renew if necessary, the dust shield (3) on companion flange. Position companion flange so it will not obstruct cotter pin hole in end of pinion shaft, slide bearing retainer oil seal on its land on companion flange, then press companion flange on pinion shaft. Install retaining nut, clamp companion flange in a vise and tighten nut to a torque of 300 ft.-lbs.

NOTE: Pressure of oil seal will generally hold bearing retainer away from bearing. If it does not, tie retainer to companion flange.

33. With pinion shaft assembled as outlined above, clamp the bearing cup in a soft jawed vise only tight enough to prevent rotation, then using an inch-pound torque wrench attached to companion flange retaining nut, check the rolling torque (bearing preload) of the pinion shaft. This rolling torque should be 13-23 in.-lbs.

If rolling torque of shaft is not as specified, it will be necessary to disassemble the shaft assembly and vary

the spacer and/or shims as required. A spacer and shim kit is available under Oliver part number 155 342-A.

With pinion shaft assembled and the rolling torque of shaft determined, press pinion shaft assembly into carrier, install bearing retainer and tighten cap screws to a torque of 25 ft.-lbs. Install cotter pin to lock nut in place.

34. Reassemble differential case assembly as follows: Place side gears, pinions and their thrust washers in differential case and install pinion pin (shaft). Secure pinion pin in place with lock pin and if lock pin is straight type, stake pin in position. If lock pin is a spring type, staking is not necessary. If bevel ring gear was removed, reinstall with bolt heads next to bevel ring gear and tighten nuts to a torque of 78-86 ft.-lbs. Press bearings on differential case with tapers facing away from case. Place bearing cups over differential bearings and place differential assembly in carrier. Position bearing adjusting nuts in carrier and install the carrier bearing caps. Tighten the bearing cap retaining cap screws until caps are snug but BE SURE threads of caps and adjusting nuts are in register. Maintain some clearance between gear teeth and tighten adjusting nuts until bearing cups are seated and all end play of differential is eliminated. Mount a dial indicator and shift differential assembly as required to obtain a backlash of 0.008-0.011 between bevel

pinion shaft gear and bevel ring gear. Differential is shifted by loosening one adjusting nut and tightening the opposite an equal amount.

Note: Fore and aft position of the bevel pinion shaft is not adjustable.

With gear backlash adjusted, tighten the bearing cap retaining cap screws to a torque of 65 ft.-lbs and secure with lock wire. Install adjusting nut lock pins and cotter pins.

NOTE: If lock pins will not enter slots of adjusting nuts after backlash adjustment has been made, tighten rather than loosen the adjusting nut, or nuts. Recheck gear backlash.

Install thrust screw and turn screw in until it contacts bevel ring gear, then back it out ¼-½ turn. Apply sealant to threads of thrust screw at surface of carrier, then install lock washer and lock nut. Secure lock nut by bending one tang of lock washer over nut and one tang over boss of carrier.

**34A. OVERHAUL (NO-SPIN).** With differential and carrier removed as outlined in paragraph 30, unit is disassembled as follows: Straighten tab of lock washer (14—Fig. O30) and remove lock nut, lock washer and thrust screw (15). Punch mark the carrier bearing caps (19) so they can be reinstalled in their original positions, then remove cotter pins and the adjusting nut lock pins (21). Cut lock wires and remove the carrier bearing caps. Lift differential from carrier and keep bearing cups (18) identified with their bearing cones (20). Bearing cones can now be removed from differential case, if necessary.

Remove cotter pin and nut (1) from pinion shaft (11), then using a puller, remove the companion flange (2) and dust shield. Remove pinion shaft bearing retainer (5) and press shaft and bearings from carrier. Use a split bearing puller to support pinion bearing cup (8) on edge nearest pinion shaft gear and press pinion shaft from

rear bearing (7) and bearing cup. Remove spacer (9) and any shims which may be present from pinion shaft. Remove front bearing (10) and inner (pilot) bearing (12) in a similar manner.

Unbolt and remove bevel ring gear from differential, if necessary. Remove differential case bolts and hold case together as last bolt is removed to keep assembly from flying apart due to the internal spring pressure. See Fig. O30A. Hold out rings (6) can be removed with snap ring spreaders.

**34B.** Clean and inspect all parts. Check splines on side gears and clutch members and remove any burrs or chipped edges with a stone or burr grinder. Renew any parts which have sections of splines broken away. Inspect springs (3) for fractures, or other damage, and renew springs which do not have a free height of 2¼-2½ inches. Center cam in central driver (7) must be free to rotate within the limits of keys in central driver. Check the weld between driven clutch (5) and cam ring on clutch by tapping lightly on cams of cam ring. If cam ring rotates in driven clutch, weld is defective (failed). Inspect teeth on central driver and driven clutches. Small defects can be dressed with a stone. If central driver or driven clutch is renewed, also renew the part it mates with. A smooth wear pattern up to 50 percent of face width is acceptable for the cams on center cam and driven clutch.

**34C.** The differential and carrier unit is assembled as follows: Place the inner (pilot) bearing on inner end of pinion shaft and stake in four places. Use a piece of pipe the size of inner race of pinion shaft bearing and press the forward pinion shaft bearing (10—Fig. O30) on shaft with taper facing threaded end of pinion shaft. Place bearing spacer, and any shims which were present during disassembly over pinion shaft, then position the bearing cup over forward bearing.

Press the rear pinion shaft bearing (7) on shaft with taper away from threaded end of pinion shaft. Check, and renew if necessary, the oil seal (4) in pinion shaft bearing retainer (5). Seal is installed with lip toward inside. Place gasket and bearing retainer over pinion shaft. Check, and renew if necessary, the dust shield (3) on companion flange. Position companion flange so it will not obstruct cotter pin hole in end of pinion shaft, slide bearing retainer oil seal on its land on companion flange, then press companion flange on pinion shaft. Install retaining nut, clamp companion flange in a vise and tighten nut to a torque of 300 ft.-lbs.

NOTE: Pressure of oil seal will generally hold bearing retainer away from bearing. If it does not, tie retainer to companion flange.

**34D.** With pinion shaft assembled as outlined above, clamp the bearing cup in a soft jawed vise only tight enough to prevent rotation, then using an inch-pound torque wrench attached to companion flange retaining nut, check the rolling torque (bearing preload) of the pinion shaft. This rolling torque should be 13-23 in.-lbs.

If rolling torque is not as specified, it will be necessary to disassemble the shaft assembly and vary the spacer and/or shims as required. A spacer and shim kit is available under Oliver part number 155 342-A.

With pinion shaft assembled and the rolling torque of shaft determined, press pinion shaft assembly into carrier, install bearing retainer and tighten cap screws to a torque of 25 ft.-lbs. Install cotter pin to lock nut (1) in place.

**34E.** Lubricate all parts and reassemble no-spin differential by reversing disassembly procedure. Be sure to position spring retainers so that spring seats inside cupped section of retainer. Be sure key in central driver is aligned with slot in holdout ring. Tighten the case bolts to 36 ft.-lbs. torque and use a single wire through holes in bolt heads to lock bolts. Place an axle shaft into each side of differential and turn axle back and forth. It should be possible to feel backlash between clutch teeth. Backlash should be about $\frac{5}{32}$-inch. If no backlash can be felt and side gear is locked solid, check differential for incorrect assembly.

If bevel ring gear was removed, reinstall with bolt heads next to bevel ring gear and tighten nuts to a torque of 78-86 ft.-lbs. Press bearings on differential case with tapers facing away from case. Place bearing cups over

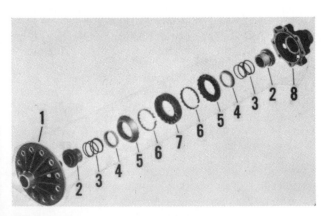

**Fig. O30A — Exploded view of the "No-Spin" differential assembly available for the Four Wheel Drive front axle.**

1. Case half
2. Side gear
3. Spring
4. Spring retainer
5. Driven clutch
6. Hold out ring
7. Central driver and center cam
8. Case half

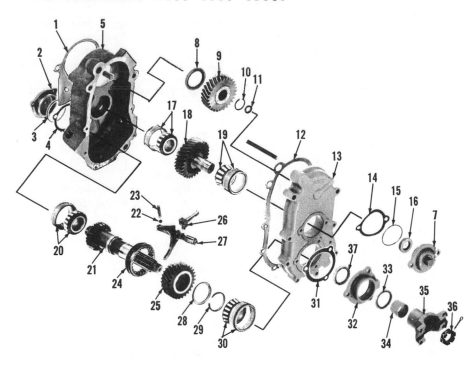

**Fig. O31 — Exploded view of the transfer drive assembly used on four-wheel drive tractors.**

1. Gasket
2. Bearing retainer
3. "O" ring
4. Shim
5. Transfer case
7. Bearing cover
8. Oil seal
9. Drive gear
10. Snap ring
11. Oil seal
12. Gasket
13. Transfer case cover
14. Shim

15. "O" ring
16. Oil seal
17. Bearing assy.
18. Idler gear and shaft
19. Bearing assy.
20. Bearing assy.
21. Output shaft
22. Detent ball
23. Detent spring
24. Coupling
25. Output gear

26. Actuator
27. Shifter fork
28. Thrust washer
29. Snap ring
30. Bearing assy.
31. Gasket
32. Bearing retainer
33. Oil seal
34. Spacer
35. Companion flange
36. Nut
37. Oil seal

differential bearings and place differential assembly in carrier. Position bearing adjusting nuts in carrier and install the carrier bearing caps. Tighten the bearing cap retaining cap screws until caps are snug but BE SURE threads of caps and adjusting nuts are in register. Maintain some clearance between gear teeth and tighten adjusting nuts until bearing cups are seated and all end play of differential is eliminated. Mount a dial indicator and shift differential assembly as required to obtain a backlash of 0.008-0.011 between bevel pinion shaft gear and bevel ring gear. Differential is shifted by loosening one adjusting nut and tightening the opposite an equal amount.

Note: Fore and aft position of the bevel pinion shaft is not adjustable.

With gear backlash adjusted, tighten the bearing cap retaining cap screws to a torque of 65 ft.-lbs. and secure with lock wire. Install adjusting nut lock pins and cotter pins.

NOTE: If lock pins will not enter slots of adjusting nuts after backlash adjustment has been made, tighten rather than loosen the adjusting nut, or nuts. Recheck gear backlash.

Install thrust screw and turn screw in until it contacts bevel ring gear, then back it out ¼-½ turn. Apply sealant to threads of thrust screw at surface of carrier, then install lock washer and lock nut. Secure lock nut by bending one tang of lock washer over nut and one tang over boss of carrier.

**TRANSFER DRIVE**

**Models So Equipped**

35. **R&R AND OVERHAUL.** To remove the transfer drive assembly, it will be necessary to split the tractor front main frame from the rear main frame. In cases where tractor is equipped with Hydra-Power drive, Reverse-O-Torc drive or Creeper drive, it will also be necessary to remove the engine, along with the auxiliary drive unit, before tractor can be split.

36. To split tractor when it is equipped with an auxiliary drive unit, remove engine as outlined in paragraph 95 or 240. Disconnect speedometer drive cable from adapter. Disconnect shifting rod from transfer drive shifter arm and pull it from hole in front

main frame. Disconnect drive shaft from companion flange of transfer drive output shaft. Unclip and disconnect power steering oil lines. Disconnect safety starting switch wires. Disconnect rear light wires and remove clips. On non-diesel tractors, unhook governor control bellcrank spring. Support tractor front frame in a manner which will prevent tipping, then support rear frame with a rolling floor jack. Remove the front frame to rear frame retaining cap screws and separate tractor.

37. To split tractor which does not have an auxiliary drive unit, remove the pto drive shaft as outlined in paragraph 356, or the hydraulic pump drive shaft as outlined in paragraph 409. Remove drive shaft shield, then remove the sprocket chain from transmission input shaft sprocket. Disconnect speedometer drive cable from adapter. Disconnect shifting rod from transfer drive shifter arm and pull it from hole in front main frame. Disconnect drive shaft from companion flange of transfer drive output shaft. Unclip and disconnect power steering lines. Disconnect safety starting switch wires. Disconnect rear light wires and remove clips. On non-diesel tractors, unhook governor control bellcrank spring. Support tractor front frame in a manner which will prevent tipping, then support rear frame with a rolling floor jack. Remove the front frame to rear frame retaining cap screws and separate tractor.

38. With tractor split as outlined in paragraph 36 or 37, drain transmission and transfer housings. Place transmission and transfer drive in gear and remove companion flange and spacer from transfer drive output shaft. Disconnect oil hose, or line, from transfer drive case, then unbolt transfer drive cover (13—Fig. O31) from case and remove cover along with idler gear (18) and output gear (25) and shaft (21). Remove shifter coupling (24) from fork if necessary. Remove snap ring (10) and gear (9) from forward end of transmission bevel pinion shaft. Discard snap ring. Unbolt and remove transfer case (5) from tractor rear frame. Shifter fork (27) and actuator (26) can be removed after shifter arm and Woodruff key are removed. Any further disassembly required will be obvious, however, save and identify shims located under bearing retainers for subsequent use.

39. Clean and inspect all parts and renew as necessary. New seals, "O" rings and gaskets should be used dur-

ing assembly and, before unit is installed on tractor rear main frame, unit should be partially assembled and the end play of the idler and output shafts checked and adjusted as follows:

Install idler shaft rear bearing cup in transfer drive case and the output shaft rear bearing cup in shaft rear bearing retainer. Be sure both bearing cups are completely seated and install the output shaft rear bearing retainer (2) along with original shim pack (4). Install the output shaft front bearing retainer (32) with new gasket (31) and the idler shaft front bearing retainer (7) with original shim pack (14). Be sure both front bearing cups are seated against the bearing retainers. Place dowels in transfer drive case, if necessary. Place output shaft and idler shaft assemblies in transfer drive case cover, then using a new gasket, secure cover to case. Use a dial indicator and check end play of both shafts. End play should measure 0.001-0.003 for both

shafts and can be adjusted by varying the number of shims under the bearing retainers. However, if shims are added, be sure bearing cups are seated against retainers before rechecking the shaft end play.

With shaft bearings adjusted, separate unit and remove shafts and all bearing retainers. Renew "O" rings on bearing retainers and install new oil seals as follows: Bevel pinion shaft oil seal (8) in case (5) with lip of seal rearward. Bevel pinion shaft seal (11) in cover (13) with lip of seal forward. Shift fork actuator (26) seal with lip inward. Idler gear shaft front oil seal (16) with lip rearward. In the output shaft front bearing retainer, install the wide seal (37) at the rear with lip facing rearward and the narrow oil seal (33) at front with lip facing forward.

Coat all seals and "O" rings with grease prior to installation, and in addition, be sure the cavity between the two oil seals in output shaft front bearing retainer is completely filled

with grease. This can be done through the grease fitting after unit is assembled.

Reassemble unit by reversing the disassembly procedure. Use a new snap ring (10) when installing drive gear (9) on transmission bevel pinion shaft. If necessary, use grease on sliding coupling to hold it in position in shifter fork. When installing companion flange, be sure not to obstruct cotter pin hole.

Complete balance of tractor assembly by reversing the disassembly procedure.

### DRIVE SHAFT

#### Models So Equipped

40. The front axle drive shaft is conventional. Removal and overhaul is obvious after an examination of the unit. Spider and bearings assemblies are available as units only. All other parts are available separately.

When installing drive shaft, yoke end is installed at rear.

# FRONT SYSTEM
## (Hydraulic Power Front Wheel System)

Series 1750, 1850 and 1950 tractors are available with a hydraulic powered front wheel drive which uses pressurized oil to furnish power for the front wheels. The power unit is mounted in front of the tractor radiator and produces two hydraulic circuits which control and drive the tractor front wheels.

The delivery (primary) circuit furnishes the high volume of pressurized oil used to power the front wheels and is produced by a variable displacement piston type pump driven by a shaft attached to the engine crankshaft pulley. This oil is routed through hoses to fixed displacement wheel motors which drive the front wheels through reduction gears.

The control (secondary) circuit furnishes control pressure to engage and disengage the front wheels clutches as well as signal (control) pressure to regulate the delivery circuit operation. The signal pressure controls the delivery circuit by actuating (varying) a stroke control valve which is mechanically connected to a trunnion (swash plate) which regulates the displacement (stroke) of the piston pump.

The system also has an electrical system consisting of interlock

switches on the transmission shift rails and the clutch release lever as well as a solenoid lock on the drive control quadrant. The interlock switches permit front wheel drive operation only when transmission is in a forward position and the clutch pedal released (clutch fully engaged). The quadrant solenoid lock prevents "on-the-go" shifting from two wheel to four wheel operation. Whenever transmission is in a forward gear position and ignition switch in "ON", the quadrant lever is locked in "OFF" position. Transmission gear shift lever must be moved to neutral position to break electrical circuit to quadrant solenoid lock before front wheel drive can be engaged.

### TROUBLE SHOOTING

When testing the hydraulic front wheel drive assembly, checks must be made in the following sequence: Control circuit, electrical circuit and delivery circuit. Any other sequence will not give valid results.

41A. CONTROL CIRCUIT. The control circuit should maintain 90-130 psi with front wheel drive engaged and engine running. If pressure is less than 90 psi, check the following:

a. Check level of tractor hydraulic lift reservoir and if low, add fluid as required.

b. Check the manually controlled valve cable for breaks or maladjustment. Renew cable and/or adjust cable as outlined in paragraph 42E.

c. If doubt exists as to condition of quadrant pressure gage, attach a gage of known condition to gage line connector at front of valve housing and compare gage readings. Renew quadrant pressure gage if necessary.

41B. ELECTRICAL CIRCUIT. A system of interlock switches on the transmission shifter rods and clutch release lever is used to provide automatic start and stop control of the hydraulic front wheel drive. Delivery circuit of the hydraulic unit will not go into operation unless an electrical signal is present at the solenoid valve.

To check the interlock electrical system, connect a test light to the solenoid valve terminal, turn ignition switch to "ON" and move gear shift lever to a forward position. With clutch pedal released, test light should be on. Repeat test with gear shift lever on other side of shift pattern.

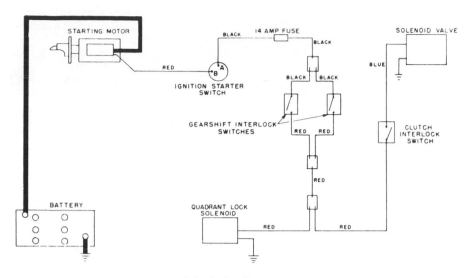

**Fig. O32—Wiring diagram of the hydraulic powered front wheel drive system.**

If test light does not react as stated, check for a blown fuse and if a blown fuse is found, locate cause before proceeding further. See interlock switch wiring diagram shown in Fig. O32. Also check clutch interlock switch adjustment as follows: Attach test light to **blue** wire terminal of clutch interlock switch, turn ignition switch to "ON" position and move gear shift lever to a forward position. Test light should now be on and should remain on until clutch pedal is depressed 1½ inches from its rest position at which time light should go off and remain off through remainder of pedal downward travel. If adjustment of clutch interlock switch is required, loosen housing retaining screws, turn housing until correct adjustment is obtained and tighten retaining screws.

If necessary, the shifter rail interlock switches can be isolated from the clutch interlock switch for testing as follows: Connect test light to **red** wire connection of clutch interlock switch and turn ignition switch on. Place shift lever in a forward position on both sides of the shift pattern and note test light which should be on in both cases. If test light is off in either case, renew the interlock switch on the side affected.

Install a new clutch interlock switch if an electrical signal is present at **red** wire connection of switch but switching action cannot be obtained. Adjust new switch as previously outlined.

**41C. DELIVERY CIRCUIT.** To test the delivery circuit, first remove test cap located on lower right side of gear pump housing and attach a 5000 psi gage. Lock tractor brakes and to assure complete safety, attach tractor drawbar to a substantial stationary

object. Start engine and place front drive quadrant lever in No. 6 position, then adjust engine speed to 1800 rpm. Check quadrant gage to be sure that control circuit pressure is at least 90 psi, then connect a jumper wire from any battery source to terminal of solenoid.

NOTE: Jumper wire must not be connected before engine is started and 90 psi pressure is showing on quadrant pressure gage.

Note reading on test gage which should be approximately 3000 psi.

If delivery circuit pressure is not as stated, adjust pressure control cable housing as outlined in paragraph 42A. If pressure cannot be adjusted by shifting cable, refer to paragraph 42C for maximum pressure adjustment.

**41D. DRIVE WHEEL.** The condition of drive wheel hydraulic clutch "O" rings and wheel motors oil seals can generally be determined as follows:

a. If oil flows from the primary gear housing breather with tractor stationary and front wheel drive engaged, the hydraulic clutch piston "O" rings are defective and should be renewed.

b. If oil flows from the primary gear housing breather with tractor moving and front wheel drive engaged, the wheel motor oil seal is defective and should be renewed.

### ADJUSTMENTS

Three adjustments can be made to control the delivery circuit pressure. They are: Pressure control cable adjustment, Minimum Stroke adjustment and Maximum Pressure adjustment. All adjustments must be made with hydraulic system at operating temperature and are made as follows.

**42A. PRESSURE CONTROL CABLE ADJUSTMENT.** Remove the test opening cap located at right bottom side of gear pump housing and attach a pressure gage of approximately 5000 psi capacity. See Fig. O32A. Lock the tractor brakes and to assure complete safety, secure drawbar to a substantial stationary object. Start engine and place quadrant control lever in No. 6 position. Adjust engine speed to 1800 rpm and note quadrant gage which must show at least 90 psi before proceeding with test.

Connect a jumper wire between a battery voltage source such as the alternator "BAT" terminal and the unit solenoid valve terminal. Note the reading on test gage which should be 1800 psi. If pressure is not as stated, loosen the jam nuts at forward end of control cable and shift cable forward and rearward as required and tighten jam nuts to hold adjustment. If the recommended pressure adjustment cannot be obtained by changing

**Fig. O32A—When making pressure control cable adjustment, install test gage as shown and refer to text.**

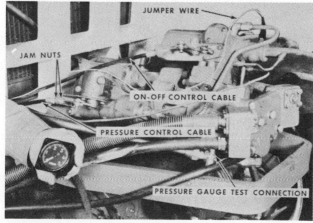

**Fig. O32B—View showing method of making minimum stroke adjustment. Refer to text for procedure.**

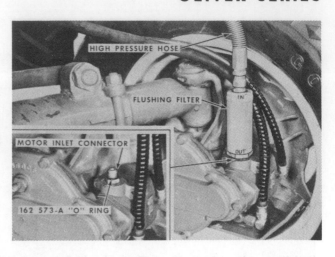

**Fig. O32C—Attach flushing filter as shown when flushing system. NO NOT operate in reverse.**

the control cable length, check and adjust, if necessary, the maximum pressure as oulined in paragraph 42C.

**42B. MINIMUM STROKE ADJUSTMENT.** With unit shroud off, raise either of the front wheels. Start engine, run at 1800 rpm and place quadrant lever in the No. 1 position. Check the raised front wheel which should turn approximately 2 rpm at the 1800 engine rpm indicating a correct minimum stroke (pump) adjustment.

If minimum stroke adjustment is required, stop engine, remove valve housing to stroke control oil line and the fitting and plug outlet on top of valve housing. Remove snap ring and stroke control valve end cap, then loosen jam nut and turn adjusting screw about half way into end cap. See Fig. O32B. Reinstall end cap and secure with snap ring. Install snap ring with chamfered O.D. next to end cap. Start engine and run at 1800 rpm then using a $\frac{3}{16}$-inch allen wrench, turn adjusting screw clockwise until the specified 2 rpm of raised front wheel is obtained. Stop engine, again remove end cap and tighten jam nut to hold the adjustment. Reinstall end cap and BE SURE chamfered O.D. of snap ring is next to end cap. Recheck adjustment and if satisfactory, reinstall plug, fitting and valve housing to stroke control oil line.

**42C. MAXIMUM PRESSURE ADJUSTMENT.** With shroud removed, remove front jam nut from pressure control cable, or remove the cable if desired. Connect a 5000 psi gage to opening in gear pump housing (Fig. O32A). Lock tractor brakes, and to insure complete safety, chain drawbar to a substantial stationary object. Start engine, place quadrant in No. 1 position and be sure quadrant gage registers at least 90 psi, then connect a jumper wire between a battery voltage source such as the alternator "BAT" terminal and terminal on unit

solenoid valve. Rotate cam control arm rearward until cam is fully bottomed and note reading of test gage. Test gage should register 3300-3500 psi.

If pressure is not as stated, remove cam housing, cam follower, shims and spring from stroke control valve housing and vary shims as required. Each shim will change pressure about 140 psi. Reassemble and recheck pressure. Adjust pressure control cable as outlined in paragraph 42A.

### FLUSH SYSTEM

**42D.** To flush (filter) the hydraulic front wheel drive system, obtain two flushing filters which are available from Oliver Corporation under part number STS-165 and proceed as follows: Raise front of tractor until both front wheels are free, then support under axle to maintain this position. Disconnect the high pressure hoses from both wheel motors, then install a 0.097 x 0.755 I.D. "O" ring (Oliver No. 162 573-A) on inlet connectors of wheel motors. Install filters with "out" end on wheel motor inlet fittings, then connect the pressure hoses to upper ends of filters. See Fig. O32C. Start engine, place quadrant control lever in No. 2 position and check quadrant pressure gage to see that it registers at least 90 psi, then connect a jumper wire between any battery source and solenoid valve.

NOTE: Be sure engine is running and quadrant pressure gage registers at least 90 psi before connecting jumper wire. To

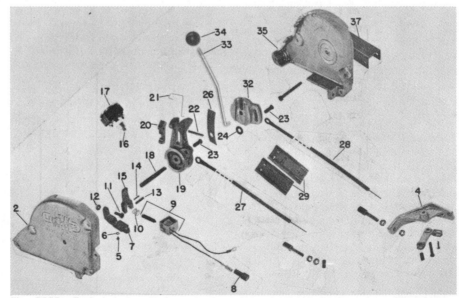

**Fig. O32D—Exploded view of the control quadrant used on hydraulic powered front wheel drive tractors.**

| | | | |
|---|---|---|---|
| 2. Housing, R.H. | 12. Lock nut | 20. Latch | 28. Pressure control cable |
| 4. Bracket | 13. Roll pin | 21. Spring | 29. Clamp |
| 7. Solenoid bracket | 14. Roll pin | 22. Roll pin | 32. Pressure control hub |
| 8. Connector | 15. Lock | 23. Headed pin | 33. Control lever |
| 9. Solenoid and plunger | 16. Flared connector | 24. Washer | 35. Housing, L.H. |
| 10. Spring | 17. Pressure gage | 26. Spring | 37. Mounting bracket |
| 11. Adjusting screw | 18. Pivot | 27. OFF-ON control cable | |
| | 19. OFF-ON control hub | | |

do otherwise will very likely damage system.

With jumper wire connected, run engine for 10 minutes at 1000 rpm, 10 minutes at 1800 rpm and 10 minutes at high idle. With flushing cycle completed, remove jumper wire and shut off engine. Remove filters and "O" rings and reinstall high pressure hoses to wheel motors. Remove blocking from under front axle.

NOTE: Always have front wheels raised when flushing. Driving the tractor with filters installed may cause the operating pressure to exceed the maximum pressure rating of the filters. In addition, should the tractor be inadvertently operated in reverse the filters will be back flushed and any contaminents present will be forced back into hydraulic system.

## CONTROL QUADRANT

Control for the hydraulic front wheel drive unit is provided by a dual cable control quadrant assembly mounted on right side of instrument panel. Included in the quadrant assembly is an electrically actuated solenoid lock and a pressure gage which registers the control circuit pressure of front wheel drive unit.

**42E. R&R AND OVERHAUL.** To remove the quadrant and cables, first remove shroud from the front wheel drive unit. Place quadrant lever in No. 6 position, then loosen clamping nipples from forward ends of cables and push cam lever and "OFF-ON" valve bellcrank from cable ends. Disconnect pressure gage line and solenoid wire from quadrant. Disconnect cable support and clip from engine and radiator grille, then unbolt and remove quadrant, mounting bracket and control cables.

To disassemble the quadrant assembly, remove cable clamp (29—Fig. O32D), loosen pressure gage bracket nuts, then remove the cap screws holding the two housings together and remove mounting bracket (37), left housing (35) and pressure gage (17). Remove control hubs (19 and 32), and pivot pin (18) from right housing (2), then remove control lever (33). Disconnect cables (27 and 28) from control hubs, then remove hubs and leaf spring (26) from pivot. Remove control hub latch (20) and spring (21). Remove solenoid (9) and bracket (7) from right housing. If solenoid or lock (15) requires renewal, disassemble solenoid and bracket by removing roll pins (13 and 14).

Inspect all parts for cracks, burrs, excessive wear or other damage and renew as necessary.

Assemble quadrant assembly as follows: Assemble and install solenoid, solenoid bracket and lock in right housing and secure solenoid ground wire to housing. Place latch (20) and spring (21) in position in control hub (19) and install roll pin (22). Place control hubs and leaf spring over pivot pin. Hubs should be centered and slot in pin should face lock notch in the "OFF-ON" control hub (19). Install control lever (33) in pressure control hub (32). Attach cables to control hubs with longest cable attached to "OFF-ON" control hub (19), then install the assembly in right housing with "OFF-ON" hub against housing. Rotate hub rearward until latch drops down behind housing ramp, then adjust screw in solenoid bracket to obtain $\frac{1}{16}$-inch clearance between hub and solenoid lock. Locate pressure gage and bracket in right housing, then with control lever in No. 1 position, install left housing and mounting bracket. Secure pressure gage bracket and install cable clamp.

Mount quadrant and bracket to instrument panel and connect pressure line and solenoid wire. Place quadrant lever in "OFF" position, secure the longer cable housing to bracket (4—Fig. O32D) and attach control cable to "OFF-ON" valve bellcrank while valve spool is "at rest" position. Place quadrant lever in No. 1 position, rotate compensating control cam arm rearward until cam contacts follower, then while holding cam arm in this position, attach shortest cable housing to the support bracket and the control cable to cam arm. Reposition cable support bracket, if necessary, to allow full travel of cam arm without control cable binding.

Secure cables and pressure gage line to engine and grille. Check pressure control cable adjustment as outlined in paragraph 42A, then install front wheel drive unit shroud.

## REMOVE AND REINSTALL (HYDRAULIC UNIT)

**42F.** To remove the hydraulic unit remove shroud, disconnect "off-on" cable and pressure control cable, then disconnect quadrant pressure gage line from hydraulic unit. Disconnect wire from solenoid valve. Disconnect oil cooler outlet and inlet lines from hydraulic unit. Drive roll pins from forward ends of both wheel motor high pressure hoses and pull hoses from hydraulic unit manifold. Do not lose the two special retaining washers. Disconnect the wheel motor return hoses and leak-off hoses at hydraulic unit end. Remove mounting guard. Attach a hoist to hydraulic unit, unbolt unit from mounting bracket, then slide unit forward until drive shaft clears piston pump drive shaft and lift unit away from tractor. Mounting bracket can now be removed, if necessary.

To reinstall hydraulic unit, proceed as follows: Install mounting bracket, if necessary. Position hydraulic unit and start drive shaft on piston pump drive shaft, then mate unit with mounting bracket and install retaining cap screws. Reconnect all hoses, quadrant pressure gage line and solenoid valve wire. Place quadrant control lever in "OFF" position, secure "OFF-ON" (longest) cable housing to to its support bracket and the cable to the manually controlled valve bellcrank with valve in the rest position. Move quadrant control lever to No. 1 position, then rotate pressure compensating cam control arm rearward until cam contacts cam follower and temporarily attach pressure control cable (shortest) to cam control arm and cable housing to support bracket. Check full travel of cable and control arm and reposition cable support bracket to prevent binding, if necessary.

## OVERHAUL

If desired, all sub-assemblies of the hydraulic power front wheel drive unit, except piston pump and drive shaft, can be removed, serviced and reinstalled without removing unit from tractor. However, because of the inter-relation of assemblies, should any damage occur within the unit the possibility exists that contaminants could be circulated throughout the entire front wheel drive system. Therefore, it would seem prudent to disassemble the complete unit for cleaning and inspection should trouble be experienced.

The following procedure is based on the unit being removed from tractor as outlined in paragraph 42F, and each sub-assembly will be removed, disassembled and reinstalled.

**42G MANIFOLD.** Remove the manifold-to-piston pump housing oil line, then unbolt and remove manifold from gear pump housing. Remove and discard the two "O" rings from recesses in rear of manifold. Remove the pressure hose port end caps, "O" rings and back-up (Teflon) rings. Discard "O" rings and back-up rings.

Clean manifold and inspect for cracks, warping and damaged threads.

2. End cap
3. "O" ring
4. Piston
5. Valve spool, hyd. controlled
6. "O" ring
7. Elbow
8. Housing
9. Socket head screw
10. Pipe plug
11. Filter element and gaskets
12. Spring seat
13. Spring
14. Shell
15. Gasket
16. Center stud
17. End caps
18. Connector
19. Oil pipe
20. Elbow

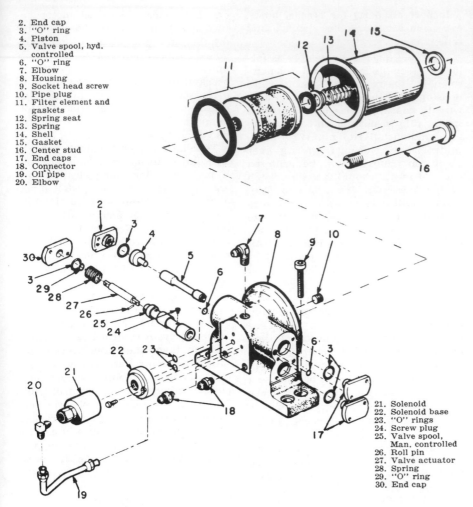

21. Solenoid
22. Solenoid base
23. "O" rings
24. Screw plug
25. Valve spool, Man. controlled
26. Roll pin
27. Valve actuator
28. Spring
29. "O" ring
30. End cap

**Fig. O32E—Exploded view of filter and valves assembly of the hydraulic powered front wheel drive unit.**

Pay particular attention to the pressure hose ports.

To reassemble and reinstall manifold, install pressure port end caps, "O" rings and back-up (Teflon) rings. The back-up rings are installed on outer sides of the smaller "O" rings. Place "O" rings in recesses in rear of manifold and install manifold on gear pump housing. Install manifold-to-piston pump housing oil line.

**42H. FILTER AND VALVES ASSEMBLY.** To remove the filter and valves assembly, first remove valve housing-to-stroke control oil line. Remove the four socket head cap screws and lift filter and valves assembly from gear pump housing. If valve housing-to-gear pump housing oil line remains with valve housing, it can now be removed. Discard all "O" rings as filter and valve assembly is removed and disassembled.

To disassemble filter and valve assembly, remove center stud (16—Fig. O32E), remove filter assembly and discard filter element (11). Remove plug (10) for cleaning purposes. Remove solenoid valve oil line (19), then remove solenoid (21), solenoid base (22) and "O" rings (23). Remove manual control valve bellcrank and bellcrank support, then remove end caps (2, 17 and 30), spring (28) and "O" rings (3). Identify end caps for reinstallation. Remove piston (4) and

the hydraulically controlled valve spool (5) from upper bore. Remove valve actuator (27) and manually controlled valve spool (25).

Clean and inspect all parts. Pay particular attention to valve spools and their bores. Spools should be a snug fit in bores yet slide freely when lubricated. Light scoring can be cleaned up with crocus cloth, however spools and/or housing should be renewed if heavy scoring is present. No test specifications are available for spring (28), however spring should be free of fractures or other defects and should be capable of returning valve spool to "OFF" position with comparative ease.

Reassemble by reversing the disassembly procedure, however keep the following points in mind. End cap (2) is thicker than end cap (17) and is installed at piston end of valve housing. Index oil holes in solenoid base (22) with similar oil holes in valve housing. Install plug (10) in right side opening of filter base as viewed from filter side. Do not block left side opening.

When reinstalling filter and valves assembly, be sure "O" rings are installed and position valve housing-to-gear pump housing oil line as assembly is installed. Tighten the valve housing retaining socket head cap screws to 70-77 Ft.-Lbs. torque.

**42J. GEAR PUMP.** To remove the gear pump assembly, first remove the manifold as in paragraph 42G and the filter and valves assembly as in paragraph 42H. Gear pump assembly can now be removed from piston pump housing plate.

With gear pump assembly removed, disassemble as follows: Remove overspeed check valve seat (10—Fig. O32F), "O" ring (9), overspeed check

**Fig. O32F — Exploded view of the hydraulic powered front wheel drive gear pump assembly.**

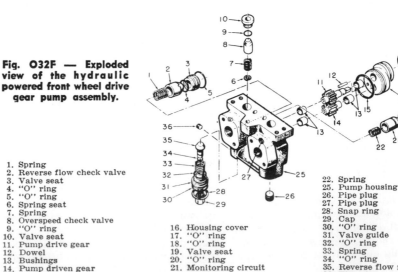

1. Spring
2. Reverse flow check valve
3. Valve seat
4. "O" ring
5. "O" ring
6. Spring seat
7. Overspeed check valve
8. Overspeed check valve
9. "O" ring
10. Valve seat
11. Pump drive gear
12. Dowel
13. Bushings
14. Pump driven gear
15. "O" ring
16. Housing cover
17. "O" ring
18. "O" ring
19. Valve seat
20. "O" ring
21. Monitoring circuit relief valve

22. Spring
25. Pump housing
26. Pipe plug
27. Pipe plug
28. Snap ring
29. Cap
30. "O" ring
31. Valve guide
32. "O" ring
33. Spring
34. "O" ring
35. Reverse flow relief valve
36. Pipe plug

valve (8), spring (7) and spring seat (6). Remove valve seat (3), reverse flow check valve (2) and spring (1). Remove pressure test port cap (29), then snap ring (28), guide (31), spring (33) and reverse flow relief valve poppet (35). Remove cover (16), alignment dowel (12) and pump gears and shafts (11 and 14). Thread a ¼-20 inch cap screw into tapped hole in seat (19) and remove seat, monitoring circuit relief valve plunger (21) and spring (22). Removal of pipe plugs will be dictated by the need for cleaning.

Clean and inspect all parts. Bushings (13) in housing (25) and cover (16) can be renewed if necessary. None of the valves in the gear pump assembly are adjustable, however all springs, as well as valves and valve seats, should be closely examined. Renew any parts which are cracked, worn or otherwise damaged.

Assemble gear pump assembly as follows: Install spring (22), monitoring circuit relief valve plunger (21) with "O" ring (20) and seat (19) with "O" ring (18). Lubricate pump gears liberally and place gears in housing with drive gear on top side, then install alignment dowel (12) and housing cover (16) with "O" rings (15 and 17). Install reverse flow relief valve poppet (35) and "O" ring (34), spring (33), guide (31) with "O" ring (30 and 32) and secure with snap ring (28). Install spring (1), reverse flow check valve (2) and valve seat (3) with "O" rings (4 and 5). Install spring seat (6) with flat side down, spring (7), overspeed check valve (8) and valve seat (10) with "O" ring (9).

After gear pump unit is assembled, secure unit to piston pump housing plate and tighten retaining cap screws to 42-48 Ft.-Lbs. Install filter and valves assembly as outlined in paragraph 42H and manifold as outlined in paragraph 42G.

**42K. STROKE CONTROL.** With control cable and valve housing-to-stroke control line removed, stroke control unit can be disassembled as follows:

Turn cylinder (29—Fig. O32G) counterclockwise and remove cylinder assembly (items 26 through 37) from housing (14). Pull piston rod (38) from housing (14). Remove snap ring (34) and end cap (33) from cylinder. Remove retaining screw (32), then remove piston (30), spring (27) and spring guide (26) from cylinder. Discard "O" rings (28 and 31), Remove compensating cylinder end cap (47), then remove spring (45), screen cap (44), piston (39) and piston rod (38)

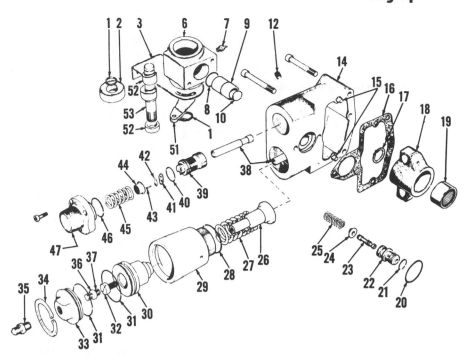

**Fig. O32G—Exploded view of stroke control used on hydraulic powered front wheel drive unit.**

| | | |
|---|---|---|
| 1. Snap ring | 17. Snap ring | 29. Control cylinder | 40. "O" ring |
| 2. Bearing support | 18. Rocker arm | 30. Control piston | 41. Reed valve |
| 3. Cable bracket | 19. Needle bearing | 31. "O" ring | 42. Back-up plate |
| 6. Cam housing | 20. "O" ring | 32. Screw | 43. Screw |
| 7. Grease fitting | 21. "O" ring | 33. End cap | 44. Screen cap |
| 8. "O" ring | 22. Valve sleeve | 34. Snap ring | 45. Spring |
| 9. Cam follower | 23. Valve pilot | 35. Connector | 46. "O" ring |
| 10. Shim | 24. Spring seat | 36. Lock nut | 47. End cap |
| 12. Pipe plug | 25. Spring | 37. Min. stroke adj. | 51. Control arm |
| 14. Housing | 26. Spring guide | screw | 52. Needle bearing |
| 15. Dowels | 27. Spring | 38. Piston rods | 53. Camshaft |
| 16. Gasket | 28. "O" ring | 39. Compensating piston | |

from housing (14). Remove reed stop (42) and reed valve (41) from piston (39), if necessary. Discard "O" rings (40 and 46). Unbolt cam housing (6) from stroke control arm (51) and camshaft (53), then press camshaft and bearing support (2) from cam housing. Remove bearing support (2) from camshaft. Remove cable support bracket (3). Remove camshaft bearings (52) if renewal is required. Remove cam follower (9), shims (10), spring (25) and spring seat (24) from stroke control housing, then remove stroke control housing from piston pump housing. Discard "O" rings (3 and 20) and gasket (16). Remove valve pilot (23) and valve sleeve (22) from bore in piston pump housing. Discard "O" ring (21). Rocker arm (18) and bearing (19) can be removed from control shaft if necessary by removing snap ring (17).

Clean and inspect all parts. Pay particular attention to springs (25, 27 and 45) and renew if any doubt exists as to their condition. Camshaft bearings (52) and rocker arm bearing (19) can be renewed, if necessary.

To reassemble stroke control, proceed as follows: Install rocker arm and bearing on control shaft so that

rocker arm is parallel with mounting surface of piston pump housing as shown in Fig. O32H. Secure rocker arm with snap ring (17—Fig. O32G). Place new "O" ring (20) in counterbore, and new gasket (16) on dowels of housing (14), then install housing on piston pump housing and tighten retaining cap screws to 20-27 Ft.-Lbs. torque. Install spring seat (24), spring (25), shims (10) and cam follower (9) with new "O" ring (8). Install lower bearing (52) in cam housing (6) and upper bearing (52) in bearing support (2), then install bearing and bearing support on upper end of

**Fig. O32H — Rocker arm is parallel with pump mounting surface when correctly installed.**

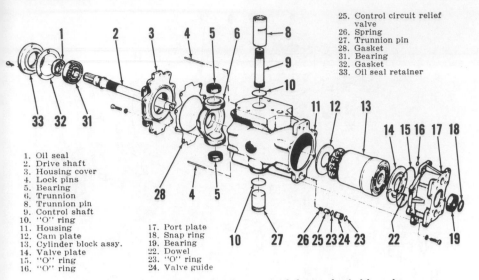

1. Oil seal
2. Drive shaft
3. Housing cover
4. Lock pins
5. Bearing
6. Trunnion
8. Trunnion pin
9. Control shaft
10. "O" ring
11. Housing
12. Cam plate
13. Cylinder block assy.
14. Valve plate
15. "O" ring
16. "O" ring

17. Port plate
18. Snap ring
19. Bearing
22. Dowel
23. "O" ring
24. Valve guide

25. Control circuit relief valve
26. Spring
27. Trunnion pin
28. Gasket
31. Bearing
32. Gasket
33. Oil seal retainer

Fig. O32J—Exploded view of the hydraulic powered front wheel drive piston pump assembly showing component parts and their arrangement.

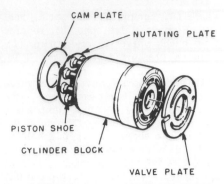

Fig. O32K—View of cylinder block (piston pump) assembly. None of the piston pump parts are available separately.

camshaft and secure with snap ring (1). Place camshaft and support and bearing assembly in cam housing, install control cable support bracket and control arm and align the previously affixed match marks on control arm and camshaft.

NOTE: If match marks were not affixed to control arm and control shaft prior to disassembly, place control arm on control shaft so that centerline of control arm is 37½ degrees clockwise from flat on end of control shaft.

Install retaining snap ring (1) to secure control arm. Tighten cam housing retaining cap screws to 20-27 Ft.-Lbs. torque. Be sure orifice in reed valve (41) is clear, then install reed valve and back-up plate (42) on compensating piston (39). Place connecting rod (38) in upper bore of housing (14) with end having annular groove next to rocker arm (18), then install compensating piston assembly screen cap (44), spring (45) and end cap (47) with "O" ring (46). Tighten end cap retaining cap screws to 10-12 Ft.-Lbs. torque. Place "O" ring (31) on piston (30) and install piston in cylinder (29). Assemble spring (27) and spring guide (26) insert assembly in rear of cylinder (29) and secure to piston (30) with the retaining screw (32). Install connecting rod (38) in lower bore of housing (14), with end having annular groove next to rocker arm (18), then using new "O" ring (28), install cylinder assembly on housing (14). If not already done, loosen the minimum stroke adjusting screw lock nut (36), install connector (35) in end cap, then install end cap in cylinder. Secure end cap with snap ring (34) and be sure chamfered side of snap ring faces end cap.

NOTE: A minimum stroke adjustment must be made after hydraulic front wheel drive unit is completely assembled and installed. Refer to paragraph 42B.

**42L. PISTON PUMP.** To disassemble and service the piston pump assembly, the hydraulic front wheel drive unit must be removed from tractor and the manifold, filter and valves assembly, gear pump assembly and stroke control valve assembly removed.

To disassemble piston pump assembly, proceed as follows: Remove snap ring and rocker arm from control shaft if not already done. Remove oil seal retainer (33—Fig. O32J) and gasket (32). Oil seal (1) can now be renewed. Support pump housing drive end down, remove all but two diagonally opposed port plate (17) retaining cap screws, then slowly loosen the remaining tow cap screws to relieve pumping element spring pressure. Carefully remove port plate (17), cylinder block (13) and drive shaft (2) as an assembly. Remove snap ring (18) and pull port plate assembly from drive shaft. Valve plate (14), "O" rings (15 and 16), dowels (22) and bearing can now be removed from port plate. Pull cylinder block (13) from drive shaft and stand it on a clean soft surface. Pull control circuit relief valve assembly (items 23 through 26) from bore in housing. Remove cam plate (12) from trunnion (6). Remove housing cover (3) and gasket (28) from pump housing and remove bearing (31) from cover. Remove both lock pins (4). Thread a ¼-inch cap screw in tapped hole in lower trunnion pin (27) and pull trunnion pin from housing. Insert a drift through lower trunnion pin hole and

bump control shaft (9) and upper trunnion pin (8) from housing. Remove "O" rings (10) from pump housing. Needle bearings (5) can now be renewed if necessary.

With piston pump assembly disassembled, thoroughly clean all parts and inspect as follows:

**42M. CYLINDER BLOCK.** Use a straight edge to check the front face of cylinder block. There should be no signs of wear or scoring of the sealing lands around the kidney-shaped slots. Any difference in height between outer lands and the two inner sealing lands will require renewal of the complete cylinder block assembly. Axial play of nutating plate (Fig. O32K) in piston shoes, or axial play of piston shoes in ball sockets must not exceed 0.005. If clearance exceeds that stated, renew complete cylinder block assembly. Examine surfaces of piston shoes that slide on cam plate and if any wear (more than 0.0001) exists, renew complete cylinder block assembly.

**42N. CAM PLATE.** Carefully examine cam plate. Fine scratches in running surface are permissible but if any wear or scoring (more than 0.0001) is present, renew the cam plate. **DO NOT** attempt to refinish cam plate.

**42P. VALVE PLATE.** Examine surface of valve plate that mates with cylinder block. Renew valve plate if running surface is 0.0005 or more lower than original surface. Also, carefully check the areas between the kidney-shaped slots. Any wear or scoring in these areas will result in poor efficiency even though other surfaces are satisfactory. DO NOT attempt to refinish valve plate.

**42R. CONTROL SHAFT.** Check condition of control shaft splines. Splines should have an interference

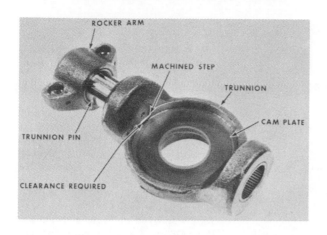

Fig. O32L—View showing relationship of rocker arm, control shaft and trunnion pin. Note position of machined step on end of control shaft.

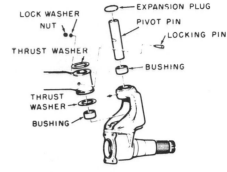

Fig. O33—Steering knuckle and pivot pin component parts. Refer to text for removal procedure.

fit in trunnion and not more than 0.005 backlash with rocker arm.

**42S. ASSEMBLY.** If necessary, install new needle bearings (5—Fig. O32J) in trunnion (6). Bearings are installed flush, or slightly below, outer edges of bearing bores. Install trunnion pin "O" rings (10) in bores of pump housing, locate trunnion (6) in housing, then install lower trunnion pin (27) with tapped hole toward outside and install lock pin (4). Place upper trunnion pin (8) on control shaft (9) with lock pin groove toward smallest end of control shaft. Position control shaft so machined step on trunnion end will provide clearance for cam plate when control shaft is installed in trunnion as shown in Fig. O32L, then install rocker arm so it is parallel with the machined step and has stroke control connecting rod sockets toward cam plate side of trunnion. Secure rocker arm with snap ring. Align lock pin groove of upper trunnion pin with lock pin hole, support lower trunnion pin, then drive control, shaft and upper trunnion pin into trunnion and housing and install lock pin. Install front bearing (9) in port plate (17). Install dowels (22), then place valve plate (14) over dowels and be sure the arrow on side

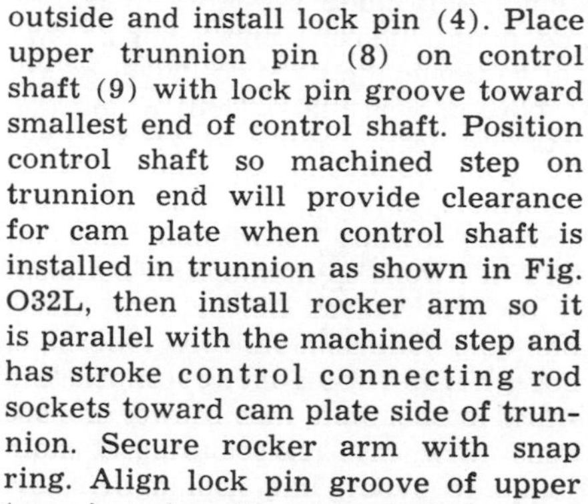

of valve plate is on the same side of port plate as the small "O" ring (16). See Fig. O32M. Lubricate valve plate liberally, set cylinder block assembly, port end down on valve plate and align the drive shaft bores. Insert drive shaft (2—Fig. O32J) through cylinder block assembly, port plate and front bearing and secure shaft with snap ring (18). Place cam plate (12) in trunnion (6) with tapered surface of inside diameter facing rearward. Place "O" rings (23) on valve guide (24), then install spring (26) control circuit relief valve (25) and guide assembly in bore of pump housing. Install "O" rings (15 and 16) to port plate, then carefully install port plate and cylinder block assembly to pump housing. Tighten the retaining cap screws to 100-115 Ft.-Lbs. torque. Install gasket (28) and cover (3) so that lock pin holes are covered and tighten retaining cap screws to 42-48 Ft.-Lbs. torque. Install rear bearing (33) with lip facing forward. Lubricate seal, then install gasket (32) and retainer on cover (3) and tighten retaining cap screws to 10-12 Ft.-Lbs. torque.

Reinstall stroke control as in paragraph 42K, gear pump as in paragraph 42J, filter and valves as in paragraph 42H and manifold as in paragraph 42G. Install the assembled unit on tractor as outlined in paragraph 42F.

## SPINDLES AND DRIVE WHEELS

**43A. PIVOT PINS.** Front drive wheels spindles are fitted with renewable pivot pins and bushings which can be renewed as follows: Raise front of tractor and support front axle in raised position. Disconnect hoses from wheel motor and cap all openings. Remove wheel and axle outer member from axle center member.

Refer to Fig. O33 and remove expansion plug and pivot pin locking pin. Thread a ¾-inch puller in tapped hole in upper end of pivot pin and remove pivot pin. Remove axle outer member and thrust washers. Remove grease fittings and bushings from spindle.

When installing bushings, index holes in bushings with lube holes in spindle and install grease fittings. Place a thrust washer on each side of pivot pin hole of axle outer member, position axle member in spindle and install pivot pin with tapped hole toward top and locking pin groove next to locking pin hole. Install locking pin and a new expansion plug. Lubricate bushings with gun grease. Reinstall wheel and outer axle assembly, reconnect hoses to wheel motor and remove blocking.

**43B. DRIVE WHEEL MOTOR.** To remove drive wheel motor, first remove rock guard, then disconnect hoses from wheel motor and remove motor-to-clutch line. Drain lubricant from primary housing then remove housing cover (26—Fig. O33C). Turn wheel hub until the rim attaching lug is in line with wheel motor, then pry primary drive gear away from wheel motor until bearing (48) contacts driven gear (28). Disconnect and pull motor from primary drive gear and housing. Remove spacer (46).

With wheel motor removed, refer to Fig. O33A and proceed as follows: Remove line fittings from motor, then clamp port plate (8) in a vise. Remove housing (27) and ramp (22) assembly and use caution not to allow thrust bearing (17) and thrust plates (16 and 18) to drop. Mark cylinder block (12) and drive shaft (5) so they can be reinstalled in the same position and remove cylinder block from drive shaft. Identify pistons (15) and their bores and remove pistons, spring seats (14) and springs

Fig. O32M—When assembling piston pump, install valve plate as shown. Refer to text.

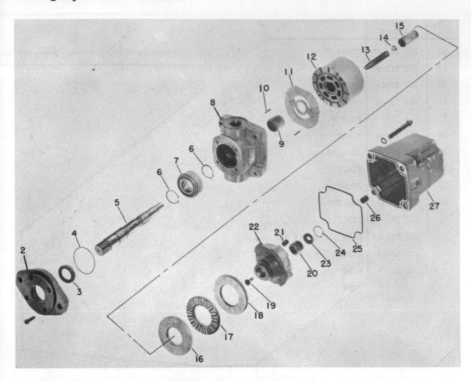

**Fig. O33A—Exploded view of front wheel drive motor. Refer to text when installing valve plate (11) and ramp (22).**

|  |  |  |  |
|---|---|---|---|
| 2. Mounting flange | 9. Bearing | 15. Piston | 21. Dowel |
| 3. Oil seal | 10. Dowels | 16. Piston plate | 22. Ramp |
| 4. "O" ring | 11. Valve plate | 17. Thrust bearing | 23. Screw seat |
| 5. Drive shaft | 12. Cylinder block | 18. Thrust plate | 24. Snap ring |
| 6. Snap ring | 13. Spring | 19. Retaining screw | 25. Seal ring |
| 7. Bearing | 14. Spring seat | 20. Bearing | 26. Dowel (sleeve) |
| 8. Port plate |  |  | 27. Housing |

(13). Mark valve plate (11) and port plate (8) so they can be reinstalled in the same position and remove valve plate and dowels (10) from port plate. Remove mounting flange (2) with oil seal (3) and "O" ring (4). Pull drive shaft (5) from port plate, then remove snap rings (5) and bearing (7). Lift piston plate from ramp (22). Remove seal ring (25) from housing. Remove screw (19) and lift ramp assembly from housing. Needle bearing (20) can be removed from ramp after removing snap ring (24) and screw seat (23).

Clean and inspect all parts. Bearing (9) can be removed from port plate (8) if necessary. Pay particular attention to pistons (15) and cylinder block (12). Pistons and/or cylinder block should not have any heavy scoring or wear. Renew valve plate if it shows signs of wear or scoring. Thrust bearing (17), plates (16 and 18) and ramp (22) should be in good condition with no wear, scoring or other damage.

Reassemble wheel motor as follows: Install ramp locating dowel, in hole adjacent to "M" location marking as shown in Fig O33B. If removed, press needle bearing (20—Fig. O33A) into ramp (22) until end of bearing is flush with bearing bore, then install

screw seat (23) with tapered side next to bearing and install snap ring (24). Place ramp in housing (27) so thickest side is toward the leakoff opening, insert drive shaft in ramp bearing and center drive shaft and ramp in housing, then install ramp retaining screw (19). Place thrust plate (18) on ramp with largest chamfer on inside diameter away from thrust bearing, then install thrust bearing (17) and piston plate (16). If necessary, install new bearing (9) in port plate so rear of bearing is flush with rear of bearing bore. Install snap rings (6)

and bearing (7) on drive shaft (5) and install assembly in port plate. Install new seal (3) in mounting flange so lip of seal is facing bearing (7). Lubricate lip of seal and tape shaft splines, then install flange on port plate so that mounting hole lugs are 180 degrees from port plate inlet and outlet ports. Tighten mounting flange retaining cap screws to 12 Ft.-Lbs. torque. Install valve plate dowels (10) in port plate, then align previously affixed match marks and install valve plate (11) on dowels and be sure the open notches in valve plate are counterclockwise from the dowels as shown in Fig. O33B. Install piston springs (13—Fig. O33A) in piston bores of cylinder block (12), tapered end first and hook springs to pins at bottom of bores. Place spring seats (14) on springs and install pistons (15). Align the previously affixed cylinder block and drive shaft marks and install cylinder block on drive shaft. If removed, install sleeve dowel (26) in housing at location shown in Fig. O33B. Install seal ring (25—Fig. O33A) in seal ring groove of housing. Clamp port plate assembly in a vise and install housing and ramp assembly so dowel (26) enters dowel counterbore of port plate. Tighten housing retaining cap screws to 37 Ft.-Lbs. torque. Reinstall all line fittings.

NOTE: When correctly assembled, wheel motor will have pressure port at top and leak-off port forward when unit is installed on tractor.

To install wheel motor on tractor, install spacer and loosely secure motor to primary gear housing. Install motor to clutch line. Bump primary reduction gear into housing until inner bearing bottoms, then reinstall primary gear housing cover and rock guard. Tighten housing cover cap screws to 30-35 Ft.-Lbs. torque. Tighten wheel motor cap screw and stud nut to 50-60 Ft.-Lbs. torque. Fill

**Fig. O33B — When assembling front wheel drive motor, install dowels and valve plate as shown.**

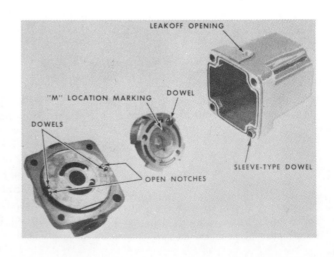

primary gear housing with lubricant, connect all hoses and flush systems as outlined in paragraph 42D.

### DRIVE WHEEL

43C. To disassemble drive wheel assembly, proceed as follows: Remove wheel weights if so equipped. Raise front of tractor and support axle in raised position, then drain primary and planetary gear housings. Remove tire and rims. Remove planet carrier (16—Fig. O33C) from wheel hub (7). Straighten tabs of lock (10), then remove adjusting nut (11) and lock. Pull internal gear (9) from spindle and take care not to drop outer bearing (6) which generally is a loose fit on gear hub. If bearing is stuck, it can be forced from internal gear hub by using two $\frac{7}{16}$-inch cap screws in the tapped holes provided in gear. Remove wheel hub (7) from spindle and remove oil seal (4) and inner bearing (5). Bearing cups can be bumped from hub if necessary. Disassemble planet carrier by driving roll pins (12) into pinion shafts (18), driving pinion inward until roll pinholes are exposed and driving roll pins from pinion shafts. Pinion shafts can now be driven out toward outside of carrier and the pinions (20), pinion bearings (21) and thrust washers (19) removed from carrier. Discard pinion shaft "O" rings (17).

Remove rock guard and disconnect inlet, outlet and leak-off lines from wheel motor. Remove wheel motor to clutch line and cap all openings. Remove primary gear housing cover (26), then remove primary gear housing (32) and gears (29 and 49) from steering knuckle. Remove gears from housing. Remove spring guide (36), spring (37) and washer (38), then pull on end of shaft (3) to release the thrust washer halves (39). Remove the thrust washer halves and pull shaft (3) from outer end of spindle. Clutch piston (41) and "O" rings (40 and 42) can now be removed from spindle. Remove bearings (48 and 50) from primary drive gear if necessary. Bearings (27 and 30) and bushing (51) can also be removed from primary reduction gear (29).

Clean and inspect all parts. Discard all "O" rings and gaskets. Renew any parts which show undue wear or scoring, or any other damage. If service is required on wheel motor, refer to paragraph 43B.

Reassemble drive wheel as follows: Install "O" rings (40 and 42) in knuckle (2). Insert drive shaft (3), small end first, through spindle and install thrust washer halves (39) in

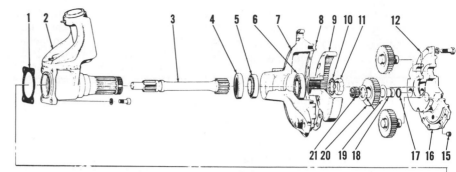
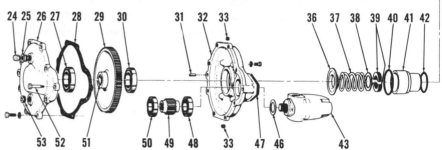

**Fig. O33C—Exploded view of front wheel reduction drive assembly. Refer to Fig. O33A for exploded view of drive motor (43).**

| | | |
|---|---|---|
| 1. Gasket | 15. Pipe plug | 29. Primary reduction gear |
| 2. Steering knuckle | 16. Planet carrier | 30. Bearing |
| 3. Drive shaft | 17. "O" ring | 31. Dowel |
| 4. Oil seal | 18. Pinion pin | 32. Primary gear housing |
| 5. Bearing assy. | 19. Thrust washer | 33. Pipe plug |
| 6. Bearing assy. | 20. Pinion | 36. Spring guide |
| 7. Wheel hub | 21. Bearing | 37. Spring |
| 8. Gasket | 24. Breather | 38. Washer |
| 9. Internal gear | 25. Bushing | 39. Thrust washer halves |
| 10. Nut lock | 26. Housing cover | 40. "O" ring |
| 11. Adjusting nut | 27. Bearing | 41. Clutch piston |
| 12. Roll pin | 28. Gasket | 42. "O" ring |
| | | 43. Drive motor |
| | | 46. Spacer |
| | | 47. Gasket |
| | | 48. Bearing |
| | | 49. Primary drive gear |
| | | 50. Bearing |
| | | 51. Bushing |
| | | 52. Dowel bolt |
| | | 53. Pipe plug |

shaft groove so chamfered sides are toward clutch cylinder, then push shaft outward to seat thrust washer halves in clutch cylinder. Install flat washer (38), spring (37), spring guide (36) and gear housing (32). Be sure gear housing is installed so motor mounting is straight forward. If new bushing (51) is required, install bushing so outer end is flush with the fully splined end of gear hub. Install primary reduction gear inner bearing (30) so shielded side is toward clutch cylinder and on side opposite bushing. Install bearing (27) on opposite hub of primary reduction gear and bearings (48 and 50) on primary drive gear (49). Install both gear and bearing assemblies in housing and be sure driven gear (29) is installed with bushing (fully splined gear hub end) away from clutch housing. Be sure breather (24) is clean, then loosely attach cover (26) to housing (32). Install dowel (31) and dowel bolt (52). Install spacer (46) and motor (43) and be sure pressure port is on top and leak-off port is toward front. Install motor to clutch oil line, then tighten the wheel motor retaining cap screw and dowel bolt nut to 37 Ft.-Lbs. torque. Install rock guard, then tighten cover retaining cap screws to 30-35 Ft.-Lbs.

If new bearing cups are being installed in hub (7), install them with smaller I.D's to inside and be sure they are bottomed. Place bearing cone (5) in hub and install oil seal (4). Install bearing cone (6) on hub of internal gear (9). Drive hub assembly on spindle, place internal gear assembly on spindle then install lock (10) and adjusting nut (11) and adjust bearings as follows: Tighten adjusting nut until snug, then back-off until all pressure is released from bearings. Use a spring scale attached to rim lug hole and check effort required to keep hub rotating; then tighten adjusting nut until an additional 3½-6 pounds pull is required to keep hub

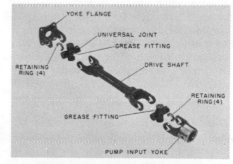

**Fig. O33D—Drive shaft for the hydraulic powered front wheel drive unit.**

rotating. Bend down tabs of lock (10) to hold nut in this position.

Place "O" rings (17) in grooves of pinion shafts (18). Place bearings (21) in bores of pinions (20), place a thrust washer (19) on each side of pinion gear, then insert pinion shaft in bore with roll pin hole aligned with roll pin hole in carrier web. Drive pinion shafts into position and install roll pins so outer ends are flush. Install planet carrier on hub and tighten retaining cap screws to 70-80 Ft.-Lbs. torque.

Connect hoses to wheel motor, then fill primary gear housing and planetary gear housing to recommended level with correct lubricant. Install tire and rim and tighten attaching bolts to 160-170 Ft.-Lbs. torque. If so equipped, install wheel weights and tighten attaching bolts to 170-180 Ft.-Lbs. torque.

Flush system as outlined in paragraph 42D.

### DRIVE SHAFT

43D. Drive shaft for the hydraulic powered front wheel drive unit can be removed after disconnecting yoke flange from crankshaft pulley adapter. Remove adapter from crankshaft after drive shaft is removed, if necessary.

Drive shaft universal joints can be disassembled after removing retaining rings, and are serviced with kits which include spider, bearing cups and retaining rings.

When reinstalling drive shaft, DO NOT secure the drive shaft to the front wheel drive piston pump shaft. To do so may limit engine crankshaft end play and could result in crankshaft bearing failure.

# POWER STEERING

44. All row crop and four-wheel drive tractors have power steering as standard equipment; other models have power steering as optional equipment.

All two-wheel drive tractors prior to tractor serial number 134 684-000 were equipped with Gemmer steering unit shown in Fig. 040; two-wheel drive tractors with tractor serial number 134 684-000 and up are equipped with Saginaw Hydramotor as shown in Fig. 060 and steering cylinder as shown in Fig. 080.

Four-wheel drive tractors prior to tractor serial number 134 684-000 were equipped with Char Lynn control unit shown in Fig. 045; serial number 134 684-000 and up are equipped with Saginaw Hydramotor shown in Fig. 060.

The power steering systems will be discussed under the brand name as there will be no difficulty in determining which system is being used.

On tractors equipped with a hydraulic lift system, the pressurized operating oil is supplied by the hydraulic lift system through a priority type flow divider valve which is bolted to the hydraulic lift housing. On tractors which are not equipped with a hydraulic lift system, the pressurized operating oil is supplied by a separate engine mounted, belt driven pump.

## LUBRICATION AND BLEEDING

### Models With Gemmer Steering Unit

45. Air trapped in the power steering system will cause erratic operation and shimmy of the front wheels. When these symptoms are encountered, bleed the power steering system as follows: Check, and fill if necessary, the hydraulic lift reservoir or the belt driven pump reservoir. Start engine and turn front wheels to the right as far as they will go. Loosen the bleed screw (5—Fig. 040) on right end of cylinder housing. When the oil escaping from the screw is free of air bubbles, tighten the screw. Refill reservoir.

NOTE: When performing the bleeding operation, use caution to avoid injury from revolving parts.

## SYSTEM PRESSURE TEST
### Models With Gemmer Steering Unit

46. If steering performance is not normal, first bleed system as outlined in paragraph 45. If bleeding does not correct the trouble, check the system pressure to determine the location of the trouble as follows:

On models with belt driven pump, install a pressure gage capable of registering at least 1100 psi, along with a shut-off valve, in the pump pressure line. Shut-off valve must be between gage and the power steering unit.

On models with flow divider, install pressure gage in the plug hole on left front side of flow divider valve as shown in Fig. 036.

With power steering oil warm, shut-off valve open and engine running at low idle speed, turn the steering wheel from one extreme to the other and note the pressure gage reading. Pressure gage reading should be approximately 1000 psi.

If pressure is less than specified on those models with belt driven pump, check the pump by closing the shut-off valve only long enough to obtain a reading. If pressure is now approximately 1000 psi, the pump can be considered satisfactory and the trouble is located in the power steering unit. If the pressure is considerably less than the 1000 psi, the pump is at fault and should be removed for repair as outlined in paragraph 47.

If the pressure is less than specified on those models with flow divider valve, add shims (4—Fig. 037) under relief valve (5). One shim should change the pressure about 50 psi. If the addition of shims will not restore pressure, check the hydraulic lift pump pressure as outlined in paragraph 403 to insure that sufficient pressure is reaching the flow divider valve. If hydraulic lift pump pressure is satisfactory, and the pressure from flow divider valve is still low, remove and overhaul flow divider valve as outlined in paragraph 49.

## PUMP
### Models With Gemmer Steering Unit

47. R & R AND OVERHAUL. This section concerns the belt driven pump used on those models without hydraulic lift.

Procedure for removal of the power steering pump is self-evident after an examination of the unit.

48. As exploded view of the pump is shown in Fig. 035 and disassembly procedure is as follows: Drain reservoir, remove the orifice fitting (28) cap screws (23) and spacers (24). Pull

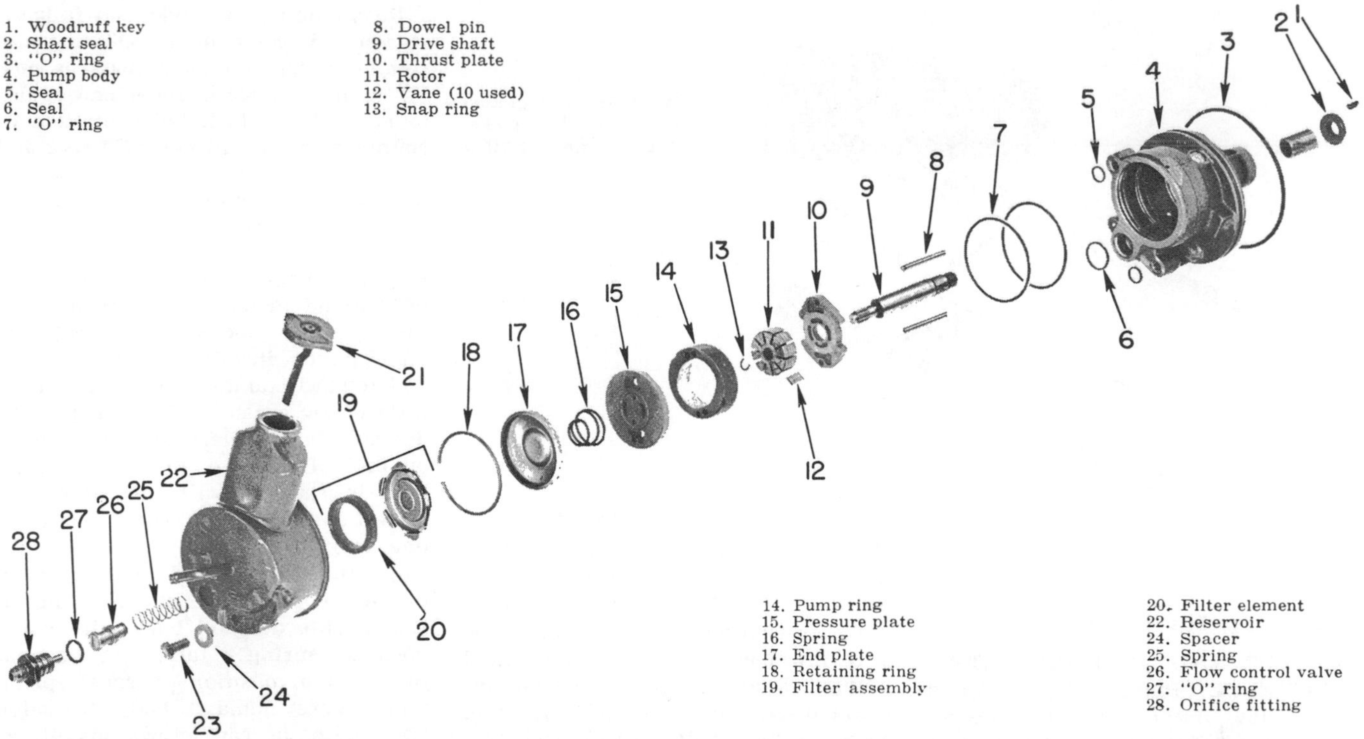

1. Woodruff key
2. Shaft seal
3. "O" ring
4. Pump body
5. Seal
6. Seal
7. "O" ring
8. Dowel pin
9. Drive shaft
10. Thrust plate
11. Rotor
12. Vane (10 used)
13. Snap ring

14. Pump ring
15. Pressure plate
16. Spring
17. End plate
18. Retaining ring
19. Filter assembly
20. Filter element
22. Reservoir
24. Spacer
25. Spring
26. Flow control valve
27. "O" ring
28. Orifice fitting

**Fig. O35 — Exploded view of the belt driven power steering pump used on tractors without a hydraulic lift system and equipped with Gemmer steering unit or Char-Lynn control unit. Refer to Fig. O50 or Fig. O52 for pump used with Saginaw steering unit.**

reservoir (22) from pump body (4), then remove "O" ring (3) from pump body. Flow control valve (26), spring (25) and filter assembly (19) are now free. NOTE: Flow control valve can be removed prior to removing reservoir, if desired. Place small punch through hole in pump body, compress retaining ring (18) and remove same. Remove end plate (15) and pump ring (14) from dowels (8). Note location and direction of arrow on pump ring prior to removal. Remove vanes (12) from rotor (11), then remove snap ring (13) and pull rotor from shaft (9). Remove thrust plate (10) and the pump shaft from pump body. Dowels (8) can be removed from

pump body, if necessary. Seal (2) can be driven from body using a soft drift.

Check I.D. of pump ring for chatter marks, wear or scoring. Check fit of vanes in rotor and for undue wear. Vanes should fit slots of rotor snugly yet be free enough to slide easily. If pump ring, vanes or rotor prove to be unserviceable, it is recommended that they all be renewed and a kit is available for renewing these parts.

Inspect wear surfaces of thrust plate and pressure plate for scoring or undue wear. Light scoring can be corrected on a fine surface grinder or by lapping, provided care is exercised. Renew parts which are excessively scored.

Inspect shaft bushing in pump body and if found to be defective, renew entire assembly as bushing is not available separately.

Except for spring (25), no parts are catalogued separately for flow control valve. However, the flow control valve can be disassembled and cleaned.

When reassembling pump, observe the following points. Install seal (2) with lip facing reservoir end of pump. Use all new seals, which are available in a kit, and coat all parts with power steering fluid prior to assembly. Rounded ends of vanes face outward and counterbore of rotor is toward reservoir end of pump. Pump

ring is installed with embossed arrow pointing in direction of pump rotation. Avoid excessive tightening of orifice fitting and cap screws when installing reservoir as distortion of reservoir could result.

## FLOW DIVIDER VALVE

Tractors equipped with power steering and hydraulic lift receive pressurized oil for operation of the power steering via a flow divider attached to the hydraulic lift housing. Tractors prior to Series C, has flow divider valve attached to front side of hydraulic lift housing. Series C, and all later tractors, have the flow divider valve attached to left front side of hydraulic lift housing.

Beginning in March 1968, a demand type flow divider valve was installed during production which gives a pilot flow of 2½-3 gallons and a demand flow of 4½-5 gallons. This dual action is obtained by adding a demand piston which reacts to increased pressure in the steering system circuit causing an increased flow from the valve spool by reducing the amount of oil being returned to hydraulic reservoir and by forcing more oil to flow through metering holes of spool. Refer to Figs. O36 and O37 for views of a flow divider valve prior to the demand type and to Fig. O37A for a view of the demand type.

**Fig. O36 — View showing pressure gage installed in flow divider valve to check relief pressure. Refer to Fig. O37 for an exploded view of flow divider valve. Early tractor is shown. Series C and later tractors have valve installed on left side of hydraulic housing.**

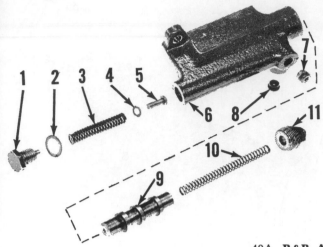

**Fig. O37 — Exploded view of the power steering flow divider valve.**

1. Plug
2. Crush washer
3. Spring
4. Shim
5. Plunger
6. Body and seat
7. Plug
8. Insert
9. Spool
10. Spring
11. Plug

## Models So Equipped

**49. R&R AND OVERHAUL. (PRIORITY TYPE).** Refer to Figs. O36 and O37. If flow divider valve is to be removed, the procedure for doing so is obvious.

To disassemble, remove plugs (1 and 11—Fig. 037) and withdraw parts from body (6).

Spool (9) should slide smoothly in its bore and be free of any nicks or burrs. Free length of spring (10) is $4\frac{25}{32}$ inches and should test 13-15 pounds when compressed to a length of $3\frac{3}{8}$ inches. Spring (3) has a free length of $2\frac{3}{16}$ inches and should test 44-54 pounds when compressed to a length of $2\frac{7}{32}$ inches. Check seating surfaces of relief valve plunger (5) and its seat. Relief valve seat and body are not available separately. The addition, or substraction of one shim (4) will change the relief valve pressure approximately 50 psi.

**49A. R&R AND OVERHAUL (DEMAND TYPE),** refer to Fig. O37A. If flow divider is to be removed, the procedure for doing so is obvious.

To disassemble, remove relief valve plug, crush washer, spring, shims and plunger. Remove cap screws from demand body, then remove body, valve spool spring, valve spool and "O" rings. Remove demand piston stop cap, crush washer, spring retainer and piston from demand body.

Clean and inspect all parts and renew as necessary. Discard all "O" rings including "O" ring and back-up washer from demand piston. Refer to the following specification data:

Pilot flow-gpm ................................2½-3
Demand flow-gpm ........................4½-5
Relief Valve
    Pressure setting full flow psi ....1600
    Pressure change-each shim ....31 psi
    Spring color ............................Purple
    Spring free length-in. .................2¾
    Test—lbs. @ in. ............69-79 @ 2¼
Spool Spring
    Spring color ................................None
    Free length—in. ..........................$4\frac{25}{32}$
    Test—lbs. @ in. ............13-15 @ $3\frac{3}{8}$

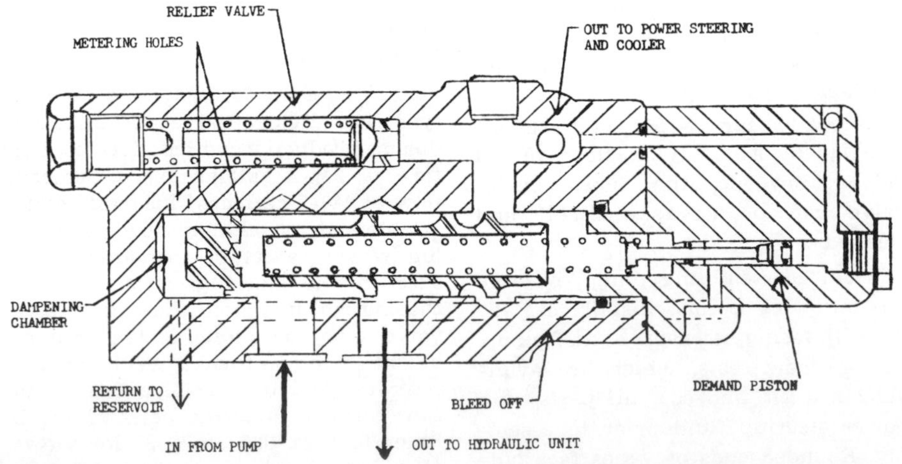

Reassemble flow divider as follows: If demand piston inner "O" ring and retainers were removed, install a **new** "O" ring retainer in inner end of demand body ½-inch below bottom of counterbore. Install new "O" ring and be sure it has no twists. Install second **new** "O" ring retainer in inner end of demand body so it is flush with bottom of counterbore. Place new "O" ring and back-up ring on outer end of demand piston with back-up ring toward valve spool side. Lubricate demand piston, install in bore with stem end toward counterbore (spool), then install new crush washer and plug in demand body. Place spring retainer on end of demand piston and index hole in retainer with end of piston. Install spool in body. Then install the one large, and the two small, "O" rings in demand body end of flow divider body. Place spool spring in spool, mate demand body with valve body so spring retainer on demand piston is in position on spool spring and secure demand body to valve body with the cap screws. Install relief valve plunger, shims and spring, then use new crush washer and install retaining plug.

Oil line insert can be renewed if necessary and old insert can be pulled by using a thread tap.

## POWER STEERING GEAR UNIT
### Models With Gemmer Steering Unit

**50.** The Gemmer power steering gear unit is mounted in the top side of the front main frame and is secured to the vertical spindle on tricycle models, or pitman shaft on other models, by the puller bolt of the unit. While the mounting position and the configuration of housing may differ between Rowcrop and other tractors, the working parts for the power steering unit remain the same.

**51. REMOVE AND REINSTALL.** To remove the power steering gear unit, first remove precleaner, muffler, fuel tank cap, side panels and hood. Remove fan blades and drain cooling system. Disconnect radiator top supports, unbolt radiator from front frame, then disconnect upper and lower radiator hoses and lift radiator from tractor.

On 1800 models, remove fan belt, then using a suitable puller, remove the crankshaft pulley. On all models, loosen steering shaft support at clutch housing, then remove steering shaft front universal joint from steering

**Fig. O37A—Cross-sectional view of the late demand type flow divider valve. Demand piston body is bolted to spool body. Note "O" rings between spool body and demand piston body.**

RELIEF VALVE
METERING HOLES
OUT TO POWER STEERING AND COOLER
DAMPENING CHAMBER
RETURN TO RESERVOIR
IN FROM PUMP
BLEED OFF
OUT TO HYDRAULIC UNIT
DEMAND PISTON

column of power steering unit. Disconnect torque link, hydraulic lines and remove protective cap from unit; then unscrew puller bolt and lift unit from tractor.

NOTE: Unit can also be removed without removing the crankshaft pulley as follows: Remove radiator and disconnect steering shaft from unit. Disconnect torque link from unit housing, then remove control valve. Rotate front wheels until unit will clear crankshaft pulley, then unscrew puller bolt and remove unit. BE SURE not to damage the control valve mounting surface.

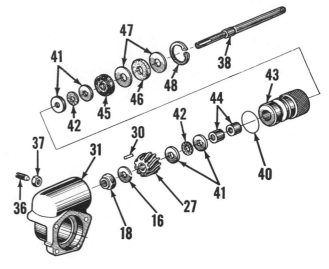

Fig. O39—Exploded view of power steering unit helical gear housing and associated parts. Refer to Fig. O40 for legend.

**52. DISASSEMBLE AND OVERHAUL.** Because of the relative complexity of the power steering gear, the written procedure for overhauling it becomes rather lengthy. It is advisable, therefore, to break down the assembly into arbitrarily assumed sub-assemblies. In presenting this overhaul procedure, each such sub-assembly will be removed from the gear, disassembled, overhauled and reassembled completely, before proceeding to the next sub-assembly. The procedures assume the gear to be off the tractor. Reinstallation of the sub-assemblies to the gear will be covered in a separate section beginning with paragraph 58. An exploded view of the complete assembly, showing the relative location of all parts, is shown in Fig. O40.

53. SECTOR SHAFT AND PULLER SUB-ASSEMBLY. To remove shaft and puller, loosen the large threaded lock nut (59—Fig. O40) with a spanner wrench, then turn the inner seal retainer lock nut (57) out until the adjusting sleeve (51) is fully unthreaded from the housing. Rotation of the lock nut (57) imparts rotation to the

Fig. O38—Sector shaft and puller bolt assembly removed from the power steering unit housing.

threaded sleeve because the lock nut is staked at 4 places to the sleeve. If the staking has broken loose, the sleeve can be removed by using two lock nuts (59) as jam nuts. Refer to Fig. O38.

To overhaul the removed unit, extract the outer end oil seal by first carefully unstaking the lock nut (57—Fig. O40) from the aluminum adjusting sleeve and turning the nut out of the sleeve. A pin spanner or equivalent is required for nut (57). It may be necessary to use two lock nuts (59) on the sleeve to hold sleeve from rotating while removing the seal lock nut. Bump end of shaft on wood block to dislodge seal and seal adaptor. Oil seal (94) and the lathe cut seal (99) at opposite end can be removed at any time. Puller bolt is available only as an assembly with shaft, but can be removed after removing retaining and "C" washers.

Install new seals when reassembling and be sure to restake the lock nut by peening the aluminum adjusting sleeve into the four slots in the nut.

The procedure for reinstalling the sector shaft and puller assembly is covered in paragraph 60.

54. HELICAL GEAR HOUSING. An exploded view of the helical gear housing is shown in Fig. O39. To remove the unit, loosen retainer screw (36), then turn the knurled trunnion until slot in outer face of trunnion is pointed toward mounting flange or is at high point of eccentricity. Remove the four mounting cap screws and pull the helical gear housing (31) and components off the main housing as shown in Fig. O41.

To overhaul the unit, remove retainer screw (36—Fig. O39) from housing and withdraw the column and

attached parts. With column mounted in soft jaws of vise, unstake and remove nut (18) and slide gear, trunnion, bearings and seals off the column. Remove snap ring (48) from trunnion and remove seal retaining washers, dust seal and column oil seal (45).

Install all new oil seals and "O" rings. If needle bearings (44) are to be renewed, they should be installed by pressing against lettered end of bearing shell and should be flush at outer ends. Lip of column oil seal (45) should be toward bearings. Nut (18) should be tightened to produce just a slight drag and then staked in position. When reassembling trunnion to housing, align slot in end of same with mounting flange.

The procedure for installing the helical gear housing is covered in paragraph 62.

55. VALVE ASSEMBLY. An exploded view of the valve and components is shown in Fig. O42. Assembly is removed after removing the attaching cap screws. Lift body and contained parts from main housing. Lift out the two small "O" rings (1) and valve actuating lever (68). Refer to Fig. O43.

Order of disassembly and reassembly is shown in Fig. O42. The valve link (81) is threaded into the spool; use wooden vise jaws to hold spool when unscrewing link. The inverted flared tubing connectors in body are pressed in but can be extracted using a 10-24 tap for small one and $\frac{5}{16}$-18 tap for the other. Run a nut well up on each tap, add a washer under it. Coat tap threads with grease, then after threading tap into connector about three turns, screw down on nut to pull connector out of body.

1. "O" ring
2. Housing
3. Housing
5. Bleed screw
6. Worm and ball nut assy.
7. Adapter and pin assy.
8. Seal retaining washers
9. "O" ring
10. Retaining ring
11. Adapter seal
12. Retaining ring
13. Thrust bearing bushing
14. Thrust bearing and spring assy. (early)
14A. Race and spring
14B. Thrust bearing
14C. Thrust bearing race
15. Centering spring
16. Lock nut washer
17. Lock nut washer
18. Lock nut
19. Piston rack
20. Thrust ring & seal assy.

21. Seal spacer
22. Snap ring
23. Piston ring
24. Retaining screw
25. Housing adapter (early)
26. "O" ring
27. Helical gear (12 or 14 teeth)
28. Helical gear (12 or 11 teeth)
30. Square key
31. Gear housing
35. Gear housing
36. Retaining screw
37. Jam nut
38. Column

39. Column
40. "O" ring
41. Thrust washer
42. Thrust bearing
43. Trunnion
44. Needle bearing
45. Oil seal
46. Dust seal
47. Retaining washer
48. Snap ring
49. Shaft & puller assy.
50. Seal washer
51. Adjusting sleeve
52. "C" washer
53. Retaining ring
54. Seal
55. Seal adapter & seal assy.
56. "O" ring
57. Lock nut
58. "O" ring
59. Lock ring
60. End cover
62. End cover assy.
65. "O" ring
66. Plug
67. "O" ring
68. Actuating lever
69. Linkage cover
70. Valve assy.
71. Valve body

72 & 73. Connector
74. Valve spool
75. Reaction spool
76. Spring
77. Thrust washer
78. Centering spring
79. "O" ring
80. Annulus
81. Link
82 & 83. "O" ring
84. Plug
85. Retaining ring
86. Retaining washer
87. Retaining ring
88. Cap
89 & 90. Shipping plugs
91. "O" ring
92. Snap ring
94. Oil seal
95. Shim
96. "O" ring
97. Spring
98. Retaining washer
99. Ring (lathe cut)
.00. Tab washer
.01. "O" ring
.02. Housing adapter (late)

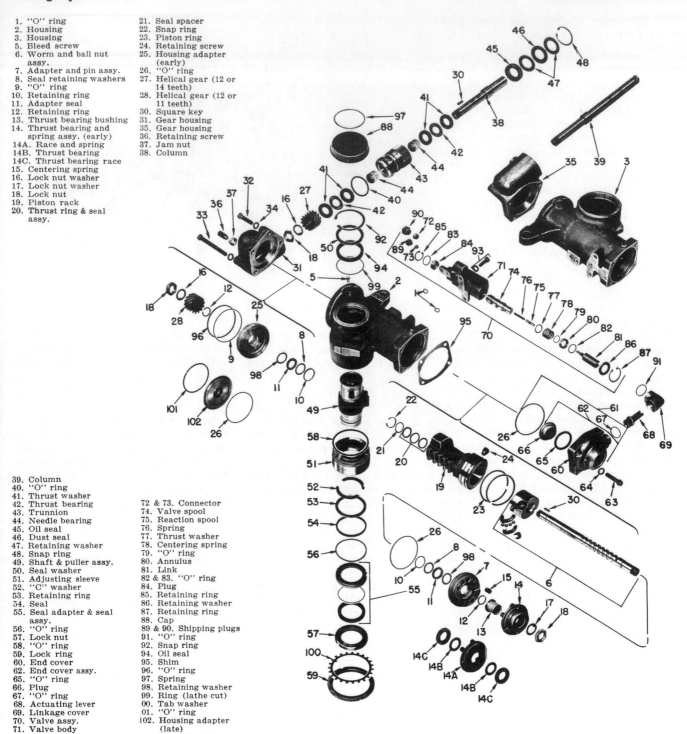

**Fig. O40 — Exploded view of the Gemmer power steering unit. Housings (2 and 3) differ slightly in appearance and are mounted differently but internal parts for both housings are similar.**

To install the unit, refer to paragraph 61.

56. COVER ASSEMBLY. The cover assembly (61—Fig. 040) can be removed after the valve assembly is off and the procedure is obvious.

57. RACK PISTON ASSEMBLY. An exploded view of the rack piston assembly and components is shown in Fig. 040. The shaft and puller, helical gear housing, valve assembly and cover assembly must be removed before removing rack piston. After the mentioned parts have been removed, unstake and remove the nut (18) from helical gear end of worm screw shaft. Remove gear (28), square key (30) and retaining ring (12). Slide the housing adapter (25 or 102) from the shaft. Push on end of worm shaft which will move the rack piston (19) and attached parts out the cover end of the main housing. Four of the loose thrust bearing centering springs (15) will drop out at this time.

Unstake and remove nut (18) from opposite end of the worm screw shaft, then remove the center race and springs (14), or race and spring (14A), thrust bearings (14B and 14C) and bushing (13). Remove retaining snap ring (12) from end of shaft and lift the adapter and pin assembly (7) from the rack piston. Mount rack piston in soft jawed vise and then un-stake and remove the ball nut retaining screw (24). Carefully push worm screw out of rack piston and when ball nut is partially exposed, hold ball guides firmly with one hand to prevent balls from falling out as nut and shaft are withdrawn from rack piston. Hold guides in position with tape or wire as shown in Fig. O44.

The nut balls are matched and cannot be obtained separately. The worm screw, ball nut, balls, guide and retainers are furnished only as a ball nut assembly.

Remove the rack snap ring (22—Fig. O40), spacer (21) and three-piece seal (20) from bore of rack piston. If piston rings (23) are scored or otherwise damaged, use a suitable ring expander to install new rings.

When reassembling, install all new oil seals and "O" rings. Do not assemble the housing adapter or helical gear until rack is reinstalled to main housing.

Assemble the worm screw and nut, long end first, into the piston ring end of the rack piston, being extremely careful to avoid cutting the oil seal in opposite end of rack piston. Bring retaining screw hole in nut into register with mating hole in rack piston, then install a new retaining screw (24). Tighten the new screw to a torque of 30-35 ft.-lbs., then stake screw into position.

Remove retaining ring (10) and install new seal assembly (11), with lip out, into adapter bore.

Reassemble the adapter and pin assembly (7), retaining ring (12), sleeve bushing (13) and thrust bearing assembly (14 or 14A, 14B and 14C) to shaft. Counterbore in flange end of bushing should cover the retaining ring. Reassemble washer (17) and nut (18) and with string wound around center race circumference of thrust bearing (14), tighten nut (18) until a scale pull of 1½ to 3 (ball type bearings) or 0 to 3 (needle type bearings) pounds is required to rotate the thrust bearing. Stake lip edge of nut into keyway. Align pin in adapter with hole in thrust bearing washer and

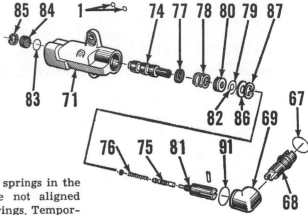

Fig. O42—Exploded view of power steering unit control valve. Refer to Fig. O40 for legend.

assemble the four loose springs in the adapter holes that are not aligned with the permanent springs. Temporarily secure adapter to thrust bearing with tape.

Note: Some early units had ball type thrust bearings (14B and 14C) which were subsequently changed to needle type and nut (18) was changed from a straight to an interference thread and copper plated for identification. Replace early parts with late parts when servicing unit.

Reinstall the assembly as in paragraph 59.

**58. REASSEMBLE.** After the sub-assemblies have been removed, disassembled, overhauled and reassembled, as outlined in paragraphs 53 through 57 they should be reinstalled to the main housing in the following order.

**59. RACK PISTON, WORM AND BEARING.** When installing this sub-assembly to main body, coat piston rings liberally with 10W oil and apply coating of light grease to main housing counterbore. Install a new "O" ring (26—Fig. O40) to adapter and pin assembly (7). Align flat on worm thrust bearing and spring assembly (14 or 14A) with the valve mounting surface of main housing, then gently push the assembly into the housing.

The cover assembly (61) can be installed only one way. Make sure that the pin on the flat surface of the bearing adapter enters the recess in the cover.

Place new "O" rings on outside of housing adapter (25 or 102), lubricate the housing bore; then, carefully slide the adapter , hub end out, on to the

Fig. O41 — Helical gear housing assembly removed from power steering unit.

worm screw shaft and into the housing. Install the snap ring (12) to shaft groove, then the gear key and helical gear (28) with counterbore of same over the snap ring. Mount assembly in a vise, place brass rod through housing hole and wedge rack piston to keep it from turning. Then install lockwasher (16) and nut (18). Tighten the nut to a torque of 25-30 ft.-lbs.

**60. SECTOR SHAFT AND PULLER.** Position the rack piston with center tooth of same in register with center of sector shaft hole in housing. The outside (upper) end of the sector shaft (49) has a slot or etched mark indicating center tooth of sector. Carefully enter marked sector tooth into mesh with marked tooth on piston, being extremely careful not to cut the housing oil seal. Start the adjusting sleeve (51) into the housing threads. If sector has been correctly installed the helical gear (28) will rotate 4¾ (row crop), or 5¾ (other) revolutions in moving the rack piston from one extreme to the other.

Using an inch-pound torque wrench, measure and record the drag required to rotate the worm screw shaft (6). Continue threading adjusting sleeve (51) into the housing and at the same time rotate worm screw shaft back and forth through about 90 degrees, so as to move it through the center "high spot" on worm screw shaft.

When an increase in drag is felt when passing through the high spot, again attach the inch-pound torque wrench to nut and tighten adjusting sleeve until torque wrench registers an amount of drag 4½ to 9 pounds greater than the previously recorded drag. Lock the adjusting sleeve in this position by installing and tightening the adjusting sleeve ring type lock nut (59).

**61. VALVE AND BODY.** Assemble valve linkage cover (69—Fig. O42) to body (71). Lower the valve actuating lever (68) into hole in end cover (60—Fig. 040) making sure the lever slot engages flat edge of worm thrust bearings (14 or 14A). Install new "O" ring between actuating lever cover and housing end cover and two new small "O" rings into counterbores on mounting face of main housing. Place valve body on main housing so pin in slot of valve link is engaged in slotted end of actuating lever. Bolt the valve body to the main housing.

62. HELICAL GEAR HOUSING. Bolt housing assembly to main housing with slot in trunnion (43—Fig. 040) parallel to mounting face of housing (31). Before installing the steering gear to tractor, adjust the column as outlined in paragraph 63.

63. ADJUSTMENT. The gear portion of the steering system is correctly adjusted when not more than 3 inches of angular movement at the end of 9-inch radius (18-inch diameter steering wheel rim) produces movement at the puller shaft or moves the front wheels.

If arc of movement (at 9-inch radius) required to move the puller shaft or front wheels is more than 5 inches, the gear needs adjustment.

With the steering gear assembly removed from the tractor, and with the gear completely assembled, rotate the trunnion retainer screw (36—Fig. 040) out (counter-clockwise) three full turns. While rotating the steering column (38 or 39) in either direction, with one hand, turn the knurled adjusting trunnion (43) either way until an increase in drag is felt in rotating the column. Now back off the knurled trunnion until column drag is just removed. Tighten and lock the trunnion retainer screw.

### LUBRICATION AND BLEEDING
## Models With Char Lynn Control Valve

65. To bleed power steering system, check and fill, if necessary, the hy-

Fig. O43—Control valve and link removed from power steering unit.

Fig. O44—Worm ball nut guides can be held in position with wire. Component parts of ball nut are not available separately.

draulic lift system reservoir, or the belt driven pump reservoir. Start engine and cycle steering system full left and right several times until system is free of air and operates smoothly. Recheck and refill reservoir, if necessary.

### SYSTEM PRESSURE TEST
## Models With Char Lynn Control Unit

66. If steering performance is not normal, first bleed system as outlined in paragraph 65. If bleeding does not correct the trouble, check the system pressure to determine the location of trouble as follows:

On models with belt driven pump, install a pressure gage capable of registering at least 1100 psi, along with a shut-off valve, in the pump pressure line. Shut-off valve must be between gage and the power steering unit.

On models with flow divider, install the pressure gage in the plug hole on left front side of flow divider valve as shown in Fig. 036.

With power steering oil warm, shut-off valve open (on models with belt driven pump) and engine running at low idle speed, turn the steering wheel from one extreme to the other and note the pressure gage reading. Pressure gage reading should be approximately 935 psi for models with belt driven pump, or 1000 psi for models with flow divider valve.

If pressure is less than specified on those models with belt driven pump, check the pump by closing the shut-off valve only long enough to obtain a reading. If pressure is now approximately 935 psi, the pump can be considered satisfactory and the trouble is located in the power steering unit. If the pressure is considerable less than the 935 psi, the pump is at fault and should be removed for repair as outlined in paragraph 47.

If the pressure is less than specified on those models with flow divider valve, add shims (4—Fig. 037) under relief valve (5). One shim should change pressure about 50 psi. If the addition of shims will not restore pressure, check the hydraulic lift

pump as outlined in paragraph 403 to insure that sufficient pressure is reaching the flow divider valve. If hydraulic lift pump pressure is satisfactory, and the pressure from flow control valve is still low, remove valve and overhaul as outlined in paragraph 49.

### PUMP
## Models With Char Lynn Control Unit

67. The belt driven power steering pump used on models not having a hydraulic lift system fitted with Char Lynn control unit is the same as the pump used on models equipped with Gemmer steering unit.

For information concerning this pump, refer to paragraphs 47 and 48.

### FLOW DIVIDER VALVE
## Models With Char Lynn Control Unit

68. The flow divider valve used on tractors fitted with Char Lynn control unit is the same as the valve used on tractors fitted with Gemmer steering unit.

For information concerning this valve, refer to paragraph 49.

### POWER STEERING CONTROL UNIT
## Models With Char Lynn Control Unit

69. REMOVE AND REINSTALL. Thoroughly clean the control unit and

Fig. O45—The Char-Lynn control unit can be removed from steering column as shown.

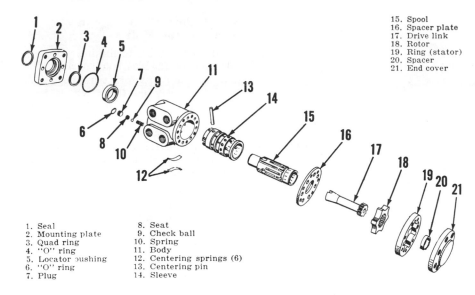

15. Spool
16. Spacer plate
17. Drive link
18. Rotor
19. Ring (stator)
20. Spacer
21. End cover

1. Seal
2. Mounting plate
3. Quad ring
4. "O" ring
5. Locator bushing
6. "O" ring
7. Plug
8. Seat
9. Check ball
10. Spring
11. Body
12. Centering springs (6)
13. Centering pin
14. Sleeve

**Fig. O45A — Exploded view of the Char-Lynn control unit used on early four-wheel drive tractors.**

lines, then identify lines and disconnect from control unit. Plug the open lines and ports in control unit. Control unit can now be unbolted from steering column and removed as shown in Fig. 045.

Install by reversing the removal procedure and tighten the control unit attaching cap screws to a torque of 23-24 ft.-lbs. Check, and refill reservoir if necessary.

70. **OVERHAUL.** With unit removed as outlined in paragraph 69, remove the seven 12-point screws from gerotor end of unit and remove cap (21—Fig. O45A), gerotor set (items 18 and 19), spacer (20), drive link (17) and spacer plate (16). Pull sleeve and spool assembly from bore of valve body. Remove plate (2) from body at which time locator bushing (5)

can also be removed. If check valve assembly (items 6 through 10) is to be removed, first remove the upper left (out) connector from valve body, then using a bent wire through the outlet hole, push plug (7) from hole. Unscrew seat (8) and remove ball (9) and spring (10). Spool (15) can be removed from sleeve (14) after pin (13) is removed. Centering springs (12) can also now be removed. Remove seal (1), quad ring (3) and "O" ring (4) from plate (2).

Clean all parts, dry with lint free wipers, and check as follows: Use a micrometer and measure width of rotor (18) and stator (19). If stator is 0.001, or more, wider than the rotor, renew the complete gerotor set. Position the rotor and stator as shown in Fig. 046 and measure between tooth of rotor and tooth (lobe) of stator. If clearance exceeds 0.005 at closest

point, renew the complete gerotor set. Place each of the six centering springs as shown in Fig. 047 and depress arch of each spring to a minimum of $\frac{1}{16}$-inch. When released, spring arch should return to a height of at least $\frac{7}{32}$-inch. Oil seal (1—Fig. O45A) is installed in plate (2) with lip facing away from valve body.

Use the following procedure to assemble the control unit. Install the centering springs in spool in two sets of three springs with arches of each set back to back and cut-outs in spring ends toward gerotor end of spool. Install spool and centering springs in sleeve, press ends of spring sets together so they will enter slot in end of sleeve and be sure ends of springs do not protrude beyond outside diameter of sleeve. Install centering pin in sleeve and spool and push it in until it is flush, or below, outside surface of sleeve. Place check valve spring in its bore with largest end of spring toward inside of body. Place check ball in bore, then install seat with counterbore toward ball and tighten to 12-13 ft.-lbs. torque. Place new "O" ring on plug and install plug. Place locator bushing in counterbore of body with largest chamfer facing outward, then with seal, quad ring and "O" ring in plate, install plate and tighten retaining cap screws to a torque of 20-21 ft.-lbs. Lubricate the spool and sleeve assembly, then carefully push spool and sleeve into body until assembly bottoms and oil seal land of spool is through seal. Set stator and rotor on end plate and drop spacer in bore of rotor. Place drive link in rotor so slot of drive link is aligned with one of the valleys between the rotor teeth as shown in Fig. 048.

**Fig. O46 — Clearance between tooth of rotor and lobe of stator must not exceed 0.005.**

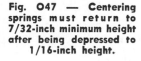

**Fig. O47 — Centering springs must return to 7/32-inch minimum height after being depressed to 1/16-inch height.**

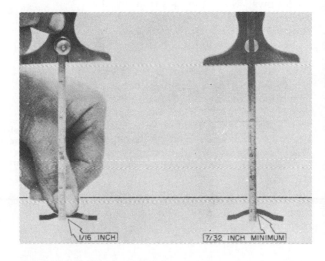

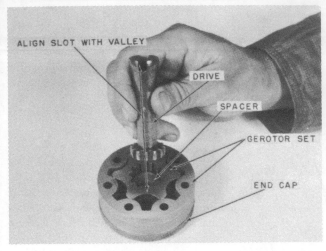

ALIGN SLOT WITH VALLEY

DRIVE

SPACER

GEROTOR SET

END CAP

**Fig. O48 — When installing drive link in rotor, be sure to align centering pin slot of drive link with a tooth valley of the rotor.**

NOTE: It is imperative that the slot in drive link be aligned with a tooth valley of the rotor as this times the control unit.

Place spacer plate over drive link, align the bolt holes, then insert two cap screws through the assembly to hold parts in position. Place assembly on body and make certain slot in drive link aligns with centering pin. Install remaining cap screws and tighten all cap screws to a torque of 12-13 ft.-lbs. Reinstall connectors, if necessary.

Reinstall control unit on tractor and bleed power steering system as outlined in paragraph 65.

### STEERING CYLINDERS
### Four-Wheel Drive Models

71. **R & R AND OVERHAUL.** To remove either steering cylinder, disconnect hoses from the three-way connectors and plug openings. Disconnect cylinder from spindle support and front axle and remove cylinder and hoses from tractor.

With cylinder removed, disassemble as follows: Note the position of the street elbow then remove the hoses and street elbow from cylinder. Remove bearing nut, then push bearing into barrel slightly and remove snap ring from I. D. of barrel. Bearing, piston rod and piston assembly can now be pulled from barrel. See Fig. O49. If bearing assembly is to be removed from piston rod, remove the piston assembly and slide bearing assembly off piston end of piston rod. Any further disassembly required will be obvious.

While reassembly is the reverse of disassembly, keep the following points in mind. Wiper seal in bearing is installed with lip facing outward.

Tighten piston retaining nut to a torque of 50-60 ft.-lbs. Pistons may be fitted with two cast iron rings (2-inch cylinder), or "O" rings with Teflon sealing rings (1¾-inch cylinder). When installing the non-metallic piston seals, place "O" rings in grooves and install Teflon seal rings over "O" rings. If piston has the cast iron rings, stagger the ring end gaps and do not let the ring end gaps align with elbow hole in cylinder barrel during installation. When piston is fitted with the Teflon seals, use a blunt tool to work the seals past elbow hole in cylinder barrel. The same blunt tool can be used to work the bearing "O" ring past elbow hole. If cylinder ends were removed, adjust them during installation until the distance between centerlines of holes in cylinder ends is 18½ inches with cylinder completely retracted. New self aligning bushings can be installed in cylinder ends, if necessary.

When installing steering cylinder on tractor, be sure the piston rod end is attached to the spindle support. Bleed power steering system as outlined in paragraph 65 and refill reservoir, if necessary.

### All Models With Saginaw
### Hydramotor Unit

All 1800 tractors beginning with serial number 134 684-000 (Series "C") and all later tractors equipped with power steering are fitted with a Saginaw Hydramotor unit. The Hydramotor directs metered, pressurized oil to the steering cylinder (two used on four-wheel drive tractors) as well as providing directional control of steering cylinder.

On tractors equipped with a hydraulic lift system, the pressurized oil for the power steering system is furnished by the hydraulic lift system pump via a priority type flow divider valve which is mounted on the hydraulic lift housing.

On tractors which do not have a hydraulic lift system, the pressurized oil for the power steering system is furnished by a pump which is mounted on engine.

All Hydramotor units are basically similar, however, units used on tractors prior to serial number 141 228-000 used a separate steel mounting bracket which also served to retain the stub shaft housing in lower housing unit. Tractors with serial number 141 228-000 and up, use a Hydramotor unit which has the stub shaft housing retained in the lower housing by a snap ring and the mounting bosses are cast into the lower housing.

Operating pressures of the early and late units differ and will be discussed in the System Pressure Test section.

### LUBRICATION AND BLEEDING
### All Models With Saginaw
### Hydramotor Unit

75. On models having an engine mounted pump, the power steering fluid reservoir is separate from pump and is located on a bracket attached to engine. Maintain fluid level at "FULL" mark on dipstick attached to reservoir filler cap. Recommended

**Fig. O49 — View showing piston, piston rod and bearing assembly removed from cylinder barrel.**

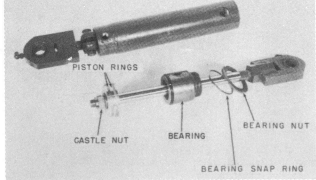

PISTON RINGS

CASTLE NUT

BEARING

BEARING NUT

BEARING SNAP RING

power steering fluid is "Type A" automatic transmission fluid.

On models with a flow divider valve, the hydraulic lift fluid is also the power steering fluid.

Steering system is self-bleeding and any air trapped in lines, cylinder or Hydramotor should be eliminated by turning steering wheel to move cylinder, or cylinders, through full range of travel several times in each direction.

## SYSTEM PRESSURE TEST
### All Models With Saginaw Hydramotor Unit

76. Factory setting for the power steering system operating pressure was 1000 psi on all tractors prior to serial number 141 228-000 (with early type Hydramotor having separate mounting bracket) and 1500 psi on tractors after that serial number. However, the system operating pressure on tractors prior to serial number 141 228-000 can be raised to 1200 psi to improve steering performance; a kit is available for the separate pump or the flow divider valve to raise operating pressues on these tractors.

To check power steering system operating pressure, proceed as follows:

On models with engine mounted pump, install a pressure gage of about 2000 psi capacity, along with a shut-off valve, in the pump pressure line. Shut-off valve must be between gage and Hydramotor.

On models with flow divider valve, install gage in flow divider valve as shown in Fig. 036. Note: Flow divider valve is located on left side of hydraulic housing on late model tractors.

With power steering oil warm, shut-off valve open (on models with engine mounted pump) and engine running at low idle rpm, turn steering wheel in either direction until steering cylinder, or cylinders, are at extreme end of travel and hold steering wheel in this position only long enough to observe the pressure gage reading. Gage reading on models with the early Hydramotor should be 1000 psi, or 1200 psi if steering improvement package has been installed. Gage reading on models with the late Hydramotor should be 1500 psi.

If pressure is less than specified on models with engine mounted pump, check pump by closing shut-off valve only long enough to obtain a gage reading. If pressure is now that specified, pump can be considered satisfactory and any trouble present will be located in Hydramotor or steering cylinder. If pump pressure remains

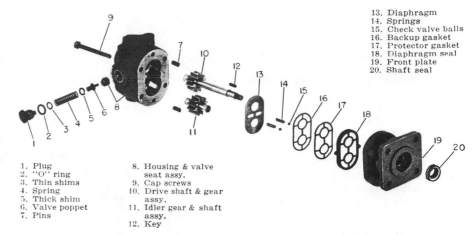

13. Diaphragm
14. Springs
15. Check valve balls
16. Backup gasket
17. Protector gasket
18. Diaphragm seal
19. Front plate
20. Shaft seal

1. Plug
2. "O" ring
3. Thin shims
4. Spring
5. Thick shim
6. Valve poppet
7. Pins
8. Housing & valve seat assy.
9. Cap screws
10. Drive shaft & gear assy.
11. Idler gear & shaft assy.
12. Key

Fig. O50 — Exploded view of power steering pump used on all models except 1850 diesel that are not equipped with hydraulic lift system. Pressure relief is adjusted by varying number of thin shims (3); a thick shim (5) is located between spring (4) and valve poppet (6).

considerably less than specified after shut-off valve is closed, adjust pump pressure as follows:

Remove plug (1—Fig. 050 or 23—Fig. 052) from pump and add shims (3 or 21) as required. Adding one shim should increase approximately 75 psi. If pump relief valve pressure cannot be brought to specifications by adding relief valve shims, remove pump as outlined in paragraph 78 and overhaul pump as outlined in paragraph 79 or 79A.

77. If pressure is less than specified on models with flow divider valve, remove plug (1—Fig. 037) and add shims (4); adding one shim should increase pressure about 50 psi. If adding shims will not restore specified pressure, check the hydraulic lift pump as outlined in paragraph 403 to insure that sufficient pressure is reaching the flow divider valve. If the hydraulic lift pump pressure is satisfactory, and the pressure from flow divider valve is still low, remove and overhaul the flow divider valve as outlined in paragraph 49.

## PUMP
### All Models With Saginaw Hydramotor Unit

78. **REMOVE AND REINSTALL.** To remove pump on all tractors except series 1850 diesel, remove hood side panel, disconnect reservoir to pump line from pump and drain pump reservoir. Disconnect pressure line from pump, then unbolt pump mounting bracket from support and remove pump and mounting bracket from tractor. Place a scribe line across bracket, pump front plate and pump housing, then remove pump pulley,

drive key and mounting bracket from pump, if necessary.

To remove pump from series 1850 diesel tractors, disconnect reservoir to pump line from pump and drain pump reservoir. Disconnect pressure line from pump. Remove pump drive coupling guard (shield) and the set screw from pump rear drive hub. Unbolt pump from mounting bracket, then pull pump from bracket, rear hub and insert and remove pump from tractor.

### All Models With Saginaw Hydramotor Unit Except Series 1850 Diesel

79. **OVERHAUL.** Prior to disassembly, thoroughly clean exterior of pump and scribe assembly marks across mounting bracket, front plate (19—Fig. 050) and housing (8). Remove pulley retaining nut, pulley and key (12) from pump drive shaft and remove mounting bracket from front plate. Remove housing retaining capscrews (9) and bump end of drive shaft against wood block to separate the housing from the front plate. Remove gear and shaft assemblies (10 and 11), diaphragm (13), gaskets (16 and 17), diaphragm seal (18), check springs (14) and check balls (15) from front plate. Using suitable tool, remove drive shaft seal (20). Remove plug (1) and relief valve components (2, 3, 4, 5, & 6) from housing.

Thoroughly clean, dry and inspect all parts. Drive shaft and gear assembly (10) or idler gear and shaft assembly (11) should be renewed if shaft bearing diameter measures less than 0.4300, or gear width measures less than 0.566. (Gears are not renewable separately from shafts, although one gear and shaft assembly may be renewed without renewing the other.) Remove minor burrs from gear teeth with fine emery cloth.

Renew back plate if gear pocket diameter exceeds 1.1695, or wear at bottom of gear pocket is more than 0.001. Renew the housing and/or front plate if bushing inside diameter is more than 0.4375. Housing and relief valve seat (8) are available as an assembly only; renew housing if seat is excessively worn or damaged. Relief valve components (items 1 through 6) are available as a repair kit which includes pins (7), or pins are available separately.

Reassemble as follows: Using a dull pointed tool, work diaphragm seal (18) into grooves in front plate with open part of "V" sections down. Drop the steel balls (15) into their bores and insert springs (14) on top of balls. Install protector gasket (17) and back-up gasket (16) into diaphragm seal and install diaphragm over backup gasket with bronze face up and grooves toward low pressure side of pump as shown in Fig. 051. Install gear and shaft assemblies (10 and 11—Fig. 050) in bores of front plate. Apply thick coat of heavy grease to milled faces of housing and front plate and install housing to front plate with previously affixed scribe marks aligned. Install retaining cap-screws and tighten to torque of 7-10 ft.-lbs. Lubricate lip of seal (20), work seal over the drive shaft with lip inward and then drive the seal into front plate with suitable tools. Install relief valve poppet (6), thick shim (5), spring (4) and plug (1) with new "O" ring (2) and same number of thin (adjusting) shims as were removed. Install mounting bracket to front plate with previously affixed

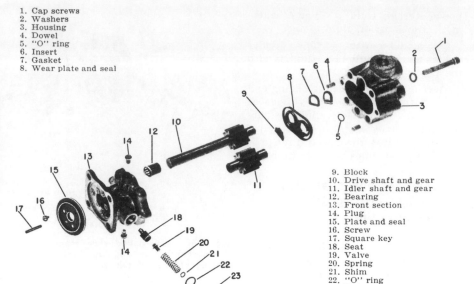

1. Cap screws
2. Washers
3. Housing
4. Dowel
5. "O" ring
6. Insert
7. Gasket
8. Wear plate and seal

9. Block
10. Drive shaft and gear
11. Idler shaft and gear
12. Bearing
13. Front section
14. Plug
15. Plate and seal
16. Screw
17. Square key
18. Seat
19. Valve
20. Spring
21. Shim
22. "O" ring
23. Plug

Fig. O52 — Exploded view of the power steering pump used on 1850 diesel tractors which are not equipped with a hydraulic lift system.

scribe marks aligned, then install key, pulley and pulley retaining nut.

Reinstall pump on tractor with test gage and shut-off valve installed, fill reservoir with proper fluid and, if necessary, adjust pump relief pressure as outlined in paragraph 76.

### Series 1850 Diesel Models

**79A. OVERHAUL.** Prior to disassembly of pump, thoroughly clean exterior of pump and place a scribe line across pump sections. Remove key (17—Fig. 052) and the housing retaining cap screws, then bump end of drive shaft against a wood block to separate the housing from front section. Remove shaft and gear assemblies (10 and 11), block (9), wear plate and seal assembly (8), gasket (7), insert (6) and "O" ring (5). Remove the retaining screws (16), then remove seal and plate assembly from recess in front section. Remove plug (23) and relief valve components (items 18, 19, 20 and 21). The four bearings (12), dowel pins (4) and plug (14) can also be removed if necessary.

Clean and inspect all parts. Renew bearings (12) and/or shafts and gears (10 and 11) if damaged or excessively worn. Gears are not available separately from shafts although one shaft and gear assembly may be renewed without renewing the other. Remove minor burrs from gear teeth with fine emery cloth. Relief valve seat (18) is available separately from housing.

Lubricate seals prior to assembly and reassemble by reversing the dis-

assembly procedure. Tighten retaining cap screws (1) to 7-10 ft.-lbs. torque.

### HYDRAMOTOR STEERING UNIT

The Saginaw Hydramotor steering units used on all tractor models are similar. Early production Hydramotor steering units were fitted with a separate steel pivot (mounting) plate and have a maximum working pressure of 1200 psi. Late production units have pivot holes in the unit covering casting and have a maximum working pressure of 1600 psi. Service on both the early and late production units is similar; any differences will be noted in the following text.

### All Models With Saginaw Hydramotor Unit

**80. R&R HYDRAMOTOR STEERING UNIT.** First, disconnect the battery ground strap, then proceed as follows: Remove emblem from steering wheel adjuster lock knob, unscrew jam nut and the lock knob from adjuster bolt, remove flat washer from bolt and then pull steering wheel and inner shaft from steering column. Remove control panel hood and the instrument panel retaining screws; then, disconnect tachourmeter cable, snap-out the instrument panel lights, disconnect wiring from gages and remove the instrument panel. Loosen the outer nuts on steering column clamp and remove instrument panel slot "door" from column. Tag the four Hydramotor hoses for ease in reconnecting same, then disconnect hoses from unit and immediately cap all openings.

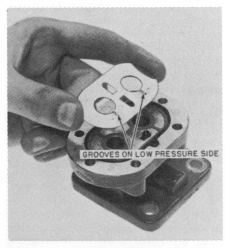

GROOVES ON LOW PRESSURE SIDE

Fig. O51 — Install diaphragm with bronze side up and grooves towards low pressure side of pump as shown.

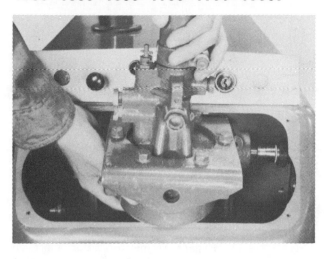

Fig. O57 — Removing early production type Hydramotor unit out through instrument panel opening.

On late production units, loosen pivot pin lock nuts and unscrew the pivot pins until clear of the Hydramotor cover casting. Then, lift the unit out from opening in instrument panel support.

On early production units, drive the spring (roll) pin from governor control shaft arm, remove control shaft nut from steering unit R.H. support and pull control shaft from arm and the L.H. steering unit support. Be careful not to lose bushing from L.H. support as shaft is withdrawn. Loosen lock nuts on pivot pins and unscrew the pins until they are clear of the Hydramotor pivot bracket. Lift Hydramotor, turn it ¼-turn counterclockwise, slip pivot bracket over front of the instrument panel opening and lift unit up out of opening as shown in Fig. O57. Also see Fig. O58.

Reverse removal procedures to reinstall either the early or late production unit. On models without hydraulic system, refill power steering reservoir as outlined in paragraph 75.

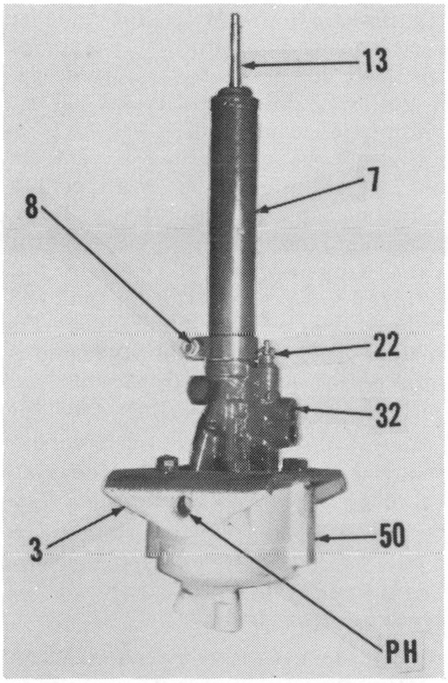

## All Models With Saginaw Hydramotor Unit

81. **R&R BLOCKING SPOOL VALVE.** The blocking spool valve and related parts can be removed and reinstalled after the Hydramotor steering unit has been removed as outlined in paragraph 80. Refer to Fig. O59 and proceed as follows:

Remove the lockout adjuster nut (22). Plug (24) and spool valve (25) may now be removed by pushing the plug into bore against spring pressure with screwdriver or other tool, then quickly releasing the plug to allow spring to pop it out of the bore. Remove plug and if spool sticks in bore, invert the unit and tap housing (32) with soft faced mallet to jar spool out. Invert unit and allow spring (26) and the spring and guide assembly (27) to drop from bore.

If spool (25) is excessively worn or badly nicked, it should be renewed. A brightly polished wear pattern is normal. Spool should slide and rotate freely in bore.

To reassemble, install parts in bore of housing (32) as shown in exploded view, renewing the "O" ring (23) on plug (24) and tightening adjuster nut to a torque of 10-15 ft.-lbs.

## All Models With Saginaw Steering Unit

82. **R&R STEERING COLUMN JACKET AND SHAFT ASSEMBLIES.** With Hydramotor unit removed as outlined in paragraph 80, proceed as follows:

Loosen clamp (8—Fig. O60) and pull column jacket assembly (7) from control valve housing (32). Hold steering shaft with large screwdriver inserted through slot in outer shaft (pull adjuster bolt up out of way) and unscrew the hex nut (28) until it nearly contacts control valve housing. Drive the tapered collar (19) towards nut

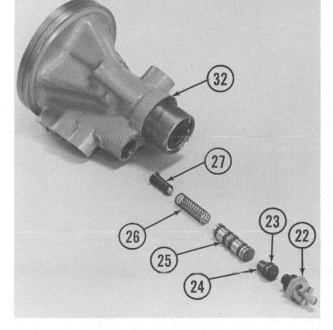

Fig. O58 — View of early production Hydramotor steering unit removed from tractor. Pivot holes (PH) for unit are in separate steel bracket (3), while pivot holes for late production units are in the cover (50) casting; see (PH—Fig. O62).

PH. Pivot holes
3. Pivot bracket
7. Column jacket
8. Jacket clamp
13. Adjuster bolt
22. Blocking valve lockout
32. Housing
50. Cover

Fig. O59—Exploded view of Hydramotor housing and blocking valve components. Blocking valve can be removed without disassembling Hydramotor steering unit.

22. Lockout
23. "O" ring
24. Plug
25. Valve spool
26. Spring
27. Spring & guide assy.
32. Housing

until collar is loose, turn collar until hole in collar is over locking ball hole in outer shaft (17), then shake the ball (20) out of hole. The outer shaft, tapered collar, hex nut and adjusting bolt (13) with nut (14) can then be removed from the stub shaft (37).

If necessary to disassemble the steering shaft or column jacket assemblies, procedure for doing so is obvious. When installing new needle bearing (6), drive or press on lettered side of bearing cage only. Bearing race (16) is used on early production units only. Reassemble the unit before reinstalling on stub shaft as follows:

Engage splines of outer shaft on splines of stub shaft being sure that locking ball hole in outer shaft is away from flat on stub shaft. Align hole in tapered collar, hole in stub shaft and groove around the stub shaft, then drop locking ball in hole and groove and turn tapered collar ¼-turn. Tighten the hex nut to a torque of 20-30 ft.-lbs. and stake nut into slot in outer shaft as shown in Fig. 061.

**Early Production Hydramotor Units**

83. **R&R PIVOT (MOUNTING) BRACKET.** The pivot bracket (3—Fig. 060) retains cover (50) to control valve housing on early production Hydramotor units. With unit removed from tractor as outlined in paragraph 80, the pivot bracket can be removed as follows: Loosen capscrews (1) retaining pivot plate to cover and remove all the cap screws except the one where interference with control valve housing is encountered. Bump pivot plate loose and rotate the control valve housing within the cover so that remaining cap screw can be removed. Then remove the pivot plate. For further service, clamp the Hydramotor unit in vise as shown in Fig. 062 and refer to paragraph 86.

Reinstall pivot plate by reversing removal procedure. Tighten the plate retaining capscrews to a torque of 45-60 ft.-lbs.

**Late Production Hydramotor Units**

84. **R&R COVER RETAINING SNAP RING.** On late production units, a snap ring (52—Fig. 060) is used to retain cover (50) to control valve housing (32). To remove the snap ring, first remove unit from tractor as outlined in paragraph 80, then proceed as follows: Refer to Fig. 063. If end gap of snap ring is not near hole in cover as shown, bump the snap

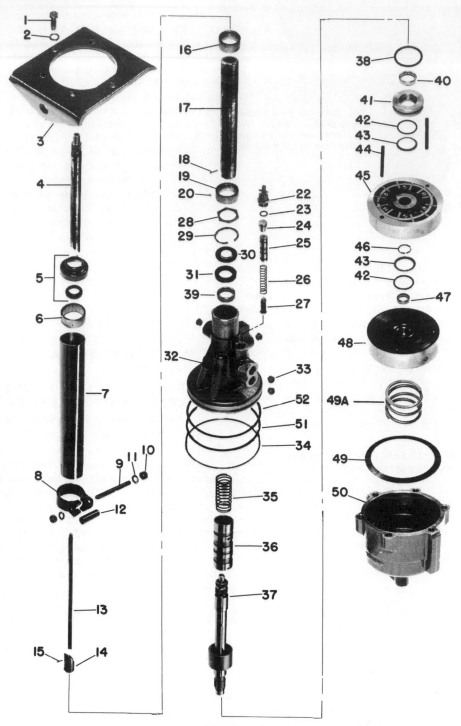

**Fig. O60 — Exploded view of composite early and late production Hydramotor steering unit. On early production units, cover (50) is retained to housing (32) by pivot bracket (3) and cap screws (1); "O" ring (51) is located in groove of housing and a Belleville washer (49) pressure plate spring is used. On late production models, cover is retained to housing by snap ring (52); "O" ring (51) and backup ring (34) are located in groove in I.D. of cover and a coil type pressure plate spring (49A) is used.**

3. Pivot bracket (early)
4. Inner shaft
5. Dust cap & scraper
6. Needle bearing (not serviced)
7. Column jacket
8. Jacket clamp
12. Spacer
13. Adjuster bolt
14. Adjuster nut
15. Pin
16. Bearing race (not serviced)
17. Outer shaft
18. Pin
19. Tapered collar
20. Locking ball

22. Lockout assembly
23. "O" ring
24. Plug
25. Valve spool
26. Spring
27. Spring & guide assy.
28. Hex nut
29. Snap ring
30. Dust seal
31. Oil seal
32. Housing
33. Connector seats
34. Backup ring
35. Spring
36. Valve spool
37. Stub shaft assembly
38. "O" ring

39. Needle bearing
40. Needle bearing
41. Bearing support
42. "O" rings
43. Teflon seals
44. Dowel pins
45. Pump ring & rotor assy.
46. Snap ring
47. Needle bearing
48. Pressure plate
49. Spring (early production)
49A. Spring (late production)
50. Cover (early production shown)
51. "O" ring
52. Snap ring

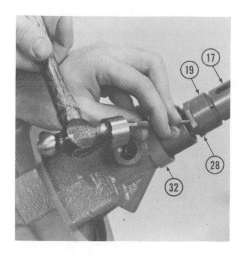

**Fig. O61 — After installing steering shaft on Hydramotor stub shaft, stake hex nut to slot in steering shaft (17) with center punch.**

17. Outer shaft
19. Tapered collar
28. Hex nut
32. Housing

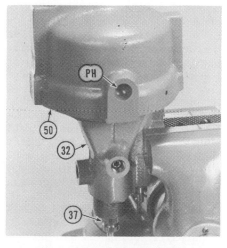

**Fig. O62 — Late production Hydramotor mounted in vise for overhaul; note pivot hole (PH) in cover (50).**

PH. Pivot holes
32. Housing
37. Stub shaft
50. Cover

ring into this position with hammer and punch. Insert a pin or punch into hole and drive pin or punch inward to dislodge snap ring end from groove. With the pin or punch under the snap ring, use screwdrivers to pry ring from cover. After removing the snap ring, place Hydramotor unit in vise as shown in Fig. 062 and refer to paragraph 86 for further service procedure.

NOTE: Usually, spring (49A—Fig. 060) will push control valve housing from cover. However, if burrs or binding condition exists it may be necessary to bump cover loose by tapping around edge of cover with mallet as shown in Fig. 064.

85. To reinstall the cover retaining snap ring, control valve housing must be held in cover against pressure from spring (49A—Fig. 060). It is recommended that the unit be placed in an arbor press and the housing be pushed into cover by a sleeve as shown in Fig. 065. **CAUTION:** Do not push against end of stub shaft (37—Fig. 060). Place snap ring over housing before placing unit in press. Carefully apply pressure on housing with sleeve until flange on housing is below snap ring groove in cover. Note that lug on housing must enter slot in cover If housing binds in cover, **do not** apply heavy pressure; remove unit from press and bump cover loose with mallet as shown in Fig. 064. When housing has been pushed sufficiently far into cover, install snap ring in groove with end gap near hole in cover as shown in Fig. 063.

### All Models With Saginaw Hydramotor Unit

86. **OVERHAUL HYDRAMOTOR STEERING UNIT.** With steering shaft and column unit removed as in paragraph 80 and the pivot bracket or cover retaining snap ring removed as outlined in paragraph 83 or 84. proceed as follows:

Remove cover (50—Fig. 062) by pulling upward with twisting motion. Remove the pressure plate spring, then lift off pressure plate as in Fig. 066. (NOTE: Tractors, serial number 194 491-000 and up, have pressure plate retained by two cap screws). Remove dowel pins as in Fig. O67, then remove snap ring (46) from stub shaft (37) with suitable snap ring pliers and screwdriver; discard snap ring. Pull pump ring and rotor assembly (45) up off of stub shaft as shown in Fig. 068. Tap end of stub shaft with soft faced mallet as shown in Fig. 069 until bearing support (41) can be removed, then carefully remove actuator assembly from housing as shown in Fig. 070.

Note: It is recommended that the actuator assembly not be disassembled as it is a factory balanced unit and serviced as an assembly only. See Fig. 071 for details of bearing support seals and bearings.

Housing (32—Fig. 060), blocking valve spool (25) and actuator assembly, which includes spring (35), spool (36) and stub shaft assembly (37) are serviced as a matched set only. If these parts are otherwise serviceable,

**Fig. O63 — To remove cover retaining snap (52) on late production units, drive pin or punch through hole (dotted lines) in cover (50) to disengage snap ring in groove. Housing is (32).**

**Fig. O64 — In event coil pressure plate spring does not push cover from housing, tap around edge of cover with soft faced mallet.**

**Fig. O65 — To reinstall cover retaining snap ring (52), place unit in arbor press and push housing (32) into cover (50) with sleeve as shown.**

**Fig. O66 — Removing pressure plate (48) from dowel pins (44).**

| | |
|---|---|
| 32. Housing | 45. Pump ring & rotor assy. |
| 44. Dowel pins | 48. Pressure plate |

**Fig. O67 — Removing dowel pins (44) from pump ring and rotor assembly (45) and housing (32). Discard snap ring (46) after it is removed from stub shaft (37) and install new snap ring on reassembly.**

needle bearing (39), seals (30 and 31), connector seats (33) and blocking valve components except the valve spool (25) may be renewed as necessary. Install new needle bearing by pressing on lettered side of bearing cage only. Connectors (33) may be removed by threading inside diameter with tap as shown in Fig. 072, then removing connector with puller bolt. Refer to paragraph 81 for information on blocking valve unit.

Needle bearings (40 and 47—Fig. 060) in bearing support (41) and pressure plate (48) may be renewed if support and/or plate are otherwise serviceable. Install new bearings by pressing on lettered side of bearing cage only.

Rotor, ring, vanes and vane springs are serviced as a complete assembly (45) only; however, the unit may be disassembled for cleaning and inspection. Reassemble by placing rotor in ring on flat surface. Insert vanes (rounded side out) in rotor slots aligned with large diameter of ring, turn rotor ¼-turn and insert remaining vanes. Hook the vane springs behind each vane with screwdriver as shown in Fig. 073; be sure that vane springs are in proper place on both sides of rotor.

87. To reassemble Hydramotor unit, place housing, with needle bearing, seals and snap ring installed, in a vise with flat (bottom) side up. Check to be sure that pin in actuator is engaged in valve spool; if spool can be pulled away from actuator as shown in Fig. 074, push spool back into actuator and

be sure that the pin is engaged into hole in spool. Then, carefully insert actuator assembly into bore of housing. Place bearing support assembly on stub shaft and carefully push the assembly in flush with housing as shown in Fig. 075. Place the pump ring and rotor assembly on stub shaft and housing; on late production units, chamfered outer edge of ring must be away from housing (up). Install a **new** rotor retaining ring and insert the dowel pins through ring into housing. Stick the "O" ring and Teflon seals into pressure plate with heavy grease, then install pressure plate over stub shaft, pump ring and rotor assembly and the dowel pins. NOTE: On tractors, serial number 194 491-000 and up, which have the bolted on pressure plate, install pressure plate and tighten the two retaining cap screws to 96-120 in.-lbs.; then, back-off the two cap screws ¼-½ turn and be sure cap screw heads are within the outer circumference of pressure plate. With pressure plate and ring and rotor assemblies final positioned and cap screws backed off, about 10 in.-lbs. of torque should remain on the retaining cap screws. Place pressure plate spring on pressure plate. Note: On early production units a Belleville washer type spring is used; cup side of spring must be away from pressure plate. A coil spring is used on all late production units. On early production units, install new sealing "O" ring in groove on outside diameter of housing. On late production units, install new "O" ring and backup ring in groove in cover (refer

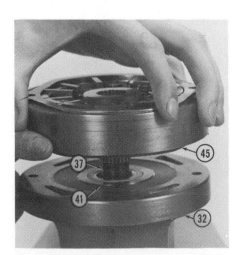

**Fig. O68 — Lifting the pump ring and rotor assembly (45) from stub shaft (37) and housing (32). Bearing support is (41).**

**Fig. O69 — Tap on end of stub shaft (37) with soft faced mallet to bump bearing support (41) out of housing (37).**

**Fig. O70 — Removing the actuator assembly (stub shaft and control valve spool) from housing. Be careful not to cock spool in bore of housing.**

to Fig. 076). Then, install cover over the assembled steering unit and install pivot bracket on early production units as in paragraph 83, or cover retaining snap ring on late production units as in paragraph 84.

### STEERING CYLINDER

### All Models With Saginaw Hydramotor Unit

88. **REMOVE AND REINSTALL (TWO-WHEEL DRIVE).** Remove hood side panels and hood. Disconnect oil cooler from radiator. On all models, remove grille and radiator from main frame. Loosen line clamps and disconnect power steering lines from power steering cylinder and cap all lines and openings. Refer to Fig. 080, remove plug (4) from cap (1) and remove cap from steering cylinder. Pry plug (7) from pinion (11). Pull pinion from splines of front assembly

shaft by turning puller bolt (9—Fig. 081) counter-clockwise. Remove nut and washer from tapered stud at end of torque link and drive the stud upward out of main frame, then lift power steering cylinder from front main frame. NOTE: On Row Crop models, removal of tapered stud will require removal of dust shield from lower side of main frame.

To reinstall the power steering cylinder, reverse the removal procedure. Use new "O" rings and seals during installation. Be sure to align punch marked tooth of pinion with center tooth groove in piston as shown in Fig. 082. Tighten the puller bolt to a torque of 200 ft.-lbs. Install plug (7—Fig. 080) and fill the cavity around pinion with power steering fluid before installing cap (1).

89. **OVERHAUL (TWO-WHEEL DRIVE)** With cylinder removed as outlined in paragraph 88, proceed as follows: Pull pinion assembly from cylinder, if necessary. Puller bolt can be removed from pinion after snap ring (8—Fig. 080) is removed. To remove end plug retaining snap ring (24), insert pin or punch in hole (H—Fig. 081) and push snap ring from groove in cylinder. Grasp one of the lugs on end plug with vise-grip pliers and with a twisting motion, pull end plug from cylinder. Remove plug sealing "O" ring (22—Fig. 080) from end of cylinder and remove the piston (21). Remove the Teflon seals (16) and "O" rings (17) from groove at each end of piston and remove pinion sealing quad ring (12) from bottom of cylinder. If necessary to remove the check valve assembly (18, 19 and

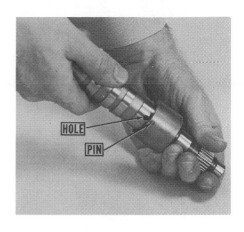

Fig. O74 — Pin in actuator sleeve must be engaged in hole in end of spool valve before actuator assembly is installed. If spool cannot be pulled out of sleeve, pin is engaged.

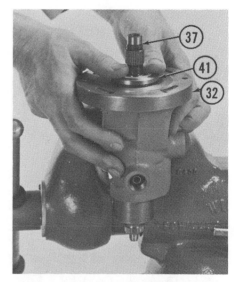

Fig. O75 — When pushing bearing support (41) for stub shaft (37) into housing (32), take care not to damage sealing "O" ring.

Fig. O71 — Lifting the Teflon seal (43) and "O" ring (42) from bearing support (41). Needle bearing (40) is serviced separately. Install new "O" ring in groove (G) before installing bearing support.

Fig. O72—To remove connector seats (33), thread inside diameter with tap as shown, then use puller bolts to remove seats.

Fig. O73 — Be sure all vane springs are engaged behind the rotor vanes. Springs can be pried into place with screwdriver as shown.

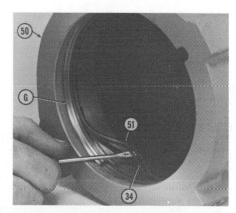

Fig. O76 — On late production units, "O" ring (51) and backup ring (34) are installed in cover (50); be sure backup ring is to outside (open side of cover). Groove (G) is for cover retaining snap ring.

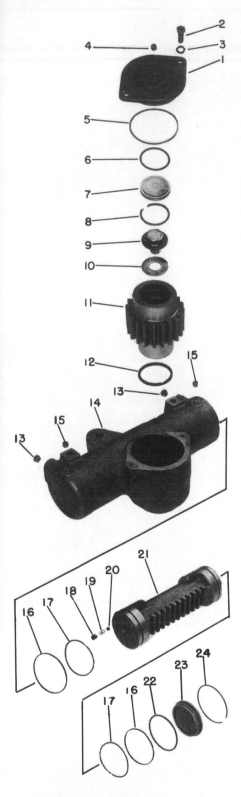

Fig. O82—View of pinion (11) and piston (21) removed from cylinder to show timing marks (TM). Puller bolt (9) is retained in pinion snap ring. White ring (16) is Teflon seal ring.

20) from either end of piston, insert a punch in hole from inner side of piston head and drive valve ball, spring and spring pin from piston. If the hydraulic line connector seats (13) are damaged, they may be removed in a manner similar to that shown in Fig. 072. Clean and inspect all parts and renew any that are worn, scored or damaged.

Use all new "O" rings and seals and reassemble cylinder as follows: Insert check ball (20) into bore in end of piston (21), insert spring (19) with small end next to ball and drive the retaining spring pin (18) in flush with end of piston. Install an "O" ring (17) in bottom of groove in each end of piston, then install a Teflon ring (16) on top of each "O" ring. Lubricate piston and install in bore of cylinder, install "O" ring (22), lubricate and install end plug (23) and install plug retaining snap ring (24). Install quad ring (12) in bottom of pinion bore in cylinder casting and lubricate ring and bore. Install washer (10), puller bolt (9) and retaining snap ring (8) in top of pinion.

Reinstall cylinder as outlined in paragraph 88.

90. **R&R AND OVERHAUL (FOUR-WHEEL DRIVE).** All four-wheel drive models are equipped with two steering cylinders of the double acting type. All models prior to serial number 137 613-000 were equipped with 2-inch cylinders, all models after this serial number were equipped with 1¾-inch cylinders. In some cases, the two inch cylinders may have been replaced by the 1¾-inch cylinders to improve steering performance. Refer to paragraph 71 for cylinder information.

## OIL COOLER

### All Models With Saginaw Hydramotor Unit

91. All series 1800 and 1900 series C, and earlier series 1850 and 1950 tractors equipped with power steering were fitted with an oil cooler of the type shown in Fig. O82A. This type cooler can be removed by removing hood lower front side panels, disconnecting lines at lower end of cooler unit and unbolting unit from radiator. Cleaning is the only service recommended and faulty units should be renewed.

91A. Beginning in February 1968, a demand type power steering flow divider valve was installed and the change in cooling requirements resulted in the installation of the type cooler shown in Fig. O82B. This oil cooler is mounted to front left side of radiator and can be removed by removing hood lower front side panels, disconnecting lines at bottom of

Fig. O81 — View showing power steering cylinder on Row Crop models with cylinder cap and pinion plug removed. On Wheatland and Industrial models, cylinder is mounted to front side of front assembly shaft instead of to rear as shown.

H. Hole for removing snap ring
9. Puller bolt
11. Pinion
14. Cylinder

Fig. O80 — Exploded view of power steering cylinder assembly used on all models except those equipped with four-wheel drive.

| | |
|---|---|
| 1. Cap | 14. Cylinder |
| 4. Plug | 15. Plugs |
| 5. "O" ring | 16. Teflon rings |
| 6. "O" ring | 17. "O" rings |
| 7. Plug | 18. Spring pins |
| 8. Snap ring | 19. Springs |
| 9. Puller bolt | 20. Check valves |
| 10. Washer | 21. Piston |
| 11. Pinion | 22. "O" ring |
| 12. Quad ring | 23. End plug |
| 13. Connectors | 24. Snap ring |

cooler and unbolting unit from radiator support. In addition to cleaning, unit can be repaired if care is exercised.

### COOLER BY-PASS VALVE

91C. When tractor is fitted with oil cooler shown in Fig. O82B, a by-pass valve (6) is located between power steering Hydramotor and the oil cooler. On series 1750, valve is mounted on left rear of front main frame. On series 1950-T, valve is located on lower front side of instrument panel support. Valve is used to prevent high pressure oil from entering oil cooler during cold weather. Valve is set to open at 65-85 psi and can be adjusted by adding or subtracting shims behind valve spring (5 —Fig. O82C).

91D. **R&R AND OVERHAUL.** By-pass valve can be removed after disconnecting lines and removing mounting cap screws. To disassemble, remove plug (1) with "O" ring (2), then remove spring (3), spring retainer (4) and ball (5) from body (6). Insert shown at top of body is renewable, however valve seat (not shown) for ball (5) is not.

Clean and inspect all parts. If seat for ball is damaged renew body (6). Spring (3) has a free length of 2 1/64 inches and should test 2.9-3.5 lbs. when compressed to a length of 1 34/64 inches.

If necessary, valve can be checked and adjusted prior to installation as follows: Look through valve ports and note that relief valve spring is visible in one of the ports. Install a plug in

the port that is NOT obstructed with valve spring. Connect an injector nozzle test pump, or a hand hydraulic pump in the port directly across from the plug previously installed. Actuate pump to build pressure and note pressure at time valve cracks. If valve cracking pressure is not 65-85 psi, add or subtract shims behind spring as necessary. Shims are available in thickness of 0.0359.

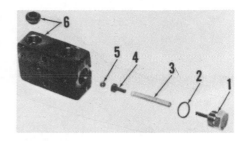

**Fig. O82C—Exploded view of oil cooler by-pass valve. No adjusting shims are shown.**

1. Plug
2. "O" ring
3. Spring
4. Spring retainer
5. Steel ball
6. Body and insert

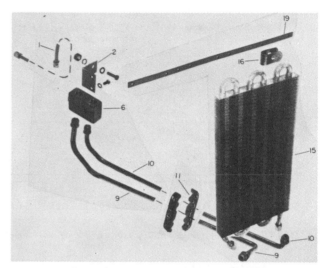

**Fig. O82B — Oil cooler unit used on models which have demand flow type of flow divider valve. Note by-pass valve (6).**

1. Oil line
2. Bracket
6. By-pass valve
9. Inlet line
10. Outlet line
15. Oil cooler
16. Clamp
19. Mounting strip

**Fig. O82A — View showing mounting location of the oil cooler used on tractors fitted with Saginaw power steering.**

# ENGINE AND COMPONENTS
## (Six Cylinder)

The Oliver series 1750, 1800, 1850 and 1950-T tractors are fitted with six cylinder, overhead valve engines. All engines except those used on series 1850 diesel tractors are equipped with wet sleeves. The series 1850 diesel engines are equipped with dry type sleeves.

Series A, 1800 tractors with serial numbers 90 525-000 through 124 395-000 may be equipped with a non-diesel engine having a bore and stroke of 3¾ x 4 inches and a displacement of 265 cubic inches; or a diesel engine having a bore and stroke of 3⅞ x 4 inches and a displacement of 283 cubic inches.

Series B and C, 1800 tractors with serial numbers 124 396-000 through 150 420-000 may be equipped with a non-diesel engine having a bore and stroke of 3⅞ x 4 inches and a dis-

placement of 283 cubic inches; or a diesel engine having a bore and stroke of 3⅞ x 4⅜ inches and a displacement of 310 cubic inches.

Series 1850 tractors with serial numbers 150 421-000 and up may be equipped with a non-diesel engine having a bore and stroke of 3⅞ x 4⅜ inches and a displacement of 310 cubic inches; or a diesel engine having a bore and stroke of 3⅞ x 5 inches and a displacement of 354 cubic inches.

Series 1750 tractors with serial number 180 537-000 and up may be equipped with a non-diesel engine having a bore and stroke of 3⅞ x 4 inches and a displacement of 283 cubic inches; or a diesel engine having a bore and stroke of 3⅞ x 4⅜ inches and a displacement of 310 cubic inches.

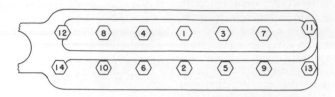

Fig. O83—Tighten all series 1750, 1800 and 1850 non-diesel engine cylinder head cap screws in the sequence shown.

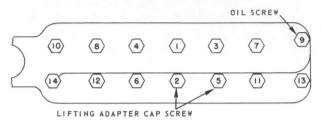

OIL SCREW

LIFTING ADAPTER CAP SCREW

Fig. O83A—Tighten all series 1750 and 1950-T diesel engine cylinder head cap screws in the sequence shown.

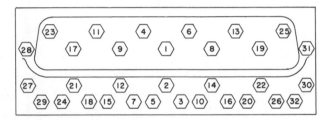

Fig. O84—Tighten all series 1850 diesel engine cylinder head stud nuts in the sequence shown.

Series 1950-T tractors with serial number 194 080-000 and up are equipped with a diesel engine having a bore and stroke of 3⅞ x 4⅜ inches and a displacement of 310 cubic inches.

All of the engines, except the series 1850 diesel engine, are basically similar.

## R & R ENGINE WITH CLUTCH
### Series 1750-1800-1850-1950-T

95. On tractors equipped with Hydra-Power drive, Creeper drive or Reverse-O-Torc, the auxiliary drive unit must be removed along with the engine. The auxiliary drive unit and clutch housing can then be removed from engine to service the clutch and/or flywheel.

To remove engine, first drain cooling system and if engine is to be disassembled, drain oil pan. Remove both hood side panels and hood.

On all series 1800 A and B and series 1800 C and 1850 diesel tractors, remove battery or batteries, hood supports and battery tray.

Disconnect clutch spring from tank support. Unclip wiring loom and control cables from bottom of fuel tank and on diesel models, disconnect leak-off and water trap lines from tank. Disconnect the fuel supply line from tank. Attach hoist to tank and take up slack, then unbolt tank cradle, drive out hinge pin and remove fuel tank and cradle.

Remove the pto drive shaft as outlined in paragraph 356, or the hydraulic pump drive shaft as outlined in paragraph 409 if tractor has no pto. Disconnect clutch rod from clutch shaft lever.

On models with no auxiliary drive unit, remove the center drive shaft shield and disconnect coupling chain. Loosen clutch shaft coupler, slide coupler onto center drive shaft and remove center shaft. Do not lose spacer located between coupling sprockets. Remove clutch shaft from clutch housing.

On all models, remove fan blades and radiator. Disconnect the necessary wiring and controls from engine and on models equipped with auxiliary

drive, disconnect the unit control linkage and oil cooler lines. Disconnect power steering lines from clips on bottom of engine, or from engine mounted power steering pump, if so equipped. On models with manual steering, disconnect steering shaft from clutch housing and separate shaft. Remove engine mounting bolts and on all models except 1850 diesel identify the left front and right rear bolts as they are a machine fit and act as dowels. Install lifting eyebolt to one of the extended (lifting) cylinder head capscrews of all models except 1850 diesel, or to brackets on series 1850 diesel, attach hoist and lift engine from tractor.

Reinstall engine by reversing the removal procedure and bleed the power steering system.

### Series 1950-T

95A. On tractors equipped with an auxiliary drive unit, the auxiliary drive unit must be removed along with the engine. The auxiliary drive unit and clutch housing can then be removed from engine to service clutch and/or flywheel.

To remove engine, first drain cooling system and if engine is to be disassembled, drain oil pan. If auxiliary drive unit is to be serviced, also drain unit. Remove both hood side panels, front and rear lower side panels and hood.

Shut off fuel and disconnect main tank fuel supply line, injector leak-off line, and if equipped with wheel guard fuel tanks, the wheel guard tank supply line. Disconnect wiring from bottom of main tank. Disconnect wire from fuel gage sending unit and restriction indicator sending unit. Remove switch panel hood (cowl). Attach a sling to main fuel tank, unbolt rear of tank, remove front tank strap and lift tank from tractor.

Remove pto shaft as outlined in paragraph 356, or the hydraulic pump drive shaft as outlined in paragraph 409 if tractor has no pto. Remove sprocket coupling chain. Disconnect clutch control rod and if so equipped, the auxiliary drive unit control rod.

On models with no auxiliary drive unit, remove the center drive shaft shield. Loosen clutch shaft coupler, slide coupler onto center drive shaft and remove center drive shaft. Do not lose spacer located between coupling sprockets. Remove clutch shaft from clutch housing.

On all models, remove fan blades, disconnect oil coolers from radiator and remove radiator. Disconnect the necessary lines, wiring and controls from engine and pull wiring rearward from fuel tank support. Disconnect auxiliary drive unit coolant lines at unit and line clip from front frame. Remove engine mounting bolts and identify left front and right rear bolts as they are a machine fit and act as dowels. Install eyebolt to cylinder head lifting cap screws, attach hoist and lift engine from tractor.

Reinstall engine by reversing the removal procedure and bleed fuel system.

## R&R CYLINDER HEAD
### All Series 1750—Non Diesel Series 1800-1850

96. To remove cylinder head, remove hood side panels and hood, then drain cooling system. Unclip wiring and control cables from bottom of fuel tank. Disconnect fuel supply line and on diesels, disconnect leak-off (drip) line and water trap line from fuel tank. Remove batteries, hood supports and battery tray from instrument panel support on models so equipped. Unbolt fuel tank bracket, swing tank upward and retain in position with a piece of wire, or rope, attached to steering wheel. Remove upper radiator hose, air cleaner and cleaner hose. On tractors equipped with separate power steering pump, disconnect the pump reservoir and move it out of the way. Disconnect breather tube from tappet cover and on diesels, disconnect wiring from preheater solenoid. On non-diesels, unbolt ignition coil from rocker cover, disconnect spark plug wires and remove carburetor from manifold. On diesels, disconnect the injection pump to injector lines at injectors and immediately cap off all openings. Remove rocker cover, rocker arm shaft oil line, rocker arms and shaft assembly, then remove cylinder head cap screws and remove cylinder head from engine.

When reinstalling, refer to Figs. O83 and O83A and tighten all screws on series 1800, except drilled oil screws, to 129-133 ft.-lbs. Tighten drilled screw to 113-117 ft.-lbs. On series 1750 and 1850 non-diesel tractors, tighten all screws to 129-133 ft.-lbs. On series 1750 diesel tractors with 9/16 inch cap screws, tighten screws 1 through 6 to 150 ft.-lbs. and cap screws 7 through 14 to 133 ft.-lbs. On 1750 Diesel tractors, if 5/8-inch cap screws are used, tighten screws 1 through 8, 11 and 12 to 180 ft.-lbs. and screws 9, 10, 13 and 14 to 133 ft.-lbs. Torque values given are for oiled threads. Note that oil screw is at rear of right hand row of cap screws. Rocker arm shaft bracket nuts and

manifold nuts are tightened to 25-27 ft.-lbs. Adjust tappet gaps as in paragraph 98.

### Series 1950-T

96A. To remove cylinder head, first remove both side panels and hood, then drain cooling system. Disconnect wires from restriction indicator sending unit and inlet hose from turbocharger. Disconnect wiring from preheater solenoid, then remove filters and support bracket. Disconnect ground cable from tappet cover stud and remove turbocharger outlet pipe hoses and preheater as a unit. Unbolt breather tube flange from tappet cover and remove flange and breather tube. Remove tappet cover, rocker shaft oil line, rocker arms and shaft assembly and push rods. Identify push rods so they can be reinstalled in their original position. Remove the small bleed line from thermostat housing and cylinder head. Disconnect engine oil cooler lines from cylinder head and thermostat housing. Disconnect wire from engine temperature sending unit, then remove by-pass hose from water pump and thermostat housing. Remove thermostat housing from cylinder head. Remove turbocharger inlet and outlet oil lines. Remove exhaust manifold, head shield and turbocharger as a unit. Remove injector leak-off line and disconnect pressure lines from injectors and immediately cap all openings. If equipped with wheel guard fuel tanks, disconnect fuel supply line from tee at top front of main fuel tank and move line out of the way. Disconnect supply line from secondary fuel filter and injection pump and remove line. Remove cylinder head cap screws and lift off cylinder head.

When reinstalling cylinder head, refer to Fig. O83A and tighten cap screws 1 through 8, 11 and 12 to 180 ft.-lbs. torque and cap screws 9, 10, 13 and 14 to 133 ft.-lbs. torque. Torque values are for oiled threads.

Note that oil screw is at rear of right hand row of cap screws. Rocker arm shaft bracket nuts and manifold nuts are tightened to 25-27 ft.-lbs. torque. Adjust valve tappet gaps as outlined in paragraph 98.

NOTE: In the event of engine overhaul, or any other service involving turbocharger or exhaust system, it is recommended by Oliver Corporation that engine be operated for approximately one hour without the turbocharger being installed to preclude the possibility of turbocharger being damaged by engine debris. Do this by any of the following methods: Plug turbocharger oil pressure line; install a separate oil filter

in turbocharger oil pressure line; or fabricate a connector and join turbocharger oil pressure and oil return lines.

### Series 1850 Diesel

97. To remove cylinder head, remove the hood side panels and hood, then drain cooling system. Unclip wiring and control cables from bottom of fuel tank. Disconnect fuel supply line water trap line and leak-off line from fuel tank. Remove battery, hood supports and battery tray from instrument panel support. Unbolt fuel tank support, swing tank upward and retain in this position with a piece of wire, or rope, attached to steering wheel. (Drain tank if necessary). Remove upper radiator hose and air cleaner assembly. Disconnect power steering pump reservoir and move it out of the way. Disconnect wiring from pre-heater and breather pipe from tappet cover. Remove injector leak-off return line. Remove the injection pump to injector lines and immediately cap all openings. Unbolt final fuel filter from cylinder head. Remove tappet cover, rocker arms and shaft assembly and push rods. Remove cylinder head stud nuts and lift cylinder head from engine. Use caution not to damage injector tips and DO NOT let cylinder head rest on combustion side while injectors are installed.

NOTE: Cylinder head can be cleaned and inspected after removal by removing the water outlet housing and rear cover. Be sure the oil hole (Fig. 085) for rocker arm oil pipe is open and clean.

Reinstall by reversing the removal procedure. Tighten the cylinder head stud nuts in the sequence shown in Fig. 084 and to a torque of 80-85 ft.-lbs. Adjust tappets to 0.012 cold, then start engine and bring to operating temperature. Recheck the torque on cylinder head stud nuts and reset tappets to 0.010, if necessary.

## VALVES, GUIDES AND SEATS
### All Series 1750-1800-1850 Non-Diesel-1950-T

98. Both inlet and exhaust valves for all models except 1750 diesel and 1950-T diesel have a face angle of 44½ degrees and a seat angle of 45 degrees. Valve stem diameter is 0.372-0.373 for inlet valves and 0.371-0.372 for exhaust valves.

For series 1750 diesel and 1950-T diesel, valves have a face angle of 29½ degrees for inlet and 45½ degrees for exhaust. Seat angles are 30

Fig. O85—Measure depth of valve heads in 1850 diesel cylinder head as shown. Valve heads should be 0.029-0.060 below cylinder head surface when correctly serviced. Note arrow indicating oil hole for rocker arm oil pipe.

degrees for inlet and 45 degrees for exhaust. Valve stem diameter is 0.3720-0.3725 for inlet valves and 0.3715-0.3720 for exhaust valves.

Valve seat width for all models is as follows:

Series 1800-1900
  All .................0.057-0.067
Series 1850-1950-1750 Non-Diesel
  All .................0.080-0.090
Series 1750 Diesel
  Inlet ................0.046-0.052
  Exhaust ..............0.062-0.072
Series 1950-T Diesel
  Inlet ................0.087-0.097
  Exhaust ..............0.062-0.072

Exhaust valve seat inserts are standard equipment for all engines. Diesel engines for series 1800 A, B and C also have intake valve seat inserts as standard equipment. Inserts are installed with 0.002-0.0035 interference fit.

Intake valves of early non-diesel engines are provided with neoprene oil guards to prevent oil from passing into the combustion chamber via the valve stems. Intake valve guides of late production non-diesel engines are fitted with Perfect Circle seals.

Install new oil guards or seals each time the valves are reseated.

Adjust valve tappet gaps cold and to the following values:

**Intake**
  All non-diesel
    except 1750 .........0.013-0.015
  Series A & B diesel ....0.009-0.011
  Series C diesel .........0.016-0.018
  Series 1750 non-diesel ..0.019-0.021
  Series 1750-1950-T
    diesel ..............0.029-0.031

**Exhaust**
  All series 1800
    non-diesel ...........0.022-0.024
  Series 1800 A & B
    diesel ...............0.015-0.017
  Series 1800 C diesel ....0.026-0.028
  Series 1850 non-diesel ..0.023-0.025
  Series 1750 non-diesel ..0.029-0.030
  Series 1750-1950-T diesel 0.029-0.031

For all series except 1750 and 1950-T diesel the cast iron valve guides may be either shouldered or straight. The intake and exhaust valve guides are not interchangeable. New shouldered type guides have an inside diameter of 0.3745-0.3755. New straight type guides have an inside diameter of 0.374-0.375. Maximum allowable inside valve guide diameter is 0.377 for intake or 0.378 for exhaust. Length of guides is 2½ inches for intake and 2⅝ inches for exhaust.

Operating clearance of valve in guide when using shouldered type guides is 0.0015-0.0035 for intake and 0.0025-0.0045 for exhaust. When using straight type valve guides, operating clearance of valve in guide is 0.001-0.003 for intake and 0.002-0.004 for exhaust. Maximum allowable clearance of valves in guides in either case is 0.005 for the intake and 0.007 for the exhaust. Renew valves and/or guides when these clearances are exceeded.

For series 1750 and 1950-T diesel, the cast iron valve guides are the straight type with a length of 3½ inches. Guides with a spiral groove end have an inside diameter of 0.373-

0.374 while those with plain end have an inside diameter of 0.3745-0.3757. Operating clearance of inlet valves in spiral groove end guides is 0.0005-0.002 and in plain end guides 0.002-0.0037. Operating clearance of exhaust valves in spiral groove end guides is 0.001-0.0025 and in plain end guides 0.0025-0.0042. Renew valves and/or guides when clearance exceeds 0.005 for the spiral groove end guides or 0.007 for the plain end guides.

Drive guides out toward top side of cylinder head. When properly installed, the valve guide height above machined top surface of cylinder head for all series except series 1750 and 1950-T diesel is as follows: $\frac{15}{16}$-inch for non-diesel inlet valves without the Perfect Circle seals; $\frac{27}{32}$-inch for non-diesel inlet valves with Perfect Circle seals, or $\frac{15}{16}$-inch for all non-diesel exhaust and all diesel valves.

For series 1750 and 1950-T diesel, guides are correctly installed when guide height is ⅞-inch above counterbore of cylinder head.

### Series 1850 Diesel

99. Both inlet and exhaust valves have a face and seat angle of 45 degrees. Desired seat width is $\frac{3}{32}$-inches for the inlet valves and 7/64-inches for the exhaust valves. Seats may be narrowed, using 20 and 60 degree stones. Maximum allowable seat runout is 0.002. When properly serviced and installed, the valve heads should be 0.029-0.060 below surface of cylinder head as shown in Fig. O85.

Valve seat inserts are available for installation in cases where wear or damage prevent bringing the valve head into the proper relationship with the cylinder head. Shrink inserts with dry ice prior to installation and install them using a piloted driver and a press. Counterbore machining dimensions are as follows:

**Inlet**
  Inside diameter-in. ...2.0165-2.0175
  Depth-in. .............0.283-0.288

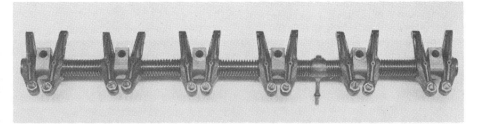

Fig. O85A — View of 1850 diesel rocker arms and shaft assembly removed. Note the offset (right and left hand) rocker arms and the seal on lower end of rocker arm oil pipe.

**Exhaust**

Inside diameter-in. . . . . .1.678-1.679
Depth-in. . . . . . . . . . . . . .0.375-0.380

Inlet valves only on engines prior to engine serial number 8512097 are fitted with neoprene oil deflectors, whereas all valves on engines after above serial number are fitted with oil deflectors and new deflectors should be installed each time inlet valves are removed. Deflectors are installed with open ends toward valve guides. Stem diameter for inlet valves is 0.3725-0.3735 and for exhaust valves is 0.372-0.373. Normal operating clearance of valves in guides is 0.0015-0.0035 for the inlet and 0.002-0.004 for the exhaust. Valve guide inside diameter is 0.375-0.376. Renew any valve which has less than $\frac{1}{32}$-inch head margin.

Adjust both inlet and exhaust valve tappet gap to 0.012 cold and reset to 0.010 after engine is brought to operating temperature, if necessary.

The straight type, cast iron valve guides are renewable and guides are not interchangeable. Guides can be identified by their length. Inlet valve guides are $2\frac{5}{16}$ inches long and the exhaust valve guides are $2\frac{7}{16}$ inches long.

To remove guides, press them out toward top side of cylinder head. When installing new guides, position with chamfered end up, then using a piloted driver, press guide into cylinder head until top end of guide is $\frac{5}{8}$-inch above machined top surface of cylinder head. Guides are presized and if carefully installed, will require no final sizing. Inside diameter of new valve guides is 0.375-0.376.

### VALVE ROTATORS
#### All Series 1750-1800-1850 Non-Diesel-1950-T

100. Exhaust valves of non-diesel engines and intake and exhaust valves of late series diesel engines are equipped with a positive type rotator. These rotators cannot be serviced but should be observed while engine is running to make sure valve rotates slightly each time valve opens. Renew rotator or any valve which fails to rotate.

### VALVE SPRINGS
#### All Series 1750-1800-1850 Non-Diesel-1950-T

101. Valve springs of four different free lengths have been used. a $2\frac{5}{16}$-inch spring, used on the intake valves of gasoline engines and intake and exhaust valves of diesel engines. A $2\frac{9}{16}$-inch spring, used for intake and exhaust valves for both diesel and LP-Gas engines and intake valves of gasoline engines. A $2\frac{11}{16}$-inch spring, used on intake and exhaust valves of diesel engines and intake valves of gasoline engines. A $2\frac{25}{32}$-inch spring, used on intake valves of gasoline engines and intake and exhaust valve of both diesel and LP-Gas engines.

Test specifications are as follows:
$2\frac{5}{16}$ in. free length. . .1.387 in. @ 82 lbs.
$2\frac{9}{16}$ in. free length. . .1.506 in. @ 95 lbs.
$2\frac{11}{16}$ in. free length. . .1.350 in. @ 90 lbs.
$2\frac{25}{32}$ in. free length. . .1.594 in. @ 71 lbs.

#### Series 1850 Diesel

102. Each valve is fitted with an outer and inner valve spring. Springs are made with dampener coils and during installation, dampener coils are positioned next to cylinder head. Renew any valve spring that has ends that are not square, is discolored, shows any sign of fracture or other damage, or does not meet the following specifications.

**Outer spring**
Test (lbs.) and
length (in.). . . . . . . . . . .38-43 @ 1.78
68.5-77.1 @ 1.358

**Inner spring**
Test (lbs.) and
length (in.). . . . . . .14.6-15.7 @ 1.563
31.8-36.2 @ 1.140

### VALVE LIFTERS (CAM FOLLOWERS)
#### All Series 1750-1800-1850 Non-Diesel-1950-T

103. The mushroom type valve lifters operate directly in machined bores in the cylinder block. Valve lifters are supplied only in standard size of 0.6240-0.6245 and should have an operating clearance of 0.0005-0.002 in their bores.

Renew valve lifters if their diameter is less than 0.619, or if they are more than 0.0003 out-of-round. If diametral clearance exceeds 0.007 with new valve lifter in bore, renew cylinder block.

Any valve lifter can be removed after camshaft has been removed as outlined in paragraph 113.

#### Series 1850 Diesel

104. The mushroom type valve lifters operate directly in machined bores in the cylinder block. Valve lifters are furnished only in standard size with a stem outside diameter of 0.7475-0.7485. Normal operating clearance of lifter in its bore is 0.0015-0.0038. Renew lifter if more than 0.001 out-of-round.

Any valve lifter can be removed after camshaft is removed as outlined in paragraph 114.

### VALVE ROCKER ARMS AND SHAFT
#### All Series 1750-1800-1850 Non-Diesel

105. The rocker arms and shaft assembly can be removed as follows: Remove both hood side panels and hood. Disconnect fuel lines, wiring and control cables from bottom of fuel tank. Remove batteries, hood support and battery tray from instrument panel support. Unbolt fuel tank cradle, swing fuel tank upward and retain in this position with a piece of wire or rope attached to steering column. On models so equipped, disconnect power steering pump reservoir and move it out of the way. Remove breather tube and on non-diesel models remove coil from rocker arm cover. On diesel models, disconnect preheater solenoid from rocker arm cover. Remove rocker arm cover and the rocker shaft oil line. Unbolt and remove the rocker arm and shaft assembly.

Disassemble the rocker arms and shaft by removing the hair pin keepers and sliding parts from shaft. Note that rocker arms (except 1750 diesel) are right and left hand assemblies and that offsets are toward the shaft supports.

Rocker arm shaft diameter is 0.742-0.743 with a reject diameter of 0.740. Inside diameter of rocker arm bushing is 0.7445-0.7455 and rocker arm should have an operating clearance of 0.0015-0.0035 on shaft. If operating clearance exceeds 0.005, renew rocker arm and/or shaft as rocker arm bushings are not available separately. Rocker arm contact button radius can be refaced providing the original radius is maintained and kept parallel with the rocker arm shaft. Shaft springs should exert 10 pounds pressure when compressed to $\frac{5}{8}$-inch length.

If the rear rocker arms are receiving an excess of oil, check location of restrictor plug located inside of rocker arm shaft. This restrictor should be $13\frac{3}{16}$-inches (non-diesel), or $13\frac{1}{2}$-inch (diesel) forward of shaft rear end. Plugs in ends of shaft must fit snug in shaft I.D.

#### Series 1950-T

105A. The rocker arms and shaft assembly are removed as follows: Remove both hood side panels and hood. Disconnect wires from restriction indicator sending unit and inlet hose from turbocharger. Unbolt preheater solenoid from filter bracket, then remove filters and support bracket. Disconnect preheater ground cable from tappet cover stud, then remove outlet pipe, hoses and preheater as a unit. Unbolt breather tube flange from tappet cover and remove flange and breather tube. Remove tappet cover and the rocker arm shaft oil line. Unbolt and remove rocker arms and shaft assembly. If push rods are removed, identify them so they can be reinstalled in their original position.

Disassemble rocker arms and shaft by removing the hair pin keepers and sliding parts off shaft. Rocker arm shaft diameter is 0.7425-0.7435 with a reject diameter of 0.740. Inside diameter of rocker arm bushing is 0.7445-0.7455 and rocker arm should have 0.001-0.003 operating clearance on shaft. If operating clearance exceeds 0.0045, renew rocker arm and/or shaft. Rocker arm bushings are not available separately. Rocker arm contact radius can be refaced providing original radius is maintained and kept parallel with rocker arm shaft. Shaft springs should exert 10 pounds when compressed to a length of $\frac{5}{8}$-inch.

### Series 1850 Diesel

106. The rocker arms and shaft assembly can be removed as follows: Remove hood side panels and hood. Disconnect fuel lines, wiring and control cables from bottom of fuel tank. Remove battery, hood supports and battery tray from instrument panel support. Unbolt fuel tank support, swing fuel tank upward, and retain in this position with a piece of wire or rope attached to steering column. Disconnect breather pipe from tappet cover, then unbolt and remove tappet cover. Gradually loosen the rocker arm shaft bracket retaining nuts to prevent distorting shaft and lift rocker arms and shaft assembly from engine. Also, use care not to damage the rocker arm oil pipe or oil pipe seal. See Fig. 085A.

NOTE: Should tappet cover leakage problems occur, use the pressed steel tappet cover along with gasket replacement package, Oliver part number 164 894-AS and tappet cover clamp package, Oliver part number 166 238-AS.

Disassemble the rocker arms and shaft by removing the two end snap rings and the oil pipe retaining cap screw and sliding parts from shaft. Note that rocker arms are right and left hand assemblies. Shaft end plugs can be removed if necessary.

Diameter of new rocker arm shaft is 0.7485-0.7495. Rocker arm bushing inside diameter should be 0.7505-0.7520 which will provide 0.001-0.0035 operating clearance. Rocker arm bushings are available separately; or rocker arm and bushing is available as an assembly. Bushings must be reamed to provide the proper clearance on shaft. Be sure oil holes in bushing and rocker arm align when installing bushings.

The four long rocker arm shaft springs have a free length of 3 inches and should test 8.55-10.05 pounds when compressed to a length of $2\frac{5}{32}$ inches. The two short rocker arm shaft springs have a free length of $1\frac{3}{32}$ inches and should test 6.73-8.23 pounds when compressed to a length of 53/64-inches.

### VALVE TIMING
### All Series 1750-1800-1850 Non-Diesel-1950-T

107. Valves are correctly timed when "C" mark on camshaft gear is meshed with an identical mark on the crankshaft gear as shown in Fig. 086.

### Series 1850 Diesel

108. Valves are correctly timed when, with number one piston at TDC, the single punch marked tooth of crankshaft gear meshes with the double punch mark of the right hand idler gear and the single punch marked tooth of right hand idler meshes with the double punch mark of the camshaft gear. See Fig. 087.

Valve timing can also be checked as follows: Remove tappet cover and turn engine until rocker arms of number six cylinder are "rocking," at which time, number one piston will be at approximately TDC on compression stroke. Adjust the clearance of number one cylinder inlet valve (front) to 0.034, then turn engine in the direction of normal rotation only far enough to take up the 0.034 clearance. NOTE: Stop rotation of engine the instant the rocker arm tightens against valve stem. Remove timing hole cover from clutch housing. If valve timing is correct, TDC mark on flywheel should be within $\frac{1}{4}$-inch of the scribe line on timing hole.

### TIMING GEAR COVER
### All Series 1750-1800-1850 Non-Diesel-1950-T

109. To remove timing gear cover, first remove side panels and hood, then drain cooling system. Remove grille and radiator. Remove fan belt, water pump and generator. On models so equipped, disconnect the belt driven power steering pump from timing gear cover. On non-diesel tractors, disconnect governor linkage at governor, then remove governor from timing gear cover and on models with two-piece sleeve and bushing use caution not to let thrust washer drop off hub of governor gear. Remove cap screw from crankshaft pulley, thread two cap screws into threaded holes of crankshaft pulley, then attach a puller to the cap screws and remove pulley. Remove the three cap screws which retain oil pan to timing gear cover. Loosen remaining oil pan cap screws, separate oil pan gasket from timing gear cover and allow oil pan to drop. Unbolt and remove timing gear cover.

Crankshaft front oil seal can now be removed and the procedure for doing so is obvious. Seal is installed with cork facing toward front on series 1800 A, B and C and series 1850 non-diesel prior to serial number 150 421-000. Lip type seal used on later tractors is installed with lip facing rearward.

When reinstalling timing gear cover, use sealant on bottom side of cover where it mates with oil pan and gasket.

### Series 1850 Diesel

110. To remove timing gear cover, first drain cooling system, shut off the fuel supply valve, then remove hood side plates and hood, grille and radiator, fan and fan belt and alternator.

Turn engine so that TDC mark on flywheel is in register with pointer in window and number one piston is on compression stroke.

Remove cover from front of camshaft gear and remove the tachometer drive adapter and gear cover plate from front of auxiliary drive shaft gear. Note: At this time, it is good

Fig. O86 — On all except 1850 diesel, valves are properly timed when marks align as shown. Non-diesel engine is shown, however, diesel engine is similar.

policy to check gear backlash to determine if timing gears should be renewed. Recommended backlash of gears is 0.003-0.006.

Remove the two retaining plates and damper assembly from crankshaft pulley. Be careful not to damage damper. Remove cap screw, retaining washer, clamping ring and crankshaft pulley. Attach hoist to engine, take weight of engine on hoist, then remove engine front support. Remove cap screws from front of oil pan, loosen remaining oil pan cap screws, separate oil pan gasket from timing gear cover and let oil pan drop.

Remove cap screws, shim, tab washer and retaining washer from front camshaft gear, attach puller to tapped holes in gear and remove gear from camshaft. Remove the three socket head cap screws, retaining plate and gear from auxiliary drive shaft. Remove remaining cap screws and remove timing gear cover from engine.

Crankshaft front oil seal in timing gear cover can now be renewed; install new seal with lip to inside of cover. To reinstall timing gear cover, proceed as follows:

110A. Remove the right hand idler gear, then reinstall the gear with timing marks on crankshaft gear, left hand idler gear and right hand idler wear aligned as shown in Fig. 087.

Tighten retaining nut to 45-50 ft.-lbs. torque. NOTE: Timing marks may not be present on left hand idler gear; however, they are not necessary and may be ignored if present.

Stick camshaft thrust washer in place with heavy grease and be sure auxiliary drive shaft and thrust washers are in place. Install timing gear cover on engine leaving the retaining cap screws loose. Place crankshaft pulley on crankshaft to align oil seal to pulley, then tighten timing gear cover retaining cap screws and remove pulley from crankshaft,

Align camshaft gear keyway with key on camshaft and pull gear on shaft with the retaining washer and cap screw, making sure camshaft gear and right hand idler gear timing marks are aligned as shown in Fig. 087. Remove the retaining cap screws, install tab washer and shim then reinstall cap screw. Tighten the cap screw to 45-50 ft.-lbs. torque and bend tab washer against cap screw head. Reinstall camshaft gear cover on timing gear cover.

Install engine front support (with pedestal offset rearward) and remove hoist from engine. Install crankshaft pulley on crankshaft, then install clamping ring with bevel toward rear. Install the pulley retaining washer and cap screw and tighten cap screw

to 270-300 ft.-lbs. torque. Install damper assembly on crankshaft pulley.

Install auxiliary drive shaft gear and time injection pump as outlined in paragraph 203. Install auxiliary drive gear cover and tachometer drive adapter on timing gear cover.

Complete reassembly of tractor and bleed fuel system as outlined in paragraph 180.

### TIMING GEARS
### All Series 1750-1800-1850 Non-Diesel-1950-T

111. Timing gear train on diesel engines consists of three helical gears and the injection pump drive gear. On non-diesel engines, two gears are used and the drive gear for the governor is driven by the camshaft gear. Gears are accessible for service after removing timing gear cover as outlined in paragraph 109.

Recommended gear backlash is 0.003-0.005. Renew all gears when backlash between any pair exceeds 0.007.

The cylinder block for all engines is designed for the installation of the idler gear, however, idler gear is used only with diesel engines. Plugs must be installed in either the idler gear shaft hole (non-diesel engines) or the governor gear hub hole (diesel engines) to block off the oil holes and maintain engine oil pressure. When servicing timing gears, note the lubrication tube which protudes from crankcase above camshaft gear. Hole in tube is toward camshaft gear.

Remove camshaft gear from camshaft by using a puller attached to two cap screws threaded into gear. Crankshaft gear is removed in a similar manner. Avoid pulling gears with pullers which clamp or pull on the gear teeth.

Before installing, heat gears in oil to facilitate installation. When installing camshaft gear, remove oil pan and buck-up camshaft at one of the lobes near front end of shaft with a heavy bar. Gears are installed with "C" mark on camshaft meshed with a similar mark on crankshaft gear as shown in Fig. 086.

Diesel engine idler gear can be removed after removing the injection pump gear. However, before removing either gear, align timing marks of camshaft and crankshaft gear, then mark injection pump drive gear and hub so gear can be reinstalled in same position.

Operating clearance of idler gear shaft (spindle ) in sleeve is 0.0015-

**Fig. O87 — View showing 1850 diesel timing gear arrangement. Gears will not normally be seen as shown as camshaft gear and auxiliary drive shaft gear must be removed before timing gear cover can be removed. Note the timing marks and their alignment.**

AS. Auxiliary drive shaft gear      CS. Camshaft gear      CR. Crankshaft gear

0.003 and if clearance exceeds 0.005, renew the sleeve and bushings assembly as a unit as parts are not available separately. Outside diameter of gear shaft (spindle) is 0.999-1.000. Bushings in new sleeve may require final sizing to provide the desired 0.0015-0.003 operating clearance. Thrust spring should test 6½-8½ pounds at $\frac{31}{32}$-inch.

### Series 1850 Diesel

112. Timing gear train consists of five helical gears; crankshaft gear, camshaft gear, auxiliary drive shaft gear and two idler gears. Recommended backlash of gears is 0.003-0.006. Lubrication for gears is supplied via the idler gear hubs.

Thrust washers for the camshaft and auxiliary drive shaft are held in counterbores in the cylinder block by the timing gear cover. Camshaft gear and auxiliary drive shaft gear must be removed before the timing gear cover can be removed. Refer to paragraph 110 for information regarding removal of timing gear cover. Crankshaft gear and the two idler gears can be serviced after removing the timing gear cover.

Idler gears and hubs can be removed after removing the self-locking nuts and crankshaft gear can be removed by using a suitable puller. Idler gear hub diameter is 1.3740-1.3745 and inside diameter of idler gear bushings is 1.3755-1.3771 which provides an operating clearance of 0.001-0.003. Bushings are available separately or gears and bushings are available as assemblies.

Idler gear hubs are installed with recessed ends toward cylinder block and dowels located in oil holes. See See Fig. O88. Be sure oil holes in idler gears and hubs, shown in Fig. O88A, are open and clean.

Install crankshaft gear with timing marks on gear teeth forward, then refer to paragraph 110A for installation of remaining timing gears and timing gear cover.

## CAMSHAFT

### All Series 1750-1800-1850 Non-Diesel-1950-T

113. To remove camshaft, first remove timing gear cover as outlined in paragraph 109. On non-diesel engines, remove ignition distributor and tachometer drive assembly. On diesel models, remove tachometer drive assembly. On all models so equipped, remove fuel pump. On tractors with separate power steering pump, disconnect and move power steering pump reservoir out of the way. Then, remove rocker arms cover, rocker arms and shaft assembly and push rods. Remove oil pan and oil pump, then block up or support valve lifters and pull camshaft and gear from cylinder block. Use caution when withdrawing camshaft not to damage lobes of shaft and the bushings in crankcase. Cam gear can be either pressed or pulled from camshaft.

Camshaft journal sizes for all models except 1750 and 1950-T are: Front, 1.749-1.750; second, third and fourth, 1.7485-1.7495. Recommended operating clearance of camshaft journals in

bushings is 0.0015-0.0035 with a maximum operating clearance of 0.005.

Camshaft journal size for models 1750 and 1950-T is 1.7485-1.7495. Recommended operating clearance of camshaft journals in bushings is 0.002-0.0035 with a maximum operating clearance of 0.005.

Camshaft bushings are pre-sized and when installing, use a closely piloted driver and be sure oil holes in bushings align with those in cylinder block. Camshaft end play is controlled by a spring loaded thrust button. Thrust button spring has a free length of $1\frac{3}{16}$-inches and should test 15.5-18.5 lbs. when compressed to a length of $\frac{25}{32}$-inches.

If gear was removed from camshaft, be sure "C" mark on gear is toward front when gear is installed.

### Series 1850 Diesel

114. To remove camshaft, first remove timing gear cover as outlined in paragraph 110. Remove tappet cover and both engine side covers. Loosen rocker arm adjusting screws, disengage push rods from rocker arms, then lift tappets and retain in raised position with spring type clothes pins, or other suitable clamps. Remove fuel pump. Remove thrust washer and carefully pull camshaft from cylinder block.

To measure camshaft bearing bores, it will be necessary to remove oil pan to gain access.

Camshaft journal diameters are:

| | |
|---|---|
| No. 1 (front) | 1.9965-1.9975 |
| No. 2 | 1.9865-1.9875 |
| No. 3 | 1.9765-1.9775 |
| No. 4 (rear) | 1.9665-1.9675 |

Camshaft journal bore diameters are:

| | |
|---|---|
| No. 1 (front) | 2.0000-2.0010 |
| No. 2 | 1.9900-1.9920 |
| No. 3 | 1.9800-1.9820 |
| No. 4 | 1.9700-1.9720 |

Normal operating clearance of camshaft in bearing bores is 0.0025-0.0045 for No. 1 journal and 0.0025-0.0055 for Nos. 2, 3 and 4 journals.

Camshaft end play is controlled by a doweled thrust washer, held in a counterbore of the cylinder block by the timing gear cover. Normal end play of camshaft is 0.004-0.016. New thrust washer is 0.216-0.218 thick. New thrust washers will be between flush and 0.004 above machined surface of cylinder block when installed.

Reinstall camshaft by reversing removal procedure and set tappet

**Fig. O88 — On 1850 diesel, idler gear hubs are installed with recessed ends toward cylinder blocks and dowels located in oil holes.**

**Fig. O88A — Oil holes are provided in 1850 diesel idler hubs and gears to lubricate gear train.**

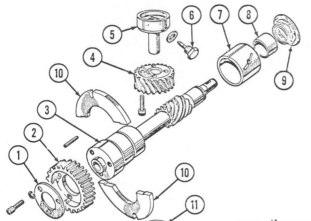

**Fig. O88B — Exploded view of 1850 diesel auxiliary drive shaft, vertical drive shaft and related parts. Although not shown, shims may be located below bearing (11). Note oil jet tube (6).**

1. Retainer plate
2. Drive gear
3. Auxiliary drive shaft
4. Worm gear
5. Vertical drive shaft
6. Oil jet tube
7. Front bushing
8. Rear bushing
9. Oil seal
10. Thrust washers
11. Bearing

gap to 0.012 cold. Reset valve tappet gap to 0.010 if necessary, after engine has been brought to operating temperature.

## IDLER GEARS
### Series 1850 Diesel

115. To remove the idler gears, first remove timing gear cover as outlined in paragraph 110. With timing gear cover off, remove the two self-locking nuts and remove retaining washers and idler gears. Idler gear hubs can now be removed from studs. Idler gear bushings are available separately, however, bushings are included with new idler gears.

NOTE: The left hand (upper) idler gear may or may not have timing marks. As long as the crankshaft gear, right hand (lower) idler gear and the camshaft gear are in their correct relationship to each other, as shown in Fig. 087, this is of no consequence as the auxiliary drive shaft is timed separately.

Inside diameter of idler gear bushings is 1.3755-1.3770. Outside diameter of idler gear hubs is 1.3740-1.3745. Normal operating clearance of gears on hubs is 0.001-0.003. Backlash between idler gears and their mating gears is 0.003-0.006.

When installing idler gears, position idler gear hubs with recessed ends next to cylinder block and align dowels with oil holes. See Fig. 088. Install gears with timing marks aligned as shown in Fig. 087 and tighten the self-locking nuts to 45-50 ft. lbs. torque.

## AUXILIARY DRIVE SHAFT
### Series 1850 Diesel

116. To remove the auxiliary drive shaft, first remove the timing gear cover as outlined in paragraph 110 and if equipped with power steering, but without hydraulic lift system, re-

move the power steering pump as outlined in paragraph 78 and remove the pump drive coupling from rear end of auxiliary drive shaft.

Then, carefully withdraw the auxiliary drive shaft from front of cylinder block, removing the split thrust washer as shaft is withdrawn. If inspection of the worm gear indicates wear, remove and check the vertical (fuel injection pump) drive shaft as outlined in paragraph 210.

With auxiliary drive shaft removed, the rear oil seal and the shaft bushings can be renewed. Oil seal is installed with lip toward front of engine. Inside diameter of front bushing is 1.9375-1.9397 and inside diameter of rear bushing is 1.250-1.2516. Shaft front journal outside diameter is 1.9355-1.9365 and should have an operating clearance of 0.001-0.0042 in front bushing. Shaft rear bearing journal outside diameter is 1.248-1.249 and should have an operating clearance of 0.001-0.0036 in rear bushing.

New thrust washers have a thickness of 0.1875-0.1905 and when installed should be between 0.0035 below and 0.0025 above the machined surface of cylinder block. Auxiliary drive shaft end play should be 0.0025-0.009.

To reassemble, insert auxiliary drive shaft carefully to avoid damage to rear oil seal and insert the thrust washer halves in groove of shaft before shaft is fully inserted. Position end of thrust washers over dowel pin in cylinder block as shaft is pushed into place. Reinstall power steering pump, if so equipped and reinstall timing gear cover as outlined in paragraph 110A.

## ROD AND PISTON UNITS
### All Series 1750-1800-1850 Non-Diesel-1950-T

117. Connecting rod and piston assemblies are removed from above after removing cylinder head and oil pan.

Cylinder numbers are stamped on front side of pistons on new engines; however, connecting rods are not marked and should be marked with the cylinder number on camshaft side prior to removal. Mark both connecting rod and rod cap.

When reinstalling units, tighten connecting rod bolt nuts to a torque of 56-58 ft.-lbs. for all diesel and non-diesel engines except series 1750 and 1950-T which should be 46-50 ft.-lbs.

## ROD AND PISTON UNITS
### Series 1850 Diesel

118. Connecting rod and piston assemblies are removed from above after removing cylinder head and oil pan.

Pistons are stamped with word "Front" on top front side of piston and should be stamped with cylinder number if they are to be reused. Connecting rods and caps are stamped with cylinder number and numbers are located on side opposite camshaft (left side). In addition to cylinder identification numbers, the connecting rods are marked with a code (weight) number which may be 11, 12 or 13. When renewing a connecting rod, be certain new rod has the same code number.

Parting line of connecting rod and cap are serrated to insure correct mating and these serrations must not be tampered with.

When installing units, tighten the self-locking rod nuts to a torque of 65-70 ft.-lbs.

## PISTONS, SLEEVES AND RINGS
### All Series 1750-1800-1850 Non-Diesel-1950-T

119. All engines are fitted with wet type sleeves which are sealed at lower end by two packing rings and at upper end by the cylinder head gasket. With cylinder head and oil pan off and the piston and connecting rod assemblies removed, the sleeves can be pulled from cylinder block using an OTC 938 puller with the proper sized sleeve attachment, or similar equipment. Sleeves and pistons are not available separately.

Specifications for sleeves are as follows:

Non-Diesel Models

Inside Diameter, Standard:
    Series 1800............3.750-3.751
    Series 1800 B & C ...3.875-3.8765
    Series 1750, 1850 .....3.875-3.8765
Inside Diameter, Max. Allowable*
    Series 1800 ...............3.761
    Series 1800 B & C..........3.886
    Series 1750, 1850 ..........3.886

Out-of-Round, Max...........0.002
Taper, Max....................0.004
Stand-Out,
    Recommended ........0.001-0.004
*At top of ring travel.
Diesel Models
    Inside Diameter,
        Standard ............3.875-3.8765
    Inside Diameter,
        Max. Allowable ............3.886
    Out-of-Round, Max. .........0.002
    Taper, Max. ...................0.004
    Stand-Out,
        Recommended ........0.001-0.004

If the side clearance of a new compression ring in piston top ring groove exceeds 0.006, or if piston and/or sleeve is scored, cracked or otherwise damaged, renew the piston and sleeve assembly.

120. Prior to sleeve installation, check the sleeve stand-out as follows: Carefully clean sleeve and sleeve bore and pay particular attention to the taper of the sleeve lower bore in crankcase as well as the sleeve flange counterbore. Insert sleeve WITHOUT seal rings in cylinder block and clamp sleeve in position using short screws and flat washers. Mount a dial indicator, place button of indicator on the outer land of sleeve flange and zero the indicator. Move indicator button from outer land of sleeve to surface of cylinder block. Dial indicator should show a sleeve stand-out of 0.001-0.004. If sleeve stand-out is less than 0.001, add a shim under sleeve flange. If sleeve stand-out is more than 0.004, check sleeve flange and its counterbore to be sure no foreign material is present and if both are perfectly clean and stand-out still exceeds 0.004, change sleeve.

NOTE: Be sure to use outer land of cylinder sleeve flange when measuring for sleeve stand-out. Do not use inner (firewall) land of sleeve flange.

121. With sleeve stand-out determined, install sleeves as follows: Coat sleeve sealing rings lightly with Lubriplate, or equivalent, and place on lower end of sleeve. Slide a small punch under seal rings and move punch around sleeve several times to eliminate all twists from sealing rings. Insert sleeves and seal rings into cylinder block and use hand pressure to position sleeves. Sleeves should go into place with only moderate hand pressure. If they do not, remove sleeve and determine cause.

122. Desired piston clearance in sleeve for series 1800 non-diesel requires 3-6 lbs. pull on a spring scale to withdraw a ½ x 0.0025 feeler gage

positioned 90 degrees from piston pin. Series 1800 diesel requires 3-5 lbs. pull on the ½ x 0.0025 feeler gage.

All series 1850 and 1750 non-diesel requires 3-5 lbs. pull on a spring scale to withdraw a ½ x 0.0035 feeler gage positioned 90 degrees from piston pin. Series 1750 and 1950-T diesel requires 3-6 lbs. on a ½ x 0.002 feeler gage.

123. RINGS. Pistons of all series 1800 engines are fitted with three compression rings and one oil control ring. Pistons of all series 1750, 1850 and 1950-T are fitted with two compression rings and one oil control ring.

Several types of rings have been used and the following information will be helpful. Rings with notch on outside diameter are installed with notch toward bottom of piston. Rings with notch on inside diameter are installed with notch toward top of piston. Rings with a tapered face are installed with taper (largest outside diameter) toward bottom of piston. Oil control rings using the coiled spring expander are installed as follows: Connect ends of expander, position the teflon sleeve over expander 180 degrees from connection, then place expander in groove and install oil control ring so expander groove is on top side and ring gap is positioned over teflon seal.

Instruction sheets are included with service ring sets.

124. Recommended ring end gap for non-diesel engines is 0.010-0.020, with a maximum allowable of 0.045 for all rings except the three piece service oil control ring which should have 0.015 end gap for the rails only with a maximum of 0.055 (rails only). For all diesel engines except 1950-T, ring end gaps are the same as for non-diesel except for the oil control ring which should have an end gap of 0.010-0.018 for production rings; 0.010-0.023 for service rings and with 0.045 maximum allowable for either ring.

For series 1950-T, recommended ring end gap is 0.014-0.020 for compression rings and 0.010-0.023 for oil rings. Maximum allowable end gap is 0.045 for all rings.

125. For all models except 1950-T, side clearance of top compression ring is 0.0025-0.004 and second and third compression ring side clearance is 0.002-0.0035 for both diesel and non-diesel engines. Oil ring side clearance is 0.001-0.003 for non-diesel production rings and 0.0015-0.003 for all diesel oil rings. Maximum side clearance for all rings including series

1950-T is 0.006. Side clearance values for the three-piece non-diesel service oil rings are not applicable.

NOTE: Top compression ring for series 1950-T is a keystone type ring.

### Series 1850 Diesel

126. Engines are fitted with cast iron, dry type sleeves which have a flange at top end along with a firewall projection. With cylinder head and oil pan off and the piston and connecting rod assemblies removed, the sleeves can be removed using a suitable puller. Sleeves and pistons are available in standard size only and are not available separately.

Specifications for sleeves are as follows:
Inside diameter .........3.878-3.879
Out-of-round, max. ............0.002
Taper, max. ...................0.005
Stand-out (firewall) ......0.030-0.035

The fit of a new piston in a new sleeve is correct when a 5-7 lb. pull on a spring scale will withdraw a ½-inch wide, 0.006 thick feeler from between piston and sleeve when feeler is positioned 90 degrees from piston pin. Piston pin bore diameter of piston is 1.37485-1.37505 and piston pin should have a palm push fit in piston when piston is heated to 100-120 degrees F.

127. RINGS. Pistons are fitted with three compression rings and two oil control rings with bottom oil ring located below the piston pin. All rings are straight faced. Compression rings with step in inside diameter are installed with step toward top of piston. All other rings may be installed with either side up.

Recommended ring end gap is 0.015-0.019 for the top compression ring and 0.011-0.016 for all other rings. Ring side clearance is 0.0019-0.0039 for the compression rings and 0.0025-0.0045 for the oil control rings.

### PISTON PINS AND BUSHINGS
### All Series 1750-1800-1850 Non-Diesel-1950-T

128. The 1.2494-1.2497 diameter full floating type piston pin is retained in the piston bosses by snap rings and in addition to standard size, is available for all except series 1750 and 1950-T in oversizes of 0.005 and 0.010. Reject diameter for piston pin is 1.2486.

Two bushings are fitted in piston pin end of connecting rod and are installed with outer edges flush with outer edges of rod bore and cut-away portions of bushings are aligned with oil hole in rod. Bushings are honed after installation to provide the piston pin with a thumb press fit at room temperature with pin and bushings dry (0.0004-0.0009). If oversize piston pin is being used, it will be necessary to hone piston bosses to provide 0.0002-0.0004 clearance for pin. Maximum allowable clearance of piston pin is 0.0019 in rod and 0.0008 in piston.

### Series 1850 Diesel

129. The 1.3748-1.3750 diameter piston pin is retained in the piston by snap rings and is available in standard size only.

Piston pin end of connecting rod is fitted with a bushing. Install bushing so oil hole in bushing is aligned with oil hole in connecting rod and ream bushing after installation to an inside diameter of 1.37575-1.3765 which will provide an operating clearance of 0.00075-0.0017 for the piston pin.

Piston pin bore in piston is 1.37485-1.37505 and piston pin should have a palm push fit in piston when piston is heated to 100-120 degrees F.

## CONNECTING RODS AND BEARINGS

### All Series 1750-1800-1850 Non-Diesel-1950-T

130. Connecting rod bearings are of the non-adjustable, insert type and can be moved after removing oil pan and connecting rod caps. When installing new bearings, make certain the bearing insert projections fit into the slot in connecting rod and cap and that cylinder numbers are in register and face toward camshaft side of engine.

NOTE: Affix cylinder numbers to connecting rod and cap prior to removal, if necessary.

Bearing inserts are available in standard sizes as well as undersizes of 0.003 and 0.020.

Check the crankshaft crankpin and the bearing inserts against the values which follow:

Crankpin diameter ......2.4365-2.4375
Rod bearing running
   clearance ...........0.0005-0.0015
Maximum running
   clearance ................0.0025
Connecting rod
   side play ...........0.0075-0.0135
Connecting rod bolt
   torque—ft.-lbs. (except 1750
   & 1950-T diesel) ...........56-58
Connecting rod bolt
   torque—ft.-lbs. (1750 &
   1950-T diesel) ..............46-50

### Series 1850 Diesel

131. Connecting rod bearings are of the non-adjustable, insert type and can be removed after removing oil pan and connecting rod caps. When installing new bearings, make certain bearing insert projections fit into slots in connecting rod and rod cap and that cylinder numbers are in register and are on side opposite camshaft. Do not attempt to alter the serrated mating surfaces of the connecting rod and cap. Bearings are available in standard size and 0.020 undersize.

NOTE: In addition to the cylinder number stamped on connecting rod assembly, a code number is also etched on the rod. This code number will be 11, 12 or 13 and indicates the weight of connecting rod, rod cap, bolts and nuts. Therefore, the code number etched on connecting rod should be indicated when ordering replacement connecting rods as rod weight of connecting rods should not vary more than two ounces. Code numbers and their corresponding weights are as follows:

Code No. 11..4 lbs. 2 oz. to 4 lbs. 4 oz.
Code No. 12..4 lbs. 4 oz. to 4 lbs. 6 oz.
Code No. 13..4 lbs. 6 oz. to 4 lbs. 8 oz.

Check the crankshaft crankpin and the bearing inserts against the values which follow:
Crankpin diameter ......2.4990-2.4995
Rod bearing running
   clearance ...........0.0015-0.0030
Connecting rod
   side play ...........0.0095-0.0130
Connecting rod nut
   torque—ft.-lbs. ...........65-70*
*These nuts to be used only once.

## CRANKSHAFT AND BEARINGS

### All Series 1750-1800-1850 Non-Diesel-1950-T

132. Crankshaft is supported in seven non-adjustable insert type bearings which can be renewed from below without removing crankshaft. The normal crankshaft end play of 0.0045-0.0095 is controlled by the flanged number five main bearing. Maximum allowable crankshaft end play is 0.011 and is corrected by renewing the number five main bearing. Bearings are available in standard size as well as undersizes of 0.002 and 0.020 for all series A 1800 and in standard as well as 0.003 and 0.020 undersizes for series 1750, 1800 B and C, 1850 non-diesel and 1950-T.

NOTE: If excessive crankshaft end play is present after installation of a new standard or 0.003 undersize thrust bearing, a 0.020 undersize, 0.017 overwidth bearing is available for tractors, serial number 124 396-000 and up, to correct this condition. To utilize this bearing grind crankshaft thrust bearing journal to 0.020 undersize and 0.017 overwidth.

To remove crankshaft, first remove engine as outlined in paragraph 95, then on all models remove clutch housing (with auxiliary drive unit if so equipped), clutch, flywheel, oil pan, rear oil seal and retainer and timing gear cover. On diesel models, remove injection pump drive gear and shaft and the idler gear assembly, then unbolt injection pump from engine front plate and let the attached fuel lines support injection pump. Unbolt and remove engine front plate which in some cases, may involve removing the dowels. Remove connecting rod caps and main bearing caps and lift crankshaft from engine.

Check the crankshaft and main bearings against the values which follow:

Crankpin diameter,
   standard ............2.4365-2.4375
Main journal diameter,
   standard .............2.624-2.625
Main bearing running
   clearance ...........0.0015-0.0045
Main bearing running
   clearance, Max. .............0.0065
Crankshaft end play,
   desired .............0.0045-0.0095
Journal out-of-round or
   taper, Max. ...............0.0003
Flywheel flange run-out........0.001
Main bearing bolt
   torque-ft.-lbs. ...........129-133

When reassembling engine, refer to paragraph 111 for information on idler gear and to paragraph 199 for information on injection pump drive shaft and gear.

### Series 1850 Diesel

133. Crankshaft is supported in seven non-adjustable, insert type bearings and the crankshaft end play is controlled by thrust washsers located at sides of the center main bearing as shown in Fig. 088C. Bearings are available in standard size as well as 0.020 undersize.

All of the main bearings, except the rear, can be renewed after oil pan and oil pump and screen assembly are removed. Removal of the rear main bearing will require removal of engine as rear bearing is located above an adapter (filler block) to which the crankshaft rear oil seal retainer is attached. Refer to paragraph 95 for information concerning engine removal.

Check the crankshaft end play, and if necessary, renew the thrust washers as follows: Use a small bar, push crankshaft forward and while holding in this position, measure between machined surface of crankshaft web and thrust washer. Normal crankshaft end play is 0.002-0.014. If shaft end play exceeds 0.014, remove the center main bearing cap and thrust washer halves, then roll the upper thrust washer halves out of their recesses in cylinder block. Note that lower thrust washer halves have lugs which fit into machined slots of the main bearing cap. When reinstalling thrust washers, position them so grooves are facing outward. A coating of grease will hold the lower thrust washers in bearing cap to ease installation. Thrust washers are available in standard size and 0.007 oversize. Tighten center main bearing cap retaining screws to a torque of 140-150 ft.-lbs.

To remove crankshaft, first remove engine as outlined in paragraph 95, then remove clutch housing (with auxiliary drive unit if so equipped), clutch, flywheel, rear oil seal and retainer, oil pan, oil pump assembly and screen and the timing gear cover. Remove connecting rod caps, adapter (rear filler block) and main bearing caps and lift crankshaft from engine.

**Fig. O88C — View showing location of 1850 diesel crankshaft thrust washers. Note that grooves in thrust washers are toward outside.**

Check crankshaft and main bearings against the values which follow:

Crankpin diameter, std...2.4990-2.4995

Main journal diameter,
standard ............2.9985-2.9990

Main bearing running
clearance ...........0.0025-0.0045

Crankshaft end play.......0.002-0.014

Main journal
out-of-round, max. .........0.0015

Main journal taper, max........0.001

Flywheel flange run-out,
max. .............0.001 at 4⅞ dia.

Main bearing bolt
torque—ft.-lbs. ...........140-150

When reassembling, be sure all the bearing cap ring dowels are in place and using all new locks, install bearing caps in their proper location with serial numbers in line and toward serial number on bottom face of cylinder block. Tighten main bearing cap screws to a torque of 140-150 ft.-lbs. Be sure adapter (filler block) is aligned with oil pan surface and rear oil seal retainer surface before tightening set screws. Use paper gaskets (shims) under adapter to align it with oil pan surface, if necessary.

## CRANKSHAFT OIL SEALS

### All Series 1750-1800-1850 Non-Diesel-1950-T

134. **FRONT SEAL.** To renew the the crankshaft front oil seal, remove the timing gear cover as outlined in paragraph 109.

NOTE: Series 1800 A, B and C, and earlier models of the 1850 series were fitted with a cork, face type front oil seal. This seal has been replaced with a lip type seal and a kit is available to convert the early face type seal to the late type lip seal.

**Fig. O89 — View showing rear oil seal typical of those used on early series 1800 engines. Later engines, except 1850 diesel, use a lip type seal which is installed in retainer.**

Kit contains new lip type seal, timing gear cover, crankshaft pulley and the necessary gaskets. To service series 1800 A gasoline and LP-Gas models, order kit number 165 486-AS; to service series 1800 B and C diesel, order kit number 165 978-AS; to service 1850 gasoline and LP-Gas models, order kit number 165 486-AS. All series 1750 and 1950-T are equipped with the lip type seal.

Cork type seals are installed with cork toward front. Lip type seals are installed with larger lip toward rear. Lubricate seal before installing the timing gear cover.

135. **REAR SEAL.** Tractors serial number 90 525-000 through 124 395-000 used a face type crankshaft rear oil seal. Tractors, serial number 124 396-000 and up, are fitted with lip type crankshaft oil seals. A rear oil seal conversion kit containing seal retainer dowels, gasket, wear sleeve and seal is available to correct leakage problems should they occur with the face type seal. A special tool (Oliver No. ST-150) must be used to align seal retainer during dowel installation when conversion kit is used.

To renew the rear oil seal, first remove the pto drive shaft as outlined in paragraph 356, or the hydraulic pump drive shaft as outlined in paragraph 409 if tractor has no pto.

On tractors with no auxiliary drive unit, remove clutch as outlined in paragraph 262. On tractors which have an auxiliary drive unit, remove engine and auxiliary drive unit as outlined in paragraph 95, then remove auxiliary drive unit and clutch housing from engine. Remove flywheel. Loosen the oil pan cap screws, separate pan gasket from retainer if necessary, let oil pan drop, then remove the oil pan to seal retainer cap screws. Remove

**Fig. O90 — View showing method of seating 1850 diesel rear oil seal in retainer.**

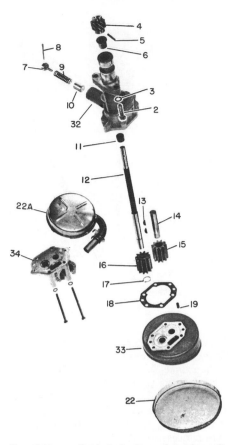

**Fig. O91 — Exploded view of engine oil pump used on all except 1850 diesel. Note items 22A and 34 which are used on late type pumps.**

| | |
|---|---|
| 4. Drive gear | 14. Idler shaft |
| 5. Spring pin | 15. Idler gear |
| 6. Bushing | 16. Driven gear |
| 7. Adjusting screw | 17. Snap ring |
| 8. Cotter pin | 18. Gasket |
| 9. Relief valve spring | 19. Dowel |
| 10. Relief valve plunger | 22. Screen (early) |
| 11. Bushing | 22A. Screen (late) |
| 12. Drive shaft | 33. Cover (early) |
| 13. Woodruff key | 34. Cover (late) |

seal retainer. Seal can now be removed from crankshaft (early models) or retainer (late models). See Fig. 089.

Lubricate new seal when installing. Face type seals are installed with cork facing rearward. Lip type seals are installed with lip facing forward.

**Series 1850 Diesel**

136. **FRONT SEAL.** To renew the crankshaft front seal, remove timing gear cover as outlined in paragraph 110, then press seal from timing gear cover. Install new seal with lip facing toward engine (rearward) and reinstall timing gear cover as outlined in paragraph 110A.

137. **REAR SEAL.** The crankshaft rear oil seal is a two-piece, asbestos rope type seal that is retained by a two-piece seal retainer bolted to rear surface of cylinder block.

To renew the rear oil seal, first remove the pto drive shaft as outlined

in paragraph 356, or the hydraulic pump drive shaft as outlined in paragraph 409, if tractor has no pto.

On tractors with no auxiliary drive unit, remove clutch as outlined in paragraph 262. On tractors which have an auxiliary drive unit, remove engine and auxiliary drive unit as outlined in paragraph 95, then remove auxiliary drive unit and clutch housing from engine. Remove flywheel. Remove the two vertical cap screws which hold the two-piece retainer together, then remove the retainer from rear face of cylinder block. Pry seal out of retainer.

New seal is installed as follows: Soak new seal pieces in clean engine oil for at least one hour prior to installing. Clean seal retainer and if necessary, true up any faces that may be nicked or distorted. Place one half of the seal retainer in a vise, start ends of seal piece into groove of retainer so that ends of seal extend 0.010-0.020 above parting line of retainer and work seal into groove of retainer.

NOTE: The 0.010-0.020 protrusion of the seal ends from the retainer half must be maintained or leaks will occur.

After seal has been seated as much as possible by hand, use a small round metal bar with a rolling and pressing motion and fully seat seal in its groove. Repeat this operation with the other half of the seal assembly and again maintain the 0.010-0.020 protrusion of seal ends from retainer half. See Fig. 090.

Use a sealing compound and install retainer gasket to face of cylinder block. Let sealant dry to secure gasket. Coat surfaces of seal with graphite grease and parting lines of retainer with sealant, then place retainer halves around crankshaft and install the vertical cap screws. Tighten these screws securely, then install retainer to cylinder block.

Complete assembly by reversing the disassembly procedure.

**FLYWHEEL**

**All Series 1750-1800-1850 Non-Diesel-1950-T**

138. Flywheel can be removed after clutch or converter and drive plate is removed and the procedure for doing so is obvious. Flywheel is doweled to crankshaft and can be installed in one position only. Tighten retaining cap screws to a torque of 66-69 ft.-lbs.

To install a new ring gear, heat gear to 450 degrees F. and install on flywheel with beveled end of teeth toward front. With flywheel installed, check flywheel run-out which should not exceed 0.005. Inside diameter of new flywheel bushing (pilot) is 0.8765-0.8795.

**Series 1850 Diesel**

139. Flywheel can be removed after clutch or converter drive plate is removed and the procedure for doing so is obvious. When installing flywheel, mate the unevenly spaced hole in flywheel with the unevenly spaced hole in crankshaft flange. Tighten cap screws to a torque of 74-80 ft.-lbs.

With flywheel installed, check face run-out which should not exceed 0.001 for each inch of radius, measured from center of flywheel.

To install a new ring gear, heat gear to approximately 475 degrees F. and install on flywheel with beveled end of teeth toward front of flywheel.

Check condition of flywheel bushing and renew if necessary. Inside diameter of a new bushing is 0.8765-0.8795.

**OIL PUMP**

**All Series 1750-1800-1850 Non-Diesel-1950-T**

140. The gear type pump is driven by the camshaft and is mounted on the underside of the crankcase. Pump removal requires removal of oil pan and the two cap screws which attach pump to cylinder block. On non-diesel, it will be necessary to remove the ignition unit whenever the oil pump is being installed. Retime ignition as outlined in paragraph 230 or 231.

Disassembly of the oil pump is obvious after an examination of the unit and reference to Fig. 091. Refer to the following information when overhauling:

Relief valve setting
(except 1950-T) ............42 psi
Relief valve setting
(1950-T) ................70 psi
Relief valve spring
free length (except
1950-T) ...............$2\frac{11}{16}$ inches
Relief valve spring free
length (1950-T) .......$1\frac{15}{16}$ inches
Idler gear bore..........0.6260-0.6265
Idler shaft diameter.....0.6220-0.6225
Idler gear operating
clearance .............0.0035-0.0045
Drive shaft lower bushing
bore, production ......0.6260-0.6265

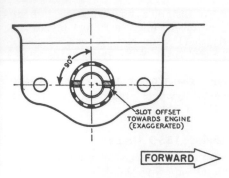

**Fig. O92 — Correct position of ignition unit drive slot when viewed from ignition unit mounting pad.**

Drive shaft lower bushing
bore, repairs . . . . . . . . .0.6255-0.6280

Drive shaft upper bushing
bore, production . . . . . .0.4950-0.4955

Drive shaft upper bushing
bore, repairs . . . . . . . . .0.4965-0.4970

Drive shaft lower
journal diameter . . . . . .0.6240-0.6245

Drive shaft upper
journal diameter . . . . . .0.4945-0.495

Drive shaft operating
clearance, lower
bushing . . . . . . . . . . . . .0.0015-0.004

Drive shaft operating
clearance, upper
bushing . . . . . . . . . . . . .0.0005-0.003

Drive shaft end play . . . . . . .0.003-0.008

Relief valve setting is controlled by adjusting screw (7).

141. Before installing oil pump to non-diesel engines, rotate engine crankshaft until number one piston is on compression stroke and flywheel mark "TDC" is indexed at inspection port. Install the oil pump so that the narrow side of the pump drive gear, as divided by the ignition unit drive slot, is on the crankshaft side and

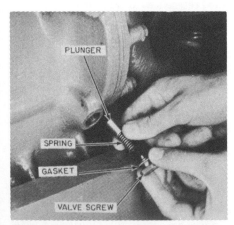

**Fig. O92A — Main oil gallery relief valve assembly removed. (Used on all 1800 and series 1850 non-diesel.)**

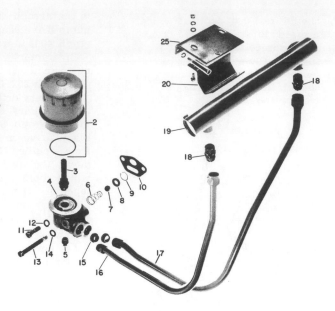

**Fig. O93 — View showing series 1850 diesel engine oil cooler and filter arrangement.**

2. Filter & gasket
3. Pipe
4. Base
6. Spring
7. Ball
8. Washer (seat)
9. Snap ring
10. Gasket
15. Seat
16. Tube, right
17. Tube, left
18. Connector
19. Oil cooler
20. Clamp
25. Bracket

parallel to the crankshaft. Refer to Fig. O92 which shows the correct position of the ignition unit drive slot when view from above through ignition unit shaft hole in cylinder block. If drive slot is not in the position as shown, remove the oil pump and re-mesh the pump drive gear.

**Series 1850 Diesel**

142. The rotor type oil pump is mounted on the underside of the crankcase and is retained in position by a set screw and lock nut located on the outside of cylinder block. The pump is driven by the vertical oil pump and injection pump drive shaft which receives its drive from the auxiliary drive shaft. After removal of oil pan, the removal of oil pump, relief valve housing and associated tubing will be obvious.

Oil pump can be disassembled, cleaned and inspected; however, if any of the pump parts are found to be defective, the complete pump (except relief valve assembly) must be renewed as parts are not catalogued separately. When reinstalling pump, align the master spline of the pump drive shaft with the master spline of the vertical drive shaft.

Refer to the following for specifications concerning the oil pump and relief valve.

Inner to outer
rotor clearance . . . . . . . .0.004-0.008
Outer rotor to body
clearance . . . . . . . . . . . .0.009-0.011
Rotors to body face
clearance . . . . . . . . . . . .0.0025-0.005
Pump drive shaft
diameter . . . . . . . . . . . .0.6855-0.6860

Drive shaft bore I. D. . . . .0.6875-0.6885
Drive shaft running
clearance . . . . . . . . . . . .0.0015-0.003
Relief valve plunger
diameter . . . . . . . . . . . . . .0.747-0.749
Relief valve spring
test . . . . . . .21.19-21.75 lbs. @ $1\frac{13}{16}$ in.
Pump pressure at
2400 rpm . . . . . . . . .25 psi or above

NOTE: Beginning with engine serial number 8516356 a six-lobe oil pump has replaced the four-lobe pumps in the series 1850 diesel tractors. Should trouble be encountered with the four-lobe pump, the following parts are required to change to the six-lobe pump which is the only pump being furnished.

**Part**

Oil pump . . . . . . . . . . . . . . . .163 942-A
Relief valve spring . . . . . . . .163 944-A
Delivery line . . . . . . . . . . . . .163 945-A
Suction line . . . . . . . . . . . . . .163 946-A
"O" rings (two) . . . . . . . . . .159 411-A

Specifications may vary slightly from the four-lobe pump but service data will remain the same.

**MAIN OIL GALLERY
RELIEF VALVE**

**All Series 1750-1800-
1850 Non-Diesel-1950-T**

143. The main oil gallery, located on right side of engine, has a pressure relief control incorporated into its forward end. Plunger and spring can be removed after removing the hex head plug. For all models except 1950-T, spring has a free length of $1\frac{3}{8}$ inches and should test 10 pounds when compressed to a length of 1 inch. For models 1950-T, spring has a free length of 1.925 inches and

should test 5.5 pounds when compressed to a length of 1⅜ inches. Spring is color coded **red** for identification. Plunger and its bore should be free of any nicks or burrs. Plunger should fit its bore snugly, yet slide freely. See Fig. 092A.

## ENGINE OIL COOLER

### Series 1850 Diesel

144. The series 1850 diesel engines are equipped with an engine oil cooler which is mounted to front of engine. Oil circulation is from base of engine oil filter and is cooled by the coolant system fluid. Oil cooler is available only as a unit and the service involved is flushing and cleaning.

To remove the oil cooler, remove the radiator panels, then identify and disconnect the oil lines and coolant hoses from oil cooler. Remove clamp from mounting bracket and pull cooler from side of tractor. See Fig. 093 for view of oil cooler parts and their arrangment and Fig. 093A for a view of the oil cooler installation.

### Series 1950-T

144A. The series 1950-T engine is fitted with an engine oil cooler mounted on lower right side of engine. See Fig. O94 for an exploded view of the unit. Also shown is the by-pass filter which is mounted on top side of clutch housing.

NOTE: Removal of engine oil cooler on Wheatland model tractors, which have a wider front frame, presents no removal problems and oil cooler can be removed after disconnecting the necessary oil and coolant lines. On Row Crop models, which have a narrower front frame, clear-

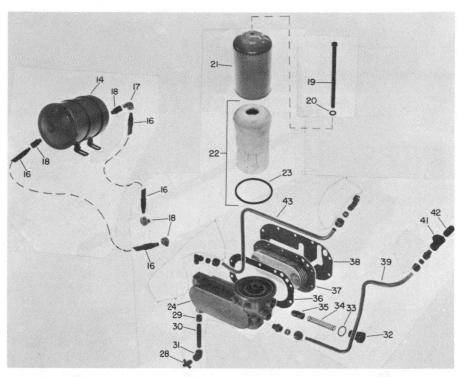

Fig. 094—Exploded view of engine oil cooler assembly used on series 1950-T. Also shown is the full flow engine oil filter and the by-pass oil filter. Cooler assembly is mounted on lower right side of engine and by-pass filter (14) is mounted on top of clutch housing.

| | | |
|---|---|---|
| 14. By-pass filter | 23. Gasket | 35. By-pass valve piston |
| 16. Hose | 24. Base | 36. Gasket |
| 17. Elbow | 28. Drain cock | 37. Cooler |
| 18. Elbow | 29. Reducer | 38. Gasket |
| 19. Bolt | 30. Nipple | 39. Coolant line |
| 20. Gasket | 31. Elbow | 41. Tee |
| 21. Shell | 32. Plug | 42. Nipple |
| 22. Filter element | 33. Washer | 43. Coolant line |
| | 34. Valve spring | |

ance between front frame and oil cooler is minimum and to assure correct installation of the two oil cooler gaskets, it is recommended that the engine be raised slightly until oil cooler is accessible.

144B. To remove engine oil cooler on Row Crop models, proceed as follows: Drain cooling system and oil cooler. Remove both hood side panels, front and rear lower side panels and hood. Turn fan until one of the blades is pointing straight up so it will help support radiator, then unbolt radiator from front frame. Disconnect the clips which retain oil lines running to coolers mounted on radiator. Remove sprocket coupling chain and disconnect lower ends of clutch control rod and auxiliary drive unit control rod. Shut off fuel and disconnect main tank fuel supply line. Disconnect injection pump control rod. Unbolt main tank front support from rear of engine and if tank is filled with fuel, place a support between tank and front frame. Remove engine mounting bolts and identify left front and right rear bolts as they are a machine fit and act as dowels. Remove rear air filter, then attach hoist

to engine and raise front of engine until engine oil cooler is accessible.

NOTE: The engine oil filter relief valve and the engine oil cooler by-pass valve are located in bores in bottom of cylinder block and can be removed, if necessary, after the engine oil pan and oil pan gasket are removed. The front valve is the oil filter relief valve and valve spring is color coded white. The rear valve is the oil cooler by-pass valve and the valve spring is color coded blue.

With oil cooler unit removed, remove plug (32), spring (34) and oil filter by-pass piston (35). Note that spring (34) is color coded black. Clean and inspect all parts and be sure all oil passages are clean and open. If cooler (37) is defective, it is recommended that cooler be renewed rather than to attempt to repair it. Spring specifications are as follows: Oil filter by-pass spring (black) has a free length of 3 11/16 inches and should test 4 pounds at 2⅝ inches. Filter relief valve spring (white) has a free length of 2.475 inches and should test 6.6 pounds at 1⅜ inches. Cooler by-pass spring (blue) has a free length of 1.8 inches and should test 1.1 pounds at 1¼ inches.

Fig. O93A — Series 1850 diesel engine oil cooler is mounted to a bracket attached to front of engine.

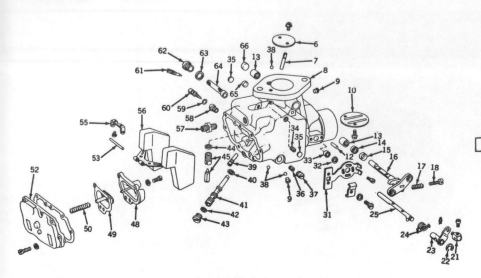

Fig. O95 — Exploded view of the model USX Marvel-Schebler carburetor used on later model tractors.

6. Throttle plate
7. Accelerator pump jet
8. Body
9. Plug
10. Choke plate
12. Stop pin
13. Needle bearings
14. Packing
15. Retainer
16. Throttle shaft
17. Spring
18. Throttle stop screw
21. Swivel
22. Retainer ring
23. Choke lever
24. Spring
25. Choke shaft

31. Bracket
32. Retainer
33. Packing
34. Economizer jet
35. Expansion plug
36. Minimum fuel restriction jet
37. Drain plug
38. Ball plugs
39. Idle jet
40. Gasket
41. Nozzle
42. Gasket
43. Plug
44. Gasket
45. Inlet needle and seat

48. Pump housing
49. Pump diaphragm
50. Spring
52. Bowl cover
53. Float shaft
55. Float bracket
56. Float
57. Connector
58. Power needle seat
59. "O" ring
60. Power adjusting needle
61. Idle mixture needle
62. Plug
63. Gasket
64. Screen
65. Choke shaft cup
66. Throttle shaft cup

Reinstall by reversing removal procedure and when installing turbocharger oil pressure line be sure check valve is installed with ball end forward and plug on top side. Ball can be seen by looking in end of check valve. Bleed fuel system.

# CARBURETOR
## (Except LP-Gas)
### Series 1750-1800-1850

145. The gasoline engines of series 1800 tractors are equipped with Marvel-Schebler carburetors and models TSX-807, USX-29, USX-32 or USX-32-1 have been used. Marvel-Schebler carburetor model USX-37 is used on series 1850 gasoline engines and model USX-44 on series 1750 gasoline engines.

### ADJUSTMENT

Carburetor adjustments are made with engine at operating temperature.

146. LOW IDLE. Set throttle stop screw to obtain approximately 350 engine rpm for series 1800, or 400 rpm for series 1850, turn idle mixture adjusting screw in until engine rpm begins to decrease, then back-out mixture adjusting screw until engine rpm stabilizes at the highest rpm. Readjust throttle stop screw if necessary, to obtain correct engine low idle rpm.

147. HIGH IDLE. With engine operating under load at approximately rated rpm, turn main (power) jet needle in until engine power starts to fall off, then back-out needle until full engine power is restored.

Note: Be sure engine is under load when making this adjustment. If a

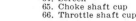

Fig. O97 — Use method shown to check model USX carburetor accelerator pump discharge jet. Refer to text.

dynamometer is not available, operate tractor with drawbar load and stop tractor to make adjustments.

### OVERHAUL

148. The overhaul and adjustment of the model TSX-807 carburetor used on tractors with serial number 90 525-000 through 124 285-000 is conventional and will be obvious after removal and an examination of the unit. Float setting and parts data are as follows:

**TSX-807**

Float setting .................. ¼-inch
Repair kit ...................286-1318
Gasket set ................. 16-594
Inlet needle and seat.......233-581
Idle jet ................... 49-203
Nozzle ................... 47-416
Power jet ............. 49-479
Economizer jet ........... 49-145

149. Beginning with tractor serial number 124 396-000 and running through 129 285-000 the model USX-29 carburetor was used and beginning with tractor serial number 129 286-000 and running through 150 420-000, either a USX-32 or a USX-32-1 carburetor was used. Refer to the following tables for parts data and to paragraph 151 for overhaul data.

**USX-29**

Float setting ................See text
Repair kit ...................286-1416
Gasket set ................. 16-704
Inlet needle and seat.......233-611
Pump diaphragm ..........237-516
Idle jet ................... 49-420
Nozzle ................... 47-A91
Main adjusting needle ..... 43-720
Main needle seat ........... 36-418
Economizer jet ............ 49-390

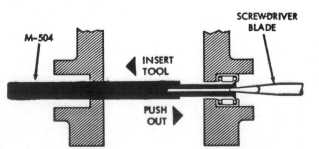

Fig. O96 — Use removal tool and screwdriver as shown to press out throttle shaft bearings from model USX carburetor.

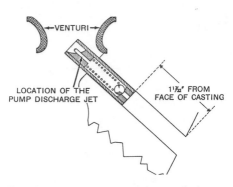

**Fig. O98 — Accelerator pump discharge jet is installed in model USX carburetor as shown.**

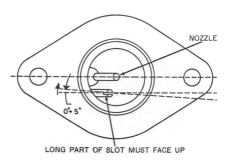

**Fig. O99 — View showing upper end of model USX carburetor accelerator pump discharge jet after installation.**

**USX-32 or USX-32-1**

| | |
|---|---|
| Float setting | See text |
| Repair kit | 286-1416 |
| Gasket set | 16-704 |
| Inlet needle and seat | 233-607 |
| Pump diaphragm | 237-550 |
| Idle jet | 49-420 |
| Nozzle | 47-A91 |
| Main adjusting needle | 43-720 |
| Main needle seat | 36-418 |
| Economizer jet | 49-390 |

150. Beginning with tractor serial number 150 421-000 a Marvel-Schebler model USX-37 carburetor is used. Refer to the following table for parts data and to paragraph 151 for overhaul data.

**USX-37**

| | |
|---|---|
| Float setting | See text |
| Repair kit | 286-1463 |
| Gasket set | 16-720 |
| Inlet needle and seat | 30-747 |
| Pump diaphragm | 237-550 |
| Idle jet | 49-420 |
| Nozzle | 47-A91 |
| Main adjusting needle assy | 43-720 |
| Main needle seat | 36-426 |
| Economizer jet | 49-390 |

**USX-44**

| | |
|---|---|
| Float setting | See Text |
| Repair kit | 286-1528 |
| Gasket set | 16-704 |
| Inlet needle and seat | 30-747 |
| Pump diaphragm | 237-550 |
| Idle jet | 49-420 |
| Nozzle | 47-471 |
| Main adjusting needle assy. | 43-720 |
| Economizer jet | 49-480 |

151. With carburetor removed, disassemble as follows: Remove throttle plate (6—Fig. 095) and pull throttle shaft (16) from body. Remove retainer (15) and packing (14). Cup (66) need not be removed unless throttle shaft bearings are to be renewed. Remove choke plate (10) and pull choke shaft (25) from body. Remove bracket (31), retainer (32) and packing (33). Remove plug (62) and screen (64). Remove cover (52), then unbolt and remove housing (48), diaphragm (49) and spring (50) from cover. Remove float shaft (53), float (56), inlet needle and needle seat (45). Remove plug (43) and nozzle (41). Remove idle tube (39), power needle (60) and seat (58). Remove plug (37) and minimum fuel jet (36). Remove idle needle (61).

The economizer jet (34) is not subject to wear and should not require service. However, if for any reason renewal is required, it will be necessary to drill expansion plug (35) and pry it out before economizer jet can be removed.

The throttle shaft needle bearings (13) can be removed from body, if necessary, by using Marvel-Schebler tool M-504 or its equivalent. Refer to Fig. 096. Insert tool through throttle shaft bore, then spread split end of tool with a small screw driver so lips of tool will engage bearing shell and press bearing from its bore. Repeat operation for opposite side.

NOTE: Use a vise as a press to remove and install the throttle shaft bearings. Bearings are hardened and could break if driven.

The accelerator pump discharge jet (7—Fig. 095) can be checked prior to removal and if satisfactory, jet need not be removed. Check jet operation as follows: Connect a vacuum gage between two short lengths of plastic or rubber tubing. Be sure bowl cover is off, then place one tube end over the end of discharge jet which protrudes into throttle bore. See Fig. 097. Apply vacuum (suck) on opposite end of tube assembly and as ball check leaves seat, note gage reading which should be 3 inches Hg. Ball

check leaving seat will be audible. If the jet check ball leaves its seat at less than 3 inches Hg. vacuum, renew the complete discharge jet as follows:

NOTE: If Marvel-Schebler tool M-505 (positioning tool) is not available, be sure to measure depth that discharge jet is installed in the drilled passage. This depth should be $1\frac{17}{32}$ inches as shown in Fig. 098.

Use a $\frac{5}{32}$-inch rod and drive discharge jet out toward throttle bore end of drilled passage. To install new discharge jet, start jet into drilled passage and position slot of discharge end so it is within 5 degrees of being paralled with the centerline of carburetor and deep end of slot is upward as shown in Fig. 099. Use Marvel-Schebler positioning tool M-505 and drive jet into position. If tool M-505 is not available, use the $\frac{5}{32}$-inch rod to drive the jet to the $1\frac{17}{32}$ inch location as shown in Fig. 098.

Clean and inspect all parts. Use compressed air to blow out all passages EXCEPT the pump discharge jet passage. Compressed air can damage discharge jet.

Reassemble by reversing the disassembly procedure and adjust float level by either adding an additional gasket under inlet needle seat or bending float lever, until float lever is parallel with surface of body mounting flange. Note: Float level can also be set by removing float needle and observing the extreme positions of float travel, then adjusting float until it is midway between these positions when float needle is installed. After carburetor is assembled, make the following initial adjustments so engine can be started and final adjustments made as outlined in paragraphs 146 and 147. Power needle 3 turns open; idle mixture adjusting needle 1½ turns open and throttle stop screw 1½ turns clockwise from closed throttle position.

**FUEL PUMP**

**Series 1750-1800-1850**

152. Beginning with tractor serial number 129 276-000, series 1800 gasoline tractors and all series 1750 and 1850 tractors are equipped with a diaphragm type fuel supply pump. The pump is mounted on right side of engine and is actuated by a lobe on the engine camshaft.

Overhaul of pump is conventional and will be obvious after an examination. Note that the pump valves are staked in position at four places.

# LP GAS SYSTEM (Zenith)

Series 1800 tractors with serial numbers 90 525-000 to 140 077-000 are fitted with a Zenith LP-Gas system which is comprised of a filter, a vaporizer and a pressure regulating carburetor. The pressure regulating carburetor serves both as a secondary regulator and a carburetor. Early tractors with the Zenith system were fitted with a carburetor having an assembly number 12719 whereas the later tractors used a carburetor with an assembly number of 12858. Service procedure is the same for both units, however when ordering parts be sure to use the correct assembly number.

## ADJUSTMENTS

153. Initial adjustments on the carburetor are 2 to 3 turns open for the idle mixture screw and 1 to 2 turns open for the main fuel adjusting screw.

Start engine and bring to operating temperature. Adjust throttle stop screw to obtain an engine low idle of 475-525 rpm. Turn the idle mixture screw in or out as required until engine runs smoothly. Correct adjustment should be obtained with idle mixture screw open approximately $2\frac{3}{8}$ turns. Recheck engine low idle rpm.

Place load on engine and run at high idle rpm. Turn the main adjusting needle inward until engine starts to lose power, then turn needle out until full power is restored. Repeat this operation until a definite setting is obtained. Correct setting should be obtained when the needle is $1\frac{3}{4}$ turns open.

If stationary loading of engine is not possible, make adjustments then

operate tractor under drawbar load to obtain final load setting, stopping tractor to make adjustments. Refer to Fig. 0101.

## FILTER

154. The LP-Gas filter, located at left side of battery tray, is fitted with a renewable element and should be cleaned monthly as follows: Turn off the vapor and liquid withdrawal valves, remove plug from bottom of filter bowl and allow any accumulations to drain. Slowly open the vapor valve and blow out any accumulations which remain.

Every year, unscrew the filter bowl and renew the element and gasket.

## VAPORIZER

155. The vaporizer used on LP-Gas burning models of the 1800 series tractors is a Zenith model number A962A-1.

156. **R&R AND OVERHAUL.** Before disconnecting any lines, be sure that all fuel is out of the lines, vaporizer, regulator and carburetor by closing the tank withdrawal valves and allowing the engine to run until it stops. Turn off ignition switch, then drain cooling system and remove hood and side panels. Disconnect and remove air cleaner and bracket assembly. Disconnect fuel lines from vaporizer, then remove the retaining cap screw at bottom of vaporizer housing and withdraw vaporizer assembly.

Refer to Fig. 0102 and proceed as follows: Remove the four caps crews and separate vaporizer coil and mounting plate assembly (18) from vaporizer body (14). Discard all three "O" rings. Remove any alternate four of the diaphragm cover screws and install in their place four aligning studs (Zenith No. C161-195, or equivalent). Maintain pressure on the diaphragm, remove the remaining four cap screws, then carefully release the pressure on diaphragm cover and remove cover (2), diaphragm springs (4 and 5) and vibration dampener (3). Remove studs, then remove diaphragm (6), baffle plate (7) baffle plate gasket (8) and fuel valve cap (9) as shown in Fig. 0103. Use a 1-inch socket wrench and remove valve seat (10—Fig. 0102), fuel valve (12) and spring (13) as shown in Fig. 0104.

Clean all parts in a grease solvent, however, DO NOT use carburetor cleaning solvents for cleaning any of the parts as carburetor cleaners will

**Fig. O101 — Points of adjustment for the Zenith LP-Gas pressure regulating carburetor.**

I. Idle mixture screw     S. Throttle stop screw
M. Main fuel needle

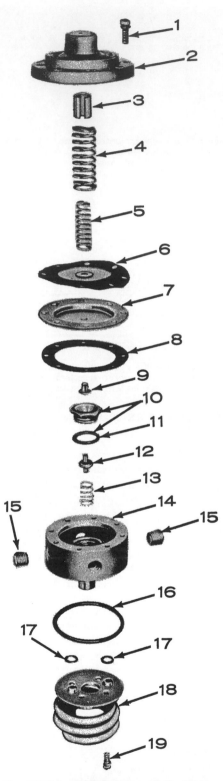

**Fig. O102 — Exploded view of the LP-Gas vaporizer assembly.**

| | |
|---|---|
| 2. Diaphragm cover | 11. "O" ring |
| 3. Vibration dampener | 12. Fuel valve |
| 4. Outer spring | 13. Valve spring |
| 5. Inner spring | 14. Body |
| 6. Diaphragm | 15. Allen plug |
| 7. Baffle plate | 16. "O" ring |
| 8. Gasket | 17. "O" ring |
| 9. Fuel valve cap | 18. Coil & mounting |
| 10. Valve seat |     plate |

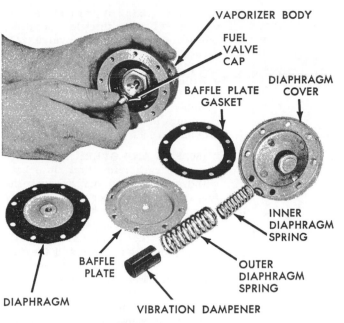

VAPORIZER BODY

FUEL
VALVE
CAP

BAFFLE PLATE
GASKET

DIAPHRAGM
COVER

**Fig. O103—Partially disassembled vaporizer assembly. Note fuel valve cap.**

INNER
DIAPHRAGM
SPRING

OUTER
DIAPHRAGM
SPRING

BAFFLE
PLATE

DIAPHRAGM

VIBRATION DAMPENER

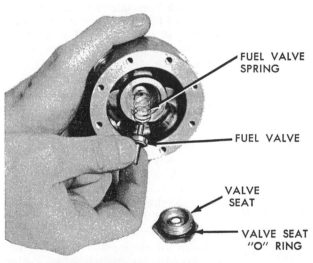

FUEL VALVE
SPRING

FUEL VALVE

VALVE
SEAT

VALVE SEAT
"O" RING

**Fig. O104—View showing the fuel valve being removed from body.**

destroy the impregnation used in the casting and the coating on the coil. Inspect all parts for undue wear or other damage and renew as necessary.

To reassemble, proceed as follows: Place fuel valve spring (13—Fig. 0102) over boss in center of vaporizer housing (14), then place fuel valve (12) on spring with shortest stem toward housing. Be sure spring is resting on machined shoulder of fuel valve, then with new "O" ring on valve seat (10), install valve seat in vaporizer body and tighten. Place fuel valve cap (9) over the protruding fuel valve stem. Reinstall the aligning studs in four alternate holes in top of vaporizer body, then install new baffle plate gasket (8), baffle plate (7), with recessed side toward body, and diaphragm (6), with flanged disc away from body. Be sure fuel valve cap enters center hole of baffle plate. Place vibration dampener (3) inside the outer diaphragm spring (4).

Place inner diaphragm spring (5) over center of diaphragm plate, then install outer spring over the inner spring. Place diaphragm cover (2) over aligning studs and depress cover until cap screws can be installed in the remaining four holes. Continue to depress the diaphragm cover and seat the cap screws lightly. Remove the aligning studs and install the remaining four cap screws. Tighten all the cap screws evenly and to a moderate tightness. Install a new "O" ring (16)

**Fig. O105—To test vaporizer assembly, install pressure gage and line as shown. Refer to text for procedure.**

on vaporizer body and two new "O" rings (17) on ends of vaporizer coil. Align mounting holes and install coil and mounting plate assembly (18) to vaporizer body.

157. **PRESSURE AND LEAKAGE TESTS.** To conduct pressure and leakage tests, proceed as follows: Install a pressure gage capable of registering at least 30 psi in the vaporizer outlet as shown in Fig. 0105. Connect the vaporizer inlet to a source of compressed air, loosen the previously installed gage until leakage occurs, then retighten gage. Gage should register a pressure of 9 to 11 psi and the reading should remain steady. If gage reading creeps up, it indicates a leaking fuel valve or fuel valve seat. If a leak is indicated, correct it by cleaning and/or renewing valve parts as necessary.

To check for leakage, use a bubble solution and proceed as follows:

Cover vent hole in diaphragm cover. If bubble forms, the diaphragm is leaking. Renew diaphragm and recheck.

Check around the diaphragm cover. If bubbles form, renew baffle plate gasket and recheck.

Check pipe plugs and if bubbles form, remove plugs; then reinstall them using a pipe plug compound and recheck.

Check the vaporizer coil mounting plate and if bubbles form, remove the vaporizer coil and mounting plate and renew "O" rings; then recheck.

### PRESSURE REGULATING
### CARBURETOR

158. **OPERATION.** The Zenith PC2 J 10 pressure regulating carburetor serves both as a secondary regulator and carburetor. See Fig. 0106 for a cross-sectional view of the carburetor.

The fuel valve seat (2) is adjustable so that the relationship of the diaphragm lever (9) and the diaphragm flange can be varied to meet the specifications for the particular

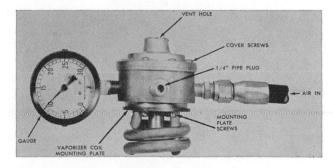

VENT HOLE

COVER SCREWS

1/4" PIPE PLUG

AIR IN

MOUNTING
PLATE
SCREWS

GAUGE

VAPORIZER COIL
MOUNTING PLATE

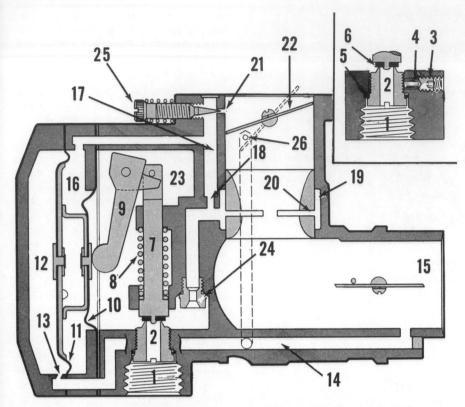

**Fig. O106 — Cross-sectional view of pressure regulating carburetor showing main components. Inset shows valve seat locking arrangement.**

| | | |
|---|---|---|
| 1. Fuel inlet | 8. Fuel valve spring | 14. Air passage | 20. Venturi |
| 2. Fuel valve seat | 9. Diaphragm lever | 15. Air intake | 21. Idle needle seat |
| 3. Locking screw | 10. Inner diaphragm | 16. Inner diaphragm | 22. Throttle fly |
| 4. Locking plug | 11. Outer diaphragm | chamber | 23. Pressure chamber |
| 5. "O" ring | 12. Outer diaphragm | 17. Idle fuel passage | 24. Main jet |
| 6. Sealing disc | chamber | 18. Idle orifice | 25. Idle needle |
| 7. Fuel valve | 13. Air passage orifice | 19. Annulus | 26. Economizer orifice |

engine operation. The fuel valve seat is locked in position in the carburetor body by means of the lock screw (3) (inset) and plug (4).

Fuel enters the carburetor at the fuel inlet (1) from the primary regulator at a pressure of 9-11 psi; but is prevented entering the carburetor by the neoprene seal (6) on the regulating (fuel) valve (7). When the engine is not running, pressures are equal in all parts of the carburetor and no fuel can flow. When the engine choke is closed and the engine turned over with the starter, a partial vacuum is built up in the carburetor air horn beyond the choke fly and transmitted to the chamber (16) between the two diaphragms by means of passage (17). The greater pressure in chamber (12) causes the large diaphragm (11) to move inward, acting through the spacer on diaphragm (10), moving the lever arm (9) to overcome the pressure of spring (8) and raise the fuel valve (7) off its seat admitting fuel to the pressure chamber (23). Fuel continues to enter the pressure chamber until pressure in chamber (23), combined with the pressure of spring (8), overcomes the pressure of

diaphragm (11) and closes the fuel valve (7). In engine idle operation, throttle valve (22) is closed, causing a high vacuum above the throttle valve, and fuel flows from the pressure chamber through orifice (18) past idle needle (25) to provide fuel for engine idle. At full throttle operation, the restriction of the venturi (20) causes a pressure drop at this point, drawing fuel into the carbu-

retor and mixing it with air at the venturi. At part throttle operation, the action of the throttle valve (22) causes a decrease of pressure at orifice (26) which is transmitted through a drilled passage to passage (14) causing a slight lowering of pressure in chamber (12), back of diaphragm (11), resulting in a progressively leaner fuel mixture at part throttle operation.

159. **R&R AND OVERHAUL.** To remove the carburetor, first close both withdrawal valves and exhaust the fuel in the regulator and lines by allowing the engine to run until it stops. Turn off the ignition switch, disconnect the choke and throttle linkage, air cleaner hose and the fuel inlet line. Unbolt and remove the carburetor assembly. Remove the six screws securing the diaphragms to the carburetor and remove the diaphragms as shown in Fig. O107. Remove the fuel inlet fitting from the bottom of the carburetor and the inlet valve seat set screw and plug; then remove the inlet valve seat while holding the valve off its seat with the diaphragm lever.

Remove the diaphragm lever shaft plug in the side of the carburetor body and remove the shaft, lever and fuel valve assembly. Remove the main and idle needle valves and the main valve body using Zenith tool No. C161-193, or equivalent. Remove the throttle and choke flys and shafts, and the venturi by removing the venturi locking screw in the side of the carburetor body.

Clean all parts, with the exception of the diaphragms and fuel valve seal, in a suitable solvent and examine for wear or damage. Examine the diaphragm for cracks, pin holes or deterioration and renew if necessary. Reassemble the carburetor by revers-

**Fig. O107 — LP-Gas pressure regulating carburetor shown with diaphragm assembly removed.**

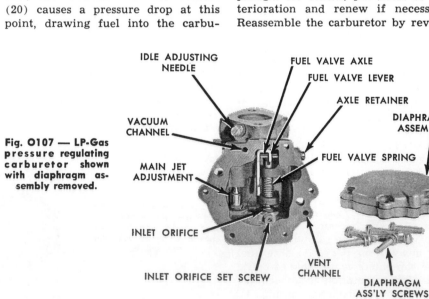

IDLE ADJUSTING NEEDLE

FUEL VALVE AXLE

FUEL VALVE LEVER

AXLE RETAINER

DIAPHRAGM ASSEMBLY

VACUUM CHANNEL

MAIN JET ADJUSTMENT

FUEL VALVE SPRING

INLET ORIFICE

VENT CHANNEL

INLET ORIFICE SET SCREW

DIAPHRAGM ASS'LY SCREWS

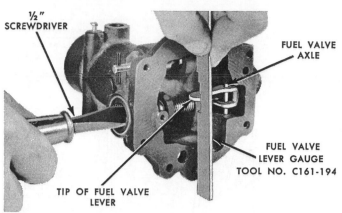

½"
SCREWDRIVER

FUEL VALVE
AXLE

FUEL VALVE
LEVER GAUGE
TOOL NO. C161-194

TIP OF FUEL VALVE
LEVER

**Fig. O108 — When adjusting diaphragm lever, use step number three of Zenith gage number C161-194 as shown.**

ing the disassembly procedure using new gaskets and a new fuel valve seal.

While holding the fuel valve open, install the fuel valve seat and adjust same until fuel valve lever just

touches the land of step number three of Zenith gage C161-194 as shown in Fig. O108. Retain fuel valve in this position by installing plug and set screw.

Note: If Zenith gage is not available, the distance between machined surface of body and tip of lever is $\frac{3}{32}$-inch.

When reinstalling the diaphragm, diaphragm cover and gasket, make sure that the parts are arranged so that the diaphragm passages are aligned with the passages in the carburetor body.

# LP GAS SYSTEM (Bosch-Ensign)

Series 1800 tractors serial number 140 077-000 and up, and all series 1850 tractors are available with an LP-Gas system manufactured by the Ensign Products Section, American Bosch Arma Corporation. Like other LP-Gas systems, this system is designed to operate with the fuel tank not more than 80% filled.

160. Bosch-Ensign Series CBX carburetors and Series RDG Regulator have three points of mixture adjustment, plus an idle stop screw. Adjustment and repair data follows:

## ADJUSTMENTS

161. **INITIAL ADJUSTMENTS.** After overhauling or installing new carburetor or regulator, make the following initial adjustments: Open idle screw on regulator 1½ turns. Open starting adjustment screw on carburetor 1 turn. Open load adjustment screw on carburetor 9 to 9½ turns. Close choke and open throttle ½ way to start engine.

162. **STARTING SCREW ADJUSTMENT.** After engine has been started and is at normal operating temperature move throttle ½ to ¾ open while leaving choke closed. Adjust starting screw to give highest engine speed, then turn screw in slightly until engine speed just starts to drop and tighten lock nut. Return throttle to slow idle position and open choke.

163. **IDLE STOP SCREW.** Adjust idle stop screw on carburetor throttle to obtain an engine slow idle speed of 500 rpm for 1800 series C tractors, or 400 rpm for series 1850 tractors.

164. **IDLE MIXTURE SCREW.** Adjust idle mixture screw on regulator to obtain smoothest engine idle performance when engine is at operating temperature. Readjust idle stop screw if necessary.

165. **LOAD SCREW ADJUSTMENT (WITH ANALYZER).** Be sure to follow the gas analyzer operating instructions and set load screw to give a reading of 12.8 on gasoline scale or 14.3 on LP-Gas scale with engine at normal operating temperature and running at high idle speed. Recheck idle mixture and stop screw adjustment.

166. **LOAD SCREW ADJUSTMENT (WITHOUT ANALYZER).** With engine running at full throttle and at normal operating temperature, apply load until governor opens carburetor throttle wide open. Find adjustment points where engine speed begins to drop from mixture being too rich, then too lean. Set adjustment midway between these two points and tighten jam nut. Recheck idle mixture and stop screw adjustment.

167. **LOAD SCREW ADJUSTMENT (WITHOUT ANALYZER OR LOAD).** Make idle adjustment carefully as outlined in paragraph 164. With engine running at high idle speed and at normal operating temperature, adjust load screw to give maximum engine rpm; then, slowly turn load screw in until engine speed begins to fall. Set load screw midway between these two positions.

### FILTER

168. The Bosch-Ensign filter (See Fig. 0110) is equipped with a felt filtering element and a magnetic ring.

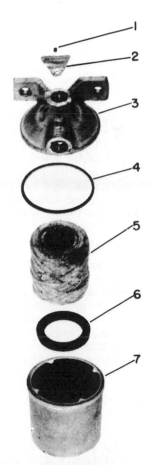

**Fig. O110 — Exploded view of the Bosch-Ensign filter.**

3. Cover
4. Gasket
5. Element
6. Magnet
7. Housing

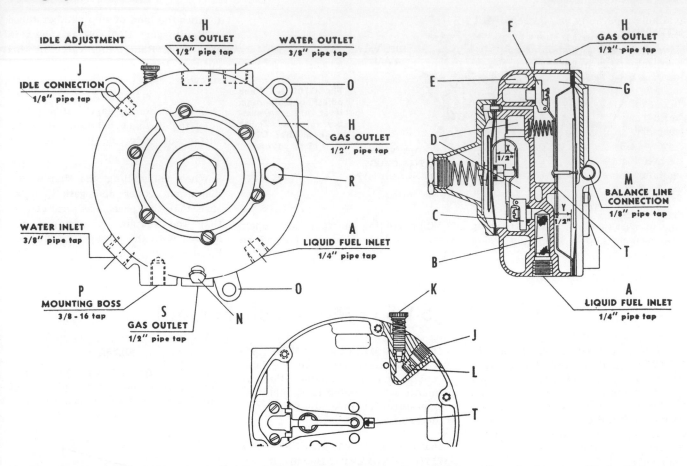

**Fig. O111 — Sectional view of a typical Bosch-Ensign LP-Gas regulator. Note inlet lever dimension in upper right hand view.**

B. Strainer
C. High pressure valve
D. Vaporizing chamber
E. Water jacket
F. Low pressure valve
G. Low pressure chamber
L. Idle orifice
N. Drain
O. Support lugs
R. Plug
T. Boss

When servicing the LP-system or on major engine overhauls, it is advisable to remove the filter housing and clean or renew the filtering element. CAUTION: Shut off both liquid and vapor valves at fuel tank and run engine until fuel is exhausted before attempting to remove the filter housing.

### REGULATOR

The Bosch-Ensign Series RDG regulator combines a heat exchanger to vaporize liquid LP-Gas with a two stage regulator to reduce fuel pressure to slightly below atmospheric. The primary regulator reduces the fuel from tank pressure to approximately 4 psi as it enters the final regulator. Heat for vaporizing the liquid fuel is obtained from coolant from the tractor cooling system. Refer to Fig. O111 for cross-sectional views of regulator.

169. **R&R REGULATOR.** Shut off both the liquid and vapor withdrawal valves at fuel tank and run engine until fuel in system is exhausted. Drain engine cooling system. Remove water connections to engine cooling system. Disconnect fuel and balance lines and

remove regulator from tractor. Reverse removal procedure to reinstall regulator.

170. **OVERHAUL REGULATOR.** Refer to Fig. O112 for an exploded view of regulator. Disassemble the regulator, clean parts with solvent and dry with air hose. Inspect and renew valve seats, valves, diaphragms and springs as necessary. Parts are

2. Cover
3. Spring
4. Diaphragm assembly
5. High pressure valve assy.
6. Plug
7. Lever
9. Valve body
10. Pin
11. "O" ring
12. Drain cock
13. Plug
14. Strainer
15. Spring
16. Low pressure valve assy.

17. Screw
18. Spring
19. Idle mixture needle
20. Body
21. Gasket
22. Plate
23. Diaphragm
24. Plate

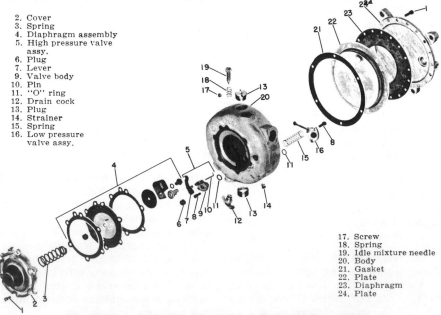

**Fig. O112 — Exploded view of Bosch-Ensign RDG regulator.**

available separately or in a repair kit.

After inlet valve (primary regulator valve) assembly is installed, open and close valve several times; then, check distance from face of housing to bottom of groove in inlet valve lever (See Fig. O113 and dimension X in Fig. O111). Bend inlet valve lever as necessary so that this measurement is exactly ½-inch.

When installing outlet valve (final regulator valve) assembly, align dimple of valve lever with arrow on reference boss (See Fig. O114) and after operating valve several times, bend valve lever as necessary so that lever is level with the two bosses on either side of lever.

**171. TROUBLE SHOOTING REGULATOR PROBLEMS.** Generally, difficulties encountered with regulator are due to leakage of gas past valves. Trouble will generally show up as excessive fuel consumption, decrease in power, inability to properly adjust fuel mixture and/or loss of gas through carburetor when engine is not running. To test regulator, remove plug (R—Fig. O111) and install a 0-10

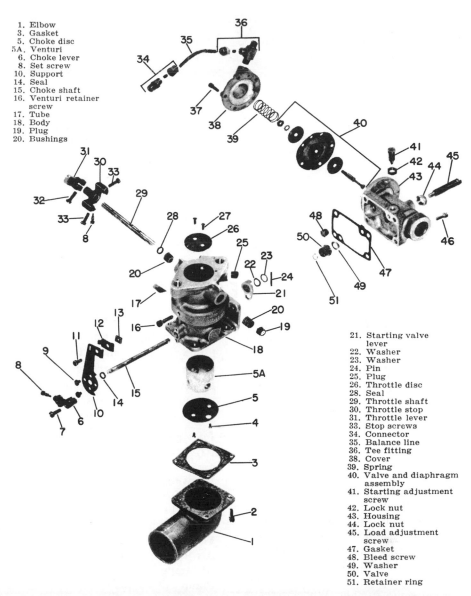

1. Elbow
3. Gasket
5. Choke disc
5A. Venturi
6. Choke lever
8. Set screw
10. Support
14. Seal
15. Choke shaft
16. Venturi retainer screw
17. Tube
18. Body
19. Plug
20. Bushings

21. Starting valve lever
22. Washer
23. Washer
24. Pin
25. Plug
26. Throttle disc
28. Seal
29. Throttle shaft
30. Throttle stop
31. Throttle lever
33. Stop screws
34. Connector
35. Balance line
36. Tee fitting
38. Cover
39. Spring
40. Valve and diaphragm assembly
41. Starting adjustment screw
42. Lock nut
43. Housing
44. Lock nut
45. Load adjustment screw
47. Gasket
48. Bleed screw
49. Washer
50. Valve
51. Retainer ring

Fig. O115 — Exploded view of the Bosch-Ensign model CBX LP-Gas carburetor.

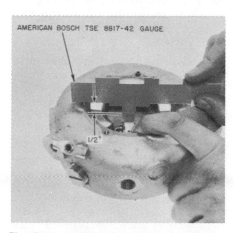

Fig. O113 — Using Bosch-Ensign gage No. 8817-42 to set high pressure (inlet) valve lever to dimension "X" (½-inch) as shown in Fig. O111.

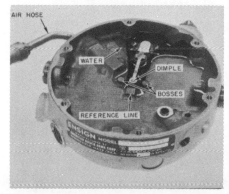

Fig. O114 — Dimple in outlet valve lever is aligned with reference line and lever should be level with the two bosses shown.

psi gage in opening. If gage pressure gradually builds up after engine is stopped, inlet valve is leaking and same should be cleaned or renewed. Remove fuel hose from regulator to carburetor. Soap bubble should hold over fuel opening at regulator. If not, fuel outlet valve or low pressure diaphragm is leaking. Clean or renew valve and check diaphragm.

### CARBURETOR

The Bosch-Ensign Series CBX carburetor is equipped with starting and load adjustment screws and with an economizer unit that richens the fuel mixture under load conditions. See Fig. O115 for exploded view of carburetor. Idle fuel mixture adjustment is provided on the regulator.

**172. OVERHAUL CARBURETOR.** Refer to Fig. O115. Repair parts are available separately, also a repair kit is available for the carburetor. Other than renewing throttle shaft and throttle shaft bushings, carburetor servicing generally concerns renewal of economizer diaphragm (40) and making sure that starting valve (21) is working properly. Check to see that economizer diaphragm will hold vacuum. Inspect starting valve for proper position when choke is closed. Valve should fit tightly against carburetor body and completely cover main fuel opening in carburetor body when choke is closed. Refer to Fig. O116 for removal and installation of fuel adjusting screw.

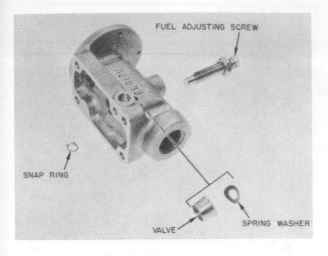

Fig. O116 — View showing installation of main fuel adjusting screw and valve of Bosch-Ensign CBX carburetor.

# DIESEL FUEL SYSTEM

## Series 1750-1800-1850-1950-T

The diesel fuel system consists of three basic units; the fuel filter, injection pump and injection nozzles. When servicing any unit associated with the fuel system, the maintenance of absolute cleanliness is of utmost importance.

Probably the most important precaution that servicing personnel can impart to owners of diesel powered tractors is to urge them to use an approved fuel that is absolutely clean and free from foreign material. Extra precaution should be taken to make certain that no water enters the fuel storage tanks. This last precaution is based on the fact that all diesel fuels contain some sulphur. When water is mixed with sulphur, an acid is formed which will quickly erode the closely fitting parts of the injection pump and fuel injection nozzles.

### TROUBLE SHOOTING

### Series 1750-1800-1850-1950-T

173. If the engine will not start or does not run properly, refer to the following paragraphs for possible cause of trouble.

174. **ENGINE WILL NOT START.** If the engine will not start, possible causes of trouble are as follows:

Fuel tank empty.

Fuel supply valve closed.

Engine stop control applied or improperly adjusted.

Water in fuel.

Inferior fuel.

Clogged fuel filter.

Air traps in system.

Low cranking speed.

Pump installed out of time.

Pump shaft broken (pump seized).

Faulty or worn fuel injection pump.

Faulty injector nozzles.

Low engine compression.

175. **ENGINE DOES NOT RUN PROPERLY.** If the engine will start but does not run properly (smokes excessively, mis-fires, etc.), possible causes of trouble are as follows:

Water in fuel.

Inferior fuel.

Fuel filter clogged.

Air cleaner element clogged.

Air traps in fuel system.

Pump out of time.

Faulty fuel injectors.

Low compression on one or more cylinders.

Faulty fuel injection pump.

### BLEEDING THE SYSTEM

### Series 1750-1800-1850-1950-T

176. Some minor changes have been made in the diesel fuel system which will vary somewhat the bleeding of the diesel fuel system. The primary fuel system filter on tractors prior to serial number 124 396-000 (Series A) utilized a pipe plug in the side of the cover assembly for bleeding, whereas filters on all tractors after serial number 124 395-000 (Series A and B) have the bleed screw incorporated into the cover nut assembly. On series 1750 and 1950-T, the bleed plugs are in header of filter assembly.

At tractor serial number 129 285-000, the hand primer was deleted from the

fuel system and a fuel pump having a built in primer lever was installed. See Fig. O122.

177. To bleed the diesel fuel system, proceed as follows. Open fuel tank shut-off valve if necessary. On Series A tractors, remove plug (1-Fig. O120) and allow fuel to flow until all bubbles disappear, then reinstall plug. On Series B and C tractors, remove bleed screw from filter cover nut and on those with no fuel pump, allow fuel to flow until all bubbles disappear, then reinstall bleed screw. On models having fuel pump, actuate the pump priming lever until bubble free fuel flows from bleeder screw hole, then install bleed screw.

Note: If priming lever (Fig. O122) of fuel pump cannot be actuated, turn engine until pump rocker arm is released by the camshaft.

178. With primary filter free of air, loosen bleed screw on final fuel filter and actuate hand primer (3—Fig. O121), or fuel pump priming lever (Fig. O122), until bubble free fuel appears, then tighten bleed screw. Continue to actuate lever for an additional fifteen or twenty strokes to force any remaining air in the low pressure system through the lines and into the tank.

179. Attempt to start engine, and if engine fails to start, or runs unevenly, loosen lines at injectors and bleed air from them either by placing fuel stop in run position and operating starting motor, or allow engine to run at low idle. Tighten connections when operation is completed.

### Series 1850 Diesel

180. To bleed the diesel fuel system, open the fuel shut-off valve, if necessary, then remove the primary fuel filter bleed screw (Fig. O123) and actuate the priming lever (P) of the fuel supply pump until bubble free fuel flows from bleed screw opening. Reinstall bleed screw.

Fig. O120 — On 1800 series A tractors, remove plug (1) to purge air from primary fuel filter.

Fig. O121 — On early 1800 diesel tractors, final fuel filter (2) can be bled by operating hand primer (3).

Fig. O122 — View showing priming lever of fuel pump and filters on 1750 diesel tractors.

Fig. O123 — Bleed screw for 1850 diesel primary fuel filter is located in cover bolt as shown. Note fuel supply pump priming lever (P).

Fig. O124 — To bleed final filter on 1850 diesel, loosen the final filter to injection pump fuel line at coupling of injection pump.

Fig. O125 — Injection pump bleed screws on 1850 diesel are located as shown. When bleeding injection pump, tighten lower bleed screw first.

Note: If priming lever of fuel supply pump cannot be actuated, turn engine until pump rocker arm is released by the camshaft.

With primary filter free of air, loosen the final filter to injection pump line at the injection pump (Fig. 0124), actuate fuel supply pump priming level until bubble free fuel flows from loosened connection, then tighten connection.

With final filter bled, loosen both bleed screws (Fig. 0125) on injection pump, actuate fuel supply pump priming lever and when bubble free fuel flows from both openings, tighten lower bleed screw first, then tighten upper bleed screw.

Attempt to start engine, and if engine fails to start, or runs unevenly, loosen lines at injectors and bleed air from them either by placing fuel stop in run position and operating starting motor, or allowing engine to run at low idle rpm. Tighten connections when operation is completed.

## INJECTOR NOZZLE

### Series 1800

Series 1800 diesel engines are equipped with throttling pintle type fuel injection nozzles. In operation, some fuel is atomized to start the combustion process but much of the fuel is emitted from the nozzle as a solid "core" which crosses the combustion chamber and enters the energy cell. As the power stroke continues, the fuel-air mixture is ejected from the energy cell into the combustion chamber where burning of the fuel is completed.

WARNING: Fuel leaves the injector nozzle with sufficient force to penetrate the skin. Keep your person clear of nozzle spray when testing.

181. **TESTING AND LOCATING A FAULTY NOZZLE.** If rough or uneven engine operation, or misfiring, indicates a faulty injector, the defective unit can usually be located as follows:

With the engine operating at low idle speed, loosen the high pressure connection at each injector in turn. As in checking spark plugs, the faulty unit is the one which least affects the running of the engine when its line is loosened.

If a faulty nozzle is found and considerable time has elapsed since the injectors have been serviced, it is recommended that all injectors be removed and new or reconditioned units installed, or the nozzles serviced as outlined in the following paragraphs. Also, energy cells should be removed and cleaned if faulty injectors are found.

182. **REMOVE AND REINSTALL.** Before loosening any fuel line connections, thoroughly clean the head surface, lines and injectors with compressed air, if available, and by washing with diesel fuel or a suitable solvent. After disconnecting the pressure and leak-off lines, cap all connections to prevent entry of dirt or dust into fuel system. Loosen pressure line connections at injection pump to prevent bending the lines. Remove the retaining cap screws and carefully withdraw the injector assembly from cylinder head, being careful not to strike the tip end of nozzle against any hard surface.

Thoroughly clean the nozzle recess in cylinder head before reinserting the injector assembly. No hard or sharp tools should be used for cleaning. A piece of wood dowel or brass stock properly shaped, or an approved nozzle bore cleaner should be used. Install injector with a new copper gasket and loosely install retaining nuts until all fuel lines are connected; then, tighten injector retaining nuts to a torque of 13-17 ft.-lbs. and bleed the injectors and lines as outlined in paragraph 179.

183. **NOZZLE TESTER.** A complete job of testing and adjusting the injector requires the use of a special tester such as that shown in Fig. 0126. Only clean approved testing oil should be used in tester tank.

The injector should be tested for spray pattern, seat leakage, back leakage and opening pressure as follows:

184. **SPRAY PATTERN.** Operate tester handle until oil flows from injector connection, then attach the injector assembly. Close the valve to tester gage and operate tester handle a few quick strokes to purge air from

injector and tester pump and to make sure injector is not plugged or inoperative.

If a straight, solid core of oil flows from nozzle tip without undue pressure on tester handle, open valve to tester gage and remove cap nut (1—Fig. 0127). Slowly depress the tester handle and observe the pressure at which core emerges. If opening pressure is not approximately 1925 psi (new spring) or 1750 psi (used spring), loosen locknut (3) and turn adjusting screw (4) in or out until opening pressure is 1925 psi (new injector assembly or new spring) or 1750 psi (used spring).

When opening pressure has been set, again close valve to tester gage and operate tester handle at approximately 100 strokes per minute while examining spray core. Fuel should emerge from nozzle opening in one solid core, in a straight line with injector body, with no branches, splits or atomization.

NOTE: The tester pump cannot duplicate the injection velocity necessary to obtain the operating spray pattern of the delay type nozzles. Also absent will be the familiar popping sound associated with the nozzle opening of conventional nozzles. Under operating velocities, the observed solid core will cross the combustion chamber and enter the energy cell. In addition, a fine conical mist surrounding the core will ignite in the combustion chamber area above the piston. The solid core cannot vary more than 7½ degrees in any direction and still enter the energy cell. While the core is the only spray characteristic which can be observed on the tester, absence of core deviation is of utmost importance.

185. SEAT LEAKAGE. The nozzle valve should not leak at pressure of 150 psi less than opening pressure (paragraph 184). To check for seat

Fig. O126 — When testing injectors, use a tester such as the one shown.

leakage, open the valve to tester gage and actuate tester handle slowly until gage pressure is 150 psi less than opening pressure. Maintain this pressure for at least 10 seconds, then observe the flat surface of nozzle body and the pintle tip for drops or undue wetness. If drops or wetness appear, the injector must be disassembled and overhauled as outlined in paragraph 188.

186. BACK LEAKAGE. A back leak test will indicate the condition of the internal sealing surfaces of the nozzle assembly. Before checking the back leakage, first check for seat leakage as outlined in paragraph 185 then proceed as follows:

Turn the adjusting screw (4—Fig. 0127) inward until nozzle opening pressure is set at 2300 psi. Release the tester handle and observe the length of time required for gage needle to drop from 2200 psi to 1500 psi. The time should be not less than 6 seconds. A faster drop would indicate wear or scoring between piston surface of needle (10) and nozzle (11), or improper sealing of pressure face surfaces of nozzle and holder (9). NOTE: Leakage at tester connections or tester check valve will show up as fast leak back in this test. If all injectors tested fail to pass this test, the tester, rather than the injector, should be suspected.

187. OPENING PRESSURE. To assure peak engine performance, it is recommended that the six injectors installed in any engine be adjusted as nearly as possible to equal opening pressures. The recommended opening pressure is 1925 psi for a new nozzle (or used nozzle with new spring); or 1750 psi for used nozzle and spring. When a new spring (7—Fig. 0127) is installed in an injector assembly, the injection pressure will drop quickly as the spring becomes seated under the constant compression. After the opening pressure has been adjusted, tighten locknut (3) to a torque of 80-85 ft.-lbs. and install cap nut (1); then recheck opening pressure to make sure adjusting screw has not moved.

188. **MINOR OVERHAUL.** The maintenance of absolute cleanliness in the overhaul of injector assemblies is of utmost importance. Of equal importance is the avoidance of nicks or scratches on any of the lapped surfaces. To avoid damage to any of the highly machined parts, only the recommended cleaning kits (Oliver Kit No. STAS—115—B, or Bacharach Kit No. 66-5034) and oil base carbon sol-

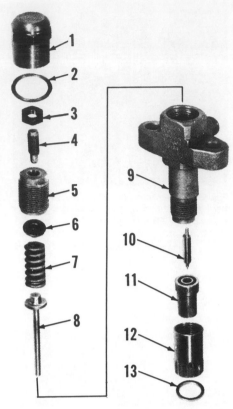

Fig. O127 — Exploded view of series 1800 throttling pintle type injector assembly.

| | |
|---|---|
| 1. Cap | 7. Pressure spring |
| 2. Gasket | 8. Spindle |
| 3. Lock nut | 9. Nozzle holder |
| 4. Adjusting screw | 10. Nozzle valve |
| 5. Spring retainer | 11. Nozzle body |
|     nut | 12. Cap nut |
| 6. Upper spring seat | 13. Gasket |

vents should be used in the injector repair sections of the shop. The nozzle valve and body are individually fit and hand lapped, and these two parts should always be kept together as mated parts.

Before disassembling a set of injectors, immerse the units in a clean carbon solvent and thoroughly clean the outer surfaces with a brass wire brush. Be extremely careful not to damage the pintle end of the nozzle valve extending out of nozzle body. Rinse the injector in clean diesel fuel.

Clamp the squared upper part of the injector holder in a soft jawed vise, tightening vise only tight enough to keep injector from slipping, or use a holding fixture. Remove cap nut (1—Fig. 0127), loosen lock nut (3) and back out the adjusting screw (4) until all tension is removed from the spring; then remove the nozzle holder nut (12). Withdraw the nozzle valve (10) from nozzle body (11) with the fingers as shown in Fig. 0128, or if valve is stuck, use special extractor such as that shown in Fig. 0129. NEVER loosen valve by tapping on exposed pintle end of valve. Remove spring retainer cap nut (5—Fig. 0127),

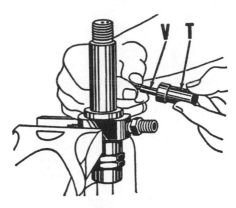

Fig. O128 — Withdraw nozzle valve (V) from body (T) as shown, or if valve is stuck, use an extractor such as that shown in Fig. O129.

spring seat (6), spring (7) and seat and spindle (8). Renew spring if rusted, cracked or distorted in any way. Carefully examine the spring seats and spindle and renew if chipped, cracked or damaged in any way.

Examine the lapped pressure faces of nozzle body (11) and holder (9) for nicks or scratches, and the piston (larger) portion of nozzle valve (10) for scratches or scoring. Clean the fuel gallery with the special hooked scraper as shown in Fig. O130, by applying side pressure while the body is rotated. Clean the valve seat with the brass seat tool as shown in Fig.

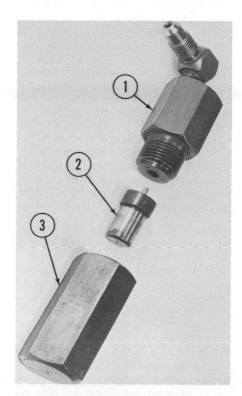

Fig. O129 — View of hydraulic nozzle valve extractor which is used in conjunction with an injector tester.

O131. Polish the set with the pointed wooden polishing stick and a small amount of tallow as shown in Fig. O132. Clean the pintle orifice from the inside, using the proper size probe. Polish the nozzle valve seat and pintle with a piece of felt and some tallow, loosening any particles of hardened carbon with a pointed piece of brass stock. Never use a hard or sharp object as any scratches will cause distortion of the injection core.

As the parts are cleaned, immerse in clean diesel fuel in a compartmented pan. Insert the nozzle valve into nozzle body while holding parts below the fuel level in pan and assemble to nozzle holder while wet. Use the center centering sleeve (Oliver Tool No. STA-115-B, or Bacharach No. 66-0064) when reassembling as shown in Fig. O133 and tighten the holder nut to a torque of 45-50 ft.-lbs.

Install spindle and seat (8—Fig. O127), spring (7), seat (6) and nut (5), tightening spring nut to a torque of 50-60 ft.-lbs. Install screw (4) and lock nut (3), adjusting the screw to obtain nozzle opening pressure of 1925 psi (new spring) or 1750 psi with used spring (7). After opening pressure is adjusted, tighten the lock nut and install cap nut (1) and sealing washer (2).

Retest the injector as previously outlined. If injector fails to meet the tests, disassemble and reclean or renew the nozzle valve and body.

### INJECTION NOZZLES

### Series 1850 Diesel

Engines of the series 1850 diesel tractors are equipped with multi-hole C. A. V. injectors which inject directly into combustion chambers located in tops of pistons.

WARNING: Fuel leaves the injector nozzle with sufficient force to penetrate the skin. Keep your person clear of nozzle spray when testing.

189. **TESTING AND LOCATING A FAULTY NOZZLE.** If rough or uneven engine operation, or misfiring indicates a faulty injector, the defective unit can usually be located as follows:

With the engine operating at low idle speed, loosen the high pressure connection at each injector in turn. As in checking spark plugs, the faulty unit is the one which least affects the running of the engine when its line is loosened.

If a faulty nozzle is found and considerable time has elapsed since the injectors have been serviced, it is rec-

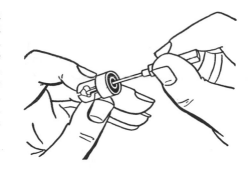

Fig. O130 — Use special scraper to clean carbon from fuel gallery of series 1800 injector.

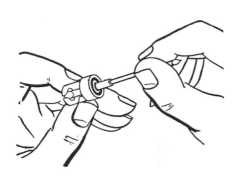

Fig. O131 — Use brass seat tool to remove carbon from valve seat of series 1800 injector.

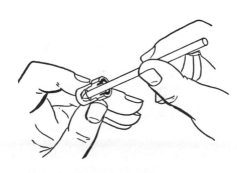

Fig. O132 — Use wooden stick and a small amount of tallow to polish valve seat of series 1800 injector.

Fig. O133 — When assembling series 1800 injector, use centering sleeve (S) to center nozzle assembly on injector body.

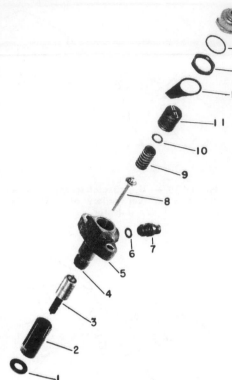

Fig. O135 — Exploded view of series 1850 multi-hole spray type injector assembly.

1. Gasket
2. Cap nut
3. Nozzle valve
4. Dowel
5. Body
6. Washer
7. Connector
8. Spindle
9. Pressure spring
10. Washer
11. Adjusting screw
12. Tab
13. Lock nut
14. Washer
15. Protective cap

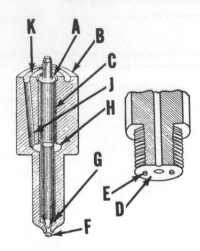

Fig. O136 — Inspect the disassembled series 1850 injector at the points indicated.

A. Nozzle body pressure face
B. Nozzle body pressure face
C. Nozzle valve piston surface
D. Nozzle holder pressure face
E. Dowel pins
F. Nozzle spray holes
G. Valve seat
H. Fuel gallery
J. Fuel passage
K. Fuel chamber

ommended that all injectors be removed and new or reconditioned units be installed, or the nozzles be serviced as outlined in the following paragraphs.

190. **REMOVE AND REINSTALL.** Before loosening any fuel line connections, thoroughly clean the head surface, lines and injectors with compressed air, if available, and by washing with diesel fuel or a suitable solvent. After disconnecting the pressure and leak-off lines, cap all connections to prevent entry of dirt into fuel system. Loosen pressure line connections at injection pump to prevent bending the lines. Disconnect fuel lines and tank support and raise fuel tank. Remove the retaining nuts and carefully withdraw the injector assembly from cylinder head, being careful not to strike the tip end of nozzle against any hard surface.

Thoroughly clean the nozzle recess in cylinder head before reinserting the injector assembly. No hard or sharp tools should be used for cleaning. A piece of wood dowel or brass stock properly shaped, or an approved nozzle bore cleaner should be used. Install a new copper gasket, tighten injector retaining nuts to a torque of 10-12 ft.-lbs. and bleed the injectors and lines as outlined in paragraph 180.

191. **NOZZLE TESTER.** A complete job of testing and adjusting the injec-

tor requires the use of a special tester such as that shown in Fig. 0126. Only clean approved testing oil should be used in tester tank.

The injector should be tested for spray pattern, seat leakage, back leakage and opening pressure as follows.

192. **SPRAY PATTERN.** Operate tester handle until oil flows from injector connection, then attach the injector assembly. Close the valve to tester gage and operate tester handle a few quick strokes to purge air from injector and tester pump and to make sure that injector is not plugged or inoperative.

If four, finely atomized sprays emerge from tip without undue pressure on tester handle, open valve to tester gage and check to see that injector opening pressure is approximately 2600 psi. If pressure is not as specified, adjust pressure by removing cap (15—Fig. 0135), loosening locknut (13) and turning adjusting screw (11) as required.

When opening pressure has been set, again close valve to tester gage and operate tester handle quickly several times while observing the spray pattern. The four conical sprays should be symmetrical, finely atomized and have no core or streaks. If spray pattern is not as described, or if one or more of the spray holes are plugged, overhaul nozzle as outlined in paragraph 196.

193. **SEAT LEAKAGE.** The nozzle valve should not leak at pressures below 2450 psi. To check for seat leakage, open valve to tester gage and actuate tester handle slowly until gage pressure aproaches 2450 psi. Maintain this pressure for at least 10 seconds, then observe nozzle tip for drops of fuel or undue wetness. If evidence of leakage appears, injector must be overhauled as outlined in paragraph 196.

194. **BACK LEAKAGE.** A back leak test will indicate the condition of internal sealing surfaces of nozzle assembly. Before checking back leakage, first test for seat leakage as outlined in paragraph 193, then proceed as follows:

Actuate tester handle slowly until gage pressure approaches 2300 psi. Release tester handle and observe the length of time required for gage needle to drop from 2200 psi to 1500 psi. The time should be not less than 6 seconds. A faster drop would indicate wear or scoring between piston surface of nozzle valve or body, improper sealing of pressure face surfaces between nozzle holder and valve body, or leakage at lines or connections. NOTE: Leakage at tester connections or tester check valve will show up as fast leak-back in this test. If all injectors tested fail to pass the leakage test, the tester rather than the injector should be suspected.

195. **OPENING PRESSURE.** To assure peak engine performance, it is recommended that the four injectors installed in any engine be adjusted as nearly as possible to equal opening pressures. The recommended opening pressure for new injectors or used injectors with new spring, is 2600 psi. Opening pressure for used nozzles is 2500 psi.

After injector has been overhauled, or has passed the other tests outlined in paragraphs 192 through 194, recheck the opening pressure of the set of injectors and adjust as outlined.

Fig. O137 — Clean the fuel gallery of 1850 injector nozzle using the special hooked scraper as shown.

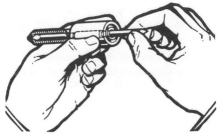

Fig. O138 — Clean nozzle valve seat of 1850 injector nozzle using the brass seat scraper.

Fig. O139 — Remove carbon from the dome cavity of 1850 injector nozzle using the small hand tool furnished in kit.

196. **OVERHAUL.** The maintenance of absolute cleanliness in the overhaul of injector assemblies is of utmost importance. Of equal importance is the avoidance of nicks, scratches or handling damage on any of the lapped or polished surfaces. To avoid damage, only the recommended cleaning kits and oil base carbon solvents should be used. The nozzle valve and body are individually fit and hand lapped, and these two parts must always be kept together and never inter-mixed.

Before disassembling a set of injectors, cap the pressure line connections with a line nut having the hole soldered shut, or with a special metal cap, and immerse the units in a clean carbon solvent. While the injectors are soaking, clean the work area and remove any accumulation of discarded parts from previous service jobs. Remove injectors one at a time from the solvent and thoroughly clean the outer surfaces with a brass wire brush. Rinse the injector in clean diesel fuel and disassemble as follows:

Clamp the injector body in a holding fixture, or soft-jawed vise, with only enough pressure to keep injector from slipping. Remove the protective cap, loosen locknut, then back off the pressure adjusting screw until all tension is removed from pressure spring. Remove the nozzle holder nut (2—Fig. O135) and withdraw the nozzle valve assembly. If valve cannot be removed from nozzle body with the fingers, return the assembly to the carbon solvent. A special nozzle valve extractor (Fig. O129) which uses hydraulic pressure from tester pump is available to remove valves which are stuck. Valve can sometimes be removed with a pair of pliers but care should be exercised not to damage the parts. Examine the lapped pressure faces (A, B & D—Fig. O136) of nozzle body and holder for nicks or scratches, and the piston surfaces (C) of nozzle valve and body for scratches, wear or scoring. Clean the fuel gallery (H)

with the special hooked scraper tool as shown in Fig. O137 by applying side pressure while body is rotated. Clean the nozzle valve seat with the brass seat scraper as shown in Fig. O138. Polish the seat with a pointed wooden polishing stick and a small amount of tallow. Use the special scraper to remove any carbon in dome cavity behind the spray holes as shown in Fig. O139, then clean the spray holes, using a 0.010 cleaning needle or broach, held in a pin vise as shown in Fig. O140. Leave only about $\frac{1}{16}$-inch of cleaning wire protruding from vise to resist bending. Clean the fuel passage with a small drill or similar tool as shown in Fig. O141.

Remove adjusting screw (11), washer (10), spring (9) and spindle (8), then clean and thoroughly inspect these parts. Renew spring if distorted, cracked or if there are rust spots on spring. Renew spindle if seat for spring or lower end of spindle are chipped, worn or cracked. Connector (7) need not be removed unless there is evidence of leakage past sealing washer (6) or if connector is damaged. Thoroughly flush all parts in clean diesel fuel and reassemble leaving adjusting screw loose until nozzle assembly is reinstalled.

As parts are cleaned, immerse in clean diesel fuel in a compartmented pan. Insert the nozzle valve into body underneath the fuel level to prevent the inclusion of dust particles between piston surfaces of nozzle valve and body. If a reverse flushing attachment is available, flush the nozzle assembly after valve is inserted, then immediately assemble the injector. Make sure the sealing surfaces (A, B & D—Fig. O136) are absolutely clean and that nozzle assembly is properly installed on dowels (E). Tighten the nozzle holder nut to a torque of 50 ft.-lbs., using the special nozzle nut socket if available. Retest the injector as previously outlined. If injector fails to meet the tests, disassemble and reclean or renew the nozzle and any other parts suspected of being faulty.

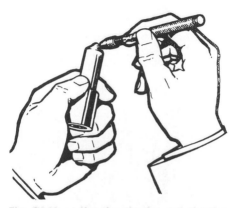

Fig. O140 — Use the pin vise and cleaning needle to clean the nozzle spray holes of 1850 injector nozzle.

Fig. O141 — Clean carbon and gum from the fuel feed hole of 1850 injector nozzle using the small drill as shown.

## INJECTOR NOZZLES

### Series 1750-1950-T

WARNING: Fuel leaves the injection nozzles with sufficient force to penetrate the skin. When testing, keep your person clear of the nozzle spray.

196A. **TESTING AND LOCATING A FAULTY NOZZLE.** If one engine cylinder is misfiring it is reasonable to suspect a faulty injector. Generally, a faulty injector can be located by running the engine at low idle speed and loosening, one at a time, each high pressure line at injector. As in checking spark plugs in a spark ignition engine, the faulty unit is the one that least affects the engine operation when its line is loosened.

Remove the suspected injector as outlined in paragraph 196B. If a suitable nozzle tester is available, test injector as outlined in paragraphs 196D through 196G. If a tester is not available, reconnect pressure line to injector and with nozzle tip directed where spray will do no harm, crank engine and observe spray pattern. The conical spray pattern should be symmetrical and without heavy lines. If spray pattern is ragged or wet, or

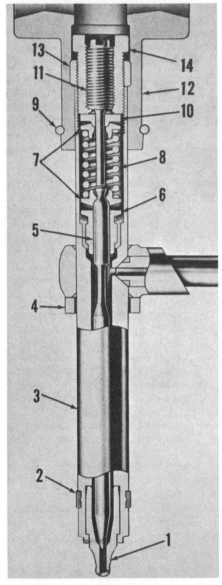

Fig. O141A—Cross sectional view of the "pencil" type Roosa-Master injector. Nozzle tip (1) and valve guide (6) are parts of finished body and are not serviced separately.

| | |
|---|---|
| 1. Nozzle tip | 9. Boot clamp |
| 2. Carbon seal | 10. Ball washer |
| 3. Nozzle body | 11. Lift adjusting |
| 4. Seal washer |     screw |
| 5. Nozzle valve | 12. Boot |
| 6. Valve guide | 13. Lock nut |
| 7. Spring seat | 14. Pressure adjusting |
| 8. Pressure spring |     screw |

if nozzle dribbles, the nozzle valve is not seating properly and the unit should be overhauled or renewed.

**196B. REMOVE AND REINSTALL.** To remove an injector, remove hood and wash injector, lines and surrounding area with clean diesel fuel. On series 1950-T, remove exhaust manifold and turbocharger. Use hose clamp pliers to expand clamp and pull leak-off boot from injector. Disconnect high pressure line, then cap all openings. Remove cap screw from nozzle clamp and remove clamp and spacer. Pull injector from cylinder head.

NOTE: Unless the carbon stop seal has failed causing injector to stick, the injectors can be easily removed by hand. If injectors cannot be removed by hand, use Oliver nozzle puller JDE-38 and be sure to pull injector straight out of bore. DO NOT attempt to pry injector from cylinder head or damage to injector could result.

When installing injector, be sure nozzle bore and seal washer seat are clean and free of carbon or other foreign material. Install new seal washer and carbon seal on injector and insert injector into its bore using a slight twisting motion. Install and align locating clamp then install holddown clamp and spacer and tighten cap screw to 20 ft.-lbs. torque.

Bleed system as outlined in paragraph 176 if necessary.

**196C. TESTING.** A complete job of nozzle testing and adjusting requires the use of an approved nozzle tester. Only clean, approved testing oil should be used in the tester tank. The nozzle should be tested for spray pattern, opening pressure, seat leakage and back leakage (leak-off). Injector should produce a distinct audible chatter when being tested and cut off quickly at end of injection with a minimum of seat leakage.

NOTE: When checking spray pattern, turn nozzle about 30 degrees from vertical position. Spray is emitted from nozzle tip at an angle to the centerline of nozzle body and unless injector is angled, the spray may not be completely contained by the beaker. Keep your person clear of the nozzle spray.

**196D. SPRAY PATTERN.** Attach injector to tester and operate tester at approximately 60 strokes per minute and observe the spray pattern. A finely atomized spray should emerge at each of the four nozzle holes and a distinct chatter should be heard as tester is operated. If spray is not symmetrical and is streaky, or if injector does not chatter, overhaul injector as outlined in paragraph 196H.

**196E. OPENING PRESSURE.** The correct opening pressure for series 1750 is 2550-2650 psi for used nozzles, or 2750-2850 psi for new nozzles or when new spring has been installed. The correct opening pressure for series 1950-T is 2750-2850 psi for used nozzles, or 2950-3050 psi for new nozzles or when new spring has been installed. There should not be more than 100 psi difference between the injectors used in any engine except when intermixing new and used nozzle assemblies.

To check nozzle opening pressure, attach injector to tester and pump tester several times to actuate injector and allow valve to seat normally. Now observe tester gage and quickly pump up pressure until gage falls off rapidly. This indicates the nozzle opening pressure and should be within the pressure ranges given above.

If opening pressure is not correct but nozzle will pass all other tests, adjust opening pressure as follows: Loosen the pressure adjusting screw lock nut (13—Fig. O141A), then hold the pressure adjusting screw (14) and

Fig. O141B — Disassembled view of the "pencil" type Roosa-Master injector. Refer to Fig. O141A for parts identification.

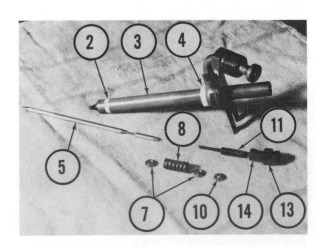

back out the valve lift adjusting screw (11) at least one full turn. Actuate tester and adjust nozzle pressure by turning adjusting screw as required. With the correct nozzle opening pressure set, gently turn the valve lift adjusting screw in until it bottoms, then back it out ½-turn which will provide the correct valve lift of 0.009. Hold pressure adjusting screw and tighten lock nut to a torque of 70-75 in.-lbs.

NOTE: A positive check can be made to see that the lift adjusting screw is bottomed by actuating tester until a pressure of 250 psi above nozzle opening pressure is obtained. Nozzle valve should not open.

**196F. SEAT LEAKAGE.** To check nozzle seat leakage, proceed as follows: Attach injector on tester in a horizontal position. Raise pressure to approximately 2400 psi (1750), or 2600 psi (1950-T), hold for 10 seconds and observe nozzle tip. A slight dampness is permissible but should a drop form in the 10 seconds, renew the injector or overhaul as outlined in paragraph 196H.

**196G. BACK LEAKAGE.** Attach injector to tester with tip slightly above horizontal. Raise and maintain pressure at approximately 1500-1700 psi and observe leakage from return (top) end of injector. After first drop falls, the back leakage should be 5 to 8 drops every 30 seconds. If back leakage is excessive, renew injector or overhaul as outlined in paragraph 196H.

**196H. OVERHAUL.** First wash the unit in clean diesel fuel and blow off with clean, dry compressed air. Remove carbon stop seal and sealing washer. Clean carbon from spray tip using a brass wire brush. Also, clean carbon or other deposits from carbon seal groove in injector body. DO NOT use wire brush or other abrasives on the Teflon coating on outside of nozzle body between the seals. Teflon coating can be cleaned with a soft cloth and solvent. Coating may discolor from use, but discoloration is not harmful.

Clamp the nozzle in a soft jawed vise, loosen lock nut (13—Fig. O141A) and remove pressure adjusting screw (14), ball washer (10), upper spring seat (7), spring (8) and lower spring seat (7).

If nozzle valve (5) will not slide from body when body is inverted, use a suitable valve extractor; or reinstall on nozzle tester with spring and lift adjusting screw removed, and use hydraulic pressure to remove the valve. See Fig. O141B.

Nozzle valve and body are a matched set and should never be intermixed. Keep parts for one injector separate and immerse in clean diesel fuel in a compartmented pan, as unit is disassembled.

Clean all parts thoroughly in clean diesel fuel using a brass wire brush and lint-free wiping towels. Hard carbon or varnish can be loosened with a suitable, non-corrosive solvent.

Clean the spray tip orifices first with an 0.008 cleaning needle held in a pin vise; then with a 0.011 (1750) or a 0.012 (1950-T) needle.

Clean the valve seat using a Valve Tip Scraper and light pressure while rotating scraper. Use a Sac Hole Drill to remove carbon from sac hole.

Piston area of valve and guide can be lightly polished by hand, if necessary, using Roosa Master No. 16489 lapping compound. Use the valve retractor to turn valve. Move valve in and out slightly while rotating, but do not apply down pressure while valve tip and seat are in contact.

If back leakage was not correct and polishing of valve does not bring unit within specifications, a new valve can be installed.

Valve and seat are ground to a slight interference angle. Seating areas may be cleaned up, if necessary, using a small amount of 16489 lapping compound, very light pressure and no more than 3-5 turns of valve on seat. Thoroughly flush all compound from valve body after polishing.

NOTE: Never lap a new valve to an old injector body seat. If lapping with the old valve did not restore seat, lapping with a new valve will only destroy angle of the new valve.

When assembling the nozzle, back lift adjusting screw (11—Fig. O141A) several turns out of pressure adjusting screw (14), and reverse disassembly procedure using Fig. O141A as a guide.

Connect nozzle to tester and turn pressure adjusting screw (14) into nozzle body until opening pressure is 2550-2650 psi with a used spring, or 2750-2850 psi with a new spring for series 1750, or 2750-2850 psi with a used spring, or 2950-3050 psi with a new spring for series 1950-T.

After spring pressure has been adjusted and before tightening lock nut (13), valve lift must be adjusted as follows: Hold pressure adjusting screw from turning and using a small screwdriver, thread lift adjusting screw (11) into nozzle unit until it bottoms on valve (5). To be sure screw is bottomed, actuate tester

handle until a pressure 250 psi above opening pressure is obtained. Nozzle valve should not open.

With lift adjusting screw bottomed, turn screw counter-clockwise ½-turn to establish the recommended 0.009 valve lift. Tighten the lock nut while holding pressure adjusting screw from turning. Recheck the opening pressure; then recheck spray pattern, seat leakage and back leakage. Install the unit as outlined in paragraph 196B.

NOTE: When adjusting opening pressure and injector has not been disassembled, back out the lift adjusting screw at least one full turn before moving pressure adjusting screw. This will prevent accidental bottoming of lift screw while attempting to set the pressure.

## FUEL INJECTION PUMP
### Series 1750-1800-1950-T

197. **PUMP TIMING.** To check injector pump timing, shut off the fuel supply valve and remove timing window cover from fuel injection pump. Turn engine so that No. 1 piston is coming up on compression stroke; then, continue to turn engine slowly until the 8 degree BTDC timing mark (Series A tractors) or the 2 degree BTDC timing mark (Series B and C tractors) or the 4 degree mark (Series 1750 and 1950-T) on engine flywheel is aligned with pointer. Note: Early model tractors may have "FP" (firing point) stamped on flywheel instead of the timing mark tape. Use "FP" mark when timing these tractors. The timing marks in pump timing window should then be exactly aligned as shown in Fig. O142. If the pump timing marks are not in register as shown, loosen the three injection pump mounting stud nuts and rotate the pump in the slotted mounting holes as necessary to align the timing marks. While holding pump in this position tighten the three mounting stud nuts securely. Then, either turn the engine through two revolutions or turn backward about ¼-turn, then forward until flywheel timing mark is again aligned with pointer and recheck timing marks in pump timing window. Readjust timing if necessary. CAUTION: **Do not** loosen the pump mounting stud nuts while engine is running if setting timing by dynamometer method.

NOTE: When aligning flywheel timing mark with pointer, engine must be turned in normal direction of rotation to align mark with pointer. If mark is turned past pointer, continue to turn the engine through two revolutions to again align mark and pointer.

or back engine up at least ¼-turn and then turn forward to timing mark. Turning the engine backwards to just the point where timing mark is aligned with pointer will result in incorrect timing due to gear backlash.

198. **REMOVE AND REINSTALL.** Thoroughly clean outside of pump, lines and connections. Shut off the fuel supply valve, remove timing window cover from fuel injection pump and turn engine so that both timing marks are visible in timing window and the 8 degree BTDC (Series A) or 2 degree BTDC (Series B and C) or 4 degree BTDC (Series 1750 and 1950-T) flywheel timing mark is aligned with pointer. Then reinstall injection pump timing window cover.

Disconnect battery ground strap to prevent engine from being turned accidently while pump is removed; then, proceed as follows: Disconnect throttle rod, stop control wire and cable, fuel supply line and excess fuel return line from fuel injection pump. Remove the injector pressure line clamps and disconnect the lines at connections near pump end of lines. Loosen the injector pressure line connections at injectors so lines can be spread to remove pump and immediately cap all open lines and connectors. Protective caps and plugs are available and are listed in Oliver parts catalog. Remove the three pump mounting stud nuts and remove pump from mounting studs and drive shaft, spreading the injector lines to allow pump to be moved rearward. After removing pump, remove seals from pump drive shaft and carefully inspect the shaft. If shaft shows any sign of damage, renew shaft as outlined in paragraph 199.

NOTE: Aligning flywheel and pump timing marks prior to removing pump is not required procedure, but is an aid in reinstalling pump. If pump drive shaft is removed or renewed in conjunction with pump re-

moval, either reinstall pump without regard to timing procedure and reinstall drive shaft as outlined in paragraph 200, or install shaft in pump and reinstall pump and shaft as an assembly as outlined in paragraph 201.

Before reinstalling pump, check to be sure engine was not turned while pump was removed. If engine was turned, or pump was removed without aligning timing marks as outlined in removal procedure, proceed as follows: Either remove the rocker arm cover and turn engine until intake valve of No. 1 cylinder just closes, or loosen the No. 1 cylinder injector mounting stud nuts and turn engine until compression leak occurs around the loosened injector. Then, continue to turn engine until the 8 degree BTDC (Series A) or 2 degree BTDC (Series B and C) or 4 degrees BTDC (Series 1750 and 1950-T) flywheel timing mark is aligned with pointer and reinstall the rocker arm cover, or remove the No. 1 cylinder injector and reinstall same with new sealing washer. Note: Early model tractors may have "FP" (firing point) stamped on flywheel instead of the timing mark tape. Use "FP" mark when timing these tractors.

Remove the timing window cover from outer side (throttle control rod side) of fuel injection pump and with a screwdriver inserted in rotor drive slot, turn pump rotor so that pump timing marks are aligned. Install two new pump drive shaft seals on shaft with lips of seals facing away from each other. Lubricate the seals and shaft area between seals with Lubriplate or equivalent grease and carefully install the pump over drive shaft and seals. Take care not to roll lip of rear seal back as pump is being installed; if this happens, remove pump and renew the seal. (Note: Use of Roosa-Master seal installation tool 13369 is recommended to install seals on shaft and use of Roosa-Master seal compressor tool 13371 is recommended

when installing pump over the drive shaft seals.) When pump is installed over drive shaft far enough to engage drive shaft tang with slot in pump rotor, it may be necessary to rock the rotor slightly with pencil eraser tip inserted through timing window or to rock the pump assembly slightly to engage the drive shaft tang and rotor slot. When pump mounting slots are engaged with the mounting studs and pump body contacts mounting plate, rotate the pump assembly in the mounting slots as required to bring the pump timing marks into exact alignment, then securely tighten the pump mounting nuts. Turn the engine through two revolutions, or back engine up at least ¼-turn, then slowly turn forward until flywheel timing mark is again aligned with pointer and recheck pump timing marks. Loosen the pump mounting stud nuts and adjust timing if necessary.

Bleed the fuel system as outlined in paragraph 176.

199. **RENEW PUMP DRIVE SHAFT.** To remove the pump drive shaft, gear and hub assembly, remove the gear cover from front side of timing gear cover. If injection pump is installed, shut off the fuel supply valve and remove the timing window cover from outer side of injection pump to drain fuel; otherwise, diesel fuel will run into engine when shaft is removed. Then, withdraw the drive gear, hub and shaft assembly from opening in front of timing gear cover. Cut the locking wire and remove the drilled head cap screws retaining gear to hub. Clamp tang of drive shaft in soft jawed vise and remove the hub retaining nut. Pull or press hub from shaft and Woodruff key. If fuel injection pump was not removed, install new drive shaft as outlined in paragraph 200. If pump is removed, new shaft can be installed in pump and pump with shaft installed as outlined in paragraph 201, or if desired, pump can be reinstalled without shaft and the shaft can then be installed as in paragraph 200.

200. To install new drive shaft with pump installed on tractor, proceed as follows: Check to see that injection pump is mounted so that mounting studs are centered in slots; if not, loosen the stud nuts, turn pump so that studs are centered in slots and tighten the nuts. Turn engine until No. 1 piston starts up on compression stroke, then continue to turn engine

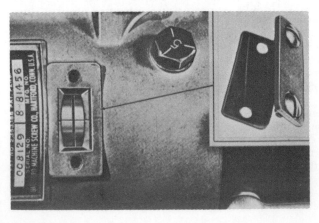

Fig. O142 — When No. 1 cylinder is on compression stroke and proper flywheel timing mark is aligned with pointer, injection pump timing marks should be aligned as shown on series 1750, 1800 and 1950-T fuel injection pump.

slowly until 8 degree BTDC (Series A) or 2 degree BTDC (Series B and C) or 4 degree BTDC (Series 1750 and 1950-T) flywheel timing mark is aligned with pointer. Note: Early tractors may have "FP" (firing point) stamped on flywheel instead of the timing mark tape. Use "FP" mark when timing these tractors. Clamp tang of drive shaft in soft jawed vise, install Woodruff key in slot and install drive hub with retaining nut and washer. Tighten the nut to a torque of 35-40 ft.-lbs. Remove shaft from vise and install new drive shaft seals with seal lips facing away from each other. Use of Roosa-Master seal installation tool 13369 is recommended. Lubricate the seals and shaft area between seals with Lubriplate or similar grease. Note the offset dimple in end of drive tang on shaft and in slot of pump rotor. Carefully insert shaft into pump with offset dimples on shaft tang and rotor slot aligned and work the rear seal lip into bore of pump body with a dull pointed tool. When drive tang is engaged in rotor slot, remove timing window cover from fuel injection pump and turn the drive shaft and hub so that pump timing marks are aligned as shown in Fig. 0142. While holding pump rotor from turning with screwdriver inserted in slot below the timing marks in timing window, install drive gear on hub and turn gear counter-clockwise as far as possible to eliminate gear backlash. Install the two drilled head capscrews in set of two holes that are aligned in gear and hub, tighten the capscrews securely and install locking wire. Insert thrust spring and thrust button in front end of drive shaft and install drive gear cover to timing gear cover with new gasket. Recheck pump timing marks and adjust if necessary as outlined in paragraph 197. Bleed diesel fuel system as in paragraph 176.

201. To install new drive shaft when fuel injection pump is removed, either reinstall pump and then install shaft as outlined in paragraph 200, or proceed as follows:

Remove timing window cover from outer side of fuel injection pump. Install new seals on drive shaft with lips of seals facing away from each other. Use of Roosa-Master seal installation tool 13369 is recommended. Lubricate seals and shaft area between seals with Lubriplate or similar grease, then carefully insert shaft into pump so that offset dimple in end of shaft tang is aligned with offset dimple in slot of pump rotor. Using a dull pointed tool, carefully work

lip of rear seal into bore of pump housing, or compress seal with Roosa-Master seal compressor tool 13371 when installing shaft. With drive tang of shaft engaged in slot of pump rotor, install the pump on tractor with mounting hole slots centered on the mounting studs. Insert Woodruff key in slot in pump shaft. Be sure that drive gear in a free fit on hub, then install hub, washer and retaining nut on shaft. Temporarily install drive gear on hub in any position and with the drilled head capscrews finger tight. Then, tighten the hub retaining nut to a torque of 35-40 ft.lbs. and remove the drive gear. Turn engine so that No. 1 piston is coming up on compression stroke, then continue to turn engine slowly until the 8 degree BTDC (Series A) or 2 degree BTDC (Series B and C) or 4 degree BTDC (Series 1750 and 1950-T) flywheel timing mark is aligned with pointer. Note: Early tractors may have "FP" (firing point) stamped on flywheel instead of timing mark tape. Use "FP" mark when timing these tractors. Turn injection pump drive shaft so that the pump timing marks are aligned as shown in Fig. 0142. While holding pump from turning with screwdriver inserted in slot below the pump timing marks, install pump drive gear on hub and turn gear as far counter-clockwise as possible to eliminate all gear backlash. Install the two drilled head capscrews into a set of holes in drive gear and hub that are aligned, tighten the capscrews and secure with locking wire. Insert the thrust spring and thrust button in front end of pump drive shaft and install drive gear cover to timing gear cover with new gasket. Recheck pump timing as outlined in paragraph 197 and readjust if necessary. Bleed the diesel fuel system as outlined in paragraph 176.

202. **GOVERNOR ADJUSTMENT.** Both the high idle speed and low idle speed adjustments are preset at the factory and normally should not be disturbed. However, should it become necessary, adjust the governed speeds as follows:

Start engine and bring to normal operating temperature. Break the seals on the high idle (rear) adjusting screw and/or low idle adjusting screw as necessary (refer to Fig. 0143); then, proceed as follows:

To adjust high idle speed, place throttle lever in high idle position, loosen lock nut and turn adjusting screw in or out as required to obtain a high idle speed of 2200 RPM on 1800 series A models, 2425 RPM on

1800 series B and C models, 2640 RPM on series 1750 models, or 2650 RPM on 1950-T models. Tighten the lock nut and install new wire seal.

To adjust low idle speed, position throttle lever at low idle position, loosen lock nut and turn adjusting screw (front screw) in or out to obtain a low idle speed of 625-675 RPM on all 1800 models or 800 RPM for 1750 and 1950-T models. Tighten lock nut and install new wire seal.

NOTE: The throttle control lever on fuel injection pump is spring loaded and should slightly over-travel the high and low speed stops on the injection pump governor control shaft. If necessary to adjust the throttle linkage, proceed as follows: With engine not running, disconnect throttle control rod at fuel injection pump and place hand throttle lever in low idle position. Loosen the lock nut on front end of control rod. While holding the injection pump throttle shaft against the low idle speed stop, adjust length of control rod so that injection pump throttle lever must be moved $\frac{1}{16}$-$\frac{1}{8}$ inch against spring pressure to reinstall the linkage pin. Then tighten lock nut on control rod and move the hand throttle lever to high idle speed position. Loosen the lock nut on stop screw in stop arm (Series A and B), or left hand steering gear support (series C and 1750), or bellcrank (series 1950-T), adjust stop screw until the injection pump lever over-travels high idle speed stop $\frac{1}{16}$ to $\frac{1}{8}$-inch, then tighen lock nut.

If hand throttle lever tends to creep at high idle speed, tighten the friction assembly located at top end of hand lever shaft (series A and B) or at right hand end of lever shaft (series C, 1750 and 1950-T).

Fig. O143 — Adjusting the series 1750, 1800 and 1950-T high speed screw. Low idle speed is adjusted by front screw.

## Series 1850 Diesel

The C.A.V. fuel injection pump used on series 1850 diesel tractors is a self-contained distributor type unit which includes the engine governor as well as all the components required for properly metering and delivering the fuel. The pump is vertically mounted on left side of engine and is driven by a vertical shaft which receives its drive from the auxiliary drive shaft.

This section will include removal, reinstallation and timing of the injection pump as well as information on the injection pump adapter plate, pump vertical drive shaft, transfer pump pressure check and pump advance check.

203. **TIMING.** Injection pump to engine timing may be checked by either aligning the injection pump timing marks with number one piston at 28° BTDC on compression stroke as outlined in paragraph 203A or by removing the fuel injection pump as outlined in paragraph 204, then aligning the timing marks on fuel injection pump drive shaft and injection pump adapter as outlined in paragraph 203B.

CAUTION: Injection pumps are sealed at high idle no-load adjusting screw, inspection hole cover and governor control linkage cover. Oliver Corporation states: "SEALS BROKEN BY UNAUTHORIZED PERSONEL DURING WARRRANTY PERIOD WILL VOID WARRANTY ON ENGINE AND FUEL INJECTION EQUIPMENT". Timing the fuel injection pump at 28° BTDC involves removing the inspection hole cover; therefore, on tractors in warranty, timing should be checked with inspection pump removed as outlined in paragraph 203B.

203A. TIMING WITH PUMP INSTALLED. **Before proceeding with the following, refer to CAUTION preceding this paragraph.**

Shut off the fuel supply valve and remove cover from timing hole in right front side of flywheel housing and remove inspection window from fuel injection pump. Turn engine so that the number one piston is coming up on compression stroke, then continue turning engine slowly until the "STATIC 28" mark on flywheel is aligned with pointer in timing hole. Note: In event that flywheel marking is turned past pointer, reverse crankshaft rotation for about ¼-turn, then turn forward until mark is aligned. This assures that backlash in timing gear train is taken up.

At this time, the "F" mark on injection pump rotor should be aligned with scribe line on snap ring, or if snap ring does not have a scribe line, with square end of snap ring as shown in Fig. 0145; if not, proceed as follows: Check to see if scribe mark on pump mounting flange is aligned with similar mark on pump adapter; loosen the pump mounting stud nuts and turn pump to align scribe marks if they are not aligned, then retighten nuts. If "F" mark is not yet aligned with square end of snap ring, remove tachometer drive adapter and auxiliary drive gear cover from front side of timing gear cover. Loosen the three socket head cap screws and turn auxiliary drive shaft so that "F" mark on pump rotor is aligned with scribe line on snap ring or square end of snap ring. While holding the auxiliary drive shaft in this position, turn auxiliary drive shaft gear as far as possible in clockwise direction (as viewed from front of engine) to remove all gear backlash, then tighten the socket head cap screws. Bleed diesel fuel system as outlined in paragraph 180 after reinstalling removed covers and tachometer drive adapter.

203B. TIMING BY REMOVING FUEL INJECTION PUMP. First remove fuel injection pump as outlined in paragraph 204; then, proceed as follows:

Turn engine so that number one piston is coming up on compression stroke. Remove cover from timing hole in front side of flywheel housing, then continue turning engine slowly until TDC mark on flywheel is aligned with pointer in timing hole. The slots in adapter plate and vertical shaft (see Fig. 0151A) should then be in exact alignment. Note: In event that flywheel TDC mark is turned past the timing pointer, reverse crankshaft rotation for about ¼-turn, then turn forward until mark is aligned. This will assure that backlash in timing gear train is taken up.

If the slots in pump adapter and vertical drive shaft are not exactly aligned when the TDC mark on flywheel is aligned with pointer, proceed as follows: Remove the tachometer drive adapter and auxiliary drive gear cover from front of timing cover. Loosen the three socket head cap screws and turn auxiliary drive shaft until the slots in adapter and vertical shaft are aligned. While holding the auxiliary drive shaft in this position, turn auxiliary drive gear as far as possible in a clockwise direction (as viewed from front of engine) to remove all gear backlash, then tighten the three socket head cap screws. Reinstall the gear cover, tachometer drive adapter, flywheel timing hole cover and reinstall fuel injection pump as outlined in paragraph 204.

203C. **TRANSFER PUMP PRESSURE AND INJECTION PUMP ADVANCE.** The transfer pump pressure

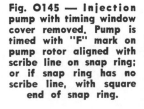

**Fig. O145 — Injection pump with timing window cover removed. Pump is timed with "F" mark on pump rotor aligned with scribe line on snap ring; or if snap ring has no scribe line, with square end of snap ring.**

**Fig. O144 — When checking injection pump transfer pressure and injection pump automatic advance, install pressure gage and advance gage as shown. Refer to text for specifications.**

and automatic advance of injection pump can both be checked during the same operation. Prior to making these checks, inspect all points between fuel tank and injection pump to be sure there are no air leaks, fuel leaks or restrictions.

Refer to Fig. 0144. Remove lower (outside) bleed screw assembly and install a length of hose connected to a gage capable of registering at least 100-150 psi in the opening. Remove the small plug from the automatic advance piston spring cap and install C. A. V. Advance Gage (No. 7044-896). Start engine and run until fuel is at operating temperature.

With the equipment installed as outlined above and the fuel at operating temperature, the following conditions should exist:

Transfer pump pressure at
150 engine rpm
(cranking) . . . . . . . . . .10 psi min.
2400 engine no-load rpm. .70-80 psi
Advance movement degrees @
engine rpm
Start . . . . . . . . . . . . . . .600-1000 rpm
2¼-3 degrees, total. . . . . . .2400 rpm

**204. REMOVE AND REINSTALL.** When injection pump is originally installed and timed, scribe lines are placed on the injection pump mounting flange and the pump mounting adapter as shown in Fig. 0147. If pump is removed for routine service, it may be reinstalled without regard to crankshaft position by aligning the master spline of injection pump rotor with master spline of pump quill

shaft and then aligning the mark on injection pump mounting flange with the mark on pump mounting adapter.

To remove the injection pump, thoroughly clean injection pump and connections with clean diesel fuel or solvent. Shut off fuel and disconnect throttle and fuel stop linkage. Disconnect fuel supply, bleed-back and injector lines from injection pump and immediately cap off all openings. Remove the three mounting stud nuts and lift injection pump from engine. If pump quill shaft is lifted out with injection pump, be sure not to drop it.

Prior to installing pump, check timing as outlined in paragraph 203B. Install by reversing the removal procedure and be sure master splines of quill shaft and injection pump rotor are aligned when installing pump on mounting studs. Align scribe lines of pump mounting flange and adapter before tightening stud nuts.

Complete installation be reconnecting all lines and linkage and bleed fuel system as outlined in paragraph 180.

**205. GOVERNOR.** The governor is integral with the injection pump; hence, only the speed and linkage adjustments will be covered in this section.

**206. LOW IDLE.** Start engine and run until it reaches operating temperature then, disconnect the top end of governor control rod (link) (GR— Fig. 0147) from the bellcrank (BC) located above injection pump. Loosen idle speed adjusting screw lock nut, hold throttle in idle position and ad-

just idle speed screw (LS) to obtain the recommended low idle speed of 650 rpm. and tighten lock nut. Push hand throttle toward low idle position until throttle shaft arm under instrument panel butts against its stop, then if necessary, adjust the clevis on rear of rear (horizontal) control rod (CR) until the governor control rod (link) (GR) will freely enter its hole in the bellcrank (BC) when injection pump throttle lever is held in the idle position. Hand throttle shaft arm and injection pump idle speed adjusting screw must meet their stops at the same time.

**207. HIGH IDLE.** Refer to **CAUTION** in paragraph 203 before adjusting the engine high idle speed.

Be sure engine low idle speed is adjusted before adjusting the engine high idle speed.

To adjust the engine high idle speed, first lift the battery door at rear of hood and turn the stop screw, located in the left steering support, to its minimum position (height). Break seal and remove cap from high idle adjusting screw (HS), then loosen adjusting screw lock nut. Disconnect governor control rod (link) (GR) from the bellcrank (BC) located above injection pump. Be sure engine is at operating temperature, then hold throttle in high idle position and adjust the injection pump high idle adjusting screw until engine is operating at the recommended 2565 rpm. Tighten the adjusting screw lock nut. Hold the injection pump throttle lever in the high idle position, then move throttle hand lever as required until governor control rod (link) (GR) will enter hole in bellcrank freely. With throttle hand lever positioned as outlined, back-out the stop screw in the left hand steering gear support until it contacts stop on throttle shaft arm and tighten lock nut. Throttle shaft arm and injection pump high idle adjusting screw must meet their stops at the same time.

NOTE: Do not change the length of the rear control rod (CR) as it will alter the engine low idle speed adjustment.

**208. INJECTOR PUMP ADAPTER PLATE AND DRIVE SHAFT.** The injection pump receives its drive from a quill shaft (13—Fig. 0150) which connects the injection pump rotor to the vertical drive shaft which in addition to driving the injection pump, also drives the engine oil pump. The vertical drive shaft is driven by the auxiliary drive shaft which receives its drive from a gear that is part of

**Fig. O147 — View of series 1850 diesel engine injection pump showing governor adjusting screws, timing marks and control linkage. Note seal on timing window cover and the cap and seal on high idle speed adjusting screw.**

BC. Bellcrank
BS. Bleed screws
CR. Rear control rod
FS. Fuel stop lever
GR. Governor control rod
HS. High idle adjusting screw
LS. Low idle adjusting screw
OJ. Oil jet tube
TM. Timing marks
TW. Timing window cover

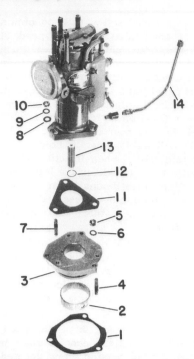

**Fig. O150 — Exploded view of adapter plate. Hole in bushing (2) must align with hole in adapter (3).**

| | |
|---|---|
| 1. Gasket | 11. Gasket |
| 2. Bushing | 12. Snap ring |
| 3. Adapter plate | 13. Quill shaft |

**Fig. O151 — Exploded view of auxiliary drive shaft and vertical drive shaft assemblies. Shims may be located under bearing (11).**

1. Retainer plate
2. Drive gear
3. Auxiliary drive shaft
4. Worm gear
5. Vertical drive shaft
6. Oil tube
7. Front bushing
8. Rear bushing
9. Oil seal
10. Thrust washers
11. Bearing

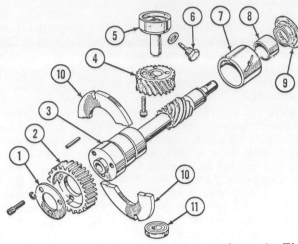

the engine timing gear train. The injection pump adapter plate can be serviced after injection pump is removed and the vertical drive shaft can be serviced after auxiliary drive shaft is removed.

209. ADAPTER PLATE. To remove the injection pump adapter plate, remove injection pump as outlined in paragraph 204, then remove the three stud nuts and lift adapter plate from engine. Bushing (2—Fig. O150) can now be removed, if necessary. Inside diameter of new bushing is 2.3125-2.3143 and should provide 0.001-0.0048 operating clearance for upper bearing journal of vertical drive shaft. When installing a new bushing, use a piloted driver and be sure hole in bushing mates with hole in adapter plate.

210. VERTICAL DRIVE SHAFT. To remove the vertical drive shaft, first remove injection pump as outlined in paragraph 204, then remove the auxiliary drive shaft as outlined in paragraph 116. If a suitable withdrawal tool is available, the vertical shaft, gear and bearing can be removed without removing the oil pump. However, if tool is not available, or if bearing is tight in its bore, remove oil pan and oil pump assembly and carefully bump vertical drive shaft from crankcase. Should there be shims located below vertical shaft bearing

(11—Fig. O151) be sure to save them for subsequent reinstallation.

With shaft removed from tractor it can be disassembled by pulling bearing, removing gear retaining cap screws and pressing shaft from worm gear.

All parts are available for service. Upper bearing journal of vertical shaft (5) has an outside diameter of 2.3095-2.3115 and should have 0.001-0.0048 operating clearance in injection pump adapter plate bushing. Pay particular attention to the bronze worm gear (4) and renew gear if it shows signs of damage or if backlash between the worm gear and auxiliary drive shaft exceeds 0.003 (0.001-0.003 recommended). When assembling shaft assembly, press worm gear on shaft with flanged side toward bearing end of shaft, secure with the retaining cap screws, then press bearing on shaft until it butts against gear.

To install vertical shaft assembly in crankcase, reinstall any shims that were present, align master spline in I.D. of drive shaft with master spline of oil pump drive shaft and press assembly into crankcase until it bottoms. Install auxiliary drive shaft and timing gear cover, but prior to installing the auxiliary drive shaft gear, align the injection pump drive gear and adapter plate markings (slots) as follows: Turn engine, if necessary, until number one cylinder is at TDC on the compression stroke. Place injection pump adapter plate on mounting studs so scribe line is nearest the outer mounting stud, then turn auxiliary drive shaft until notch in upper bearing journal of vertical shaft aligns with the vertical slot in the adapter plate. See Fig. O151A. Install auxiliary shaft drive gear retaining cap screws being careful not to move the slots of vertical drive shaft and adapter plate out of alignment. Check snap ring on injection pump quill shaft,

align master spline as shown in Fig. O151B and place quill shaft in the vertical drive shaft, then install injection pump and bleed fuel system.

## ENERGY CELLS

### Series 1800

These assemblies are mounted directly opposite from the fuel injectors in the cylinder head, as shown in Fig. O152. Oliver Corporation catalogs the cell body (100) separately. It is suggested, however, that both of these parts be renewed if either one requires renewal.

In almost every instance where a carbon-fouled or burned energy cell is encountered, the cause is traceable either to a malfunctioning injector, incorrect fuel or incorrect installation of the energy cell. Manifestations of a fouled or burned unit are misfiring, exhaust smoke, loss of power or pronounced detonation (knock).

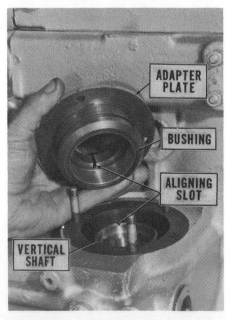

**Fig. O151A — View showing adapter plate positioned for installation. Note aligning slots which must mate. Refer to text.**

**211. REMOVE AND REINSTALL.** Any energy cell can be removed without removing any engine component. To remove an energy cell, first remove the threaded cell holder plug (97) and take out the cell holder spacer (98). With a pair of thin nosed pliers, remove cell cap (99). To remove the cell body, screw a $\frac{15}{16}$-20 NF bolt into the threaded end of the cell body. A nut and collar on the bolt will make it function as a puller. If puller is not available, remove the fuel injector; and use a brass rod to drift the cell body out of the cylinder head. Clean unit as outlined in paragraph 212 and reinstall by reversing the removal procedure.

**212. CLEANING.** Clean all carbon from front and rear crater of cell body using a brass scraper or a shaped piece of hard wood. Clean the exterior of the energy cell with a brass wire brush, and soak the parts in carbon solvent. Reject any part which shows signs of leakage or burning. If parts are not burned, re-lap sealing surfaces (99 and 100) by using a figure 8 motion on a lapping plate coated with fine compound.

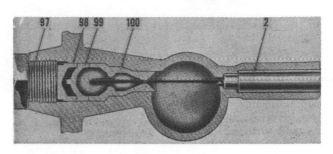

Fig. O152—Top sectional view showing location of 1800 series energy cell which is directly opposite injector.

> 2. Injector
> 97. Cell holder
> 98. Spacer
> 99. Cell cap
> 100. Cell body

Refer to Fig. O153. Loosen bail clamp from plunger, then disengage ends of bail wire from primer pump body. Remove plunger, plunger guide and piston seal from body. Place pump body in vise, then remove valve nut. Seals, valves and spring can now be removed.

Examine all parts for tears, undue wear, or other damage and renew as necessary. If plunger guide is to be renewed, old guide can be cut. Lubricate new guide and slide over plunger with chamfer toward inside of pump body. Install valves so that they open in the same direction as the arrow embossed on the primer pump body.

Bleed system as outlined in paragraph 176.

### PRIMER PUMP

### Series 1800 (Prior Ser. No. 129 285-000)

**213. R&R AND OVERHAUL.** Shut off fuel, disconnect inlet and outlet lines from primer pump and cap off open ends of fuel lines. Remove the two mounting screws and remove pump from tractor.

### FUEL PUMP
### Series 1750-1800

**214.** Beginning with tractor serial number 129 286-000, 1800 series B, all series C and series 1750 diesel tractors were equipped with a diaphragm type fuel supply pump. The pump is mounted on right side of engine and is actuated by a lobe on the engine camshaft. The pump has a priming lever incorporated into the pump.

Overhaul of pump is conventional and will be obvious after an examination. Note that the pump valves are staked in position at four places.

### Series 1850

**215.** Series 1850 diesel tractors are equipped with a diaphragm type fuel supply pump as shown in Fig. O154. Pump is mounted on right side of engine and is actuated by a lobe on the engine camshaft. The pump has

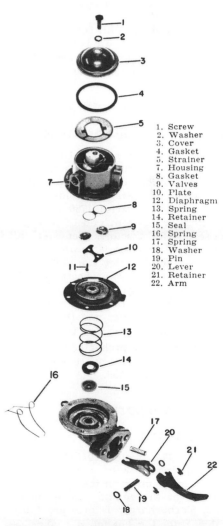

1. Screw
2. Washer
3. Cover
4. Gasket
5. Strainer
7. Housing
8. Gasket
9. Valves
10. Plate
12. Diaphragm
13. Spring
14. Retainer
15. Seal
16. Spring
17. Spring
18. Washer
19. Pin
20. Lever
21. Retainer
22. Arm

Fig. O151B — When installing quill shaft, align master spline with master spline of vertical drive shaft.

MASTER SPLINE

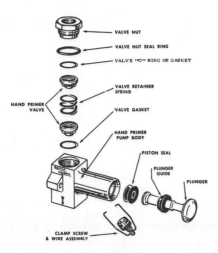

VALVE NUT
VALVE NUT SEAL RING
VALVE "O" RING OR GASKET
VALVE RETAINER SPRING
HAND PRIMER VALVE
VALVE GASKET
HAND PRIMER PUMP BODY
PISTON SEAL
PLUNGER GUIDE
PLUNGER
CLAMP SCREW & WIRE ASSEMBLY

Fig. O153 — Disassembled view of hand primer used on early 1800 tractors.

Fig. O154 — Exploded view of fuel supply pump used on series 1850 diesel tractors. Valves (9) are retained in housing (7) by plate (10).

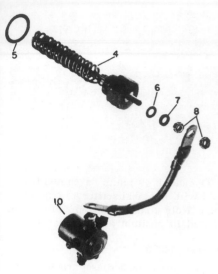

**Fig. O155—View showing component parts of the manifold preheater typical of those used on diesel tractors.**

| | |
|---|---|
| 4. Heating coil | 7. Washer |
| 5. Gasket | 8. Nuts |
| 6. Insulator | 10. Switch (solenoid) |

a priming lever incorporated into the pump.

Overhaul of the unit is conventional and will be obvious after an examination. Note that valves (9) are held in place by a retaining plate (10).

## MANIFOLD PREHEATER

### Series 1750-1800-1850-1950-T

216. The inlet manifold (except 1950-T) of diesel engines is fitted with a solenoid controlled preheater unit which is installed in lower side of manifold of series 1750 and 1800 tractors, or in front end of manifold on series 1850 tractors. Preheater unit for 1950-T tractors is located in the turbocharger to inlet manifold pipe.

Operation of the preheater can be checked by depressing the control switch located on the instrument panel for about 30 seconds and then touching manifold to see if it is warm.

Service on the preheater and solenoid is accomplished by renewing the units. See Fig. 0155.

## TURBOCHARGER

### Series 1950-T

216A. The series 1950-T tractors are equipped with a turbocharger which is mounted on the top forward portion of the engine exhaust manifold. Turbocharger is pressure lubricated from the engine oil cooler assembly and with engine running at governed speed, the oil pressure to turbocharger should be within 10 psi

of the engine oil pressure and a minimum of 10 psi should be maintained at engine low idle rpm. In addition to proper lubrication, it is also extremely important that no leaks or restrictions exist in the air induction system. Because of the critical tolerances machined into the turbocharger, and the high speeds of the rotating members, any dirt or other debris entering the turbocharger, either in the incoming air or the lubricating oil, can quickly cause the unit to fail.

Regular inspections should be made of the turbocharger and related components. Run engine and check for leaks in the air inlet system and exhaust system. Leaks can often produce noise. If noise is present and no leaks are evident, check rotating members of turbocharger for interference with housing and for faulty bearings. If oil is present on compressor wheel or in housing, oil is probably being pulled or pushed from center housing and faulty seals and/or bearings are indicated. Turbocharger rotating members must spin freely with no drag. Be sure unit is being properly lubricated as indicated in preceding paragraph; also be sure the check valve of oil inlet line is operating correctly to prevent oil drain back of the oil inlet line.

Certain operating procedures must be observed with a turbocharged engine that are not normally considered with a naturally aspirated engine and the following precautions should be taken.

1. If engine oil filters have been changed, oil lines disconnected, service performed on turbocharger, or if engine has been shut down for a considerable length of time, pull fuel shut-off (no fuel position) and crank engine until oil pressure registers on gage. Start engine and run at about 1000 rpm until oil pressure is normal. Do not run engine at high rpm until gage registers at least 20 psi.

2. Prior to engine shut-down, let engine idle for 3 to 5 minutes

to allow turbocharger to settle out. Immediate shut-down of engine could result in oil starvation of turbocharger bearings.

3. If engine has been serviced, or turbocharger has been removed, the Oliver Corporation recommends that engine be run for about one hour before installing turbocharger. Remove oil pressure line from engine oil cooler and plug hole. Clean line and check valve. This will assure cleanliness of lubricating oil and exhaust system.

NOTE: At this time, no parts are being furnished to service turbocharger and any service required will be done through a factory rebuild exchange program. Rebuilt units will carry new parts warranty. Refer to paragraph 216B for removal and reinstallation procedure.

216B. **REMOVE AND REINSTALL.** To remove turbocharger, first remove hood side panels and hood, then disconnect air inlet hose from turbocharger. Unbolt and remove muffler elbow and adapter ring. Disconnect oil inlet and outlet lines from turbocharger and cap lines. Unbolt and remove turbocharger from adapter block. Remove and discard gasket.

Prior to installing turbocharger, refer to item #3 of paragraph 216A. Install turbocharger as follows: Invert turbocharger and pour oil into return line port to fill center housing. Turn rotating member to coat bearings and thrust washer with oil. Be sure oil pressure line is clean, then fill line with oil. Install new gasket on adapter block, coat threads of mounting studs with high temperature thread lubricant, then install turbocharger and connect oil pressure and return lines. Connect air inlet hose. Place adapter ring in muffler elbow and install elbow and ring with ring in groove of turbine housing. Install hood and hood side plates.

Prior to starting engine, refer to item 1 of paragraph 216A.

# NON-DIESEL GOVERNOR

All series 1750, 1800 and 1850 non-diesel engines are fitted with a flyweight type governor, mounted on front face of the timing gear cover and driven by the engine camshaft gear. Some changes were made in the governors used on Series A (prior serial number 124 396-000) tractors

and the changes are obvious after an examination of Figs. 0159 and 0160. Beginning with Series B (serial number 124 396-000 and up) tractors, a governor having two levers (throttle lever and governor speed control lever) is used on all non-diesel engines. See Fig. 0161.

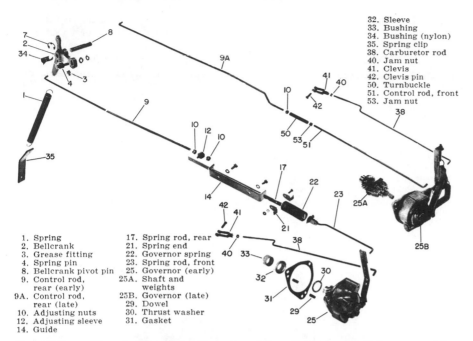

32. Sleeve
33. Bushing
34. Bushing (nylon)
35. Spring clip
38. Carburetor rod
40. Jam nut
41. Clevis
42. Clevis pin
50. Turnbuckle
51. Control rod, front
53. Jam nut

1. Spring
2. Bellcrank
3. Grease fitting
4. Spring pin
8. Bellcrank pivot pin
9. Control rod, rear (early)
9A. Control rod, rear (late)
10. Adjusting nuts
12. Adjusting sleeve
14. Guide
17. Spring rod, rear
21. Spring end
22. Governor spring
23. Spring rod, front
25. Governor (early)
25A. Shaft and weights
25B. Governor (late)
29. Dowel
30. Thrust washer
31. Gasket

Fig. O156 — View showing early (lower) and late (upper) type linkage assemblies. Refer to text for adjustment.

Throttle linkage travel stops are used to limit travel of linkage at the high idle position. Series A tractors have an adjustable arm attached to lower end of throttle hand lever shaft. Series B tractors have the stop incorporated into the throttle hand lever lower bearing and stop is not adjustable. Series C and series 1750 and 1850 tractors have a stop screw located in the left hand steering gear support.

Overhaul procedure for all governors is similar and any differences which may occur will be obvious.

### ADJUSTMENT
### Series 1750-1800-1850

217. **LINKAGE.** Normally linkage will not need adjustment unless linkage has been disassembled or damaged.

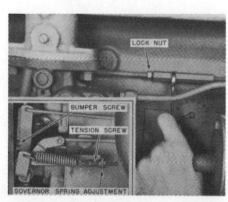

Fig. O157 — Make sure spring arm is horizontal as shown when adjusting engine high idle rpm on late type governors. Regulate spring tension to control engine rpm.

If linkage has been disassembled, use Fig. 0156 as a guide for assembly and adjust linkage as follows: Place hand throttle lever in high idle position, loosen lock nut (40) on the governor to carburetor rod (38), then disconnect the rod from governor lever. Hold governor lever in high idle position and turn rod as required until rod end is $\frac{1}{16}$-inch forward of hole in governor lever which on single lever governors will be the top hole. Reinstall rod in governor lever and tighten lock nut.

218. **LOW IDLE.** With linkage adjusted as outlined in paragraph 217, start engine and run until it reaches operating temperature. Place hand throttle lever in low idle position and adjust the carburetor stop screw to obtain the following low idle engine speeds.

1800 Series A Gas...........325 rpm
1800 Series A LPG...........500 rpm
1800 Series B & C Gas......350 rpm
1800 Series B & C LPG......500 rpm
1850 Series Gas & LPG......400 rpm
1750 Series Gas ...........400 rpm

219. **HIGH IDLE.** With engine at operating temperature and low idle speed adjusted, check and adjust the engine high idle speed as follows:

Place hand throttle lever in high idle position and check engine high idle rpm which should be 2175-2225 (2200 desired) for 1800 series A (serial number 90 525-000—12 395-000); 2400-2450 (2425 desired) for 1800 series B and C (serial number 124 396-000—150 420-000) or 2615-2665 (2640 de-

sired) for 1850 series (serial number 150 421-000—) and series 1750.

If adjustment is required, stop engine and place throttle in low idle position, then loosen lock nut and back out bumper screw until governor operating lever is released. See Fig. 0158.

On 1800 Series A tractors, start engine and place hand throttle lever in high idle position, loosen and adjust nuts (10—Fig. 0156) as required to obtain the stated engine high idle rpm. Tighten lock nuts and check to see that stop screw in stop arm is against the steering column when engine is operating at high idle. Adjust bumper screw as outlined in paragraph 220.

On 1800 series B tractors, pull the hand throttle lever toward the high idle position until arm on lower end of hand lever shaft butts against stop pin of the lower bearing assembly, then if necessary, loosen lock nuts and turn turnbuckle (50) until the governor spring arm is in a horizontal position as shown in Fig. 0157. Start engine and adjust governor spring eye screw to obtain the stated engine high idle rpm. Tighten all lock nuts and adjust bumper screw as outlined in paragraph 220.

On 1800 series C and series 1750 and 1850 tractors, push hand lever forward until stop of arm on left end of hand throttle shaft butts against stop screw located in left hand steering gear support. Check governor spring arm which should be in a horizontal position. If governor spring arm is not set a horizontal position, lift battery door and adjust the stop screw in left hand steering gear support as necessary. Start engine and adjust governor spring eye screw to obtain the stated engine high idle rpm. Tighten all lock nuts and adjust bumper screw as outlined in paragraph 220.

220. **BUMPER SCREW.** Engine surging at high idle rpm can be eliminated by adjusting the bumper screw as follows. Stop engine, refer to Fig. 0158, loosen bumper screw lock nut and turn bumper screw inward slightly. Restart engine and recheck for surging. Continue this operation until engine surging is eliminated. DO NOT turn bumper screw in further than necessary as the low idle rpm of engine may be increased. Tighten lock nut. Refer to paragraphs 218 and recheck the engine low idle rpm.

BUMPER SCREW

LOCK NUT

Fig. O158 — Bumper screw is located as shown. Refer to text for adjustment.

## R & R AND OVERHAUL

### Series 1750-1800-1850

221. Governor can be removed after disconnecting linkage and removing the governor to timing gear cover retaining cap screws. Carefully withdraw the unit from timing gear cover.

NOTE: Do not allow the governor gear thrust washer to drop off hub of governor gear. To do so may require removal of timing gear cover or oil pan.

Refer to Figs. O159, O160 and O161 for exploded views of the governors used. Governor gear can be removed by using a suitable puller. Disassembly of weight unit and thrust bearing assemblies is evident. On early governors (Fig. O159), pin (13) can be driven from governor fork (12) after removing soft plug (5). All bushings are presized and should need no final sizing if carefully installed. All shafts should have an operating clearance of 0.0015-0.002 in the housing bushings.

Governor gear hub rotates in a sleeve and bushing assembly which is pressed into front face of crankcase. The sleeve and bushing assembly can be renewed by removing timing gear cover and using a suitable puller. Inside diameter of governor gear bushing is 1.0015-1.002. Outside diameter of governor gear hub is 0.999-0.9995 and should have 0.002-0.003 operating clearance in bushing. Renew bushing and/or gear when operating clearance exceeds 0.005.

NOTE: About March 1, 1966, a one-piece governor bushing was installed in all production non-diesel engines replacing the previously used two-piece governor sleeve and bushing. The governor gear thrust washer (3—Fig. O161) is NOT used with the one-piece governor bushing.

When installing the one-piece bushing, oil groove is toward rear of engine and front of bushing must protrude $\frac{3}{16}$-inch from front

Fig. O159 — Exploded view of the early governor used on 1800 series A tractors. Refer also to Fig. O160.

1. Lever
2. Oil seal
3. Bushing
4. Housing
5. Expansion plug
6. Expansion plug
7. Expansion plug
8. Bushing
9. Bumper screw
10. Jam nut
11. Bumper spring
12. Fork
13. Groove pin
14. Thrust bearing
15. Thrust sleeve
16. Shaft and carrier
17. Weight
18. Weight pin
18A. Weight pin
19. Retainer
19A. Retainer
20. Spacer
21. Gear

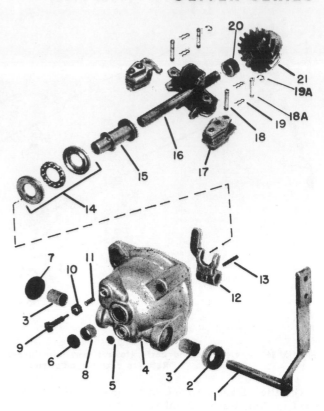

Fig. O160 — Exploded view of late governor used on 1800 series A tractors. Some units may have spacer (4) brazed to shaft and carrier (8). Refer also to Fig. O159.

3. Gear
4. Spacer
5. Retainer
6. Weight pin
7. Weight
8. Shaft and carrier
9. Thrust sleeve
10. Thrust bearing
11. Lever and shaft
12. Spring pin
13. Retainer
14. Oil seal
15. Ball bearing
16. Housing
17. Fork
19. Bushing
20. Bumper screw
21. Jam nut
22. Bumper spring
23. Bushing

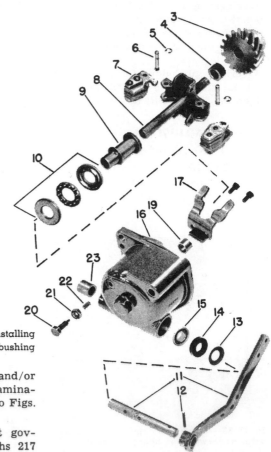

of cylinder block. Be careful when installing the one-piece bushing as depth of bushing is critical.

Any further disassembly and/or overhaul is obvious after an examination of the unit and reference to Figs. O159, O160 and O161.

After reinstallation, readjust governor as outlined in paragraphs 217 through 220.

# COOLING SYSTEM

### RADIATOR

#### Series 1750-1800-1850-1950-T

222. To remove the radiator, first drain cooling system, then remove hood side panels and hood and front lower panels. On tractors fitted with an oil cooler (or coolers) unbolt same from radiator (Fig. 082A). Unbolt fan blades from hub and let fan rest in fan shroud. Disconnect upper and lower hoses, unbolt radiator from tractor front frame and lift same from tractor. Grille need not be disturbed.

### WATER PUMP

#### All Series 1750-1800-1850 Non-Diesel-1950-T

223. R&R AND OVERHAUL. To remove water pump, drain cooling system and remove left side panel. Unbolt fan blades from hub and let fan blades rest in fan shroud. Remove fan belt, disconnect lower radiator hose from pump, then unbolt and remove pump from left side of tractor.

To disassemble the water pump, refer to Fig. 0162 and proceed as follows: Remove pump pulley and pump cover. Remove snap ring from front of shaft, then using a press, push the shaft and bearing assembly out of impeller and pump body. Seal, seal seat and slinger can now be renewed. Shaft and bearing assembly are available as a unit only.

When reinstalling impeller, press same on shaft until aft side is flush with aft end of pump shaft.

Note: Pumps used prior to engine serial number 1126152 do not have seal seat (12). If impeller is being renewed, be sure impeller hub is same length as original, or be sure to install seal seat if a later impeller with shorter hub is installed.

#### Series 1850 Diesel

224. R&R AND OVERHAUL. To remove water pump, first drain cooling system and remove hood side panels. Remove the oil cooler to filter oil lines and the fan belt. Remove fan blades and let blades rest in the fan shroud. Disconnect oil cooler lines, and if so equipped, the Hydra-Power Drive heat exchanger lines, from water pump. Disconnect water pump inlet hose, then unbolt and remove water pump and spacer from cylinder block.

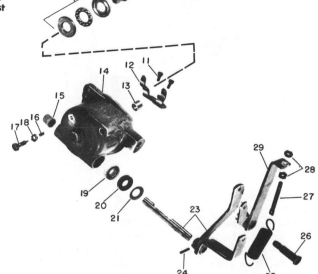

Fig. 0161 — Exploded view of the latest two lever governor used on 1800 series B and C and 1850 tractors. Units for series 1750 are similar except that thrust washer (3) is not used.

3. Thrust washer
4. Gear
5. Retainer ring
6. Weight pin
7. Weight
8. Shaft and spider
9. Thrust sleeve
10. Thrust bearing
12. Fork
13. Bushing
14. Housing
15. Bushing
16. Bumper spring
17. Bumper screw
18. Jam nut
19. Ball bearing
20. Oil seal
21. Seal retainer
23. Throttle lever
24. Spring pin
25. Governor spring
26. Shoulder screw
27. Adjusting eye
28. Jam nuts
29. Speed change lever
30. Shaft

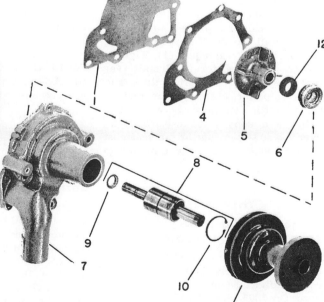

Fig. 0162 — Exploded view of water pump used on all series 1800 and non-diesel series 1850 tractors. Water pump for series 1750 and 1950-T has different body and pulley but internal parts are the same.

1. Cover
4. Gasket
5. Impeller
6. Seal
7. Body
8. Shaft assembly
9. Slinger
10. Snap ring
11. Pulley
12. Seal seat

To disassemble water pump, refer to Fig. 0163 and proceed as follows: Remove lock nut (1) and washer (2), then remove pulley (3) and key (4) from front end of pump shaft. Remove snap ring (5). Support body (11), then press on rear of shaft (12) and push shaft, bearings, spacer, seal flange, seal and seal retainer from impeller and pump body. Rear seal (13) can now be removed from pump body.

To reassemble water pump, first install seal retainer (10) with cup side forward and lip flush with edge of bearing bore. Install seal (9) and flange (8) with offset of flange inner

**Fig. O163 — Exploded view of water pump used on series 1850 diesel tractors. A spacer (not shown) is located behind body (11).**

1. Lock nut
2. Washer
3. Pulley
4. Key
5. Snap ring
6. Bearings
7. Spacer
8. Flange
9. Seal
10. Retainer
11. Body
12. Pump shaft
13. Seal
14. Impeller

### Series 1850 Diesel

226. To remove thermostat, remove hood side panels and hood. Drain coolant and disconnect upper radiator hose. Disconnect wire from temperature indicator sending unit. Disconnect engine oil cooler bracket from thermostat housing and remove air cleaner assembly. Complete removal of retaining cap screws, hold oil cooler forward and remove thermostat housing from front of cylinder head.

Thermostat can be removed from housing after removing the retaining snap ring. Be sure to check condition of the jiggle pin while thermostat housing is off.

Thermostat should start to open at 175 degrees F. and be fully open at 190 degrees F.

diameter toward seal. Install inner bearing (6) in body with sealed side toward seal (rear). Place bearing spacer (7) in body with chamfered end next to rear bearing, fill the cavity between spacer and body with water pump lubricant, then install front bearing (6) with shielded side of bearing forward and install snap ring (5). Support front bearing inner race, start pump shaft (12) into rear side of body and press shaft through bearings and spacer until it bottoms. Place left center cap screw in its hole, then install woodruff key (4), pulley (3), washer (2) and nut (1). Tighten nut to 55 ft.-lbs. torque. Install rear seal (13), then support pulley and press impeller (14) on rear end of shaft until clearance between impeller and body is 0.028-0.035; however, be sure bearings are in rearmost position when measuring impeller to body clearance.

### THERMOSTAT

### All Series 1750-1800-1850 Non-Diesel-1950-T

225. To remove thermostat, remove hood side panels and hood. Drain coolant and disconnect upper radiator hose. On 1800 series C, 1750, 1850 and 1950-T tractors also disconnect bypass hose and bleed line from thermostat housing. Remove cap screws from air cleaner bracket and remove air cleaner, thermostat housing and thermostat on all models except 1950-T which does not have air cleaner bracket on thermostat housing.

Thermostat may be either a bellows or a pellet type. Bellows type should start to open at 167-172 degrees F. and be fully open at 200 degrees F.

The pellet type should start to open at 175 degrees F. and be fully open at 190 degrees F.

# IGNITION AND ELECTRICAL SYSTEM

### SPARK PLUGS
### Series 1800

227. For normal or heavy duty service, the following spark plugs are recommended for use in the series 1800 gasoline engines. AC 44XLS, Autolite AG-42 or Champion N-11-Y. For LP-Gas engines, AC C44XL, Autolite AG4 or Champion N6 spark plugs are recommended.

Spark plug electrode gap is 0.025 for gasoline engines, or 0.020 for LP-Gas engines. Tighten plugs to 30 ft.-lbs. torque when installing.

### Series 1750-1850

228. For normal or heavy duty service, the following spark plugs are recommended for use in the series 1750 and 1850 gasoline engines. AC 44XLS, Autolite AG-42, Champion N-11-Y or Prestolite 14GT42. For LP-Gas engines, AC C44XL, Autolite AG4, Champion N6 or Prestolite 14G3 spark plugs are recommended.

Spark plug electrode gap is 0.025 for all plugs. Tighten spark plugs to 30 ft.-lbs. torque when installing.

### DISTRIBUTOR

### Series 1800

229. Non-diesel tractors prior to serial number 124 396-000 (Series A) are equipped with a Delco-Remy

model 1112603 distributor. Tractors serial number 124 396-000 and up (Series B and C) are equipped with a model 1112632 Delco-Remy distributor. Specification data follows:

**Distributor 1112603**
Breaker contact gap..........0.022
Breaker arm spring
   tension .................19-23 oz.
Cam angle degrees...........31-37
Start advance ..........,....0.5 @ 275
Intermediate advance.....2-4 @ 375
Maximum advance ....10-12 @ 1100
**Distributor 1112632**
Breaker contact gap..........0.016
Breaker arm spring
   tension .................19-23 oz.
Cam angle degrees...........31-34
Start advance ........0.7-3.5 @ 300
Intermediate advance..7.5-9.5 @ 850
Maximum advance ....11-13 @ 1200

Advance data is in distributor degrees and distributor rpm; double the listed values for flywheel degrees and rpm. Breaker arm spring tension is taken at rear of and at 90 degrees to, the contact point.

230. **IGNITION TIMING.** To set static timing, remove flywheel timing hole cover, then crank engine until number one piston is coming up on compression stroke. Continue cranking engine until the correct following timing marks align with the timing hole pointer: O degrees for gasoline, or 4 degrees BTDC for LP-Gas tractors prior to serial number 124 396-000

(Series A); 2 degrees BTDC for gasoline, or 7 degrees BTDC for LP-Gas tractors serial number 124 396-000 and up (Series B and C). Be sure distributor breaker points are set at 0.022 for Series A tractors, or 0.016 for Series B and C tractors, then loosen distributor clamp screw and rotate distributor in the direction opposite rotor rotation until breaker points just start to open. Tighten distributor clamp screw to retain distributor in this position.

NOTE: The foregoing procedure sets static timing; however, the preferred method is to use a timing light as outlined below. If a choice exists, always use a timing light.

In addition, LP-Gas mixtures may vary from place to place and consequently it may be necessary to deviate somewhat from both the static and timing light settings to reach optimum performance of the LP-Gas engines.

To set running timing, remove flywheel timing hole cover and install timing light. Start and run engine at high idle rpm. With timing light held close to the timing hole, the following BTDC timing marks should register with the timing pointer: 22 degree mark for gasoline, or 26 degree mark for LP-Gas engines of the Series A tractors; 24 degree mark for gasoline, or 28 degree mark for LP-Gas engines of the Series B and C tractors. If the proper timing mark does not register with the timing pointer, loosen distributor clamp screw and rotate distributor as necessary. Tighten distributor clamp screw and remove timing light.

### Series 1750-1850

231. Series 1750 and 1850 non-diesel tractors are equipped with a Holley D-2563AA ignition distributor which incorporates a vacuum advance unit. Refer to Fig. O164 for an exploded view.

Specifications are as follows:
Breaker contact gap............0.025
Cam angle @ idle speed......35°-38°
Centrifugal advance curve:

| Distributor rpm. | advance degrees |
|---|---|
| 225 | 0-1½ |
| 275 | ½-2¾ |
| 325 | 2-4 |
| 900 | 8-10 |
| 1200 | 11-13 |

Vacuum advance @ 1000 distributor rpm.:

| Inches Hg. | advance degrees |
|---|---|
| 7 | 0 |
| 10 | 0-3 |
| 13 | 3-6 |
| 16 | 6-8 |
| 20 | 6-8 |

**232. IGNITION TIMING.** To set static timing, remove flywheel timing hole cover, then crank engine until number one piston is coming up on compression stroke. Continue cranking engine until flywheel 0 degree mark for 1850 gasoline, or 2 degree BTDC mark for 1850 LP-Gas and 1750 gasoline tractors, aligns with timing hole pointer, then loosen distributor clamp screws and turn distributor in direction opposite to rotor rotation until breaker points just start to open. Tighten distributor clamp screws to retain distributor in this position.

NOTE: The foregoing procedure sets static timing; however, the preferred method is to use a timing light as outlined below. If a choice exists, always use a timing light.

In addition, LP-Gas mixtures may vary from place to place and consequently it may be necessary to deviate somewhat from both the static and timing light settings to reach optimum performance of the LP-Gas engines.

To check running timing, remove flywheel timing hole cover and connect timing light. Disconnect the vacuum advance line from distributor and plug open end of line. Start engine and run at 2400 rpm. With timing light held close to the timing hole, the 24 degree BTDC mark for 1850 gasoline or 26 degree BTDC mark for 1850 LP-Gas and 1750 gasoline on flywheel should register with the timing hole pointer. If the proper timing mark does not register with the timing pointer, loosen distributor clamp screws and rotate distributor as necessary. Tighten clamp screws, remove light and reconnect the distributor vacuum advance line.

### GENERATOR AND REGULATOR
### Series 1800

233. A Delco-Remy generator, model 1100400 or 1100419 and a Delco-Remy model 1118997 regulator are used on Series 1800 tractors. Specification data follows.

#### Generator 1100400-1100419
Brush spring tension.........28 oz.
Field draw @ 80° F.
  Volts ........................12
  Amperes ................1.5-1.62
Output (cold)
  Amperes (max.) ...........25
  Volts ......................14
  RPM ....................2710

#### Regulator 1118997
Cut-out relay
  Air gap ..................0.020
  Point gap ................0.020
  Closing voltage range....11.8-14.0
  Adjust to ................12.8

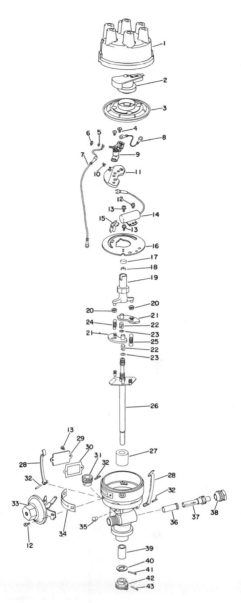

Fig. O164 — Exploded view of Holley distributor used on Series 1850 non-diesel tractors. The Holley distributor incorporates a vacuum spark advance as well as a centrifugal advance. Timing is checked with vacuum line disconnected and plugged. To remove shaft (26), it is first necessary to remove pin (32) from Tachourmeter drive gear (31) and pins (41 and 43) from lower end of shaft. Pin (32) is accessible after removing plate (29).

| | |
|---|---|
| 1. Cap | 26. Shaft and weight plate |
| 2. Rotor | 27. Bushing |
| 3. Dust cover | 28. Clamps |
| 7. Low tension lead | 29. Cover plate |
| 8. Ground wire | 30. Gasket |
| 9. Contact set | 31. Tachourmeter drive gear |
| 11. Breaker plate | 32. Pin |
| 14. Condensor | 33. Diaphragm assy. |
| 15. Spring | 34. Gasket |
| 16. Base plate | 35. Plug |
| 17. Wick | 36. Bushing |
| 18. Retainer | 37. Gear and sleeve |
| 19. Cam | 38. Bushing |
| 20. Bushings | 39. Bushing |
| 21. Weights | 40. Washer |
| 22. Bearing strip | 41. Pin |
| 23. Washer | 42. Coupling |
| 24. Primary spring | 43. Pin |
| 25. Secondary spring | |

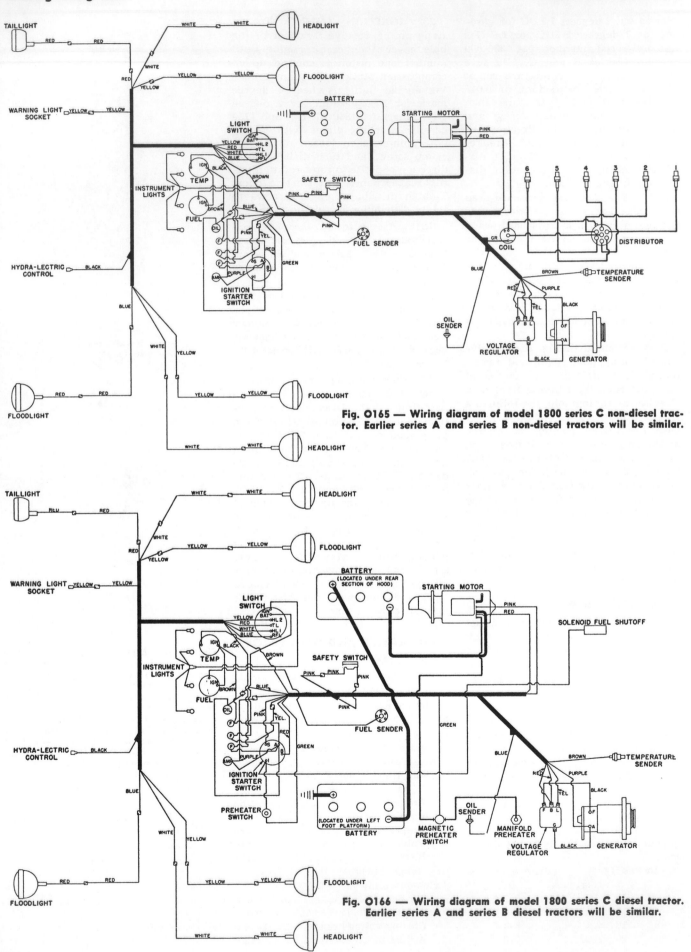

**Fig. O165 — Wiring diagram of model 1800 series C non-diesel tractor. Earlier series A and series B non-diesel tractors will be similar.**

**Fig. O166 — Wiring diagram of model 1800 series C diesel tractor. Earlier series A and series B diesel tractors will be similar.**

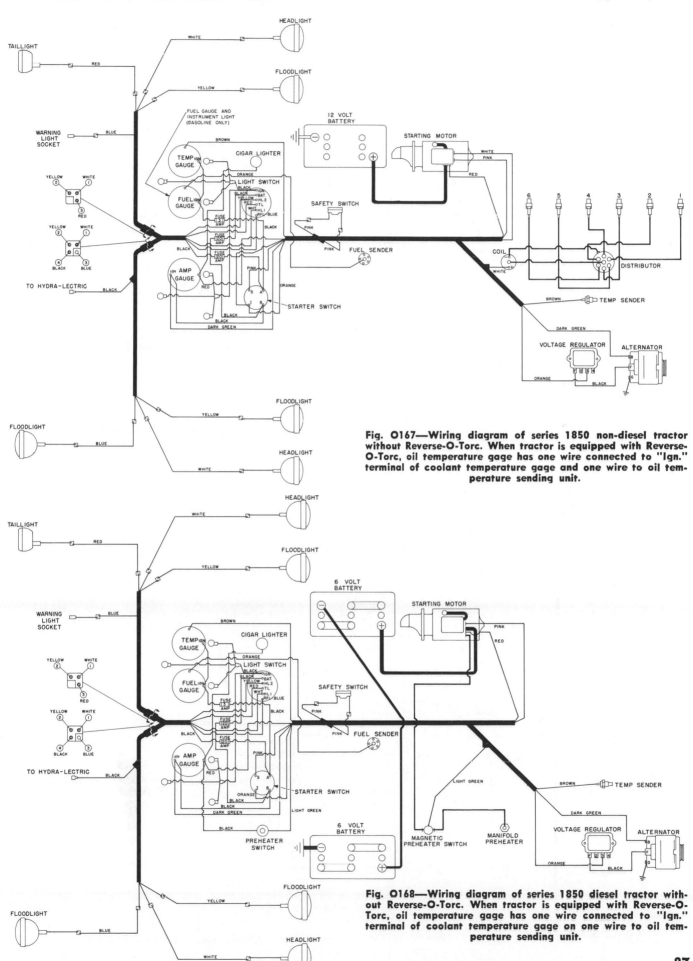

Fig. O167—Wiring diagram of series 1850 non-diesel tractor without Reverse-O-Torc. When tractor is equipped with Reverse-O-Torc, oil temperature gage has one wire connected to "Ign." terminal of coolant temperature gage and one wire to oil temperature sending unit.

Fig. O168—Wiring diagram of series 1850 diesel tractor without Reverse-O-Torc. When tractor is equipped with Reverse-O-Torc, oil temperature gage has one wire connected to "Ign." terminal of coolant temperature gage on one wire to oil temperature sending unit.

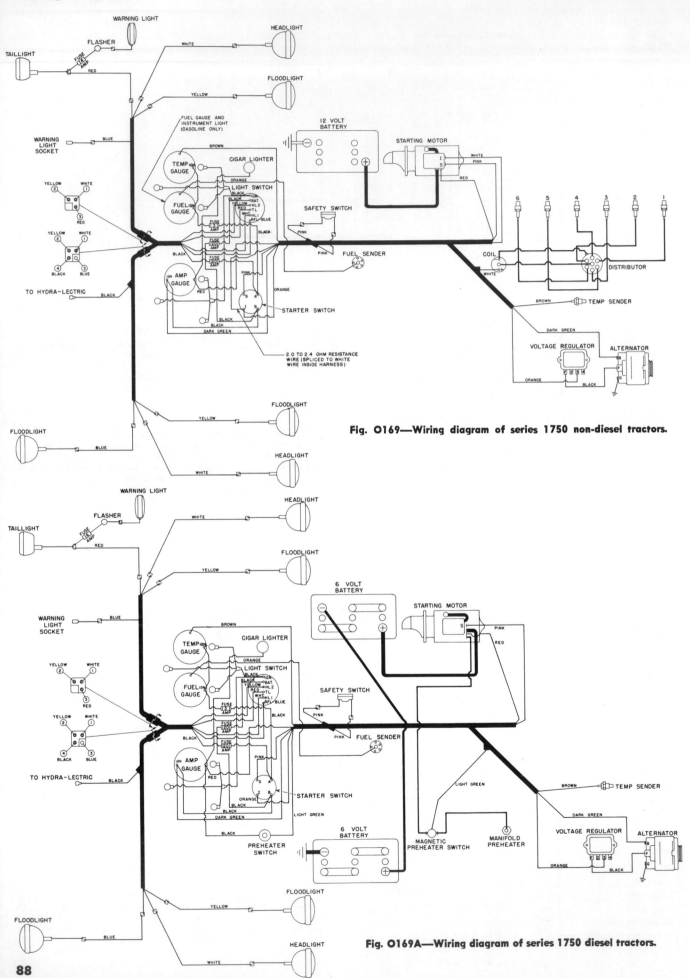

Fig. O169—Wiring diagram of series 1750 non-diesel tractors.

Fig. O169A—Wiring diagram of series 1750 diesel tractors.

Voltage regulator
  Air gap . . . . . . . . . . . . . . . . .0.075
  Voltage range . . . . . . . . .13.6-14.5
  Adjust to . . . . . . . . . . . . . . . .14.0
  Ground polarity . . . . . . . . .Positive

## ALTERNATOR AND REGULATOR

### Series 1750-1850-1950-T

CAUTION: An alternator (AC generator) is used to supply the charging current for the Series 1750, 1850 and 1950-T tractors. Because certain components of the alternator can be damaged by procedures that would not affect a DC generator, the following precautions must be observed.

1. When installing batteries or connecting a booster battery, be sure that the negative post of all batteries are grounded.
2. Never short across any of the alternator or regulator terminals.
3. Never attempt to polarize the alternator.
4. Always disconnect all battery straps before removing or installing any electrical unit.
5. Never operate alternator on an open circuit; be sure all leads are properly connected and tightened before starting engine.

234. Series 1850 tractors are equipped with a Delco-Remy model 1100725 (non-diesel) or 1100731 (diesel) alternator and a model 1119517 regulator. Series 1750 and 1950-T tractors are equipped with a Delco-Remy model 1100720 or 1100777 alternator and a model 1119517 regulator. Specification data follows:

**Alternator 1100720-1100725-1100731**
Ground . . . . . . . . . . . . . . .Negative
Field current @ 80°F.
  Amperes . . . . . . . . . . . . . . .2.2-2.6
  Volts . . . . . . . . . . . . . . . . . .12.0
Cold output @ specified voltage
  Specified volts . . . . . . . . . . .14.0
  Amps. @ 2000 rpm . . . . . . .21.0
  Amps. @ 5000 rpm . . . . . . . .30.0
  Rated output—hot . . . . . . . .32.0

**Alternator 1100777**
Ground . . . . . . . . . . . . . . . .Negative
Field current @ 80°F.
  Amperes . . . . . . . . . . . . . . .2.2-2.6
  Volts . . . . . . . . . . . . . . . . . .12.0
Cold output @ specified voltage
  Specified volts . . . . . . . . . .14.0
  Amps. @ 2000 rpm . . . . . . .32.00
  Amps. @ 5000 rpm . . . . . . .50.0
  Rated output-hot . . . . . . . .55.0

**Regulator 1119517**
Polarity . . . . . . . . . . . . . . .Negative
Field relay
  Air gap . . . . . . . . . . . . . . . .0.015
  Point opening . . . . . . . . . . .0.030
  Closing voltage range. . . . .1.5-3.2

Voltage regulator
  Air gap . . . . . . . . . . . . . . .0.067 (1)
  Point opening . . . . . . . . . . .0.015
  Voltage setting @ degree F.
    13.9-15.0 @ 65
    13.8-14.8 @ 85
    13.7-14.6 @ 105
    13.5-14.4 @ 125
    13.4-14.2 @ 145
    13.2-14.0 @ 165
    13.1-13.9 @ 185

(1) Starting point for point gap adjustment after bench repair. Subsequent adjustment is made if necessary so that operation on lower contacts is 0.05-0.40 volts lower than on upper contacts.

## STARTING MOTOR

### Series 1800

235. Delco-Remy starting motors, model numbers 1107682, 1113088, 1113098 and 1113165 have been used on series 1800 tractors. Specification data follows:

**Starter 1107682**
Brush spring tension (min.) . . .35 oz.
No-load test
  Volts . . . . . . . . . . . . . . . . . .10.6
  Max. amps. . . . . . . . . . . . . . .94
  RPM . . . . . . . . . . . . . . . . . .3240
Resistance test
  Volts . . . . . . . . . . . . . . . . . .3.5
  Min. amps. . . . . . . . . . . . . . .325
  Max. amps. . . . . . . . . . . . . . .390

**Starter 1113088-1113098-1113165**
Brush spring tension (min.) . . .48 oz.
No-load test
  Volts . . . . . . . . . . . . . . . . . .11.5
  Min. amps.* . . . . . . . . . . . . . .57
  Max. amps.* . . . . . . . . . . . . . .70
  Min. RPM . . . . . . . . . . . . . .5000
  Max. RPM . . . . . . . . . . . . . .7400
Lock test
  Volts . . . . . . . . . . . . . . . . . .3.4
  Amperes . . . . . . . . . . . . . . .500
  Torque (ft.-lbs.) . . . . . . . . . . .22
* Includes solenoid.

## STARTING MOTOR

### Series 1750-1850

236. The series 1850 tractors are equipped with a Delco-Remy model 1107358 (non-diesel) or a 1113656 (diesel) starting motor. The series 1750 tractors are equipped with a

Delco-Remy 1107358 (non-diesel) or a 1113098 or 1113139 (diesel) starting motor. The series 1950-T tractors are equipped with a Delco-Remy 1113656 starting motor. Specification data follows.

**Starter 1107358**
Brush spring tension (min.) . . .35 oz.
No-load test:
  Volts . . . . . . . . . . . . . . . . . .10.6
  Amps. (max.) . . . . . . . . . . . .94.0
  RPM (min.) . . . . . . . . . . . . .3240
Resistance test:
  Volts . . . . . . . . . . . . . . . . . .3.5
  Amps. (min.) . . . . . . . . . . . .325
  Amps. (max.) . . . . . . . . . . . .390

**Starter 1113098**
Brush spring tension (min.) . .48 oz.
No-load test
  Volts . . . . . . . . . . . . . . . . . .11.5
  Min. amps.* . . . . . . . . . . . . . .57
  Max. amps.* . . . . . . . . . . . . . .70
  Min. RPM . . . . . . . . . . . . . .5000
  Max. RPM . . . . . . . . . . . . . .7400
Lock test
  Volts . . . . . . . . . . . . . . . . . .3.4
  Amperes . . . . . . . . . . . . . . .500
  Torque (ft.-lbs.) . . . . . . . . . . .22
*Includes solenoid.

**Starter 1113139**
Brush spring tension (min.) . .80 oz.
No-load test
  Volts . . . . . . . . . . . . . . . . . .11.5
  Min. amps* . . . . . . . . . . . . . .57
  Max. amps.* . . . . . . . . . . . . . .70
  Min. RPM . . . . . . . . . . . . . .5000
  Max. RPM . . . . . . . . . . . . . .7400
Lock test
  Volts . . . . . . . . . . . . . . . . . .3.4
  Amperes . . . . . . . . . . . . . . .500
  Torque (ft.-lbs.) . . . . . . . . . . .22
*Includes solenoid.

**Starter 1113656**
Brush spring tension (min.) . . . . .80
No-load test:
  Volts . . . . . . . . . . . . . . . . . .11.6
  Amps. (min.) . . . . . . . . . . . .85.0*
  Amps. (max.) . . . . . . . . . . .125.0*
  RPM (min.) . . . . . . . . . . . . .5900
  RPM (max.) . . . . . . . . . . . . .8100
Lock test:
  Volts . . . . . . . . . . . . . . . . . .2.35
  Amps. . . . . . . . . . . . . . . . . .600.0
  Torque ft.-lbs. . . . . . . . . . . . .17.0
* Includes solenoid

# ENGINE AND COMPONENTS
## (Four Cylinder)

Oliver 1900 and 1950 tractors are fitted with a General Motors (Detroit Diesel) series 4-53 diesel engine. This four cylinder, two-cycle engine has a bore and stroke of 3⅞x4½ inches, with a displacement of 212.4 cubic inches and a 17 to 1 compression ratio. A Roots type blower unit is mounted on left side of engine and provides air for scavenging and combustion via an air box and ports in the cylinder liners.

Oliver Corporation does not catalogue repair parts for the engine and any required parts must be ordered from an authorized Detroit Diesel source.

Engine service information in this manual is limited to engine removal and installation and tune-up which includes governor adjustments. For other service information on the engine, consult the General Motors series 53 Maintenance Manual.

## R & R ENGINE WITH CLUTCH

### Series 1900-1950

240. On tractors equipped with Hydra-Power drive, Creeper drive or Reverse-O-Torc, the auxiliary drive unit must be removed along with the engine. The auxiliary drive unit and clutch housing can then be removed from engine to service the clutch and/or flywheel.

To remove engine, first drain cooling system and remove both hood side panels and hood.

Remove battery (or batteries), hood supports and battery tray from instrument panel support. Disconnect clutch spring from tank support. Unclip wiring loom and control cables from bottom of fuel tank, then disconnect leak-off line, water trap line and fuel supply line from tank. Attach hoist to tank and take up slack, then unbolt tank cradle, drive out hinge pin and remove fuel tank and cradle. Disconnect upper and lower radiator hoses from engine. Remove air cleaners and bracket. Remove fan shaft and fan shaft support as an assembly. Remove air intake manifold from blower and plug the blower opening. On models so equipped, disconnect steering shaft bearing from its supports on right side of clutch housing, then disconnect front steering shaft from steering gear and rear steering shaft and remove the front steering shaft from tractor. Disconnect emergency stop wire from blower and governor control rod from extension (cross) shaft arm. Disconnect fuel shut-off wire. Disconnect the necessary wires and cables from starting motor solenoid. Disconnect wires from coolant temperature and oil pressure sending units. Disconnect wires from generator and tachometer drive cable from adapter. Disengage all retaining clips and pull wires and cables rearward. Disconnect power steering lines from the clips beneath engine, if equipped with a hydraulic system, or from power steering pump, if so equipped.

Remove the pto drive shaft as outlined in paragraph 356, or the hydraulic pump drive shaft as outlined in paragraph 409 if tractor has no pto.

On models with no auxiliary drive unit, remove the center drive shaft shield and disconnect the coupling chain. Loosen clutch shaft coupler, slide coupler onto center drive shaft and remove center drive shaft. Do not lose spacer located between coupling sprockets. Remove clutch shaft from clutch housing.

On models equipped with an auxiliary drive unit, remove sprocket coupling chain and disconnect control linkage from drive unit.

Attach hoist to engine lifting brackets, remove engine mounting bolts and lift engine from tractor. Identify the two rear mounting bolts so they can be installed in their original positions as they act as engine aligning dowels.

Reinstall engine by reversing the removal procedure and bleed the power steering system and fuel system.

## TUNE-UP

### Series 1900-1950

241. Normally a tune-up on an engine in service will only involve making certain that the various adjustments have not changed. However, if cylinder head, injectors or governor have been serviced, it will be necessary to make a preliminary (cold) tune-up which will allow the engine to be started and brought to operating temperature at which time the valve clearance, injector timing and other linkage adjustments rechecked.

Engine tune-up should be performed in the following sequence.

1. Adjust valve clearance.
2. Time fuel injectors.
3. Adjust governor gap.
4. Adjust injector rack levers.
5. Adjust engine high idle.
6. Adjust engine low idle.
7. Adjust governor buffer screw.
8. Adjust governor booster spring.

When making the preliminary (cold) tune-up, perform the first four of the items listed above, then start engine and bring to operating temperature at which time all eight of the above listed items should be performed. Restart engine, if necessary, to maintain engine temperature during the tune-up operation.

Refer to the following paragraphs.

242. **ADJUST EXHAUST VALVES.** To adjust the clearance (cold) for exhaust valves on number one cylinder, remove hood, raise fuel tank, and remove tappet cover; then proceed as follows: Pull fuel shut-off control to the "no-fuel" position and turn engine in the normal direction of rotation until the number one injector follower is fully depressed. Loosen lock nuts on push rods and rotate push rods as required until clearance between valve stems and rocker arms is 0.011 cold for 1900 series A, B and early C with two valve cylinder heads, or 0.026 cold for late series 1900C and 1950 with four valve cylinder heads. Repeat the above operation for the remaining cylinders. NOTE: Exhaust valve clearance must be rechecked, and adjusted if necessary, to 0.009 (two valve head) or 0.024 (four valve head) after engine has been started and brought to operating temperature.

243. **TIME FUEL INJECTORS.** To time the fuel injector on number one cylinder, proceed as follows: Pull fuel shut-off control to the "no-fuel" position and turn engine in the normal direction of rotation until exhaust valves for the number one cylinder are fully depressed. Place the small end of the injector timing gage (General Motors tool J1853) in the hole provided in top of injector body. Loosen jam nut on push rod and rotate push rod until shoulder of tim-

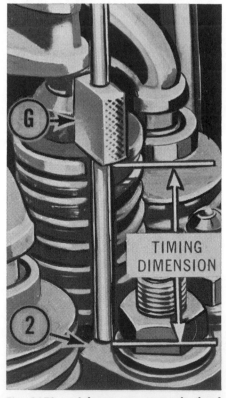

Fig. O170 — Injectors are correctly timed when shoulder of timing gage (G) will just pass over top of injector follower as shown. Refer to text.

ing gage will just pass over the top of injector follower as shown in Fig. O170. With injector set to this dimension (1.460 inches), hold push rod and tighten jam nut. Repeat the above operation for the remaining cylinders.

**244. ADJUST THE GOVERNOR GAP.** After timing the injectors and adjusting the exhaust valves, check and adjust, if necessary, the governor gap prior to adjusting the injector rack controls. Disconnect the fuel shut-off wire. Remove cover from governor control housing and place the speed control lever in high idle position. Measure gap between spring plunger and plunger guide as shown in Fig. O171. This gap should measure 0.006 and if adjustment is necessary, loosen lock nut and turn gap adjusting screw in or out as required. Tighten lock nut and recheck gap.

When installing cover on housing, be sure pin of throttle control lever fits into slot of the differential lever. Pull cover assembly away from engine to take up slack in screw holes and while holding cover in this position, tighten screws. Reconnect fuel shut-off wire.

Check and adjust the injector racks as outlined in paragraph 245.

**245. ADJUST INJECTOR RACK CONTROLS.** Disconnect clevis from governor speed control cross shaft, then loosen lock nut and back out buffer screw approximately ⅝-inch. Disconnect linkage from stop lever. Loosen all injector rack adjusting screws and be certain that all control levers are free on the control tube. Place governor control (shut-off)

lever in run position and the speed control (throttle) lever in the high idle position, then starting with rear injector, turn the inner adjusting screw of rack control lever downward until a slight increase in turning effort is noted. See Fig. O172. This will position the injector rack in full fuel position. Now turn outer adjusting screw of rack control lever downward until it bottoms lightly on the control tube; then alternately tighten the two adjusting screws.

With adjusting screws tight, check to be sure that governor control lever is in run position and the speed control lever is at high idle position; then check the adjustment as follows: Press down on injector rack lightly with a screw driver which will cause the injector rack to rotate slightly. Release screw driver and if rack returns to its original position, adjustment can be considered satisfactory. If rack does not return to its original position, it is too loose and adjustment is corrected by loosening the outer adjusting screw slightly and tightening the inner adjusting screw.

The adjustment is too tight if when moving the governor control (stop) lever from the stop to the run position, the injector rack becomes tight before the stop lever reaches the end of its travel. This can be noted by the increase in the effort required to move the stop lever and by a deflection of the fuel (control) rod. Correct this condition by loosing the inner adjusting screw slightly and tightening the outer adjusting screw.

With rear injector rack adjusted, manually hold the control tube in the

full fuel position and adjust the remaining rack control levers in the same manner.

When all injector racks have been adjusted, place control tube in full fuel position and push down on the injector racks with a screw driver. Racks should return to their original positions when screw driver is removed. When settings are correct, the racks of all injectors will be snug on the ball end of their respective rack control levers.

Refer to paragraph 246 to adjust governor.

**246. ADJUST GOVERNOR.** Prior to adjusting the governed engine speeds, be sure governor gap is set as outlined in paragraph 244; then proceed as outlined in paragraphs 247 through 250.

**247. ADJUST HIGH IDLE SPEED.** Start engine and run until it reaches operating temperature, then place hand throttle in high idle position and check the engine speed which should be 2125 rpm for series A tractors, 2375 rpm for series B and C tractors, or 2575 rpm for series 1950 tractors. If engine rpm is not as specified, disconnect booster spring and cross shaft coupling from spring housing (H—Fig. O173), then unbolt and remove spring housing. Add or subtract shims (102—Fig. O174) between spring and spring retainer (104) as necessary. Shims are available in thicknesses of approximately 0.010 and 0.078.

NOTE: If the high idle speed of engine is changed more than 50 rpm by the addi-

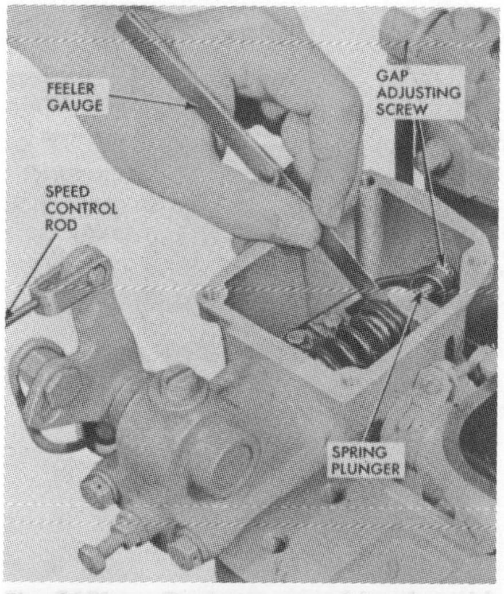

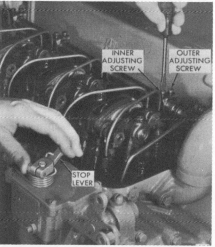

Fig. O172 — View showing injector rack control tube being adjusted. Refer to text for procedure. Note that governor is mounted on left side of engine instead of right as it is on Oliver tractors.

Fig. O171 — Governor control housing with cover removed. Adjust gap to 0.006. Note that unit shown is mounted on left side of engine instead of right as it is on Oliver tractors.

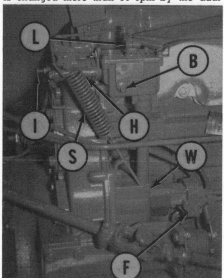

Fig. O173 — Engine governor showing the points of adjustments.

B. Buffer screw
F. Fuel pump
H. Spring housing
I. Idle screw
L. Stop lever
S. Booster spring
W. Weight housing

tion or subtraction of shims, the governor gap must be rechecked. If the governor gap is readjusted, then the injector racks adjustment must also be rechecked.

**248. ADJUST LOW IDLE SPEED.** Loosen lock nut and back-out buffer screw (B—Fig. 0173). Start engine and place hand throttle in low idle position and check the engine low idle speed which should be 500 rpm. If engine rpm is not as specified, loosen lock nut and turn idle adjusting screw (I—Fig. 0173) in or out as required to obtain the specified 500 rpm. Tighten lock nut and recheck. Adjust buffer screw as outlined in paragraph 249.

**249. ADJUST BUFFER SCREW.** With low idle speed adjusted, turn buffer screw (B—Fig. 0173) in until it contacts the governor differential lever as lightly as possible and still eliminates any engine roll or surge. DO NOT raise the engine low idle speed more than 15 rpm with the buffer screw.

**250. ADJUST BOOSTER SPRING.** With low idle speed adjusted, the booster spring (S—Fig. 0173) can be adjusted, if necessary, as follows: Disconnect clevis from speed control

cross shaft. Loosen nut of spring retainer bolt on speed control lever and the adjusting nuts on eye bolt at lower end. Move the spring retainer bolt in the slot of the speed control lever until the center of the bolt is on, or slightly below, an imaginary line running through the center of the bolt, speed control lever shaft and eye bolt. Tighten the spring retainer bolt at

this point. Start engine and tighten adjusting nut on eye bolt until the point is reached where the speed control lever, when released, will not return to the low idle position from the high idle position. Now, reduce the spring tension until the speed control lever will return to low idle from high idle position and tighten jam nut. Reconnect linkage.

# COOLING SYSTEM

## RADIATOR

### Series 1900-1950

**251.** To remove radiator, first drain cooling system, then remove hood side panels and hood. Unbolt fan blades from fan drive hub and let fan rest in fan shroud. Disconnect upper and lower radiator hoses, then unbolt and remove both air cleaners and bracket. On series 1950, disconnect the fuel and hydraulic oil coolers from radiator. On all models, remove cap screws from radiator supports, lift radiator from tractor and remove fan blades from fan shroud.

Reinstall by reversing removal procedure.

## FAN AND FAN DRIVE

### Series 1900-1950

**252. R&R AND OVERHAUL.** To remove the fan and fan drive assembly, first remove fan blades from fan drive hub and allow fan to rest in fan shroud. Loosen bolts retaining fan drive assembly in support and remove fan belt. Complete removal of retaining bolts and remove fan drive assembly from support. Remove fan from fan shroud. If necessary, support can now be removed.

**253.** With fan drive assembly removed, remove nut (14—Fig. 0175) from aft end of shaft and remove pulley (12) and Woodruff key (13).

Note: On tractors prior to serial

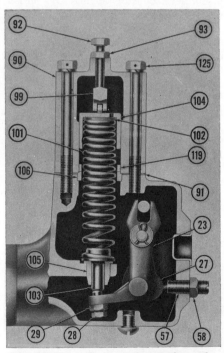

Fig. 0174 — Cross-sectional view of a typical control housing and spring housing assembly. Unit shown is for mounting on left side of engine. Item (106) not used on GM 4-53 engine.

| | |
|---|---|
| 23. Differential lever | 92. Low idle |
| 27. Operating lever | adjusting screw |
|     shaft | 101. Spring |
| 28. Governor gap | 102. Shims |
|     adjusting screw | 103. Plunger |
| 57. Buffer screw | 104. Spring retainer |
| 90. Spring housing | 105. Plunger guide |
| | 119. Stop |

| | | | |
|---|---|---|---|
| 1. Fan | 7. Shaft | 13. Woodruff key |
| 4. Fan hub | 8. Carrier | 14. Conelock nut |
| 5. Snap ring | 12. Pulley | 19. Support |
| 6. Bearing | | |

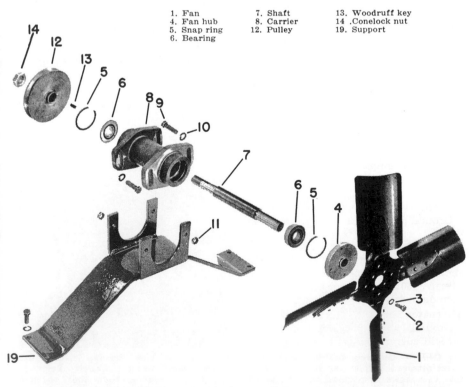

Fig. O175 — Exploded view of fan drive assembly used on Oliver 1900 tractor.

number 94 156-000, pulley was a press fit with no nut.

Use a suitable puller and remove fan hub (4). Remove snap rings (5), place assembly in a press and press shaft from carrier. The bearing that was removed along with the shaft can now be removed from shaft. Remove other bearing from carrier.

Inspect parts and renew as necessary.

254. To reassemble fan drive assembly, proceed as follows: Press a bearing on shaft until it bottoms against shoulder of shaft. Press a bearing into carrier until it bottoms and install a snap ring. Place carrier in a press, and with a hollow sleeve, support the installed bearing under inner race, then press the shaft and bearing into position and install snap ring. Install Woodruff key and pulley, with flange next to bearing, on aft (threaded) end of shaft and tighten

nut securely. Press hub on shaft until measured distance between rear of pulley and front of fan hub is $8\frac{5}{32}$ inches.

Balance of reassembly and installation is obvious. Adjust fan belt to allow ½-inch deflection midway between pulleys.

## THERMOSTAT

### Series 1900-1950

255. The cooling system is equipped with a pellet type thermostat which starts to open between 167 and 172 degrees F. and should be fully opened at 190 to 192 degrees F.

The thermostat is located in a housing at the front of the cylinder head. To renew the thermostat, remove hood, drain coolant to a point below the thermostat housing, then unbolt water outlet elbow and remove thermostat.

## ELECTRICAL SYSTEM

### GENERATOR AND REGULATOR

#### Series 1900

256. Delco-Remy generator 1100400 or 1100419 and regulator 1118997 are used on the Oliver 1900 model tractor. Specification data follows:

**Generator 1100400-1100419**
Brush spring tension .........28 oz.
Field draw @ 80° F.
  Volts .......................12
  Amperes .............1.5-1.62
Output (cold)
  Amperes (max.) ..............25
  Volts ....................14.0
  RPM .....................2710
**Regulator 1118997**
Cut-out relay
  Air gap ...................0.020
  Point gap .................0.020
  Closing voltage range ...11.8-14.0
  Adjust to ..................12.8
Voltage regulator
  Air gap ..................0.075
  Voltage range .........13.6-14.5
  Adjust to ..................14.0
  Ground polarity ..........Positive

#### Series 1950

257. Delco-Remy alternator 1100725 and regulator 1119517 are used on the Oliver 1950 model tractors. Specification data follows:

**Alternator 1100725**
Ground .................Negative
Field current @ 80° F.
  Amps. .................2.2-2.6
  Volts ...................12.0
Cold output @ specified voltage
  Specified volts ...........14.0

Amps. @ 2000 rpm..........21.0
Amps. @ 5000 rpm..........30.0
Rated output—hot ............32.0

**Regulator 1119517**
Polarity ................Negative
Field relap
  Air gap .................0.015
  Point opening .............0.030
  Closing voltage range......1.5-3.2
Voltage regulator
  Air gap ..................0.067 (1)
  Point opening .............0.015
  Voltage setting @ degrees F.
    13.9-15.0 @ 65
    13.8-14.8 @ 85
    13.7-14.6 @ 105
    13.5-14.4 @ 125
    13.4-14.2 @ 145
    13.2-14.0 @ 165
    13.1-13.9 @ 185

(1) Starting point for point gap adjustment after bench repair. Subsequent adjustment is made if necessary so that operation on lower contacts is 0.05-0.40 volts lower than on upper contacts.

## STARTING MOTOR

### Series 1900-1950

258. A Delco-Remy, model number 1113100 or 1113136 starting motor is used on the Oliver model 1900 and 1950 tractors. Specification data are as follows:

**Starter 1113100-1113136**
Brush spring tension (min.)..80 oz.

No-load test
  Volts .....................11.5
  Min. Amps.
    (includes solenoid) .......57.0
  Max. Amps.
    (includes solenoid) .......70.0
  RPM (min.) ...............5000
  RPM (max.) ..............7400
Lock test
  Volts .....................3.4
  Amps. .....................500
  Torque (Ft.-Lbs.) ............22

# C L U T C H

260. Engine clutch for all tractors is a dry, single plate, spring loaded type. The series 1800 tractors have a 12 or 13 inch clutch, series 1750, 1850 and 1950-T have a 13 inch clutch and the 1900 and 1950 tractors have a 14 inch clutch. Service procedure is similar for all clutches.

NOTE: Beginning with tractor serial number 191 062-000 for series 1750, and with September 1967 production for series 1850, a ceramic metallic button type clutch disc is being installed. All series 1950-T tractors are fitted with the ceramic button type clutch.

This button type clutch assembly can be installed in early series 1750 and 1850 tractors except those which were equipped with asbestos faced clutch discs. Substitution on the asbestos faced clutch models is not feasible because of the number of parts which must be changed.

An instruction sheet is included with button type clutch package when package is ordered to replace earlier clutches. Package includes button type disc, cover assembly and spacers. Package part number is 165 477-AS.

### ADJUSTMENT

#### All Models

261. Clutch pedal free travel is ¾-inch, measured vertically at the clutch pedal pad.

To adjust pedal free travel, refer to Fig. O184 or O184A, loosen the clevis jam nut at clutch lever end of the clutch control rod, remove cotter pin and clevis pin and rotate clevis as required to obtain correct free travel. Reinstall clevis pin and cotter pin, then tighten jam nut.

NOTE: In some cases on early 1800 series tractors, interference between clutch rod and fuel tank may be encountered. This condition may be corrected either by removing clutch rod and bending it to provide ¾-inch deflection at a point 9¾ inches from threaded end of clutch rod; or by drilling a 25/64-inch hole 2 inches below existing hole in clutch release lever and installing clutch rod in the new hole. Readjust clutch pedal free travel in either case.

On late series 1800 and 1900 tractors and all 1850 and 1950 tractors, the clutch release lever has two holes; however, only the top hole is to be used on series 1900 and 1950 tractors.

## REMOVE AND REINSTALL

### All Models

262. On tractors equipped with auxiliary drive unit, remove engine as outlined in paragraph 95, 95A or 240, then remove clutch housing and proceed as outlined in paragraph 263.

On tractors with no auxiliary drive, proceed as follows: First remove the pto drive shaft as outlined in paragraph 356, or the hydraulic pump drive shaft as outlined in paragraph 409 for those tractors not equipped with pto. Remove the drive shaft shield, then remove master link from sprocket chain and remove chain. Remove bolts from clutch shaft coupling, slide coupling rearward until it is free from clutch shaft and lift out drive shaft. Do not lose the sprocket spacer which will drop out as the clutch shaft is removed. Sprockets are also free to be removed, if desired.

Remove the clutch shaft by pulling it rearward.

On diesel tractors, disconnect the water trap drain line from clutch housing. Disconnect clutch control rod from clutch lever. On series 1750, 1800, 1850 and 1950-T disconnect necessary wiring from starter, then re-

move starter and on all series 1750, 1800 and 1850 non-diesel, remove the dust shield from lower front side of clutch housing. On all models, unbolt clutch cover and slide it rearward. See Fig. O185.

**TABLE No. 1**

| SERIES 1800—12 inch clutch | |
|---|---|
| Cover assy. part no..... | 105 621-AS |
| Serial number range... | 90 525-115 346 |
| Cover assy. part no..... | 107 175-AS |
| Serial number range.... | 115 347-150 420 |
| Spring Data—105 621-AS | |
| Number used........ | 12 |
| Color .............. | Light blue |
| Test (lbs. @ in.) ..... | 160-170 @ $1\frac{11}{16}$ |
| Spring Data—107 175-AS | |
| Number used ....... | 12 |
| Color .............. | 6 Pink |
| .............. | 6 Dark blue |
| Test (lbs. @ in.) Pink ... | 209-255 @ $1\frac{11}{16}$ |
| Dark blue........... | 183-196 @ $1\frac{11}{16}$ |

| SERIES 1800-1850-13 inch clutch | |
|---|---|
| Cover assy. part. no. (opt. 1800) ......... | 105 627-AS |
| Cover assy. part no..... | 105 627-ASA |
| Serial number range.... | 90 525-150 420 |
| Spring Data-105 627-AS | |
| Number used........ | 8 |
| Color .............. | Dark gray |
| Test (lbs. @ in.) ..... | 243-257 @ $1\frac{11}{16}$ |
| Spring Data—105 627-ASA | |
| Number used........ | 12 |
| Color .............. | 4 Brown |
| .............. | 8 Dark blue |
| Test (lbs. @ in.) Brown .. | 224-236 @ $1\frac{11}{16}$ |
| Dark blue ........ | 183-196 @ $1\frac{11}{16}$ |

| SERIES 1750-13 inch clutch | |
|---|---|
| Cover assy. part no..... | 160 973-AS |
| Serial number range.... | 180 537 up |
| Spring Data—160 973-AS | |
| Number used........ | 12 |
| Color .............. | 4 Brown |
| .............. | 8 Pink |
| Test (lbs. @ in.) Brown .. | 224-236 @ $1\frac{11}{16}$ |
| Pink .............. | 209-225 @ $1\frac{11}{16}$ |

| SERIES 1950-T-13 inch clutch | |
|---|---|
| Cover assy. part no..... | 164 005-AS |
| Serial number range.... | 194 080 up |
| Spring Data—164 005-AS | |
| Number used........ | 12 |
| Color .............. | 4 Pink |
| .............. | 8 Green |
| Test (lbs. @ in.) Pink ... | 209-225 @ $1\frac{11}{16}$ |
| Green ........... | 170-180 @ $1\frac{11}{16}$ |

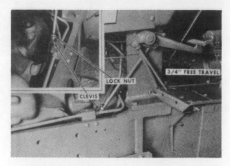

**Fig. O184A—Clutch rod and adjusting clevis for series 1950-T is located as shown.**

**TABLE No. 2**

| SERIES 1900-1950—14 inch clutch | |
|---|---|
| Cover assy. part no..... | 7AS-3569-A |
| Serial number range.... | 90 525-189 007 |
| Cover assy. part no..... | 165 305-AS |
| Serial number range.... | 189 008 up |
| Spring Data—7AS-3569-A & 165 305-AS | |
| Number used........ | 15 |
| Color .............. | Yellow |
| Test (lbs. @ in.)..... | 176-186 @ $1\frac{11}{16}$ |

263. On series 1750, 1800, 1850 and 1950-T tractors, loosen clutch cover screws approximately one turn each in rotation to prevent distortion of clutch cover and be sure not to lose the spacers which are located between clutch cover and flywheel on series 1800.

On series 1900 and 1950 tractors, install three ⅜ x 2½ inch cap screws in the holes provided and unload (compress) clutch prior to unbolting it from flywheel.

Reinstall by reversing the removal procedure. Position driven disc so widest hub flange is toward rear and use clutch shaft as a pilot to align disc.

## OVERHAUL

### Series 1750-1800-1850-1950-T (12 & 13 Inch)

264. Remove clutch as outlined in paragraphs 262 and 263, then on the

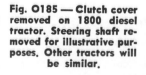

**Fig. O184 — Adjust clevis shown to obtain ¾-inch clutch pedal free travel. Series 1850 tractor is shown. On 1900 and 1950 tractors, clutch rod is installed in top hole of clutch release lever.**

**Fig. O185 — Clutch cover removed on 1800 diesel tractor. Steering shaft removed for illustrative purposes. Other tractors will be similar.**

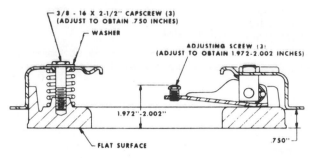

3/8 - 16 X 2-1/2" CAPSCREW (3)
(ADJUST TO OBTAIN .750 INCHES)

WASHER

ADJUSTING SCREW (3)
(ADJUST TO OBTAIN 1.972-2.002 INCHES)

1.972"-2.002"

FLAT SURFACE

.750"

Fig. O185A—Cross sectional view of the clutch cover assembly used on late 1950 tractors. Note adjustment measurements and location of adjusting screws.

13 inch clutches, remove the drive straps and ferrules. Place clutch cover assembly in a press with a block under pressure plate that will permit movement of clutch cover. Place a bar across top of cover assembly, compress the unit slightly and remove nuts from the eye bolts. Slowly release pressure and lift off clutch cover. All parts are now available for inspection and/or servicing. Note location and color of springs prior to removal. Remove levers as follows: Grasp inner end of lever and threaded end of eye bolt between thumb and fingers and while holding them as close together as possible, lift strut up and over end of lever. Any further disassembly is obvious.

Renew parts as necessary and refer to the following Table No. 1 for spring data. Clutch facings are available for relining the clutch driven disc, however carefully inspect driven disc for cracks, worn splines or loose hub rivets prior to relining.

265. ADJUST LEVERS. Reassemble clutch cover by reversing the disassembly procedure and adjust clutch fingers as follows: Measure, and record, the distance between friction face of flywheel and inner race of pilot bearing. Use three pieces of shim stock 0.365 thick for 12 inch clutches; or 0.355 thick for 13 inch clutches, equally spaced between clutch pressure plate and flywheel and bolt clutch cover in position. Actuate clutch fingers several times to insure proper functioning, then turn adjusting nuts on eyebolts until the measured distance between lever and inside pilot bearing race is the total of the previously recorded pilot bearing to flywheel face distance plus $2\frac{3}{16}$ inches, for the 12 inch clutch; or $2\frac{1}{16}$ inches for the 13 inch clutch. Stake nuts into eye bolts.

Remove clutch cover and shim stock. Reinstall clutch disc and the clutch cover to flywheel and use clutch shaft as a pilot to align clutch driven disc. Clutch disc is installed with widest hub flange toward rear.

### Series 1900-1950 (14 Inch) Prior Ser. No. 189 008-000

266. Remove clutch as outlined in paragraphs 262 and 263. Place clutch on a press and position a block under pressure plate that will permit movement of the clutch cover. Place a bar across top of clutch cover, compress the unit slightly and remove the three cap screws which were installed to unload clutch, as well as the nuts from the adjusting screws. Turn the adjusting screws clockwise (downward), then release the press pressure a slight amount. Continue this operation until adjusting screws are freed from cover. Slowly release pressure and lift off clutch cover. All parts are now available for inspection and/or service. Note location and color of springs prior to removal. Any further disassembly required is obvious. Renew parts as necessary and refer to the following Table No. 2 for spring data. Clutch facings are available for relining the clutch driven disc, however carefully inspect driven disc for cracks, worn splines or loose hub rivets prior to relining.

267. ADJUST LEVERS. Reassemble clutch by reversing disassembly procedure and do not use too much pressure when starting and threading adjusting screws into cover as stripped threads could result. Screw adjusting screws into cover until they protrude about ½-inch and install lock nuts finger tight. Install the three unloading cap screws before releasing the pressure on clutch cover.

Mount clutch cover assembly on flywheel with three pieces of 0.438 keystock (in place of driven disc) between pressure plate and flywheel and under clutch fingers. Remove the three unloading cap screws. Actuate each lever several times to insure proper functioning, then place a straight edge across clutch cover and turn adjusting screws as required to obtain a measurement of $\frac{27}{32}$-inch between inner end of finger and bottom of straight edge. Tighten lock nuts and recheck.

### Series 1950 (14-Inch) Ser. No. 189 008-000 And Up

267A. Remove clutch as outlined in paragraphs 262 and 263. Place clutch on a press and position a block under pressure plate that will permit movement of the clutch cover. Place a bar across top of clutch cover, compress unit until levers are slack, then remove lever pins. Remove the three unloading cap screws previously installed, then release press slowly and lift off clutch cover. Parts are now available for inspection and/or service. Renew parts as necessary and refer to Table No. 2 for spring data. Clutch facings are available for relining clutch, however, carefully inspect driver disc for cracks, worn splines or loose hub rivets prior to relining. Reassemble clutch by reversing disassembly procedure.

627B. ADJUST LEVERS. To adjust levers, place cover assembly on a flat surface and using the three unloading cap screws adjust cover so that distance between lower edge of cover and the flat surface is 0.750 inches as shown in Fig. O185A. Check this dimension at several points around cover to insure accuracy, then turn each lever adjusting screw as required until distance between contact surface of adjusting screw head and flat surface is 1.972-2.002. Be sure all lever adjusting screws are adjusted equally. See Fig. O185A. Tighten the unloading cap screws, reinstall clutch driven disc and clutch cover on flywheel and use clutch shaft as a pilot to align clutch driver disc. Clutch driven disc is installed with widest flange hub toward rear. Remove the three clutch unloading cap screws.

Reinstall the three cap screws to unload clutch, then remove clutch cover and shim stock. Reinstall the clutch driven disc and clutch cover

Fig. O186 — View of pilot bearing and cage with clutch removed.

to flywheel and use clutch shaft as a pilot to align clutch driven disc. Clutch disc is installed with widest hub flange toward rear. Remove the three clutch unloading cap screws.

### PILOT BEARING AND DAMPER

#### All Models

268. To remove the pilot bearing and damper plate, first remove clutch assembly as outlined in paragraphs 262 and 263. See Fig. O186. Remove cap screws from pilot bearing cage and remove cage and damper disc. See Fig. O187.

Remove snap ring from front side of pilot bearing cage and press bearing from cage.

Reinstall by reversing removal procedure and install damper plate with longest flange of hub toward front.

NOTE: The damper shown in Fig. O187 was replaced in series 1850 at tractor serial number 193 981-000 with a solid pto drive hub, and in series 1950 at tractor serial number 189 009-000 with a new bearing cage and a solid pto drive hub. Conversion packages are available to convert tractors built prior to the above serial numbers. However, flywheel must be reworked to install the solid damper and rework instructions are included with the conversion package.

Order conversion package, part number 166 080-AS, for series 1850 diesel tractors, or conversion package 166 100-AS for series 1850 non-diesel and 1950 tractors.

### CLUTCH RELEASE BEARING

#### All Models

269. To renew the clutch release bearing, remove the clutch housing as outlined in paragraph 262. On 1900,

1950 and 1850 diesel tractors, remove the clutch housing assembly from left side of tractor. On all 1750, 1800, 1950-T and 1850 non-diesel tractors, allow clutch housing to rest in tractor frame.

Remove fitting from outer end of grease tube, remove rubber grommet from clutch housing, then pull grease tube inside the clutch housing. Pull release bearing, bearing sleeve and grease tube from release fork and tube. Remove grease tube and elbow from bearing sleeve and press bearing from sleeve.

When installing release bearing on bearing sleeve, be sure grease apertures are aligned and thrust surface of bearing faces forward. Pack groove in sleeve with proper grease prior to installation. Reinstall by reversing the removal procedure and adjust clutch pedal free travel to ¾-inch.

### RELEASE FORK AND SHAFT

#### All Models

270. Release fork and shaft can be removed as follows: Disconnect clutch control rod from clutch lever and remove top inspection cover. Remove cap screws and the round key from release fork, then hold fork and pull shaft from left side of clutch cover. Remove fork from top side of housing.

When reinstalling, be sure tapped hole of release fork faces forward. Adjust clutch pedal free travel to ¾-inch.

### RELEASE BEARING TUBE

#### All Models

271. Should it become necessary to renew the release bearing tube, remove the clutch housing as outlined in paragraph 262. Clutch housing for 1900, 1950 and 1850 diesel tractors can be removed from tractor frame with no further disassembly. On 1800 and 1850 non-diesel tractors, and all 1750 and 1950-T tractors, proceed as follows: Remove left hood side plate and the left lower rear plate. Remove throttle rod. Remove clutch fork shaft, fork and throw-out bearing assembly. Loosen steering shaft bearing supports and disconnect steering shaft at rear of engine on tractors with manual steering. Shut off fuel and remove fuel supply line as well as the sediment bowl. Disconnect wire from temperature sending unit, then pull all wires from grommet at left rear of engine and move out of the way. Disconnect stop wire from injection pump; or choke wire from carburetor and move out of the way. Rotate clutch housing in tractor frame until open side is toward top and the top side of housing is toward rear of tractor, then remove clutch housing from left side of tractor.

Remove the retainer and felt seal from rear of clutch housing and press release bearing tube from housing.

To install new release bearing tube, press same from inside housing until end is within $\frac{1}{16}$-inch of the felt seal counterbore. Use caution not to damage tube and be sure it is installed parallel with center line of clutch housing. Install new felt seal and retainer and reinstall clutch housing.

**Fig. O187 — With pilot bearing cage removed, damper plate can be removed. Series 1800 tractor shown.**

# CREEPER DRIVE

Some models of the 1800 and 1900 series C tractors and series 1750, 1850, 1950 and 1950-T tractors may be equipped with a Creeper Drive unit which provides direct drive and creeper drive speeds. The Creeper Drive operation can be used only in conjunction with 1st and 2nd and 4th speed positions of the tractor main transmission as a lock-out system prevents creeper drive operation in the other three speeds of the tractor main transmission. The Creeper Drive unit must not be shifted while tractor is in motion.

### REMOVE AND REINSTALL

#### Models So Equipped

275. To remove the Creeper Drive unit, first remove engine with Creeper Drive unit attached as outlined in paragraph 95 or 240. Support engine on wood blocks so oil pan will not be damaged. Remove starting motor and on models so equipped, remove dust shield from lower front of clutch housing. Disconnect throwout bearing grease tube, then remove throw-out bearing, clutch cross shaft and release fork. Creeper Drive unit can

now be removed from clutch housing.

Reinstall by reversing the removal procedure.

## OVERHAUL

### Models So Equipped

276. With unit removed as outlined in paragraph 275, drain unit, if necessary, then refer to the following appropriate paragraphs for disassembly and reassembly of the cover assembly, input shaft assembly, output shaft assembly and the countershaft assembly.

277. COVER ASSEMBLY. To disassemble the cover assembly, remove the countershaft support shaft oil line, then unbolt and remove cover assembly from housing. Drive out roll pin (65—Fig. 0190), pull shift lever (64) from shifter arm (81), then pull arm from cover. Seal (63) can now be renewed. Remove baffle plate (75) and gasket (74). Remove one plug (59), lift snap rings (80) from their grooves, then bump out shifter rail (73) and catch detent ball (79) and spring (78) as shift rail clears shifter hub (82). Note: Remaining plug (59) will be removed by shift rail as it is bumped out. Remove shifter fork (83) if necessary, then drive out roll pin (71) and remove interlock (lock-out) rod (70) and hub (82) from cover. "O" ring (72) can now be renewed.

Clean and inspect all parts and renew as necessary. Use new "O" ring and seal and reassemble by reversing disassembly procedure.

278. INPUT SHAFT ASSEMBLY. To remove input shaft assembly, first remove the cover assembly as outlined in paragraph 277. Remove shaft retainer (98—Fig. O190 or 94—Fig. O190A), pull countershaft support shaft (101 or 97) and let countershaft assembly rest on bottom of housing. On the models shown in Fig. O190, remove throw-out bearing tube and support (43) and bearing retainer (48), then pull input shaft assembly from bearing support (38) and housing. Bump off bearing support (38). Remove snap ring (50) and press input shaft (40) from bearing (41).

On the models shown in Fig. O190A, unbolt support assembly (41) and pull support and input shaft assembly from housing. Remove snap ring (45), remove input shaft (40) and bearing (44) from support, then remove snap ring (47) and press input shaft from bearing.

On either model, push gear (36 or 38) and coupling (35 or 37) forward on input shaft to ease removal of snap ring, then remove snap ring, coupling

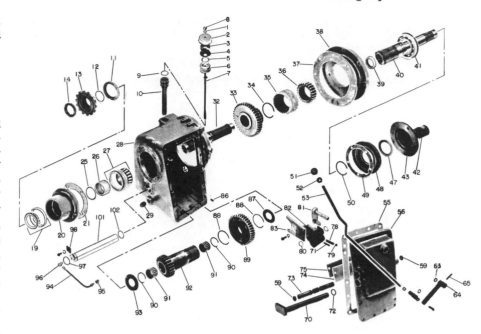

**Fig. 0190—Exploded view of early type Creeper Drive unit. Refer to Fig. 0190A for view of later type.**

| | | | |
|---|---|---|---|
| 1. Washer | 26. Spacer | 50. Snap ring | 81. Shifter arm |
| 2. Cap | 27. Front bearing | 51. Knob | 82. Hub |
| 3. Element | 28. Housing | 52. Bushing | 83. Shifter fork |
| 4. Cup | 29. Drain plug | 53. Control rod | 86. Dowel |
| 5. Gasket | 32. Output shaft | 55. Gasket | 87. Thrust washer |
| 6. Cap | 33. Output gear | 56. Cover | 88. Snap rings |
| 7. Dip stick | 34. Snap ring | 59. Plug | 89. Countershaft gear |
| 8. Nut | 35. Coupling | 63. Oil seal | 90. Snap rings |
| 9. "O" ring | 36. Input gear | 64. Shifter lever | 91. Bearings |
| 10. Filler tube | 37. Bearing support | 65. Groove pin | 92. Countershaft |
| 11. Oil seal | 38. Bearing support | 70. Lock-out rod | 93. Thrust washer |
| 12. "O" ring | 39. Oil seal | 71. Roll pin | 94. Oil line |
| 13. Coupling sprocket | 40. Input shaft | 72. "O" ring | 95. Connector |
| 14. Lock nut | 41. Ball bearing | 73. Shifter rail | 96. Elbow |
| 19. Rear bearing | 42. Tube | 74. Gasket | 97. "O" ring |
| 20. Retainer | 43. Tube support | 75. Plate | 98. Retainer |
| 21. Gasket | 47. Oil seal | 78. Detent spring | 101. Support shaft |
| 25. Shim | 48. Retainer | 79. Detent ball | 102. "O" ring |
| | 49. Gasket | 80. Snap rings | |

and gear from input shaft. Seal (39) can now be removed from rear of input shaft and seal (47 or 43) can be removed from retainer (48—Fig. O190) or support (41—Fig. O190A).

Clean and inspect all parts and renew as necessary. Use new gaskets and seals during assembly. Seal (39) is installed with lip facing toward front of input shaft. Seal (47 or 43) is installed with lip toward rear.

Reassemble the input shaft assembly by reversing the disassembly procedure.

279. OUTPUT SHAFT ASSEMBLY. To remove the output shaft assembly, remove the cover as outlined in paragraph 277, then unbolt bearing retainer (20), pull shaft assembly rearward and remove gear (33) from side of housing.

With shaft assembly removed, unstake lock nut (14), then using a spanner wrench (Oliver No. ST-149, or equivalent), remove nut, coupling sprocket (13) and "O" ring (12). Press output shaft from rear bearing and retainer, remove shims and spacer, then press output shaft from front

bearing. Remove oil seal (11) and rear bearing from retainer. If necessary, bearing cups can now be removed from retainer.

Clean and inspect all parts and pay particular attention to the bearings and bearing cups. Use new "O" ring, seal and gasket during assembly. Assemble output shaft components and adjust bearings as outlined in paragraph 280 prior to installing shaft assembly in housing.

280. To assemble shaft and adjust bearings, proceed as follows: Install bearing cups in retainer (20), if necessary, with smaller diameters toward inside. Press front bearing on shaft with smaller diameter toward rear. Place spacer and the original shims on shaft, insert shaft in retainer and press on rear bearing with smaller diameter toward front. Sprocket coupling can be used as a driver. Install a new lock nut (14) on shaft and tighten nut to a torque of 200 ft.-lbs. Check the shaft bearing adjustment which should be between 0.001 shaft end play and 5 inch pounds rolling torque. If adjustment is not as stated,

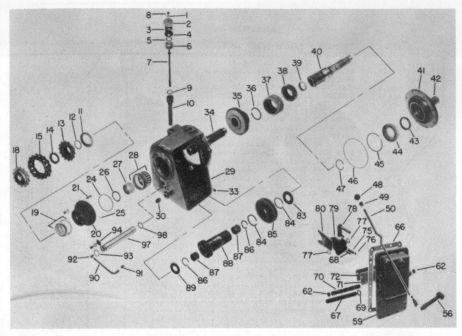

**Fig. O190A—Exploded view of the late type Creeper Drive unit. Note difference between support (41) and the three-piece assembly shown in Fig. O190. "O" rings (24, 25 and 46) are now used instead of gaskets.**

| | | | |
|---|---|---|---|
| 1. Washer | 21. Stud | 43. Oil seal | 76. Detent ball |
| 2. Cap | 24. "O" ring | 44. Bearing | 77. Snap ring |
| 3. Filter | 25. "O" ring | 45. Snap ring | 78. Shifter arm |
| 4. Cup | 26. Shims | 46. "O" ring | 79. Hub |
| 5. Gasket | 27. Spacer | 47. Snap ring | 80. Shifter fork |
| 6. Cap | 28. Front bearing | 48. Knob | 83. Thrust washer |
| 7. Dipstick | 29. Housing | 49. Bushing | 84. Snap ring |
| 8. Nut | 30. Drain plug | 50. Control rod | 85. Countershaft gear |
| 9. "O" ring | 33. Dowel | 56. Shifter lever | 86. Snap ring |
| 10. Filler tube | 34. Output shaft | 59. Cover | 87. Bearing |
| 11. Oil seal | 35. Output gear | 62. Plug | 88. Countershaft |
| 12. "O" ring | 36. Snap ring | 66. Gasket | 89. Thrust washer |
| 13. Coupling sprocket | 37. Coupling | 67. Lock-out rod | 90. Oil line |
| 14. Lock nut | 38. Input gear | 69. "O" ring | 91. Connector |
| 15. Coupling chain | 39. Oil seal | 70. Shifter rail | 92. Elbow |
| 18. Coupling sprocket | 40. Input shaft | 71. Gasket | 93. "O" ring |
| 19. Rear bearing | 41. Tube support | 72. Plate | 94. Shaft retainer |
| 20. Retainer | 42. Tube | 75. Detent spring | 97. Support shaft |
| | | | 98. "O" ring |

remove nut, coupling sprocket and rear bearing and vary shims as required. Shims are available in thicknesses of 0.004, 0.007 and 0.015.

With shaft adjustment made, remove nut and coupling sprocket and install new "O" ring (12) and oil seal (11) with lip toward front. Reinstall coupling sprocket and lock nut, tighten nut to a torque of 200 ft.-lbs. and stake nut in position.

**281. COUNTERSHAFT ASSEMBLY.** With cover assembly off, remove retainer (98—Fig. O190 or 94—Fig. O190A) and pull countershaft support shaft (101 or 97). Lift snap ring (88 or 84) from its groove and move it about ¼-inch rearward. Thrust washers and countershaft assembly can now be removed from housing. Removal of countershaft gear and bearings is obvious.

Clean and inspect all parts and renew as necessary.

When reassembling, install countershaft gear on shaft with hub rearward. Install needle bearings only far enough to install the retaining snap rings.

**282.** To reassemble Creeper Drive unit, proceed as follows: On models shown in Fig. O190, place new gasket (21) on retainer (20) and align oil drain-back notch in gasket with oil drain-back notch of retainer. On mod-

els shown in Fig. O190A, place new "O" ring (24) on pilot of retainer (20) and use grease to stick small "O" ring (25) in counterbore of oil drain-back hole. On either unit, start output shaft into housing and place output gear (33 or 35) on shaft with shifter fork groove to front. Align the oil drain-back hole in retainer with oil drain-back hole in housing (5 o'clock) and secure retainer to housing.

On models shown in Fig. O190, use new gasket (37) and place bearing support (38) on housing with cut-out in O.D. of support at the 9 o'clock position. Insert input shaft assembly into housing and over pilot of output shaft. Align notches of gasket (49) and bearing retainer (48) with oil drain-back hole in bearing support and install bearing retainer and throw-out bearing tube and support (43). On models shown in Fig. O190A, place new "O" ring (46) on pilot of support, insert input shaft assembly into housing and over pilot of output shaft. Secure support to housing.

Place new "O" ring in countershaft support shaft front bore and in groove at rear of shaft. Use grease to hold countershaft thrust washers in place. If necessary, lift rear countershaft gear snap ring from its groove and move it about ¼-inch rearward. Place countershaft assembly in housing and align the support shaft bores, then install support shaft and shaft retainer.

Align shifter fork with shifter fork groove of output shaft gear, then install cover and oil line. Disassemble breather assembly and clean and/or renew the breather filter element and fill unit with 5 quarts of SAE 80 multi-purpose lubricant.

# REVERSE-O-TORC DRIVE

The Reverse-O-Torc drive is a forward and reverse auxiliary transmission interposed between engine and tractor transmission providing a 1:1 ratio to the tractor transmission input shaft. The unit consists of two hydraulically controlled wet type multiple disc clutches which are coaxially mounted, along with a countershaft and idler gear, mounted in a cast iron housing. Pressurized oil to actuate clutches is furnished by an externally mounted pump and oil is controlled and directed by valving mounted on top side of unit. A 12 inch, single stage torque converter is

mounted on the engine flywheel and provides 2.15 maximum torque multiplication.

Reverse-O-Torc is available only for Industrial model tractors and when so equipped, tractors are not fitted with engine clutch or reverse gears in the tractor transmission as the forward and reverse clutches in the Reverse-O-Torc unit are used to start and stop tractor in both forward and reverse directions in any selected gear.

Power take-off or hydraulic system cannot be installed on Reverse-O-Torc equipped tractors.

## Models So Equipped

**283. PERFORMANCE TEST.** Run engine at 1000-1200 rpm with Reverse-O-Torc in neutral until engine reaches normal operating temperature, then with brakes locked and transmission in neutral, depress forward pedal (or lever) to full throttle position and note engine rpm. Engine rpm should be 2640 for all 1750 and for 1850 non-diesel; 2565 for 1850 diesel; 2550 for 1950; or 2650 for 1950-T tractors. Repeat test for reverse pedal (or lever) and correct any deviation from the above stated engine rpm.

With engine high idle established, place tractor transmission in sixth gear and be sure tractor brakes are fully locked. Slowly depress forward pedal (or lever) to full throttle position and note engine rpm (converter stall speed) which should be 1450-1750 rpm for series 1750, 1850 and 1950-T or 1500-1750 for series 1950. Repeat test with reverse pedal (or lever).

NOTE: Do not allow torque converter to overheat. If converter temperature goes beyond normal, cool by running engine at 1000-1200 rpm with Reverse-O-Torc in neutral.

If engine speeds (converter stall speeds) are not as stated above and engine is known to be in satisfactory condition, check pressure of regulator valves as outlined in paragraph 283A.

**283A. REGULATOR VALVES.** A pressure check can be performed to check the clutch pressure regulator valve and the converter charging pressure relief valve. In addition, an oil flow check can be made for the converter regulator valve.

To check clutch pressure regulator valve, attach a pressure gage of about 500-600 psi capacity, to pressure test port in top of control valve cover. Bring Reverse-O-Torc oil temperature up to normal range, if necessary, then run engine at 1200 rpm and note the gage reading which should be 145-175 psi. If pressure is not as stated, refer to paragraph 289 for service information on control valve.

To check converter charging pressure relief valve, plug hose running from converter regulator valve to oil cooler at the valve (A—Fig. O191A). Remove spring from clutch pressure regulator valve but leave the guide pin in valve bore. See Fig. O191. With gage still connected to test port and oil at operating temperature run engine at 1200 rpm and note gage pressure which should be 65-95 psi. If pressure is not as stated, refer to paragraph 289 for service information on control valve.

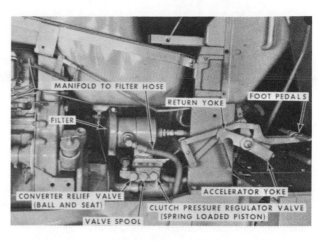

Fig. O191 — Left side of control valve showing location of clutch pressure regulator valve and converter relief valve.

MANIFOLD TO FILTER HOSE
RETURN YOKE
FOOT PEDALS
FILTER
CONVERTER RELIEF VALVE (BALL AND SEAT)
VALVE SPOOL
ACCELERATOR YOKE
CLUTCH PRESSURE REGULATOR VALVE (SPRING LOADED PISTON)

Condition of converter regulator can be approximated by checking oil flow as follows: Disconnect oil cooler to sump hose (B—Fig. O191A) at control valve and plug valve port. Make provision to catch oil, then run engine at 1200 rpm and note oil flow from open end of hose. A substantial amount of oil should flow from hose. If little, or no oil, flows from open end of hose, refer to paragraph 289 for service information on control valve.

**284. TROUBLE SHOOTING.** The following will serve as a guide when trouble shooting the Reverse-O-Torc unit.

1. LOW ENGINE RPM AT CONVERTER STALL.
   Could be caused by:
   a. Low engine output
   b. Converter element interference
   c. Incorrect converter assembly
2. HIGH ENGINE RPM AT CONVERTER STALL.
   Could be caused by:
   a. Clutch slippage
   b. High oil temperature
   c. Low oil level
   d. Low converter out pressure
3. LOSS OF POWER.
   Could be caused by:
   a. Clutch slippage

b. Converter charging pressure low
c. Converter stall speed low
d. Incorrect converter assembly

4. NO POWER IN EITHER DIRECTION.
   Could be caused by:
   a. Low clutch pressure
   b. Faulty selector valve
   c. Maladjusted control linkage
   d. Faulty charging pump
5. POWER IN ONE DIRECTION ONLY
   Could be caused by:
   a. Broken seal rings
   b. Faulty clutch
   c. Faulty selector valve
   d. Maladjusted control linkage
6. TRACTOR DRIVES IN ONE DIRECTION AND STALLS IN OPPOSITE DIRECTION.
   Could be caused by:
   a. Faulty clutch
7. IRREGULAR OIL PRESSURE.
   Could be caused by:
   a. Faulty clutch pressure valve
   b. Plugged oil filter
   c. Faulty charging pump
   d. Plugged oil lines
   e. Oil and/or air leaks
   f. Low oil level
8. HIGH OIL TEMPERATURE.
   Could be caused by:
   a. Clogged oil cooler

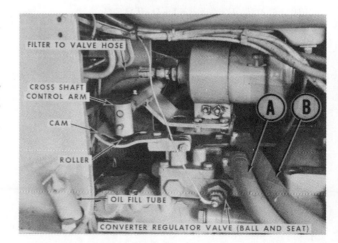

Fig. O191A — Right side of control valve showing location of converter regulator valve. Item (A) is converter to oil cooler hose. Item (B) is oil cooler to sump hose.

FILTER TO VALVE HOSE
CROSS SHAFT CONTROL ARM
CAM
ROLLER
A
B
OIL FILL TUBE
CONVERTER REGULATOR VALVE (BALL AND SEAT)

Fig. O191B—View of Reverse-O-Torc hand lever. Refer to text for adjustment procedure.

b. Low converter stall speed
c. Low converter out pressure
d. Low oil level

## LINKAGE ADJUSTMENT

**285. FOOT PEDALS.** To adjust foot pedals, have both pedals touching return yoke and the control valve arm in centered position, then adjust the control arm on the cross shaft until control arm roller is centered in slot of cam plate. Check pedal operation to see that control arm is fully shifted before pedal strikes accelerator yoke. Roller on control arm must not strike either end of cam slot when either pedal is fully depressed. Recheck engine governor adjustment as outlined in paragraph 283.

**285A. HAND LEVER.** To check hand lever adjustment, place hand lever in neutral position (Fig. O191B) and check that control valve arm is in centered position. If necessary, adjust link located below lower control rod pivot. Push hand lever forward in both positions and recheck engine governor adjustment as outlined in paragraph 283.

**286. R&R REVERSE-O-TORC DRIVE AND CONVERTER.** To remove Reverse-O-Torc drive unit, first disconnect oil cooler hoses from control valve, then remove drain plug and drain Reverse-O-Torc, if necessary. Disconnect hoses and remove filter from control valve, then unscrew and remove oil filler tube. Remove engine and Reverse-O-Torc unit as outlined in paragraph 95, 95A or 240. Support reversing unit with a hoist, unbolt converter housing from flywheel housing and pull assembly from engine. Remove oil manifold and pickup tube from oil pump, then separate converter housing from reversing gear case.

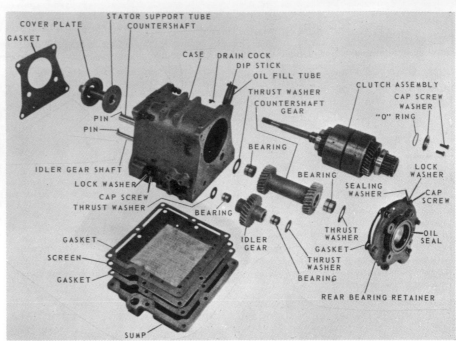

Fig. O192—View showing component parts comprising the Reverse-O-Torc unit. Note location of sealing washer on rear bearing retainer. Sump screen has a gasket located on both sides.

Remove cap screws retaining converter to drive plate and remove converter from engine.

Reinstall reverser unit and engine by reversing removal procedure.

**287. OVERHAUL REVERSE-O-TORC UNIT.** To disassemble Reverse-O-Torc unit, remove shift linkage, bracket and control valve from top of reverser case. Discard "O" rings located between control valve and reverser case. Refer to Fig. O192, turn reverser case bottom side up and remove sump, screen and both gaskets. Discard both baskets. Push idler gear shaft out front of reverser housing and lift idler gear assembly and thrust washers from housing. If idler gear bearings must be removed break bearing cages and extract rollers, then use a suitable puller to remove bearing races. Remove the two cap screws from end of output shaft, then remove washer (retainer), sprocket and

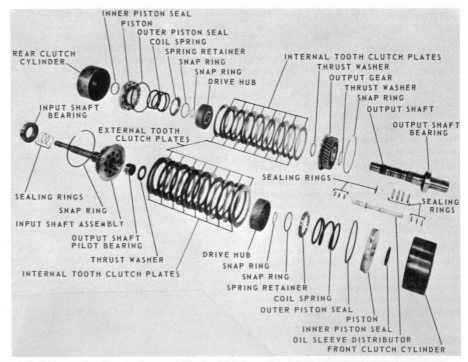

Fig. O192A—Exploded view of clutch assembly. Note the cast iron seal rings used on input shaft, output shaft and oil sleeve distributor.

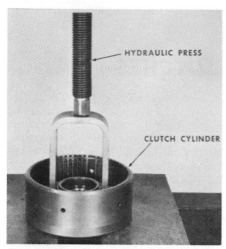

Fig. O192B — Use a tool similar to that shown to compress clutch piston return spring during disassembly and reassembly.

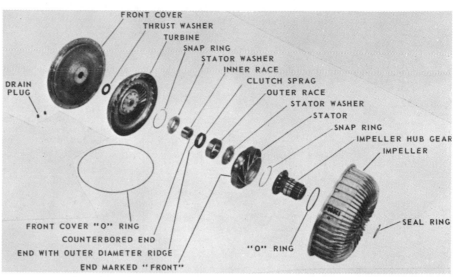

Fig. O192D—Exploded view of torque converter unit. Drain plugs in front cover must mate with holes in torque converter drive plate when converter is mounted to engine fly-wheel.

"O" ring. Remove cover and stator support tube. Unbolt rear bearing retainer and use caution during this operation as clutch assembly will drop. Also note the seal washer on one of the retaining cap screws located at upper right (one o'clock) position. Press oil seal from retainer and discard retainer gasket. Push countershaft out front of housing and lift out countershaft gear and thrust washers. If countershaft gear bearings are to be removed, press them both out together. Remove clutch assembly from housing.

Refer to appropriate subsequent paragraphs for information on clutch assembly, control valve, charging pump and charging pump idler gear. Refer to paragraph 288A for reassembly procedure.

**288. CLUTCH OVERHAUL.** To service clutch assembly, use Fig. O192A as a guide and proceed as follows: Remove the large snap ring from inside front edge of front clutch cylinder and remove input shaft. Do not remove the three cast iron sealing rings unless they have lost tension, are broken, stuck in grooves, or will

Fig. O192C—Compressed air may be used to remove piston from cylinder, if necessary. Plug one hole and carefully apply compressed air to opposite hole.

allow insertion of a 0.015 feeler gage between side of ring and groove. If sealing rings are not removed, use caution not to damage rings during disassembly and reassembly. If renewal of input shaft front bearing is required, input shaft can be pressed from bearing. If necessary to renew the output shaft pilot bearing from bore at rear if input shaft, break bearing cage and extract rollers and bearing race.

Pull oil distributor sleeve from output shaft, examine the seal rings and remove if necessary. Remove thrust washer, clutch plates and drive hub from output shaft and front clutch cylinder, then remove snap ring and pull front clutch cylinder from output shaft. Remove the large snap ring from rear inside edge of rear clutch cylinder, then remove the clutch cylinder and clutch plates from output gear and drive hub. Remove snap ring, drive hub, thrust washer, output gear and rear thrust washer from output shaft. Remove sealing rings and bearing from output shaft, if necessary.

Clutch cylinder and piston assemblies are identical and either can be disassembled as follows: Use a "U" shaped pressing tool as shown in Fig. O192B placed on spring retainer and slightly compress clutch spring. Remove spring retainer snap ring, slowly release press and remove spring retainer and clutch spring. Remove piston by bumping open end of cylinder on a wood block. If piston is liquid locked, plug one oil hole in I.D. of cylinder hub and cautiously apply compressed air to the remaining oil hole as shown in Fig. O192C. Remove and discard piston inner and outer seals after piston is removed.

Refer to the following data as a guide for parts renewal.

Clutch Plate Thickness:
  Internal tooth ........0.0605-0.0630
  External tooth ........0.0665-0.0710
Thrust washer thickness 0.0615-0.0635

**288A.** To reassemble clutch assembly, proceed as follows: Press output shaft bearing on rear of input shaft until it butts against shaft collar. then place output gear over front of output shaft with countershaft gear rearward and against thrust washer. Place next thrust washer over output shaft, install drive hub with thrust face rearward and secure with snap ring. If necessary, install the four cast iron sealing rings on output shaft. Start with an internal toothed clutch plate and alternately install the eight internal and eight external toothed clutch plates over clutch drive hub.

Assemble both clutch cylinders at this time. Place new seals on O.D. of piston and cylinder hub. Lubricate seals, then with piston positioned flat side inward, push piston to bottom of cylinder. Place piston return spring and spring retainer on top of piston, compress spring until snap ring groove is exposed and install retaining snap ring.

Place rear clutch cylinder assembly over rear clutch plates and secure clutch cylinder to output gear with large snap ring. Place front clutch cylinder on output shaft and secure with snap ring. Install drive hub with thrust face forward. Start with an external toothed clutch plate and alternately install the eight external and eight internal toothed clutch plates. If necessary, install seal rings on oil distributor sleeve, place oil distributor sleeve in I.D. of output shaft with

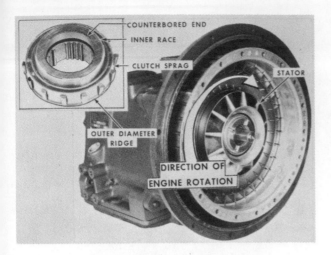

Fig. O192E—With Sprag clutch assembled as shown, stator will turn only in the direction of engine rotation.

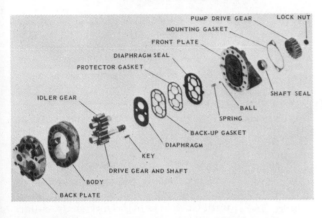

Fig. O193 — Exploded view of charging (primary) pump. Protector gasket, back-up gasket and diaphragm must fit under lip of diaphragm seal.

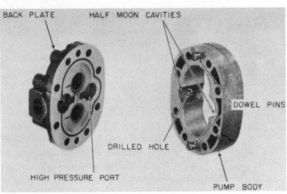

Fig. O193A — When assembling charging pump, be sure cavity with drilled hole is toward pressure side of pump.

splines. Thickness of thrust washer is 0.061-0.063 and thickness of stator washers is 0.768-0.778.

289A. Reassemble torque converter as follows: Place "O" ring and seal ring on impeller hub, install hub gear in impeller and secure with cap screws. Place snap ring in rear of stator, then install a stator washer against snap ring with counterbore toward inside of stator. Place outer race in stator against stator washer. Install Sprag clutch in outer race so that the ridge on outside diameter of clutch is on same side as the word "FRONT" marked on stator. Place inner race in Sprag clutch so counterbored end of inner race is on side opposite to the ridge on outside diameter of Sprag clutch. See Fig. O192E. Place remaining stator washer in stator with counterbored side next to Sprag clutch and install remaining snap ring.

NOTE: At this time, correct assembly of Sprag clutch can be checked, if desired. While facing the word "FRONT" marked on stator, either place assembly on stator support, or hold inner race. Stator will turn in a clockwise direction but will lock up when an attempt is made to turn stator counterclockwise. Stator must be able to turn in direction of engine rotation when converter is assembled. See Fig. O192E.

If stator was placed on stator support, remove it and place in the impeller with word "FRONT" upward and facing front cover. Place turbine, vane side down, on stator and install thrust washer. Install new "O" ring on front cover and secure cover to impeller.

Mount converter on drive plate in original position. BE SURE drain plugs are located behind access holes in drive plate.

289B. CHARGING PUMP AND IDLER GEAR. The gear type charging (primary) pump is externally mounted on the torque converter housing and is driven via an idler gear by the converter impeller hub gear. Pump can be removed from torque converter housing after engine and Reverse-O-Torc unit have been removed, however, if service is required for the pump idler gear assembly, it will be necessary to remove Reverse-O-Torc unit from converter housing.

To remove pump, proceed as follows: Remove the pump to filter hose, then unbolt oil manifold from rear of pump and remove manifold and oil pickup tube. Discard all "O" rings. Remove mounting cap screws and pull

short end forward. Place front thrust washer on output shaft. If necessary, install pilot bearing, front bearing and the cast iron sealing rings on input shaft. Front bearing is bottomed against input shaft gear. Install input shaft gear in front clutch cylinder and install the large snap ring.

Refer to paragraph 289E for installation information.

289. TORQUE CONVERTER. Mark location of converter on drive plate so access holes in drive plate will be over drain plugs when converter is reinstalled, then remove converter from drive plate on engine flywheel. Refer to Fig. O192D, set converter on bench with impeller side down, remove front cover and discard "O"

ring. Remove thrust washer and lift turbine and stator assembly from impeller. Remove snap rings from each side of stator assembly and remove stator washers, inner race, Sprag clutch and outer race from stator. Remove cap screws, pull impeller hub gear from impeller and discard "O" ring and seal ring.

Clean and inspect all parts. Pay particular attention to the Sprag clutch and the inner and outer races. If any of these parts show signs of scoring or wear, renew all three parts. Measure thickness of thrust and stator washers carefully as they maintain operating clearances within the converter. Impeller and turbine must have no cracks or damaged vanes or

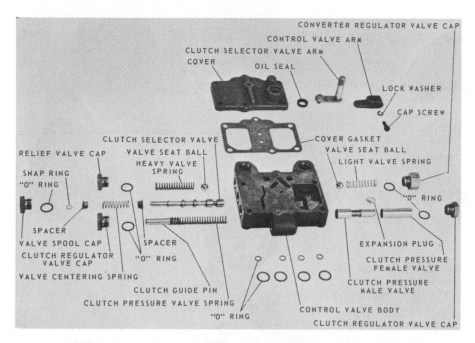

**Fig. O193B—Exploded view of Reverse-O-Torc control valve assembly. Refer also to Figs O191 and O191A. Note location of heavy spring for converter relief valve.**

pump from converter housing. Refer to paragraph 289C for service information.

If pump idler gear assembly is to be serviced, remove Reverse-O-Torc unit from converter housing, then remove the expansion plug at front of idler gear shaft and bump shaft rearward out of converter housing. Idler gear and bearing and both thrust washers can now be removed from converter housing. Bearing, inner race, shaft and pin and thrust washers are available for repairs. New thrust washers are 0.0615-0.0635 thick.

289C. To overhaul charging pump, use Fig. O193 as a guide. Remove and discard drive gear lock nut, then remove drive gear and key. Place a scribe line across pump assembly to assist in reassembly, then remove the four retaining cap screws and separate pump. Bump pump drive shaft against a wood block if pump sections stick together. Remove pump diaphragm and remove the two small springs and balls. Pry back-up gasket, protector gasket and diaphragm seal from pump front plate. Inspect shaft seal and if it is satisfactory do not attempt removal as seal will be damaged when removed.

Clean and inspect all parts. Small nicks or burrs can be removed with emery cloth. Drive gear and shaft, or idler gear and shaft are available separately, however shafts and gears are not. If bearings in front plate or back plate require renewal the complete front plate or back plate

must be renewed as bearings are not available separately.

Refer to the following data as a guide for parts renewal.

Gear Shaft Dia., Min. . . . . . . . . . .0.685
Gear Width, Min. . . . . . . . . . . . . .1.107
Bearing I.D., Max. . . . . . . . . . . . .0.691
Gear Pocket I.D., Max. . . . . . . . .1.719
Back Plate Wear, Max. . . . . . . . .0.0015

A seal kit is available and should be used when reassembling pump. Kit contains diaphragm, diaphragm seal, back-up gasket and protector gasket. Assemble pump as follows: If installing new shaft seal, press seal into front plate with lip of seal facing in. Lubricate seal lip. With open side of diaphragm seal down, work diaphragm seal into grooves of front plate. Place protector gasket, then back-up gasket, in diaphragm seal, push them down as far as possible and be sure they are under lip of diaphragm seal. Place the two steel balls in their bores and place springs on top of balls. Position diaphragm with bronze side up, then work diaphragm under lip of diaphragm seal. Lubricate gears and shafts, install in front plate and be careful not to damage shaft seal. Apply a thin coat of heavy grease to both sides of pump body, align scribe line on body with scribe line on front plate and install pump body over dowels of front plate. Body is correctly installed if half-moon cavities of body are towards back plate and the small drilled hole in one cavity is toward pressure side of pump. See Fig. O193A. Install back plate so scribe line is aligned with scribe line

on body, install cap screws and tighten to 27-30 ft.-lbs. torque. Install key and drive gear and secure with new lock nut.

289D. CONTROL VALVE. To disassemble control valve, use Fig. O193B as a guide and proceed as follows: Loosen cap screw and remove control valve arm and cam plate (not shown) from clutch selector valve arm. Remove the remaining two cap screws from valve cover, remove cover and discard gasket. Remove and discard selector valve arm oil seal. Remove clutch regulator valve cap, guide pin and spring, then remove cap from opposite end of bore and remove the two-piece pressure valve. Remove relief valve cap, spring (heavy) and ball. Remove converter regulator valve cap, spring (light) and ball. Remove control valve spool cap, pull spool assembly outward as far as possible, then remove snap ring, centering spring and spacers (spring seats). Remove expansion plug from opposite end of spool bore and push spool from valve body.

Discard all "O" rings, then clean and inspect all parts. Renew any parts which show undue wear or scoring. Refer to the following data as a guide for parts renewal.

Clutch Pressure Valve Spring:
    Free length . . . . . . . . . . .4.230-4.270
    Test lbs. @ in. . .16.2-19.8 @ 2.750
Relief Valve Spring:
    Free length . . . . . . . . . . .2.980-3.020
    Test lbs. @ in. . . . .17.1-20.9 @ 1.750
Converter Regulator Valve Spring:
    Free length . . . . . . . . . . . . . . . .2.500
    Test lbs. @ in. . . . . . .5.4-6.6 @ 1.50
Control Spool Centering Spring:
    Free Length . . . . . . . . . . .1.950-2.000
    Test lbs. @ in. . . . . . .4.5-5.5 @ 0.88
Clutch pressure
    valve bore . . . . . . . . . .0.7495-0.7500
Control spool bore . . . . . .0.6250-0.6255

Oil parts prior to assembly and reassemble control valve by reversing the disassembly procedure. Male half of the two-piece clutch pressure valve should face to inside of valve body.

289E. REASSEMBLE REVERSE-O-TORC. To reassemble Reverse-O-Torc, proceed as follows: Turn housing bottom side up and place clutch assembly in housing with input (long) shaft forward. Use heavy grease to stick countershaft gear thrust washers in housing, place countershaft gear in housing with larger gear rearward, then install countershaft from front of housing and be sure countershaft pin is in its slot. Use same procedure and install idler gear assembly and shaft. Place stator support tube and cover plate over input shaft and secure to housing. Use new gasket and install rear bearing retainer.

At this time, use a dial indicator and check end play of clutch assembly. Correct end play of clutch assembly is 0.020-0.035 and can be adjusted, if necessary, by adding or subtracting bearing retainer gaskets.

Place new "O" ring on output shaft, then install sprocket, retainer washer and cap screws and secure cap screws with lock wire after tightening. Place sump screen between two new gaskets, then place screen and gaskets on housing so indentation in screen and gaskets align with indentation in housing. Place sump cover on housing so oil pick-up tube hole aligns with indentation in housing and secure with cap screws. Install drain plug, if necessary.

Turn reversing unit over and install control valve and shift linkage on top side of housing. Install charging pump and oil manifold. Install reversing unit on converter housing, then mount complete assembly on engine. Install Reverse-O-Torc oil filter after engine is installed.

# HYDRA-POWER DRIVE

All Series tractors have available as optional equipment, a Hydra-Power drive unit interposed between clutch and transmission, which provides both direct drive and underdrive operation. This auxiliary drive is a self-contained unit and includes a hydraulically operated multiple disc clutch as well as a pump and a filter. The gear type pump mounted on inner side of housing cover furnishes pressurized oil to operate the multiple disc clutch as well as providing lubrication for the unit. In direct drive position the multiple disc clutch is engaged and a "straight through" power flow is obtained, whereas, in the under-drive position, the multiple disc clutch is disengaged and the power flow is routed to the countershaft where an approximate 26 percent speed reduction is produced. Unit can be shifted while tractor is in motion.

Some modifications have been made in the unit since its introduction and these will be noted as necessary.

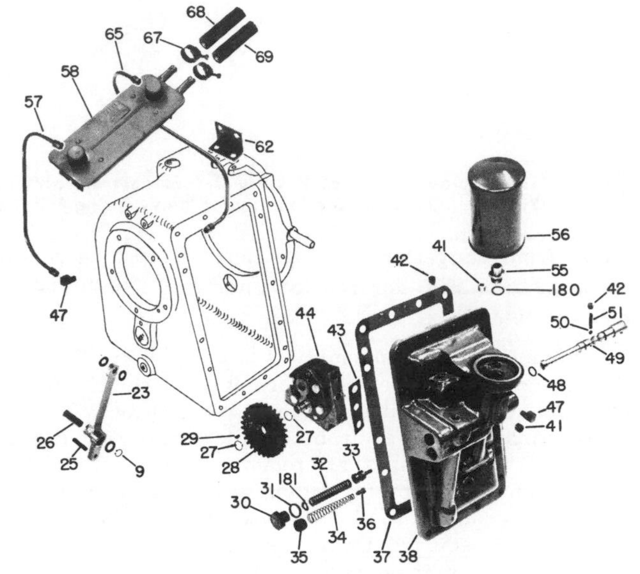

**Fig. O196 — Exploded view of the early Hydra-Power drive cover showing component parts. Later cover assemblies are similar. See Fig. 196A.**

| | |
|---|---|
| 9. | Snap ring |
| 23. | Control lever |
| 25. | Pin |
| 26. | Bushing |
| 27. | Snap ring |
| 28. | Drive gear |
| 29. | Woodruff key |
| 30. | Plug |
| 31. | Crush washer |
| 32. | Regulator spring |
| 33. | Regulator spool |
| 34. | By-pass spring |
| 35. | Plug |
| 36. | By-pass valve plunger |
| 37. | Gasket |
| 38. | Cover |
| 41. | Plug |
| 42. | Plug |
| 43. | Gasket |
| 44. | Oil pump |
| 47. | Elbow |
| 48. | "O" ring |
| 49. | Control valve spool |
| 50. | Detent ball |
| 51. | Detent spring |
| 55. | Adapter |
| 56. | Oil filter |
| 57. | Oil line |

| | |
|---|---|
| 58. | Heat exchanger |
| 62. | Bracket |
| 65. | Oil line |
| 67. | Hose clamp |

| | |
|---|---|
| 68. | Inlet hose |
| 69. | Outlet hose |
| 180. | "O" ring |
| 181. | Shim |

## TROUBLE SHOOTING
### All Models So Equipped

290. The following symptoms and their causes will be helpful in trouble shooting the Hydra-Power unit.

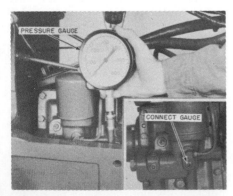

**Fig. O195 — When checking operating pressure of the Hydra-Power drive unit, install pressure gage as shown. Early model is shown.**

1. OPERATING TEMPERATURE TOO HIGH.
   a. Coolant hoses, heat exchanger or oil lines obstructed.
   b. Multiple disc clutch slipping.
   c. Engine cooling system overheated.
   d. Low pump pressure.
   e. Defective or improperly adjusted bearings.

2. INOPERATIVE OR HESTITATES HYDRA-POWER DRIVE.
   a. Defective over-running clutch.
   b. Defective clutch.
   c. Weak retractor spring.

3. INOPERATIVE OR HESITATES IN DIRECT DRIVE.
   a. Multiple disc clutch slipping or defective.
   b. Clutch operating pressure low.

4. MULTIPLE DISC CLUTCH SLIPPING.
   a. Clutch operating pressure low.

   b. Oil passages obstructed.
   c. Defective clutch.
   d. Oil collector sealing rings worn or broken.
   e. Defective oil collector "O" ring.
   f. Foreign material in clutch.

5. ABNORMAL LUBRICATION CIRCUIT PRESSURES (TOO HIGH OR TOO LOW).
   a. Defective pump.
   b. Pump inlet clogged.
   c. Defective by-pass spring.
   d. By-pass valve plunger not seating properly.
   e. Low oil level.

6. LOW CLUTCH OPERATING PRESSURE.
   a. Low oil level.
   b. Defective pump.
   c. Pump inlet clogged.
   d. Oil collector sealing rings worn or broken.
   e. Defective clutch "O" ring.
   f. Defective clutch piston ring.

g. Defective regulator spring.

h. Regulator valve spool stuck open.

i. Clogged oil passage.

7. HIGH CLUTCH OPERATING PRESSURE.

a. Defective regulator spring.

b. Regulator valve spool stuck closed.

c. Clogged oil passage.

8. ERRATIC CLUTCH OPERATING PRESSURE.

a. Regulator valve spool installed backward.

## OPERATING PRESSURE

### All Models So Equipped

291. Hydra-Power drive unit has two oil pressure circuits; a clutch operating circuit and a lubrication circuit. Both circuits can be checked at the same point using a single gage.

To check the Hydra-Power drive operating pressures, proceed as follows: Remove the pipe plug (41—Fig. O196 or 30—Fig. O196A), and install a pressure gage capable of registering at least 300 psi as shown in Fig. O195. Be sure oil level in Hydra-Power drive is at the proper level, then start engine and run until oil in Hydra-Power drive is at operating temperature. With oil warmed, operate engine at approximately 2000 rpm, place Hydra-Drive unit in direct drive and observe pressure gage. Gage should read 140-190 psi. If pressure is not as specified, and no malfunctions of the unit are evident, add shims (181—Fig. O196) behind the regulator valve spool spring to raise pressure. Shims are 0.0359 thick and each shim will raise the pressure approximately 5 psi.

Place Hydra-Power unit in the under-drive position and with engine running at 2000 rpm, observe the gage pressure which should be 20-60 psi. If gage reading is not as specified, refer to the Trouble Shooting section or the section pertaining to overhaul.

## REMOVE AND REINSTALL

### All Models So Equipped

292. To remove the Hydra-Power drive unit it is necessary to remove the engine and Hydra-Power unit as outlined in paragraph 95, 95A or 240. With engine removed support it on wood blocks so oil pan will not be damaged, then on Series 1750, 1800 and 1850 non-diesel, remove clutch

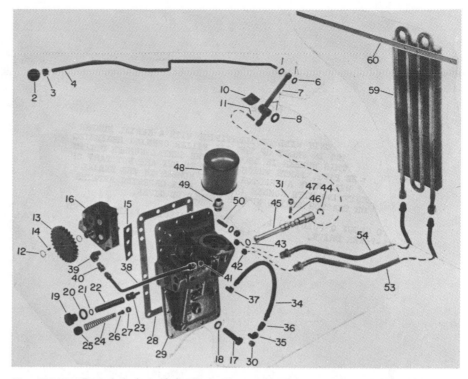

**Fig. O196A—Exploded view of the Hydra-Power drive cover used on Series 1950-T tractors. Note oil cooler (59) used on these models which mounts on front of radiator.**

| | | |
|---|---|---|
| 3. Bushing | 18. Lock washer | 30. Plug | 44. Snap ring |
| 4. Control rod | 19. Plug | 31. Plug | 45. Control spool |
| 7. Control lever | 20. Washer | 34. Hose | 46. Detent ball |
| 10. Bushing | 21. Shim | 35. Elbow | 47. Spring |
| 11. Lever pin | 22. Spring | 36. Connector | 48. Filter |
| 12. Snap ring | 23. Spool | 37. Elbow | 49. Adapter |
| 13. Pump gear | 24. Spring | 38. Oil line | 50. Stud |
| 14. Key | 25. Plug | 39. Elbow | 53. Coolant line |
| 15. Gasket | 26. Plunger | 40. Connector | 54. Coolant line |
| 16. Pump | 27. Seat | 41. Seat | 59. Oil cooler |
| 17. Pump bolts | 28. Gasket | 42. Seat | 60. Support strip |
| | 29. Cover | 43. "O" ring | |

housing lower dust shield and starter. Disconnect coolant lines from heat exchanger (early models). Unbolt and remove clutch housing and Hydra-Power unit from engine. Remove clutch release bearing and lubrication hose. Remove cap screw and key from release bearing fork and pull clutch shaft from fork and clutch housing. Unbolt and remove the clutch housing from Hydra-Power unit.

Reinstall by reversing the removal procedure.

## OVERHAUL

### All Models So Equipped

293. With unit removed as outlined in paragraph 292, drain unit and on early models so equipped, remove the front control rod. Remove valve control lever and if necessary, renew lever bushing. On early units, remove filter to heat exchanger and heat exchanger to countershaft oil lines. On late units, remove filter to countershaft oil line. Remove filter to oil collector stem hose. Remove cover and attached pump from housing.

293A. On early units, remove clutch release bearing tube support (142—

Fig. O197), and retainer (140). Remove snap ring from outer race of input shaft bearing (137) and bump off input shaft bearing support plate (118). Remove support shaft lock (90), pull support shaft (96) and allow countershaft assembly to lower to bottom of housing. Remove and discard support shaft "O" ring from shaft and bore in housing. Pull input shaft (136), bearing (137), input gear (120) and clutch hub (121) as an assembly from housing. Remove snap ring (123) from front end of output shaft (104), then unbolt output shaft bearing retainer (81). Pull output shaft rearward and remove the clutch and oil collector assembly and the output gear (105) from housing as the shaft is withdrawn. Countershaft assembly can now be removed from housing. If necessary, the heat exchanger, filler neck and valve spool control lever shaft can also be removed.

At this time all of the components of the Hydra-Power drive unit are available for inspection and/or overhaul. Refer to the appropriate following paragraphs.

293B. On late units, remove support shaft lock (95—Fig. O197A), pull

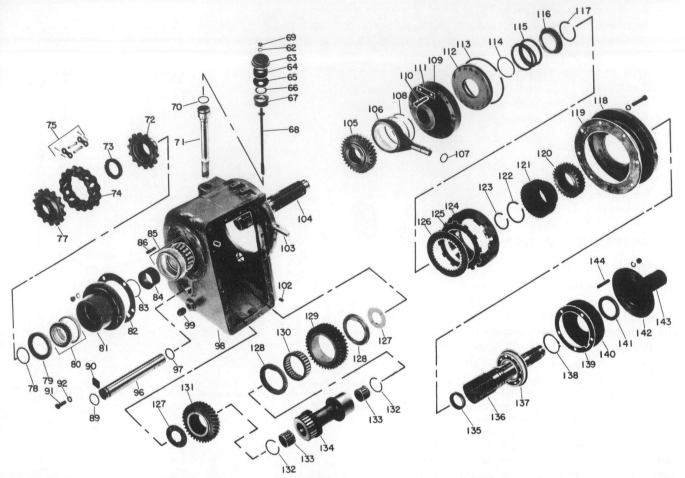

**Fig. O197 — Exploded view of the Hydra-Power drive unit used on Series 1750 and 1850. Earlier units did not use bearing spacer (84), clutch gear bearings (128) were retained by snap rings (not shown) and pump was driven by double gear (131) (see Fig. 203) instead of gear integral with countershaft (134). Refer to text for difference in service procedure on early and late output shaft assemblies. Also refer to Fig. O197A.**

| | | | |
|---|---|---|---|
| 62. Washer | 79. Oil seal | 102. Dowel | 117. Snap ring | 130. Sprag (over-|
| 63. Cap | 80. Rear bearing assy. | 103. Lever shaft | 118. Bearing support | running) clutch |
| 64. Breather element | 81. Bearing retainer | 104. Output shaft | 119. Gasket | 131. Gear |
| 65. Cup | 82. Gasket | 105. Output gear | 120. Input gear | 132. Snap rings |
| 66. Gasket | 83. Shim | 106. Oil collector | 121. Clutch hub | 133. Bearings |
| 67. Cap | 84. Spacer | 107. "O" ring | 122. Snap ring | 134. Countershaft |
| 68. Dip stick | 85. Front bearing | 108. Sealing rings | 123. Snap ring | 135. Oil seal |
| 69. Nut | assy. | 109. Clutch spider | 124. Plate retainer | 136. Input shaft |
| 70. "O" ring | 86. Stud | 110. Cap screw | 125. Driven plate | 137. Ball bearing |
| 71. Filler tube | 89. "O" ring | 111. Lock plate | 126. Drive plate | 138. Snap ring |
| 72. Sprocket, front | 90. Retainer | 112. Piston | 127. Thrust washer | 139. Gasket |
| 73. Nut | 96. Support shaft | 113. Piston ring | 128. Clutch gear | 140. Retainer |
| 74. Coupling chain | 97. "O" ring | 114. Seal ring | bearing | 141. Oil seal |
| 75. Master link | 98. Housing | 115. Release spring | 129. Clutch gear | 142. Support |
| 77. Sprocket, rear | 99. Drain plug | 116. Spring retainer | | 143. Tube |
| 78. "O" ring | | | | 144. Stud |

support shaft (98) and allow counter-shaft to rest on bottom of housing. Remove retaining cap screws from support (136) and pull support and input shaft (122) assembly from housing. Remove snap ring (120) from front of output shaft (104) and remove clutch assembly, oil collector and output gear (105) from shaft and housing. Unbolt retainer (85) and remove retainer and output shaft assembly from housing. Remove countershaft assembly from housing.

At this time, all components of the Hydra-Power drive unit are available for inspection and/or overhaul. Refer to the appropriate following paragraphs.

**294. CONTROL VALVE SPOOL, REGULATOR VALVE AND BY-PASS VALVE.**

The control valving can be removed from housing cover as follows: Remove the allen plug (42—Fig. O196 or 31—Fig. O196A) on top side of control valve bore and remove detent spring and ball. Pull control valve spool forward out of its bore, then remove and discard "O" ring. Remove plugs, then remove regulator valve (upper) and by-pass valve (lower) assemblies. Save any shims (181) which might be present behind the regulator valve spring.

NOTE: Removal of the regulator and by-pass valves can be accom-plished with Hydra-Power drive unit on tractor.

Check springs against the values which follow:

Detent spring
   Free length ..............$1\frac{9}{32}$ in.
   Test lbs. @ inches..15.3-18.7 @ $\frac{7}{8}$

Regulator spring
   Free length .............$3\frac{5}{8}$ in.
   Test lbs. @ inches....55-67 @ $2\frac{3}{4}$

By-pass spring
   Free length .............$3\frac{1}{2}$ in.
   Test lbs. @ inches..$5\frac{1}{2}$-$6\frac{1}{2}$ @ $2\frac{7}{16}$

Outside diameter of regulator valve spool and land of control valve spool is 0.6865-0.6869 and their bore inside diameters are 0.6875-0.6883.

Fig. O197A—Exploded view of Hydra-Power unit used in Series 1950-T. Service procedure will not vary greatly from earlier Hydra-Power units although some parts are constructed differently. Note that clutch spider now contains clutch plates as well as piston and plate retainer is not used. Clutch hub and input gear are now integral with input shaft. Support (136) now incorporates bearing retainer and bearing support used on earlier models.

| | | | |
|---|---|---|---|
| 67. Cap | 79. Master link | 105. Output gear | 116. Drive plate | 126. Sprag clutch |
| 68. Element | 81. Rear sprocket | 106. Oil collector | 117. Driven plate | 127. Gear |
| 69. Cup | 82. "O" ring | 107. "O" ring | 118. Retainer | 128. Snap ring |
| 70. Gasket | 83. Oil seal | 108. Sealing rings | 119. Snap ring | 129. Bearing |
| 71. Cap | 84. Rear bearing | 109. Clutch spider | 120. Snap ring | 130. Countershaft |
| 72. Dipstick | 85. Retainer | 110. Piston | 121. Oil seal | 131. Snap ring |
| 74. "O" ring | 86. "O" ring | 111. Piston ring | 122. Input shaft | 132. Bearing |
| 75. Filler tube | 87. "O" ring | 112. Seal ring | 123. Thrust washer | 133. Snap ring |
| 76. Front sprocket | 88. Shim (.004, | 113. Release spring | 124. Clutch gear | 134. Oil seal |
| 77. Nut | .007, .015) | 114. Spring retainer | bearing | 135. "O" ring |
| 78. Coupling chain | 89. Spacer | 115. Snap ring | 125. Clutch gear | 136. Support |
| | 90. Front bearing | | | 139. Tube |
| | 91. Stud | | | |
| | 94. "O" ring | | | |
| | 95. Retainer | | | |
| | 98. Support shaft | | | |
| | 99. "O" ring | | | |
| | 100. Housing | | | |
| | 101. Drain plug | | | |
| | 102. Dowel | | | |
| | 103. Lever shaft | | | |
| | 104. Output shaft | | | |

In some cases, trouble may be encountered in that the by-pass valve has damaged the seat which is machined in the by-pass plunger bore in cover. This results in reduced lubrication oil flow and pressure. Beginning with Hydra-Power unit serial number 44200, a hardened seat has been incorporated into housing cover and a field package, Oliver part number 166 372-AS, is available to correct this trouble on units prior to unit serial number 44200. Field package contains a by-pass valve seat, set screw, Grade AV Loctite and an instruction sheet.

A new valve seat can be installed as follows: Refer to Fig. 0197B and center punch cover in location shown. Coat seat area with heavy grease,

then drill a hole in cover with a No. 7 drill. Thread drilled hole with a ¼-20 tap and be sure to clean all cuttings from passageways. Be sure surfaces of old valve seat and new hardened seat are clean, then apply a coating of Loctite to new valve seat. Install new seat in bore with outer diameter chamfer facing inward, use a soft drift and drive new seat firmly against old seat. See Fig. 0197C. Apply Loctite to set screw, install in tapped hole and tighten to retain seat in position. Check original by-pass valve spring to insure it meets the specifications already given, then reassemble unit and allow at least four hours at a minimum of 72 degrees F. for Loctite to set before filling unit with oil.

When reassembling, use new "O" ring on control valve spool and install regulator valve with pin inward. If any shims were present behind regulator spring during disassembly, be sure to reinstall them. Prior to install-

Fig. O197B—When installing new by-pass valve seat, drill unit cover at position indicated. Refer also to Fig. O197C.

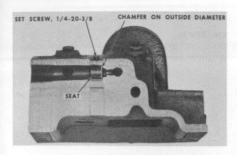

Fig. O197C—Cut-away view showing new by-pass valve seat installation. Refer to text for procedure.

ing cover on housing, renew "O" ring on stem of oil collector and in oil collector stem bore of housing cover, if not already done (early models not converted).

NOTE: A field service package, Oliver part number 159 346-AS, is available to correct any oil leakage problems which may occur between the oil collector stem and housing cover on Hydra-Power units prior to unit serial number 16234. By using this service package and reworking the housing cover, the clutch operating oil is rerouted and oil collector stem "O" rings are no longer pressurized.

Service package contains a new oil collector ring along with the necessary oil lines and fittings but does not include any seals and gaskets. When reworking housing cover, an "O" ring and gasket package, Oliver part number 155 234-AS, must be ordered in addition to the service package.

295. To rework the housing cover, remove Hydra-Power unit from tractor as outlined in paragraph 292, then remove housing cover, input shaft assembly and the output shaft assembly.

Remove the oil collector ring from clutch spider and discard. Remove oil filter and if so equipped, the control valve front control rod. Remove control valve spool lever, then remove the allen plug, detent spring and ball and pull control valve spool from its bore.

Remove the allen plug at top of oil collector stem bore and drive the tapered brass plug furnished into oil passage in bottom of stem bore until top of plug is $\frac{1}{16}$-inch below bore surface, then reinstall the allen plug. See Fig. O198. Center punch the two locations indicated in Fig. O199 and use a 37/64-inch drill for top hole and a 21/64-inch drill for bottom hole.

NOTE: Be accurate when locating and drilling these holes. The bottom hole intersects with the control valve spool bore and a depth stop should be used on drill when drilling bottom hole to prevent drill from scoring opposite side of spool bore. Carefully remove all burrs from spool bore.

Use a ⅜-inch NPTF tap and tap the top hole from outside of cover and during threading operation, alternately remove tap and install the furnished adapter. Continue this operation until inner end of adapter is approximately flush with inner surface of cover when tightened. Tap bottom hole from inside of cover using a ⅛-inch NPTF tap. Install the two elbows and oil line on inner surface of cover as shown in Fig. O200.

Place the new oil collector on clutch spider with word "FRONT" toward spider. Reinstall output shaft assembly and input shaft assembly. Install cover, then install elbow in adapter in top hole and connector in oil collector stem and install oil hose as shown in Fig. O201. If oil hose shown in Fig. O201 rubs against tractor main

frame, install a 90 degree elbow in bottom hole.

Complete balance of reassembly and reinstall by reversing the removal procedure.

296. **PUMP.** Unbolt pump from housing cover and discard gasket. Remove drive gear outer retainer snap ring, remove gear and key from pump drive shaft, then remove inner gear retainer snap ring.

Remove through bolts and separate pump. Dowels can be driven out and screen removed, if necessary. See Fig. O202.

NOTE: If same pump gears and shafts are to be reinstalled, mark idler gear so it will be reinstalled in original position.

Pump dimensional data are as follows:

| | |
|---|---|
| Gear width | 0.9975-0.9980 |
| Pump body width | 0.9995-1.0000 |
| Gear shaft diameter | 0.6247-0.6250 |
| Shaft bore diameter | 0.6265-0.6275 |
| Shaft operating clearance | 0.0015-0.0028 |

Reassemble by reversing disassembly procedure and use new gasket when mounting pump on housing cover. Heads of through bolts should be on drive gear side of pump and dowels should be installed prior to tightening through bolts.

Fig. O200 — View showing installation of interior oil line.

Fig. O198 — Drive tapered plug, furnished in package, to 1/16-inch below flush of oil hole.

Fig. O199 — Use illustration to locate the holes to be drilled when reworking early Hydra-Power drive cover. Also refer to text.

Fig. O201 — View showing installation of exterior oil hose.

**297. INPUT SHAFT ASSEMBLY (EARLY UNITS).** To disassemble input shaft assembly, remove snap ring (122—Fig. 0197) and pull clutch hub (121) and input gear (120) from input shaft. Remove snap ring (138) and press bearing from shaft. Oil seal (135) can be removed from aft end of input shaft at any time.

Inspect input shaft, input gear and clutch hub for damaged or worn splines or teeth. Check condition of input shaft bearing. Renew parts as necessary.

Reassemble input shaft assembly as follows: Install oil seal in aft end of input shaft with lip facing toward front of shaft. Press bearing on shaft with snap ring groove in bearing outer race toward front of shaft, install retaining snap ring, then press bearing toward snap ring to insure that bearing inner race is tight against snap ring. Install input gear on shaft with hub next to bearing, then install the clutch drive hub on shaft with its hub toward input gear and install the retaining snap ring.

**297A. INPUT SHAFT ASSEMBLY (LATE UNITS).** To disassemble input shaft assembly, remove snap ring (131—Fig. 0197A) and press shaft (122) and bearing (132) from support (136). Remove snap ring (133) and press shaft from bearing. Remove oil seal (134) from support. Oil seal (121) can be removed at any time. Tube (139) can be renewed if necessary.

Inspect input shaft, gear and clutch hub for damaged or worn splines or teeth. Check condition of input shaft bearing. Renew oil seals (121 and 134) if necessary.

Reassemble input shaft assembly as follows: Install oil seal in rear of input shaft with lip of seal facing toward front of shaft. Install oil seal in support with lip of seal facing toward rear. Press bearing on input shaft and install retaining snap ring. Press input shaft and bearing in front support and install bearing retaining snap ring.

**298. CLUTCH ASSEMBLY.** To disassemble the clutch assembly, first pull oil collector (106) from clutch spider hub, then remove the two sealing rings (108). On early models or models which do not have cover reworked, remove plug from outer end of oil collector stem if oil passage requires cleaning. Straighten tabs of lock plates (111), remove clutch plate retaining cap screws and lift clutch spider (109) from plate retainer (124). Clutch plates can now be removed

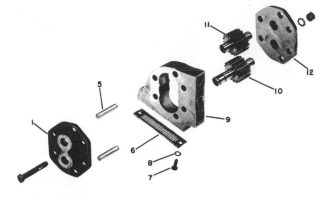

**Fig. O202 — Exploded view of the Hydra-Power drive oil pump.**

1. Rear plate
5. Dowel
6. Screen
9. Center plate
10. Drive shaft and gear
11. Idler shaft and gear
12. Front plate

from the plate retainer. Place clutch spider in a press, depress the spring retainer (116) slightly and remove retainer snap ring (117). Release press and remove retainer and retractor spring (115), then remove piston (112) from clutch spider either by bumping spider hub on a wood block, or by carefully applying air pressure to the oil hole located in land between oil collector sealing rings.

Clean and inspect all parts. Use the following specification data as a guide for renewing parts.

Clutch reaction spring
　test ........155-185 lbs. @ $1\frac{3}{16}$ in.
Driven plate thickness...0.112-0.120
Driving plate thickness..0.064-0.067
　Max. allowable cone.......0.010
Clutch piston ring
　Width ...............0.123-0.124
　End gap .............0.005-0.015
　　Max. allowable .........0.025
　Side clearance .......0.003-0.006
　　Max. allowable .........0.025
Oil collector hub O.D....3.544-3.546
Oil collector I.D.........3.550-3.552
Oil collector seal ring
　Width .............0.0930-0.0935
　End gap .............0.003-0.008
　　Max. allowable .........0.020

Reassemble clutch as follows: Install new seal ring on O.D. of spider piston hub. Install piston ring on O.D. of clutch piston. Start piston into spider with flat side toward oil cavity and install by compressing piston ring with fingers. Place clutch spider in a press, position retractor spring and retainer and compress retainer until snap ring can be installed. Start with an external lug clutch plate (driven) and alternate with an internal spline plate (drive) and install the six driven plates and five drive plates. Install the clutch spider on plate retainer, install lock plates and cap screws and tighten cap screws to a torque of 14-15 ft.-lbs. Secure cap screws with lock plates. Install sealing rings on oil collector hub and compress rings with soft wire. Start

oil collector over hub with word "Front" toward clutch spider and remove wire as oil collector goes over sealing rings. Lubricate clutch plates and if necessary, reinstall plug in outer end of oil collector stem.

**298A.** To disassemble the clutch assembly, first pull oil collector (106) from rear hub of clutch spider (109) and remove the two sealing rings (108). Remove snap ring (119), retainer (118), driven plates (117) and drive plates (116) from clutch spider. Use a straddle tool on spring retainer (114), compress release spring slightly and remove snap ring (115). Slowly release pressure and remove retainer and release spring from clutch spider. Remove piston (110) assembly from clutch spider by bumping open end of clutch spider against a wood block, or by applying compressed air to the oil hole located in land between oil collector sealing rings.

Clean and inspect all parts. Use the following specifications as a guide for renewing parts.

Clutch release
　spring test ..155-185 lbs. @ $1\frac{3}{16}$ in.
Driven plate thickness
　(ext. tooth) ..........0.0665-0.0715
Drive plate thickness
　(int. tooth) ............0.112-0.120
　Max. allowable cone .......0.010
Clutch piston ring
　Width ................0.123-0.124
　End gap ................0.005-0.015
　　Max. allowable .............0.025
　Side clearance ........0.003-0.006
　　Max. allowable .............0.025
Oil collector hub O.D. ....3.544-3.546
Oil collector I.D. ........3.550-3.552
Oil collector seal rings
　Width ................0.0939-0.0935
　End gap ................0.003-0.008
　　Max. allowable .............0.020

Reassemble clutch as follows: Install new seal ring on O.D. of piston hub in spider. Install piston ring on O. D. of clutch piston. Start piston into spider with flat side toward oil cavity and install by compressing

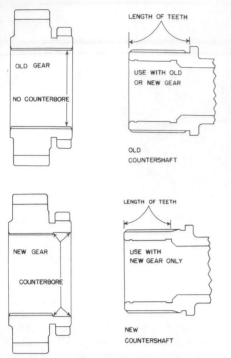

**Fig. O203 — When servicing Hydra-Power drive on series 1800 and 1900 be sure to use correct countershafts and double gears as shown.**

piston ring with fingers. Place clutch spider in a press, position release spring and retainer and compress retainer until snap ring can be installed. Start with an external splined clutch plate and alternate with an internal splined clutch plate and install the 10 clutch plates. Install retainer (pressure plate) and retaining snap ring. Install sealing rings on oil collector hub and compress rings with soft wire. Start collector over hub with word "Front" toward clutch spider and remove wire as collector goes over sealing rings. Lubricate clutch plates.

**299. OUTPUT SHAFT ASSEMBLY.** Very early Hydra-Power drive units were equipped with an output shaft which was not shouldered. Subsequently, a field service package, Oliver part number 158 601-AS, was made available which used a shouldered input shaft, along with a spacer and shims to provide for the bearing adjustment. In addition to the shouldered output shaft, spacer and shims, the package also includes a new output gear, one new bearing cone and a new lock nut. However, package does not include any seals, "O" rings or gaskets, so when installing the output shaft package, a seal and gasket package, Oliver part number 155 234-AS must also be ordered.

To disassemble output shaft assembly, unstake lock nut and using a spanner wrench (Oliver No. ST-149) or its equivalent, remove nut. Remove sprocket and "O" ring from shaft, then press shaft from rear bearing and retainer. On early units with shaft having no shoulder, note location of front bearing on shaft, then press bearing from shaft. On later units having the shouldered shaft, remove shims and spacer, then press front bearing from shaft. Remove oil seal and rear shaft bearing from retainer and if necessary, pull bearing cups from retainer.

Clean and inspect all parts and renew as necessary. Always use new "O" ring and oil seal.

NOTE: Beginning with Hydra-Power unit serial number 44200, an output shaft gear with a larger inside diameter and full length splines, an output shaft with a larger outside diameter and a clutch spider with a larger inside diameter to accommodate the larger shaft were installed. Both old and new parts are supplied for service and although old and new parts are not interchangeable indiivdually, the three late parts can be installed in prior units if operating conditions require it.

In adition, a Field Package, Oliver part number 165 649-AS, is available to convert the very early units having output shaft with no shoulder to the later type having the shouldered output shaft. Field Package also contains an instruction sheet giving information on the ⅞-inch counterbore required in the clutch spider.

**300. Assemble early output shaft** without shoulder as follows: Start front bearing on shaft with smaller diameter toward threaded (rear) end of shaft and press bearing on shaft until inner race is even with rear edge of grind (polished surface) on shaft.

NOTE: This is to insure sufficient room for installation of spider retaining snap ring.

If necessary, install bearing cups in bearing retainer with smaller diameters toward center, then insert shaft and bearing in retainer. Use coupling sprocket as a driver and press rear bearing on shaft. Remove coupling sprocket, install new "O" ring and oil seal (lip inward), then reinstall coupling sprocket. Install new lock nut and tighten only finger tight at this time. Refer to paragraph 304 in assembly procedure for information concerning bearing adjustment.

**301. Assemble later output shaft** assembly as follows: Place front bearing on rear of output shaft with smaller diameter toward rear and press bearing on shaft until it bottoms

against shoulder. If necessary, install bearing cups in retainer with smaller diameters toward inside. Place spacer on output shaft with counter-bored end next to front bearing cone, then install shims that were previously removed (all of the shims if service package is being installed). Place shaft in bearing retainer, install rear bearing, sprocket and lock nut and tighten nut to a torque of 200 ft.-lbs. Check shaft bearing adjustment which should be from 0.001 shaft end play to 5 inch pounds rolling torque. If bearing adjustment is not as stated, remove nut, sprocket and rear bearing and vary shims as required. Shims are available in thicknesses of 0.004, 0.007 and 0.015. With bearing adjustment made, remove lock nut and sprocket, then install new "O" ring and oil seal. Seal is installed with lip forward. Install sprocket and lock nut, tighten nut to a torque of 200 ft.-lbs. and stake in position.

**302. COUNTERSHAFT ASSEMBLY.** Lift countershaft from housing, if necessary, and retrieve both thrust washers. Pull gear (131—Fig. O197 or 127—Fig. O197A) from rear of shaft. Note: On series 1800, a double gear is used as shown in Fig. O203. Remove retaining ring (series 1800 and 1900) and front bearing from front end of shaft and pull clutch gear from the sprag clutch. Note position of sprag clutch, then remove clutch from shaft. Remove snap ring (series 1800) and rear bearing from shaft. Remove snap rings and needle bearings from I. D. of countershaft.

Clean and inspect all parts and renew as necessary. Use the following data as a guide for renewing parts.

| | |
|---|---|
| Front thrust washer thicknesses | 0.060-0.062 |
| Rear thrust washer thicknesses | 0.060-0.062 |
| Countershaft support shaft O.D. | 1.3745-1.3750 |
| Countershaft O.D., clutch end | 2.8429-2.8434 |
| Max. allowable taper | 0.0002 |
| Clutch gear I. D. | 3.4988-3.4998 |
| Max. allowable I. D. taper | 0.0003 |
| Front and rear clutch gear bearing surface O.D. | 3.4986-3.4978 |

To reassemble countershaft, proceed as follows: Install needle bearings in I. D. of countershaft only far enough to allow installation of snap rings. Place rear clutch gear bearing on shaft with bearing surface toward front and install rear snap ring. Note: On series 1800 and 1900, rear clutch gear bearing is notched, on series 1850 and 1950 front and rear clutch bearings are identical and no snap rings are used. Install the sprag clutch on countershaft so that drag strips point toward rear bearing. Install clutch gear over the sprag clutch and rotate clutch gear clockwise around the sprag clutch to assist in installing gear. After installation, clutch should overrun when gear is turned clockwise but lock up when counter-clockwise gear rotation is attempted. Install the front gear bearing and, on series 1800 and 1900 install retaining ring and the output double gear with smaller gear next to shaft shoulder. On series 1850 and 1950 install rear gear next to pump drive teeth on shaft.

NOTE: On 1800 and 1900 series, some differences exists in countershafts and the countershaft double gears as shown in Fig. O203. Note that the later double gear has a counterbore in front I. D. whereas the early gear is not counterbored. Early countershaft will work with either gear but late countershaft must be used only with late double gear. Part numbers remain the same so identification of parts must be visual. Therefore, when ordering new parts, be sure the correct combination of double gear and countershaft is obtained. On 1850 and 1950 series, a single gear is used at rear of countershaft and pump is driven from gear which is integral with shaft.

#### ASSEMBLY

303. Install oil filler tube, heat exchanger and bracket on models so equipped, control valve spool lever shaft or input shaft support studs if any were previously removed.

Use heavy grease on the countershaft thrust washers and position in housing with their tangs in the slots provided. On early models, the thrust washer with the largest outside diameter is the rear washer; late model thrust washers are alike. Place assembled countershaft in case with clutch gear forward and allow shaft assembly to rest on bottom of case. Do not dislodge the thrust washers. Delay installation of countershaft support shaft until input shaft assembly has been installed.

Use a new bearing retainer gasket, (82—Fig. O197), or a new "O" ring (87—Fig. O197A), start output shaft in housing and as shaft is moved forward, install output gear, with hub rearward, and the clutch assembly, with oil collector rearward, on output shaft. Align notches of gasket and retainer with oil drain-back notch in retainer (Fig. O197); or use grease

to hold small "O" ring (86—Fig. O197A) in counterbore of retainer. Align drain-back opening of retainer with hole in housing, then slide retainer over studs and secure with stud nuts. Install clutch spider retaining snap ring on front of output shaft.

304. Adjust output shaft bearings on units which do not have a shouldered shaft at this time. Bump front end of output shaft to insure that snap ring is seated against clutch spider. Install "O" ring and coupling sprocket on output shaft, if not already done, then install and tighten lock nut until only a slight amount of shaft end play exists. Now use an adapter such as Oliver No. ST-152 and an inch-pound torque wrench and check and record the amount of oil seal and bearing drag. Continue to tighten lock nut until output shaft has not more than 0.001 end play, or shaft bearings have not more than 5 inch pounds preload (in addition to seal and bearing drag). Bump output shaft both ways to be sure bearings are seated and end operation by bumping forward end of output shaft to insure that snap ring is seated against clutch spider. Recheck shaft setting and readjust if necessary. Stake lock nut into shaft slot and at 180 degrees from shaft slot to maintain bearing adjustment.

305. On units shown in Fig. O197, align splines of clutch drive (internal spline) plates and insert input shaft assembly. All five of the driven discs will rotate if drive hub is properly positioned. Use new gasket, install bearing support so notch in support aligns with notch in housing, then install snap ring on outer race of input shaft bearing. Install new seal in bearing retainer with lip toward front and install the bearing retainer and the release bearing tube and support. Align notches in gasket and bearing retainer with oil drain back hole in bearing support.

305A. On units shown in Fig. O197A, install new "O" ring (135) on

support and lubricate freely. Align splines of clutch drive (internal spline) plates and start clutch hub portion of input shaft into clutch. Align notch in support with notch in housing and push input shaft into position and support it against housing. If resistance of "O" ring entering housing bore cannot be overcome by hand, use two mounting cap screws 180 degrees apart and carefully pull support into place. However, BE SURE clutch plate splines are in place on clutch hub or the inner plates will be damaged. Install and tighten remaining cap screws.

NOTE: A new release bearing tube can be installed in the tube support, if necessary. Install tube so it protudes 4 inches from front flange of tube support and be sure tube is not cocked during assembly. Any misalignment of release bearing tube will result in rapid wear of the clutch release bearing.

306. Install the countershaft support shaft as follows: Install new "O" ring in shaft bore in front of housing. Install new "O" ring on rear of shaft, and if necessary, install oil line elbow in rear of shaft. Lubricate both "O" rings, make certain both thrust washers are in position, then lift countershaft assembly to align bores and install support shaft from the rear. Install shaft retainer and the support shaft to heat exchanger oil line (early units).

307. To install housing cover, install "O" ring on stem of oil collector and on early units which have not been modified with the field conversion package, install "O" ring in oil collector stem bore in housing. Use a new cover gasket, then lift stem of oil collector, start stem in cover bore and carefully install cover by rotating slightly as stem "O" ring enters cover bore. Tighten cover cap screws securely. On late units (Fig. O197A), install filter to countershaft line.

Fill unit with 6 quarts (with filter) of Type A Automatic Transmission Fluid.

# HYDRAUL SHIFT

The Hydraul Shift is a three-speed auxiliary unit interposed between engine clutch and transmission and provides on-the-go shifting in over, under or direct drive. This auxiliary unit is a self-contained unit and includes its own reservoir, lubrication and pressure systems and control valving. Oil operating temperature is controlled by a cooler located in front

of the radiator. Gears are in constant mesh and are helical cut. Drive selection is accomplished by engaging or disengaging multiple disc clutches and the operation of a planetary gear set and an overrunning (Sprag) clutch. The pressurized oil used to operate clutches, as well as oil for lubrication, is furnished by a gear type pump mounted on inner side of

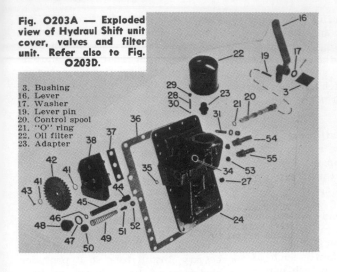

**Fig. O203A — Exploded view of Hydraul Shift unit cover, valves and filter unit. Refer also to Fig. O203D.**

3. Bushing
16. Lever
17. Washer
19. Lever pin
20. Control spool
21. "O" ring
22. Oil filter
23. Adapter

24. Cover
27. Plug (ctsk)
28. Spring
29. Plug
30. Detent ball
31. Stud
34. Insert
35. "O" ring
36. Gasket
37. Gasket
38. Pump
41. Snap ring
42. Pump gear
43. Key
44. Regulator valve spool
45. Spring
46. Shim (.035)
47. Washer
48. Spring
50. Plug (ctsk)
51. By-pass valve plunger
52. Seat
53. Insert
54. Oil cooler line, right
55. Oil cooler line, left

housing cover.

In overdrive position, oil pressure engages overdrive (rear) clutch and power flow goes through countershaft which drives the planet gear carrier. Overdrive clutch holds sun gear shaft stationary and the speed increase is obtained by the planet gears driving output shaft through the ring gear.

In underdrive position, both direct and overdrive clutches are disengaged and power flow is through the countershaft which drives the planet gear carrier. The overrunning clutch locks sun gear shaft and output shaft together causing them to rotate as a unit. This makes the planet drive inactive and results in a speed reduction.

In direct drive, the direct (front) clutch is engaged. This locks input shaft and output shaft together and power flow is straight through the unit. Overrunning clutch over-runs and permits free rotation of sun gear shaft, planet gears and countershaft.

### TROUBLE SHOOTING

#### All Models So Equipped

308. The following symptoms and their causes will be helpful in trouble shooting the Hydraul Shift unit.

1. OPERATING TEMPERATURE TOO HIGH.
   a. Coolant hoses, oil cooler or oil lines obstructed.
   b. Low pump pressure or pump intake clogged.
   c. One or both clutches slipping.
   d. Engine cooling system overheated.
   e. Defective or improperly adjusted bearings.
   f. Reservoir oil level low.

2. INOPERATIVE OR HESITATES IN OVERDRIVE.
   a. Overdrive clutch slipping or defective.

   b. Low pump pressure or pump intake clogged.
   c. Oil passage obstructed.

3. INOPERATIVE OR HESITATES IN UNDERDRIVE.
   a. Overrunning clutch worn or defective.
   b. Damaged or sticking underdrive and/or overdrive clutch.

4. INOPERATIVE OR HESITATES IN DIRECT DRIVE.
   a. Low pump pressure.
   b. Direct drive clutch faulty or slipping.
   c. Oil passage obstructed.

5. LUBRICATION CIRCUIT PRESSURE ABNORMAL (TOO HIGH OR TOO LOW).
   a. Defective by-pass valve spring.
   b. By-pass valve plunger sticking or not seating.
   c. Pump defective or pump intake clogged.
   d. Oil passage obstructed.

6. LOW CLUTCH OPERATING PRESSURE.
   a. Pump defective or pump intake clogged.
   b. Reservoir oil level low.
   c. Defective regulator valve spring.
   d. Regulator valve spool stuck open.

7. HIGH CLUTCH OPERATING PRESSURE.
   a. Defective regulator valve spring.
   b. Regulator valve spool stuck closed or sticking in bore.
   c. Oil too heavy or too cold.

8. ERRATIC CLUTCH OPERATING PRESSURE.
   a. Regulator valve spool incorrectly installed.
   b. Sticky regulator valve spool.

### SYSTEM PRESSURE CHECKS

#### All Models So Equipped

308A. The Hydraul Shift unit has two oil pressure circuits; a lubrica-

tion circuit and a clutch operating pressure circuit. When making any system pressure check, be sure reservoir oil is at correct level, oil is at operating temperature and engine is operating at rated rpm.

To check lubrication circuit, proceed as follows: Install a pressure gage either in place of the cover-to-countershaft oil line, or in place of heat exchanger (oil cooler) line, at the housing cover. Location will depend on space available as either position will suffice.

NOTE: If desired, a hole can be cut in the top of an oil filter and a gage welded in place. This assembly can be installed in place of the regular oil filter and the lubrication circuit pressure checked.

Start engine, run at rated rpm and observe gage. Gage should register 20-60 psi with control valve in any drive position. If lubrication pressure is not as stated, refer to item 5 in Trouble Shooting section. Also, be sure hose on pressure gage has at least ½-inch I. D. as a smaller hose might restrict oil flow and give a false reading.

To check clutch operating pressure circuit, remove pipe plug (T—Fig. O203D) from bore below oil filter assembly and install gage. Start engine, run at rated rpm and observe gage. Gage should register 140-190 psi with control valve in either overdrive or direct drive (clutch pressurized). If clutch pressure is not as stated, refer to items 6 and 7 in Trouble Shooting section. Shims (46—Fig. O203A) may be used behind the pressure regulator valve spring (45), if necessary. Shims are 0.0359 thick and each shim will vary pressure approximately 5 psi.

### REMOVE AND REINSTALL

#### All Models So Equipped

309. To remove the Hydraul Shift drive unit it is necessary to remove the engine and Hydraul Shift unit as outlined in paragraph 95 or 240. With engine removed, support it on wood blocks so oil pan will not be damaged, then on series 1750, 1800 and 1850 non-diesel, remove lower dust shield and starter. On all models, unbolt and remove clutch housing and Hydraul Shift unit from engine. Remove clutch release bearing and lubrication hose. Remove cap screw and key from release bearing fork and pull clutch shaft from fork and clutch housing. Unbolt and remove the clutch housing from Hydraul Shift unit.

Reinstall by reversing removal procedure.

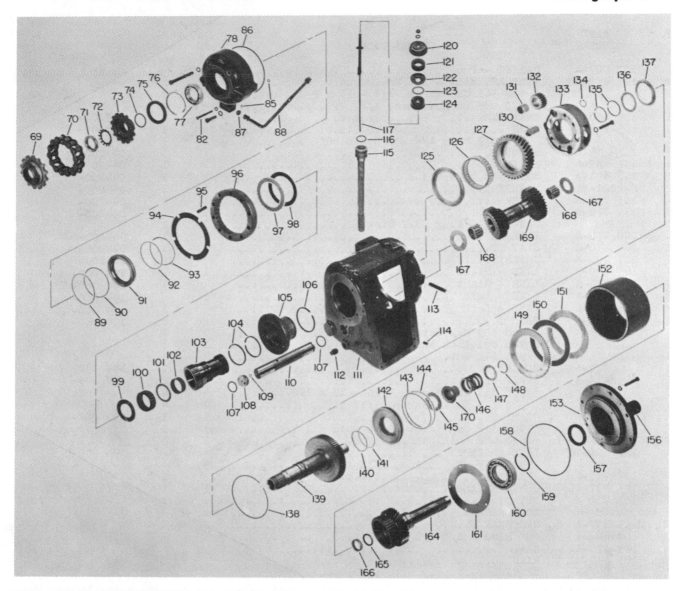

**Fig. O203B—Exploded view of Hydraul Shift unit showing component parts and their relative positions.**

| | | | | | |
|---|---|---|---|---|---|
| 69. Sprocket, rear | 89. Seal ring | 101. Spacer | 117. Dipstick | 137. Bushing | 152. Internal ring |
| 70. Coupling chain | 90. "O" ring | 102. Bearing | 120. Cap | 138. Retaining ring | gear |
| 71. Spanner nut | 91. Piston | 103. Sun gear shaft | 121. Filter element | 139. Output shaft | 153. Support |
| 72. Lock washer | 92. "O" ring | 104. Seal rings | 122. Cup | 140. Seal ring | 156. Tube |
| 73. Sprocket, front | 93. Seal ring | 105. Support | 123. Gasket | 141. "O"ring | 157. Oil seal |
| 74. "O" ring | 94. Spring plate | 106. Seal ring | 124. Cap | 142. Piston | 158. "O" ring |
| 75. Oil seal | 95. Return springs | 107. "O" ring | 125. Bushing | 143. "O" ring | 159. Snap ring |
| 76. Snap ring | (5 used) | 108. Plug | 126. Bearing | 144. Seal ring | 160. Bearing |
| 77. Bearing | 96. Retainer | 109. "O" ring | 127. Drive gear | 145. Thrust washer | 161. Bearing retainer |
| 78. Clutch housing | 97. Separator plates | 110. Support | 130. Planet gear pin | 146. Return spring | 164. Input shaft |
| 82. Cap screw | (6 used) | 111. Housing | 131. Bearing | 147. Retainer | 165. Oil seal |
| (¼ x 1½) | 98. Clutch plates | 112. Plug, magnetic | 132. Planet gear | 148. Snap ring | 166. Bearing |
| 85. "O" ring | (5 used) | 113. Lever shaft | 133. Planet carrier | 149. Pressure plate | 167. Thrust washer |
| 86. "O" ring | 99. Thrust washer | 114. Dowel | 134. Snap ring | 150. Clutch plate | 168. Bearing |
| 87. Insert | 100. Overrunning | 115. Filler tube | 135. Seal rings | (6 used) | 169. Countershaft |
| 88. Oil line | (Sprag) clutch | 116. "O" ring | 136. Thrust washer | 151. Separator plate | 170. Spring guide |
| | | | | (5 used) | |

## OVERHAUL

### All Models So Equipped

309A. **DISASSEMBLE HYDRAUL SHIFT UNIT.** With Hydraul Shift unit removed as outlined in paragraph 309, drain unit and remove dipstick and filler cap assembly. Straighten tabs of lock washer (72—Fig. O203B), then use a spanner wrench (Oliver No. ST-149, or equiv-

alent) and remove nut (71). Remove sprocket (73) and "O" ring (74) from output shaft. Unscrew filter element (22—Fig. O203A) from adapter (23), then remove cover (24) and pump (38) assembly from unit housing (111—Fig. O203B). Pull input shaft support (153) and input shaft (164) from housing and direct drive clutch. See Fig. O203E. Remove cap screws which retain overdrive clutch housing (78—Fig. O203B), then rotate

housing slightly to allow removal of countershaft support (110). See Fig. O203F. Support the countershaft (169—Fig. O203B) to prevent binding the support, bump on front end of support until sealing plug (108) is removed, then remove support (110) from front of housing. Allow countershaft to rest on bottom of housing. Remove output shaft (139), direct drive clutch and internal ring gear (152) assembly from housing. Remove

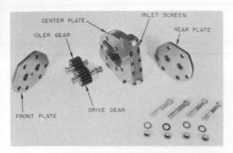

Fig. O203C—Hydraul Shift oil pump disassembled. Refer to text for dimensional data.

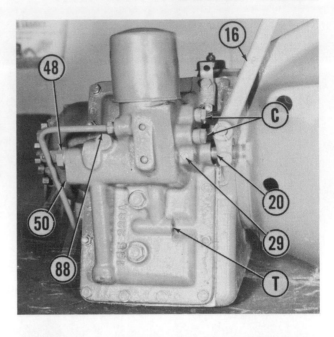

Fig. O203D — Hydraul Shift unit cover installed. Refer to Fig. O203A for legend. (T) is test port for checking clutch pressure. Shipping plugs (C) are installed in oil cooler line ports.

Fig. O203E — Hudraul Shift unit with input shaft assembly removed. Refer to Fig. O203B for legend.

planet gear carrier (133), drive gear (127) and bearing (126) if not removed with output shaft, clutch and internal ring gear assembly. Remove thrust bushing (125) from housing. Remove overdrive clutch housing (78), clutch and sun gear shaft (103) from housing by bumping lightly on front end of sun gear shaft. Remove support (105) from housing if it was not removed with the overdrive clutch and housing assembly. Remove countershaft and countershaft thrust washers (167) from housing. Remove filler neck (115), studs, dowels or control lever shaft (113), as necessary.

With Hydraul Shift unit disassembled, discard all "O" rings, seals and gaskets and refer to the appropriate following paragraphs to service subassemblies. Refer to paragraph 309H for reassembly procedure.

**309B. COVER AND VALVES. NOTE:** If only pressure regulator valve, by-pass valve or oil filter are to be serviced, it can be accomplished without removing cover and with Hydraul Shift unit in tractor.

Remove plug (48—Fig. O203A or O203D), washer (47), spring (45) and pressure regulator spool (44) from bore in cover. DO NOT lose any shims (46) which may be used. A $\frac{5}{16}$-inch cap screw can be used to remove spool if it is stuck. Remove countersunk plug (50), spring (49) and bypass plunger (51). Seat (52) is renewable, if necessary. Remove plug (29), spring (28) and detent ball (30), then pull control spool (20) from bore and discard "O" ring (21). Also See Fig. O203D. DO NOT attempt to remove spool (20—Fig. O203B) until detent assembly is removed or damage to spool could result. Remove oil filter adapter (23) so oil passages can be inspected and cleaned.

Clean and inspect all parts. Control spool (20) and pressure regulator spool (44) should be free of nicks or scoring and should slide freely in their bores when lubricated. By-pass plunger (51) and seat (52) should be free of nicks, burrs or grooves. Springs (28, 45 and 49) should be free of rust, bends or fractures.

Refer to the following data as a guide for parts renewal.

Pressure Regulator Valve Spring
    Free length—in. ...........................$3\frac{5}{8}$
    Spring test lbs. @ in. 55-57 @ $2\frac{3}{4}$
By-Pass Valve Spring
    Free length—in. ...........................$3\frac{1}{2}$
    Spring test lbs.
      @ in. .......................$5\frac{1}{2}$-$6\frac{1}{2}$ @ $2\frac{7}{16}$
Detent Ball Spring
    Free length—in. ...........................$1\frac{9}{32}$
    Spring test lbs @ in. 15.3-18.7 @ $\frac{7}{8}$
Pressure Regulator Spool
    Diameter ........................0.6865-0.6869
Control Spool Land—
    Diameter ........................0.6865-0.6869
Pressure Regulator Spool
    Bore—I.D. ......................0.6875-0.6883
Control Spool Bore—I.D. 0.6875-0.6883

Reassemble by reversing the disassembly procedure.

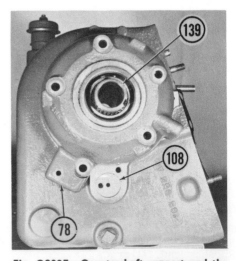

Fig. O203F—Countershaft support and the plug (108) must have cap screw hole positioned as shown to allow installation of support retaining cap screw.

309C. PUMP. Unbolt pump from housing cover and discard gasket. Remove drive gear outer retainer snap ring, remove gear and key, then remove gear inner retainer snap ring.

Remove through bolts and separate pump. Dowels can be driven out and screen removed, if necessary. See Fig. O203C.

NOTE: If same pump gears and shafts are to be reinstalled, mark idler gear so it will be reinstalled in original position. Do not remove gears from shafts as they cannot be serviced separately.

Pump dimensional data are as follows:

Gear width ....................0.9975-0.9980
Pump body width ..........0.9995-1.000
Gear shaft diameter ....0.6247-0.6250
Shaft bore diameter ....0.6265-0.6275
Shaft operating
　　clearance ....................0.0015-0.0028

Reassemble by reversing the disassembly procedure and use new gasket when mounting pump on housing cover. Heads of bolts should be on drive gear side of pump and dowels should be installed prior to tightening through bolts.

309D. INPUT SHAFT AND SUPPORT. Remove "O" ring (158—Fig. O203B) from support (153), then remove bearing (166) and seal (165) from bore in rear of input shaft. Unbolt bearing retainer (161) and press input shaft (164) and bearing (160) from support. Oil seal (157) can now be removed from support, and if renewal is required, clutch release bearing tube (156) can also be removed at this time. Do not remove bearing (160) from input shaft unless renewal of bearing or bearing retainer (161) is indicated. If either part is to be renewed, remove snap ring (159) and press input shaft from bearing.

To reassemble input shaft and support, place bearing retainer on input shaft, then press bearing on shaft and install snap ring. Press bearing tube (if renewed) into support and be sure tube is parallel with bore. Install oil seal (157) in support with lip of seal toward rear. Press input shaft and bearing into support and secure with bearing retainer. Install oil seal (165) in rear bore of input shaft with lip facing toward rear, then install bearing (166). Install "O" ring (158) in groove of support.

Refer to paragraph 309H for installation procedure.

309E. OUTPUT SHAFT, INTERNAL RING GEAR AND DIRECT DRIVE CLUTCH. To disassemble the output shaft, internal ring gear and direct drive clutch, proceed as follows: If planet gear carrier remained with output shaft assembly during disassembly, separate it from output shaft, then remove snap ring (138) and output shaft (139) from internal ring gear (152). With a press, compress spring (146), remove snap ring (148), retainer (147), return spring (146), guide (170) and thrust washer (145) from output shaft. Remove clutch piston (142) from output shaft (gear), either by bumping end of output shaft on a block of wood, or by directing compressed air into oil passage of output shaft, then remove seal rings (140 and 144) and "O" rings (141 and 143) from I.D. and O.D. of piston. Remove seal rings (135), thrust washer (136) and bushing (137) from rear of output shaft. Remove pressure plate (149), clutch plates (150) and separator plates (151) from internal ring gear.

Clean and inspect all parts. Sprag (overrunning) clutch surface of output shaft must be smooth and show no signs of wear. Clutch plates must have no broken lugs, splines or warpage. Refer to the following data as a guide for parts renewal.

Return Spring
　Free length ...............................2.270
　Spring test lbs. @ in. ....230 @ 1.750
Clutch Plate Thickness ......0.076-0.084
Separator Plate Thickness 0.087-0.093
Pressure Plate Thickness ..0.248-0.252
Piston Ring Groove—Outside
　Width ..................................0.156-0.162
　Depth ..................................0.193-0.194
Piston Ring Groove—Inside
　Width ..................................0.128-0.132
　Depth ..................................0.127-0.128

Reassemble output shaft, internal ring gear and direct drive clutch assembly as follows: Start with a clutch plate (150) and alternately install the six clutch plates and five separator plates (151) in internal ring gear (152), then install pressure plate (149). Install "O" rings (141 and 143) and seals (140 and 144) in O.D. and I.D. grooves of piston (142), grease seals and press piston in front side of output shaft gear. Use caution when installing piston not to damage seals. Install washer (145), guide (170), return spring (146) and retainer (147) on front end of output shaft, compress spring and install snap ring (148). Install output shaft assembly in internal ring gear with clutch piston next to clutch plates and install snap ring (138). Install thrust washer (136) over splined end of output shaft, then install the two seal rings (135) in grooves of output shaft.

NOTE: Bushing (137) is catalogued as a separate part, however, if renewal is not indicated do not attempt to remove it from output shaft.

309F. PLANET GEAR CARRIER. To disassemble the planet gear carrier assembly, first remove drive gear (127) from carrier (133) by removing the four retaining cap screws. Remove the planet gear pin retaining rings (134) and drive panet gear pins (130) from carrier. Remove planet gears (132) from carrier and bearings (131) from planet gears.

Inspect planet gears, bearings and pins for wear, scoring, or other damage and renew as necessary. Inspect drive gear teeth and inside diameter. Inside diameter of gear should not be unduly worn or scored. Check condition of bearing (126) and with bearing in gear check fit of gear and bearing on support (105). Gear and bearing should have a free fit on support without excessive clearance. At this time, also inspect bushing (125) for wear or scoring.

To reassemble planet carrier, place planet gear pin retainer in groove at front of pin bore. Place bearing in planet gear, position gear in carrier, then insert planet gear pin from rear of carrier until pin is against retainer and step on rear of shaft is flush with lower surface of carrier. Repeat operation for the three remaining planet gears. Place drive gear on rear of carrier, align cap screw holes, then install retaining cap screws and tighten them to 48-52 ft.-lbs. torque. Bushing (125) and bearing (126) are placed on support (105) when unit is being assembled.

309G. OVERDRIVE CLUTCH AND SUN GEAR. If removed with overdrive clutch assembly, remove sun gear shaft (103) from overdrive clutch. Remove overrunning clutch (100), spacer (101), bearing (102) and seal rings (104) from sun gear shaft. Bump support (105) from housing, if necessary, and remove seal ring (106). Remove clutch plates (98), separator plates (97), retainer (96), return springs (95) and return spring plate (94) from housing (78). See Fig. O203G. Remove clutch piston (91—Fig. O203B) from housing by bumping housing on a wood block. Remove seal rings (89 and 93) and "O" rings (90 and 92) from O.D. and I.D. of piston. Remove oil seal (75), snap ring (76) and bearing (77) from bore of housing.

With overdrive clutch disassembled, discard oil seal (75), seals (89 and 93) and "O" rings (90 and 92), then clean and inspect all parts. Clutch plates and separator plates must not be warped, scored or unduly worn.

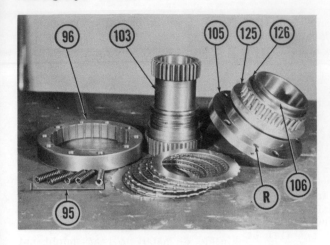

Fig. O203G—Component parts of overdrive clutch, sun gear shaft and support except clutch housing and spring plate. Note small "O" ring (R) in front side of support (105).

Fig. O203H—Install bearing (102), spacer (101) and Sprag clutch (100) in sequence shown. Drag strips (bronze) of Sprag clutch are toward front of sun gear shaft (103).

Clutch splines of sun gear shaft must not have deep grooves worn by clutch plates and bore for overrunning clutch must be smooth and show no wear. Refer to the following data as a guide for parts renewal.

Return Springs
    Free length .................1.320
    Spring test lbs. @ in. ..10 @ 1.159
Clutch Plate Thickness ...0.073-0.078
Separator Plate Thickness 0.077-0.082
Piston Ring Groove—Outside
    Width .................0.128-0.132
    Depth .................0.195-0.197
Piston Ring Groove—Inside
    Width .................0.128-0.132
    Depth .................0.151-0.152

Partially assemble overdrive assembly as follows: Install bearing (77) in housing (78) and secure with snap ring (76). Install oil seal (75) with lip toward bearing. Install "O" rings in outer and inner grooves of piston (91), install seals over both "O" rings, then grease seals and install piston in housing (78). Install seals (104) in grooves of sun gear shaft (103). Place bearing (102) and spacer (101) in bore of sun gear shaft, then install the overrunning clutch so the drag strips (bronze) are toward bearing (102). See Fig. O203H. Delay any further assembly of the overdrive clutch until the Hydraul Shift unit is assembled.

**309H. REASSEMBLE HYDRAUL SHIFT UNIT.** When reassembling the Hydraul Shift unit, proceed as follows: Install seal ring (106—Fig. O203G) on hub of support and the small "O" ring (R) in oil hole counterbore on front side of support flange, then install support in rear bore of housing so oil hole in support flange (with "O" ring) is aligned with the oil hole in housing which is on right side of lower cap screw hole. Temporarily install one of the short cover retaining cap screws finger tight to keep support from shifting

when planet carrier and output shaft assembly are installed.

Coat countershaft thrust washers (167—Fig. O203B) with grease and stick them to inner sides of countershaft support bores and be sure tangs of thrust washers are in slots. With bearings (168) installed, carefully place countershaft in housing, with double gear rearward, so thrust washers are not dislodged, then pull countershaft toward cover opening until gear teeth clear front bore of housing. Place bushing (125) on support with cupped side rearward, then install bearing (126) on support and use fingers to compress seal ring, if necessary. Install planet carrier and drive gear (gear rearward) in housing with gear (127) over bearing (126). Be sure bushing (137), thrust washer (136) and seal rings (135) are installed on splined (long) end of output shaft, start output shaft through planet carrier and support and install internal ring gear over the planet gears.

Place new "O" ring (107) on front end of countershaft support (110). Move countershaft into position, align thrust washers, lift countershaft so support will not bind, then slide support rearward until support is ready to enter rear bore. At this time, position tapped hole in rear end of support at the nine o'clock position and push shaft rearward until front of support is flush with housing.

NOTE: When overdrive clutch housing (78) is finally installed, holes in housing, plug (108) and support (110) must be aligned to allow installation of the ¼ x 1½ inch support retaining cap screw, as well as to place the oil hole in support on top side. Support is slotted at forward end if subsequent positioning is required.

With countershaft support positioned, place new "O" ring (107) on

O.D. of plug (108), then use grease to stick the small "O" ring (109) in counterbore of cap screw hole in front of plug. Insert cap screw (82) through plug and small "O" ring (109) and thread a few turns into support, then bump plug into housing bore. Remove cap screw after plug is installed. See Fig. O203F.

Obtain a small ½-inch thick wood block about 3x3 inches square, place block against forward end of output shaft, then tip assembly forward so output shaft is supported by the wood block. With small end forward, place sun gear shaft assembly over output shaft and push sun gear shaft on output shaft until it bottoms.

NOTE: Installation of overrunning clutch can be checked at this time. Sun gear shaft will turn (clutch overrun) clockwise but will lock up when attempt is made to turn shaft counterclockwise.

Remove cap screw that was installed earlier in flange of support (105—Fig. O203B). To ease assembly of overdrive clutch, install two guide studs in cap screw holes of housing (111). Install retainer (96) over guide studs with chamfer in I.D. of retainer on top side. Start with a separator plate (97) and alternately install the six separator plates and the five clutch plates (98). Place thrust washer (99) over end of output shaft and be sure it does not hang up on shoulder of output shaft. Place return springs (95) in blind holes of retainer (96), then place plate (94) over return springs. Place new "O" ring (86) over flange of support (105) and small "O" rings (85) in counterbores on front face of housing (78), then carefully install housing (78) over guide studs. Remove guide studs and secure clutch housing with cap screws. Install countershaft support retaining cap screw (82). Place new "O" ring (74), sprocket (73), lock washer (72) and nut (71) on end of output shaft and using a spanner wrench (Oliver No. ST-149, or equivalent), tighten nut (71). Secure nut by bending a tab of lock washer into slot in nut.

Set unit in an upright position, align splines and center direct drive clutch plates as nearly as possible. Install new "O" ring (158) on pilot of front support (153), then insert input shaft hub into direct drive clutch and rotate input shaft as necessary to engage the clutch plates. NOTE: Two cap screws can be temporarily installed to help in overcoming resistance of "O" ring; however, BE SURE input shaft can be rotated while drawing support into housing to prevent damage to inner clutch plates. Cut-out (C—Fig. O203E) in O.D. of support flange must mate with a similar cut-out (C) at right side of housing.

Place the wood block used earlier under front of housing to tip it slightly rearward, then remove the two temporarily installed cap screws from support. Place engine clutch housing on support and secure clutch housing and support to Hydraul Shift unit housing. Install clutch throw-out bearing and lubrication hose, throw-out bearing fork and clutch shaft. Use new gasket and install cover and pump assembly. Install cover-to-overdrive clutch oil line and the control spool lever. When installing new oil filter, tighten filter not more than ½-turn after gasket contacts base. Install drain plug if necessary and fill unit with 3½ quarts of Automatic Transmission Fluid, Type "A".

# TRANSMISSION

## OVERHAUL

### All Models

All series tractors are equipped with a constant mesh, helical gear transmission. Several changes have been made and in some cases, will require different service procedures. When this occurs, it will be noted in the appropriate section. Be specific in regard to tractor serial numbers when ordering replacement parts. Tractors equipped with Reverse-O-Torc will not have any reversing gears as this function is accomplished by the Reverse-O-Torc unit. Spacers are substituted for the reversing gears.

**310. TRANSMISSION TOP COVER.** On models with batteries under platforms, remove battery inspection plate, disconnect battery cables and protect cable ends to prevent possible shorting. Loosen battery case bolts and remove battery case and battery. Disconnect brake springs and on series 1950-T, also disconnect clutch spring. Remove all platforms. Disconnect power steering lines from flow divider (early models with flow divider on front of lift housing) and hydraulic housing, if so equipped. On late models with hydraulic system, remove return line. Disconnect starter safety switch wires at the connectors. Unbolt and remove the gear shift lever and lift the two springs from top cover. Unbolt and remove transmission top cover.

Before installing transmission top cover, check the safety switch and if

necessary, adjust same until switch opens when the roller is about one-half way up the side of the interlock ramp.

Use new gasket and reinstall top cover by reversing the removal procedure.

**311. R&R BULL GEAR COVER.** Unhook brake return springs and remove both platforms. On tractors with wheel guard extensions, remove brake pedal cover and the splash panels. Remove seat assembly, then unbolt and remove the bull gear cover.

**312. TRACTOR SPLIT.** Except for the mechanical four wheel drive models, all tractors can be split without removing the fuel tank and instrument panel providing proper tools are available. A special heavy duty ¾-inch drive 11/16-inch socket, proper extensions and a break-over handle are required. To make a split on mechanical four wheel drive tractors will require removal of engine to gain access to the front frame to rear frame bolts.

To make tractor split, remove pto drive shaft or the hydraulic pump drive shaft. Remove sprocket coupling chain and on models with no auxiliary drive unit, remove clutch shaft shield. Disconnect speedometer drive cable from pinion shaft front housing. Disconnect tail light wires and Hydra-Lectric wires at connectors under support. Disconnect power steering oil line from flow divider and oil cooler return line from front of

lift housing. Block between front axle and main frame, or support front frame so it will not tip, then support rear main frame with a rolling floor jack. Work from top and bottom sides of front frame and remove the front frame to rear frame cap screws and separate tractor.

Reassemble by reversing the disassembly procedure and tighten the front frame to rear frame cap screws to 280-300 ft.-lbs. torque.

**315. SHIFTER RAILS AND FORKS.** Remove transmission top cover as outlined in paragraph 310. Remove the hydraulic lift as in paragraph 414, or the bull gear cover as outlined in paragraph 311. Disconnect the bevel pinion shaft oil hose, and on models equipped with Creeper Drive, remove locknut guide and shaft from rear frame, then remove the cap screws which retain shifter forks to shifter rails. Push the two outer rails forward and push out the expansion plugs located at forward ends of shifter rails. Be careful not to lose poppet balls and springs as rails clear rear bore. Prior to removing the center shifter rail, remove shifter lug and if so equipped, the cotter pin.

NOTE: On tractors equipped with an auxiliary drive that do not have sufficient clearance to allow removal of shifter rails, the engine must be removed from the mechanical four wheel drive models, or tractor must be split as outlined in paragraph 312.

Reinstall as follows: Place poppet spring in center poppet hole and place poppet ball on top of spring. Insert center shifter rail part way into rear frame, and if so equipped, install the cotter pin at inner end of shifter rail. Turn shifter rail so flat on inner end is downward, then use a small punch to depress poppet ball and spring and push shifter rail rearward until rail goes over poppet ball. Turn shifter rail until poppet ball seats in detent, then install shifter lug and fork. Install right and left shifter rail in the same manner. Use sealing compound and install new expansion plugs in shifter rail front bores.

Complete balance of reassembly by reversing disassembly procedure.

**316. LUBRICATION PUMP.** Transmissions of all 1800 and 1900 series C and later tractors are pressure lubricated by a pump mounted on rear end of transmission countershaft. An input shaft conversion package was made available for all 1800 and 1900 series A and B tractors, except those with

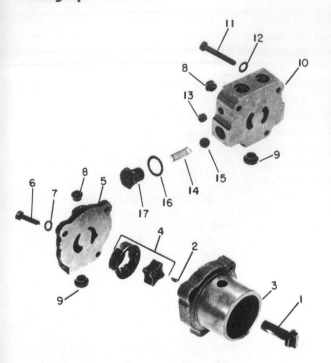

**Fig. O204 — Exploded view of transmission lubrication pump. Cover (5) is used on all models except those equipped with Reverse-O-Torc Drive. On Reverse-O-Torc equipped models, valves (15) in pump cover (10) allow pump to pressurize the transmission lubrication circuit when rotating in either direction.**

1. Shaft
2. Key
3. Body
4. Rotor set
5. Cover (except R.O.T)
8. Tubing seats
9. Tubing seats
10. Cover (R-O-T only)
13. Plug
14. Valve springs (4)
15. Valve balls (4)
16. Sealing washers (4)
17. Plugs (4)

Reverse-O-Torc, which provides lubrication for the transmission from a pump mounted on front end of countershaft. The front mounted pump is driven by a slotted drive stud screwed into the tapped hole in front end of countershaft and the pump body functions as the countershaft front bearing retainer (cover).

To remove the front mounted pump, drain transmission, disconnect filter line from pump cover connector and pump inlet outer pipe from inlet line plug, then unbolt and remove pump cover assembly. Pump body can now be removed from rear frame, however, be careful not to lose or damage shims located back of the pump body.

To remove the rear mounted pump, remove the hydraulic lift housing, or the bull gear cover. Disconnect suction tube from pump and remove the pressure tube (pump to oil filter tube). Depress the pump housing lock with a small rod or punch inserted through hole in main frame casting at front of pump and withdraw pump assembly from bore in main frame.

Refer to Fig. O204 for an exploded view of the pump which is mounted on rear end of transmission countershaft. Pumps which are mounted on front of transmission countershaft are very similar except for the pump housing (body).

Inner rotor should be a free fit on pump shaft (1) with drive key (2) installed. Polish shaft with crocus cloth, or file drive key, if binding of rotor on shaft occurs. Pump shaft (1) has two key slots and in some cases, binding conditions can be cor-

rected by moving key to opposite slot. On tractors equipped with Reverse-O-Torc, pump cover (10) is fitted with four check balls (15), springs (14) and retainers (17). If damaged, tubing seats (8 and 9) can be removed by threading with tap and pulling the seats from pump cover.

NOTE: Pumps which mount on front of countershaft may have poppet type check valve plungers in pump cover instead of the ½-inch steel balls (15) used in the rear mounted pumps.

When installing rear mounted pumps, be sure to insert lock spring and plunger in pump body (See Fig. O205), then depress the lock and insert pump in bore of main frame.

**317. R&R INPUT SHAFT.** Several input shaft modifications have been made since the introduction of the series 1800 and 1900 tractors. The original input shaft assembly used is shown by numbers 86 through 99 in Fig. O206. At tractor serial number 94 271-000, the input shaft assembly was changed to the type shown by numbers 14 through 27 in Fig. O206 and a conversion package (Oliver No. 106 446-AS) was made available to convert the original assembly to the later type.

Beginning with tractor serial number 134 684-000, the input shaft assembly was pressure lubricated and included parts numbered 13 through 28 as shown in Fig. O207 and a conversion package (Oliver No. 158 290-AS) was made available to modify the earlier input shaft assemblies which used the nut on aft end of input shaft. (Note: If tractor is a four wheel drive

model a supplemental transfer case package, Oliver No. 158 264-AS, must also be used which provides installation room for the pressure pump assembly). The conversion package contains the pressure pump and the necessary input shaft components required.

All of the conversion packages include instruction sheets that contain the necessary procedure information.

Refer to the following appropriate paragraph for information pertaining to the input shaft assemblies.

**318. INPUT SHAFT (PRIOR SERIAL NO. 94 271-000).** To remove the input shaft, split tractor as outlined in paragraph 312.

NOTE: If tractor does not have an auxiliary drive unit it is not absolutely necessary to split the tractor, however, most mechanics prefer to do so because of the limited working room.

Drain rear frame, then refer to the appropriate sections and remove pto drive shaft or hydraulic pump drive shaft, transmission top cover and hydraulic lift unit or bull gear cover.

Remove the input shaft front bearing retainer (7—Fig. O206) and shims and remove oil seal (11) from retainer. Remove the snap ring (97) from spacers (96) by sliding it forward onto thrust washer (94). Remove the two spacers. Remove input shaft rear snap ring (95) from its groove, then slide snap ring, thrust washer (94) and spacer snap ring (97) rearward on shaft. Place a spacer between front input shaft gear (89) and housing so gears cannot move forward. Use an Oliver No. ST-188 driver, or equivalent, bump shaft forward until front bearing (13) is free of housing and input shaft is free of rear bearing (98).

Note: The spacer used to block the gears can be made from a piece of 3½ inch I. D. pipe, 1½ inches long

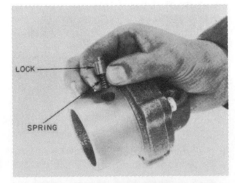

**Fig. O205 — Lock assembly of transmission lubrication pump is installed as shown.**

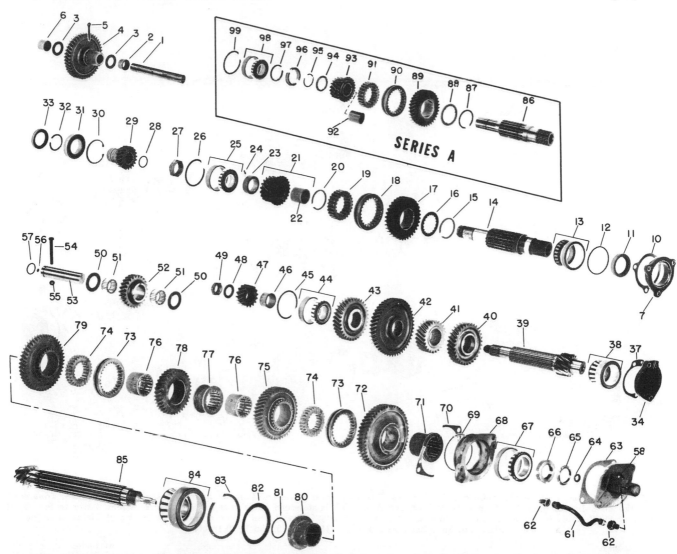

**Fig. O206 — Exploded view of the 1800 and 1900 series B transmission gears and shafts. Note also the input shaft components shown in inset which were used in early 1800 and 1900 series A tractors. Refer also to Figs. O207 and O208.**

| | | | | |
|---|---|---|---|---|
| 1. Idler gear shaft | 22. Bushing | 43. 4th & 6th speed gear | 63. Gasket | 82. Oil seal |
| 2. Bearing | 23. Spacer | 44. Bearing assembly | 64. Oil seal | 83. Snap ring |
| 3. Thrust washers | 24. Spring pin | 45. Snap ring | 65. Lock nut | 84. Bearing assembly |
| 4. Idler gear | 25. Bearing assembly | 46. Locating sleeve | 66. Adjusting nut | 85. Countershaft |
| 5. Cotter pin | 26. Snap ring | 47. Reverse gear | 67. Bearing assembly | 86. Input shaft |
| 6. Bearing | 27. Nut | 48. Washer | 68. Bearing cage | 87. Snap ring |
| 7. Bearing retainer | 28. "O" ring | 49. Nut | 69. "O" ring | 88. Thrust washer |
| 10. Shims | 29. Pump drive pinion | 50. Thrust washers | 70. Shims | 89. High range gear |
| 11. Oil seal | 30. Snap ring | 51. Bearings | 71. Mounting sleeve | 90. Shift coupling |
| 12. "O" ring | 31. Bearing | 52. Reverse idler gear | 72. 1st & 3rd speed gear | 91. Drive collar |
| 13. Bearing assembly | 32. Snap ring | 53. Idler gear shaft | 73. Shift couplings | 92. Bushing |
| 14. Input shaft | 33. Oil seal | 54. Cap screw | 74. Drive collars | 93. Low range gear |
| 15. Snap ring | 34. Cover | 55. Self-locking nut | 75. 2nd & 5th speed gear | 94. Thrust washer |
| 16. Thrust washer | 37. Shims | 56. Plug | 76. Mounting sleeves | 95. Snap ring |
| 17. High range gear | 38. Bearing assembly | 57. "O" ring | 77. Spacer | 96. Spacer |
| 18. Shift coupling | 39. Countershaft | 58. Speedometer drive | 78. 4th & 6th speed gear | 97. Snap ring |
| 19. Drive collar | 40. High range gear | housing | 79. Reverse gear | 98. Bearing assembly |
| 20. Snap ring | 41. 2nd & 5th speed gear | 61. Oil hose | 80. Mounting sleeve | 99. Snap ring |
| 21. Low range gear | 42. Low range gear | 62. Connectors | 81. "O" ring | |

with a longitudinal section cut out to permit it to slip over the input shaft.

Shaft can now be disassembled.

When installing new bushing (92— Fig. O206), install same from helical tooth end of low range input gear (93) and position so it is $\frac{3}{32}$-inch below end surface of gear. Should it be that bushing fits loosely in gear, installation procedure will be aided by installing bushing on input shaft

instead of in bore of gear. Vary shims (10) under bearing retainer (7) to provide between 0.001 shaft end play and 5 inch pounds bearing preload. Shims are available in thicknesses of 0.004, 0.007 and 0.0149.

319. INPUT SHAFT (SERIAL NO. 94 271-000 THROUGH 134 683-000). To remove the input shaft, split tractor as outlined in paragraph 312.

NOTE: If tractor does not have an auxiliary drive unit it is not absolutely necessary to split the tractor, however, most mechanics prefer to do so because of the limited working room.

Drain rear frame, then refer to the appropriate sections and remove pto drive shaft or hydraulic pump drive shaft, transmission top cover and hydraulic lift unit or bull gear cover.

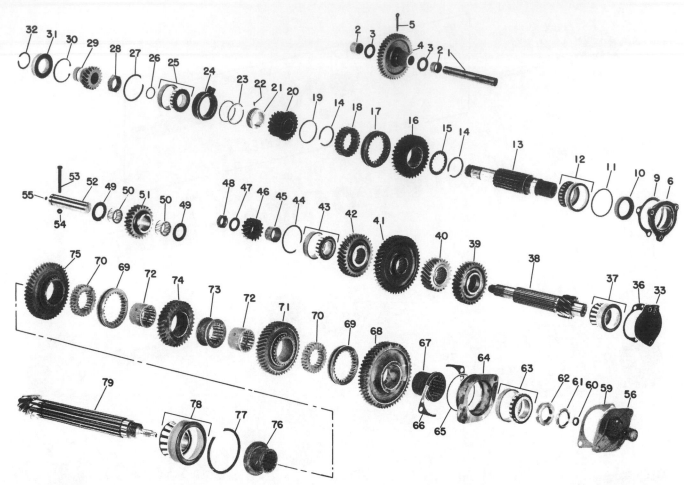

Fig. O207 — Exploded view of the 1800 and 1900 series C and all later series transmission gears and shafts for two-wheel drive models. Note the oil collector assembly (items 21, 22, 23, and 24) which is standard equipment on these transmissions. Refer to Fig. O206 for 1800 and 1900 series A and B transmission gears. Refer to Fig. O208 for exploded views of the different bevel pinion shaft assemblies that have been used in 1800, 1850, 1900 and 1950 tractors.

| | | | | |
|---|---|---|---|---|
| 1. Idler gear shaft | 18. Drive collar | 33. Cover | 50. Bearings | 66. Shims |
| 2. Bearings | 19. Sealing ring | 36. Shims | 51. Reverse idler gear | 67. Mounting sleeve |
| 3. Thrust washers | 20. Low range gear | 37. Bearing assembly | 52. Idler gear shaft | 68. 1st & 3rd speed gear |
| 4. Idler gear | 21. Spacer | 38. Countershaft | 53. Cap screw | 69. Shift coupling |
| 5. Cotter pin | 22. Spring pin | 39. High range gear | 54. Self-locking nut | 70. Drive collars |
| 6. Bearing retainer | 23. Sealing rings | 40. 2nd & 5th speed gear | 55. Plug | 71. 2nd & 5th speed gear |
| 9. Shims | 24. Oil collector | 41. Low range gear | 56. Speedometer drive | 72. Mounting sleeves |
| 10. Oil seal | 25. Bearing assembly | 42. 4th & 6th speed gear | housing | 73. Spacer |
| 11. "O" ring | 26. "O" ring | 43. Bearing assembly | 59. Gasket | 74. 4th & 6th speed gear |
| 12. Bearing assembly | 27. Snap ring | 44. Snap ring | 60. Oil seal | 75. Reverse gear |
| 13. Input shaft | 28. Nut | 45. Locating sleeve | 61. Lock nut | 76. Mounting sleeve |
| 14. Snap rings | 29. Pump drive pinion | 46. Reverse gear | 62. Adjusting nut | 77. Snap ring |
| 15. Thrust washer | 30. Snap ring | 47. Washer | 63. Bearing assembly | 78. Bearing assembly |
| 16. High range gear | 31. Bearing | 48. Nut | 64. Bearing cage | 79. Pinion shaft |
| 17. Shift coupling | 32. Snap ring | 49. Thrust washers | 65. "O" ring | |

Remove the input shaft front bearing retainer (7—Fig. 0206) and shims and remove oil seal (11) from retainer. Remove cotter pin from idler gear (4), pull shaft (1) forward and remove gear (4) and thrust washers (3). Remove "O" ring (28) from drive pinion (29), then remove snap ring (30) from front side of bearing (31) and bump pinion assembly forward out of bore. Unstake and remove nut (27) from aft end of input shaft and move the sliding coupling (18) onto the forward (high range) input gear. Remove snap ring (26), then remove snap ring (15) from its groove and slide it forward on input shaft. Use driver ST-138, or equivalent, and

bump input shaft rearward until rear (low range) input gear (21) and drive collar (19) can be separated far enough to allow removal of snap ring (20) which is hidden by the drive collar. Remove snap ring (20) from its groove and slide it rearward on shaft. Place a spacer between front input shaft gear and housing so gears cannot move forward, then using an Oliver STS-100 push-pull assembly and an Oliver ST-138 driver, or equivalent units, push input shaft forward until front bearing cup is free of housing and input shaft is free of rear bearing.

Note: The spacer used to block the

gears can be made from a piece of 3½ inch I. D. pipe, 1½ inches long with a longitudinal section cut out to permit it to slip over the input shaft.

Shaft can now be disassembled.

When reinstalling new bushing (22), install same from helical tooth end of low range input gear (21) and position so it is $\frac{3}{32}$-inch below end surface of gear. Should it be that bushing fits loosely in gear, installation procedure will be aided by installing bushing on input shaft instead of in bore of gear. Shift center shifter rail to its forward position, insert input shaft through front bearing bore in housing and install component parts

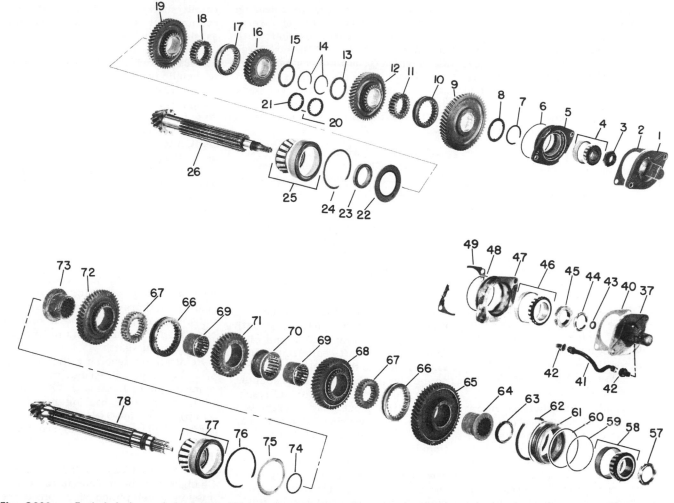

Fig. O208 — Exploded views of the bevel pinion shaft assemblies that have been used. Upper shaft and gears were used in 1800 and 1900 series A tractors. Lower shaft and gears are used in 1800 and 1900 series B and C and all later tractors. Center bearing cage and bearing assembly (items 37 through 49) are used with two-wheel drive tractors. Lower bearing cage and bearing assembly (items 57 through 63) are used on four-wheel drive tractors.

| | | |
|---|---|---|
| 1. Speedometer drive housing | 12. 2nd & 5th speed gear | 22. Oil seal |
| 2. Gasket | 13. Thrust washer | 23. Spacer |
| 3. Nut (self-locking) | 14. Snap rings | 24. Snap ring |
| 4. Bearing assembly | 15. Thrust washer | 25. Bearing assembly |
| 5. Bearing cage | 16. 4th & 6th speed gear | 26. Bevel pinion shaft |
| 6. "O" ring | 17. Shift coupling | 37. Speedometer drive housing |
| 7. Snap ring | 18. Drive collar | 40. Gasket |
| 8. Thrust washer | 19. Reverse gear | 41. Oil hose |
| 9. 1st & 3rd speed gear | 20. Spring (6 used) | 42. Connectors |
| 10. Shift coupling | 21. Snap ring retainers | 43. Oil seal |
| 11. Drive collar | | 44. Lock nut |
| | | 45. Adjusting nut |
| | | 46. Bearing assembly |

| | |
|---|---|
| 47. Bearing cage | 67. Drive collars |
| 48. "O" ring | 68. 2nd & 5th speed gear |
| 49. Shims | 69. Mounting sleeves |
| 57. Nut | 70. Spacer |
| 58. Bearing assembly | 71. 4th & 6th speed gear |
| 59. "O" ring | 72. Reverse gear |
| 60. "O" ring | 73. Mounting sleeve |
| 61. Bearing cage | 74. "O" ring |
| 62. Snap ring | 75. Oil seal (series B) |
| 63. Spacer (selective) | 76. Snap ring |
| 64. Mounting sleeve | 77. Bearing assembly |
| 65. 1st & 3rd speed gear | 78. Bevel pinion shaft |
| 66. Shift couplings | |

of input shaft as shown in Fig. 0206. Be sure spring pin (24) is next to bearing and engages keyway. Tap spring pin to seat it in keyway when the spacer is in place. With bearing retainer in place, drive the rear bearing into position. Install nut on aft end of input shaft, tighten to a torque of 90-100 ft.-lbs. and stake nut into slot of input shaft. Install rear bearing cup and snap ring. Bump front of input shaft to seat bearings, retighten bearing retainer and check input shaft bearing adjustment which should be between 0.001 shaft end play and 5 inch pounds bearing preload. Add or subtract shims behind bearing re-

tainer, if necessary, to obtain specified end play. Shims are available in thicknesses of 0.004, 0.007 and 0.0149.

320. INPUT SHAFT (SERIAL NO. 134 685-000 AND UP). To remove the input shaft, split tractor as outlined in paragraph 312.

NOTE: If tractor does not have an auxiliary drive unit it is not absolutely necessary to split the tractor, however, most mechanics prefer to do so because of the limited working room.

Drain rear frame, then refer to the appropriate sections and remove pto drive shaft or hydraulic pump drive

shaft, transmission top cover and hydraulic lift unit or bull gear cover.

Remove the input shaft front bearing retainer (6—Fig. 0207) and remove oil seal (10) and "O" ring (11) from retainer. Remove cotter pin (5) from idler gear (4), pull shaft (1) forward and remove gear and thrust washers (3). Remove snap ring (30) from front side of bearing (31) and bump pinion assembly (29) forward out of bore. Remove the inlet oil line from oil collector (24) and tee at housing wall. Unstake and remove nut (28) from aft end of input shaft and move sliding coupling (17) onto the front (high range) input gear. Remove snap ring (27), then lift front snap ring (14) from its groove and slide it forward on input shaft. Use driver ST-138, or its equivalent, and drive input shaft rearward until rear (low range) input gear (20) and drive collar (18) can be separated far enough to allow access to rear snap ring (14) and Teflon ring (19), which are hidden by the drive collar. Cut and remove Teflon ring, then lift snap ring from its groove and slide it rearward on shaft. Place a spacer between front input shaft gear and housing so gears cannot move forward, then using driver, Oliver ST-138 or equivalent, bump input shaft forward until front bearing cup is free of housing and input shaft is free of rear bearing. Pull shaft from front of housing and remove gears from top of housing as shaft is removed.

NOTE: The spacer used to block the gears can be made from a piece of 3½ inch I. D. pipe, 1½ inches long with a longitudinal section cut out to permit it to slip over the input shaft.

Further disassembly of shaft and components is obvious.

321. Reinstall input shaft assembly as follows: Press front bearing cone on input shaft with smaller diameter forward. Place front snap ring (14) on shaft about ½-inch forward of its groove, then place thrust washer (15) next to the snap ring. Place seal rings (23) in grooves of spacer (21), oil seal rings freely, then install spacer and rings in oil collector (24) from flat side of oil collector using fingers to compress sealing rings. Push center shift rail to its forward position. Place shift coupling (17) on its hub of high range gear (16), start rear of input shaft in front bearing bore and install gear and collar on shaft with collar rearward and mated with its shifter fork. Install drive collar (18) with counterbore rearward. Start rear snap ring (14) on shaft spline with sharp ends forward but do not slide it into its groove. Place Teflon seal ring (19) on shaft at rear of snap ring (14). Install low range gear (20) on shaft with helical teeth toward rear. Place pin (22) in spacer, align it with keyway of input shaft and install the oil collector and spacer assembly on shaft with counterbored side of oil collector forward. Install the rear snap ring (14) in its groove and be sure Teflon sealing ring (19) is on splined portion of shaft. Install front bearing cup in its bore, install front bearing retainer (without seal and "O" ring) and shims and tighten retaining cap screws finger tight. Lubricate "O" ring (26) and install in groove next to threads

on aft end of shaft. Start rear bearing on shaft with smaller diameter rearward and bump bearing on shaft until nut (28) can be started. Be sure large diameter flange of nut is against bearing, tighten nut until bearing is tight against spacer (21), then complete tightening of nut to a torque of 90-100 ft.-lbs. Stake nut into keyway of shaft.

NOTE: While tightening nut be sure sealing ring (19) stays on splined portion of input shaft or damage to sealing ring will occur.

Install rear bearing cup and snap ring (27), then bump on forward end of shaft to seat cup against snap ring. Slide drive collar (18), high range input gear (16) and thrust washer (15) rearward and install front snap ring (14) in its groove.

Further tighten the front bearing retainer cap screws, if necessary, until input shaft has only a slight end play. Demesh input shaft gears from countershaft gears, then using an inch-pound torque wrench and adapter, check and record the rotational resistance of the input shaft. Completely tighten the bearing retainer cap screws and recheck the shaft rotational drag which should be not more than 5 inch-pounds greater than the previously recorded value. Vary shims (9) as required to obtain this adjustment. Shims are available in thicknesses of 0.004, 0.007 and 0.0149. Note: If shims are varied be sure shaft is bumped slightly forward after each check to insure that bearing cup contacts flange of bearing retainer.

With input shaft adjustment made, remove front bearing retainer and install oil seal (10) with lip rearward and "O" ring (11) in retainer. Lubricate oil seal and "O" ring and install retainer. Install input shaft oil line between oil collector and tee. Reinstall hydraulic pump drive pinion assembly and the idler gear and shaft assembly. Shift transmission to neutral.

322. R&R COUNTERSHAFT. To remove the countershaft, first remove input shaft as outlined in paragraph 318, 319 or 320. Unstake and remove nut (48) and washer (47). Use two ⅜x1⅛ inch cap screws, fitted with nuts, between gear (39) and front wall of transmission housing to prevent loading of gears during shaft removal.

On early model tractors which do not have a lubrication pump installed, remove front bearing cover (33—Fig. 0207) and shims (36), attach a puller to the threaded hole in front of

countershaft and pull shaft. Remove countershaft gears from top of housing as shaft is withdrawn.

On models prior to serial number 134 684-000 which have been modified to include the transmission lubrication pump, disconnect lines, remove the pump, shims (36) and pump drive stud from front of countershaft, then attach puller and remove countershaft. Lift gears from top of housing as shaft is withdrawn.

On models serial number 134 684-000 and up, disconnect filter line from pump, depress lock and pull pump from rear of countershaft. Use a tool which will fit into pump drive slot in rear of countershaft and bump shaft forward until front bearing cup clears housing and rear bearing cone is free of countershaft. Remove countershaft gears from top of housing as countershaft is withdrawn. If necessary, remove snap ring (44) and rear bearing cup (43) from housing and press front bearing cone from shaft.

Reinstall by reversing the removal procedure and tighten nut to a minimum torque of 150 ft.-lbs. Vary shims (36) under front bearing cover (or pump) to provide 0.001-0.003 countershaft end play. Shims (36) are available in thicknesses of 0.004, 0.007 and 0.0149.

323. R&R BEVEL PINION SHAFT. Changes have been made in the transmission bevel pinion shaft since the introduction of the series 1800 and 1900 tractors. Refer to Fig. O208 which shows the difference in bevel pinion shaft assemblies and the manner in which the shaft bearings are adjusted.

324. TWO-WHEEL DRIVE. Two-wheel drive tractors prior to tractor serial number 124 396-000 used the bevel pinion shaft assembly shown at top of Fig. 0208. Two-wheel drive tractors serial number 124 396-000 and up use the bevel pinion shaft components (items 64 through 78) shown at bottom of Fig. 0208 along with bearing cage, shims and nuts (items 37 through 49) as shown at center of Fig. 0208. Bottom bevel pinion shaft shown in Fig. 0208 (including items 57 through 63) is used in four-wheel drive tractors; however, some two-wheel drive tractors may be equipped with the four-wheel drive pinion shaft assembly. Refer to paragraph 329 if this situation is encountered.

325. To remove the early type bevel pinion shaft (26—Fig. 0208), first remove the input shaft as outlined in paragraph 318 or 319, the countershaft as outlined in paragraph 322 and the differential as outlined in paragraph 335.

Note: While it is not mandatory that the shifter rails and forks be removed, their removal will simplify procedure.

Disconnect oil hose from speedometer housing (1), if so equipped, then unbolt and remove speedometer housing and gasket (2) from pinion shaft bearing cage (5). Wrap forward end of shaft to prevent damage to speedometer drive, then remove nut (3) from forward end of shaft. Install two ⅜-16 cap screws in the tapped holes provided in bearing cage and remove bearing cage; then remove snap ring (7) and thrust washer (8). Lift center and rear snap rings (14) from their grooves, move snap rings forward on shaft and, at the same time, move shaft toward rear. Continue this operation until shaft is removed from housing. Remove pinion shaft component parts from top of housing. BE SURE to renew oil seal (22) whenever pinion shaft is removed.

326. Reinstall by reversing removal procedure and use a wooden block approximately 2½ inches thick under reverse gear (19) to prevent shaft resting on oil seal (22). Prior to complete reassembly however, pinion shaft preload must be set as follows: Use two, short ½-13 cap screws to secure bearing cage to housing. Block aft end of pinion shaft and drive front bearing on the shaft. Install shaft nut and tighten to a point where there is only a slight end play of shaft. Shift the front sliding coupling rearward onto output gear (12) and place the rear sliding coupling in neutral position. Wrap a length of cord around output gear (12), attach a spring scale to string and while pulling on the scale, note and record the reading when the pinion shaft starts to turn. This reading indicates the amount of seal drag. Now tighten the pinion shaft nut (3), bump both ends of shaft to seat bearings and recheck spring scale reading required to rotate pinion shaft. Continue this operation until the spring scale reads 5 to 6 pounds greater than the seal drag reading.

Note: When bumping ends of shaft, use a piece of pipe over speedometer drive on front of shaft to preclude any possibility of damage.

Adjust differential bearings and backlash as outlined in paragraph 337.

NOTE: Tractors with serial numbers 122 513-000 through 124 395-000 have snap ring retainers (20 and 21) installed to assist in keeping snap rings (14) in their grooves. To provide access to snap rings (14), remove the six small springs (20) and push retainers (21) off snap rings.

Should the problem of snap rings coming out of their grooves be encountered on tractors prior to serial number 122 513-000, two snap ring retainer service packages are available. On tractors with serial numbers 90 525-000 through 95 275-000, having snap rings 0.107-0.111 or 0.108-0.110 thick, order packge part number 156 040-AS. On tractors with serial numbers 94 276-000 through 122 512-000, having snap rings 0.126-0.128 thick, order package part number 155 958-AS.

When installing snap ring retainers, be sure the shallower counterbore of retainers fits over snap rings and install a retainer spring in every third spline. Be careful not to overstress (warp) snap rings during installation or difficulty will be encountered in sliding retainers over snap rings. The projections (points) of snap ring ends should be away from gears when installed.

327. To remove the bevel pinion shaft from tractors serial number 124 396-000 and up, first remove the input shaft as outlined in paragraph 319 or 320, the countershaft as outlined in paragraph 322 and the differential as outlined in paragraph 335.

NOTE: While it is not mandatory that the shifter rails and forks be removed, their removal will simplify procedure.

Disconnect oil line (41—Fig. 0208) from speedometer housing (37), then unbolt and remove speedometer housing from bearing cage (47). Wrap forward end of shaft to prevent damage to speedometer drive, then remove lock nut (44) and adjusting nut (45). Remove and save shims (49). Install two ⅜-16 cap screws in the tapped holes provided in bearing cage (47) and remove bearing cage and front bearing. Pull shaft rearward and remove gears and spacers from top of housing as shaft is withdrawn. On 1800 and 1900 series "B" tractors, BE SURE to renew oil seal (75) whenever pinion shaft is removed.

328. Reinstall by reversing removal procedure and on 1800 and 1900 series "B" tractors, use a wood block approximately 2½ inches thick under reverse gear (72) to prevent pinion shaft from resting on seal (75). Note: Series C, and later tractors do not use seal (75). Prior to complete reassembly, bevel pinion shaft bearing preload must be adjusted as follows: Tight adjusting nut (45) until a slight amount of pinion shaft end play remains, then bump shaft both ways to seat bearings and bearing

cups. Slide front coupling (66) rearward onto gear (68) and place rear coupling (66) in neutral position. Wrap a length of cord, with a spring scale attached, around gear (68) and, on series B tractors check reading on scale required to keep shaft in motion. This is oil seal drag and must be used when setting bearing preload on series B tractors.

Tighten the bevel pinion shaft adjusting nut until the spring scale reads 3-5 pounds with shaft rotating or, on series B tractors, until the spring scale reads 3-5 pounds more than the oil seal drag reading. If the bevel pinion shaft adjusting nut tightens before correct preload is obtained, add an equal number of shims (49) behind flanges of bearing retainer (47) until the correct bearing preload can be obtained. Shims are 0.0149 thick.

With bearing preload adjusted, mount a dial indicator so a reading can be taken from spacer (70), then move gears back and forth by prying behind gears (65) and (72) and check indicator which should show a spacer spacer end play.

With correct bearing preload and end play of 0.001-0.017. If end play of gears on shaft is not as stated, vary number of shims (49) behind bearing retainer (47) and reset the bearing preload; adding shims will increase gear end play determined, loosen adjusting nut aproximately ¼-turn, install lock nut (44) and while holding adjusting nut (45), tighten lock nut securely. Recheck the bearing preload and if necessary, readjust nut (45). Stake lock nut when adjustment is complete.

329. FOUR-WHEEL DRIVE. Four-wheel drive tractors use a bevel pinion shaft which is similar, except for front bearing cage assembly (items 57 through 63), to the bevel pinion shaft used in the late two-wheel drive tractors. The four-wheel drive pinion shaft assembly may be installed in some two-wheel drive tractors. See lower bevel pinion shaft assembly in Fig. 0208.

To remove bevel pinion shaft, first remove the input shaft as outlined in paragraph 319 or 320, the countershaft as outlined in paragraph 322, the differential as outlined in paragraph 335 and the four-wheel drive transfer case as outlined in paragraph 38. Tape seal contact surface of bevel pinion shaft to prevent damage to shaft.

Note: While it is not mandatory that the shifter rails and forks be removed, their removal will simplify procedure.

Unstake and remove the nut (57—Fig. 0208); then using driving sleeve over oil tube on front end of shaft, bump shaft (78) rearward and remove front bearing, bearing cage (61) and spacer (63). Pull shaft rearward and lift gears and spacers from housing as shaft is moved rearward. On series B tractors, BE SURE to renew oil seal (75) whenever bevel pinion shaft is removed.

330. Reinstall by reversing the removal procedure and on series B tractors, use a wood block approximately 2½ inches thick under reverse gear (72) to prevent pinion shaft from resting on seal (75). Note: Series C and later tractors do not use seal (75). Prior to complete assembly however, bevel pinion shaft preload must be adjusted as follows: Use the original spacer (63) and tighten nut (57) until

end play is removed from shaft. Bump shaft both ways to seat bearings and cups. Slide front shift collar (66) rearward onto gear (68) and place rear shift collar (66) in neutral position. Wrap a cord around gear (68) and attach a spring pull scale to cord.

On series B tractors be sure bearings are not preloaded, then check and record reading on scale required to keep shaft in motion; this is oil seal drag.

Tighten nut until a scale reading of 3-5 pounds is required to keep shaft in motion or, on series B tractors, the reading is 3-5 pounds more than the previously determined oil seal drag. If nut tightens before correct bearing preload is reached, remove nut, front bearing and spacer (63) and install the next smaller spacer. Spacers are available in thicknesses of 0.288, 0.314 and 0.340.

With bearing preload adjusted, mount a dial indicator so a reading can be taken from spacer (70), then move gears back and forth by prying behind gears (65) and (72) and check indicator which should show a spacer end play of 0.001-0.027. If end play of gears on shaft is not as stated, change spacer (63) as required and readjust bearing preload. Spacers are available in thicknesses of 0.288, 0.314 and 0.340.

331. **R&R REVERSE IDLER GEAR.** To remove the reverse idler gear, first remove the bevel pinion shaft as outlined in paragraphs 323 through 329, then remove the bolt and nut which retain idler gear shaft (52—Fig. 0207) in housing and lift out shaft, gear (51) and thrust washers (49). Note: Models with Reverse-O-Torc drive are not equipped with a reverse idler.

# DIFFERENTIAL, BEVEL GEARS, FINAL DRIVE
# & REAR AXLE

## DIFFERENTIAL

### All Models

335. **REMOVE AND REINSTALL.** To remove the differential, first drain transmission and final drive comparment, then remove the pto drive shaft, or the hydraulic pump drive shaft. Remove the hydraulic lift unit or the bull gear cover. Attach a hoist to pto unit, then unbolt and remove unit from rear frame. Disconnect wiring and remove both fenders, then support tractor under rear frame and remove both wheel and tire units. Remove cap screws from pto actuating arms and remove pto actuating shaft and arms. Remove the draft control spring and support assembly and spacers. On either side of tractor, remove snap ring from inner end of axle and the cap screws retaining axle

carrier to rear frame; then using a pusher, push axle from bull gear and remove axle and carrier. Repeat operation on opposite side of tractor using a piece of bar stock to support pusher. Lift bull gears from rear frame.

336. On all models except 1950-T, disconnect right hand brake pull rod from brake lever, then remove cotter pin from right end of brake cross shaft and remove right brake pedal and lock assembly. Remove cap screw from left brake pedal and remove pedal and lock assembly and Woodruff key. Disconnect left brake pull rod from brake lever and pull brake cross shaft from rear frame. On all models, remove jam nuts, adjusting nuts and spherical washers from brake actuating rods. Remove brake housing covers and lift out brake discs and actuating assemblies.

On all models except 1950 serial number 189 007-000 and up, and 1950-T, proceed as follows.

Remove cap screws from brake housings and wedge them out far enough to remove the two sets of shim packs from behind each brake housing. Identify shims as to their locations and tie them together to prevent intermixing. Pull the right hand brake housing and bull pinion out until the right differential side gear can be grasped, then hold gear and bull pinion in place and pull assembly out of rear frame. Remove left brake housing and bull pinion assembly in same manner.

Turn differential so left shaft points toward left front corner of housing and lift out differential.

On series 1950, serial number 189 007-000 and up, and series 1950-T, in-

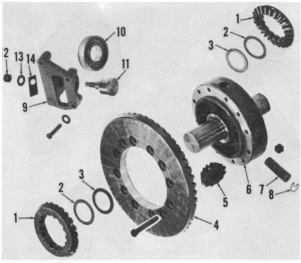

Fig. O210A—View showing component parts of the differential assembly used in series 1950-T tractors. Bearing shaft (11) is eccentric. Refer to text for adjustment of thrust bearing (10).

1. Side gear
2. Thrust bearing
3. Thrust washer
4. Bevel gear
5. Pinion gear
6. Shaft and spider
7. Pinion pin
8. Retaining ring
9. Support
10. Bearing (thrust)
11. Bearing shaft
12. Jam nut
13. Washer
14. Clip (lock)

stall a lifting eye in the tapped hole of one of the differential pinion pins, attach a hoist to lifting eye and take slack from hoist. Remove thrust bearing assembly. Remove cap screws from brake housings and wedge them out far enough to remove the two sets of shim packs from behind each brake housing. Identify shims as to their location and tie them together to prevent intermixing. Carefully pull either brake housing and bull pinion out until differential side gear can be grasped, then hold gear and bull pinion in place and pull entire assembly from rear frame and brake housing support. Remove opposite brake housing and bull pinion in the same manner. Unbolt supports (31 and 32—Fig. O210B) from rear frame and save shims (33) located under supports. Remove supports and differential from rear frame.

336A. If any of the loose roller bearings drop out of bull pinion inner bearing, refer to paragraph 344 for reassembly. Reinstall differential by reversing the removal procedure and using the correct thickness of shim packs between brake housings and tractor rear main frame as outlined in paragraph 337. When installing brake housing and bull pinion assemblies on 1800 and 1900 series A tractors, install the cone head cap screws first and tighten to a torque of 75-80 ft. lbs. (1800 and 1900 series A brake housings are retained by two conical head cap screws and four hex head cap screws and lock washers; 1800 and 1900 series B and later brake housings are retained by a dowel pin and six hex head cap screws and lock washers). Adjust bearings and backlash as outlined in paragraph 337.

NOTE: When reinstalling differential assembly in the late series 1950 and the series 1950-T, be sure the shims located under brake housing supports are the same thickness as the values stamped on each side of the rear main frame. See Fig. O210B. Values may not be the same for both sides. Shims are available in thicknesses of 0.004, 0.007 and 0.0149.

337. BEARINGS ADJUST. Differential end play and gear backlash are controlled by the two (split) shim packs interposed between the brake housings and the rear main frame. When adjusting bearings, do not depend on a shim count or prior assembly. Use a micrometer to measure shim pack thickness to be sure that both shim packs installed between brake housing and tractor rear frame are identical in thickness. Shims are available in thicknesses of 0.004, 0.007 and 0.0149.

Mount a dial indicator and check the differential end play which should be 0.001-0.003. Add or subtract shims as necessary to obtain the specified end play, however, BE SURE to maintain some gear backlash during this operation.

With differential end play adjusted, use the dial indicator to check the backlash between bevel pinion and bevel drive gear at four points around drive gear; any large differences in backlash readings will indicate runout of drive gear due to improper tightening of gear mounting bolts or foreign material between gear and spider. Average the four readings, if no large differences exist, to determine gear backlash. Transfer shims, in equal amounts from one brake housing to the other until the average of four dial indicator readings agrees with the backlash specifications etched on bevel drive (ring) gear.

Mesh position of bevel gears is fixed and non-adjustable.

After correct bearing and backlash adjustments have been made on late series 1950 and series 1950-T differential, loosen jam nut on bearing shaft (11—Fig. O211D) and turn shaft until bearing (10) contacts bevel gear (4), then tighten jam nut and lock with clip (14).

338. OVERHAUL. Remove differential as outlined in paragraph 335. Remove pinion pin retaining snap rings, then with puller Oliver No. STS-100, or equivalent, threaded into tapped hole of pinion pin, pull pinion pins and remove pinions.

When reassembling, install pinions with taper facing inward and tapped hole of pinion pin facing outward. When mounting bevel drive gear to spider, be sure there is no foreign material between gear and spider and install bolts with heads next to bevel drive gear. Tighten nuts alternately and evenly to a torque of 110-115 ft.-lbs.

Note: When installing new differential spider, ream holes in spider to 0.530-0.532 as follows: Place bevel

Fig. O210B — View of series 1950-T differential installed. Note brake housing supports (31). Shim thickness required under each support is stamped at the locations (S) shown.

11. Brake housing
18. Bull pinion
31. Brake housing supports
P. Transmission oil pump
S. Shim thickness stamp

drive gear on spider, ream one hole and install bolt. Now ream the hole which is directly opposite and install bolt. Tighten nuts to hold bevel drive gear in position and ream balance of holes. Install remainder of bolts and tighten nuts alternately and evenly to a torque of 110-115 ft.-lbs.

### BEVEL GEARS

339. Data for removal and overhaul of bevel pinion shaft is given in paragraph 323 and for differential in paragraph 338. Adjust backlash as outlined in paragraph 337.

## FINAL DRIVE AND REAR AXLES

### All Models

340. For the purposes of this section, the final drive will include those components of the drive line which follow the differential. That is, the bull pinion shafts, bull gears and the wheel axle shafts. Due to their interrelation, they will be treated in the order of disassembly.

341. **WHEEL AXLE SHAFTS.** To remove a wheel axle shaft, first drain final drive compartment, remove the pto drive shaft, or the hydraulic unit drive shaft, then remove the hydraulic lift unit as outlined in paragraph 414, or the bull gear cover as outlined in paragraph 311. Attach hoist to pto unit, unbolt and remove same from rear frame, Remove fenders, then support tractor rear frame and remove the wheel and tire assemblies. Remove the draft control spring and support assembly and spacers. Remove cap screws from axle outer bearing cap, then remove snap ring from inner end of axle. Use a suitable pusher and push axle from bull gear. Remove wheel hub, bearing retainer, shims and bearing assembly from outer end of axle. Remove seals from bearing retainers. Remove bearing and spacer from inner end of axle.

Note: If desired, the axle and axle carrier can be removed as a unit by unbolting the axle carrier from the rear main frame and leaving the outer bearing retainer attached to axle carrier. Because of the weight involved, use a hoist for this operation.

If necessary to check and/or adjust the axle bearing preload, proceed as follows: Remove the axle carrier from tractor, if not already done, then install axle in carrier and bull gear on inner end of axle. Strike both ends of axle with a soft faced hammer to seat bearings. Wrap a length of strong cord around bull gear outer diameter and attach a

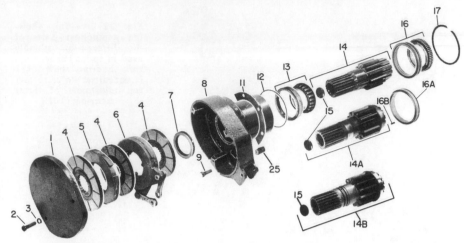

Fig. O211 — Exploded view of brake housing, brake and bull pinion assembly. Different bull pinions used are 14, 14A and 14B; bearing (16) is used with bull pinion (14) only. Bull pinions (14A and 14B) use outer bearing race (16A) and 57 loose bearing rollers (16B). Industrial models use three brake discs (4) and plate (5); other models have only two brake discs and do not use plate (5).

| | | |
|---|---|---|
| 1. Brake cover | 12. "O" ring | 16A. Bearing outer |
| 4. Brake discs | 13. Outer bull pinion | race (all except |
| (2 or 3) | bearing | early 1800 and |
| 5. Brake plate | 14. Bull pinion | 1900 series A) |
| (Industrial models | (early 1800 and | 16B. Bearing rollers |
| only) | 1900 series A) | (57 count) (all |
| 6. Brake actuating | 14A. Bull pinion (1800 | except early 1800 & |
| assembly | and 1900 series B | 1900 series A) |
| 7. Oil seal | and late series A) | 17. Snap ring |
| 8. Brake housing | 14B. Bull pinion (1850, | (1800 and 1900 |
| 9. Conical head cap | 1950 and 1800, 1900 | series A only) |
| screws (2) | series C) | 25. Dowel pin (all |
| (1800 & 1900 | 15. Expansion plug | except 1800 and |
| series A only) | 16. Inner bearing | 1900 series A) |
| 11. Shims | (early 1800 and | |
| | 1900 series A only) | |

spring scale to cord. Pull on scale and check the reading at which axle starts to rotate. This reading should be 15 pounds (120 inch pounds) for the Agricultural tractors; or 19½ pounds (156 inch pounds) for the Industrial tractors. Add or subtract shims under outer bearing retainer to obtain this value. Shims are available in thicknesses of 0.004, 0.007 and 0.0149.

342. **AXLE CARRIERS.** To R&R and overhaul the axle carriers, first remove the wheel axle shafts as outlined in paragraph 341. On early tractors, remove the lower link supports to provide removal clearance, then unbolt and remove axle carrier.

Note: Late model tractors have the axle carriers notched to provide removal clearance.

Remove the "O" ring seal from pilot of axle carrier and the snap ring or retainer which retains the axle inner bearing cup. Use a suitable puller and pull bearing cup.

Note. Two bearing cup retaining snap rings are used on models 1800 and 1900 Industrial tractors prior to tractor serial number 124 395-000.

When reassembling, adjust axle bearing preload as outlined in paragraph 341, and the draft control spring as outlined in paragraph 421.

343. **BULL GEARS.** To remove the bull gears, first remove axle shafts as outlined in paragraph 341, then remove pto actuator levers and shaft. Bull gears can now be lifted from rear frame.

NOTE: Whenever necessary to renew a bull gear in an 1800 or 1900 series A tractor, note part number stamped on old gear; if part number is 105 073-A or 105 074-A, it will also be necessary to renew mating bull pinion, pinion inner bearing and differential side gear. It is also recommended, but not mandatory, that opposite set of bull gears be renewed.

344. **BULL PINIONS.** To remove the bull pinions, proceed as outlined in paragraph 335 except that the brake cross shaft need not be removed and the differential can rest in the bottom of the housing.

Four different bull pinions have been used in production since the introduction of 1800 and 1900 series tractors. Refer to the following appropriate paragraph, or paragraphs, for bull pinion overhaul information:

344A. **EARLY 1800 AND 1900 SERIES A.** Early series A tractors were fitted with a bull pinion (14—Fig. O211) and inner bearing assembly (16) with the bearing rollers retained to inner race by a band en-

**Fig. O211A—Exploded view of the brake housings, brakes and bull pinions used in series 1950-T tractors. Note that brakes use three disc assemblies (4). Sleeves (36) are also used on bull pinion shafts. Refer to text for adjustment of supports (31 and 32).**

| | | |
|---|---|---|
| 1. Cover | 14. Shims | 22. Roller (114 |
| 4. Disc | 15. Dowel | used) |
| 5. Intermediate disc | 16. "O" ring | 23. Bearing race |
| 6. Actuating disc | 17. Bearing assy. | 31. Support, R. H. |
| 7. Oil seal | 18. Bull pinion | 32. Support, L. H. |
| 8. Brake housing, | 19. Expansion plug | 33. Shims |
| R. H. | 20. Teflon ring | 34. Copper washer |
| 11. Brake housing, | 21. "O" ring | 35. Cap screw |
| L. H. | | 36. Sleeve |

circling the notched rollers and race. Inner race and roller assembly was pressed onto inner end of bull pinion. Snap ring (17) in groove on outside diameter of outer race located the race in brake housing (8).

To disassemble this early type unit remove differential side gear from inner end of bull pinion and withdraw the bull pinion from inner end of brake housing. Carefully inspect the bull pinion, mating bull gear and the bull pinion inner bearing (16); if necessary to renew any of these parts, refer to Oliver parts catalog and obtain the following parts: Bull pinion

gear, outer race for bull pinion inner bearing, 57 bearing rollers, differential side gear and bull gear. It is also recommended, but not mandatory, that like new parts be installed in opposite final drive. Assemble as outlined in paragraph 344B.

If all parts are in useable condition, install new seal (7) in brake housing with lip to inside. Install new plug (15) in outer end of bull pinion if leakage is indicated. Lubricate seal and carefully insert bull pinion and bearing assembly into brake housing to avoid damaging seal. Place differential side gear on inner end of bull pinion and refer to paragraphs 336A and 337 for reassembly procedure. Note: New bull pinion outer bearing

(13) and/or differential side gear can be renewed on early units without renewing other parts of the unit.

**344B. LATE 1800 AND 1900 SERIES A.** Late series A tractors were fitted with a bull pinion (14A) with inner bearing consisting of 57 loose rollers (16B), separate outer race (16A) and with inner race an integral part of the bull pinion. However, pressure angle of bull pinion teeth and bull gear teeth can be either 20° or 25° and if necessary to renew bull pinion with 20° pressure angle teeth, the following parts must be renewed: Bull gear, differential side gear, inner bearing rollers and outer race and the bull pinion. If bull pinion and mating bull gear with 20° pressure angle teeth are useable, the differential side gear, inner bearing rollers and outer race, or the tapered roller outer bull pinion bearing may be renewed if necessary. Bull pinion with 20° pressure angle teeth may be identified as it has no identifying number stamped on it; bull pinion with 25° pressure angle teeth will have part number stamped on the gear.

NOTE: When renewing one set of bull pinion and mating bull gear having 20° pressure angle teeth with bull pinion and bull gear having 25° pressure angle teeth, it is recommended, but not mandatory, that the opposite bull pinion and bull gear be renewed also. CAUTION: Under no circumstances try to install a bull pinion and mating bull gear with different pressure angle teeth.

To disassemble the bull pinion and brake housing unit, remove differential side gear from inner end of bull pinion and remove the 57 loose bearing rollers. Then withdraw bull pinion from the brake housing. Remove outer bearing cone from bull pinion and outer race for inner bearing and outer bearing cup from brake housing if

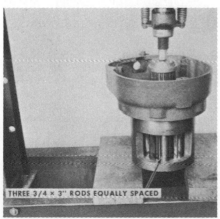

THREE 3/4 × 3" RODS EQUALLY SPACED

**Fig. O211B — Removing bull pinion from models having loose rollers in bull pinion inner bearing. Refer to text for procedure.**

**Fig. O211C — On 1850, 1950 and 1800, 1900 series C, left bull pinion is fitted with two "O" rings in grooves next to outer bearing as shown, then Teflon rings are installed on top of the "O" rings as shown in inset at upper left.**

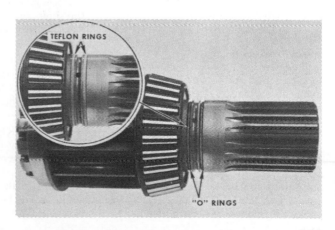

TEFLON RINGS

"O" RINGS

necessary. Note: Do not attempt to remove inner bearing race from inner end of bull pinion; this race is ground after being installed on the bull pinion, is not renewable separately and will be damaged beyond further use if removed.

To reassemble, install new oil seal (7) in brake housing with lip of seal inward. Drive new outer bearing cup into brake housing until seated and drive or press new bearing cone on outer end of bull pinion shaft until seated. Install new expansion plug (15) in outer end of bull pinion if there is evidence of oil leakage. Be sure snap ring is seated in groove on outside diameter of inner bearing outer race, then insert race into brake housing with snap ring side towards differential end of brake housing and drive the race into housing until snap ring is tight against shoulder. Lubricate lip of seal and carefully insert bull pinion into brake housing to avoid damage to seal. Insert the 57 loose bearing rollers, then place differential side gear on inner end of bull pinion. Refer to paragraphs 336A and 337 for reinstalling assembly.

344C. ALL 1800 AND 1900 SERIES B AND LATER TRACTORS. Series B and later tractors, bull pinion outer

bearing (13) is larger than on prior models. After removing the brake housing and bull pinion unit, disassemble as follows:

Remove differential side gear from inner end of bull pinion and remove the 57 loose bearing rollers. Equally space three ¾-inch diameter rods of 3 inches in length (be sure all three rods are exactly same length) around the bull pinion gear as shown in Fig. O211B, with the rods between the outer bearing cone and the inner bearing outer race. Be sure housing is supported so that inner bearing outer race will clear and press on spline (outer) end of bull pinion shaft to remove shaft, outer bearing cone and inner bearing race from the brake housing. With bull pinion removed, remove outer bearing cone from the bull pinion and outer bearing cup from the brake housing if necessary. Remove plug (15) from outer end of bull pinion only if there is evidence of oil leakage at this point.

NOTE: The inner bearing race for the 57 loose bearing rollers is pressed on the bull pinion, then ground during manufacture and must not be removed from the bull pinion shaft. If necessary to renew the inner bearing race, install new bull pinion and inner bearing race assembly.

On all tractors series C and up, the bull pinion (14B) has two seal ring grooves. On the left bull pinion only, install new "O" rings in the grooves as shown in Fig. O211C, then install Teflon rings on top of the "O" rings as shown in inset. Note: Stretch the Teflon rings only enough to allow them to be installed over the shaft and "O" rings and allow the rings to return to normal size before attempting to install bull pinion in brake housing. **Do not** install the "O" rings or Teflon rings in the grooves on bull pinion to be installed on right hand side of tractor.

Install new bearing cup in brake housing with taper facing inward, and install new oil seal with lip inward. Install new outer bearing cone on bull pinion. Install new plug (15) if plug was removed on disassembly. Be sure bearing cup and bearing cone are seated, then lubricate lip of oil seal and the sealing rings on left bull pinion and carefully install bull pinion taking care not to damage the seals. Drive inner bearing outer race into brake housing until seated against shoulder in housing, then insert the 57 loose bearing rollers and place differential side gear on bull pinion. Refer to paragraphs 336A and 337 for reinstalling assembly.

# BRAKES

The brakes, which are mounted on outer ends of bull pinions are the self-energizing disc type. Agricultural model tractors (except late series 1950 and 1950-T) use a double disc brake while the Industrial models (late series 1950 and series 1950-T) use a triple disc brake fitted with an intermediate disc which provides the required two additional friction surfaces. The actuating assembly and intermediate disc have a slot which fits over a boss in the brake housing to prevent rotation.

## All Models

345. **ADJUSTMENT.** To adjust brakes, loosen jam nut on brake actuating rod and turn adjusting nut until the third or fourth notch of brake lock will engage when the brake pedal is depressed moderately. Adjust both brakes the same.

On tractors, except the early series A, and 1950-T, brake pedals have stop screws which provide pedal pad alignment for brake pedals in the released position.

346. **R&R AND OVERHAUL.** Unhook brake return springs and remove step assembly from left side of tractor (early models). Remove jam nut, adjusting nut and the spherical washer from brake actuating rods. Remove brake housing cover and the brake assemblies.

Disassembly of the actuating assembly is accomplished by removing actuating rod from links and disengaging the three extension springs. If necessary, polish the steel balls and the ball ramps. A small amount of Lubriplate may be used on steel balls during reassembly.

Linings are bonded to brake discs and renewal requires that a complete new disc be used.

Adjust brakes as outlined in paragraph 345.

347. **BRAKE CROSS SHAFT.** If only the brake cross shaft seals are to be renewed, it can be accomplished by sliding the cross shaft from the seals. However, if the cross shaft is to be renewed, it may be

necessary to remove the left wheel and tire assembly to provide room enough to completely remove cross shaft.

Renew seals and/or cross shaft as follows: Unhook brake return springs. Remove right platform and on tractors with wheel guard extensions, also remove brake pedal cover and right platform splash panel. Disconnect the brake pull rod from right brake pedal, remove cotter pin from right end of cross shaft and remove right brake pedal and lock assembly. Remove cap screws from left brake pedal and remove pedal and lock assembly and Woodruff key. Disconnect brake pull rod from left brake arm, then remove cap screw from brake arm and the brake arm and Woodruff key from cross shaft.

At this point, seals can be renewed by sliding shaft out of the seal to be renewed. Lip of seal faces inward.

Reassemble by reversing the disassembly procedure and adjust brakes as outlined in paragraph 345.

# BELT PULLEY UNIT

Both 540 and 1000 rpm belt pulleys are available to use in conjunction with 540 and 100 rpm pto assemblies. The belt pulleys are controlled by the pto and are mounted on aft side of pto housing in place of the pto shaft housing assembly.

## All Models

**350. REMOVE AND REINSTALL.** Removal of belt pulley is obvious. However, remember that pulley must be on left side when installed.

If an original installation of the belt pulley unit is being made, proceed as follows. Remove the button plug, or pto shaft housing and housing spacer from lower hole in pto housing. If tractor has three-point hitch, remove the left lower link and lift link or leveling gear assembly. Mount belt pulley to pto housing so pulley is on the left side.

**351. OVERHAUL. (540 RPM).** With belt pulley removed, drain oil and pull connecting (stub) shaft (42—Fig. 0212) from carrier (32). Remove pulley from pulley shaft (4), then unbolt and remove the drive shaft (28) and carrier (32) assembly from housing (14). Remove "O" ring (26), shims (25) and oil seal (40) from

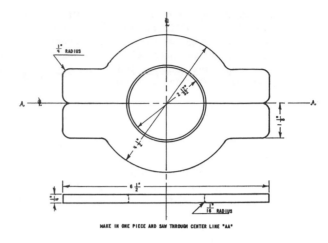

Fig. O213 — Seal installation tool can be fabricated locally, using dimensions shown.

carrier, then using a spanner wrench ST-142, or equivalent, remove adjuster nut (39). Press drive shaft (28) and gear (30) from carrier. Remove snap ring (27) and pull bevel drive gear from drive shaft. Press inner bearing from drive shaft and bearing cups from carrier. Remove expansion plugs (29 and 35) from drive shaft and carrier only if they show signs of leakage.

Unbolt pulley shaft carrier (8) from housing and remove carrier and pulley shaft (4) assembly from housing. Remove shims (13) from carrier, unstake adjuster nut (12) and use a puller to remove pinion gear (11) and bearing cone from pulley shaft.

Press pinion gear from bearing. Press pulley shaft (4) from carrier, then remove outer bearing and oil seal (6). Remove bearing cups from carrier. Remove expansion plug (16) from pulley housing only if it shows signs of leakage.

Remove cap screw (19) and disassemble breather assembly.

Inspect all parts for excessive wear, scoring, chipping or other damage and reassemble as follows: Place new oil seal (6) on pulley shaft with lip facing inward, then install bearing cone with bevel toward housing. Install bearing cone on bevel pinion with bevel away from gear. Install bearing cups in carrier, then with an

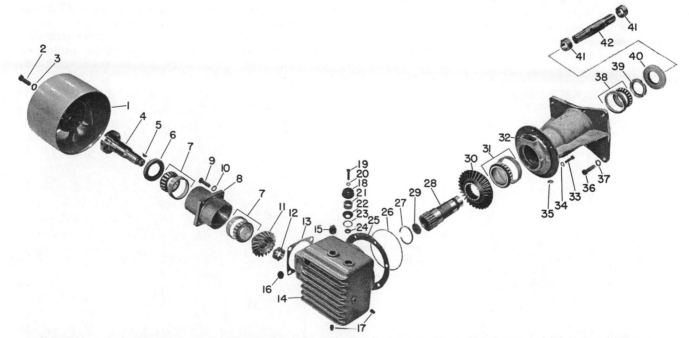

**Fig. 0212 — Exploded view of belt pulley for use with 540 rpm pto. Refer to Fig. O217 for those used with 1000 rpm pto.**

| | | | |
|---|---|---|---|
| 1. Pulley | 12. Nut | 25. Shims | 32. Carrier |
| 4. Pulley shaft | 13. Shims | 26. "O" ring | 35. Expansion plug |
| 5. Woodruff key | 14. Housing | 27. Snap ring | 38. Bearing assy. |
| 6. Oil seal | 16. Expansion plug | 28. Drive gear shaft | 39. Spanner nut |
| 7. Bearing assy. | 18. Breather cap | 29. Expansion plug | 40. Oil seal |
| 8. Carrier | 21. Breather element | 30. Bevel drive gear | 41. Spacer |
| 11. Bevel pinion | 22. Cup | 31. Bearing assy. | 42. Stub shaft |

oil seal installation tool similar to that shown in Fig. 0213, positioned between pulley flange and oil seal as shown in Fig. 0214, press oil seal and pulley shaft into carrier. Install the pinion and bearing on pulley shaft and screw nut on shaft until only a slight end play of shaft remains. Use a spring scale and note the effort required to rotate carrier. Tighten adjuster nut until the spring scale reads 5 to 6 pounds more than the reading taken prior to tightening nut. This will give the desired 10-12 inch pounds bearing preload.

With the bearing preload set, the pulley shaft location in housing must be determined as follows. Measure the distance between flat surface of pinion gear and the mounting flange of carrier as shown in Fig. 0215. Take this value and subtract from it the value found stamped on housing as shown in Fig. 0216. To this result add 1.157. If a plus, or a minus value in thousandths is shown on the pinion gear, add or subtract accordingly and this will give the thickness of the shims required between carrier and housing. Shims are available in thicknesses of 0.005, 0.007 and 0.0149.

Assemble drive shaft and install in drive shaft carrier. Tighten adjusting nut until bearings have 10-12 inch pounds preload and stake nut. Install oil seal with lip facing inward. Note the backlash value stamped on bevel drive gear and vary shims between carrier and housing to obtain this value. Shims are available in thicknesses of 0.004, 0.007 and 0.0149. In-

stall stub shaft and spacers. Use new breather element and install breather assembly. Fill unit with 6½ pints of proper lubricant.

352. OVERHAUL (1000 RPM). With belt pulley removed, drain oil and remove pulley from pulley shaft. Unbolt carrier (29—Fig. 0217) from housing (14) and remove the carrier and drive shaft assembly. Remove "O" ring (26) and shims (25) from carrier, then unstake and remove nut (11) from drive shaft. Pull bevel drive gear (27) and bearing from drive shaft and press gear from bearing. Press drive shaft (36) from carrier, then remove oil seal (37) and bearing from shaft. Pull bearing cups from carrier.

Unbolt pulley shaft carrier (7) from housing and remove the carrier and pulley shaft assembly. Remove "O" ring (12) and shims (13) from carrier, unstake adjuster nut (11) and use a puller to remove pinion gear (10) and bearing cone from pulley shaft. Press pinion gear from bearing. Press pulley shaft (4) from carrier, then remove outer bearing and oil seal (5). Remove bearing cups from carrier. Remove expansion plug (17) from housing only if it shows signs of leakage.

Remove cap screw (19) and disassemble breather assembly.

353. Inspect all parts for excessive wear, scoring, chipping or other damage and reassemble as follows: Place new oil seal (5) on pulley shaft with lip facing inward, then install bearing cone with bevel toward housing. Install bearing cone on pinion gear

Fig. O215 — Measure between points shown when locating carrier in belt pulley housing. Refer to text for procedure.

with bevel away from gear. Install bearing cups in carrier, then with an oil seal installation tool similar to that shown in Fig. 0213 positioned between pulley flange and oil seal as shown in Fig. 0214, press oil seal and pulley shaft into carrier. Install the pinion and bearing on pulley shaft and screw nut on shaft until only a

Fig. O214 — Use tool positioned as shown when installing oil seal.

Fig. O216 — Assembly dimension is stamped on belt pulley housing as shown.

DIMENSION

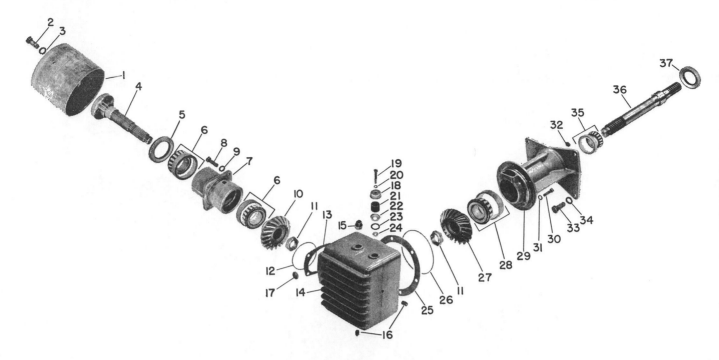

Fig. O217 — Exploded view of belt pulley for use with 1000 rpm pto. Refer to Fig. O212 for those used with 540 rpm pto.

| | | | |
|---|---|---|---|
| 1. Pulley | 11. Nut | 21. Breather element | 29. Carrier |
| 4. Pulley shaft | 12. "O" ring | 22. Cup | 32. Expansion plug |
| 5. Oil seal | 13. Shims | 25. Shims | 35. Bearing assy. |
| 6. Bearing assy. | 14. Housing | 26. "O" ring | 36. Drive shaft |
| 7. Carrier | 17. Expansion plug | 27. Bevel drive gear | 37. Oil seal |
| 10. Bevel pinion | 18. Breather cap | 28. Bearing assy. | |

slight end play of shaft remains. Use a spring scale and note the effort required to rotate carrier. Tighten the adjuster nut until the spring scale reads 5 to 6 pounds more than the reading taken prior to tightening nut. This will give the desired 10-12 inch pounds bearing preload.

With the bearing preload set, the pulley shaft location must be determined as follows: Measure the distance between the flat surface of pinion gear and mounting flange of carrier as shown in Fig. O215. Take this value and subtract from it the value found stamped on housing as shown in Fig. O216. To this result add 1.788. If a plus, or a minus value in thousandths is shown on the pinion gear, add or subtract accordingly and this will give the thicknesses of the shims required between carrier and housing. Shims are available in thicknesses of 0.004, 0.007 and 0.0149.

Install bearings on drive shaft and bevel drive gear so bevel of bearings face inward. Install bearing cups in carrier. Install shaft in carrier, install drive bevel gear and bearing on drive shaft, then install adjuster nut. Tighten adjuster nut until 10-12 inch pounds preload is placed on bearings. Stake adjuster nut in place. Install new oil seal with lip facing inward.

Note the backlash value stamped on bevel drive gear and vary the shims between carrier and housing to obtain this backlash. Shims are available in thicknesses of 0.004, 0.007 and 0.0149. Use new breather element and install the breather assembly. Fill unit with 6½ pints of proper lubricant.

# POWER TAKE-OFF

Both 540 and 1000 rpm pto units are available for all models as well as a dual speed 540 and 1000 rpm pto for 1750, 1850, 1950 and 1950-T series tractors. In addition, an engine speed pto may be used in conjunction with any unit. Rotation of engine speed pto is counter-clockwise, viewed from the rear. Refer to Fig. O218 for an exploded view of the single speed pto unit and to Fig. O218A for the dual speed pto.

NOTE: Use of the 540 rpm pto must be limited to not more than 60 horsepower.

## All Models

355. **ADJUST PTO CLUTCH.** To adjust the pto clutch, remove the large pipe plug (42—Fig. O219), then rotate clutch assembly until the adjusting lock pin is centered in hole. Depress lock pin and rotate adjusting ring, one notch at a time, clockwise to increase, or counter-clockwise to decrease the clutch engaging pressure. Clutch is properly adjusted when a spring scale reading of 40-45 pounds, taken just below control lever knob, is required to engage clutch.

355A. **ADJUST PTO BRAKE.** To adjust pto brake on models having left hand lever, start engine and be sure pto lever is latched in disengaged position, then adjust latch stop screw until pto output shaft rotation is eliminated. See Fig. O219A.

To adjust pto brake on models with right hand lever, loosen lock nut, remove cotter key and pin and rotate clevis until lever locks securely in disengaged position. See Fig. O219A.

356. **R&R PTO DRIVE SHAFT.** To remove the pto drive shaft, remove the button plug, or the engine speed

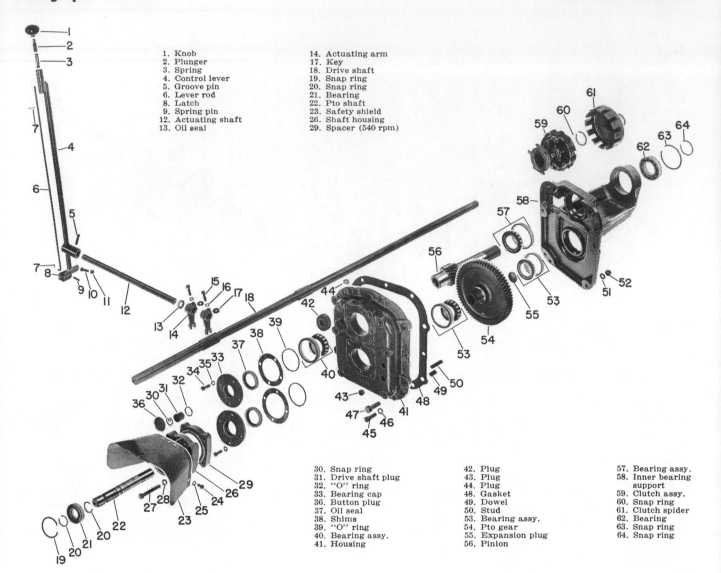

1. Knob
2. Plunger
3. Spring
4. Control lever
5. Groove pin
6. Lever rod
8. Latch
9. Spring pin
12. Actuating shaft
13. Oil seal
14. Actuating arm
17. Key
18. Drive shaft
19. Snap ring
20. Snap ring
21. Bearing
22. Pto shaft
23. Safety shield
26. Shaft housing
29. Spacer (540 rpm)

30. Snap ring
31. Drive shaft plug
32. "O" ring
33. Bearing cap
36. Button plug
37. Oil seal
38. Shims
39. "O" ring
40. Bearing assy.
41. Housing

42. Plug
43. Plug
44. Plug
48. Gasket
49. Dowel
50. Stud
53. Bearing assy.
54. Pto gear
55. Expansion plug
56. Pinion

57. Bearing assy.
58. Inner bearing
    support
59. Clutch assy.
60. Snap ring
61. Clutch spider
62. Bearing
63. Snap ring
64. Snap ring

Fig. O218 — Exploded view showing single speed pto unit and component parts. Drive shaft (18) is driven from engine flywheel.

pto shaft, if so equipped, from upper bearing bore of pto housing. Remove snap ring (30—Fig. O218) from inside diameter of pinion (56). Install puller assembly Oliver No. STS-100, or its equivalent, as shown in Fig. O219, and pull plug (31—Fig. O218) from pinion. Remove "O" ring (32), then reinstall puller to pto shaft (18), be sure clutch is disengaged, and pull pto shaft from tractor.

357. R&R PTO UNIT. To remove the pto unit, first remove the pto shaft as outlined in paragraph 356. Drain final drive compartment, attach hoist to pto, then unbolt and remove unit from rear frame. When reinstalling, be sure oil slots in sleeve collar are on top side.

Should it be necessary to remove the control lever and shaft assembly, remove the left (or right) wheel, tire and fender, then remove cap screws

and keys from actuating arms (14—Fig. O218) and slide shaft and lever assembly from rear frame. Any overhaul to the control lever and shaft will be obvious after an examination of the unit and reference to Fig. O218, O218A or O219A.

358. DISASSEMBLE SINGLE SPEED PTO UNIT. Remove the pto unit as outlined in paragraph 357. Remove safety shield (23—Fig. O218), then unbolt and remove the pto shaft and housing assembly (26) from pto housing. On 540 rpm models, also remove spacer (29). Remove snap ring (19) and press shaft from housing. (26). Remove bearing caps (33), shims (38) and "O" rings (39), then remove seals (37) from bearing caps. Unbolt and separate inner bearing support (58) from pto housing (41). Pull gear (54) from housing and renew expansion plug (55) in gear if it shows signs of leakage.

Engage clutch and pull pinion (56) and clutch assembly (59) from clutch spider (61) as far as possible. Remove snap ring (60) from aft end of pinion shaft and remove clutch assembly and pinion shaft from inner bearing support. See Fig. O220.

Note: If additional room is required for removal of snap ring (60—Fig. O218), remove snap ring (63) and bump spider and bearing slightly forward.

The need for further disassembly will be dictated by the need for renewal of parts and the procedure is obvious. Refer to paragraph 362 for information on clutch overhaul.

359. REASSEMBLE SINGLE SPEED PTO UNIT. Reassemble the pto by reversing the disassembly procedure. Use a dial indicator to check the end play of the pinion (56) and gear (54) which should be 0.003-0.005.

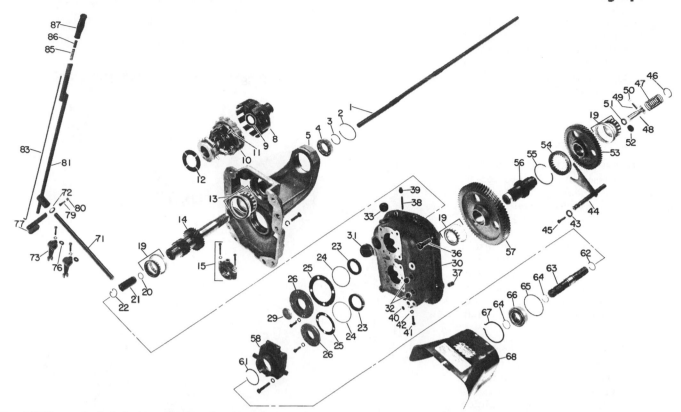

Fig. O218A — Exploded view showing the component parts and their relative positions of the dual speed 540 and 1000 rpm pto. The single speed (540 or 1000) pto does not include all the parts shown, however, basic construction remains similar.

| | | |
|---|---|---|
| 1. Drive shaft | 21. Plug | 41. Lock screw (dual speed) |
| 2. Snap ring | 22. Snap ring | 42. Lock nut |
| 3. Snap ring | 23. Oil seal | 43. Oil seal (dual speed) |
| 4. Bearing | 24. "O" ring | 44. Shift rail and fork (dual speed) |
| 5. Front bearing support | 25. Shim | 45. Cap screw |
| 8. Spider | 26. Bearing cap | 46. Snap ring (dual speed) |
| 9. Oil seal | 30. PTO housing | 47. Spring (dual speed) |
| 10. Clutch assembly | 31. Plug (adjusting hole) | 48. Locking cam (dual speed) |
| 11. Groove pin | 32. Plugs | 49. Lock pin (dual speed) |
| 12. Brake washer | 33. Plug | 50. Spring (dual speed) |
| 13. Bearing | 36. Cone head screw | |
| 14. Pinion shaft | 38. Collector sleeve | |
| 15. Oil collector | 39. Plug | |
| 19. Bearing | 40. Plug (single speed) | |
| 20. "O" ring | | |

| | |
|---|---|
| 51. Oil seal (dual speed) | 65. "O" ring |
| 52. Plug (single speed) | 66. Bearing |
| 53. Output gear (1000 rpm) | 67. Snap ring |
| 54. Drive collar | 68. Safety shield |
| 55. Snap ring | 71. Shaft |
| 56. Output gear shaft | 72. Oil seal |
| 57. Output gear (540 rpm) | 73. Lever arm |
| 58. Housing | 76. Key |
| 61. Snap ring | 77. Latch |
| 62. Snap ring | 79. Stop screw |
| 63. Output shaft | 80. Lock nut |
| 64. Lock ring | 81. Lever |
| | 83. Rod |
| | 85. Spring |
| | 86. Plunger |
| | 87. Grip |

Vary shims behind bearing caps to obtain this reading. Shims (38) are available in thicknesses of 0.004, 0.007 and 0.0149.

**360. DISASSEMBLE DUAL SPEED PTO.** Remove the pto unit as out-

Fig. O219 — Install Oliver puller STS-100 as shown. Item (42) is plug to be removed for adjusting clutch.

lined in paragraph 357. Remove safety shield and the output shaft and housing (58—Fig. O218A) assembly. Remove snap ring (67) and pull shaft (63) and bearing (66) from housing. Remove snap ring and lock ring from front of bearing (66) and press shaft (63) from bearing. Remove remaining lock ring from shaft. Remove "O" ring (65) from housing bore, and if necessary, also remove snap ring (61).

Remove bearing caps (26), shims (25), "O" rings (24) and seals (23) from pto housing (30), then remove plug (39) and using a ¼ - 20 inch cap screw, pull oil sleeve (38) from housing. Unbolt and separate inner bearing support (5) from pto housing (30). Remove shifter rod lock screw (41), then remove output gears (53 and 57), coupling (54), output gear shaft (56) and the shifter rail and fork (44). Remove snap ring (46) from front end of output gear shaft (56) and remove shifter lock assembly (items 47, 48,

49 and 50) from bore of output gear shaft. Remove seal (51) from bore of output gear shaft and shifter rail seal (43) from pto housing. Drive clutch hub groove pin (11) into bore of pinion shaft (14), remove pinion shaft and brake washer (12), then remove bearings (13 and 19) and oil collector (15) from pinion shaft.

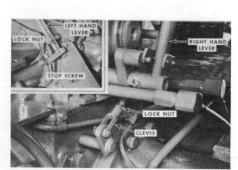

Fig. O219A — Adjust pto stop screw as shown to eliminate output shaft rotation. Refer to text.

Remove snap rings (2 and 3) at front of clutch spider (8), bump spider forward and remove clutch assembly (10) from spider. Remove spider from bearing (4), then remove bearing from front bearing support.

Drive bearing cups from pto housing and bearing front support.

Refer to paragraph 362 for pto clutch overhaul information.

**361. REASSEMBLE DUAL SPEED PTO.** To reassemble pto, proceed as follows: Install bearing cups in bearing front support (5). Install bearing (4) in front support and install snap ring (2). Install oil seal (9) in spider (8) with lip toward front, then install spider in bearing (4) and install snap ring (3) on spider. Now remove snap ring (2), bump spider forward and install clutch assembly (10), then bump assembly rearward and reinstall snap ring (2).

Note: At this time, some mechanics prefer to temporarily secure clutch in spider to prevent it shifting and possibly damaging the clutch discs.

Install oil collector (15) on pinion shaft (14) with machined face forward. Install bearing cones (13 and 19) on pinion shaft with tapers of bearing cones toward ends of shaft. Hold brake washer (12) in position behind clutch sleeve, insert pinion shaft from rear and align groove pin hole in pinion shaft with groove pin hole in clutch hub. Install groove pin (11) and be sure inner end does not extend into pinion bore.

NOTE: In some cases on Series 1850 and 1950 tractors, prior to serial number 163 115-000, the pto output shaft continued to turn with the pto clutch disengaged. This is the result of excessive flow of lubricant to the pto clutch causing the clutch plates to drag.

This condition can be corrected by drilling four $\frac{3}{16}$-inch holes equidistantly around shaft $4\frac{5}{8}$ inches from splined (forward) end of pinion shaft (14—Fig. O218A); or, installing a new shaft having the holes. Shaft hav-

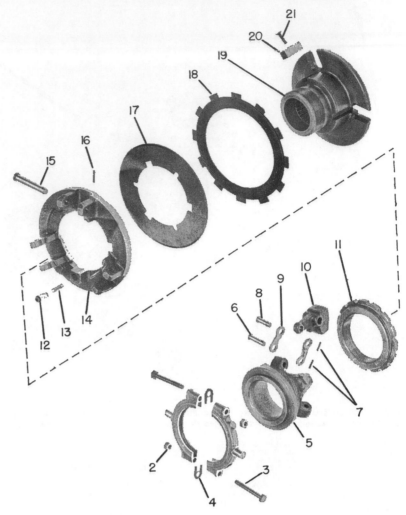

**Fig. O221 — Exploded view of pto clutch assembly.**

| | | | |
|---|---|---|---|
| 3. Bolt | 8. Link pin | 13. Spring | 18. Driving plate |
| 4. Shims | 9. Link | 14. Floating plate | 19. Hub |
| 5. Sliding sleeve | 10. Lever | 15. Lever pin | 20. Drive key |
| 6. Link pin | 11. Adjusting ring | 16. Spring pin | 21. Flat head screw |
| 7. Spring pin | 12. Lock pin | 17. Driven plate | |

ing holes was installed in Series 1850 and 1950 at tractor serial number 163 115-000. If shaft is drilled, be sure to remove any burrs or sharp edges.

Install shifter rail seal (43) in pto housing with lip facing forward. Install lock cam seal (51) in bore of output gear shaft (56) with lip facing

**Fig. O220—With pto unit disassembled as shown, all parts are available for servicing.**

forward. Place spring (50) on lock pin (49), insert pin in its hole in inner bore of output gear shaft, then hold pin to compress spring and install cam lock (48), return spring (47) and snap ring (46). Place gear (53) on front end of output gear shaft, then install bearing cone (19) on shaft with taper facing forward. Place coupling (54) on shaft (56) and slide it forward onto hub of gear (53). Place gear (57) on rear of output gear shaft and install bearing cone (19) with taper facing rearward. Position the output gear shaft and gears assembly and the shifter fork assembly in pto housing, hold parts in place and install the front bearing support assembly on the pto housing. Be sure clutch (plates) are not dislodged from spider. Install the pinion shaft and output gear shaft bearing cups in pto housing, then without oil seals (23) and "O" rings (24), install bearing caps (26) along with the originally removed shims

# HYDRAULIC LIFT

(25). Use a dial indicator and check the end play of the pinion shaft and output gear shaft. End play for both shafts is 0.003-0.005 and if adjustment is required, vary shims (25). However, be sure bearing cups are in contact with bearing caps to preclude a false reading. Shims are available in thicknesses of 0.004, 0.007 and 0.0149.

With the end play of both shafts adjusted, remove both bearing caps, install oil seals (23) with lips toward front, and the "O" rings, then reinstall the bearing caps. Align collector sleeve bore in oil collector with bore in top of pto housing, then install collector sleeve (38) and plug (39).

Install rear lock ring (64) in rear groove of output shaft (63), press shaft into bearing (66), install front lock ring (64) and secure with snap ring (62). Install snap ring (61) and "O" ring (65) in housing (58), then install output shaft and bearing assembly in housing and secure with snap ring (67). Position shift rail (44) to accommodate the output shaft being used (rearward for 540 rpm, forward for 1000 rpm) and tighten lock screw (41) and lock nut (42). Mount output shaft and housing assembly to pto housing. Delay installation of safety shield (68) until pto unit is mounted on tractor.

Install pto unit on tractor as outlined in paragraph 357 and adjust unit as outlined in paragraphs 355 and 355A.

362. PTO CLUTCH OVERHAUL Remove clutch assembly a outlined in paragraph 358 or 360 with the exception that it is not necessary to remove and disassemble the pto shaft and housing. Drive out roll pins (16—Fig. O221) and remove pins (15). Remove the sliding collar (5), levers and sleeve collar (throw-out) assembly from floating plate (14). Depress locking pin (12), unscrew adjusting ring (11) and remove adjusting ring, floating plate and clutch discs from clutch hub (19). Any further disassembly which might be required is obvious.

When reassembling, keep the following points in mind. If clutch drive keys (20) are renewed, be sure to stake the retaining screws. Renew clutch plates (17 and 18) if coned more than 0.005. Install bolts (3) in sleeve collar and any link pins that were removed so that heads point in the direction of engine rotation. Oil plates liberally prior to installation.

Adjust clutch as outlined in paragraph 355.

The hydraulic lift system can be any of several variations as follows:

A Hydra-Lectric system for operation of one or two remote electric cylinders.

A Hydra-Lectric system in conjunction with a draft control system for operation of a three-point hitch and remote cylinders.

An internal valve Depth-Stop system for operation of one or two remote non-electric cylinders.

An internal valve Depth-Stop system in conjunction with a draft control system for operation of a three-point hitch and remote cylinders.

An external valve Depth-Stop system for operation of remote non-electric cylinders. Two control valves are normally installed, however four valves can be stacked to operate four remote cylinders, if desired. Control valves are available for either single acting or double acting cylinders.

Non-electric remote cylinders may be used with Hydra-Lectric systems, however levers must be held in place and released when desired position is obtained. When using single acting cylinders with internal valve Depth-Stop system, control lever must be manually returned to neutral.

On Hydra-Lectric systems, single acting cylinders may be connected to either side of unit but can be used only one one side at a time. If single and double acting cylinders are used at the same time, the single acting cylinder must be connected to right side of unit. Single acting cylinder must be disconnected for three-point hitch operation.

On internal valve Depth-Stop systems, single acting cylinders may be connected to either side of unit and may be used two at a time. If single and double acting cylinders are being used together, they may be connected to whichever side desired.

As the combined draft control and Hydra-Lectric system is the one most generally used; and because of the similarity of the internal valve Depth-Stop and the Hydra-Lectric control systems, these units will be discussed in the following paragraphs. For information on the external valve Depth-Stop system, refer to paragraph 430.

## TROUBLE SHOOTING

365. The following are troubles and causes which may be encountered when operating the hydraulic system. When tests indicate that trouble is in the electrical portion of the hydraulic lift system, refer to paragraphs 385 through 393.

366. INTERNAL OIL LEAKS. Could be caused by: Defective pump mounting gasket, pump shaft seal, housing gasket, draft control rod seal or rod guide gasket.

367. UNIT OVERHEATS. Could be caused by: Unit overloaded. Relief valve improperly adjusted. Low oil level, oil of wrong viscosity being used or oil contaminated.

368. NOISY UNIT. Could be caused by: Worn or damaged pump or pump drive gear. Internal high pressure leaks. Low oil level or oil too heavy. Oil filter plugged. By-pass valve plugged or defective manifold.

369. OIL FOAMS OUT OF BREATHER. Could be caused by: Oil level too high or too low. Oil of wrong viscosity.

370. UNIT WON'T LIFT. Could be caused by: Pressure relief valve improperly adjusted or defective. Pump or pump drive failure. Pump not primed. Dirty or faulty servo valve. Bent or broken internal linkage. Low oil level or excessive leakage. Broken housing, cylinder, piston or defective piston seal. Faulty remote cylinder. Mechanical failure of lifting mechanism parts.

371. UNIT WILL NOT LOWER. Could be caused by: Servo valve pilot spool sticking. Lowering valve worn or sticking. Lowering valve check valve not opening.

372. RELIEF VALVE BLOWS AFTER LIFTING. Could be caused by: Improper adjustment of unit. Servo valve pilot spool not returning to neutral.

373. UNIT GOES OUT OF ADJUSTMENT. Could be caused by: Broken linkage spring inside pilot spool in servo valve. Loose servo valve mounting cap screws. Draft control linkage broken or worn. Lifting mechanism worn or loose.

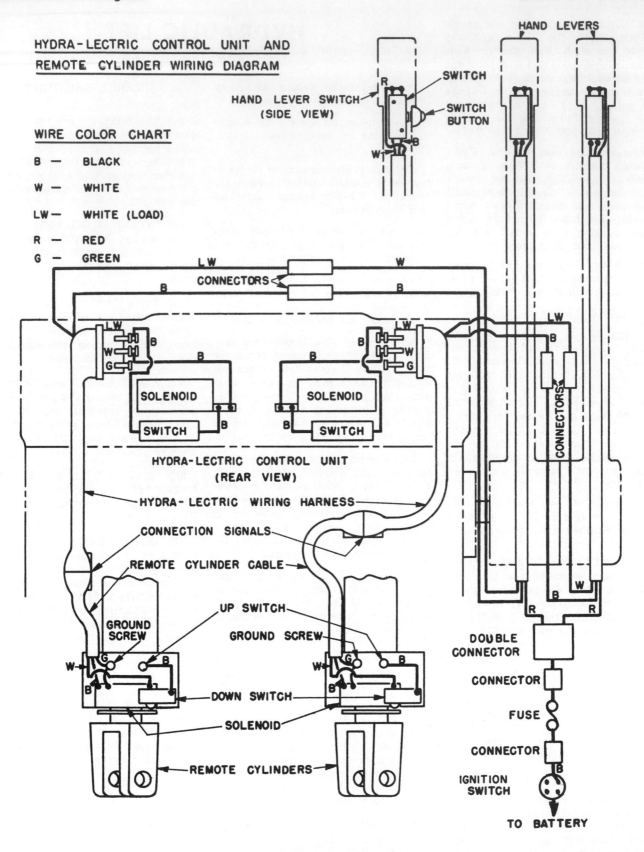

HAND LEVERS

HYDRA-LECTRIC CONTROL UNIT AND
REMOTE CYLINDER WIRING DIAGRAM

HAND LEVER SWITCH
(SIDE VIEW)

SWITCH
SWITCH
BUTTON

WIRE COLOR CHART

B — BLACK

W — WHITE

LW— WHITE (LOAD)

R — RED

G — GREEN

LW        CONNECTORS        W

SOLENOID        SOLENOID

SWITCH        SWITCH

HYDRA-LECTRIC CONTROL UNIT
(REAR VIEW)

CONNECTORS

HYDRA-LECTRIC WIRING HARNESS

CONNECTION SIGNALS

REMOTE CYLINDER CABLE

UP SWITCH

GROUND SCREW        GROUND SCREW

DOWN SWITCH

SOLENOID

REMOTE CYLINDERS

DOUBLE
CONNECTOR

CONNECTOR

FUSE

CONNECTOR

IGNITION
SWITCH

TO BATTERY

Fig. O222 — Schematic view of the wiring diagram for the Hydra-Lectric control unit and remote cylinders. Some tractors may have only one remote cylinder.

**374. ERRATIC DRAFT CONTROL OPERATION.** Could be caused by: Servo valve sticking. Worn or binding linkage or lift mechanism. Unit out of adjustment. Broken or loose draft control spring.

**375. RELIEF VALVE BLOWS WHEN REMOTE CYLINDER IS EXTENDED.** Could be caused by: Control lever being held too long in up position. Selector valve sticking. Shorted "up" switch. Contact arm of "up" switch not opening. Hand control lever not returning to neutral due to defective or maladjusted solenoid or solenoid switch, valve centering springs or remote cylinder switches.

**376. RELIEF VALVE BLOWS WHEN REMOTE CYLINDER IS CONTRACTED.** Could be caused by: Control lever held too long in "down" position. Selector valve sticking. Defective or maladjusted "down" switch. Remote cylinder friction collar improperly adjusted. Hand control lever not returning to neutral due to defective or maladjusted solenoid or solenoid switch, valve centering springs or remote cylinder switches.

**377. RELIEF VALVE BLOWS DURING REMOTE CYLINDER EXTENDING STROKE.** Could be caused by: Excessive load. Restricted piston stroke. Binding lift linkage. Couplings improperly connected. Restrictor valve stuck or passageway plugged.

**378. RELIEF VALVE BLOWS DURING REMOTE CYLINDER CONTRACTING STROKE.** Could be caused by: Restrictor valve improperly adjusted, stuck or oil line blocked. Couplings improperly connected. Bleed screw not properly adjusted when single acting cylinder is used.

**379. UNIT WILL NOT HOLD LOAD.** Could be caused by: Fittings or couplings leaking. Thermal relief valve leaking. Cylinder scored or worn or piston seal leaking. Interlock seals or seats leaking. Oil passage connectors leaking.

**380. ERRATIC REMOTE CYLINDER OPERATION.** Could be caused by: Loose or corroded electrical connections. Defective or bent cylinder stop collar. Bent "up" switch contact arm. Defective solenoid assemblies. Weak or broken valve centering spring.

**381. REMOTE CYLINDER STOP COLLAR FAILURE.** Could be caused by: Excessive air gap between collar and magnet. Cylinder solenoid not energized.

## ELECTRICAL TESTS

Electrical malfunctions fall into two classes; a dead circuit and a shorted circuit. When making electrical tests, determine first which kind of malfunction is involved.

It is necessary to have a test light assembly to perform the electrical tests and if one is not available, make one up as follows: Use two lengths of insulated wire and attach battery clips to one end of each wire. Place a 12-volt bulb and socket in series in one of the two wires. On remaining ends of the two wires, affix stiff, insulated probes.

Before initiating any electrical tests, be sure the trouble is not caused by a dead battery, blown fuses, loose or corroded connections or broken wiring. Use wiring diagram shown in Fig. O222 as a guide.

**385. TESTING ELECTRICAL CIRCUITS.** Before starting test, disconnect battery connections and remove battery from instrument panel support on those models so equipped; however, place battery so test light assembly can be connected. Disconnect all wire connectors except the connector for the fuse holder in the instrument panel support. Unplug wiring harness from Hydra-Lectric control unit and the cables from remote cylinders.

**386.** Attach test lamp assembly to battery, remove fuse holder cap and remove the fuse from cap. Put fuse back into holder, place one probe on exposed end of fuse and the other on the Hydra-Lectric terminal of ignition switch. If bulb lights, circuit is satisfactory from ignition switch to fuse.

**387.** Replace fuse cap and fuse in holder, then move same probe that was used on fuse, to the double connector located below the hand control levers. If the bulb lights the circuit between ignition switch and the double connector is satisfactory.

**388.** Check wiring and switches in each hand control lever by placing one probe on red wire and the other on black wire. Light should come on when switch is depressed. Now move the probe from the black wire to the white wire (other probe still on red wire). The light should come on but will go off when the switch is depressed. If both these conditions are met, the wiring and switches in hand control levers can be considered satisfactory.

**389.** Check the wiring, solenoids and switches in the Hydra-Lectric unit by placing a probe on each of the solenoid terminals on right side of tractor. Bulb should light when left hand control lever is pushed forward or rearward. Note: The solenoid switches are in off position when levers are in neutral position and bulb will not light. Repeat this operation for opposite side of tractor, using other control lever, if unit utilizes two remote cylinders. If bulb does not light as already indicated, remove the Hydra-Lectric control unit and check the solenoids and switches individually for complete circuits. Bulb should light when probes are placed on solenoid leads.

**390.** Test continuity of Hydra-Lectric wiring harness as follows: Place one probe on back pig tail and the other on the "B" terminal of the male plug. Bulb should light. Place one test probe on white pig tail and the other on "LW" terminal of 90 degree plug. Bulb should light. Test "G" and "W" circuits by placing a probe on corresponding letters of each plug. Bulb should light on both tests.

**391.** Test continuity of remote cylinder wiring harness by placing a test probe on same lettered terminals of each plug (W, B & G). Bulb should light in each case.

**392.** Check circuits of remote cylinder as follows: Remove cylinder head cover. Place one probe on "B" terminal of cable female connector and ground other probe to an unpainted surface of cylinder. Bulb should light indicating solenoid circuit is satisfactory. Move probe from "B" terminal to "G" terminal of cable. Bulb should light indicating ground circuit is satisfactory. Move probe from "G" terminal to "W" terminal of cable. Bulb should light indicating circuit through "down" switch and "up" switch is satisfactory. Make a further check of down switch by actuating switch several times. Bulb should go off and on each time switch is operated, indicating satisfactory operation of "down" switch. Also move piston rod to the completely extended position at which time the bulb should go off. Now move piston rod in and out a short distance. Bulb should go off and on each time rod moves in and out, indicating that the "up" switch is operating satisfactorily.

**393.** If all the conditions outlined under electrical tests are met, the electrical system can be considered satisfactory. However, if bulb does not react as indicated, the circuit is probably short circuited and a search should be made for loose or corroded connections, or broken, frayed or bare wires.

## LUBRICATION AND MAINTENANCE

### All Models

394. When performing service or maintenance of any kind on the hydraulic lift system, the practice of cleanliness is of the utmost importance. Pump, valves and cylinders are manufactured to close tolerances and are mirror finished. The induction of dirt and foreign material is most detrimental and should be avoided at all times.

Capacity of hydraulic unit is 20 quarts when unit has draft control or 24 quarts when only the Hydra-Lectric control unit is used. Add ¾ quart for each 3 inch external cylinder, 1½ quarts for each 4 inch external cylinder and ½ pint for each 5 feet of hose. Recommended fluid is SAE 10W Supplement 1 engine oil or Type A Automatic Transmission Fluid.

395. **DRAIN AND REFILL.** Drain and refill the hydraulic lift system each year, or 1000 hours, as follows: Place lift arms in down position and remove drain plug which, on the series A and B tractors is the middle (square headed) plug on upper left side of hydraulic lift housing, or on series C, early 1850 and 1950 tractors is the square headed plug located just below the flow divider valve on left side of hydraulic lift housing. On series 1750, 1950-T and late series 1850 and 1950, the drain plug is the countersunk plug located at upper left

side of housing as shown in Fig. O223A. Attach a length of hose in plug opening and place open end in a suitable container as shown in Fig. O223. Start engine, operate at about 1000 rpm and pump oil from reservoir. Stop engine as soon as oil stops flowing. DO NOT operate engine when the hydraulic system is without oil. If lift system is equipped with remote cylinders, remove hoses from cylinders and move piston in and out to clear cylinders of oil.

Reconnect hoses, if necessary, and reinstall drain plug. Fill reservoir to "FULL" mark on the dip stick. Start engine and cycle the hydraulic equipment several times. Recheck fluid level and add as necessary to bring level to "FULL" mark.

396. **BREATHER.** A breather fitted with a micronic element is located on the top side of the hydraulic housing near the right front corner on early model tractors, or combined with dip stick on late model tractors. Under normal conditions, this element is renewed each time the hydraulic fluid is renewed. However, under abnormal conditions, more frequent renewal is required.

Renewal is obvious after removing cap screw, or nut, from top side. Use new gaskets and crush washers when renewing element.

397. **OIL FILTER.** The hydraulic lift system is equipped with a full flow oil filter located on left side of lift housing on 1800 and 1900 series A and B tractors, or on front of lift housing on 1800 and 1900 series C, 1750, 1950-T, 1850 and 1950 tractors. Also incorporated in the filter circuit is a by-pass valve which opens at a 20 pound pressure differential should the filter become plugged. Service of the by-pass valve on series A and B tractors requires removal of hydraulic

housing from tractor rear frame and removal of work cylinder from housing. See Fig. O240. On 1800 series C, 1750, 1950-T, 1850 and 1950 tractors, by-pass valve has been moved to upper left front of housing and valve can be removed if necessary by removing the square headed plug which is second in line at upper left front of hydraulic lift housing.

Renewal of filter element is obvious after removing the filter body. Use new "O" ring and lubricate same prior to installing filter body. Tighten body to 150-200 ft.-lbs. torque.

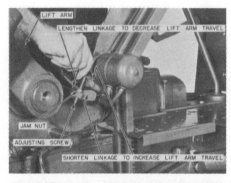

Fig. O224 — To adjust the ½ - 1 inch lift arm free travel, adjust linkage as shown.

Fig. O225 — Refer to text and adjust restrictor valve as shown. An identical valve is located on opposite side of housing.

Fig. O223 — Drain hydraulic system as shown. Early model tractor is shown, however, late models are similar. Refer to text for drain plug location.

Fig. O223A—Hydraulic system drain plug for series 1750, 1950-T, and late 1850 and 1950 tractors, except those with external valve Depth-Stop hydraulic systems, is located as shown.

Fig. O226 — Auxiliary restrictor valve installed on early model tractors. Refer to text for adjustment.

## SYSTEM ADJUSTMENTS

The adjustments in this section will include those external adjustments which can be accomplished with no disassembly. For those internal adjustments which require disassembly, refer to the appropriate section.

**Fig. O226A—Auxiliary restrictor valve installed on late model tractors. Refer to text for adjustment.**

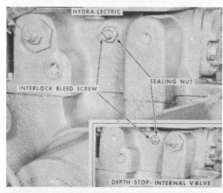

**Fig. O227 — Refer to text and adjust interlock bleed screw as shown. An identical bleed screw is located on opposite side of housing.**

398. **LIFT ARM ADJUSTMENT.** Place the three-point hitch (inner) control lever in down (forward) position. Start engine and run at approximately 1000 rpm. Move control lever to up (rear) position and allow lift arms to raise to upper limit. Grasp the lift links and check the free travel which should be ½-1 inch, measured at ends of lift arms. If free travel is not as specified, refer to Fig. O224 and loosen jam nut of the adjusting link, which connects the control lever to the servo lever, and adjust as necessary. Shorten the linkage to increase, or lengthen the linkage to decrease the lift arm free travel. Tighten jam nut and recheck.

399. **RESTRICTOR VALVE.** Loosen jam nut and turn adjusting screw (Fig. O225) as required until implement (load) will lower at the maximum rate without bouncing. Clockwise rotation of adjusting screw will increase restriction and slow the lowering rate. An extreme change in loading will require readjustment of restrictor valve. An identical valve is located on opposite side. Note different location of the restrictor adjusting screw on internal valve Depth-Stop housing as shown in inset.

400. **AUXILIARY RESTRICTOR VALVE.** This externally mounted valve is designed to provide restriction to the return flow of oil from rod end of a remote cylinder and is primarily used when a roll-over plow or similar equipment is attached to tractor and when valve is used in conjunction with restrictor valve in Hydra-Lectric or internal valve Depth-Stop control unit, restriction is provided for oil return flow from both base and rod end of remote cylinder.

This valve is installed on the right side of lift housing as shown in Fig.

O226 or O226A. This is the free flow port for the right side of the control unit. Be sure that the arrow which is embossed on the valve body points in direction of oil flow when unit is installed.

The control unit restrictor valve and the auxiliary restrictor valve should be adjusted until restriction of both valves is as nearly equal as possible. Adjustment should prevent implement bounce or reduce it to a minimum. Turning auxiliary restrictor valve adjusting screw inward increases the return oil flow restriction.

401. **INTERLOCK BLEED SCREW.** The interlock bleed screw (Fig. O227) must be backed out three or more turns when a single acting remote cylinder is used in order to provide a passageway to the reservoir. An identical valve is located on opposite side.

Note. The bleed screw must be all the way in when a double acting cylinder is used.

402. **BY-PASS VALVE (SINGLE ACTING CYLINDER).** This valve is mounted at the right front corner of the Hydra-Lectric control unit on early models (Fig. O228) or is a steel ball retained by a snap ring in lift cylinder in late models (Fig. O229) and is used when a single acting remote cylinder is used. When using a single acting cylinder, loosen jam nut and back out adjusting needle at least 4 turns (Fig. O228), or the cap screw (21—Fig. O241) five or six turns. Screw (21) is the by-pass valve shown in Fig. O229.

When a single acting cylinder is not being used, valve must be completely closed; however, do not exceed 10 ft.-lbs. torque when seating needle on external valve or damage may occur.

**Fig. O228 — View showing installation of external by-pass valve used on early model tractors. Refer to text.**

**Fig. O229 — By-pass valve used on late tractors is located as shown. Refer to text for adjustment.**

**Fig. O230 — Install pressure gage in drain hole to check pressure relief valve setting. Early tractor is shown but late tractors will be similar. Refer to text for drain plug location.**

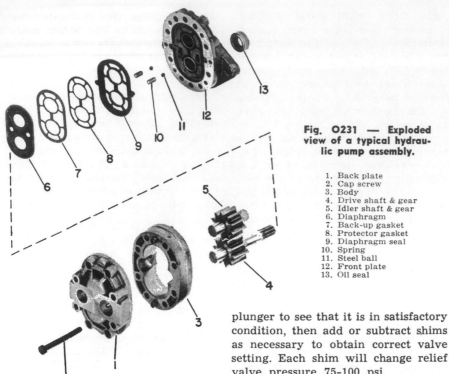

**Fig. O231 — Exploded view of a typical hydraulic pump assembly.**

1. Back plate
2. Cap screw
3. Body
4. Drive shaft & gear
5. Idler shaft & gear
6. Diaphragm
7. Back-up gasket
8. Protector gasket
9. Diaphragm seal
10. Spring
11. Steel ball
12. Front plate
13. Oil seal

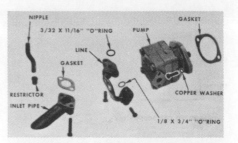

**Fig. O231A—Component parts which make up the vane type hydraulic pump conversion package.**

### SYSTEM OPERATING PRESSURE

403. The hydraulic system pressure relief valve is preset at the factory to open at 1750 psi for Series A and B tractors; 1900 psi for Series C tractors and 2050 psi for series 1750, 1850, 1950 and 1950-T tractors, except those with external valve Depth Stop systems which should be 2100 psi. Pressure setting can be checked as follows: Remove hydraulic unit drain plug and install in the drain plug hole, a gage capable of registering at least 3000 psi. See Fig. O230. Start engine and bring hydraulic fluid to operating temperature. Run engine at about 2000 rpm, then overload hydraulic system by placing a control lever in a power position and note gage reading. The gage reading should be approximately 250 psi lower than specified valve opening pressure.

NOTE: The gage reading will be a full flow reading and is equal to the cracking pressure setting of the valve. It is generally impossible to read the cracking pressure during this operation and hence the full flow pressure reading, which is about 250 psi lower, must be used.

If relief valve adjustment is necessary, record the pressure gage reading for subsequent reference, then on models prior to tractor serial number 134 684-000 remove the hydraulic lift unit as outlined in paragraph 414. Remove plug (V—Fig. O240), spring, shims and valve plunger. Inspect valve

plunger to see that it is in satisfactory condition, then add or subtract shims as necessary to obtain correct valve setting. Each shim will change relief valve pressure 75-100 psi.

Tractors serial number 134 684-000 and up have the relief valve located at front right of lift housing and housing need not be removed for servicing valve. Valve plunger (49—Fig. O238), shims (50) and valve spring (51) are identical to those used in earlier tractors.

### PUMP

Beginning with tractor serial number 191 121-000, a vane type hydraulic system pump replaced the gear type pump used in all series 1750, 1850, and 1950 tractors and is available in kit form to service series C 1800 tractors, serial number 136 501-000 and up as well as all series 1750, 1850 and 1950 tractors prior to serial number 191 121-000.

Tractors prior to the above serial numbers can be converted from gear pump to vane pump by ordering conversion package, Oliver part number 165 529-000 for series 1800 C, 1750, 1850 and 1950 with Hydra-Lectric or internal valve Depth-Stop hydraulic systems. For series 1750, 1850 and 1950 tractors with external valve Depth-Stop hydraulic systems, order conversion package, Oliver part number 165 742-000. The conversion packages contain all of the necessary parts as well as instruction sheets. Refer to Figs. O231A and O231B.

404. **REMOVE AND REINSTALL. (GEAR TYPE).** The hydraulic pump and the pump manifold are removed as a unit as follows: Remove the hydraulic lift unit as outlined in paragraph 414. On late model tractors,

back-out the by-pass valve cap screw until it clears piston housing. Remove the cylinder and piston assembly, then remove the two cap screws which retain manifold to hydraulic housing and the two nuts from pump mounting studs. Lift pump and manifold from hydraulic housing.

404A. **REMOVE AND REINSTALL (VANE TYPE).** Remove the hydraulic lift unit as outlined in paragraph 414. Back out the by-pass valve cap screw until it clears cylinder housing, then unbolt and remove cylinder and piston assembly. Remove inlet pipe and gasket. Disconnect line from pump body, then remove the pump mounting cap screws and pull pump from lift housing.

Coat both sides of pump mounting gasket with gasket cement (Permatex No. 3 or equivalent), use new "O" ring in pump line and reinstall pump by reversing removal procedure. Use new gasket when installing inlet pipe.

405. **OVERHAUL. (GEAR TYPE).** With the pump and manifold removed as outlined in paragraph 404, remove cap screws and separate pump and manifold.

NOTE: Pumps of different capacities have been used and the pump housing configuration may differ, but internal parts remain basically the same.

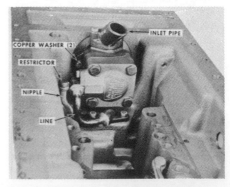

**Fig. O231B—View showing installation of the vane type hydraulic pump. Note position of the "S" shaped nipple and restrictor assembly.**

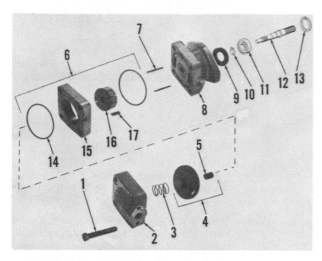

**Fig. O231C — Exploded view of a typical vane type hydraulic pump assembly.**

1. Cap screw
2. Cover
3. Spring
4. Pressure plate
5. Bushing
6. Cartridge kit
7. Pin (dowel)
8. Body
9. Oil seal
10. Snap ring
11. Bearing
12. Shaft
13. Snap ring
14. "O" ring
15. Center ring
16. Rotor
17. Vanes

Refer to Fig. O231 and proceed as follows: Place a scribe mark across front plate, body and back plate to insure proper assembly. Place pump in a vise and remove cap screws (2) which hold pump together. Remove pump from vise and while holding pump in hands, bump drive shaft against wood block and separate pump. If body (3) remains with either front plate (12) or back plate (1), remove by tapping with a soft faced hammer, or by installing drive shaft in bearing and bumping shaft. Removal of diaphragm (6), springs (10), balls (11), gaskets and oil seal (13) is obvious.

Clean all parts and inspect as follows: Check pump shafts for wear or scoring and if diameter of shafts at bearing area is less than 0.685, renew shafts and gears as they are not available separately. Measure inside diameter of bearings and if greater than 0.691, renew complete front or back plate as bearings are not available separately. Bearings should be flush with pattern islands. If wear or scoring in face of back plate exceeds 0.0015, renew back plate. Measure gear pockets of body (3) and if diameter exceeds 1.719, renew body.

When reassembling, use new diaphragm, diaphragm seal, back-up gasket, protector gasket, shaft oil seal

and manifold "O" rings. A kit is available which includes the diaphragm, diaphragm gasket and seals. Coat parts with engine oil and install diaphragm seal in front plate with open part of "V" section down. Be sure both gaskets and diaphragm are under lip of diaphragm seal and bronze face of diaphragm is toward gears. If necessary, use a dull tool to work diaphragm seal into front plate and the gaskets and diaphragm under lip of diaphragm seal. Half-moon cavities of pump body face toward the pump back plate. Shaft oil seal is installed with lip facing inside of pump. Torque the retaining cap screws to 25 ft-lbs.

Use new "O" rings and mount pump to manifold; then install assembly in hydraulic housing.

**405A. OVERHAUL (VANE TYPE).** With pump removed as outlined in paragraph 404A, place a scribe line across pump body, center ring and cover, then remove the four retaining cap screws and cover (2—Fig. O231C), spring (3) and plate (4). Bushing (5) is available separately and can be renewed if necessary. Remove center ring and dowel pins (7). Remove rotor and vanes from pump shaft. Remove snap ring (13) and remove shaft (12), bearing (11) and seal (9) from body (8). Remove snap ring (10) and bearing from shaft.

Clean and inspect all parts. Pay particular attention to rotor vanes and inside surface of center ring. Vanes should be a snug fit in rotor slots yet be able to slide freely. Inside diameter of center ring should not be grooved or show chatter marks. If rotor, rotor vanes or center ring are defective, all three parts must be renewed and a cartridge kit is available from Oliver Corporation. A seal kit containing shaft seal (9) and the cover (2) and body (8) sealing "O" rings is also available.

Reassemble by reversing disassembly procedure and lubricate all moving parts during assembly. Oil seal (9) is installed with lip facing rotor (rearward).

## PUMP MANIFOLD

**406. R&R AND OVERHAUL. (PRIOR SERIAL NO. 191 121-000).** Remove the pump and manifold assembly as outlined in paragraph 404, then unbolt and remove manifold from hydraulic pump. On tractors prior to serial number 134 684-000, remove relief valve plug, spring and plunger. Do not lose any of the shims which may be interposed between plunger and spring. Use a ½-inch socket, with extension, and remove plunger seat and crush washer. Remove "O" rings if not already done.

NOTE: On tractors prior to serial number 94 156-000, relief valve plunger seat was retained in housing by Type AV Loctite sealant. On tractors serial number 94 156-000 and up, a set screw is used to retain the plunger seat. Should the early type manifold be encountered, modify same as follows: Locate and drill a hole as shown in Fig. O232, using a number 7 (0.2010) drill bit. Use a ¼-20 tap and tap the hole. Clean all metal chips from manifold, then install plunger seat with new crush washer and tighten seat to a torque of 25 ft.-lbs. With a $\frac{3}{16}$-inch drill bit, and using set screw hole as a guide, drill (countersink) the valve seat $\frac{1}{16}$ to $\frac{3}{32}$-inch deep. Clean all metal chips from hole and install a ¼-20x⅝-inch socket head set screw. Tighten securely, then install an additional, similar set screw to lock the first set screw in place.

Reassemble manifold and refer to paragraph 407 to set the pressure relief valve before assembling pump and pump manifold.

407. The hydraulic system pressure relief valve on tractors prior to serial number 134 684-000 is located in the pump manifold and can be tested and adjusted, if necessary, before any reassembly is done.

To test and adjust the pressure relief valve when hydraulic pump and

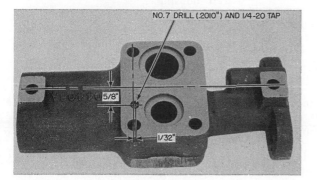

NO.7 DRILL (.2010") AND 1/4-20 TAP

5/8"

1/32"

**Fig. O232 — When reworking early type manifolds, locate hole as shown. Refer to text.**

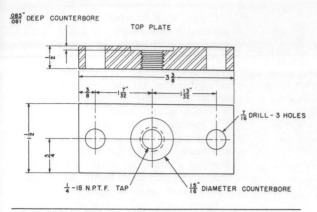

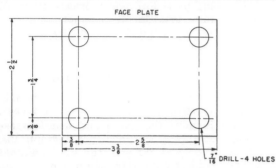

**Fig. O233 — On early type manifolds use adapter plates, made to the dimensions shown to test pressure relief valve when hydraulic unit is off tractor.**

manifold are disassembled, a diesel injection nozzle tester and two adapter plates are required. The adapter plates can be fabricated locally, using the dimensions shown in Fig. O233. When testing relief valve with this equipment the cracking pressure setting of 1750 psi for series A and B tractors, or 1900 psi for series C tractors is obtained.

To test and adjust the pressure relief valve, use "O" ring and install the adapter plates, making certain the top adapter plate is aligned with hole in top of manifold. Attach nozzle tester to top plate, then raise pressure slowly and note gage reading when relief valve opens (cracks). Repeat this operation several times in order to be accurate. If valve cracking pressure is not as stated above, remove relief valve and add or subtract shims as necessary from the relief valve plunger stem. Each shim will change relief valve pressure 75-100 psi.

## HYDRAULIC PUMP DRIVE SHAFT AND PINION

408. The hydraulic pump is driven from an idler gear which is in turn driven from a gear mounted in the wall which separates the final drive and transmission compartments. This internally splined drive pinion (66—Fig. O238) is driven by the pto shaft on tractors so equipped, or by a separate shaft (70) on those tractors having no pto. Refer to pto section for information on pto shaft and to paragraph 410 for information on pinion.

409. **R&R PUMP DRIVE SHAFT.** Remove cap screws which retain bearing carrier (82) to rear frame rear cover, pull carrier and shaft assembly from tractor rear frame. Remove snap ring (78) from bearing carrier and pull shaft and bearing from carrier. Remove snap ring (80) from aft end of shaft and remove bearing from shaft.

When reinstalling shaft assembly in tractor, it may be necessary to engage starter momentarily to engage splines of shaft and flywheel.

410. **R&R PUMP DRIVE PINION.** Remove the pto shaft as outlined in paragraph 356, or the hydraulic pump drive shaft as outlined in paragraph 409. Remove hydraulic lift unit as outlined in paragraph 414. Remove cotter pin from hub of idler gear and remove idler gear shaft, idler gear and thrust washers. Remove "O" ring (early models so equipped) from front of drive gear pinion, then remove snap ring from front side of pinion bearing and bump pinion and bearing forward out of housing wall. Remove snap ring from rear of pinion hub and press bearing from pinion. On early models, remove oil seal from bearing bore in housing.

When reinstalling, be sure lip of seal used in early models is toward front of tractor and do not install seal so deep that lip will catch on pinion snap ring. Install idler gear with long hub on front side.

## HYDRA-LECTRIC CONTROL UNIT

The Hydra-Lectric control unit serves as a top cover for the hydraulic unit and contained within it are the control valve spools, interlock assemblies, restrictor valves, interlock bleed screws and the Hydra-Lectric switches and solenoids. Refer to Fig. O234 for an exploded view of the unit.

411. **R&R AND OVERHAUL.** Remove the seat and tool box, then unbolt and remove the Hydra-Lectric control unit from hydraulic lift housing.

Remove the interlock valve guide retainers (45) and turn unit bottom side up. Note: Early guide retainers, prior to tractor serial number 120 995-000, were one-piece. Indentify and disconnect the wires from terminals in housing. Remove solenoid cap (67), then disconnect the wires from switches and remove switches from bracket and tube assembly (55). Remove the solenoid springs (66) and the solenoid plungers (61). Overtravel springs are riveted to plungers and should not be removed. If spring is defective, renew complete plunger assembly. Also, do not remove the small positioning roll pins from solenoid spring unless necessary. Remove the bracket and tube assembly (55), solenoids (54), spacers (late models) and the two small ¼-inch steel balls (62) from their recesses in the housing.

Drive roll pins (11) from inner ends of spools, remove end plate (1), then remove snap rings from spool bores and push spools out rear of housing. See Fig. O235. Centering spring (8—Fig. O234), washers (7) and travel limiter (9) can be removed from spool, if necessary, by removing snap ring (6). Grasp stem of rear interlock valve guides (12) with a pair of pliers and pull from housing. Remove spring (15) and check ball (13). Use a soft drift of not more than ½-inch diameter and drive front valve guide (12), spring (15), check ball (13), valve seat (16) and plunger (18) forward out of housing. Work from front of housing and bump valve seat (16) from housing.

Restrictor valves and interlock bleed screws can be removed at any time and the procedure for doing so is obvious.

The thermal relief valves shown in Fig. O236 can be removed and inspected if necessary. Thermal relief valves are preset at the factory to relieve at 4400-5100 psi and are non-adjustable. A faulty valve is renewed as an assembly.

NOTE: Beginning with tractor serial

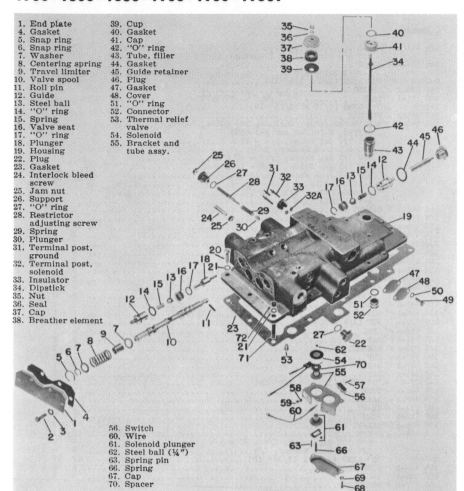

1. End plate
4. Gasket
5. Snap ring
6. Snap ring
7. Washer
8. Centering spring
9. Travel limiter
10. Valve spool
11. Roll pin
12. Guide
13. Steel ball
14. "O" ring
15. Spring
16. Valve seat
17. "O" ring
18. Plunger
19. Housing
22. Plug
23. Gasket
24. Interlock bleed screw
25. Jam nut
26. Support
27. "O" ring
28. Restrictor adjusting screw
29. Spring
30. Plunger
31. Terminal post, ground
32. Terminal post, solenoid
33. Insulator
34. Dipstick
35. Nut
36. Seal
37. Cap
38. Breather element

39. Cup
40. Gasket
41. Cap
42. "O" ring
43. Tube, filler
44. Gasket
45. Guide retainer
46. Plug
47. Gasket
48. Cover
51. "O" ring
52. Connector
53. Thermal relief valve
54. Solenoid
55. Bracket and tube assy.

56. Switch
60. Wire
61. Solenoid plunger
62. Steel ball (¼")
63. Spring pin
66. Spring
67. Cap
70. Spacer

**Fig. O234—Exploded view of late type Hydra-Lectric control unit. Unit serves as a top cover for the hydraulic housing. Parts may vary somewhat depending upon what type hydraulic system is used. Earlier type units are similar. A washer (not shown) is used between jam nut (25) and support (26).**

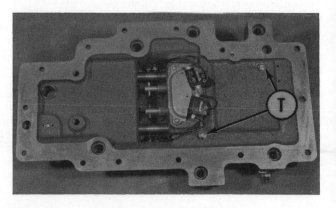

**Fig. O235 — View of valve spools and interlock assemblies removed from housing.**

number 135 193-000, four thermal relief valves are used.

Clean and inspect housing bores, valve spools and all other parts for scoring, nicks or other damage. Spools and interlock plungers should be a snug fit in bores, yet slide freely. Check all springs for fractures or other damage. Check wiring, solenoids and terminals for damage. Check switch action and continuity. If deemed necessary, resistance of solenoid can be checked with an ohmmeter. Solenoid should show approximately 4 ohms resistance.

412. Reassemble by reversing the disassembly procedure. Coat all working parts with engine oil to provide initial lubrication. Coat all "O" rings and seals with heavy grease to prevent damage during installation and be sure to use new "O" rings, gaskets and seals.

NOTE: If a new control housing is being installed, be sure the two ¼-inch steel sealing balls are installed. One is located in right top side of housing directly above the right interlock bleed screw and the other

in left lower mounting surface directly below left interlock bleed screw. If necessary, drive the steel balls into housing and peen around the outside area to retain balls in position.

In addition, at tractor serial number 120 995-000, interlock balls (13—Fig. O234) were reduced in diameter from ¾-inch to ⅝-inch. Be sure to use correct parts when renewing.

After assembly, be sure to check operation of solenoid switches by actuating valve spools. A click should

be heard when spools are moved either way. Be sure that all six passage connectors (52—Fig. O234) are in place before installing control unit to hydraulic housing.

### INTERNAL VALVE DEPTH-STOP CONTROL UNIT

413. The control unit for internal valve Depth-Stop unit is shown in Fig. O237 and note that except for housing (19), piston assemblies (34, 35 and 36) and tube (31) assembly, the unit is almost identical to the control unit used for Hydra-Lectric units shown in Fig. O234.

System adjustments are the same as those given for the Hydra-Lectric control unit and are given in paragraphs 399 through 402.

To service the internal valve Depth-Stop unit, use the service information given in paragraphs 411 and 412 except that any reference to electrical units should be disregarded. Pistons (34) and springs (35) can be removed after cap (36) is off. See Fig. O236A.

### HYDRAULIC HOUSING

The hydraulic housing (10—Fig. O238) contains the hydraulic pump, servo valve and control lever linkage along with the rockshaft and its op-

**Fig. O236 — Thermal relief valves (T) are located in bottom of control housing as shown. Late units have four thermal relief valves instead of two as shown for early units.**

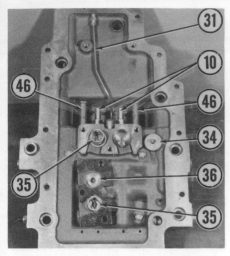

**Fig. O236A—Bottom side of Depth-Stop internal valve system control housing. Note spring counterbores in cap (36) and detent pin on top side of piston (34).**

| | |
|---|---|
| 10. Control valves | 35. Spring |
| 31. Oil line | 36. Cap |
| 34. Piston | 46. Guide retainer |

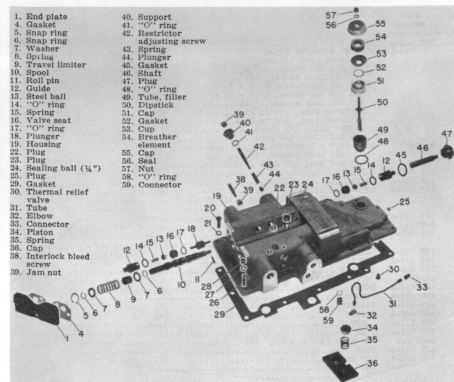

| | |
|---|---|
| 1. End plate | 40. Support |
| 4. Gasket | 41. "O" ring |
| 5. Snap ring | 42. Restrictor |
| 6. Snap ring | adjusting screw |
| 7. Washer | 43. Spring |
| 8. Spring | 44. Plunger |
| 9. Travel limiter | 45. Gasket |
| 10. Spool | 46. Shaft |
| 11. Roll pin | 47. Plug |
| 12. Guide | 48. "O" ring |
| 13. Steel ball | 49. Tube, filler |
| 14. "O" ring | 50. Dipstick |
| 15. Spring | 51. Cap |
| 16. Valve seat | 52. Gasket |
| 17. "O" ring | 53. Cup |
| 18. Plunger | 54. Breather |
| 19. Housing | element |
| 22. Plug | 55. Cap |
| 23. Plug | 56. Seal |
| 24. Sealing ball (¼") | 57. Nut |
| 25. Plug | 58. "O" ring |
| 29. Gasket | 59. Connector |
| 30. Thermal relief | |
| valve | |
| 31. Tube | |
| 32. Elbow | |
| 33. Connector | |
| 34. Piston | |
| 35. Spring | |
| 36. Cap | |
| 38. Interlock bleed | |
| screw | |
| 39. Jam nut | |

**Fig. O237—Exploded view of Hydra-Lectric and Depth Stop internal valve system control unit as applied to series 1750 or 1950-T. Note that except for the solenoid assemblies, this control unit is very similar to the control unit used for Hydra-Lectric units which is shown in Fig. O234. A washer (not shown) is used between jam nut (39) and support (40).**

erating mechanism. The type of hydraulic system will determine what components are included in the hydraulic housing.

Everything except the hydraulic pump, manifold (when used), piston rod and the piston and cylinder assembly can be serviced without removing the hydraulic unit from tractor.

**414. R&R HYDRAULIC HOUSING.** Drain hydraulic unit as outlined in paragraph 395. Remove seat and tool box. Disconnect lift links and the upper link. Disconnect the electrical connections. Disconnect hoses and/or piping and plug the openings. Unhook the brake return springs (brake rods on 1950-T) and remove both platforms. On tractors equipped with wheel guard extensions, also remove the brake pedal cover and the platform splash panels. If tractor is equipped with power steering, disconnect lines from flow divider valve and housing and plug openings. Clean the entire unit thoroughly, then remove the hydraulic pump drive cover (39—Fig. O238) and to provide a lifting point, install two ⅜ x 3 inch cap screws into the two tapped holes in housing. Place a chain under these two cap screws and the upper link support at rear, then attach a hoist. Unbolt hydraulic unit from rear frame and lift from tractor. With unit still supported, drive dowels (24) from bottom left front and right rear of hydraulic housing, remove the actuating rod guide, or the hole cover (74) and remove oil pan (71). Now

set unit on a bench, or preferably, mount unit on an engine stand if complete overhaul is anticipated. See Figs. O239 and O240.

Reinstall by reversing the removal procedure.

## CONTROL LEVERS AND LINKAGE

**415. R&R AND OVERHAUL.** The Hydra-Lectric and draft control levers assembly can be removed as follows: Right fender must be in outermost position (or removed), then remove the Hydra-Lectric control unit as outlined in paragraph 411. Remove the cap screws and keys from control arms at inner ends of Hydra-Lectric lever shafts and remove control arms. Pull retaining rings from swivels of adjusting link (screw) and remove link from draft control lever and external servo valve control arm (draft control arm). Remove cap screws which retain the lever support to hydraulic housing and pull support and levers assembly from hydraulic housing. Drive roll pin from draft control shaft inner arm, pull shaft from support and remove arm. Remaining internal linkage can be removed after removing the actuating rocker arm spring support and spring and servo valve. Draft control shaft support and actuating rocker shaft

can be removed from hydraulic housing if desired.

When installing new draft control shaft support, drive support into housing until it protudes about ⅜-inch inside housing. When installing new actuating rocker arm pivot, install with ring groove toward inside and drive pivot into housing until rocker arm operates freely on pivot with the retaining ring installed.

Any further disassembly and/or overhaul of the levers assemblies or linkage is obvious.

## SERVO VALVE

**416. R&R AND OVERHAUL.** To remove and overhaul the servo valve, first drain hydraulic unit, then remove the Hydra-Lectric control unit as outlined in paragraph 411. Remove the actuating rocker support and spring, then unbolt and remove servo valve.

NOTE: Production changes have been made in the servo valve. The early valve shown in Fig. O242 has two check valve assemblies (items 18, 20, 21, 22 and 23) whereas late valves have only one. In addition, snap spool (3) and snap spool spring (2) have been changed. However, service procedure remains the same.

In some cases, excessive cycling of the three point hitch might occur

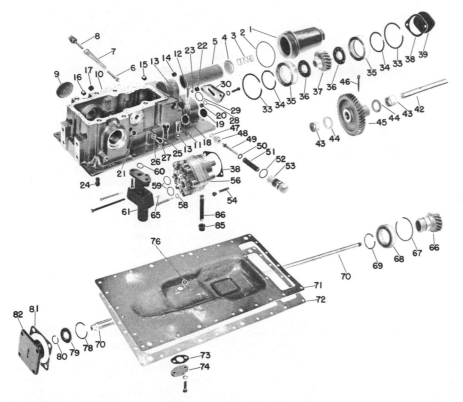

**Fig. O239 — Hydraulic housing (early) with Hydra-Lectric control unit removed.**

A. Rocker arm
C. Control arms
F. Filter
L. Linkage

R. Rockshaft
S. Support
V. Servo valve

**Fig. O238 — Exploded view of late type hydraulic lift housing and related parts. Early units differ in that the system pressure relief valve assembly is located in pump manifold, oil filter assembly is on left side of housing and filter by-pass valve is in lower right rear (inside) of housing. Tractors after serial number 191 121-000 do not use manifold (61) and have vane type pump instead of the gear type shown.**

| | | |
|---|---|---|
| 1. Filter body | 24. Dowel | 46. Cotter pin |
| 2. "O" ring | 28. "O" ring | 47. Washer |
| 3. Spring | 29. "O" ring | 48. Valve seat |
| 4. Retainer | 30. Cover | 49. Relief valve |
| 5. Filter element | 33. Snap rings |     plunger |
| 6. By-pass plunger | 34. Snap rings | 50. Shim(s) |
| 7. Spring | 35. Carrier | 51. Spring |
| 8. Plug and pin | 36. Bearings | 52. "O" ring |
| 9. Expansion plug | 37. Drive gear | 53. Retainer |
| 10. Hydraulic housing | 38. Gasket | 54. Stud |
| 18. Plug | 39. Cover | 56. Pump |
| 19. Plug | 42. Idler gear shaft | 59. "O" rings |
| 20. Plug | 43. Bearings | 60. "O" ring |
| 21. Plug | 44. Thrust washers | 61. Manifold |
| 22. Plug | 45. Idler gear | 66. Pinion |
| 23. Flared insert | | |

| | |
|---|---|
| 67. Snap ring | |
| 68. Bearing | |
| 69. Snap ring | |
| 70. Pump drive | |
|     shaft | |
| 71. Reservoir | |
| 72. Gasket | |
| 73. Gasket | |
| 74. Cover | |
| 78. Snap ring | |
| 79. Bearing | |
| 80. Snap ring | |
| 81. Gasket | |
| 82. Housing | |
| 85. Restrictor | |
| 86. Nipple | |

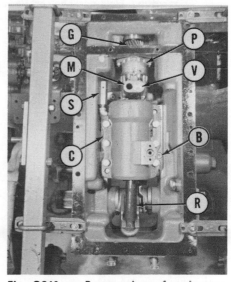

**Fig. O240 — Bottom view of early type hydraulic housing. Cylinder (C) must be off to remove filter by-pass valve. Some early tractors may not have safety valve (S).**

B. Filter by-pass
   valve
C. Cylinder
G. Pinion gear
M. Manifold

P. Pump
R. Piston rod
S. Safety valve
V. Relief valve

when engine is running; or hitch might lower when engine is stopped. This could result from lowering valve ball (17—Fig. O242) being jammed into spring (26) and not allowing ball to seat. In addition, some early servo valves, identified by having part number 160 665-A stamped on body, may have check ball pin (25) bore misaligned which can also cause the hitch to malfunction. A field package, part number 165 686-AS, is available to service servo valves on tractors prior to serial number 197 799-000.

For repairs on tractors prior to tractor serial number 168 536-000, install spring (P/N 160 215-A), cap (P/N 160 216-000) and spring guide (P/N 165 637-A). For repairs on tractors between serial numbers 168 536-000 and 189 768-000, install spring guide (P/N 165 637-A). For repairs

on tractors between serial numbers 189 768-000 and 191 799-000, remove stop pin which was added to cap (27) at tractor serial number 189 768-000, and install spring guide (P/N 165 637-A).

When servicing the 160 665-A valves, also check the check ball pin bore as follows: Obtain a rod ⅛-inch in diameter and at least six inches long and insert it in check ball pin bore. Measure at check ball cap bore and if end of rod can be moved more than $\frac{1}{16}$-inch from true center, renew the servo valve.

With servo valve removed it can be disassembled as follows: Refer to Fig. O242 and push pilot spool (4) in as far as possible; then tap lightly to bump out retainer (31). Remove pilot spool assembly from body (13), then remove nuts and retainer (31), drive out roll pin (7) and remove spring

(5) from pilot spool. Remove plug (1), spring (2) and snap spool (3). Remove snap ring (8), install a cap screw in plug (9) and pull plug. Remove lowering spool (11) and spring (12). Remove the two check valve caps (20), springs (22) and balls (18). Remove cap (27), then remove spring (26), ball (17) and lowering valve pin (25) from body. Remove cap (28), spring (29) and flow control spool (30).

145

cap screw (21—Fig. O241) on tractors serial number 134 684-000 and up. Piston can be removed from cylinder assembly by bumping open end of cylinder on a wood block.

Note. At this time, the oil filter by-pass valve can be serviced on tractors prior to serial number 134 684-000, if necessary. Valve is disassembled by removing the plug located in the left rear of the cylinder mounting flange. Filter by-pass valve on tractors serial number 134 685-000 and up is located on left front side of hydraulic lift housing.

Piston rod can be removed after removing the piston rod retainer.

When reassembling, be sure piston "O" ring is on pressure side, that is, toward closed end of piston. Grease "O" ring prior to installation into cylinder.

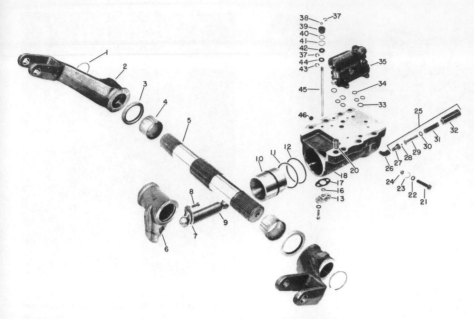

Fig. O241 — Exploded view of rockshaft, rockshaft cylinder and piston used in late model tractors. Early model tractors are similar except for dashpot (39) and by-pass (items 21, 22, 23 and 24). For early models without dashpot, a field package is available which is designed to dampen draft reaction and reduce draft control cycling in cases where necessary. Field package contains necessary parts and instruction sheet.

| | | | |
|---|---|---|---|
| 1. Snap ring | 13. Guide | 26. Elbow | 37. Retaining ring |
| 2. Lift arm, left | 16. Seal | 27. Valve seat | 38. "O" ring |
| 3. Oil seal | 17. Gasket | 28. Ball | 39. Piston (dash pot) |
| 4. Bushing | 18. Cylinder | 29. Plunger | 40. "O" ring |
| 5. Rockshaft | 20. Dowel | 30. Shim(s) | 41. Sealing ring |
| 6. Rocker arm | 21. Cap screw (special) | 31. Spring | 42. Seal |
| 7. Retainer | 22. Sealing nut | 32. Body | 43. Retaining ring |
| 9. Piston rod | 23. Snap ring | 33. "O" rings | 44. Washer |
| 10. Piston | 24. Ball (by-pass) | 34. "O" rings | 45. Actuating rod |
| 11. Back-up washer | 25. Safety valve assy. | 35. Servo valve | 46. Plug |
| 12. "O" ring | | | |

Any further removal of plugs, or seat (24), necessary for inspection of bores and passages is obvious. Check valve seats (23) are not available separately.

Clean all parts and inspect for excessive wear or scoring. Spools should be a snug fit in bores yet move freely. Renew all parts showing excessive wear or damage.

Use new "O" rings and gaskets when reassembling. Lubricate all working parts with engine oil prior to assembly to provide initial lubrication. Use heavy grease on all steel balls and springs so they will stay in position during assembly.

## ROCKSHAFT

**417. R&R AND OVERHAUL.** The rockshaft can be removed from hydraulic housing without removing housing from tractor as follows: Remove the Hydra-Lectric control unit, or cover, as outlined in paragraph 411. To provide removal clearance, remove either fender and the tire and wheel. Disconnect lift links, then remove snap rings from outer ends of rockshaft and remove lift arms. Use a soft faced hammer and bump rockshaft out of housing and rocker arm.

Note: This operation will force oil seal and bushing out on the side rockshaft is being removed from. The seal will be damaged and will require renewal, however, bushing can be salvaged if caution is used. Opposite oil seal and bushing can now be removed, if necessary. Rocker arm can be removed after disconnecting piston rod and the rocker arm feed-back link.

When reinstalling, insert rockshaft from side of housing that has no bushing, align blind splines of rockshaft and rocker arm and slide rockshaft into position. Use Oliver driver No. ST-144, or equivalent, and install bushing. Lubricate lip of oil seal and coat outer edge with sealant. Slide seal over rockshaft and drive into counterbore.

## ROCKSHAFT CYLINDER AND PISTON

**418. R&R AND OVERHAUL.** The work cylinder and piston assembly can be removed after removing the hydraulic housing as outlined in paragraph 414.

With hydraulic housing and oil pan removed, unbolt and remove cylinder and piston assembly which will require backing out the by-pass valve

## DRAFT CONTROL ACTUATING ASSEMBLY

This portion of the hydraulic lift assembly is located in the bottom rear of the tractor rear main frame and is composed of the lower link supports, actuating lever assembly and the lift draft control spring and spring support assembly.

To service these parts, refer to the appropriate sections and remove the pto, or rear frame cover and the hydraulic lift unit. See Fig. O243.

**420. R&R AND OVERHAUL.** With pto and hydraulic lift units removed, refer to Fig. O244 and proceed as follows: Loosen jam nuts (21), then remove adjusting nuts (20) and (22). Remove retainer (19) and spring (18) from support (15). Unbolt and remove support and spacers (14). Remove cap screws (8) and keys (10) and pull lower link support (1) and thrust washers (2) from rear frame. Remove actuating arm (7) from housing, remove pin (12) and remove rod (13) from actuating arm. Pry out oil seals (3), remove snap rings (4), then using a suitable driver, bump bearings (5) from housing.

Note: Seals will be damaged during removal. Do not remove seals and/or bearings unless necessary. If seals (3) are being renewed use the late type seal which has a double lip.

Reassemble by reversing the disassembly procedure and adjust draft control spring as outlined in paragraph 421.

**421. ADJUST DRAFT CONTROL SPRING.** After installation, draft control spring (18) is adjusted as follows: Install nut (22) and turn it

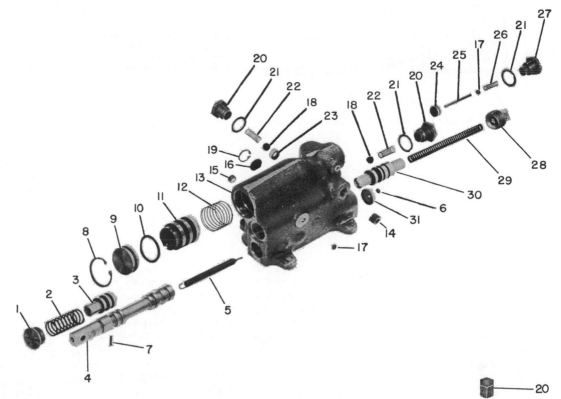

**Fig. O242 — Exploded view of servo valve assembly.**

1. Plug
2. Spring
3. Snap spool
4. Pilot spool
5. Return spring
7. Roll pin
8. Snap ring
9. Plug
10. "O" ring
11. Lowering spool
12. Spring
13. Body
14. Plug
15. Plug
16. Expansion plug
17. Steel ball (¼ in.)
18. Steel ball (7/16 in.)
19. Snap ring
20. Cap
21. Crush washer
22. Spring
23. Seat
24. Seat
25. Pin
26. Spring
27. Cap
28. Plug
29. Spring
30. Flow control spool
31. Retainer

down on rod (13) until it contacts retainer (19) and all slack is removed, then tighten nut an additional 2½ turns. This will compress spring the required ⅛-inch. Install and tighten one of the jam nuts (21) to retain adjusting nut in this position.

Place a straight edge across top surface of tractor rear frame (no gasket) and use a depth gage to measure the distance from top of adjusting nut (20) and straight edge. This distance should be 2⁹⁄₁₆ inches. Adjust nut (20) to obtain this measurement and tighten remaining jam nut (21) to hold adjusting nut in this position.

**Fig. O243 — View showing location of draft control actuating assembly.**

A. Actuating lever
L. Lower link support
N. Adjusting nut
S. Spring

## HYDRA-LECTRIC (REMOTE) CYLINDERS

Hydra-Lectric cylinders are available in three and four inch sizes.

Both size cylinders are similar in construction, with the exception that the cylinder head is bolted to the barrel on three inch cylinders while the four inch cylinder head is retained to the barrel with a snap ring. Any differences in service procedure will be noted. See Fig. O245.

**422. R&R AND OVERHAUL.** Removal of cylinder for service work is obvious.

**Fig. O244 — Exploded view of the draft control actuating assembly.**

1. Left lower link support
2. Thrust washer
3. Oil seal
4. Snap ring
5. Needle bearing
7. Actuating lever
8. Cap screw
9. Lock washer
10. Key
11. Retaining ring
12. Pin
13. Rod
14. Spacer
15. Spring support
16. Cap screw
17. Lock washer
18. Spring
19. Spring retainer
20. Adjusting nut
21. Jam nuts
22. Nut

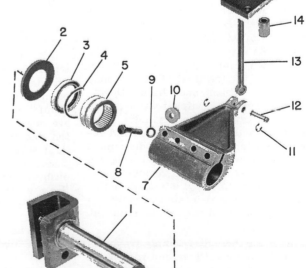

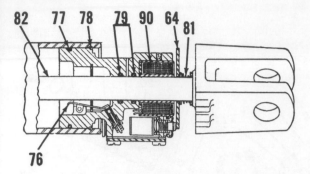

64. Stop collar
76. Contact arm
77. "O" ring
78. Snap ring
79. "O" ring
81. Spring
82. Piston rod
90. Solenoid

Clear fluid from cylinder by moving piston in and out, then thoroughly clean assembly. Refer to Fig. O246 and remove the 45 degree elbow, pipe nipple and the cable clamp from cylinder. On three inch cylinders, unbolt cylinder head. On four inch cylinders, insert small punch through hole in cylinder barrel and push snap ring out of its groove. Remove snap ring using a shall screw driver. Note: It may be necessary to bump cylinder head into barrel slightly to take pressure off snap ring. Use a series of sharp jerks on piston rod and bump cylinder head from barrel. Remove piston lock nut, then remove piston and cylinder head from piston rod. If desired, stop collar and spring can also be removed. "O" rings and back-up washers can now be removed from cylinder head and piston. Remove "Up" switch contact arm outer shaft from cylinder head by pushing on inner end. The four inch cylinder outer shaft has an "O" ring. Remove the "Up" switch contact arm, spring and inner shaft from inside cylinder head. Remove cover from cylinder head, identify and disconnect wire from "Up" switch terminal, then remove the set screw and nylon locking pin which retain "Up" switch terminal and remove terminal. Remove snap ring and rubber boot from "Down" switch, remove switch retaining nut, then pull switch from cylinder head. Identify and disconnect wires and remove switch. Identify and disconnect remaining wires, then loosen solenoid retaining set screw and remove solenoid. On three inch cylinders, the cylinder rod dust seal located directly behind the solenoid can also be removed, if necessary.

Any further disassembly required will be obvious. Clean and inspect all parts and renew any which show signs of undue wear, scoring, or other

damage. Use all new "O" rings, gaskets and back-up washers when reassembling. Pay close attention to wiring and renew any that show signs of fraying or other deterioration. Adjust cylinder as outlined in paragraph 423.

**423. CYLINDER ADJUSTMENT.** Adjust the Hydra-Lectric "Down" switch as follows: Extend the piston rod a short distance and remove cover from cylinder head. Now push stop collar toward "Down" switch until a gap of 0.030-0.050 for three inch cylinders, or 0.030-0.070 for four inch cylinders, exists between stop collar and cylinder magnet as shown in Fig. O247. Loosen jam nut and adjust "Down" switch until it just breaks the electrical circuit with stop collar in the position stated above. This point can be determined by an audible click of the switch, or by the light going off if a test light is used. Tighten switch jam nut and recheck the adjustment.

424. To check the adjustment of the "Up" switch, a test light is required and is accomplished as follows: Connect one test light probe to the "Up" switch terminal, or the "W" terminal of plug connector, and ground the other probe to an unpainted surface of the cylinder. Light should come on as points of switch are normally closed. Now extend piston rod to its limit at which time light should go out. Measure the exposed portion of the piston rod, then push the piston rod inward ⅛-inch. See Fig. O248. Light should now come on. If light does not come on, loosen the "Up" switch terminal locking set screw and adjust terminal until the ⅛-inch piston rod movement described above will actuate switch. Tighten set screw and recheck the adjustment.

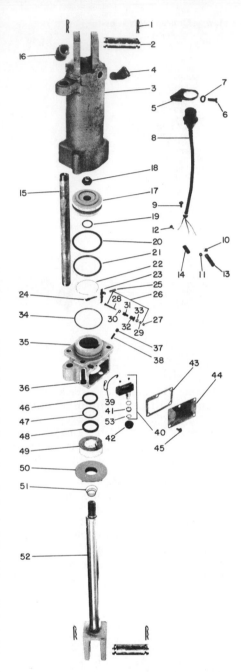

**Fig. O246 — Exploded view of 3 inch Hydra-Lectric remote double acting cylinder. Piston and piston seals will differ slightly on late units. The 4-inch cylinder is similar.**

| | |
|---|---|
| 3. Barrel | 32. Sleeve |
| 5. Bracket | 33. Washer |
| 8. Cable | 34. "O" ring |
| 13. Tubing | 35. Head |
| 14. Tubing | 37. Set screw |
| 15. Pipe | 38. Pin |
| 17. Piston | 39. Wire |
| 18. Nut | 40. Switch assy. |
| 19. "O" ring | 42. Boot |
| 20. Washer, backup | 43. Gasket |
| 21. "O" ring | 44. Cover |
| 22. Spring | 46. Packing |
| 23. Arm | 47. Washer |
| 24. Pivot | 48. Seal |
| 25. Pivot | 49. Coil |
| 26. Terminal assy. | 50. Stop |
| 28. Terminal | 51. Spring |
| 30. "O" ring | 52. Rod |
| 31. Insulator | |

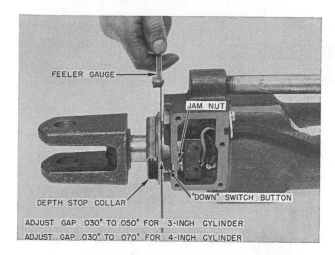

Fig. O247—Position stop collar as shown when adjusting "Down" switch.

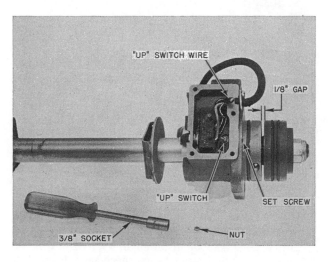

Fig. O248 — "Up" switch must be actuated when piston is 1/8-inch from cylinder head as shown. Refer to text.

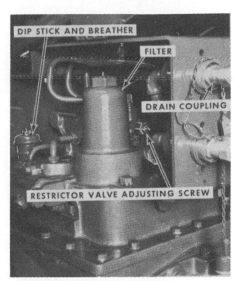

Fig. O250 — View of left rear corner of depth stop cylinder hydraulic housing (with shroud removed) showing location of hydraulic fluid dipstick, filter, drain coupling and restrictor valve adjusting screw on left control valve unit.

# DEPTH STOP HYDRAULIC SYSTEM (EXTERNAL VALVE)

Late production 1850 and 1950, and all 1750 and 1950-T tractors are available with an External Valve Depth-Stop hydraulic system for use with remote cylinders. Two valves are normally used, however, an additional one or two valves may be "stacked" to operate either three or four remote cylinders.

The external control valves can be either for single acting or double acting cylinders.

System is equipped with a pump having 20.4 gpm capacity at 2400 rpm of which 5 gpm is diverted by the flow divider to the power steering, leaving 15.4 gpm for operation of remote cylinders.

## SYSTEM ADJUSTMENTS

430. SYSTEM RELIEF PRESSURE. System relief pressure is controlled by the plunger type relief valve (24—Fig. O251) and is adjusted by varying the number of shims (25) located between the valve and spring (26). To check relief pressure, connect a 3000 psi capacity hydraulic gage to one of the remote cylinder quick disconnect couplings and pressurize that port by holding the control valve lever in proper position. Note: Hold lever in pressurized position only long enough to observe pressure gage reading. With engine running at 2200 RPM and hydraulic fluid at normal operating temperature, system relief pressure should be 2050-2150 psi. If not within the range of 2050-2150 psi, stop engine, remove the valve plug (28), spring (26), shims (25) and plunger (24). Add or remove shims as necessary to obtain correct relief pressure and renew the sealing "O" ring (27) when reinstalling plug. Note: If adding shims does not increase the system relief pressure, a faulty hydraulic system pump should be suspected.

431. FLOW DIVIDER RELIEF PRESSURE. A relief valve within the flow divider valve assembly should limit pressure within the power steering system to aproximately 1500 psi. Refer to paragraph 46 for checking and adjusting the flow divider relief valve pressure. Other than minor differences, the flow divider valve used with the depth stop cylinder hydraulic system is identical to that used with the Hydra-Lectric and/or 3-point hitch hydraulic system.

432. RESTRICTOR VALVE ADJUSTMENT. A restrictor valve on the rear of each control valve unit (See Fig. O250) controls the flow of fluid returning to the hydraulic pump from the base of the remote cylinder connected to that valve unit. An implement will have a tendency to bounce as it is lowered if flow is not sufficiently restricted. If bouncing is encountered, loosen lock nut (18—Fig. O253) on restrictor valve adjusting screw (17) and turn screw in (clockwise) to increase restriction. Note: Remove load on remote cylinder before attempting to adjust the restrictor valve. If adjusting screw is turned in too far resulting in too much restriction, the relief valve will blow causing a buzzing noise whenever remote cylinder is retracted. If this occurs, turn the screw out (counter-clockwise) to correct the condition. Tighten lock nut to maintain restrictor valve adjustment.

## FLUID AND SYSTEM FILTER

Recommended fluid for the system is SAE 10W Supplement 1 motor oil or Type "A" Automatic Transmission

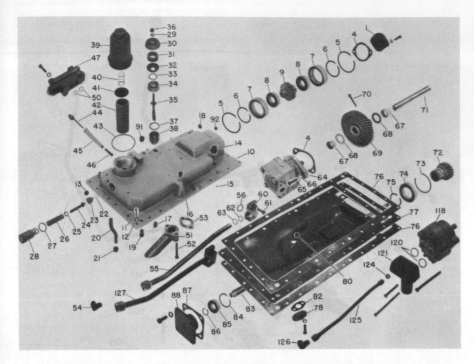

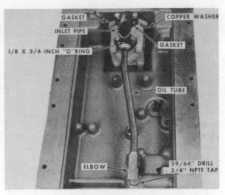

Fig. O251A—Bottom side of external Depth-Stop hydraulic housing after vane type pump conversion package has been installed.

Fig. O251—Exploded view of late type external valve Depth-Stop hydraulic system housing and related parts. Refer to Figs. O252, O253 and O253A for exploded views of the control assembly and external control valve units. Although above housing and parts are the late type, the earlier series 1850 and 1950 housings are basically similar and the above view will suffice for informational purposes. Vane type hydraulic system pump (64) has replaced gear type pump (118) in production and a kit is available to convert early model tractors, refer to text.

| | | |
|---|---|---|
| 1. Cover | 25. Shim | 45. Spring | 75. Snap ring |
| 4. Gasket | 26. Spring | 46. Plunger | 76. Gasket |
| 5. Snap ring | 27. "O" ring | 47. Flow divider | 77. Reservoir |
| 6. Snap ring | 28. Retainer | 50. "O" ring | 78. Cover |
| 7. Carrier | 29. Seal | 51. Pipe | 82. Gasket |
| 8. Bearing | 30. Cap | 53. Gasket | 83. Drive shaft |
| 9. Pinion | 31. Element | 54. Elbow | 84. Snap ring |
| 10. Housing | 32. Cup | 55. Tube | 85. Bearing |
| 13. Plug | 33. Gasket | 56. "O" ring | 86. Snap ring |
| 14. Plug | 34. Cap | 60. Flange | 87. Gasket |
| 15. Steel ball(¼") | 35. Dipstick | 62. Washer | 88. Housing |
| 16. Plug | 36. Nut | 63. "O" ring | 91. Plug |
| 17. Guide | 37. "O" ring | 64. Pump (vane type) | 92. Plug |
| 18. Insert | 38. Neck | 67. Bearing | 118. Pump (gear type) |
| 19. Dowel | 39. Body | 68. Washer | 120. "O" ring |
| 20. Nipple | 40. Spring | 69. Idler gear | 121. Manifold |
| 21. Restrictor | 41. Retainer | 70. Cotter pin | 124. Insert |
| 22. Washer | 42. Element | 71. Shaft | 125. Tube |
| 23. Seat | 43. "O" ring | 72. Drive pinion | 126. Elbow |
| 24. Plunger | 44. Pin (stop) | 73. Snap ring | 127. Tube |
| | | 74. Bearing | |

Fluid. Reservoir capacity is 4 gallons. System fluid should be changed yearly or after each 1000 hours of operation and the filter should be renewed whenever the fluid is changed.

433. To drain the hydraulic system, connect an open end hose to the upper left quick disconnect coupling ("Drain Coupling"—Fig. O250), start engine and move inner (left) control lever forward to pump fluid from the unit. CAUTION: Stop engine as soon as flow has diminished. Remove hose from quick disconnect port. Disconnect hoses from cylinders and work piston rods in and out to expel all fluid. Renew the filter element, connect hoses and refill unit with new, clean recommended fluid. Start engine, extend and contract cylinders

several times to bleed all air trapped in system and recheck fluid level. Add fluid as necessary to bring fluid to full level on dipstick.

434. To renew the filter, unscrew the cast iron filter body (See Fig. O250) and remove the old element. Clean the filter body, spring and retainer with clean solvent. Install new "O" ring on filter body, lubricate "O" ring and "O" ring seat in hydraulic housing with grease and assemble as follows: Hold element in upright position, place retainer on upper end of element and place spring on top of retainer. Position filter body over spring, retainer and element and carefully install the unit on hydraulic housing. Tighten the filter body to a minimum of 150 ft.-lbs. torque.

## OVERHAUL

435. FLOW DIVIDER VALVE. Except for tubing seats, flow divider valve used on "depth stop cylinder hydraulic system" is same as that used on models with Hydra-Lectric and/or three-point hitch hydraulic systems. Refer to paragraph 49 for information on servicing the flow divider valve.

436. HYDRAULIC PUMP. After removing the hydraulic housing as outlined in paragraph 414, the pump can then be removed and overhauled as outlined in paragraph 405 or 405A.

436A. HYDRAULIC PUMP (VANE TYPE). A vane type hydraulic pump, shown installed in Fig. O251A, is being used when necessary to replace the gear type pump used with Depth Stop hydraulic systems. The parts required to convert from gear to vane pump are available in a package consisting of oil tube, inlet pipe, pump and filter element along with the necessary "O" rings, cap screws and gaskets as well as two copper washers for the pump mounting cap screws.

To convert from gear pump to vane pump, proceed as follows: Remove housing from rear main frame as outlined in paragraph 441, remove oil pan, then turn hydraulic unit over to gain access to pump. Remove manifold to housing oil tube and tube elbow. Remove pump and manifold from housing.

Before installing vane pump assembly, the oil tube elbow hole in housing must be enlarged and tapped as follows: Remove by-pass and pressure relief valves and using heavy grease, plug both holes from inside of oil tube elbow hole. Use a 59/64-inch drill and drill 1-inch deep in the oil tube elbow hole. Clean cuttings from hole, then using a ¾-inch tap, alternately thread hole and install elbow (from package) until ½-

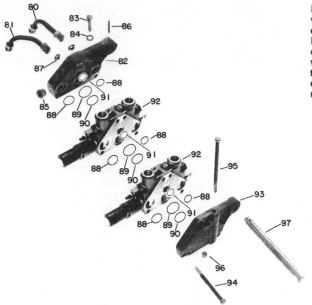

Fig. O252 — Exploded view of double acting control valve assembly. Refer to Fig. O253 for exploded view of control valve unit (92). Two control valve units are standard; one or more units may be added as an option.

80. Return line
81. Pressure line
82. Manifold
85. Plug
86. Roll pin
87. Tube inserts
88. "O" rings
89. "O" rings
90. Backup rings
91. "O" rings
92. Valve units
93. Plate
97. Lever shaft

inch of the elbow threads are in housing when elbow is tightened. CAUTION: Be careful during tapping operation. If hole is tapped too deep a tight fit of elbow cannot be obtained. Remove elbow and clean elbow, elbow hole and valve passages of grease and cuttings, then coat threads of elbow with sealant and install elbow. Elbow must be in position shown in Fig. O251A when tight.

To install pump assembly, coat pump gasket with gasket cement (Permatex No. 3 or equivalent) and install gasket. Coat pump shaft splines with grease. Place copper washers on pump mounting cap screws, install pump in housing and tighten retaining cap screws to 26-28 ft.-lbs. torque. Place "O" ring in counterbore of oil tube flange and install oil tube to elbow and pump. At this point, check to see that pump will rotate freely by hand. If pump binds, loosen pump retaining cap screws, reposition pump as necessary, and re-torque pump retaining cap screws. Install inlet pipe and gasket. Install oil pan and turn unit over. Install new filter element and if filter includes retainer (41—Fig. O251), it should be discarded.

Install hydraulic unit on rear main frame and refill with proper lubricant.

For overhaul information on vane type pump, refer to paragraph 405A.

**437. PUMP DRIVE SHAFT AND PINION.** After removing the hydraulic housing as outlined in paragraph 441, service procedure for the pump drive shaft and the pump drive pinion is the same as outlined in paragraphs 409 and 410.

**438. CONTROL VALVE UNITS AND ASSEMBLY.** Two, three or four control valve units are mounted on top of hydraulic housing in an assembly as shown in Fig. O252. To remove the valve assembly, first remove the operator's seat and the shroud from the seat support bracket. Disconnect the remote cylinder tube and hose assemblies from the valve ports, then unbolt and remove the seat support bracket, hose and quick disconnect coupling units from the tractor as an assembly. Drive the roll pin (86—Fig. O252) from valve assembly manifold (82) and lever pilot shaft (97), then remove the shaft and levers. Disconnect the pressure line (81) and return line (80); then, remove the cap screws (83 and 95) securing valve assembly to hydraulic housing and remove the assembly from tractor.

NOTE: Late model tractors have the return port located in bottom of manifold (82) and return line (80) is not used. In addition, back-up washers are now used behind all "O" rings (88 and 89) and if leakage is encountered in early units, these back-up rings should be installed.

**439. DOUBLE ACTING.** To disassemble the control valve assembly and individual units, use Figs. O252 and O253 as guides and proceed as follows: Remove the capscrews (94—Fig. O252) retaining the end plate (93) and valve units (92) to manifold (82) and carefully separate the parts. "O" rings (88, 89 and 91) are used

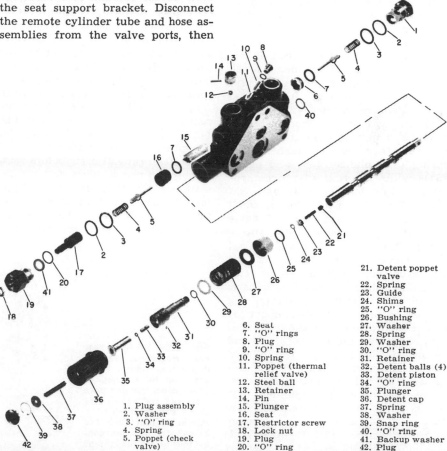

6. Seat
7. "O" rings
8. Plug
9. "O" ring
10. Spring
11. Poppet (thermal relief valve)
12. Steel ball
13. Retainer
14. Pin
15. Plunger
16. Seat
17. Restrictor screw
18. Lock nut
19. Plug
20. "O" ring

1. Plug assembly
2. Washer
3. "O" ring
4. Spring
5. Poppet (check valve)

21. Detent poppet valve
22. Spring
23. Guide
24. Shims
25. "O" ring
26. Bushing
27. Washer
28. Spring
29. Washer
30. "O" ring
31. Retainer
32. Detent balls (4)
33. Detent piston
34. "O" ring
35. Plunger
36. Detent cap
37. Spring
38. Washer
39. Snap ring
40. "O" ring
41. Backup washer
42. Plug

Fig. O253 — Exploded view of one double acting control valve unit; tractor may be equipped with two, three or four units. Refer to Fig. O252 for view showing assembly of two control units. The restrictor valve for each control unit must be individually adjusted; adjusting screw is (17).

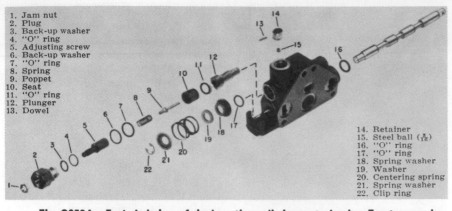

1. Jam nut
2. Plug
3. Back-up washer
4. "O" ring
5. Adjusting screw
6. Back-up washer
7. "O" ring
8. Spring
9. Poppet
10. Seat
11. "O" ring
12. Plunger
13. Dowel

14. Retainer
15. Steel ball (5/16)
16. "O" ring
17. "O" ring
18. Spring washer
19. Washer
20. Centering spring
21. Spring washer
22. Clip ring

**Fig. O253A—Exploded view of single acting cylinder control valve. Tractor may be equipped with up to four valves. Spool and body are not available separately.**

to seal passages between valve units, end plate and manifold. A backup ring (90) is used with "O" ring (89) on early models and with "O" rings (88 and 89) on late models.

Thoroughly clean outside of control valve unit in suitable solvent, refer to Fig. O253 and proceed as follows: Unscrew detent cap (36) and remove complete detent and valve spool assembly from valve body. Clamp flat (lever) end of valve in vise and remove detent cap (36) from assembly by driving against hex nut portion with brass drift; this unseats detent balls from groove inside cap. Catch the four detent balls as cap is removed. Place small spanner wrench in detent ball holes of retainer (31) and unscrew retainer from valve spool. Remove washer (27), spring (28) and washer (29) from retainer. Remove snap ring (39) from detent retainer, then remove washer (38), spring (37), plunger (35) and piston (33), with "O" ring (34), from retainer. Slide bushing (26) and "O" ring (25) from spool. Run a small (tag) wire into open end of valve spool through the shims (24), spring guide (23), spring (22) and poppet (21); then remove spool from vise and tip lever end up to let shims, etc., slide down the wire. Keep these parts on wire for cleaning. Unscrew plug (1) and remove spring (4), poppet (5) and seat (6). Remove lockout plunger (15). Unscrew plug assembly (19) and remove spring (4), poppet (5) and seat (16). Remove steel ball (12) and retainer (13); it is not necessary to remove pin (14) from retainer (13) unless pin is to be renewed. Unscrew lock nut (18) and remove restrictor adjusting screw (17). Remove both thermal relief plugs (8) and shims (not shown), springs (10) and poppets (11). NOTE: Identify each thermal relief plug, shim, spring and poppet unit so that they may be reinstalled

in same location from which they were removed, using the same number of shims between plug and spring as were removed. Remove and discard all "O" rings and back-up washers.

Thoroughly clean all parts in solvent and dry thoroughly. Small nicks and burrs should be removed with fine emery cloth. The control valve spool and valve body are not serviceable items: if either of these parts are damaged, a new valve unit assembly must be installed. Inspect point of poppet (21) and mating seat in valve spool for excessive wear. Check to be sure small orifice hole in valve spool is open and clean, and be sure small drilled holes in thermal relief bores are clean. Inspect detent grooves inside cap (36) for excessive wear. Inspect large hole in side of lockout seat (16) for excessive wear from ball (12). Inspect all springs for excessive weakness, wear or breakage.

440. Lubricate all parts before reassembly and reassemble as follows: Install new "O" ring (34) on detent piston (33) and install piston in the detent retainer (31). Place detent release plunger (35) in detent retainer, place

spring (37) inside plunger and washer (38) on spring, then install snap ring (39) to retain washer, spring and plunger in detent retainer. Install new "O" ring (30) on outside of detent retainer, then place washer (29), centering spring (28) and washer (27) over retainer. Place shims (24), spring guide (23), spring (22) and poppet (21) on tag wire and slide parts into position inside valve spool. Be sure parts are seated correctly, then clamp valve spool flat end in vise with open end up. Install new "O" ring (25) and bushing (26) over spool, then thread detent sub-assembly into spool and, using small spanner wrench, tighten to a torque of 5-8 ft.-lbs. Stick detent balls in holes of detent retainer with heavy grease. Carefully lower detent cap (36) over detent retainer until cap comes in contact with the steel balls, then use plastic hammer to drive cap down over the balls until the balls engage the second detent groove in cap. Install new "O" ring (40) in spool bore of valve body, then install the detent and spool assembly tightening the detent cap snugly. NOTE: Be sure arrow on flat (lever) end of valve spool points toward ports of the valve unit. Install new back-up washer (41) in bore of plug (19) and place new "O" ring (20) in plug at inner side of back-up washer. Thread restrictor adjusting screw (17) into plug (19), then install lock nut (18) on adjusting screw. Install back-up washer (2) and "O" ring (3) on outside of plugs (19 and 1). Install new "O" rings (7) on seats (16 and 6). Install seat (16) in lockout bore being sure large hole in seat is aligned with port to allow seating of ball (12). Install poppet (5), spring (4) and plug assembly (19). Install lockout plunger (15) in bore, then install remaining seat (6), poppet (5), spring (4) and

**Fig. O254 — Note that hose from base end of depth stop control cylinder must be coupled to lower connector and hose from rod end coupled to upper connector. That is, the hose from base end of cylinder must be connected to restrictor port of control valve unit.**

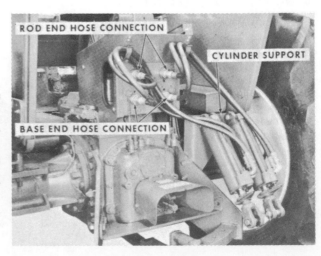

ROD END HOSE CONNECTION

CYLINDER SUPPORT

BASE END HOSE CONNECTION

plug (1). Install steel ball (12) and retainer (13) in port towards the restrictor adjusting screw (17). Install the thermal relief poppets (11), springs (10), shims and plugs (8), with new "O" rings (9), in bores from which they were removed. Install button plug (42) in detent cap.

**440A. SINGLE ACTING.** To disassemble the control valve assembly and individual units, use Figs. O252 and O253A as guides and proceed as follows: Remove cap screws (94—Fig. O252) retaining end plate (93) and valve units (92) to manifold (82) and carefully separate parts. "O" rings (88, 89 and 91) are used to seal passages between valve units, end plate and manifold. A back-up ring (90) is used with "O" ring (89) on early models and with all three "O" rings (88, 89 and 91) on late models. If leakage was present prior to disassembly, install back-up rings during reassembly.

Thoroughly clean outside of control valve unit in a suitable solvent, refer to Fig. O253A and proceed as follows: Disengage clip ring (22) from end of valve spool and pull spool from front of valve body. Remove shallow washer (21), spring (20), spacer washer (19) and deep washer (18), then remove "O" rings (16 and 17). Remove lockout plug (2) with adjusting screw (5), back-up washer (6) and "O" ring (7). Remove spring (8), poppet (9), poppet seat (10) with "O" ring (11) and lockout plunger (12). Remove jam nut (1), then remove adjusting screw (5) from lockout plug (2) and remove "O" ring (4) and back-up washer (3) from bore of lockout plug. Lift out check ball retainer (14) and check ball (15). Remove dowel (13) from retainer.

Clean and inspect all parts for scoring or excessive wear. Small nicks or burrs can be removed with fine emery cloth. Spool should be a snug fit in its bore with no noticeable side play. Spool and body are not available separately. Check large hole in poppet seat for excessive wear from check ball and see that small hole is clean and clear.

Reassemble control valve as follows: Place back-up ring (3) in bore of lockout plug (2) and install "O" ring (4) on top of back-up ring. Thread adjusting screw (5) into lockout plug and install jam nut (1). Install lockout plunger (12) small end first in upper bore of valve body, then install "O" ring (11) on poppet seat (10) and install seat so large hole of seat is aligned with port in top

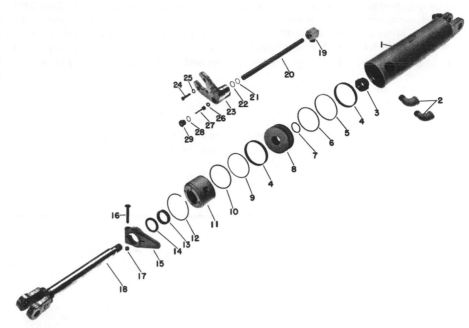

**Fig. O255 — Exploded view of depth stop control cylinder. Stop arm (15) can be clamped in different positions on piston rod (18) to adjust length of cylinder stroke. When stop arm contacts valve (27), pressure build up in control valve opens poppet valve (21—Fig. O253) in valve spool releasing detent and allowing valve to return to neutral position.**

| | | | |
|---|---|---|---|
| 1. Cylinder tube | 8. Piston | 14. Seal | 22. "O" ring |
| 2. Elbows | 9. "O" ring | 15. Stop arm | 23. Stop valve |
| 3. Nut | 10. Washer | 18. Piston rod | housing |
| 4. Wear rings | 11. Bearing | 19. Elbow | 26. Seal |
| 5. Ring | 12. Snap ring | 20. Tube | 27. Valve |
| 6. "O" ring | 13. Seal | 21. Washer | 28. "O" ring |
| 7. "O" ring | | | 29. Guide |

side of body and will permit check ball to seat. Install poppet (9), long end first, into seat and install spring (8). Place back-up washer (6) on threaded end of lockout plug, install "O" ring (7), then screw lockout plug assembly in valve body. Drop check ball (15) into cylinder port and into large hole of poppet valve seat. Insert dowel (13) in retainer (14) and install retainer in cylinder port on top of check ball.

Install "O" ring (16)' at front of spool bore and be sure it is seated with no twists. Install spool from rear side of valve body and push spool through front "O" ring only far enough to expose rear "O" ring groove. Install rear "O" ring (17) and be sure it is seated with no twists. Push spool rearward through rear "O" ring, install deep washer (18), spacer washer (19), spring (20) and shallow washer (21) and secure with clip ring (22).

NOTE: Single acting control valves have restrictor valve adjustment only and adjustment is made the same as it is for double acting control valves. See paragraph 432.

**441. R&R HYDRAULIC HOUSING.** First, drain hydraulic system fluid as outlined in paragraph 433; then, proceed as follows:

If removing the housing for access to the hydraulic pump, bull gears, etc., it will not be necessary to remove the seat support, control valve assembly or remote hose and connectors from the hydraulic housing. Remove the operators's seat assembly, shroud from around seat support and the platform panels. Thoroughly clean the unit. Disconnect pressure line to power steering and/or oil cooler from flow divider valve and disconnect return line from hydraulic housing. Hook a chain sling to seat support, unbolt hydraulic housing from rear main frame and lift unit from tractor. While unit is still supported (if necessary to gain access to hydraulic pump), drive the dowels from hydraulic housing and remove the oil pan.

The seat support and control valve assembly can be removed from the hydraulic housing as outlined in paragraph 438 either before or after removing the housing from tractor. If removing hydraulic housing with seat bracket and hydraulic control valve assembly already removed, attach lift chain with the forward seat support bracket retaining capscrews.

Reverse removal procedure to re-install unit on tractor and refill the system as in paragraph 433.

442. **OVERHAUL CYLINDER.** Refer to exploded view of cylinder in Fig. O255 and proceed as follows:

Loosen nut (17) and slide stop arm (15) to yoke end of piston rod (18). Remove cap screws (24) retaining stop valve housing (23) to bearing (11) and withdraw valve housing from end of tube (20). Unscrew nut (29) and remove stop valve (27).

Push bearing (11) into cylinder tube (1) far enough to allow removal of snap ring (12), then bump bearing from cylinder tube with piston (8) by

working piston rod in and out.

Remove nut (3), piston (8) and bearing (11) from piston rod. Need and procedure for further disassembly is evident on inspection of unit.

Install new piston rod seals (13 and 14) in bearing (11) and reassemble unit using all new "O" rings, wear rings and backup rings. Reverse disassembly procedure to reassemble the cylinder.

# NOTES

# OLIVER
## (Includes some Cockshutt
## and Minneapolis-Moline Models)

**Oliver & Cockshutt Models** ■ 1755 ■ 1855 ■ 1955
**Minneapolis-Moline Models** ■ G-850 ■ G940

Previously contained in I&T Shop Manual No. O-24

**SHOP MANUALS**

# SHOP MANUAL
# OLIVER & COCKSHUTT

## SERIES 1755, 1855, 1955

## Also Covers MINNEAPOLIS-MOLINE
## SERIES G-850, G-940

NOTE: Servicing the Minneapolis-Moline Series G-850 can be accomplished by using the information contained in this manual for Series 1755. To service Minneapolis-Marine Series G-940, use information for Series 1855.

## SERIAL NUMBER LOCATION

Tractor serial number plate is located on the rear side of the instrument panel support. Engine serial number is stamped on right flange of engine directly below the alternator.

## BUILT IN THESE VERSIONS

Rowcrop Tricycle, Rowcrop Adjustable Front Axle, Rowcrop Utility, Wheatland and Four-Wheel Drive. Wheatland models are available with non-adjustable front axles only.

## INDEX (By Starting Paragraph)

| | Series 1755, G-850 | Series 1855, G-940 | Series 1955 |
|---|---|---|---|
| **FRONT SUSPENSION CONT.** | | | |
| Front bolster . . . . . . . . . . . . | 4 | 4 | 4 |
| Pivot pin . . . . . . . . . . . . . | 1 | 1 | 1 |
| Spindle (knuckle) . . . . . . . . | 6 | 6 | 6 |
| Tie rods & toe-in . . . . . . . . | 5 | 5 | 5 |
| **GOVERNOR** | | | |
| Adjustment, non-diesel . . . . | 119 | 119 | . . . . . |
| Adjustment, diesel . . . . . . . . | 114 | 114 | 114 |
| Overhaul, non-diesel . . . . . . | 123 | 123 | . . . . . |
| R&R, non-diesel . . . . . . . . . | 123 | 123 | . . . . . |
| **HYDRA-POWER** . . . . . . . . . . | 145 | 145 | 145 |
| **HYDRAUL SHIFT** . . . . . . . . . | 162 | 162 | 162 |
| **LIFT SYSTEM, HYDRAULIC** | | | |
| Draft control spring, adjust . . | 245 | 245 | 245 |
| Draft input shaft . . . . . . . . | 266 | 266 | 266 |
| Drain and refill . . . . . . . . . | 240 | 240 | 240 |
| External lift cylinders . . . . . . | 271 | 271 | 271 |
| Hitch control lever . . . . . . . | 265 | 265 | 265 |
| Hydraulic housing . . . . . . . . | 262 | 262 | 262 |
| Lubrication and maintenance . . | 239 | 239 | 239 |
| Cylinder and piston . . . . . . . | 269 | 269 | 269 |
| Manifold . . . . . . . . . . . . | 267 | 267 | 267 |
| Oil filter . . . . . . . . . . . . . | 241 | 241 | 241 |
| Pump . . . . . . . . . . . . . . . | 256 | 256 | 256 |

| | Series 1755, G-850 | Series 1855, G-940 | Series 1955 |
|---|---|---|---|
| **LIFT SYSTEM CONT.** | | | |
| Pressure relief valve . . . . . . . | 249 | 249 | 249 |
| Quick check . . . . . . . . . . . | 237 | 237 | 237 |
| Remote cylinders . . . . . . . . | 288 | 288 | 288 |
| Remote valves . . . . . . . . . . | 280 | 280 | 280 |
| Rockshaft . . . . . . . . . . . . | 263 | 263 | 263 |
| Servo valve . . . . . . . . . . . | 267 | 267 | 267 |
| System adjustments . . . . . . . | 243 | 243 | 243 |
| Trouble shooting . . . . . . . . | 225 | 225 | 225 |
| **POWER STEERING** . . . . . . . . | 40 | 40 | 40 |
| **POWER TAKE-OFF** | | | |
| Control valve . . . . . . . . . . | 219 | 219 | 219 |
| Clutch overhaul . . . . . . . . . | 217 | 217 | 217 |
| Drive shaft, R&R . . . . . . . . | 215 | 215 | 215 |
| PTO unit, R&R . . . . . . . . . | 216 | 216 | 216 |
| **TRANSFER DRIVE** . . . . . . . . | 32 | 32 | 32 |
| **TRANSMISSION** | | | |
| Bevel pinion shaft . . . . . . . | 191 | 191 | 191 |
| Countershaft . . . . . . . . . . | 190 | 190 | 190 |
| Input shaft . . . . . . . . . . . | 188 | 188 | 188 |
| Lubrication system . . . . . . . | 176 | 176 | 176 |
| Reverse idler . . . . . . . . . . | 194 | 194 | 194 |
| Shifter rails & forks . . . . . . | 184 | 184 | 184 |
| Top cover . . . . . . . . . . . . | 183 | 183 | 183 |
| **TURBOCHARGER** . . . . . . . . . | . . . . . | 117 | 117 |

# CONDENSED SERVICE DATA

| | Series 1755, G-850 | Series 1855, G-940 | Series 1955 |
|---|---|---|---|
| **GENERAL** | | | |
| Engine Make . . . . . . . . . . . . | Own | Own | Own |
| Cylinders, No. of . . . . . . . . . | 6 | 6 | 6 |
| Cylinder Bore—Inches . . . . . . | 3 7/8 | 3 7/8 | 3 7/8 |
| Stroke—Inches | | | |
| Non-Diesel . . . . . . . . . . . | 4 | 4 | . . . . . |
| Diesel . . . . . . . . . . . . . . | 4 3/8 | 4 3/8 | 4 3/8 |
| Displacement—Cubic Inches | | | |
| Non-Diesel . . . . . . . . . . . | 283 | 283 | . . . . . |
| Diesel . . . . . . . . . . . . . . | 310 | 310 | 310 |
| Compression Ratio: | | | |
| Gasoline . . . . . . . . . . . . | 8.5:1 | 8.5:1 | . . . . . |
| Diesel . . . . . . . . . . . . . . | 16.0:1 | 16.5:1 | 16.0:1 |
| Main Bearings No. of . . . . . . . | 7 | 7 | 7 |
| Cylinder Sleeves Type . . . . . . | Wet | Wet | Wet |
| Forward Speeds, No. of . . . . . | 6 | 6 | 6 |
| Battery Terminal Grounded . . . | — Negative — | | |
| **TUNE-UP** | | | |
| Firing Order . . . . . . . . . . . | — 1-5-3-6-2-4 — | | |
| Valve Tappet Gap Inlet: | | | |
| Non-Diesel . . . . . . . . . . . | 0.020 | 0.020 | . . . . . |
| Diesel . . . . . . . . . . . . . . | 0.030 | 0.030 | 0.030 |
| Valve Tappet Gap, Exhaust: | | | |
| Non-Diesel . . . . . . . . . . . | 0.030 | 0.030 | . . . . . |
| Diesel . . . . . . . . . . . . . . | 0.030 | 0.030 | 0.030 |
| Valve Face Angle, Degrees . . . | — See Par. 58 — | | |
| Valve Seat Angle, Degrees . . . | — See Par. 58 — | | |
| Ignition Distributor Make . . . . . | — Holley — | | . . . . . |
| Ignition Distributor Model . . . . . | — D-2563AA — | | . . . . . |
| Generator and Regulator Make . . . . . . . . . . . . . . . | — Delco-Remy — | | |
| Generator Model . . . . . . . . . . | — 1100808 — | | |
| Regulator Model . . . . . . . . . . | — 1119511 — | | |
| Starting Motor Make . . . . . . . | — Delco-Remy — | | |
| Starting Motor Model: | | | |
| Non-Diesel . . . . . . . . . . . | 1108431 | 1108431 | . . . . . |
| Diesel . . . . . . . . . . . . . . | 1113139 | 1113656 | 1113656 |
| Ignition Distributor Contact Gap . . . . . . . . . . . . . . . . | 0.025 | 0.025 | . . . . . |
| Ignition Distributor Timing (Static): | | | |
| Gasoline . . . . . . . . . . . . | 0.025 | 0.025 | . . . . . |
| Injection Pump Make . . . . . . . | — Roosa-Master — | | |
| Injection Pump Timing (Static) . . . . . . . . . . . . . | 4° | 4° | 2° |
| Carburetor . . . . . . . . . . . . | – Marvel Schebler – | | . . . . . |

| | Series 1755, G-850 | Series 1855, G-940 | Series 1955 |
|---|---|---|---|
| **TUNE-UP (Cont.)** | | | |
| Carburetor Model . . . . . . . . . | USX-44 | USX-37 | . . . . . |
| Engine Low Idle RPM . . . . . . | 800 | 800 | 800 |
| Engine High Idle RPM . . . . . . | 2640 | 2640 | 2640 |
| Engine Rated RPM . . . . . . . . | 2400 | 2400 | 2400 |
| PTO RPM @ Engine Rated RPM: | | | |
| 540 rpm pto . . . . . . . . . . . | 549 | 549 | 549 |
| 1000 rpm pto . . . . . . . . . . | 984 | 984 | 984 |
| **SIZES—CLEARANCE—CAPACITIES** | | | |
| Crankshaft Journal Dia. . . . . . | — 2.624-2.625 — | | |
| Crankshaft Diameter . . . . . . . | — 2.4365-2.4375 — | | |
| Rod Length . . . . . . . . . . . . | — 6.749-6.750 — | | |
| Camshaft Journal Diameter: | | | |
| Front . . . . . . . . . . . . . . | — 1.749-1.750 — | | |
| Others . . . . . . . . . . . . . | — 1.7485-1.7495 — | | |
| Piston Pin Diameter . . . . . . . . | — 1.2494-1.2497 — | | |
| Valve Stem Diameter: | | | |
| Inlet . . . . . . . . . . . . . . | — 0.372-0.373 — | | |
| Exhaust . . . . . . . . . . . . | — 0.371-0.372 — | | |
| Piston Ring Side Clearance . . . . . . . . . . . . . | — See Par. 71 — | | |
| Main Bearing Clearance . . . . . | — 0.0015-0.0045 — | | |
| Rod Bearings Clearance . . . . . | — 0.0005-0.0015 — | | |
| Piston Skirt Clearance . . . . . . | — See Par. 70 — | | |
| Crankshaft End Play . . . . . . . | — 0.0045-0.0095 — | | |
| Camshaft Bearing Clearance . . . . . . . . . . . . . | — See Par. 66 — | | |
| Cooling System—Gallons . . . . | 5 | 5 | 5 |
| Engine Crankcase w/filter—Qts . . . . . . . . . . | 9 | 9 | 9 |
| Fuel Tank, Gasoline—Gals. . . . | 35 | 35 | 35 |
| Fuel Tank Diesel—Gallons. . . . | 35 | 35 | 35 |
| Transmission and Final Drive—Qts . . . . . . . . . . . | 43 | 43 | 43 |
| Transfer Case—Quarts . . . . . . | 1 | 1 | 1 |
| Hydra-Power Drive—Quarts . . | 6 | 6 | 6 |
| Hydraulic Lift—Quarts . . . . . . | 27 | 27 | 27 |
| Diff. Housing (4WD)—Quarts . . . . . . . . . . . . . . | 8 | 8 | 8 |
| Planetary Drive (4WD)—Qts . . | 2.5 | 2.5 | 2.5 |
| **TIGHTENING TORQUES—Ft.-Lbs.** | | | |
| Cylinder Head: | | | |
| Non-Diesel . . . . . . . . . . . | — 129-133 — | | . . . . |
| Diesel . . . . . . . . . . . . . . | — See Par. 57 — | | . . . . |
| Main Bearings . . . . . . . . . . . | — 129-133 — | | |
| Connecting Rods . . . . . . . . . | — 44-46 — | | |
| Flywheel . . . . . . . . . . . . . | — 67-69 — | | |

# FRONT SYSTEM (AXLE TYPE)

### AXLE MAIN MEMBER AND PIVOT PIN

#### Rowcrop Models

1. Rowcrop tractors may be equipped with an adjustable front axle as shown in Fig. 1. The center main member (16) is carried on a pivot pin which is retained in front support (carrier) by a pin.

To remove the front axle, stay rod (26) and wheels as an assembly, support front of tractor, then unbolt stay rod support (32) from front main frame. Disconnect inner ends of tie rods from center steering arm. Remove pivot pin from front support (Fig. 3) and axle, raise front of tractor and roll front axle and wheels assembly from under tractor.

Remove nut from rear of stay rod and remove stay rod support (32—Fig. 1). Bushings (31) and (17) should be reamed, if necessary, to provide 0.001-0.002 diametral clearance on pins.

#### Short Wheel Base Models

2. Short wheel base model tractors are equipped with the axle shown in Fig. 2. Axle rear pivot pin (33) is separate and is retained in rear support (32) by a set screw. The axle front pivot pin is integral with axle center member and is carried in a bushing (11—Fig. 4) fitted in axle carrier (3). Bushings for pivot pins are presized and should not require reaming if carefully installed.

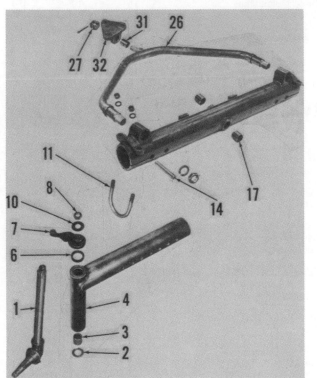

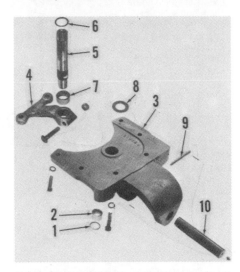

Fig. 1–Exploded view of Rowcrop adjustable front axle assembly used on Series 1755, 1855 and 1955.

| | |
|---|---|
| 1. | Steering knuckle |
| 2. | Thrust washer |
| 3. | Bushing |
| 4. | Axle extension |
| 6. | Felt washer |
| 7. | Steering arm |
| 8. | Nut |
| 10. | Washer |
| 11. | "U" bolt |
| 14. | Lock pin |
| 16. | Center member |
| 17. | Pivot bushing |
| 26. | Stay rod |
| 27. | Nut |
| 31. | Bushing |
| 32. | Support |

Fig. 3–Carrier (bolster) used when tractor is equipped with adjustable axle shown in Fig. 1.

| | | | |
|---|---|---|---|
| 1. | Seal | 6. | Seal |
| 2. | Bearing | 7. | Bearing |
| 3. | Carrier | 8. | Washer |
| 4. | Steering arm | 9. | Lock pin |
| 5. | Shaft | 10. | Pivot shaft |

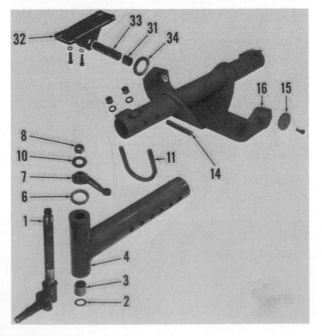

Fig. 2–Exploded view of Short Wheel Base (Utility) front axle assembly.

| | |
|---|---|
| 1. | Steering knuckle |
| 2. | Thrust washer |
| 3. | Bushing |
| 4. | Axle extension |
| 6. | Felt washer |
| 7. | Steering arm |
| 8. | Nut |
| 10. | Washer |
| 11. | "U" bolt |
| 14. | Lock pin |
| 15. | Thrust washer |
| 16. | Center member |
| 31. | Bushing |
| 32. | Support |
| 33. | Pivot pin |
| 34. | Washer (shim) |

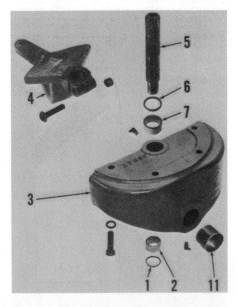

Fig. 4–Carrier used on Short Wheel Base model tractors.

| | | | |
|---|---|---|---|
| 1. | Seal | 5. | Shaft |
| 2. | Bearing | 6. | Seal |
| 3. | Carrier | 7. | Bearing |
| 4. | Steering arm | 11. | Bushing |

To remove front axle and wheels as an assembly, support front end of tractor and disconnect tie rods from center steering arm (4). Remove rear support (32—Fig. 2) and thrust washer (15) and slide axle rearward out of front support (3—Fig. 4). Bushings (11 and 31—Fig. 2) can now be renewed.

When installing bushings, be sure grease holes in bushings are aligned with grease fittings.

Be sure to install shim washer (34) during reassembly.

### Wheatland Models

3. Wheatland model tractors have a non-adjustable front axle which is fitted with rotating spindles as shown in Fig. 5.

To remove axle, stay rod (19) and wheels as an assembly, support front end of tractor, remove stay rod support (17) or stay rod nut (22), and disconnect tie rods from center steering arm (4—Fig. 6). Remove cap screw which retains pivot pin (10) in front support (3), drive pivot pin from front support and axle, then raise front of tractor and roll front axle and wheels assembly from under tractor.

Stay rod bushing and pivot pin bushing can be renewed and should be reamed, if necessary, to provide 0.001-0.002 diametral clearance.

### FRONT AXLE CARRIER
### Series 1755-1855-1955.

4. Front axle carriers (bolsters) are shown in Figs. 3, 4 and 6.

To remove front carrier, remove

front axle as outlined in paragraph 1, 2 or 3 and the power steering cylinder as outlined in paragraph 48, then unbolt carrier from tractor main frame.

Remove clamp bolt from steering arm, then remove steering shaft and thrust washer (when used) from carrier. Remove oil seals, then note location of bearings prior to driving them out. When installing new bearings, drive or press on lettered end of bearing cage only and be sure oil holes in bearing cages are aligned with lubrication fittings when bearings are installed. Seals are installed with lips toward ends of steering shaft (outward).

### TIE RODS and TOE-IN
### All Axle Models

5. Tie rod ends are the non-adjustable type and are serviced either by renewing the end assembly or the complete tie rod assembly.

Toe-in for all axle type models is 3/16-inch. Obtain correct toe-in by adjusting each tie rod an equal amount.

### STEERING KNUCKLES AND BUSHINGS
### Rowcrop and Short Wheel Base Models

6. To remove steering knuckles and renew bushings (3—Fig. 1 or Fig. 2), support front end of tractor and remove tire and wheel assembly. Straighten tab of washer (10) and remove nut (8). Remove steering arm (7) from knuckle and withdraw knuckle from axle extension (4).

Drive bushings (3) from axle extension and install new bushings flush with bore. Bushings are presized and reaming should not be necessary if bearings are installed with a suitable driver.

Thrust washer (2) can be installed either side up. Insert knuckle into axle extension, install new felt (6), then on Rowcrop models, install steering arm (7) extending to rear and at approximately 90 degrees to wheel spindle. On Short Wheel Base models, install steering arm (7) extending to front at approximately 90 degrees to wheel spindle.

### LIVE SPINDLES AND SPINDLE CARRIERS
### Wheatland Models

7. **SPINDLES.** Front wheels on Wheatland models are attached to live (rotating) spindles (1—Fig. 5) which are carried in tapered roller bearings mounted in the spindle carriers (5). Spindles can be removed by removing the cap (14) from inner side of spindle carrier and removing the cotter pin (13) and slotted nut (12) from inner end of spindle. Refer to paragraph 9 for bearing and seal service information.

8. **SPINDLE CARRIERS.** To renew the spindle carrier pivot pin (15 —Fig. 5) and/or pivot pin needle bearings (7), first remove wheel and disconnect tie rod outer end from carrier. Remove nut from bottom end of pivot pin and drive the pin up out of axle (20) and carrier. Drive bearings (7) from carrier and install new bearings by driving or pressing on lettered end of cage only. To renew carrier, remove

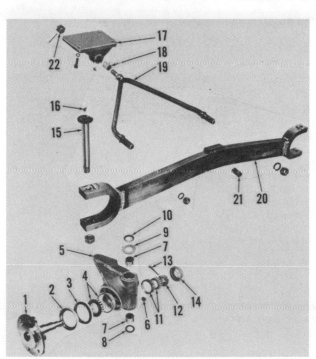

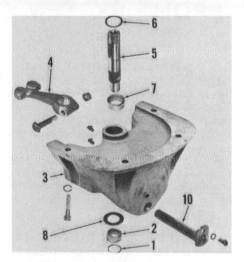

Fig. 5–Exploded view of front axle assembly used on Wheatland model tractors.

1. Spindle
2. Wear cup
3. Seal
4. Outer bearing
5. Spindle carrier
6. Plug
7. Bearing
8. Seal
9. Thrust washer
10. Thrust bearing
11. Inner bearing
12. Nut
13. Cotter pin
14. Cap
15. Pivot pin
16. Grease fitting
17. Support
18. Bushing
19. Stay rod
20. Main member
21. Bushing
22. Nut

Fig. 6–View of carrier used on Wheatland model tractors.

1. Seal　　　　　　　　6. Seal
2. Bearing　　　　　　7. Bearing
3. Carrier　　　　　　8. Washer
4. Steering arm　　　10. Pivot pin
5. Shaft

spindle as outlined in paragraph 7. When reassembling, renew seal (8) and, if damaged, renew thrust bearing (10) and washer (9). Refer to paragraph 9 for wheel bearing and seal information.

### WHEEL BEARINGS AND SEAL
### Series 1755-1855-1955

9. Refer to Fig. 7 for exploded view of Rowcrop and Short Wheel Base model front wheel bearing and seal installation and to Fig. 5 (items 1 through 14) for Wheatland models.

To install new seal (3—Fig. 5, or 2—Fig. 7), first apply thin coat of gasket sealer to outer metal rim of seal. On

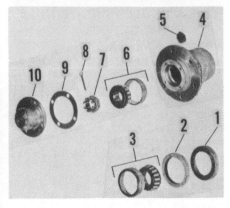

**Fig. 7–Exploded view of front wheel hub, bearing and seal assembly used on Rowcrop and Short Wheel Base tractors.**

| | |
|---|---|
| 1. Wear cup | 6. Outer bearing |
| 2. Seal | 7. Nut |
| 3. Inner bearing | 8. Cotter pin |
| 4. Hub | 9. Gasket |
| 5. Plug | 10. Hub cap |

Rowcrop or Short Wheel Base model, install seal in wheel hub (4—Fig. 7) with Oliver ST-97 or ST-145 Seal Driver and ST-125 Mandrel or equivalent tool. On Wheatland model, install seal in spindle carrier (5—Fig. 5) with Oliver ST-123 or ST-146 Seal Driver and ST-125 Mandrel or equivalent tool. Apply thin coat of gasket sealer to inside of seal wearing cup (1—Fig. 7 or 2—Fig. 5) and install wearing cup on spindle using Oliver ST-98 Driver on Rowcrop or Short Wheel Base model, or ST-124 Driver on Wheatland model. Face of cup must be smooth and flat after installation.

When reassembling wheel hub or spindle carrier the bearings should be packed with ½-pound of lithium base multi-purpose grease. Threads of plug (5—Fig. 7) in wheel hub or (6—Fig. 5) in spindle carrier should be coated with gasket sealer or white lead if plug has been removed.

To adjust wheel bearings, tighten nut until definite drag is felt when rotating wheel, then back nut off to remove preload. Continue to tighten and loosen nut until a point is located where torque required to tighten nut definitely increases, then tighten nut from this point to where cotter pin can be installed. Bearing adjustment is correct when torque of 20 inch-pounds is required to rotate wheel. Renew gasket (9—Fig. 7) when installing cap on Rowcrop or Short Wheel Base model hub.

Install grease retainer (4) with ridge upward and drive upper bearing cup in firmly against retainer. Drive lower bearing cup in firmly against shoulder in bottom of bore. Pack cone and roller assemblies of both bearings with No. 1 multi-purpose grease and pack ½-pound of same type grease in cavity above lower bearing cup in column. Install new felt (2) and retainer (3) on axle post, then drive cone and roller of lower bearing firmly against shoulder of axle post. Insert axle post in column and install upper bearing cone and roller assembly. Install seal (1) with lip up (away from bearing), lubricate seal contact surface of nut (5) and install nut for proper bearing adjustment as outlined in paragraph 10.

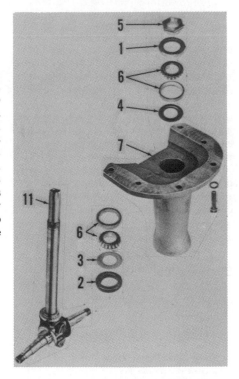

**Fig. 8–Exploded view of support (column), bearings and axle post used on tricycle model tractors.**

| | |
|---|---|
| 1. Seal | 5. Nut |
| 2. Felt washer | 6. Bearing assy. |
| 3. Felt retainer | 7. Support |
| 4. Grease retainer | 11. Axle post |

# FRONT SYSTEM (TRICYCLE TYPE)

10. **ADJUST AXLE POST BEARINGS.** To adjust the axle post bearings it will be necessary to remove the power steering cylinder as outlined in paragraph 48. If unit has been disassembled, be sure that bearing cups and cones are fully seated, then adjust nut (5—Fig. 8) so that axle post bearings have a preload of 70-80 inch-pounds, including the seal drag. Stake nut to slot in axle post when bearings are properly adjusted.

11. **R&R FRONT SUPPORT.** To remove the tricycle front support (pedestal), it will first be necessary to remove the power steering cylinder as outlined in paragraph 48. Support tractor and remove cap screws retaining support (column) to tractor main frame. Raise front of tractor and

remove unit.

12. **OVERHAUL FRONT ASSEMBLY.** Remove assembly from tractor as outlined in paragraph 11. Refer to Fig. 8 for exploded view of unit. Unstake and remove nut (5), then bump or push axle post (11) from upper bearing (6) and column (7). Remove cone and roller assembly of lower bearing (6), felt retainer (3) and felt (2) from axle post. Remove seal (1), bearing assembly (6) and grease retainer (4) from upper bore of column and drive cup of lower bearing (6) from bottom of bore.

Carefully inspect all parts and renew any that are excessively worn, scored or damaged. Reassemble unit using new seal (1), felt (2) and grease retainer (4) as follows:

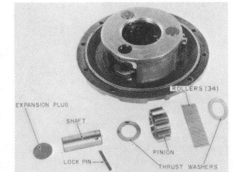

**Fig. 10–View showing one pinion assembly removed from planet spider.**

# FRONT SYSTEM
## (Front-Wheel Drive Axle)

Series 1755, 1855 and 1955 tractors are available with a front drive axle which is driven from the transmission bevel pinion shaft via a transfer case and a drive shaft fitted with two universal joints. A shifting mechanism in the transfer case allows connecting or disconnecting power to the front drive axle.

All four-wheel drive tractors are equipped with power steering. All models are equipped with a Saginaw Hydramotor steering unit. Refer to POWER STEERING section for information on the Saginaw Hydramotor and the two steering cylinders.

### FRONT AXLE AND CARRIER
### All Models So Equipped

13. **R&R AXLE ASSEMBLY.** The complete front axle assembly can be removed from tractor as follows: Remove drive shaft shield and shield front support. Disconnect drive shaft from companion flange of differential pinion shaft. Disconnect both power steering cylinders from axle and spindle supports and lay cylinders on top of axle carrier. Remove bolts retaining axle to axle carrier, then raise tractor and roll the complete axle and wheels unit forward and away from tractor.

Note: A rolling floor jack can be placed under differential pinion shaft to keep axle from rotating as tractor is lifted from axle.

If necessary, wheels and tie-rod can now be removed and procedure for doing so is obvious.

Reinstall axle by reversing the removal procedure and be sure piston rod ends of steering cylinders are attached to steering spindle supports. Tighten the cylinder attaching bolt lock nuts until they just contact the mounting flanges. Further tightening may distort mounting flanges and cause cylinder to bind.

14. **R&R AXLE CARRIER.** To remove axle carrier, first remove the front axle as outlined in paragraph 13, then, secure steering cylinders to tractor frame. Place a rolling floor jack under axle carrier and take weight of carrier. Remove pivot pin retaining cap screws, slide pivot pins from pivot supports and lower the axle carrier from tractor. If necessary, pivot supports can be removed from tractor frame.

Bushings in axle carrier can now be renewed. Bushings are pre-sized and should not require reaming if carefully installed.

15. **OVERHAUL FRONT AXLE.** Overhaul of the front drive axle assembly will be discussed as four operations; the planet spider assembly, the hub assembly, the spindle support assembly and the differential and carrier assembly. All operations except the differential and carrier overhaul can be accomplished without removing the front drive axle from tractor. Both outer ends of axle are identical, hence, only one outer end will be discussed.

16. **PLANET SPIDER.** To overhaul the planet spider assembly, support outer end of axle and remove the tire and rim. Remove relief valve from center of planet spider, remove plug from wheel hub and drain oil from planet spider. Remove capscrews that retain planet spider to wheel hub and the two puller hole cap screws. Use two of the removed retaining cap screws in the puller holes to remove planet spider assembly from wheel hub.

With unit removed, remove the three pinion shaft lock pins by driving them toward center of unit. Remove pinion shafts and expansion plugs by driving pinion shafts toward outside of planet spider. Remove planet pinions, rollers (34 in each pinion) and thrust washers. Discard the expansion plugs. Refer to Fig. 10.

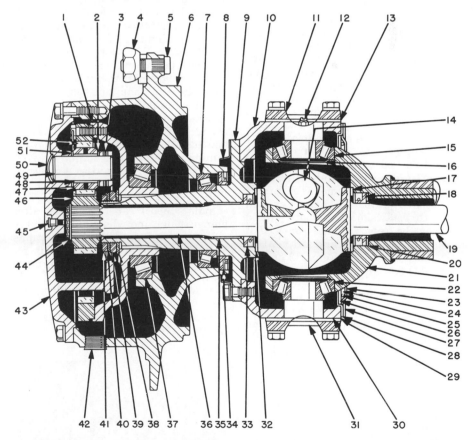

*Fig. 11–Cross-sectional view showing components of front drive axle outer end. Current production axles have a one-piece internal gear and hub instead of the two-piece assembly shown above.*

| | | |
|---|---|---|
| 1. Internal gear hub | 16. Grease retainer | 28. Seal retainer | 40. Lock washer |
| 2. Thrust washer | 17. Thrust washer | 29. Gasket | 41. Outer nut |
| 3. Pinion shaft lock pin | 18. Oil seal | 30. Shims | 42. Filler hole plug |
| 6. Wheel hub | 19. Axle shaft (inner) | 31. Lower trunnion | 43. Planet spider |
| 7. Hub inner bearing | 20. Washer | 32. Thrust washer | 44. Snap ring |
| 8. Dirt shield | 21. Axle housing | 33. Oil seal | 45. Relief valve |
| 9. Spindle | 22. Lower trunnion bearing | 34. Oil seal | 46. Sun gear |
| 10. Spindle support | 23. Dust seal retainer | 35. Bushing | 47. Pinion |
| 11. Upper trunnion | 24. Dust seal | 36. Axle shaft (outer) | 48. Pinion rollers |
| 12. Grease fitting | 25. Spring | 37. Hub outer bearing | 50. Expansion plug |
| 13. Shims | 26. Felt seal | 38. Thrust washer | 51. Thrust washer |
| 14. Universal joint | 27. Retainer ring | 39. Inner nut | 52. Internal gear |
| 15. Upper trunnion bearing | | | |

Clean and inspect all parts and renew as necessary. Pay particular attention to the pinion rollers and thrust washers.

17. When reassembling, use heavy grease to hold rollers in inner bore of pinions. Be sure tangs of thrust washers are in the slots provided for them and that holes in pinion shaft and mounting boss are aligned before pinion shafts are final positioned. Coat mating surfaces of planet spider and wheel hub with No. 2 Permatex or equivalent sealer and install planet spider in wheel hub. Tighten retaining cap screws to a torque of 52-57 ft.-lbs.

18. HUB ASSEMBLY. To overhaul the wheel hub assembly, first remove planet spider assembly as outlined in paragraph 16. With planet spider removed, remove snap ring and sun gear from outer end of axle shaft. Straighten tabs of spindle nut lockwasher, then use OTC tool JD-4 or equivalent and remove spindle outer nut and lockwasher. Now loosen but **do not** remove the spindle inner nut. Unbolt spindle from spindle support and remove wheel hub assembly.

Place hub assembly on bench with spindle nut on top side and block up assembly so spindle will be free to drop several inches. Remove the spindle inner nut, then place a wood block over end of spindle and bump spindle from internal gear hub. Lift internal gear hub and bearing from wheel hub and be careful not to allow bearing to drop from hub of internal gear hub. Complete removal of spindle from wheel hub. All bearings and seals, thrust washers and dirt shield can now be removed and renewed if necessary. Bushing and oil seal in inner bore of spindle (items 33 and 35 —Fig. 11) can also be renewed at this time.

19. Use Fig. 11 as a reference and reassemble wheel hub unit as follows: Install bearing cups (7 and 37) in hub with smallest diameters toward inside of hub. Place inner bearing in inner

bearing cup, then install oil seal (34) with lip facing bearing. Bump seal into bore until it bottoms. Place dirt shield (8) on hub so flat side is toward flange of spindle, then using caution not to damage seal, install spindle in wheel hub. Hold spindle in that position and turn unit over so threaded end of spindle shaft is on top. Place outer bearing over end of spindle and push bearing down into cup. Start bearing hub of internal gear (1) into outer bearing cone and, if necessary, tap gear lightly with a soft faced hammer to position. Install thrust washer (38) and spindle inner nut (39) and tighten nut finger tight. Coat mating surfaces of spindle and spindle support with No. 2 Permatex or equivalent sealer and install dirt shield and spindle on spindle support. Tighten retaining capscrews to a torque of 80-88 ft.-lbs.

Adjust inner nut as required until a pull of 33-38 pounds on a spring scale attached to a wheel stud is required to keep hub in motion. See Fig 12. Install lockwasher (40—Fig. 11) and outer nut (41). Tighten outer nut and re-check hub rolling torque. When adjustment is correct, bend tabs on lockwasher to secure both nuts. Install sun gear (46) and snap ring (44) on outer end of axle shaft. Coat mating surfaces of wheel hub and planet spider with No. 2 Permatex or equivalent sealer and install planet spider. Tighten retaining capscrews to a torque of 52-57 ft.-lbs. Install the puller hole capscrews and the tire and rim.

20. SPINDLE SUPPORT. The spindle support can be serviced after planet spider and wheel hub assembly are removed as outlined in paragraphs 16 and 18. However, if service is required only on the spindle support, the planet spider, wheel hub and axle shaft can be removed as a unit as follows:

Raise outer end of axle and remove tire and rim. Attach hoist to wheel

stud, then unbolt spindle from spindle support and pull complete hub assembly and axle shaft from outer end of axle. Refer to Fig. 16. **Do not** allow weight of assembly to be supported by axle shaft or damage to oil seal in axle housing outer end will result.

With the complete hub assembly and axle shaft removed, disconnect tie-rod and power steering cylinder from spindle support. Remove the capscrews from the two-piece retainer ring on inner side of spindle support and separate the retainers, seals and gasket from spindle support as shown in Fig. 13.

Note: At this time, it is desirable to remove the grease from cavity formed by spindle support and outer end of axle housing.

Remove upper trunnion, pull top of spindle support outward and remove spindle support from outer end of axle housing. Keep shims present under top trunnion tied to the trunnion for use during reassembly. Remove upper trunnion bearing from axle housing. Remove lower trunnion, shims and bearing from spindle support. Both trunnion bearing cups and upper trunnion bearing grease retainer can now be removed if necessary. If necessary to remove axle shaft thrust washer, oil seal and oil seal washer from axle outer end, a slide hammer and puller attachment can be used. Seals (Fig. 13) on outer end of axle housing can also be removed at this time.

Clean and inspect all parts and renew as necessary. It is recommended that new seals be used during reassembly.

21. To reassemble spindle support, proceed as follows: Install axle shaft seal washer and oil seal with lip toward inside and be sure oil seal is bottomed. Install axle shaft thrust washer. See Fig. 14. Install seal components over outer end of axle housing in the following order: Inner seal retainer with step toward inside of tractor; dust seal spring, rubber dust seal and felt grease seal and be sure

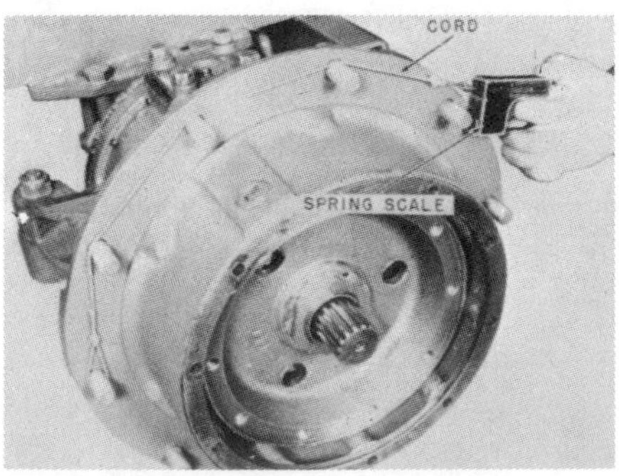

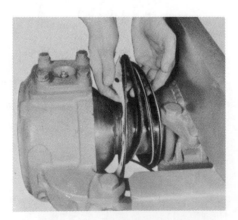

Fig. 12–Use method shown to check the wheel bearing preload which should be 33-38 lbs. pull on spring scale.

Fig. 13–Axle outer end seals and retainers are removed from spindle support as shown.

Fig. 14–Install seal washer, seal and thrust washer in the order shown.

bevel in inside diameter of both seals is toward bell of axle outer end; outer seal retainer with step toward outside of tractor and gasket. Install grease retainer (cup side up) and upper bearing cup (smallest I.D. down) in the upper trunnion bearing bore. Bolt lower trunnion to spindle support using original shims and tighten cap screws to a torque of 80-88 ft.-lbs. Place lower trunnion bearing over lower trunnion. Install lower trunnion bearing cup in outer end of axle housing with smallest I.D. of cup up. Place upper trunnion bearing in the upper trunnion bearing cup, then while tipping upper side of spindle support slightly outward, position spindle support over outer end of axle housing and install upper trunnion with original shim pack. Tighten trunnion retaining cap screws to a torque of 80-88 ft.-lbs.

Before attaching tie-rod, power steering cylinder or seal assembly to spindle support, check adjustment of trunnion bearings as follows: Connect a spring pull scale to tie-rod hole of spindle support and check pull required to rotate the spindle support. Refer to Fig. 15. Adjustment is correct when 12 to 18 pounds pull is required. To adjust bearings, vary number of shims located under the trunnions as required to obtain proper adjustment keeping the total thickness of shims under the top trunnion and lower trunnion as equal as possible. Shims are available in thicknesses of 0.003, 0.005 and 0.010.

Use grease to hold seal retainer gasket in place, then install seal assembly on spindle support, being sure that the split ends of outer seal assembly do not align. Attach tie-rod and tighten tie-rod stud nut to a torque of 200 ft.-lbs. Attach power steering cylinder and tighten attaching bolt lock nut until it just contacts mounting flange.

Place approximately four pounds of grease in the cavity of spindle support and pack universal joint of axle shaft. Coat mating surfaces of spindle and

spindle support with No. 2 Permatex or equivalent sealer and install the planet spider, wheel hub and axle assembly on the spindle support. Tighten attaching capscrews to a torque of 80-88 ft.-lbs. Then, install tire and rim.

## DIFFERENTIAL AND CARRIER

The Four Wheel Drive front axle can be equipped with either a conventional differential assembly or a "No-Spin" differential assembly. Removal procedure will be the same for both types. For service information, refer to paragraphs 23 and 27.

**22. REMOVE AND REINSTALL.** To remove the differential and carrier assembly, raise front of tractor, block axle carrier to prevent front axle assembly from rocking and remove tires and rims. Drain differential housing. Disconnect power steering cylinders and lay them on top side of axle carrier. Remove drive shaft shield and shield front support. Disconnect drive shaft from companion flange of differential pinion shaft.

Attach hoist to one of the wheel studs and take up slack of hoist. Unbolt spindle from spindle support and pull complete assembly from outer end of axle housing. Refer to Fig. 16. Remove opposite assembly in like manner.

Fig. 16–Planet spider, wheel hub and axle shaft assembly can be removed as shown.

Note: Do not allow weight of hub assembly to be supported by axle shaft or damage to oil seal in axle housing outer end will result.

Place a rolling floor jack under front axle, unbolt axle from axle carrier and lower the axle from tractor. Position axle on supports with differential pinion shaft up and secure assembly in this position with blocks. Disconnect one end of tie-rod and swing it out of way. Remove capscrews retaining differential carrier to axle housing and remove the assembly from housing.

Reinstall by reversing removal procedure and coat carrier retaining capscrews and mating surfaces of carrier and axle housing with No. 2 Permatex or equivalent sealer. Tighten capscrews to a torque of 37-41 ft.-lbs. Tighten tie-rod stud nut to a torque of 200 ft.-lbs. When joining spindle to spindle support, coat mating surfaces with No. 2 Permatex or equivalent sealer and tighten retaining capscrews to a torque of 80-88 ft.-lbs. Piston rod end of power steering cylinders are attached to spindle supports. Tighten cylinder attaching bolt lock nuts until they just contact mounting flanges.

**23. OVERHAUL. (CONVENTIONAL)** With differential and carrier removed as outlined in paragraph 22, disassemble unit as follows: Punch mark carrier bearing caps (14—Fig. 17) so they can be reinstalled in original position, then remove cotter pins and the adjusting nut lock pins (25). Cut lock wires and remove the carrier bearing caps. Lift differential from carrier and keep bearing caps (16) identified with their bearing cones (15). Bearing cones can now be removed from differential case if necessary. Unbolt and remove bevel ring gear from differential case if necessary to renew gear. Drive pinion pin (shaft) lock pin out of differential case and remove pinion pin (23), pinions (20), side gears (21) and thrust washers (18 and 19) from differential case.

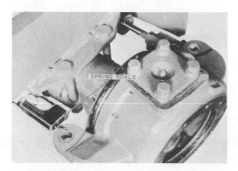

Fig. 15–Use a spring scale in the tie-rod stud hole to check trunnion bearing adjustment.

1. Nut
2. Companion flange
3. Dirt shield
4. Oil seal
5. Bearing retainer
6. Gasket
7. Bearing cone
8. Bearing cup
9. Spacer and shim kit
10. Bearing cone
11. Pinion shaft
12. Pilot bearing
13. Carrier
14. Bearing cap
15. Bearing cone
16. Bearing cup
17. Adjusting nut
18. Thrust washer
19. Thrust washer
20. Pinion gear
21. Side gear
22. Differential case
23. Pinion gear pin
24. Bevel ring gear
25. Lock pin

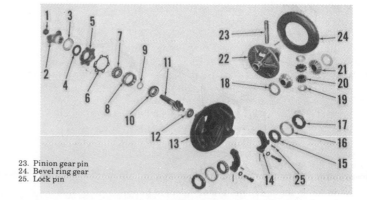

Fig. 17–Exploded view of the conventional differential and carrier assembly used on front drive axle of models equipped with four-wheel drive.

Remove cotter pin and nut (1) from pinion shaft (11), then using a puller, remove the companion flange (2) and dust shield. Remove pinion shaft bearing retainer (5) and press pinion shaft and bearing from carrier. Use a split bearing puller to support pinion bearing cup (8) on edge nearest pinion shaft gear and press pinion shaft from rear bearing and bearing cup. Remove spacer (9) and any shims which may be present from pinion shaft. Remove front bearing and inner (pilot) bearing (12) in a similar manner.

Clean and inspect all parts. Pay particular attention to bearings, bearing cups and thrust washers. If any of the differential side gears or pinions are damaged or excessively worn, renew all gears and thrust washers. Pinion shaft and bevel ring gear are available in a matched set only.

24. The differential and carrier unit is assembled as follows: Place inner (pilot) bearing on inner end of pinion shaft and stake in four places. Use a piece of pipe the size of inner pinion shaft bearing race to press forward bearing cone (10) onto shaft with taper facing threaded end of pinion shaft. Place bearing spacer and any shims which were present during disassembly over pinion shaft, then position the bearing cup over forward bearing. Press the rear pinion shaft bearing (7) on shaft with taper away from threaded end of pinion shaft. Check and, if necessary, renew the dust shield (3) on companion flange. Position companion flange so it will not obstruct cotter pin hole in end of pinion shaft, slide bearing retainer oil seal on its land on companion flange, then press companion flange on pinion shaft. Install retaining nut, clamp companion flange in a vise and tighten nut to a torque of 300 ft.-lbs.

Note: Pressure of oil seal will generally hold bearing retainer away from bearing. If it does not do so, tie retainer to companion flange.

25. With pinion shaft assembled as outlined above, clamp the bearing cup in a soft jawed vise just tight enough to prevent rotation, then using an inch-pound torque wrench on companion flange retaining nut, check torque required to rotate pinion shaft. Pinion shaft bearing adjustment is correct if 13 to 23 inch-pounds is required to rotate shaft. If rolling torque is not as specified, disassemble the pinion shaft assembly and vary thickness of spacer and/or shims as required to obtain proper rolling torque. A spacer and shim kit is available under Oliver part number 155 342-A.

With pinion shaft assembled and correct rolling torque (bearing adjustment) obtained, press pinion shaft assembly into carrier, install bearing retainer and tighten capscrews to a torque of 25 ft.-lbs. Install cotter pin to lock the nut in place.

26. Reassemble differential case assembly as follows: Place side gears, pinions and thrust washers in differential case and install pinion pin (shaft). Secure pinion pin with lock pin and, if lock pin is straight type, stake pin in position. It is not necessary to stake the spring type lock pin. If bevel ring gear was removed, reinstall with bolt heads on ring gear side of assembly and tighten the nuts to a torque of 78-86 ft.-lbs. Press bearings on differential case with tapers facing away from case. Place bearing cups over differential bearings and place differential assembly in carrier. Position bearing adjusting nuts in carrier and install the carrier bearing caps. Tighten the bearing cap screws until caps are snug but **be sure** threads of caps and adjusting nuts are in register. Maintain some clearance between gear teeth and tighten adjusting nuts until bearing cups are seated and all end play of differential is eliminated. Mount a dial indicator and shift differential assembly as required to obtain a backlash of 0.008-0.011 between bevel pinion shaft and bevel ring gear. Differential is shifted by loosening one adjusting nut and tightening the opposite nut an equal amount. Note: Mesh position of the bevel pinion shaft is not adjustable.

With gear backlash adjusted, tighten the bearing cap retaining capscrews to a torque of 65 ft.-lbs. and secure with lock wire. Install adjusting nut lock pins and cotter pins.

Note: If lock pins will not enter slots of adjusting nuts after backlash adjustment has been made, tighten rather than loosen the adjusting nut, or nuts. Recheck gear backlash.

If used, install thrust screw and turn screw in until it contacts back side of bevel ring gear, then back screw out ¼ to ½-turn. Apply sealer to threads of thrust screw at surface of carrier, then while holding screw from turning, install locking washer and nut. Secure nut by bending one tang of locking washer over nut and another tang over boss of carrier.

27. **OVERHAUL (NO-SPIN).** With differential and carrier removed as outlined in paragraph 22, unit is disassembled as follows: Punch mark the carrier bearing caps (14—Fig. 17) so they can be reinstalled in their original positions, then remove cotter pins and the adjusting nut lock pins (25). Cut lock wires and remove the carrier bearing caps. Lift differential from carrier and keep bearing cups (16) identified with their bearing cones (15). Bearing cones can now be removed from differential case, if necessary.

Remove cotter pin and nut (1) from pinion shaft (11), then using a puller, remove the companion flange (2) and dust shield. Remove pinion shaft bearing retainer (5) and press shaft and bearings from carrier. Use a split bearing puller to support pinion bearing cup (8) on edge nearest pinion shaft gear and press pinion shaft from rear bearing (7) and bearing cup. Remove spacer (9) and any shims which may be present from pinion shaft. Remove front bearing (10) and inner (pilot) bearing (12) in a similar manner.

Unbolt and remove bevel ring gear from differential, if necessary. Remove differential case bolts and hold case together as last bolt is removed to keep assembly from flying apart due to the internal spring pressure. See Fig. 18. Hold out rings (6) can be removed with snap ring spreaders.

28. Clean and inspect all parts. Check splines on side gears and clutch members and remove any burrs or

Fig. 18—Exploded view of the "no-spin" differential case assembly available for the four-wheel drive front axle.

1. Case half
2. Side gear
3. Spring
4. Spring retainer
5. Driven center
6. Hold-out ring
7. Central driver & center cam
8. Case half

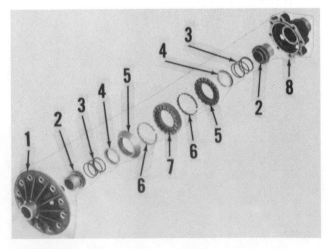

chipped edges with a stone or burr grinder. Renew any parts which have sections of splines broken away. Inspect springs (3) for fractures, or other damage, and renew springs which do not have a free height of 2¼-2½ inches. Center cam in central driver (7) must be free to rotate within the limits of keys in central driver. Check the weld between driven clutch (5) and cam ring on clutch by tapping lightly on cams of cam ring. If cam ring rotates in driven clutch, weld is defective (failed). Inspect teeth on central driver and driven clutches. Small defects can be dressed with a stone. If central driver or driven clutch is renewed, also renew the part it mates with. A smooth wear pattern up to 50 percent of face width is acceptable for the cams on center cam and driven clutch.

29. The differential and carrier unit is assembled as follows: Place the inner (pilot) bearing on inner end of pinion shaft and stake in four places. Use a piece of pipe the size of inner race of pinion shaft bearing and press

the forward pinion shaft bearing (10—Fig. 17) on shaft with taper facing threaded end of pinion shaft. Place bearing spacer, and any shims which were present during disassembly over pinion shaft, then position the bearing cup over forward bearing. Press the rear pinion shaft bearing (7) on shaft with taper away from threaded end of pinion shaft. Check, and renew if necessary, the oil seal (4) in pinion shaft bearing retainer (5). Seal is installed with lip toward inside. Place gasket and bearing retainer over pinion shaft. Check, and renew if necessary, the dust shield (3) on companion flange. Position companion flange so it will not obstruct cotter pin hole in end of pinion shaft, slide bearing retainer oil seal on its land on companion flange, then press companion flange on pinion shaft. Install retaining nut, clamp companion flange in a vise and tighten nut to a torque of 300 ft.-lbs.

NOTE: Pressure of oil seal will generally hold bearing retainer away from bearing. If it does not, tie retainer to companion flange.

30. With pinion shaft assembled as outlined above, clamp the bearing cup in a soft jawed vise only tight enough to prevent rotation, then using an inch-pound torque wrench attached to companion flange retaining nut, check the rolling torque (bearing preload) of the pinion shaft. This rolling torque should be 13-23 in.-lbs.

If rolling torque is not as specified, it will be necessary to disassemble the shaft assembly and vary the spacer and/or shims as required. A spacer and shim kit is available under Oliver part number 155 342-A.

With pinion shaft assembled and the rolling torque of shaft determined, press pinion shaft assembly into carrier, install bearing retainer and tighten cap screws to a torque of 25 ft.-lbs. Install cotter pin to lock nut (1) in place.

31. Lubricate all parts and reassemble no-spin differential by reversing disassembly procedure. Be sure to position spring retainers so that spring seats inside cupped section of retainer. Be sure key in central

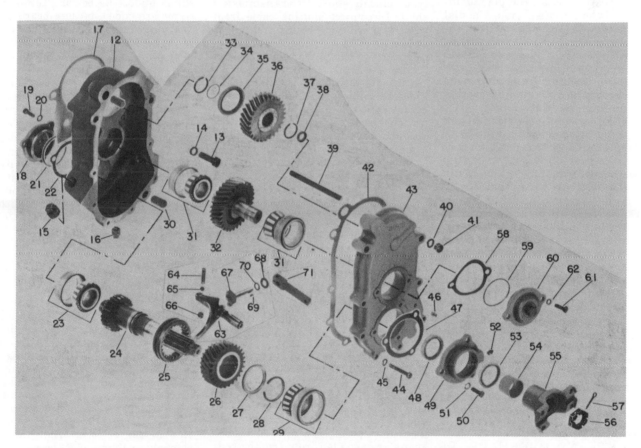

*Fig. 19—Exploded view of the four-wheel drive transfer drive assembly. Breather filter element (not shown) should be renewed yearly, or after each 1000 hours of operation.*

| | | | |
|---|---|---|---|
| 12. Case | 25. Coupling | 34. "O" ring | 47. Gasket | 60. Retainer |
| 15. Filler plug | 26. Gear (27 teeth) | 35. Oil seal | 48. Seal | 63. Shift fork & rod |
| 16. Drain plug | 27. Washer | 36. Gear | 49. Retainer | 64. Detent spring |
| 17. Gasket | 28. Snap ring | 37. Snap ring | 53. Seal | 65. Detent ball |
| 18. Bearing retainer | 29. Bearing | 38. Seal | 54. Spacer | 66. Plug |
| 21. "O" ring | 30. Dowels | 39. Stud bolt | 55. Companion flange | 67. Actuator |
| 22. Shim (0.004, 0.0075, 0.015) | 31. Bearing | 42. Gasket | 56. Castellated nut | 69. Woodruff key |
| 23. Bearing | 32. Idler gear | 43. Cover | 58. Shims (0.004, 0.0075, 0.015) | 70. Seal |
| 24. Shaft | 33. Snap ring | | 59. "O" ring | 71. Arm |

driver is aligned with slot in hold out ring. Tighten the case bolts to 36 ft.-lbs. torque and use a single wire through holes in bolt heads to lock bolts. Place an axle shaft into each side of differential and turn axle back and forth. It should be possible to feel backlash between clutch teeth. Backlash should be about 5/32-inch. If no backlash can be felt and side gear is locked solid, check differential for incorrect assembly.

If bevel ring gear was removed, reinstall with bolt heads next to bevel ring gear and tighten nuts to a torque of 78-86 ft.-lbs. Press bearings on differential case with tapers facing away from case. Place bearing cups over differential bearings and place differential assembly in carrier. Position bearing adjusting nuts in carrier and install the carrier bearing caps. Tighten the bearing cap retaining cap screws until caps are snug but BE SURE threads of caps and adjusting nuts are in register. Maintain some clearance between gear teeth and tighten adjusting nuts until bearing cups are seated and all end play of differential is eliminated. Mount a dial indicator and shift differential assembly as required to obtain a backlash of 0.008-0.011 between bevel pinion shaft gear and bevel ring gear. Differential is shifted by loosening one adjusting nut and tightening the opposite an equal amount.

Note: Fore and aft position of the bevel pinion shaft is not adjustable.

With gear backlash adjusted, tighten the bearing cap retaining cap screws to a torque of 65 ft.-lbs. and secure with lock wire. Install adjusting nut lock pins and cotter pins.

NOTE: If lock pins will not enter slots of adjusting nuts after backlash adjustment has been made, tighten rather than loosen the adjusting nut, or nuts. Recheck gear backlash.

If used, install thrust screw and turn screw in until it contacts bevel ring gear, then back it out ¼ to ½-turn. Apply sealant to threads of thrust screw at surface of carrier, then install lock washer and lock nut. Secure lock nut by bending one tang of lock washer over nut and one tang over boss of carrier.

### TRANSFER DRIVE
### All Models So Equipped

32. **R&R AND OVERHAUL.** To remove the transfer drive assembly, it will be necessary to split the tractor front main frame from the rear main frame. If tractor is equipped with an auxiliary drive, it will also be necessary to remove the engine and drive unit before tractor can be split.

33. To split tractor equipped with auxiliary drive, first remove engine and drive unit as outlined in paragraph 55, then proceed as follows: Disconnect speedometer drive cable from adapter. Disconnect shifting cable from transfer drive shifter arm and bracket and pull it from front main frame. Disconnect drive shaft from companion flange of transfer drive output shaft. Unclip and disconnect power steering oil lines. Disconnect safety starting switch wires. Disconnect rear light wires and remove clips. On non-diesel models, unhook governor control bellcrank spring. Support tractor front frame in manner which will prevent tipping, then support rear frame with rolling floor jack. Remove front frame to rear frame retaining cap screws and separate tractor.

34. To split direct drive models, proceed as follows: Remove PTO drive shaft as outlined in paragraph 180, or the hydraulic pump drive shaft as outlined in paragraph 181, if so equipped. Remove coupling or coupling chain from clutch shaft to transmission shaft. Disconnect shifting cable from transfer drive and bracket and pull it from hole in front main frame. Disconnect drive shaft from companion flange of transfer drive output shaft. Unclip and disconnect power steering lines. Disconnect rear light wires and remove clips. On non-diesel models, unhook governor control bellcrank spring. Support tractor front frame in a manner which will prevent tipping, then support rear frame with a rolling floor jack. Remove front frame to rear frame retaining capscrews and separate tractor.

35. With tractor split as outlined in paragraph 33 or 34, drain transmission and transfer housings, refer to Fig. 19 and proceed as follows: Place transmission and transfer drive in gear to unscrew nut (56), then remove companion flange and spacer from transfer drive output shaft. Disconnect oil line from transfer drive case, then unbolt transfer drive cover (43) from case and remove cover along with idler gear (32), output gear (26) and shaft (24). Remove shifter coupling (25) from fork if necessary. Remove snap ring (37) and gear (36) from forward end of transmission bevel pinion shaft; discard snap ring. Unbolt and remove transfer case (12) from tractor rear frame. Shifter fork (63) and actuator (67) can be removed after shifter arm (71) and Woodruff key (69) are removed. Any further disassembly required will be evident. Save shims (22 and 58) located under bearing retainers (18 and 60) for reassembly.

36. Clean and inspect all parts and renew as necessary. New seals, "O" rings and gaskets should be used during assembly. The unit should be partially reassembled and the end play of idler and output shafts checked and adjusted before unit is installed on tractor rear main frame.

Install idler shaft rear bearing cup in transfer drive case and install output shaft rear bearing cup in shaft rear bearing retainer. Be sure both bearing cups are firmly seated and install output shaft rear bearing retainer (18) along with original shim pack (22). Install output shaft front bearing retainer (49) and idler shaft front bearing retainer (60) with original shim pack. Be sure both bearing cups are against the bearing retainers. Install dowels (30) in transfer case if removed. Place output shaft and idler shaft assemblies in the transfer drive case cover, then using a new gasket (42), secure cover to case. Use dial indicator to check end play of both shafts; end play should be 0.001-0.003 and is adjusted by varying number of shims under the bearing retainers. If shims are added, be sure bearing cups are seated against retainers before rechecking shaft end play.

With shaft bearings (end play) adjusted, remove cover, shafts and bearing retainers. Renew the "O" rings (21 and 59) on bearing retainers and install new oil seals as follows: Bevel pinion shaft oil seal (35) in case with lip rearward. Bevel pinion shaft seal (38) in cover with lip forward. Shift fork actuator seal (70) with lip inward. Install wide seal (48) at rear with lip facing rearward and the narrow oil seal (53) at front with lip facing forward in the output shaft front bearing retainer.

Coat all seals and "O" rings with grease prior to installation and fill cavity between the two seals in output shaft front bearing retainer with grease.

Reassemble unit by reversing the disassembly procedure. Renew the "O" ring on bevel pinion shaft front bearing retainer and install case with new gasket (17). Use a new snap ring (37) and "O" ring (34) when installing drive gear (36) on transmission bevel pinion shaft. Use heavy grease on sliding coupling (25) to hold it in position in shifter fork. When installing companion flange (55), be sure not to obstruct cotter pin hole.

### DRIVE SHAFT
### All Models So Equipped

37. The front axle drive shaft is of

conventional design. Removal and overhaul procedure is evident after removal of shield and an examination of unit. Spider and bearing assemblies are available as units only. All other parts are available separately.

When installing drive shaft, yoke end is installed to rear.

# POWER STEERING

### Series 1755-1855-1955

40. All Series 1755, 1855 and 1955 tractors are equipped with power steering that utilizes a Saginaw closed center, rotary valve type Hydramotor which furnishes pressurized fluid in metered amounts to actuate the front wheel steering cylinder (Two steering cylinders used on four-wheel drive models).

Hydraulic system fluid is used as power steering fluid and is furnished by the hydraulic system pump via a priority valve located in the pressure control valve.

### LUBRICATION AND BLEEDING
### Series 1755-1855-1955

41. Steering system can be considered self-bleeding and any air trapped in lines, cylinder or Hydramotor should be eliminated by turning steering wheel to move steering cylinder, or cylinders, through full range of travel several times in each direction.

NOTE: Prior to bleeding the power steering system, be sure the hydraulic lift system is bled as outlined in paragraph 240.

### SYSTEM PRESSURE TEST
### Series 1755-1855-1955

42. Steering system operating pressure should be 1775-1825 psi when checked with engine running at 2000 rpm.

To check steering system pressure (priority valve), refer to paragraph 237 of HYDRAULIC LIFT section of this manual.

### PUMP
### Series 1755-1855-1955

43. The hydraulic lift pump also furnishes the pressurized oil for operation of the power steering system.

For information on the hydraulic lift system pump, refer to paragraph 257.

### HYDRAMOTOR STEERING UNIT

Hydramotor steering units used in Series 1755, 1855 and 1955 tractors are a rotary valve, closed center type, designed to operate in conjunction with the closed-center hydraulic lift system of the tractor. The Hydramotor receives the operating fluid from the hydraulic lift system pump via a priority valve which is incorporated in the pressure control valve located on the left side of the tractor transmission case below hydraulic lift system pump.

Hydramotors can be identified by a code number and serial number located directly beside pressure port. Code number for closed-center type Hydramotors is "CB9".

### Series 1755-1855-1955

44. **R&R HYDRAMOTOR UNIT.** First, disconnect battery ground strap, then proceed as follows: Remove emblem from steering wheel adjuster lock knob, unscrew jam nut and the lock knob from adjuster bolt, remove flat washer from bolt and then pull steering wheel and inner shaft from steering column. Unbolt boot retainer from dash panel and remove boot and retainer from steering column. Remove dash side panels. NOTE: In some cases there will be enough slack in speedometer cable, tachourmeter cable, oil pressure gage line and temperature gage wire to allow dash side panels to be moved aside. Identify and disconnect wiring from center dash gages, lights and fuses. Disconnect ignition switch from center dash and fuel shut-off control from injection pump. Remove center dash.

Tag the four Hydramotor hoses for ease in reconnecting, then disconnect hoses from unit and immediately cap all openings. Loosen pivot pin lock nuts, unscrew pivot pins until they clear Hydramotor casting, then lift the unit from opening in instrument panel support.

Reverse removal procedure to install Hydramotor and bleed unit as outlined in paragraph 41.

45. **OVERHAUL HYDRAMOTOR STEERING UNIT.** With unit removed as in paragraph 44, proceed as follows: Loosen lock nut (4—Fig. 20) and remove jacket (3) and adapter (2) from Hydramotor. Remove lock nut and press dust cover and bearing (1) from jacket. Pull upward on outer steering shaft to remove rotary valve and steering shaft assembly from housing (32). Secure outer steering shaft assembly in a vise so hole in tapered ring (11) is upward, then loosen lock nut (12) and drive tapered ring toward lock nut. Remove shaft assembly from vise and rotate tapered ring until hole in ring is over ball (13), then tap ball from shaft assembly. Separate shaft assembly from valve assembly, then remove nut (12), tapered ring (11), retaining spring (9), inner shaft (4), dowel pin (6), adjuster bolt (8) and nut (7) from outer shaft (10). Pull bearing plug assembly (20) from rotary valve assembly (23) then remove snap ring (15) and press seals (16 and 17) and bearing (18) from plug. Push valve spool and shaft out of valve body, unhook spool from shaft and slide spool from shaft. Remove "O" rings and backup rings from valve body. Remove "O" ring dampener from valve spool.

Position housing (32) with end plug (41) on top, then if necessary, position end of retaining ring (42) to within 15 degrees of hole in bottom side of housing. Use a 1/6-inch punch in hole in housing to unseat and remove retaining ring. Remove end plug (41) and spring (40). Remove backup ring (39) and "O" ring (38) from housing. Remove pressure plate (37), rotor assembly (36), thrust plate (35), drive shaft (24) and dowels (33) from housing. If necessary, push rotor through rotor ring to disassemble rotor vanes and springs. Remove snap rings (27), plugs (28), springs (30) and washers (31), then remove spool (32) from housing. Identify end position of spool (32) for reassembly.

46. Discard all "O" rings, backup rings, seal rings and oil seals. Renew bearing (1 or 18) if rollers are broken, pitted or rusted. Renew shaft (5) if bearing surface is burred or pitted. Examine valve for signs of leakage and if leakage exists, renew complete valve assembly. Also inspect valve and valve shaft pins and grooves for undue wear or other damage. Small nicks or burrs can be removed from valve assembly by using a fine hone or crocus cloth, however do not remove any metal from valve spool or body. After installing teflon ring (26) on drive shaft (24), delay installation of shaft about 30 minutes to allow the teflon ring to shrink to correct size. Renew both transfer valve (32) and housing if valve or valve bore are deeply scored, but small nicks and burrs can be removed with crocus cloth. Use an air hose to check operation of pressure plate (37) valve. Air entering angled holes will cause valve to shift so that air comes out small center hole. If valve malfunctions, or if large burrs or deep scoring exists, renew pressure plate. Be sure that vanes slide freely

in slots of rotor and bear in mind that vanes may appear to stick at extreme top and bottom of rotor slots. If nicks or burrs cannot be removed from thrust plate (35) or end plug (41) without undue honing, renew parts.

NOTE: In some early models of the rotary valve Hydramotors, a problem developed in that displacement of the transfer valve plugs (28) and/or the plug in pressure plate (37) occurred. Beginning in June of 1970, this problem was corrected at the factory by redesigning the transfer plugs to make the retaining snap rings captive and the use of dowel pins in pressure plate to prevent pressure plate plugs from blowing out. These factory corrected units can be identified by a dab

of white paint directly below hose ports on round surface of Hydramotor housing.

Should these problems be encountered in the field, an improvement package is available from the Oliver Corporation.

47. To reassemble Hydramotor, proceed as follows: Install transfer spool, spring washers (31), springs (30) and plugs (28) with "O" rings (29) installed, then secure plugs with snap rings (27). If disassembled, install rotor, vanes and springs inside of rotor ring. Vanes are installed with rounded edges next to ring and springs must be installed in the same sequence on both sides of vanes. Install "O" ring (25) on drive shaft (24), then install teflon

ring (26) over "O" ring and allow time for teflon ring to shrink to size. Place drive shaft in housing, use a 1 x 2 x 2½ inch long wood block to support shaft and turn housing bottom side up on bench. Install dowel pins (33) in housing. Install the two small "O" rings (34) in thrust plate (35), then with "O" rings toward control valve, align notches in plate with dowel pins and install plate in housing. Align dowel pin notches of rotor assembly (36) and pressure plate (37) and install in housing using care not to damage seals on drive shaft. Place "O" ring (38) and backup ring (39) in groove of housing with "O" ring next to pressure plate. Place large end of spring (40) on pressure plate, then install end plug

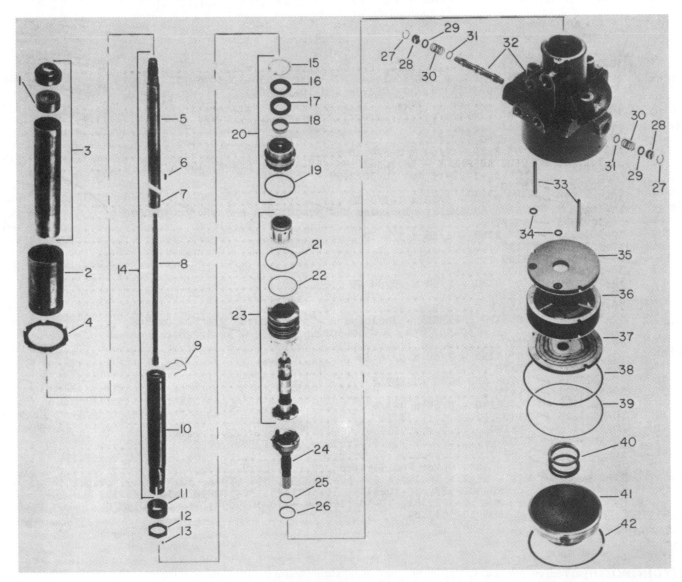

*Fig. 20–Exploded view of rotary valve Hydramotor unit used in Series 1755, 1855 and 1955 tractors.*

| | | | |
|---|---|---|---|
| 1. Bearing | 9. Retainer spring | 18. Bearing | 26. Seal ring | 35. Thrust plate |
| 2. Adapter | 10. Outer steering shaft | 19. "O" ring | 27. Snap ring | 36. Rotor assy. |
| 3. Jacket assy. | 11. Tapered ring | 20. Plug assy. | 28. Plug | 37. Pressure plate |
| 4. Lock nut | 12. Nut | 21. Seal ring | 29. "O" ring | 38. "O" ring |
| 5. Inner steering shaft | 13. Steel ball | 22. "O" ring | 30. Spring | 39. Back-up ring |
| 6. Dowel | 14. Steering shaft assy. | 23. Rotary valve assy. | 31. Spring washer | 40. Spring |
| 7. Nut | 15. Snap ring | 24. Drive shaft | 32. Transfer valve and housing | 41. End plug |
| 8. Adjuster bolt | 16. Oil seal (small) | 25. "O" ring | 33. Dowel pins | 42. Retainer ring |
| | 17. Oil seal (large) | | 34. "O" rings | |

(41) and secure with retaining ring (42).

NOTE: Face of end plug (41) has a cast recess and recess inside diameter is not concentric with outside diameter of plug. If interference with dowel pin is encountered during assembly, rotate plug until interference is eliminated.

Place "O" rings (22) on valve body and install seal rings (21) over "O" rings. Position spool so shaft engagement hole is downward and push spool in top of valve body until bottom of spool is flush with bottom of valve body. Insert spool shaft into bottom of spool and engage pin on shaft with engagement hole of spool. Align outer notch in shaft base with pin inside body and slide shaft in position.

Install bearing (18) in top of plug so that top of bearing is flush with lower edge of tapered flange. Install seals (16 and 17) with lips downward and largest seal on bottom. Install snap ring (15) and "O" ring (19), then slide the bearing plug assembly over top of valve assembly shaft.

Install adjuster nut (7) on adjuster bolt (8), insert bolt, nut, dowel pin (6) and inner steering shaft (5) into outer steering shaft (10) and install spring retainer (9). Place outer shaft in a vise with ball hole upward and loosely install tapered ring (11) and nut (12). Align hole in tapered ring with hole in outer shaft, then insert valve shaft into outer steering shaft and when ball groove in valve shaft aligns with holes in tapered ring and outer steering shaft, drop ball (13) into hole. Turn the tapered ring 180 degrees and tighten nut (12) to 40-50 ft.-lbs. torque. Install valve and steering shaft into housing and be sure notch in valve body aligns with pin on drive shaft.

If installing bearing (1), press bearing into jacket until top side is flush with end of jacket and install dust cover. Install lock nut (4), screw jacket into housing until it butts against plug assembly (20), then tighten jacket to 10 ft.-lbs. torque. Back-off jacket ⅛ to ¼-turn and tighten locknut to 50-100 ft.-lbs. torque. Reinstall tilt bracket on housing.

**POWER STEERING CYLINDER**
**All Models Except Four-Wheel Drive**

48. **R&R POWER STEERING CYLINDER.** Disconnect oil coolers from their supports and remove grille from main frame. Then, on all models, remove radiator as outlined in paragraph 124. Loosen power steering line clamps at side of engine, disconnect lines from power steering cylinder and immediately cap all openings. Refer to

Fig. 21; remove plug (4) from cap (1) and remove cap from power steering cylinder. Pry plug (7) from pinion (11).

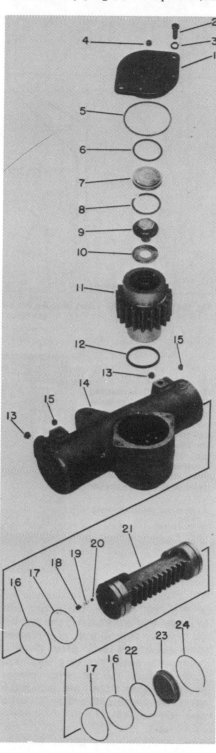

*Fig. 21—Exploded view of power steering cylinder assembly used on all models except those equipped with four wheel drive.*

| | | | |
|---|---|---|---|
| 1. Cap | | 14. Cylinder | |
| 4. Plug | | 15. Plugs | |
| 5. "O" ring | | 16. Teflon rings | |
| 6. "O" ring | | 17. "O" rings | |
| 7. Plug | | 18. Spring pins | |
| 8. Snap ring | | 19. Springs | |
| 9. Puller bolt | | 20. Check valves | |
| 10. Washer | | 21. Piston | |
| 11. Pinion | | 22. "O" ring | |
| 12. Quad ring | | 23. End plug | |
| 13. Connectors | | 24. Snap ring | |

Pull pinion from splines of front assembly shaft by turning puller bolt (9) counter-clockwise.

On Rowcrop and Short Wheel Base models, lift cylinder upward to disengage stud in cylinder from torque link, turn cylinder counter-clockwise on front assembly shaft and push cylinder downward to free pinion from shaft. Withdraw pinion from cylinder, then remove cylinder from front end of main frame.

On Wheatland models, remove nut and washer from tapered stud at end of torque link and drive the stud upward out of main frame and the link. Then remove the power steering cylinder, with pinion installed, from front end of main frame.

Reverse removal procedures to reinstall cylinder. One of the index marks on pinion (11) must be aligned with centering mark on piston (21). Tighten the puller bolt to a torque of 200 ft.-lbs., and tighten the nuts on torque link and tapered stud securely. Install plug (7—Fig. 21) with new "O" ring (6). Fill cavity around pinion with proper power steering fluid before reinstalling cap (1) with new "O" ring (5).

49. **OVERHAUL POWER STEERING CYLINDER.** After removing the cylinder assembly as outlined in paragraph 48, proceed as follows: On Wheatland models, remove pinion (11—Fig. 21). Remove snap ring (8), puller bolt (9) and washer (10). To remove end plug retaining snap ring (24) insert pin or punch in hole in end of housing (14) and push snap ring from groove in cylinder. Grasp one of the lugs on end plug with vise-grip pliers and, with a twisting motion, pull end plug from cylinder. Remove plug sealing "O" ring (22—Fig. 21) from end of cylinder and remove the piston (21). Remove the Teflon seals (16) and "O" rings (17) from groove at each end of piston and remove pinion sealing quad ring (12) from bottom of cylinder. If necessary to remove the check valve assembly (18, 19 and 20) from either end of cylinder, insert punch in hole from inner side of piston head and drive valve ball, spring and spring pin from piston. If the hydraulic line connector seats (13) are damaged, they may be removed by threading inside diameter of seats and using puller bolts. Carefully clean and inspect all parts and renew any that are excessively worn, deeply scored or damaged beyond further use.

Reassemble cylinder, using all new sealing rings as follows: Insert check ball (20) into bore in end of piston (21), insert spring (19) with small end next to ball and drive the retaining spring pin (18) in flush with end of piston. Install "O" ring (17) in bottom of groove in end of piston, then install Teflon ring (16) on top of the "O" ring. Lubricate piston and install in bore of cylinder. Install "O" ring (22) in groove in open end of cylinder, lubricate and install end plug (23) and install plug retaining snap ring (24). Install quad ring (12) in bottom of pinion bore in cylinder casting and lubricate ring and bore. Install pinion (11), and washer (10), puller bolt (9) and retaining snap ring (8) in top of pinion. On Wheatland models, insert pinion in cylinder with timing marks on pinion and piston aligned and reinstall cylinder as outlined in paragraph 48.

## POWER STEERING CYLINDERS
### Four Wheel Drive Models

50. **R&R AND OVERHAUL.** To remove either power steering cylinder, disconnect the hoses, cap all openings, then remove the bolts from cylinder end assemblies and remove the cylinder from tractor. Loosen the lock nut on rod end of assembly and the clamp on cylinder end, then remove the end assemblies. Bushings in either end assembly may be renewed if worn or damaged.

To disassemble cylinder, refer to Fig. 22 and proceed as follows: Remove end plate (19), push bearing (14) into cylinder to expose snap ring (18) and remove the snap ring. Pull the piston rod (6), bearing and piston (10) from cylinder as an assembly. Hold rod from turning at flat near outer end, then remove cotter pin and piston retaining nut (7). Slide piston and bearing from inner end of rod. Remove the sealing rings and wiper from piston and bearing. Clean and carefully inspect all parts and renew any that are excessively worn, scored or damaged.

Reassemble, using all new seals, as follows: Install "O" ring (15) and backup ring (16) in bore of bearing with backup ring at outer side of "O" ring. Install wiper (17) with lip outward, lubricate bore of bearing and install bearing on piston rod with "O" ring groove on outer diameter of bearing towards small (inner) end of rod. Install new "O" ring (13) on bearing. Install new "O" rings (11) in bottom of grooves on piston and install the Teflon rings (12) on top of the "O" rings. Install piston, washer (9) and nut (7) on piston rod and tighten nut securely. Note: If nut threads onto

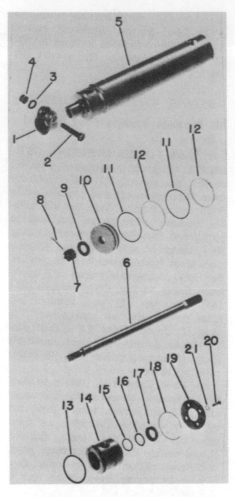

**Fig. 22–Exploded view of power steering cylinder used on four-wheel drive models.**

| | |
|---|---|
| 1. Clamp | 13. "O" rings |
| 5. Cylinder barrel | 14. Bearing |
| 6. Piston rod | 15. "O" ring |
| 7. Slotted nut | 16. Back-up ring |
| 9. Washer | 17. Wiper |
| 10. Piston | 18. Snap ring |
| 11. "O" rings | 19. Retainer plate |
| 12. Teflon rings | |

piston rod too far and slots in nut will not engage cotter pin, remove the nut and add washers (9) as necessary. Install cotter pin (8), lubricate piston and cylinder bore and carefully insert piston and bearing into bore but avoid damaging sealing rings on snap ring groove. Using a dull tool, compress the sealing rings as they pass hole for hydraulic fitting in cylinder tube. Push bearing far enough into bore to install snap ring (18), bump bearing out against snap ring with piston and install plate (19).

Before reinstalling cylinder, install the end assemblies and, with cylinder rod fully retracted, adjust position of end assemblies so that center-to-center measurement of the mounting holes is 18½ inches. Then, tighten the lock nut and clamp to secure end assemblies in this position and install cylinder on tractor.

# ENGINE AND COMPONENTS

Oliver 1755, 1855 and 1955 Series tractors are equipped with six cylinder engines. Series 1755 tractors are available with either a non-diesel engine having a bore and stroke of 3⅞ x 4 inches and a 283 cubic inch displacement; or a diesel engine having a bore and stroke of 3⅞ x 4⅜ inches and a displacement of 310 cubic inches.

The Series 1855 tractors are available with either non-diesel or diesel engines. Both engines have a bore and stroke of 3⅞ x 4⅜-inches and a displacement of 310 cubic inches.

The Series 1955 tractors are available only with a diesel engine which is the same dimensionally as the diesel engine used in the Series 1755 and 1855.

The diesel engine used in Series 1755 tractors is naturally asperated, whereas the diesel engine used in Series 1855 and 1955 is turbocharged.

## R&R ENGINE ASSEMBLY

On direct drive model tractors the clutch can be serviced without removing engine, or the engine and clutch can be removed as an assembly.

On models equipped with an auxiliary drive unit (Hydra-Power or Hydraul Shift), the engine and auxiliary drive unit must be removed as an assembly to service engine, clutch or auxiliary drive unit.

### Series 1755-1855-1955

55. To remove engine, first disconnect battery ground strap, then drain cooling system and if engine is to be disassembled, drain oil pan. On models with an auxiliary drive unit, drain drive unit if it is to be disassembled.

Remove lower side panels, hood side panels and hood. Remove the following hoses and lines: Hydramotor to power steering cylinder hoses and lines; hydraulic unit to oil cooler circulating pump line; circulating pump to cooler line; cooler to hydraulic unit line and the two auxiliary transmission to cooler lines. On all models, disconnect fuel supply line from fuel strainer and disconnect wire from fuel gage sending unit. On diesel models, also disconnect drain tube and leak-off line. Remove fuel tank straps and fuel tank.

Disconnect both radiator hoses, then remove fan from water pump pulley and allow fan to rest in fan shroud. Unbolt and remove radiator and the attached oil coolers. Disconnect all wiring and engine controls from engine and on direct drive models, remove clutch shaft shield. On all models, remove sprocket (coupling) chain. On models with an auxiliary drive unit, disconnect cable from control lever.

Remove pto shaft as outlined in paragraph 180, or the hydraulic pump drive shaft as outlined in paragraph 181 if tractor has no pto.

Unhook clutch pedal return spring and disconnect clutch rod at forward end. Remove engine mounting bolts and identify left front and right rear bolts as they are a machine fit and act as dowels. Install lifting eye-bolt (Oliver No. STS-137, or equivalent) to lifting cap screw and lift engine from tractor. NOTE: Use rear lifting cap screw if tractor is equipped with auxiliary drive unit. Save any shims which may be present for subsequent use when reinstalling engine.

Before installing engine, loosen lock nut and unthread auxiliary drive unit adjusting screw which is located in lower side of front main frame. Reinstall engine by reversing the removal procedure. Check sprocket chain to see that it can be moved freely by hand and if chain is tight, use shims on engine mounts to provide correct engine alignment. With engine installed turn auxiliary drive unit adjusting screw in until it contacts bottom of drive unit housing and tighten lock nut.

Bleed the diesel fuel system as outlined in paragraph 96.

### R&R CYLINDER HEAD
### Series 1755-1855 Non-Diesel Models

56. Drain cooling system and remove hood. Disconnect hydraulic line support from hydraulic lines and tappet cover. Disconnect outlet hose and remove air cleaner. Disconnect breather tube and ignition coil from tappet cover, then remove tappet cover.

Disconnect rocker shaft oil line from oil screw, then remove nuts from rocker arm shaft support studs and lift off rocker arms and shaft assembly. Remove push rods.

Disconnect upper radiator hose, bypass hose, the small bleed line and the temperature sending unit wire. Disconnect wires from spark plugs. Disconnect exhaust and intake manifold

(carburetor attached) from cylinder head, then unbolt and remove cylinder head.

When reinstalling cylinder head, refer to Fig. 23 and tighten all cylinder head cap screws to 129-133 ft.-lbs. torque. Torque values given are for oiled threads. Note that oil screw is at rear of right hand row of cap screws. Rocker arm shaft support bracket nuts and manifold retaining nuts are tightened to 25-27 ft.-lbs. torque. Adjust valve tappet gaps cold to clearances given in paragraph 58.

### Series 1755-1855-1955 Diesel Models

57. Drain cooling system and remove hood. Disconnect hydraulic line support from hydraulic lines and tappet cover. Disconnect wiring from preheater solenoid and restriction indicator switch (1855 and 1955). Disconnect hoses from air cleaner and remove air cleaner. Remove breather tube, then remove tappet cover, air cleaner support (1855 and 1955) and preheater solenoid as an assembly.

Disconnect rocker shaft oil line from oil screw, then remove nuts from rocker arm shaft support studs and lift off rocker arms and shaft assembly. Remove push rods.

Disconnect upper radiator hose, bypass hose, the small bleed line and the temperature sending unit wire. Remove intake manifold. On Series 1755, remove exhaust manifold and heat shield. On Series 1855 and 1955, remove turbocharger lube lines and remove exhaust manifold, turbocharger and heat shield.

Disconnect pressure lines from injectors. Remove leak-off line from injectors, fuel tank and injection pump.

Remove cylinder head cap screws and lift off cylinder head.

NOTE: Support cylinder head with wood blocks at each end to prevent damage to injectors which protrude from combustion side of cylinder head.

When reinstalling cylinder head, refer to Fig. 23 and tighten cap screws 1 through 8, 11 and 12 to 180 ft.-lbs. torque and cap screws 9, 10, 13 and 14 to 133 ft.-lbs. torque. Torque values given are for oiled threads. Note that oil screw is at rear of right hand row of cap screws. Rocker arm shaft support bracket nuts and manifold retaining

nuts are tightened to 25-27 ft.-lbs. torque. Adjust valve tappet gap to clearances given in paragraph 58.

### VALVES, GUIDES AND SEATS
### Series 1755-1855-1955

58. Valve face angle for non-diesel engines is 44½ degrees with a seat angle of 45 degrees. Diesel engine intake valves have a face angle of 29½ degrees and a seat angle of 30 degrees while exhaust valves have a face angle of 45½ degrees and a seat angle of 45 degrees. Valve stem diameter is as follows:

**Intake**

All non-diesel . . . . . . . . . 0.3720-0.3730
  Min. dia. . . . . . . . . . . . . . . . . . . 0.3690
All diesel . . . . . . . . . . . . 0.3720-0.3725
  Min. dia. . . . . . . . . . . . . . . . . . . 0.3690

**Exhaust**

All non-diesel . . . . . . . . 0.3710-0.3720
  Min. dia. . . . . . . . . . . . . . . . . . . 0.3690
All diesel . . . . . . . . . . . . 0.3715-0.3720
  Min. dia. . . . . . . . . . . . . . . . . . . 0.3690

Valve seat width for all series is as follows:

**Series 1755-1855 Non-Diesel**

Intake . . . . . . . . . . . . . . . . . . . 0.0625
Exhaust . . . . . . . . . . . . . 0.080-0.090

**Series 1755-1855-1955 Diesel**

Intake . . . . . . . . . . . . . 0.087-0.097
Exhaust . . . . . . . . . . . . . 0.062-0.072

Exhaust valve seat inserts are standard equipment for Series 1755 and 1855 non-diesel engines. Outside diameter of valve seat insert is 1.6270-1.6275. Insert counterbore in cylinder head has a diameter of 1.6240-1.6250 and a depth of 0.217-0.219. Chill inserts before installation and peen cylinder head material around insert after insert is installed.

Intake valve stems of both diesel and non-diesel engines are fitted with seals which should be renewed each time valves are reseated. Seals in non-diesel engines are Perfect Circle.

Adjust valve tappet gaps cold and to the following values:

**Intake**

All non-diesel . . . . . . . . . . . 0.019-0.021
All Diesel . . . . . . . . . . . . . . . 0.029-0.031

**Exhaust**

All models . . . . . . . . . . . . . . 0.029-0.031

Valve guides are a straight type and in diesel engines may have either a spiral groove end or a plain end. Intake valve guides in non-diesel engines are fitted with Perfect Circle valve stem seals. Valve guide length for non-diesel engines is 2 13/32 inches for intake and 2⅝ inches for exhaust. Valve guide length for all diesel engines is 3½ inches.

Valve guide inside diameter is 0.374-0.375 for non-diesel engines, or 0.373-0.374 for spiral groove end and

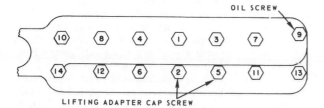

**Fig. 23—Tighten all cylinder head cap screws in the sequence shown. Refer to text for torque values on non-diesel and diesel engines.**

0.3745-0.3757 for plain end guides used in diesel engines.

Operating clearance of intake valves in guides for non-diesel engines is 0.001-0.003. Operating clearance of intake valves in spiral groove end guides for diesel engines is 0.0005-0.002, or 0.002-0.0037 in guides with plain end. Operating clearance of exhaust valve in guide for non-diesel engines is 0.002-0.004. Operating clearance of exhaust valve in spiral groove end guides for diesel engines is 0.001-0.005, or 0.0025-0.0042 in guides with plain end.

Renew valves and/or guides in non-diesel engines when clearance for intake valves exceeds 0.005 or exhaust valves exceeds 0.0065. Renew valves and/or guides in diesel engines when clearance for intake valves exceeds 0.004 or exhaust valves exceeds 0.0055.

When renewing valve guides, drive guides out toward top side of cylinder head. When properly installed, guide height above machined surface of cylinder head is 27/32-inch for non-diesel intake; 15/16-inch for non-diesel exhaust and ⅞-inch for both intake and exhaust guides of diesel engines.

## VALVE ROTATORS
### All Non-Diesel Models

59. Exhaust valves of non-diesel engines are fitted with a positive type rotator. Rotators are not serviceable but should be observed while engine is running to be sure valve rotates a slight amount each time it is opened. Renew any rotator or any valve which fails to rotate.

## VALVE SPRINGS
### All Models

60. Valve springs used for intake valves of non-diesel engines and all valves of diesel engines have a free length of 2.562 inches and should test 91 - 99 lbs. when compressed to a length of 1.506 inches. Valve springs used for exhaust valves of non-diesel engines have a free length of 2.440 inches and should test 86-94 lbs. when compressed to a length of 1.350 inches.

Renew any spring which does not have square ends, is rusted or shows any signs of fracture.

## VALVE LIFTERS
### All Models

61. Valve lifters are the mushroom type and operate directly in machined bores in the cylinder block. Lifters are supplied only in standard size of 0.6240-0.6245 and should have an op-erating clearance of 0.0005-0.002 in their bores.

Renew lifters if their diameter is less than 0.619 or if they are more than 0.0003 out-of-round. If diametral clearance exceeds 0.007 with new lifter in bore, renew cylinder block.

Any valve lifter can be removed after camshaft has been removed as outlined in paragraph 66.

## VALVE ROCKER ARMS AND SHAFT
### All Models

62. The rocker arms and shaft assembly can be removed as follows: Remove both hood side panels and hood. Disconnect oil cooler lines from center support bracket and remove bracket from tappet cover.

On diesel models, disconnect wires (except 1755) from restriction indicator sending unit, then disconnect inlet hose from turbocharger (except 1755) and remove air cleaner assembly and bracket. Unbolt preheater solenoid from fuel filters bracket, then remove filters and bracket. Disconnect preheater ground cable from tappet cover stud, then remove filter outlet pipe, hoses and preheater as a unit. Remove breather tube from tappet cover.

On all models, remove tappet cover and the rocker arm shaft oil line. Unbolt and remove rocker arms and shaft assembly. If push rods are removed, identify them so they can be reinstalled in their original position.

Disassemble rocker arms and shaft by removing the hair pin keepers and sliding parts off shaft. Rocker arm shaft diameter is 0.742-0.743 for all models. Reject diameter for all models is 0.740. Inside diameter of rocker arm bushings is 0.7445-0.7455 and rocker arms should have 0.0015-0.0035 operating clearance on shaft for all models. Renew rocker arms and/or shaft if clearance exceeds 0.0045. Rocker arm bushings are not available separately.

Rocker arm contact radius can be refaced providing original radius is maintained and kept parallel with rocker arm shaft.

Shaft springs should exert 10 pounds when compressed to a length of ⅝-inch.

## VALVE TIMING
### All Models

63. Valves are correctly timed when "C" mark on camshaft gear is meshed with an identical mark on the crankshaft gear as shown in Fig. 24.

## TIMING GEAR COVER
### All Models

64. To remove timing gear cover, first remove side panels and hood, then drain cooling system. Remove grille and radiator with oil coolers. Remove drive belts, water pump and alternator. Unbolt bracket of hydraulic oil circulating pump and move pump out of the way. Disconnect turbocharger oil return line from timing gear cover. On non-diesel engines, disconnect governor linkage at governor and remove governor. Remove cap screw from crankshaft pulley, thread two cap screws into threaded holes of crankshaft pulley, then attach a puller to the cap screws and remove pulley. Remove the cap screws which retain oil pan to timing gear cover. Loosen remaining oil pan cap screws, carefully separate oil pan gasket from timing gear cover and allow oil pan to drop. Unbolt and remove timing gear cover.

Crankshaft front oil seal can now be renewed. New seal is installed with lip facing rearward.

When reinstalling timing gear cover, use sealant on bottom side of cover where it mates with oil pan and gasket.

## TIMING GEARS
### All Models

65. Timing gear train on diesel engines consists of three helical gears and the injection pump drive gear. On non-diesel engines, two gears are used and the drive gear for the governor is driven by the camshaft gear. Gears are accessible for service after removing timing gear cover as outlined in paragraph 64.

Recommended gear backlash is 0.004-0.006. Renew all gears when backlash between any pair exceeds 0.007.

The cylinder block for all engines is designed for the installation of the idler gear, however, idler gear is used only with diesel engines. Plugs must be installed in either the idler gear shaft hole (non-diesel engines) or the governor gear hub hole (diesel engines) to block off the oil holes and maintain engine oil pressure. When servicing timing gears, note the lubrication tube which protrudes from crankcase above camshaft gear. Hole in tube is toward camshaft gear.

Remove camshaft gear from camshaft by using a puller attached to two cap screws threaded into gear. Crankshaft gear is removed in a similar manner. Avoid pulling gears with pullers which clamp or pull on the gear teeth.

Before installing, heat gears in oil to facilitate installation. When installing camshaft gear, remove oil pan and buck-up camshaft at one of the lobes near front end of shaft with a heavy bar. Gears are installed with "C" mark on camshaft meshed with a similar mark on crankshaft gear as shown in Fig. 24.

Diesel engine idler gear can be removed after removing the injection pump gear. However, before removing either gear, align timing marks of camshaft and crankshaft gear, then mark injection pump drive gear and hub so gear can be reinstalled in same position.

Operating clearance of idler gear shaft (spindle) in sleeve is 0.0015-0.003 and if clearance exceeds 0.005, renew the sleeve and bushings assembly as a unit as parts are not available separately. Outside diameter of gear shaft (spindle) is 0.999-1.000. Bushings in new sleeve may require final sizing to provide the desired 0.0015-0.003 operating clearance. Thrust spring should test 6¼-8½ pounds at 31/32-inch.

## CAMSHAFT
### All Models

66. To remove camshaft, first remove timing gear cover as outlined in paragraph 64. On non-diesel engines, remove distributor and on diesel engines, remove tachourmeter drive. On all models, remove fuel pump. Remove tappet cover, rocker arms and shaft assembly and push rods. Identify push rods so they can be reinstalled in their original position. Remove oil pan and oil pump, then support valve lifters and pull camshaft and gear from cylinder block. Use caution when withdrawing camshaft not to damage lobes of shaft and bushings in crankcase. Cam gear can be either pressed or pulled from camshaft.

Camshaft journal diameter is as follows:

**Non-Diesel**
No. 1 . . . . . . . . . . . . . . . . . . . . .1.749-1.750
No. 2-3-4 . . . . . . . . . . . . . 1.7485-1.7495
**Diesel**
No. 1-2-3-4 . . . . . . . . . . 1.7485-1.7495

Recommended operating clearance of camshaft journals in bushings is 0.0015-0.003 for No. 1 journal of non-diesel engines, and 0.002-0.0035 for Nos. 2, 3 and 4 journals of non-diesel engines and all journals of diesel engines. Maximum allowable operating clearance of non-diesel front journal is

0.0045. Maximum allowable clearance for all other journals is 0.005.

Camshaft bushings are pre-sized and when installing, use a closely piloted driver and be sure oil holes in bushings align with those in cylinder block. Bushing inside diameter is 1.7515-1.7520 with a maximum allowable inside diameter of 1.7560.

Camshaft end play is controlled by a spring loaded thrust button.

Thrust button spring has a free length of 1 3/16-inches and should test 15½-18½ lbs. when compressed to a length of 25/32-inch.

If gear was removed from camshaft, be sure "C" mark is toward front when gear is installed.

## ROD AND PISTON UNITS
### All Models

67. Connecting rod and piston assemblies are removed from above after removing cylinder head, oil pan and oil pump.

Prior to removal of rod and piston units, check to be sure that units are identified to insure correct installation. If necessary, affix cylinder number on front side of piston and on camshaft side of connecting rod and rod cap.

When reinstalling units, tighten connecting rod bolts to 44-46 ft.-lbs. torque for all models.

## PISTONS, SLEEVES AND RINGS
### All Models

68. **PISTONS AND SLEEVES.** All engines are fitted with wet type sleeves which are sealed at lower end by two packing rings and at upper end by the cylinder head gasket. With cylinder head and oil pan off and the connecting rod and piston assemblies removed, the sleeves can be pulled from cylinder block using a sleeve puller with the proper size sleeve attachment.

Specifications for sleeves are as follows:

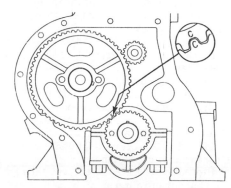

*Fig. 24–Valves are correctly timed when timing marks are aligned as shown.*

Inside diameter, std. . . . 3.8750-3.8765
　Max. allowable . . . . . . . . . . 3.8775
Out-of-round, max. . . . . . . . . . . . 0.002
Taper, max. . . . . . . . . . . . . . . . . 0.004
Projection above crankcase 0.001-0.004
Flange thickness . . . . . . . . 0.251-0.252
Counterbore depth . . . . . . .0.248-0.250

Prior to sleeve installation, check the sleeve stand-out as follows: Carefully clean sleeve and sleeve bore and pay particular attention to the taper of the sleeve lower bore in crankcase as well as the sleeve flange counterbore. Insert sleeve WITHOUT seal rings in cylinder block and clamp sleeve in position using short screws and flat washers. Mount a dial indicator, place button of indicator on the outer land of sleeve flange and zero indicator. Move indicator button from outer land of sleeve to surface of cylinder block. Dial indicator should show a sleeve stand-out of 0.001-0.004. If sleeve stand-out is less than 0.001, add a shim under sleeve flange. If sleeve stand-out is more than 0.004, check sleeve flange and its counterbore to be sure no foreign material is present and if both are perfectly clean and stand-out still exceeds 0.004, change sleeve.

69. With sleeve stand-out determined, install sleeves as follows: Coat sleeve sealing rings lightly with Lubriplate, or equivalent, and place on lower end of sleeve. Slide a small punch under seal rings and move punch around sleeve several times to eliminate all twists from sealing rings. Insert sleeve with seals into cylinder block and use hand pressure to position sleeves. Sleeves should go into position with only moderate hand pressure. If they do not, remove the sleeve and determine the cause.

70. Desired piston clearance in sleeve for all 1755 models and 1855 non-diesel models requires 3-5 lbs. pull on a spring scale to withdraw a ½ x 0.002 feeler gage positioned 90 degrees from piston pin. 1855 and 1955 diesel models require 3-6 lbs. pull on the ½ x 0.002 feeler gage.

71. **RINGS.** Pistons of all 1755, 1855 and 1955 engines are fitted with two compression rings and one oil ring. Oil ring for Series 1755 and 1855 non-diesel engines is a three-piece type being made up of two chrome faced rails and a spacer.

Side clearance for top compression ring in all Series 1755 and Series 1855 non-diesel is 0.0025-0.004. Top compression ring for Series 1855 and 1955 diesel is a keystone type.

Side clearance for second compression ring in all engines except 1955 diesel is 0.002-0.004. Side clearance for second compression ring in Series

1955 engine is 0.002-0.0038.

Maximum side clearance for all compression rings is 0.006.

Side clearance for oil rings in all engines except Series 1755 and 1855 non-diesel is 0.0015-0.003. Side clearance values for the three-piece oil rings used in Series 1755 and 1855 non-diesel are not applicable.

Maximum side clearance for all oil rings (except three-piece) is 0.006.

End gap for top compression ring for all Series 1755 and 1855 non-diesel is 0.010-0.020, or 0.014-0.020 for Series 1855 and 1955 diesel. End gap for second compression ring is 0.010-0.020 for all models except 1955 diesel which is 0.014-0.020. End gap for oil ring in 1755 and 1955 diesel is 0.010-0.023, or 0.010-0.018 for Series 1855 diesel. Oil ring rail end gap on Series 1755 and 1855 non-diesel should be 0.015-0.055.

## PISTON PINS AND BUSHINGS
**All Models**

72. The 1.2494-1.2497 diameter full floating type piston pin is retained in the piston bosses by snap rings and in addition to standard size, is available in 0.005 oversize.

Two bushings are fitted in piston pin end of connecting rod and are installed with outer edges flush with outer edges of rod bore and cut-away portions of bushings are aligned with oil hole in rod. Bushings are honed after installation to provide the piston pin with a thumb press fit at room temperature with pin and bushings dry (0.0004-0.0009). If oversize piston pin is being used, it will be necessary to also hone the piston bosses to provide 0.0002-0.0004 clearance for pin. Maximum allowable clearance of piston pin is 0.0019 in rod and 0.0016 in piston.

## CONNECTING RODS AND BEARINGS
**All Models**

73. Connecting rod bearings are of the non-adjustable, insert type and can be removed after removing oil pan and connecting rod caps. When installing new bearings, make certain the bearing insert projections fit into the slot in connecting rod and cap and that cylinder numbers are in register and face toward camshaft side of engine.

NOTE: Affix cylinder numbers to connecting rod and cap prior to removal, if necessary.

Bearing inserts are available in standard sizes as well as undersizes of 0.003 and 0.020.

Check the crankshaft crankpin and the bearing inserts against the values

which follow:

Crankpin diameter .... 2.4365-2.4375
Rod bearing running
  clearance .......... 0.0005-0.0015
Maximum running clearance .. 0.0025
Connecting rod side
  play .............. 0.0075-0.0135
Connecting rod nut
  torque—ft.-lbs. ............. 44-46

## CRANKSHAFT AND BEARINGS
**All Models**

74. Crankshaft is supported in seven non-adjustable insert type bearings which can be removed from below without removing crankshaft. The normal crankshaft end play of 0.0045-0.0095 is controlled by the flanged number five main bearing. Maximum allowable crankshaft end play is 0.011 and is corrected by renewing the number five main bearing. Bearings are available in standard size as well as undersizes of 0.003 and 0.020.

NOTE: If excessive crankshaft end play is present after installation of a new standard or 0.003 undersize thrust bearing, a 0.020 undersize bearing which is 0.017 overwidth is available. To utilize this bearing, grind crankshaft thrust bearing journal to 0.020 undersize and 0.017 overwidth.

To remove crankshaft, first remove engine as outlined in paragraph 55, then on all models remove clutch housing (with auxiliary drive if so equipped), clutch, flywheel, oil pan, rear oil seal and retainer and timing gear cover. On diesel models, remove injection pump drive gear and shaft and the idler gear assembly, then unbolt injection pump from engine front plate and let the attached fuel lines support injection pump. Unbolt and remove the engine front plate which in some cases, may involve removing the dowels. Remove connecting rod caps and main bearing caps and lift crankshaft from engine.

Check crankshaft and main bearings against the values which follow:
Crankpin diameter,
  standard .......... 2.4365-2.4375
Main journal diameter,
  standard ............. 2.624-2.625
Main bearing running
  clearance .......... 0.0015-0.0045
Main bearing running clearance,
  max. .................... 0.0055
Crankshaft end play,
  desired ............ 0.0045-0.0095
Crankshaft end play, max.
  allowable ................ 0.011
Journal out-of-round or
  taper, max. .............. 0.0003
Flywheel flange run-out ....... 0.001

Main bearing bolt torque—
  ft.-lbs. ................. 129-133

When reassembling engine, refer to paragraph 64 for information on idler gear and to paragraph 65 for information on injection pump drive shaft and gear.

## CRANKSHAFT OIL SEALS
**All Models**

75. FRONT SEAL. After removing the timing gear cover as outlined in paragraph 64, front oil seal can be pressed from timing gear cover.

New seal is installed with lip toward rear. Seal is installed 3/16-inch (non-diesel) or 3/8-inch (diesel), below outer edge of seal bore. If crankshaft pulley surface is grooved from old seal, install new seal an additional 1/32-inch deeper. Lubricate seal prior to installing crankshaft pulley.

76. REAR SEAL. On direct drive models, the crankshaft rear oil seal can be renewed without removing engine from tractor. On models with auxiliary drive, remove engine as outlined in paragraph 55.

To renew crankshaft rear oil seal, remove flywheel as outlined in paragraph 77, then drain and remove oil pan. Remove the oil seal retainer cap screws and carefully pry retainer loose from locating dowels and gasket. Carefully press new seal into retainer with lip forward. Reinstall retainer with new gasket, tighten the retaining cap screws, then install oil pan with new gasket. If crankshaft flange is grooved, add another gasket behind retainer so seal will run on a new surface.

## FLYWHEEL
**All Models**

77. On direct drive models, flywheel can be removed with engine in tractor by removing the clutch shaft, flywheel housing and clutch assembly.

On models with auxiliary drive, remove engine from tractor, then remove auxiliary drive unit and clutch housing and the clutch assembly.

With clutch removed, remove the cap screws retaining the pto drive hub and flywheel to crankshaft and lift off flywheel.

NOTE: Use caution not to allow flywheel to drop as flywheel cap screws are removed.

To install a new flywheel ring gear, heat gear evenly to a temperature of 400° F. to 450° F. and install gear with beveled end of teeth toward front of flywheel. If necessary to renew the timing tape, thoroughly clean flywheel, peel backing from tape and apply tape to flywheel with edge of tape even with edge of flywheel and the TDC marks on flywheel and tape aligned.

When installing flywheel and pto drive hub, coat flywheel cap screws with Loctite sealant C or CV and tighten the retaining cap screws to a torque of 67-69 ft.-lbs. After flywheel is installed, check flywheel run-out which should not exceed 0.005.

### OIL PUMP

#### All Models

78. The gear type oil pump shown in exploded view Fig. 25 is used in all engines. To remove pump, drain and remove oil pan, then remove the two pump retaining cap screws and withdraw pump from bottom of engine crankcase.

When disassembling pump, place alignment marks on adjusting screw (2) and pump body (8) prior to removing screw, then count the number of turns required to remove the screw. Record the number of turns to assure correct reassembly. Also, place assembly marks on the drive gear (5) and shaft (10) in the event that the original gear and shaft are reinstalled.

Refer to the following specifications when overhauling pump:

Relief valve setting . . . . . . . . . . . . 70 psi
Relief valve spring
   free length—in. . . . . . . . . . . . 1 15/16
Relief valve plunger
   diameter . . . . . . . . . . . . . . 0.745-0.747

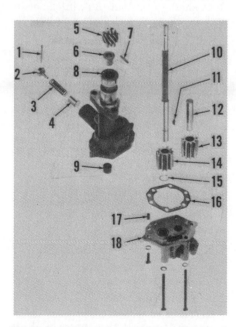

*Fig. 25—Exploded view of oil pump used on all engines.*

| | |
|---|---|
| 1. Cotter pin | 10. Drive shaft |
| 2. Adjusting screw | 11. Woodruff key |
| 3. Relief valve spring | 12. Idler shaft |
| 4. Relief valve plunger | 13. Idler gear |
| 5. Drive gear | 14. Driven gear |
| 6. Bushing | 15. Snap ring |
| 7. Spring pin | 16. Gasket |
| 8. Pump body | 17. Dowel |
| 9. Bushing | 18. Cover |

Idler gear to shaft
   clearance . . . . . . . . . . 0.0035-0.0045
   Max. allowable . . . . . . . . . . . . . 0.006
Drive shaft to bushing clearance:
   Upper bushing . . . . . . . .0.001-0.0025
   Max. Allowable . . . . . . . . . . . . . 0.005
   Lower bushing . . . . . . 0.0015-0.0025
   Max. allowable . . . . . . . . . . . . . 0.005
Drive gear to body
   clearance . . . . . . . . . . . . . .0.004-0.008

If installing new drive shaft upper bushing (6), be sure hole in bushing is aligned with hole in pump body (8). If installing new drive gear (5) and shaft (10), press gear onto shaft so that drive gear to body clearance is 0.004, then drill a 3/16-inch hole through shaft and install spring pin (7).

When assembling relief valve, thread adjusting screw in same number of turns as required to remove screw, align the previously affixed marks and install cotter pin (1). NOTE: If new screw (2) or spring (3) is being installed, or if original position of screw is not known, install screw flush with pump body, then turn screw in to the first position that cotter pin can be installed.

79. When installing oil pump on non-diesel engines, remove distributor and turn engine to where No. 1 piston is at TDC on compression stroke. Then, install oil pump so that slot in top of drive gear is positioned as shown in Fig. 26. Reinstall distributor and time ignition as outlined in paragraph 130.

80. When installing oil pump on diesel engines, remove the tachourmeter drive unit, then install oil pump. Insert tachourmeter drive shaft into slot in top of drive gear and install tachourmeter drive housing and retaining clamps.

### MAIN OIL GALLERY PRESSURE RELIEF VALVE

#### All Models

81. Oil pressure on all models is controlled by a spring loaded poppet type relief valve located at forward end of main oil gallery as shown in Fig. 26. To check engine oil pressure, remove plug from oil gallery and install a master oil pressure gage. With engine at normal operating temperature, oil pressure should be 10 (gas) or 20 (dsl) psi at idle speed and 25 (gas) or 30-35 (dsl) psi at 2400 rpm.

If pressure is not as specified, gallery oil pressure relief valve should be removed, cleaned and inspected. The 0.497-0.498 diameter plunger should slide freely in bore of cylinder block. Spring has a free length of 1⅜-inches and should exert 10 lbs. pressure when

compressed to a length of one inch.

If after reinstalling the main gallery relief valve the oil pressure is still not as specified, a faulty oil pump relief (safety) valve should be suspected.

### ENGINE OIL COOLER

#### Series 1955

82. The 1955 model tractors are equipped with an engine oil cooler which is mounted on lower right side of engine. See Fig. 27 for an exploded view of the unit. Also shown is the by-pass filter which is mounted on right rear of cylinder block.

83. To remove engine oil cooler, proceed as follows: Drain cooling system and cooler. Remove cooler to thermostat housing line. Disconnect cooler to cylinder head line and turbocharger oil supply line from cooler. Remove retaining cap screws and separate cooler and filter assembly from engine.

NOTE: The engine oil filter relief valve and the engine oil cooler by-pass valve are located in bores in bottom of cylinder block and can be removed, if necessary, after engine oil pan and oil pan gasket are removed. The front valve is the oil filter relief valve and the valve spring is color coded white. The rear valve is the oil cooler by-pass valve and the valve spring is color coded blue.

84. With oil cooler unit removed, remove plug (23), spring (25) and oil filter by-pass piston (26). Note that spring (25) is color coded black. Clean and inspect all parts and be sure all oil passages are clean and open. If cooler (28) is defective, it is recommended that cooler be renewed rather than to attempt to repair it. Spring specifications are as follows: oil filter by-pass spring (black) has a free length of 3 11/16-inches and should test 4 pounds at 2⅝-inches. Filter relief valve spring (white) has a free length of 2.279 inches and should test 11.7 pounds at 1⅜-inches. Cooler by-pass spring (blue) has a free length of 1.8 inches

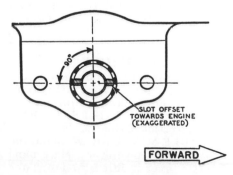

*Fig. 26—When installing oil pump in non-diesel engines, slot in pump drive gear must be positioned as shown. Refer to text.*

and should test 1.1 pounds at 1¼-inches.

85. Reinstall by reversing the removal procedure and when installing turbocharger oil pressure line, be sure check valve is installed with ball end forward and plug on top side. Ball can be seen by looking in end of check valve.

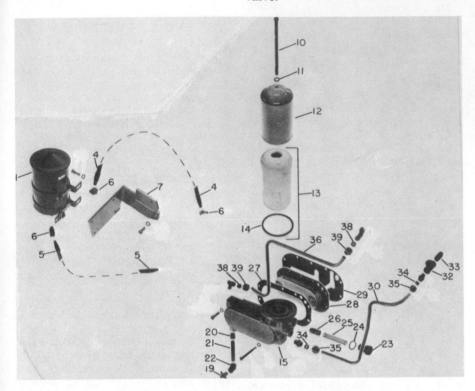

Fig. 27—Exploded view of engine oil cooler assembly used on 1955 models. Also shown is the full flow engine oil filter and the by-pass filter. Cooler assembly is mounted on lower right side of engine and by-pass filter (1) at upper right rear of cylinder block.

| | | | |
|---|---|---|---|
| 1. By-pass filter | 13. Filter element | 23. Plug | 30. Coolant line |
| 4. Hose | 14. Gasket | 24. Washer | 32. Tee |
| 5. Hose | 15. Base | 25. Valve spring | 33. Nipple |
| 6. Elbow | 19. Drain cock | 26. By-pass valve | 34. Sleeve |
| 7. Bracket | 20. Reducer | 27. Gasket | 35. Nut |
| 10. Bolt | 21. Nipple | 28. Cooler | 36. Coolant line |
| 11. Gasket | 22. Elbow | 29. Gasket | 38. Sleeve |
| 12. Shell | | | 39. Nut |

# CARBURETOR

### Series 1755-1855 Non-Diesel

86. Gasoline engines of Series 1755 and 1855 tractors are fitted with Marvel-Schebler carburetors. Series 1755 use model USX-44 while Series 1855 use model USX-37. Gasket kits and repair kits are available for service.

### ADJUSTMENTS

87. Carburetor adjustments are made with engine at operating temperature. For linkage adjustments, refer to the NON-DIESEL GOVERNOR section of this manual.

88. IDLE MIXTURE. With engine at operating temperature, place hand throttle lever in idle position. Turn idle mixture adjusting screw in until engine begins to lose speed, then turn screw out until engine again begins to lose speed. Adjust screw between these two extremes until highest engine idle speed is obtained. Use small increments when adjusting idle mixture screw, allow engine to stabilize after each adjustment and if necessary, flash engine occasionally to clear engine of excess fuel.

89. LOW IDLE. Be sure idle mixture adjustment is made before adjusting the engine low idle rpm.

Place hand throttle lever in low idle position and disconnect governor rod from governor spring lever. Hold governor spring lever in low idle position and adjust carburetor idle speed adjusting screw until engine low idle speed is 775-825 rpm. At this time, governor rod should enter hole in governor spring lever with no further

movement. If it does not, adjust turnbuckle on governor rod until governor rod enters hole in governor spring lever freely when hand throttle lever and carburetor are both in low idle position. Hand throttle lever and carburetor low idle speed screw must contact their stops simultaneously.

90. HIGH SPEED LOAD. Be sure idle mixture adjustment is made before making high speed load adjustment.

Place load on engine with a dynamometer and operate engine at about rated rpm. Turn high speed load adjusting screw in until engine begins to lose power, then back-out screw until full engine power is restored. Repeat this operation until a definite setting is established.

### OVERHAUL

91. With carburetor removed, disassemble as follows: Remove throttle plate (6—Fig. 28) and pull throttle shaft (16) from body. Remove retainer (15) and packing (14). Cup (66) need not be removed unless throttle shaft bearings are to be renewed. Remove choke plate (10) and pull choke shaft (25) from body. Remove bracket (31), retainer (32) and packing (33). Remove plug (62) and screen (64). Remove cover (52), then unbolt and remove housing (48), diaphragm (49) and spring (50) from cover. Remove float shaft (53), float (56), inlet needle and needle seat (45). Remove plug (43) and nozzle (41). Remove idle tube (39), power needle (60) and seat (58). Remove plug (37) and minimum fuel jet (36). Remove idle needle (61).

The economizer jet (34) is not subject to wear and should not require service. However, if for any reason renewal is required, it will be necessary to drill expansion plug (35) and pry it out before economizer jet can be removed.

The throttle shaft needle bearings (13) can be removed from body, if necessary, by using Marvel-Schebler tool M-504 or its equivalent. Refer to Fig. 29. Insert tool through throttle shaft bore, then spread split end of tool with a small screw driver so lips of tool will engage bearing shell and press bearing from its bore. Repeat operation for opposite side.

NOTE: Use a vise as a press to remove and install the throttle shaft bearings. Bearings are hardened and could break if driven.

The accelerator pump discharge jet (7—Fig. 28) can be checked prior to removal and if satisfactory, jet need not be removed. Check jet operation as follows: Connect a vacuum gage between two short lengths of plastic or rubber tubing. Be sure bowl cover is off, then place one tube end over the end of discharge jet which protrudes into throttle bore. See Fig. 30. Apply vacuum (suck) on opposite end of tube assembly and as ball check leaves seat, note gage reading which should be 3 inches Hg. Ball check leaving seat will be audible. If the jet check ball leaves its seat at less than 3 inches Hg. vacuum, renew the complete discharge jet as follows:

NOTE: If Marvel-Schebler tool M-505 (positioning tool) is not available, be sure to measure depth that discharge jet is installed in the drilled passage. This depth should be 1 15/32-inches.

Use a 5/32-inch rod and drive discharge jet out toward throttle bore end of drilled passage. To install new discharge jet, start jet into drilled passage and position slot of discharge end so it is within 5 degrees of being parellel with the centerline of carburetor

and deep end of slot is upward as shown in Fig. 31. Use Marvel-Schebler positioning tool M-505 and drive jet into position. If tool M-505 is not available, use the 5/32-inch rod to drive the jet to the 1 15/32-inch location.

Clean and inspect all parts. Use compressed air to blow out all passages EXCEPT the pump discharge jet passage. Compressed air can damage discharge jet.

Reassemble by reversing the disassembly procedure and adjust float level by either adding an additional gasket under inlet needle seat or bending float lever, until float lever is parallel with surface of body mounting flange. Note: Float level can also be set by removing float needle and observing the extreme positions of float travel, then adjusting float until it is midway between these positions when float needle is installed. After carburetor is assembled, make the following initial adjustments so engine

can be started and final adjustments made as outlined in paragraphs 88, 89 and 90. Power needle 3 turns open; idle mixture adjusting needle 1½ turns open and throttle stop screw 1½ turns clockwise from closed throttle position.

### FUEL PUMP
### Series 1755-1855 Non-Diesel

92. The fuel pump is mounted on right side of engine and is actuated by a lobe on the engine camshaft.

Diaphragm kits or repair kits are available for service. Overhaul of pump is conventional and will be obvious after an examination. When new pump valves are installed, be sure to stake them in position.

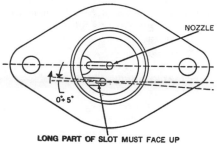

Fig. 30–Use method shown to check carburetor accelerator pump discharge jet. Refer to text.

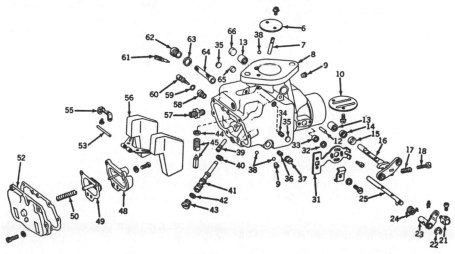

Fig. 28–Exploded view of a typical USX model Marvel-Schebler carburetor.

| | | |
|---|---|---|
| 6. Throttle plate | 31. Bracket | 48. Pump housing |
| 7. Accelerator pump jet | 32. Retainer | 49. Pump diaphragm |
| 8. Body | 33. Packing | 50. Spring |
| 9. Plug | 34. Economizer jet | 52. Bowl cover |
| 10. Choke plate | 35. Expansion plug | 53. Float shaft |
| 12. Stop pin | 36. Minimum fuel restriction | 55. Float bracket |
| 13. Needle bearings |  jet | 56. Float |
| 14. Packing | 37. Drain plug | 57. Connector |
| 15. Retainer | 38. Ball plugs | 58. Power needle seat |
| 16. Throttle shaft | 39. Idle jet | 59. "O" ring |
| 17. Spring | 40. Gasket | 60. Power adjusting needle |
| 18. Throttle stop screw | 41. Nozzle | 61. Idle mixture needle |
| 21. Swivel | 42. Gasket | 62. Plug |
| 22. Retainer ring | 43. Plug | 63. Gasket |
| 23. Choke lever | 44. Gasket | 64. Screen |
| 24. Spring | 45. Inlet needle and seat | 65. Choke shaft cup |
| 25. Choke shaft | | 66. Throttle shaft cup |

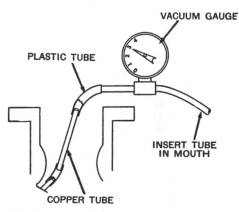

LONG PART OF SLOT MUST FACE UP

Fig. 31–View showing upper end of accelerator pump discharge jet after installation.

# DIESEL
# FUEL SYSTEM

The diesel fuel system consists of three basic units; the fuel filter, injection pump and injection nozzles. When servicing any unit associated with the fuel system, the maintenance of absolute cleanliness is of utmost importance.

Probably the most important precaution that servicing personnel can impart to owners of diesel powered tractors is to urge them to use an approved fuel that is absolutely clean

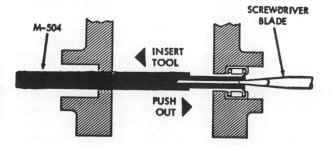

Fig. 29–Use removal tool and screwdriver as shown to press out throttle shaft bearings.

and free from foreign material. Extra precaution should be taken to make certain that no water enters the fuel storage tanks. This last precaution is based on the fact that all diesel fuels contain some sulphur. When water is mixed with sulphur, an acid is formed which will quickly erode the closely fitting parts of the injection pump and fuel injection nozzles.

## TROUBLE SHOOTING
### Series 1755-1855-1955 Diesel

93. If the engine will not start or does not run properly, refer to the following paragraphs for possible cause of trouble.

94. **ENGINE WILL NOT START.** If the engine will not start, possible causes of trouble are as follows:

Fuel tank empty.
Fuel supply valve closed.
Engine stop control applied or improperly adjusted.
Water in fuel.
Inferior fuel.
Clogged fuel filter.
Air traps in system.
Low cranking speed.
Pump installed out of time.
Pump shaft broken (pump seized).
Faulty or worn fuel injection pump.
Faulty injector nozzles.
Low engine compression.

95. **ENGINE DOES NOT RUN PROPERLY.** If the engine will start but does not run properly (smokes excessively, mis-fires, etc.), possible causes of trouble are as follows:

Water in fuel.
Inferior fuel.
Fuel filter clogged.
Air cleaner element clogged.
Air traps in fuel system.
Pump out of time.
Faulty fuel injectors.
Low compression on one or more cylinders.
Faulty fuel injection pump.

## BLEEDING THE SYSTEM
### Series 1755-1855-1955 Diesel

96. To bleed the diesel fuel system, proceed as follows: Open fuel tank shut-off valve, if necessary. Refer to Fig. 32 and remove the primary fuel filter bleed plug, then actuate priming lever of fuel supply pump until bubble free fuel flows from bleeder plug hole and reinstall bleeder plug.

NOTE: If supply pump priming lever cannot be actuated, turn engine until pump rocker arm is released by the camshaft.

With primary fuel filter free of air, loosen bleed plug on final fuel filter and actuate supply pump priming

lever until bubble free fuel appears, then tighten bleed plug. Continue to actuate supply pump priming lever for an additional fifteen or twenty strokes to force any remaining air in the low pressure system through the lines and into tank.

97. Attempt to start engine, and if engine fails to start, or runs unevenly, loosen lines at injectors and bleed air from them either by placing fuel stop in run position and operating starting motor, or allow engine to run at low idle. Tighten connections when operation is completed.

## INJECTOR NOZZLES
### Series 1755-1855-1955 Diesel

100. **LOCATING A FAULTY NOZZLE.** If rough or uneven engine operation, or misfiring, indicates a faulty injector, the defective unit can usually be located as follows:

With the engine operating at low idle speed, loosen the high pressure connection at each injector in turn. As in checking spark plugs, the faulty unit is the one which, when its line is loosened, least affects the running of the engine.

If a faulty nozzle is found and considerable time has elapsed since the injectors have been serviced, it is recommended that all injectors be removed and new or reconditioned units be installed, or the nozzles be serviced as outlined in the following paragraphs.

101. **REMOVE AND REINSTALL.** To remove an injector, remove hood and wash injector, lines and surrounding area with clean diesel fuel. On all models, remove exhaust manifold and on Series 1855 and 1955, leave turbocharger attached to manifold. Use hose clamp pliers to expand clamp and pull leak-off boot from injector. Disconnect high pressure line, then cap all openings. Remove cap screw from nozzle clamp and remove clamp and spacer. Pull injector from cylinder head.

NOTE: Unless the carbon stop seal has failed causing injector to stick, the injectors can be easily removed by

*Fig. 32—View showing fuel filter bleed plugs and primary lever located on fuel supply pump.*

hand. If injectors cannot be removed by hand use nozzle puller and be sure to pull injector straight out of bore. DO NOT attempt to pry injector from cylinder head or damage to injector could result.

When installing injector, be sure nozzle bore and seal washer seat are clean and free of carbon or other foreign material. Install new seal washer and carbon seal on injector and insert injector into its bore using a slight twisting motion. Install and align locating clamp, then install hold-down clamp and spacer and tighten cap screw to 20 ft.-lbs. torque.

Bleed system as outlined in paragraph 96 if necessary.

102. **TESTING.** A complete job of nozzle testing and adjusting requires the use of an approved nozzle tester. Only clean, approved testing oil should be used in the tester tank. The nozzle should be tested for spray pattern, opening pressure, seat leakage and back leakage (leak-off). Injector should produce a distinct audible chatter when being tested and cut off quickly at end of injection with a minimum of seat leakage.

NOTE: When checking spray pattern, turn nozzle about 30 degrees from vertical position. Spray is emitted from nozzle tip at an angle to the centerline of nozzle body and unless injector is angled, the spray may not be completely contained by the beaker. Keep your person clear of the nozzle spray.

103. **SPRAY PATTERN.** Attach injector to tester and operate tester at approximately 60 strokes per minute and observe the spray pattern. A finely atomized spray should emerge at each of the four nozzle holes and a distinct chatter should be heard as tester is operated. If spray is not symmetrical and is streaky, or if injector does not chatter, overhaul injector as outlined in paragraph 107.

104. **OPENING PRESSURE.** The correct opening pressure for series 1755 and 1855 is 2550-2650 psi for used nozzles, or 2750-2850 psi for new nozzles or when new spring has been installed. The correct opening pressure for series 1955 is 2750-2850 psi for used nozzles or 2950-3050 psi for new nozzles or when new spring has been installed. There should not be more than 100 psi difference between the injectors used in any engine except when intermixing new and used nozzle assemblies.

To check nozzle opening pressure, attach injector to tester and pump tester several times to actuate injector and allow valve to seat normally. Now observe tester gage and quickly pump up pressure until gage falls off rapidly. This indicates the nozzle opening pressure and should be within the pressure ranges given above.

If opening pressure is not correct but nozzle will pass all other tests, adjust opening pressure as follows: Loosen the pressure adjusting screw lock nut (13—Fig. 33), then hold the pressure adjusting screw (14) and back out the valve lift adjusting screw (11) at least one full turn. Actuate tester and adjust nozzle pressure by turning adjusting screw as required. With the correct nozzle opening pressure set, gently turn the valve lift adjusting

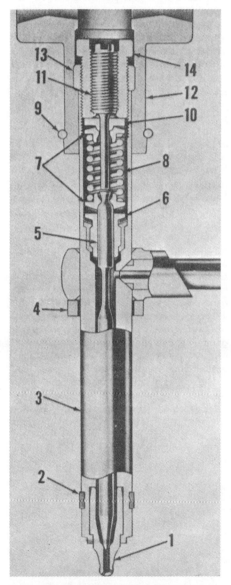

Fig. 33–Cross sectional view of injector used in Series 1755, 1855 and 1955. Nozzle tip (1) and valve guide (6) are parts of finished body and are not serviced separately.

| 1. Nozzle tip | 8. Pressure spring |
|---|---|
| 2. Carbon seal | 9. Boot clamp |
| 3. Nozzle body | 10. Ball washer |
| 4. Seal washer | 11. Lift adjusting screw |
| 5. Nozzle valve | 12. Boot |
| 6. Valve guide | 13. Lock nut |
| 7. Spring seat | 14. Pressure adjusting screw |

screw in until it bottoms, then on Series 1755 and 1855, back it out ½-turn which will provide the correct valve lift of 0.009. On Series 1955, back out the screw ¾-turn which will provide a valve lift of 0.0135. Hold pressure adjusting screw and tighten lock nut to a torque of 70-75 in.-lbs.

NOTE: A positive check can be made to see that the lift adjusting screw is bottomed by actuating tester until a pressure of 250 psi above nozzle opening pressure is obtained. Nozzle valve should not open.

**105. SEAT LEAKAGE.** To check nozzle seat leakage, proceed as follows: Attach injector on tester in a horizontal position. Raise pressure to approximately 2400 psi (1755 and 1855) or 2600 psi (1955), hold for 10 seconds and observe nozzle tip. A slight dampness is permissible but should a drop form in the 10 seconds, renew the injector or overhaul as outlined in paragraph 107.

**106. BACK LEAKAGE.** Attach injector to tester with tip slightly above horizontal. Raise and maintain pressure at approximately 1500 psi and observe leakage from return (top) end of injector. After first drop falls, the back leakage should be 3 to 10 drops every 30 seconds. If back leakage is excessive, renew injector or overhaul as outlined in paragraph 107.

**107. OVERHAUL.** First wash the unit in clean diesel fuel and blow off with clean, dry compressed air. Remove carbon stop seal and sealing washer. Clean carbon from spray tip using a brass wire brush. Also, clean carbon or other deposits from carbon seal groove in injector body. DO NOT use wire brush or other abrasives on the Teflon coating on outside of nozzle body between the seals. Teflon coating can be cleaned with a soft cloth and solvent. Coating may discolor from use, but discoloration is not harmful.

Clamp the nozzle in a soft jawed vise, loosen lock nut (13—Fig. 33) and remove pressure adjusting screw (14), ball washer (10), upper spring seat (7), spring (8) and lower spring seat (7).

If nozzle valve (5) will not slide from body when body is inverted, use a suitable valve extractor; or reinstall on nozzle tester with spring and lift adjusting screw removed, and use hydraulic pressure to remove the valve. See Fig. 34.

Nozzle valve and body are a matched set and should never be intermixed. Keep parts for one injector separate and immerse in clean diesel fuel in a compartmented pan, as unit is disassembled.

Clean all parts thoroughly in clean diesel fuel using a brass wire brush and lint-free wiping towels. Hard carbon or varnish can be loosened with a suitable, non-corrosive solvent.

Clean the spray tip orifices first with an 0.008 cleaning needle held in a pin vise; then with a 0.011 (1755 and 1855) or a 0.012 (1955) needle.

Clean the valve seat using a Valve Tip Scraper and light pressure while rotating scraper. Use a Sac Hole Drill to remove carbon from sac hole.

Piston area of valve and guide can be lightly polished by hand, if necessary, using Roosa Master No. 16489 lapping compound. Use the valve retractor to turn valve. Move valve in and out slightly while rotating, but do not apply down pressure while valve tip and seat are in contact.

If back leakage was not correct and polishing of valve does not bring unit within specifications, a new valve can be installed.

Valve and seat are ground to a slight interference angle. Seating areas may be cleaned up, if necessary, using a small amount of 16489 lapping compound, very light pressure and no more than 3-5 turns of valve on seat. Thoroughly flush all compound from valve body after polishing.

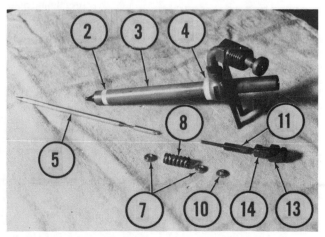

Fig. 34–Disassembled view of injector. Refer to Fig. 33 for parts identification.

NOTE: Never lap a new valve to an old injector body seat. If lapping with the old valve did not restore seat, lapping with a new valve will only destroy angle of the new valve.

When assembling the nozzle, back lift adjusting screw (11—Fig. 33) several turns out of pressure adjusting screw (14), and reverse disassembly procedure using Fig. 33 as a guide.

Connect nozzle to tester and turn pressure adjusting screw (14) into nozzle body until opening pressure is 2550-2650 psi with a used spring, or 2750-2850 psi with a new spring for series 1755 and 1855, or 2750-2850 psi with a used spring, or 2950-3050 psi with a new spring for series 1955.

After spring pressure has been adjusted and before tightening lock nut (13), valve lift must be adjusted as follows: Hold pressure adjusting screw from turning and using a small screwdriver, thread lift adjusting screw (11) into nozzle unit until it bottoms on valve (5). To be sure screw is bottomed, actuate tester handle until a pressure 250 psi above opening pressure is obtained. Nozzle valve should not open.

With lift adjusting screw bottomed, turn screw counter-clockwise ½-turn to establish the recommended 0.009 valve lift for Series 1755 and 1855 or ¾-turn to establish the recommended 0.0135 valve lift for Series 1955. Tighten the lock nut while holding pressure adjusting screw from turning. Recheck the opening pressure; then recheck spray pattern, seat leakage and back leakage. Install the unit as outlined in paragraph 101.

NOTE: When adjusting opening pressure and injector has not been disassembled, back out the lift adjusting screw at least one full turn before moving pressure adjusting screw. This will prevent accidental bottoming of lift screw while attempting to set the pressure.

## FUEL INJECTION PUMP
### Series 1755-1855-1955

108. All diesel model tractors are equipped with Roosa-Master fuel injection pumps. Following is the tractor and injection pump application. Series 1755, DBGFC633-23DH; Series 1855, DBGFC633-27DH; Series 1955, DBGFC637-12DH.

Because of the special equipment needed, and the skill required of the servicing personnel, service of injection pumps is generally beyond the scope of that which should be attempted in the average shop. Therefore, this section will include only timing of pump to engine, removal and installation and the linkage adjust-

ments which control engine speeds.

If additional service is required, the injection pump should be turned over to an Oliver facility that is equipped for injection pump service, or to an authorized Roosa-Master diesel service station. Inexperienced personnel should never attempt to service diesel fuel injection pumps.

109. **PUMP TIMING.** To check injection pump timing, shut off fuel supply valve and remove timing window cover from injection pump. Turn engine until No. 1 piston is coming up on compression stroke, then continue to turn engine slowly until the 4 degree BTDC for Series 1755 and 1855, or the 2 degree BTDC for Series 1955, timing mark on engine flywheel aligns with timing pointer.

With engine flywheel positioned as indicated above, the timing lines in injection pump timing window should be exactly aligned as shown in Fig. 35. If pump timing lines are not in register as shown, loosen the pump mounting stud nuts and turn pump as required to bring lines into register, then tighten mounting stud nuts securely. Recheck setting by turning engine two complete revolutions forward; or turn engine about one-half turn backward, then forward again, until the correct flywheel timing mark is again aligned with timing pointer and recheck the timing marks in injection pump timing window. Readjust timing if necessary.

CAUTION: Do not loosen the pump mounting stud nuts while engine is running if timing is being set by dynamometer method.

NOTE: When aligning flywheel timing mark with pointer, engine must be turned in normal direction of rotation. If mark is turned past pointer, continue to turn the engine through two revolutions to again align mark and pointer, or back engine up at least ¼-turn and then turn forward to timing mark. Turning the engine backwards to just the point where timing mark is aligned with pointer will result in incorrect timing due to gear backlash.

110. **R&R PUMP.** Thoroughly clean outside of pump, lines and connections. Shut off the fuel supply valve, remove timing window cover from fuel injection pump and turn engine so that both timing marks are visible in timing window and the flywheel timing mark given in paragraph 109 is aligned with pointer. Then reinstall injection pump timing window cover.

Disconnect battery ground strap to prevent engine from being turned accidentally while pump is removed; then, proceed as follows: Disconnect throttle rod, stop control wire and cable, fuel supply line and excess fuel return line from fuel injection pump. Remove the injector pressure line clamps and disconnect the lines at connections near pump end of lines. Loosen the injector pressure line connections at injectors so lines can be spread to remove pump and immediately cap all open lines and connectors. Remove the three pump mounting stud nuts and remove pump from mounting studs and drive shaft, spreading the injector lines to allow pump to be moved rearward. After removing pump, remove seals from pump drive shaft and carefully inspect the shaft. If shaft is broken at the safety groove or is otherwise damaged, renew shaft as outlined in paragraph 111.

NOTE: Aligning flywheel and pump timing marks prior to removing pump is not required procedure, but is an aid in reinstalling pump. If pump drive shaft is removed or renewed in conjunction with pump removal, either reinstall pump without regard to timing procedure and reinstall drive shaft as outlined in paragraph 112, or install shaft in pump and reinstall pump and shaft as assembly as outlined in paragraph 113.

Before reinstalling pump, check to be sure engine was not turned while pump was removed. If engine was turned, or pump was removed without aligning timing marks as outlined in removal procedure, proceed as follows: Either remove the rocker arm cover

*Fig. 35—When No. 1 cylinder is on compression stroke and correct flywheel timing mark is aligned with pointer, injection pump timing marks should be aligned as shown.*

and turn engine until intake valve of No. 1 cylinder just closes or loosen the No. 1 cylinder injector mounting stud nuts and turn engine until compression leak occurs around the loosened injector. Then, continue to turn engine slowly until the flywheel timing mark given in paragraph 109 is aligned with pointer and reinstall the rocker arm cover, or remove the No. 1 cylinder injector and reinstall same with new sealing washer.

Remove the timing window cover from outer side (throttle control rod side) of fuel injection pump and with a screwdriver inserted in rotor drive slot, turn pump rotor so that pump timing marks are aligned. Install two new pump drive shaft seals on shaft with lips of seals facing away from each other. Lubricate the seals and shaft area between seals with Lubriplate or equivalent grease and carefully install the pump over drive shaft and seals. Take care not to roll lip of rear seal back as pump is being installed; if this happens, remove pump and renew the seal (Note: Use of Roosa-Master seal installation tool 13369 is recommended to install seals on shaft and use of Roosa-Master seal compressor tool 13371 is recommended when installing pump over the drive shaft seals). When pump is installed over drive shaft far enough to engage drive shaft tang with slot in pump rotor, it may be necessary to rock the rotor slightly with pencil eraser tip inserted through timing window or to rock the pump assembly slightly to engage the drive shaft tang and rotor slot. When pump mounting slots are engaged with the mounting studs and pump body contacts mounting plate, rotate the pump assembly in the mounting slots as required to bring the pump timing marks into exact alignment, then securely tighten the pump mounting nuts. Turn the engine through two revolutions, or back engine up at least ¼-turn, then slowly turn forward until flywheel timing mark is again aligned with pointer and recheck pump timing marks. Loosen the pump mounting stud nuts and adjust timing if necessary.

Bleed the fuel system as outlined in paragraph 96.

**111. RENEW PUMP DRIVE SHAFT.** To remove the pump drive shaft, gear and hub assembly, remove the gear cover from front side of timing gear cover. If injection pump is installed, shut off the fuel supply valve and remove the timing window cover from outer side of injection pump to drain fuel; otherwise, diesel fuel will run into engine when shaft is

removed. Then, withdraw the drive gear, hub and shaft assembly from opening in front of timing gear cover. Cut the locking wire and remove the drilled head cap screws retaining gear to hub. Clamp tang of drive shaft in soft jawed vise and remove the hub retaining nut. Pull or press hub from shaft and Woodruff key. If fuel injection pump was not removed, install new drive shaft as outlined in paragraph 112. If pump is removed, new shaft can be installed in pump, then install pump and shaft as outlined in paragraph 113; or if desired pump can be reinstalled without shaft and the shaft can then be installed as in paragraph 112.

112. To install new drive shaft with pump installed on tractor, proceed as follows: Check to see that injection pump is mounted so that mounting studs are centered in slots; if not, loosen the stud nuts, turn pump so that studs are centered in slots and tighten the nuts. Turn engine until No. 1 piston starts up on compression stroke, then continue to turn engine slowly until flywheel timing mark given in paragraph 109 is aligned with pointer. Clamp tang of drive shaft in soft jawed vise, install Woodruff key in slot and install drive hub with retaining nut and washer. Tighten the nut to a torque of 35-40 ft.-lbs. Remove shaft from vise and install new drive shaft seals with seal lips facing away from each other. Use of Roosa-Master seal installation tool 13369 is recommended. Lubricate the seals and shaft area between seals with Lubriplate or similar grease. Note the offset dimples in end of drive tang on shaft and in slot of pump rotor. Carefully insert shaft into pump with offset dimples on shaft tang and rotor slot aligned and work the rear seal lip into bore of pump body with a dull pointed tool. When drive tang is engaged in rotor slot, remove timing window cover from fuel injection pump and turn the drive shaft and hub so that pump timing marks are aligned as shown in Fig. 35. While holding pump rotor from turning with screwdriver inserted in slot below the timing marks in timing window, install drive gear on hub and turn gear counter-clockwise as far as possible to eliminate gear backlash. Install the two drilled head cap screws in set of two holes that are aligned in gear and hub, tighten the capscrews securely and install locking wire. Insert thrust spring and thrust button in front end of drive shaft and install drive gear cover to timing gear cover with new gasket. Recheck pump timing marks and adjust if necessary

as outlined in paragraph 110. Bleed diesel fuel system as in paragraph 96.

113. To install new drive shaft when fuel injection pump is removed, either reinstall pump and then install shaft as outlined in paragraph 112, or proceed as follows:

Remove timing window cover from outer side of fuel injection pump. Install new seals on drive shaft with lips of seals facing away from each other. Use of Roosa-Master seal installation tool 13369 is recommended. Lubricate seals and shaft area between seals with Lubriplate or similar grease, then carefully insert shaft into pump so that offset dimple in end of shaft tang is aligned with offset dimple in slot of pump rotor. Using a dull pointed tool, carefully work lip of rear seal into bore of pump housing or compress seal with Roosa-Master seal compressor tool 13371 when installing shaft. With drive tang of shaft engaged in slot of pump rotor, install the pump on tractor with mounting hole slots centered on the mounting studs. Insert Woodruff key in slot in pump shaft. Be sure that drive gear is a free fit on hub, then install hub, washer and retaining nut on shaft. Temporarily install drive gear on hub in any position and with the drilled head cap screws finger tight. Then, tighten the hub retaining nut to a torque of 35-40 ft.-lbs. and remove the drive gear. Turn engine so that No. 1 piston is coming up on compression stroke, then continue to turn engine slowly until the flywheel timing mark given in paragraph 109 is aligned with pointer. Turn injection pump drive shaft so that the pump timing marks are aligned as shown in Fig. 35. While holding pump from turning with screwdriver inserted in slot below pump timing marks, install pump drive gear on hub and turn gear as far counter-clockwise as possible to eliminate all gear backlash. Install the two drilled head cap screws into a set of holes that are aligned in drive gear and hub, tighten the capscrews and secure with wire. Insert the thrust botton spring and thrust button in front end of pump drive shaft and install drive gear cover to timing gear cover using a new gasket. Recheck pump timing as outlined in paragraph 109 and readjust if necessary. Bleed the diesel fuel system as outlined in paragraph 96.

**114. GOVERNOR ADJUSTMENT.** Both the high idle speed and low idle speed adjustments are preset at the factory and normally should not be disturbed. However, should it become necessary, adjust the governed speeds as follows:

Start engine and bring to normal operating temperature. Break the seals on the high idle (rear) adjusting screw and/or low idle adjusting screw as necessary, then proceed as follows:

To adjust high idle speed, place throttle lever in high idle position, loosen lock nut and turn adjusting screw in or out as required to obtain a high idle speed of 2640 rpm for all models. Tighten the lock nut and install new wire seal.

To adjust low idle speed, position throttle lever at low idle position, loosen lock nut and turn adjusting screw (front screw) in or out to obtain a low idle speed of 800 rpm for all models. Tighten lock nut and install new wire seal.

NOTE. The throttle control lever on fuel injection pump is spring loaded and should slightly over-travel the high and low speed stops on the injection pump governor control shaft. To properly adjust the throttle linkage, proceed as follows: With engine not running, disconnect throttle control rod at fuel injection pump and place hand throttle lever in slow idle position. Loosen the lock nut on front end of control rod. While holding the injection pump throttle shaft against the low idle speed stop, adjust length of control rod so that injection pump throttle lever must be moved 1/16 to ⅛-inch against spring pressure to reinstall the linkage pin. Then tighten lock nut on control rod and move the hand throttle lever to high idle speed position. Loosen the lock nut on stop screw in fuel tank support, adjust the stop screw until the injection pump lever over-travels high idle speed stop 1/16 to ⅛-inch and then tighten the lock nut.

If throttle hand lever tends to creep at high idle speed, tighten the friction adjusting nut located at lever end of control lever shaft in instrument panel support.

### FUEL PUMP
**Series 1755-1855-1955**

115. The fuel pump is mounted on right side of engine and is actuated by a lobe on the engine camshaft.

Diaphragm kits or repair kits are available for service. Overhaul of pump is conventional and will be obvious after an examination. When new pump valves are installed, be sure to stake them in position.

### MANIFOLD PREHEATER
**Series 1755-1855-1955**

116. The inlet manifold of Series 1755 diesel engines is fitted with a solenoid controlled preheater which is installed in lower side of manifold. The preheater unit for Series 1855 and 1955 diesel engines is located in the turbocharger to inlet manifold pipe.

Operation of the preheater can be checked by depressing the control switch located on instrument panel for about 30 seconds and then touching manifold, or inlet pipe, to see if it is warm.

Service on preheater and solenoid is accomplished by renewing the units.

### TURBOCHARGER
**Series 1855-1955**

117. The Series 1855 and 1955 tractors are equipped with a turbocharger which is mounted on the top forward portion of the engine exhaust manifold. Turbocharger is pressure lubricated from the engine and with engine running at governed speed, the oil pressure to turbocharger should be within 10 psi of the engine oil pressure and a minimum of 10 psi should be maintained at engine low idle rpm. In addition to proper lubrication, it is also extremely important that no leaks or restrictions exist in the air induction system. Because of the critical tolerances machined into the turbocharger, and the high speeds of the rotating members, any dirt or other debris entering the turbocharger, either in the incoming air or the lubricating oil, can quickly cause the unit to fail.

Regular inspections should be made of the turbocharger and related components. Run engine and check for leaks in the air inlet system and exhaust system. Leaks can often produce noise. If noise is present and no leaks are evident, check rotating members of turbocharger for interference with housing and for faulty bearings. If oil is present on compressor wheel or in housing, oil is probably being pulled or pushed from center housing and faulty seals and/or bearings are indicated.

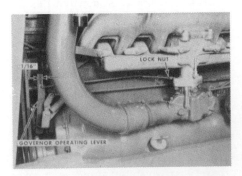

*Fig. 36—Adjust rod until end is 1/16-inch forward of hole in governor lever. Refer to text.*

Turbocharger rotating members must spin freely with no drag. Be sure unit is being properly lubricated as indicated in preceding paragraph; also be sure the check valve of oil inlet line is operating correctly to prevent oil drain back of the oil inlet line.

Certain operating procedures must be observed with a turbocharged engine that are not normally considered with a naturally aspirated engine and the following precautions should be taken.

1. If engine oil filters have been changed, oil lines disconnected, service performed on turbocharger, or if engine has been shut down for a considerable length of time, pull fuel shut-off (no fuel position) and crank engine until oil pressure registers on gage. Start engine and run at about 1000 rpm until oil pressure is normal. Do not run engine at high rpm until gauge registers at least 20 psi.

2. Prior to engine shut-down, let engine idle for 3-5 minutes to allow turbocharger to settle out. Immediate shut-down of engine could result in oil starvation of turbocharger bearings.

3. If engine has been serviced, or turbocharger has been removed, the Oliver Corporation recommends that engine be run for about one hour before installing turbocharger. Remove oil pressure line from engine and plug hole. Clean line and check valve. This will assure cleanliness of lubricating oil and exhaust system.

NOTE: At this time, no parts are being furnished to service turbocharger and any service required will be done through a factory rebuild exchange program. Rebuilt units will carry new parts warranty. Refer to paragraph 118 for removal and reinstallation procedure.

118. **REMOVE AND REINSTALL.** To remove turbocharger, first remove hood side panels and hood, then disconnect air inlet hose from turbocharger. Unbolt and remove muffler elbow and adapter ring. Disconnect oil inlet and outlet lines from turbocharger and cap lines. Unbolt and remove turbocharger from adapter block. Remove and discard gasket.

Prior to installing turbocharger, refer to item #3 of paragraph 117. Install turbocharger as follows: Invert turbocharger and pour oil into return line port to fill center housing. Turn rotating member to coat bearings and thrust washer with oil. Be sure oil pressure line is clean, then fill line with oil. Install new gasket on adapter block, coat threads of mounting studs with high temperature thread lubricant, then install turbocharger and connect oil pressure and return lines. Connect air inlet hose. Place adapter ring in muffler elbow and install elbow and ring with ring in groove of turbine housing. Install hood and hood side plates.

Prior to starting engine, refer to item 1 of paragraph 117.

# NON-DIESEL GOVERNOR

All Series 1755 and 1855 non-diesel engines are fitted with a flyweight type governor that is mounted on front face of engine timing gear cover and is driven by engine camshaft gear.

### ADJUSTMENT

119. Linkage will normally not need adjustment unless linkage has been disassembled or damaged.

To adjust linkage, place hand throttle in high idle position, loosen lock nut on the governor-to-carburetor rod, then disconnect the rod from governor operating lever. Hold governor operating lever in high idle position and adjust rod end until end is 1/16-inch forward of hole in governor operating lever. Reinstall rod in governor lever and tighten turnbuckle lock nut.

Complete governor adjustments as outlined in paragraphs 120, 121 and 122.

120. **LOW IDLE.** With linkage adjusted as outlined in paragraph 119, start engine and run until it reaches operating temperature. Place hand throttle lever in low idle position and adjust carburetor stop screw to obtain a low idle of 775-825 rpm (800 desired).

121. **HIGH IDLE.** With engine at operating temperature and engine low idle speed adjusted, check and adjust engine high idle speed as follows:

Place hand throttle lever in high speed position and check engine high idle rpm which should be 2615-2665 (2640 desired) rpm.

If adjustment is required, stop engine and place hand throttle lever in low idle position, then loosen lock nut and back-out bumper screw until governor operating lever is released. See Fig. 37.

Push hand throttle lever forward until bellcrank on hand lever shaft butts against stop screw located in fuel tank support. Check governor spring arm which should be 15 degrees from centerline of engine crankshaft. See Fig. 38. If governor spring arm is not in correct position, adjust the stop screw in fuel tank support as necessary. Start engine and adjust spring tension screw as necessary to obtain stated engine high idle rpm. Tighten all lock nuts and adjust bumper screw as outlined in paragraph 122.

122. **BUMPER SCREW.** Engine surging at high idle rpm can be eliminated by adjusting the bumper screw as follows: Stop engine, refer to Fig.

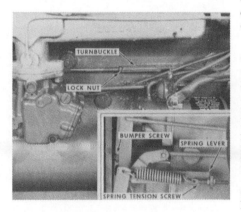

Fig. 38–Position governor spring arm as shown when making engine high idle adjustment. Refer to text.

37, loosen bumper screw lock nut and turn bumper screw inward slightly. Restart engine and recheck for surging. Continue this operation until engine surging is eliminated. DO NOT turn bumper screw in further than necessary as engine low idle rpm may be increased. Tighten bumper screw lock nut, refer to paragraph 120 and recheck the engine low idle rpm.

### R&R AND OVERHAUL
### Series 1755-1855 Non-Diesel

123. To remove governor, loosen fan belts and disconnect both rods from governor levers. Remove housing cap screws and seals and pull housing from timing gear cover. Use a clockwise motion to remove the shaft, weight and gear assembly.

NOTE: Use caution not to allow governor gear thrust washer to drop off governor gear. To do so may require removal of timing gear cover or oil pan.

Refer to Fig. 39 for an exploded view of governor assembly. To remove governor operating lever, first remove fork, then remove plug from shaft bore and remove spacers and outer retainer. Pull operating lever out of housing far enough to remove inner retainer, then remove inner retainer and complete removal of shaft from housing. Seal, bearings and shaft bushing can now be serviced. Weight shaft bushing is presized and should provide 0.0015-0.002 operating clearance for shaft. Shaft, weights and gear assembly is available as a unit only.

Fig. 37–Bumper screw is located as shown. Refer to text for adjustment.

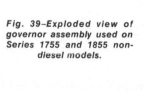

Fig. 39–Exploded view of governor assembly used on Series 1755 and 1855 non-diesel models.

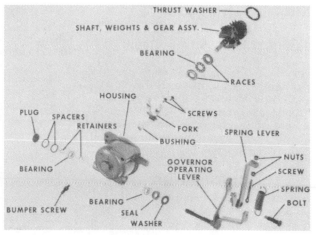

Governor gear hub rotates in a bushing which is pressed into front face of crankcase. Bushing can be renewed by removing timing gear cover and using a suitable puller. Inside diameter of bushing is 1.0022-0.0030.

Outside diameter of governor gear hub is 0.9990-0.9995 and should have 0.0027-0.0040 operating clearance in bushing. Renew bushing and/or gear when clearance exceeds 0.005.

Any further disassembly and/or

overhaul is obvious after an examination of the unit and reference to Fig. 39.

After reinstallation, readjust governor as outlined in paragraph 119 through 122.

# COOLING SYSTEM

## RADIATOR
### Series 1755-1855-1955

124. To remove radiator, drain cooling system and remove hood side panels and hood. Remove the hydraulic oil cooler and auxiliary drive unit oil lines. Disconnect both hoses from radiator, then unbolt fan blades from water pump and let fan rest in fan shroud. Remove radiator support to radiator cap screws and remove radiator and fan.

## WATER PUMP
### Series 1755-1855-1955

125. **R&R AND OVERHAUL.** To remove water pump, first drain cooling system, then remove hood, hood side panels and air cleaner assembly. Disconnect alternator adjusting link from water pump. Remove fan belts, then unbolt fan from water pump pulley and let fan rest in fan shroud. Disconnect inlet hose and bleed line from water pump and on non-diesel engines, disconnect control rods from governor. Unbolt water pump from cylinder block and remove pump from side of tractor.

126. Disassemble pump as follows: Press or pull pulley from shaft and remove rear cover and gasket. Remove snap ring (4—Fig. 40), then press shaft forward out of impeller (8) and

body (1). Remove impeller, seal seat (9) and seal (10) from body.

Inspect all parts for wear or other damage and renew as necessary. Slinger (2) can be renewed separately, if necessary.

Reassemble pump as follows: Use a short piece of pipe, or a socket, which has the same diameter as the brass cup of seal and press seal in body. Install new slinger (2) on shaft (3) if nec-

**Fig. 40–Exploded view of water pump. Pulleys (5) will vary between tractor models.**

| | |
|---|---|
| 1. Body | 6. Cover |
| 2. Slinger | 7. Gasket |
| 3. Shaft and bearing | 8. Impeller |
| 4. Snap ring | 9. Seat |
| 5. Pulley | 10. Seal |

essary and start shaft into body, slinger end first. Use a piece of pipe having same diameter as outer race of shaft bearing and press shaft and bearing into body until it bottoms. Install snap ring (4). DO NOT press on end of shaft.

Place seal seat on shaft so that rubber insert of seat faces toward impeller. Place impeller on bed of press with flat side down, then press shaft into impeller until rear face of impeller is flush with rear face of pump body. Support impeller end of shaft and press pulley on shaft until distance between front face of pulley hub and rear face of pump body is 7 31/32-inches.

Use new gasket (7) and install cover (6) on pump body.

## THERMOSTAT
### Series 1755-1855-1955

127. To remove thermostat, drain cooling system and remove hood side panels and hood. Remove air cleaner. Disconnect upper radiator hose from thermostat housing and remove bleed line. Loosen clamp on by-pass hose, then unbolt and remove thermostat housing and thermostat.

Thermostat is a pellet type and should start to open at 175 degrees F. and be fully open at 190 degrees F.

# IGNITION AND ELECTRICAL SYSTEM

## SPARK PLUGS
### Series 1755-1855 Non-Diesel

128. For normal or heavy duty service the following spark plugs are recommended: Champion, N-11-Y; AC, 44XLS (or C-44NS—1755 only); Autolite, AG42; Prestolite, 14GT42.

Spark plug electrode gap is 0.025 for all plugs. Tighten spark plugs to 30 ft.-lbs. torque when installing.

## DISTRIBUTOR
### Series 1755-1855-Non-Diesel

129. Series 1755 and 1855 non-diesel tractors are equipped with a

Holley D-2563AA ignition distributor which incorporates a vacuum advance unit. Refer to Fig. 41 for an exploded view.

Specifications are as follows:
Breaker contact gap .......... 0.025
Cam angle @ idle speed ....... 35°-38°
Centrifugal advance curve:

| Distributor rpm. | Advance degrees |
|---|---|
| 225 | 0-1½ |
| 275 | ½-2¾ |
| 325 | 2-4 |
| 900 | 8-10 |
| 1200 | 11-13 |

Vacuum advance @ 1000 distributor rpm:

| Inches Hg. | Advance degrees |
|---|---|
| 7 | 0 |
| 10 | 0-3 |
| 13 | 3-6 |
| 16 | 6-8 |
| 20 | 6-8 |

130. **IGNITION TIMING.** To set static timing, remove flywheel timing hole cover, then crank engine until number one piston is coming up on compression stroke. Continue cranking engine until flywheel 2 degree BTDC mark for Series 1755, or 0 degree BTDC mark for Series 1855 tractors, aligns with timing hole pointer, then loosen distributor clamp

screws and turn distributor in direction opposite to rotor rotation until breaker points just start to open. Tighten distributor clamp screws to retain distributor in this position.

**NOTE: The foregoing procedure sets static timing; however, the preferred method is to use a timing light as outlined below. If a choice exists, always use a timing light.**

To check running timing, remove flywheel timing hole cover and connect timing light. Disconnect the vacuum advance line from distributor and plug open end of line. Start engine and run at 2400 rpm. With timing light held close to the timing hole, the 26 degree BTDC mark for Series 1755, or 24 degree BTDC mark for Series 1855 on flywheel should register with the timing hole pointer. If the proper timing mark does not register with the timing pointer, loosen distributor clamp screws and rotate distributor as necessary. Tighten clamp screws, remove light and reconnect the distributor vacuum advance line.

## ALTERNATOR AND REGULATOR
### Series 1755-1855-1955

CAUTION: An alternator (AC generator) is used to supply the charging current for the Series 1755, 1855 and 1955 tractors. Because certain components of the alternator can be damaged by procedures that would not affect a DC generator, the following precautions must be observed.

1. When installing batteries or connecting a booster battery, be sure that the negative post of all batteries are grounded.
2. Never short across any of the alternator or regulator terminals.
3. Never attempt to polarize the alternator.
4. Always disconnect all battery straps before removing or installing any electrical unit.
5. Never operate alternator on an open circuit; be sure all leads are properly connected and tightened before starting engine.

131. Series 1755, 1855 and 1955 tractors are equipped with a Delco-Remy model 1100808 alternator and a Delco-Remy model 1119511 regulator. Specification data follows:

**Alternator 1100808**

Ground ................ Negative
Field current @ 80° F.
  Amperes ......... 2.2-2.6
  Volts ............... 12.0
Field resistance @ 80° F.
  Ohms .............. 4.6-5.5

Cold output @ specified voltage
  Specified volts ............ 14.0
  Amps @ 2000 rpm ........ 32.0
  Amps @ 5000 rpm ........ 50.0
Rated output-hot ............ 55.0

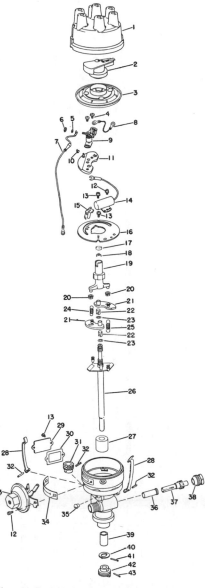

*Fig. 41–Exploded view of Holley distributor used on Series 1755 and 1855 non-diesel tractors. The Holley distributor incorporates a vacuum spark advance as well as a centrifugal advance. Timing is checked with vacuum line disconnected and plugged. To remove shaft (26), it is first necessary to remove pin (32) from tachometer drive gear (31) and pins (41 and 43) from lower end of shaft. Pin (32) is accessible after removing plate (29).*

| | |
|---|---|
| 1. Cap | 26. Shaft and weight plate |
| 2. Rotor | 27. Bushing |
| 3. Dust cover | 28. Clamps |
| 7. Low tension lead | 29. Cover plate |
| 8. Ground wire | 30. Gasket |
| 9. Contact set | 31. Tachometer drive gear |
| 11. Breaker plate | 32. Pin |
| 14. Condensor | 33. Diaphragm assy. |
| 15. Spring | 34. Gasket |
| 16. Base plate | 35. Plug |
| 17. Wick | 36. Bushing |
| 18. Retainer | 37. Gear and sleeve |
| 19. Cam | 38. Retainer |
| 20. Bushings | 39. Bushing |
| 21. Weights | 40. Washer |
| 22. Bearing strip | 41. Pin |
| 23. Washer | 42. Coupling |
| 24. Primary spring | 43. Pin |
| 25. Secondary spring | |

**Regulator 1119511**

Polarity ................ Negative
Field relay
  Air gap ................. 0.015
  Point opening ........... 0.030
  Closing voltage range .... 1.5-3.2
Voltage regulator
  Air gap .............. 0.067 (1)
  Point opening ........... 0.015
  Voltage setting @ degree F.
    13.9-15.0 @ 65
    13.8-14.8 @ 85
    13.7-14.6 @ 105
    13.5-14.4 @ 125
    13.4-14.2 @ 145
    13.2-14.0 @ 165
    13.1-13.9 @ 185

(1) Starting point for point gap adjustment after bench repair. Subsequent adjustment is made if necessary so that operation on lower contacts is 0.05-0.40 volts lower than on upper contacts.

## STARTING MOTOR
### Series 1755-1855-1955

132. Series 1755 and 1855 non-diesel engines are equipped with a Delco-Remy model 1108431 starting motor. Series 1755 diesel engines are equipped with a Delco-Remy model 1113139 starting motor. Series 1855 and 1955 diesel engines are equipped with a Delco-Remy model 1113656 starting motor. Specification data follows:

**Starter 1108431**
Brush spring tension (min.) .... 35 oz.
No-load test:
  Volts ........................9.0
  Max. amps ................. 80
  Min. rpm ................. 3500

**Starter 1113139**
Brush spring tension (min.) .... 80 oz.
No-load test:
  Volts ..................... 11.5
  Min. amps* ................. 57
  Max. amps* ................. 70
  Min. rpm ................. 5000
  Max. rpm ................. 7400
Lock test:
  Volts ......................3.4
  Amperes ................. 500
  Torque (ft.-lbs.) ............ 22
*Includes solenoid.

**Starter 1113656**
Brush spring tension (min.) .... 80 oz.
No-load test
  Volts ..................... 11.6
  Min. amps* ................. 85.0
  Max. amps* ................. 125.0
  Min. rpm ................. 5900
  Max. rpm ................. 8100
Lock test:
  Volts ..................... 2.35
  Amps ................. 600.0
  Torque (ft.-lbs.) ............ 17.0
*Includes solenoid.

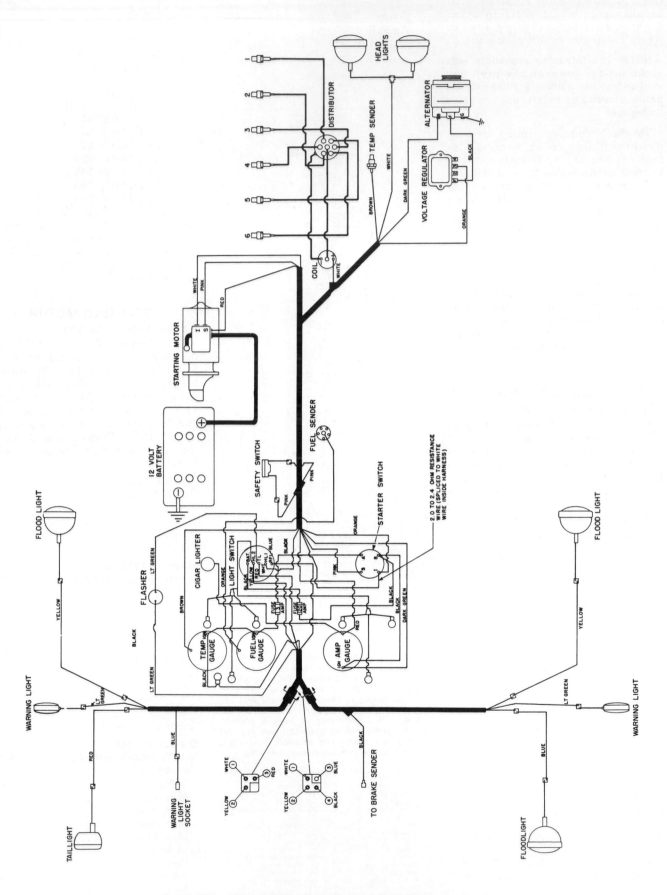

*Fig. 42–Wiring diagram for the Series 1755 and 1855 non-diesel tractors.*

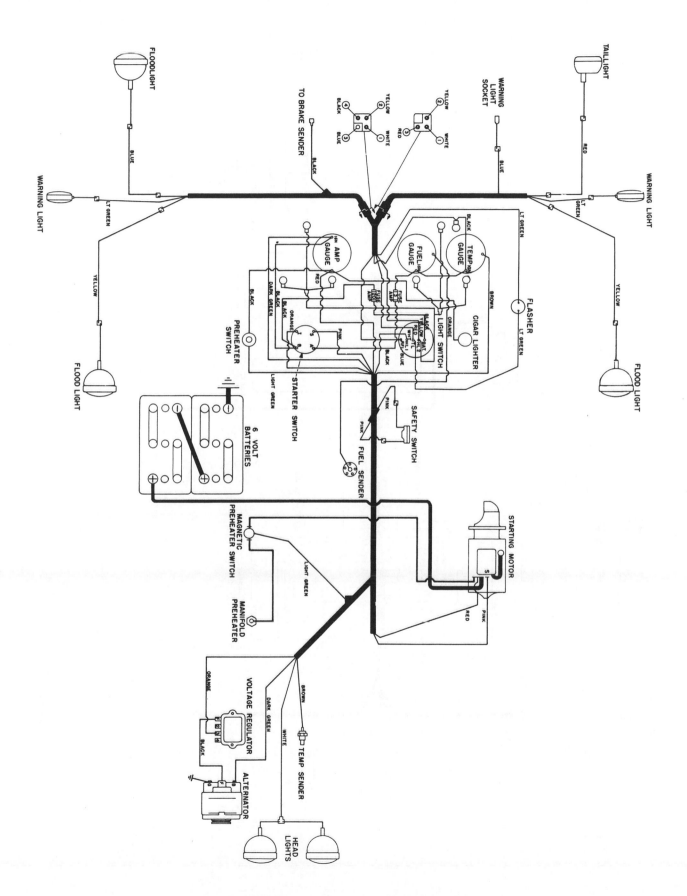

*Fig. 43–Wiring diagram for the Series 1755 diesel tractors.*

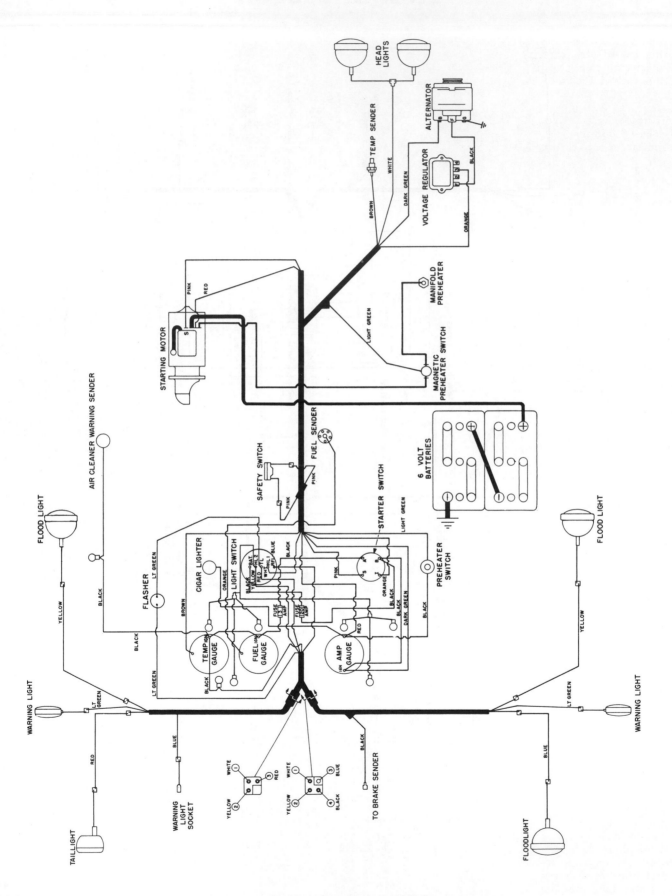

*Fig. 44–Wiring diagram for the Series 1855 and 1955 diesel tractors.*

# ENGINE CLUTCH

Engine clutch for all models is a dry, spring loaded type utilizing a button type driven plate. Series 1755 use a 12 (11.88) inch clutch whereas the Series 1855 and 1955 tractors use a 13 inch clutch. The 12 inch clutch disc has four buttons and the 13 inch clutch disc has 6 buttons. Service procedure is similar for both clutches.

## ADJUSTMENT
### Series 1755-1855-1955

133. Clutch pedal free travel is 1½-inches measured vertically at the clutch pedal pad.

To adjust pedal free travel, loosen the clevis jam nut at clutch lever end of the clutch control rod, remove cotter pin and clevis pin and rotate clevis as required to obtain the correct pedal free travel. Reinstall the clevis pin and cotter pin, then tighten jam nut.

## REMOVE AND REINSTALL
### Series 1755-1855-1955

134. On tractors equipped with an auxiliary drive unit, or a one-piece clutch shaft, remove engine as outlined in paragraph 55, then remove clutch housing.

On models with no auxiliary drive unit which are equipped with a two-piece clutch shaft, proceed as follows: Remove the pto shaft, or the hydraulic pump drive shaft if tractor has no pto. Remove clutch shaft shield, remove master link from sprocket chain and remove chain. Remove bolts from clutch shaft coupling, slide coupling rearward until it is free from front clutch shaft and lift out rear clutch shaft. Do not lose sprocket spacer which will drop out as rear shaft is removed. Sprockets are also free to be removed if desired.

Remove front clutch shaft by pulling shaft rearward.

On diesel models, disconnect water trap drain line from clutch housing. Disconnect clutch control rod from clutch shaft lever. Disconnect wiring from starter and remove starter. Remove dust shield from lower front side of clutch housing, then unbolt clutch cover and slide it rearward.

Remove clutch cover by loosening the retaining cap screws approximately one turn at a time to prevent distortion of the clutch cover.

Reinstall by reversing the removal procedure. Position driven disc so widest hub flange is toward rear and use clutch shaft as a pilot to align disc. Tighten cover retaining cap screws in the same sequence as they were removed.

## OVERHAUL
### Series 1755-1855-1955

135. Remove clutch as outlined in paragraph 134, then on the 13 inch clutches, remove the drive straps and ferrules. Place clutch cover assembly in a press with a block under pressure plate that will permit movement of clutch cover. Place a bar across top of cover assembly, compress the unit slightly and remove nuts from the eye bolts. Slowly release pressure and lift off clutch cover. All parts are now available for inspection and/or servicing. Note location and color of springs prior to removal. Remove levers as follows: Grasp inner end of lever and threaded end of eyebolt between thumb and fingers and while holding them as close together as possible, lift strut up and over end of lever. Any further disassembly is obvious.

Renew parts as necessary and refer to the following for spring data.

NOTE: Driven plate buttons are not renewable. Renew driven plate if buttons are excessively worn, cracked or otherwise damaged.

The 12 inch clutch used in Series 1755 has a total of 9 springs of which 6 are color coded black and three are unpainted (plain). The black springs should test 223-227 lbs. when compressed to a length of 1½-inches. The unpainted springs should test 189-201 lbs. when compressed to a length of 1½-inches.

The 13 inch clutch used in Series 1855 and 1955 has a total of 12 springs which are color coded pink. Springs should test 209-225 lbs. when compressed to a length of 1 11/16-inches.

Refer to Fig. 45 for spring location when reassembling.

136. **ADJUST LEVERS.** Reassemble clutch cover by reversing the disassembly procedure and adjust clutch fingers as follows: Measure, and record, the distance between friction face of flywheel and inner race of pilot bearing. Use three pieces of shim stock 0.380 thick for 12 inch clutches; or 0.398 thick for 13 inch clutches, equally spaced between clutch pressure plate and flywheel and bolt clutch cover in position. Actuate clutch fingers several times to insure proper functioning, then turn adjusting nuts on eyebolts until the measured distance between lever and inside pilot bearing race is the total of the previously recorded pilot bearing to flywheel face distance plus 2.10-2.16 inches, for the 12 inch clutch; or 2.30-2.36 inches for the 13 inch clutch. Stake nuts into eyebolts.

Remove clutch cover and shim stock. Reinstall clutch disc and the clutch cover to flywheel and use clutch shaft as a pilot to align clutch driven disc. Clutch disc is installed with widest hub flange toward rear.

## PILOT BEARING AND PTO HUB
### Series 1755-1855-1955

137. To remove pilot bearing and pto hub, remove clutch as outlined in paragraph 134, then remove snap ring and press pilot bearing from pto hub.

Pilot bearing is a sealed ball bearing.

## CLUTCH RELEASE BEARING
### Series 1755-1855-1955

138. To renew the clutch release bearing, remove clutch housing as outlined in 134, then pull release bearing and release bearing sleeve from release bearing tube. Press release bearing from sleeve.

Clutch release bearing is a sealed ball bearing.

*Fig. 45–View showing location of clutch springs.*

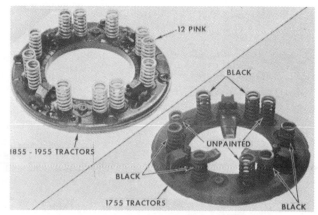

When reassembling, be sure lugs in bearing sleeve engage release bearing fork. When assembly is complete, adjust clutch pedal free travel to 1½-inches.

### RELEASE FORK AND SHAFT
### Series 1755-1855-1955

139. Release fork and shaft can be removed as follows: Disconnect clutch control rod from clutch lever and remove top inspection cover. Remove cap screws and the round key from release fork, then hold fork and pull shaft from left side of clutch cover. Remove fork from top side of housing.

When reinstalling, be sure tapped hole of release fork faces forward. Adjust clutch pedal free travel to 1½-inches.

### RELEASE BEARING TUBE
### Series 1755-1855-1955

140. Should it become necessary to renew the release bearing tube, remove the clutch housing as outlined in paragraph 134.

On direct drive models, press out old release bearing tube and when installing new tube, press it into clutch housing until inner end is within 1/16-inch of inside edge of counterbore for felt seal.

NOTE: On models equipped with an auxiliary drive transmission, the release bearing tube is a part of the auxiliary transmission input shaft support. For installation dimension, refer to appropriate section of this manual.

# HYDRA-POWER DRIVE

All Series tractors have available as optional equipment, a Hydra-Power drive unit interposed between clutch and transmission, which provides both direct drive and underdrive operation. This auxiliary drive is a self-contained unit and includes a hydraulically operated multiple disc clutch as well as a pump and a filter. The gear type pump mounted on inner side of housing cover furnishes pressurized oil to operate the multiple disc clutch as well as providing lubrication for the unit. In direct drive position the multiple disc clutch is engaged and a "straight through" power flow is obtained, whereas, in the underdrive position, the multiple disc clutch is disengaged and the power flow is routed to the countershaft where an approximate 26 per-cent speed reduction is produced. Unit can be shifted while tractor is in motion.

### TROUBLE SHOOTING
### All Models So Equipped

145. The following symptoms and their causes will be helpful in trouble shooting the Hydra-Power unit.
1. OPERATING TEMPERATURE TOO HIGH.
   a. Coolant hoses, heat exchanger or oil lines obstructed.
   b. Multiple disc clutch slipping.
   c. Engine cooling system overheated.
   d. Low pump pressure.
   e. Defective or improperly adjusted bearings.
2. INOPERATIVE OR HESITATES IN HYDRA-POWER DRIVE
   a. Defective over-running clutch.
   b. Defective clutch.
   c. Weak retractor spring.
3. INOPERATIVE OR HESITATES IN DIRECT DRIVE.
   a. Multiple disc clutch slipping or defective.
   b. Clutch operating pressure low.

4. MULTIPLE DISC CLUTCH SLIPPING.
   a. Clutch operating pressure low.
   b. Oil passages obstructed.
   c. Defective clutch.
   d. Oil collector sealing rings worn or broken.
   e. Defective oil collector "O" ring.
   f. Foreign material in clutch.
5. ABNORMAL LUBRICATION CIRCUIT PRESSURES (TOO HIGH OR TOO LOW).
   a. Defective pump.
   b. Pump inlet clogged.
   c. Defective by-pass spring.
   d. By-pass valve plunger not seating properly.
   e. Low oil level.
6. LOW CLUTCH OPERATING PRESSURE.
   a. Low oil level.
   b. Defective pump.
   c. Pump inlet clogged.
   d. Oil collector sealing rings worn or broken.
   e. Defective clutch "O" ring.
   f. Defective clutch piston ring.
   g. Defective regulator spring.
   h. Regulator valve spool stuck open.
   i. Clogged oil passage.
7. HIGH CLUTCH OPERATING PRESSURE.
   a. Defective regulator spring.
   b. Regulator valve spool stuck closed.
   c. Clogged oil passage.
8. ERRATIC CLUTCH OPERATING PRESSURE.
   a. Regulator valve spool installed backward.

### OPERATING PRESSURE
### All Models So Equipped

146. Hydra-Power drive unit has two oil pressure circuits; a clutch operating circuit and a lubrication circuit. Either circuit can be checked in two places and selection will depend on equipment and the space available for its use. See Figs. 50 and 51.

147. **CLUTCH PRESSURE.** Use a gage capable of registering at least 300 psi and install gage in the control valve to oil collector line (Fig. 50), or in the port located at front of boss under oil filter (Fig. 51). Be sure oil in

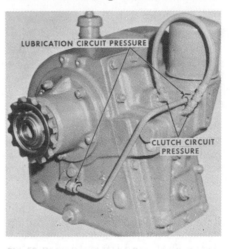

Fig. 50–Rear view of Hydra-Power unit showing pressure check points. Refer to text and Fig. 51.

Fig. 51–Front view of Hydra-Power unit showing pressure check points. Refer to text and Fig. 50.

Hydra-Power unit is at correct level, then start and run engine until oil is at operating temperature. With oil warmed, run engine at 2400 rpm, place unit in Direct Drive position and check gage which should read 140-160 psi. If pressure is not as specified, and no malfunctions of the unit are evident, vary shims (45—Fig. 52) behind regulator valve spring (44) as required. Shims are 0.0395 thick and each shim will vary pressure about 5 psi.

148. **LUBRICATION PRESSURE.** To check lubrication circuit pressure, install gage in cover to countershaft line (Fig. 50), or in place of the oil cooler lines (Fig. 51). Start and run engine at 2400 rpm, place unit in Under-Drive position and observe gage which should read 20-60 psi. If pressure is not as specified, refer to TROUBLE SHOOTING Section or to section pertaining to overhaul.

## REMOVE AND REINSTALL
### All Models So Equipped

149. To remove the Hydra-Power drive unit it is necessary to remove the engine and Hydra-Power unit as outlined in paragraph 55. With engine removed support it on wood blocks so oil pan will not be damaged, then remove clutch housing lower dust shield and starter. Unbolt and remove clutch housing and Hydra-Power unit from engine. Remove clutch release bearing. Remove cap screw and key from release bearing fork and pull clutch shaft from fork and clutch housing. Unbolt and remove the clutch housing from Hydra-Power unit.

Reinstall by reversing the removal procedure.

## OVERHAUL
### All Models So Equipped

150. **DISASSEMBLY PROCEDURE.** With unit removed as outlined in paragraph 149, drain unit if necessary. Remove valve control lever and, if necessary, lever bushing can be renewed at this time. Remove filter to countershaft oil line, then disconnect oil hose from oil collector stem and remove fitting. Remove filter, if desired, then unbolt and remove housing cover and attached pump.

Remove support shaft retainer (30—Fig. 53), pull support shaft (34) and allow countershaft to rest on bottom of housing. Remove and discard "O" rings from support shaft and shaft front bore in housing.

Unbolt support (64) from housing and pull input shaft (58) and support (64) as an assembly from housing.

Remove snap ring (56) from front end of output shaft (40), then unbolt output shaft bearing retainer (21). Pull output shaft rearward and remove the clutch and oil collector assembly and the output gear (41) from housing as the shaft is withdrawn. Countershaft assembly can now be removed from housing.

If necessary, filler neck and valve spool control lever shaft can also be removed.

At this time all of the components of the Hydra-Power drive unit are accessible for inspection and/or overhaul. Refer to the appropriate following paragraphs.

151. **CONTROL VALVE SPOOL, REGULATOR VALVE AND BY-PASS VALVE.** The control valving can be removed from housing cover as follows: Remove the Allen plug (23—Fig. 52) on top side of control valve bore and remove detent spring (24) and ball (25). Pull control valve spool (22) forward out of its bore, then remove and discard "O" ring. Remove plugs, then remove regulator valve (43) and by-pass valve (40) assemblies.

Save any shims (45) which may be present behind the regulator valve spring. Seat (39) is also renewable.

NOTE: Removal of the regulator and by-pass valves can be accomplished with Hydra-Power drive unit on tractor.

Check springs against the values which follow:
Detent spring
    Free length . . . . . . . . . . . . . . 1 9/32 in.
    Test lbs. @ inches . . . 15.3-18.7 @ ⅞
Regulator spring
    Free length . . . . . . . . . . . . . . . .3⅝ in.
    Test lbs. @ inches . . . . . .55-67 @ 2¾
By-pass spring
    Free length . . . . . . . . . . . . . . . .3½ in.
    Test lbs. @ inches . . 5½-6½ @ 2 7/16

Outside diameter of regulator valve spool and land of control valve spool is 0.6865-0.6869 and their bore inside diameters are 0.6875-0.6883.

When reassembling, use new "O" ring on control valve spool and install regulator valve with pin inward. If any shims were present behind regulator spring during disassembly, be sure to reinstall them. Prior to installing cover on housing, renew "O"

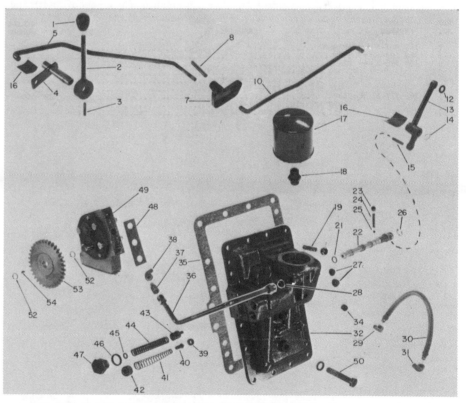

*Fig. 52—Exploded view of Hydra-Power cover assembly showing component parts used.*

| | | | |
|---|---|---|---|
| 1. Knob | 16. Bushing | 29. Elbow | 42. Plug |
| 2. Lever | 17. Filter | 30. Hose | 43. Spool |
| 3. Pin | 18. Adapter | 31. Elbow | 44. Spring |
| 4. Shaft | 19. Stud | 32. Cover | 45. Shim |
| 5. Rod | 21. "O" ring | 34. Plug | 46. Washer |
| 7. Bellcrank | 22. Control spool | 35. Gasket | 47. Cap |
| 8. Pin | 23. Plug | 36. Oil line | 48. Gasket |
| 10. Rod | 24. Spring | 37. Connector | 49. Pump |
| 12. Washer | 25. Detent ball | 38. Elbow | 50. Pump screws |
| 13. Lever | 26. Snap ring | 39. Seat | 52. Snap ring |
| 14. Snap ring | 27. Inserts | 40. Plunger | 53. Pump gear |
| 15. Pin | 28. Seat | 41. Spring | 54. Key |

ring on stem of oil collector.

**152. PUMP OVERHAUL.** After removing cover from Hydra-Power drive unit as outlined in paragraph 150, unbolt pump from cover and discard gasket. Remove drive gear outer retainer snap ring, remove gear and key from pump drive shaft, then remove inner gear retainer snap ring. Remove the through bolts and separate pump. Dowels can be driven out and the screen removed if necessary. See Fig. 54.

NOTE: If the pump gears and shafts are to be reinstalled, mark idler gear so that it will be reinstalled in its original position.

Pump dimensional data are as follows:

Gear width . . . . . . . . . . 0.9975-0.9980
Pump body width . . . . . 0.9995-1.0000
Gear shaft diameter . . . 0.6247-0.6250
Shaft bore diameter . . . 0.6265-0.6275
Shaft operating
 clearance . . . . . . . . . . 0.0015-0.0028

Reassemble by reversing disassembly procedure and use new gasket when mounting pump on housing cover. Heads of through bolts should be on drive gear side of pump and dowels should be installed prior to tightening through bolts.

**153. INPUT SHAFT ASSEMBLY.** Remove input shaft assembly as outlined in paragraph 150. To disassemble input shaft assembly, disengage snap ring (59—Fig. 53) from support (64) and pull input shaft and bearing (60) from support. Remove snap ring (61), then remove bearing and snap ring (59) from input shaft. Remove oil seal (62) from support and oil seal (57) from bore at inner end of input shaft. If clutch release bearing tube (67) is damaged, press it from support.

Inspect input shaft for damaged or worn splines or teeth. Check condition of input shaft bearing. Renew parts as necessary.

Reassemble input shaft assembly as follows: If new release bearing tube is being installed, press it in support until end of tube protrudes 3 5/8-inches from flange on front of support. Install oil seal (62) in support with lip facing rearward. Install oil seal (57) in end of input shaft with lip of seal facing forward. Place large snap ring (59) over end of input shaft, install bearing (60) and secure with snap ring (61). Carefully install input shaft and bearing in support and secure with the large snap ring (59).

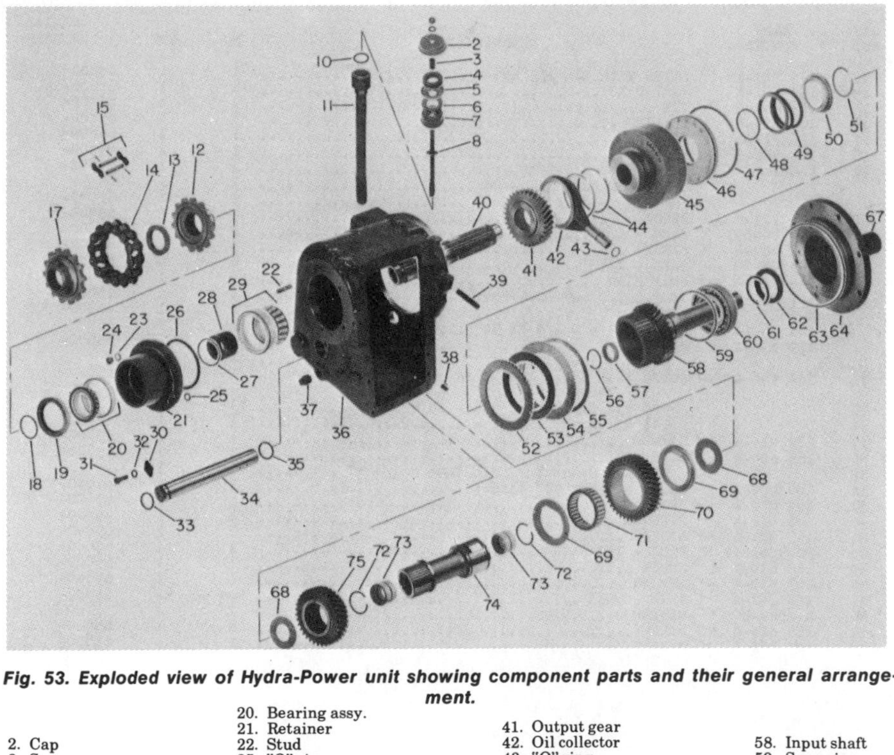

**154. CLUTCH ASSEMBLY.** Remove clutch assembly as outlined in paragraph 150. To disassemble the clutch assembly, first pull oil collector (42—Fig. 53) from clutch spider (45) hub and remove the two sealing rings (44). Remove snap ring (55), retainer (54) and clutch plates (52 and 53). Place clutch spider in a press, depress spring retainer (50) slightly and remove retaining snap ring (51). Release press and remove retainer and retractor spring (49). Use ⅛-inch brazing rod and fabricate four U-pins to the dimensions shown in Fig. 55. Place the U-pins equidistantly around edge of clutch spider so inner leg will prevent the piston outer seal ring from expanding. Hold pins in position, turn spider over and bump spider gently on bench until piston comes out of bore.

Clean and inspect all parts. Use the following specifications data as a guide for renewing parts.

Clutch retractor spring
 test . . . . . . . 155-185 lbs. @ 1 3/16 in.
Driven plate thickness . . . . 0.066-0.072
Driving plate thickness . . . 0.066-0.072
Clutch piston ring:
 Width . . . . . . . . . . . . . . . . 0.123-0.124
 End gap (desired) . . . . . . 0.005-0.015
 Max. allowable . . . . . . . . . . . 0.025
 Side clearance (desired) . 0.003-0.006
 Max. allowable . . . . . . . . . . . 0.025
Oil collector hub O.D. . . . . . 3.544-3.546
Oil collector I.D. . . . . . . . . 3.550-3.552
Oil collector seal ring:
 Width . . . . . . . . . . . . . . 0.0930-0.0935
 End gap (desired) . . . . . . 0.003-0.008
 Max. allowable . . . . . . . . . . . 0.020

Reassemble clutch by reversing the disassembly procedure. Clutch piston is installed with flat side inward. When installing clutch plates, start with a driven plate (external teeth) and alternately install the five drive plates and five driven plates, ending with a drive plate (internal teeth). Compress oil collector seal rings with soft wire, start oil collector over hub with word "Front" toward clutch spider and remove wire as oil collector goes over sealing rings. Lubricate clutch plates.

**155. OUTPUT SHAFT ASSEMBLY.** To disassemble the output shaft assembly, unstake lock nut (13—Fig. 53) and using a spanner wrench (Oliver No. ST-149 or equivalent), remove nut. Remove sprocket (12) and "O" ring (18) from shaft, then press output shaft from rear bearing (20) and retainer (21). Remove shims (27) and spacer (28), then press front bearing (29) from output shaft (40). Remove oil seal (19) and rear bearing (20) from retainer and, if necessary, remove bearing cups from retainer.

**Fig. 53. Exploded view of Hydra-Power unit showing component parts and their general arrangement.**

| | | | |
|---|---|---|---|
| 2. Cap | 20. Bearing assy. | 41. Output gear | 58. Input shaft |
| 3. Spacer | 21. Retainer | 42. Oil collector | 59. Snap ring |
| 4. Element | 22. Stud | 43. "O" ring | 60. Bearing |
| 5. Cup | 25. "O" ring | 44. Sealing rings | 61. Snap ring |
| 6. Gasket | 26. "O" ring | 45. Clutch spider | 62. Oil seal |
| 7. Cap | 27. Shim (.004, .007, | 46. Piston | 63. "O" ring |
| 8. Dip stick | .015) | 47. Piston ring | 64. Support |
| 10. "O" ring | 28. Spacer | 48. Seal ring | 67. Tube |
| 11. Filler tube | 29. Bearing assy. | 49. Release spring | 68. Thrust washer |
| 12. Front sprocket | 30. Retainer | 50. Spring retainer | 69. Gear bearing |
| 13. Nut | 33. "O" ring | 51. Snap ring | 70. Clutch gear |
| 14. Coupler chain | 34. Support shaft | 52. Drive plate | 71. Sprag clutch |
| 15. Master link | 35. "O" ring | 53. Driven plate | 72. Snap ring |
| 17. Rear sprocket | 36. Housing | 54. Retainer plate | 73. Bearing |
| 18. "O" ring | 37. Drain plug | 55. Snap ring | 74. Countershaft |
| 19. Oil seal | 38. Dowel | 56. Snap ring | 75. Gear |
| | 39. Lower shaft | 57. Oil seal | |
| | 40. Output shaft | | |

Clean and inspect all parts and renew as necessary. Always use new "O" ring (26) and oil seal (19) when reassembling unit.

156. Assemble output shaft assembly as follows: Place front bearing on rear of output shaft with smaller diameter toward rear and press bearing on shaft until it bottoms against shoulder. If necessary, install bearing cups in retainer with smaller diameters towards inside. Place spacer (28) on output shaft with the counterbored end next to front bearing cone, then install shims (27) that were previously removed. Place shaft in bearing retainer (21), install rear bearing (20), sprocket (12) and the lock nut (13); tighten lock nut to a torque of 200 ft.-lbs. Check shaft bearing adjustment which should be from a maximum of 0.001 shaft end play to a maximum shaft rolling torque of 5 inch pounds. If bearing adjustment is not within these limits, remove lock nut, sprocket and rear bearing and vary the number of shims (27) as necessary. Shims are available in thicknesses of 0.004, 0.007 and 0.015. When bearing adjustment is within the desired limits remove the lock nut and sprocket and install new "O" ring (18) and oil seal (19). Seal is installed with lip forward. Install sprocket and lock nut, tightening the lock nut to a torque of 200 ft.-lbs. Stake nut in position.

157. **COUNTERSHAFT ASSEMBLY.** Lift the countershaft from housing if necessary and retrieve both thrust washers (68—Fig. 53). Remove gear (75) from rear of countershaft. Remove front bearing (69), clutch gear (70), over-running (Sprag) clutch (71) and rear bearing (69) from front end of countershaft. Remove snap rings (72) and roller bearings (73) from bore of countershaft.

Clean and inspect all parts and renew as necessary.

Use the following data as a guide for renewing parts:

Front thrust washer
  thickness . . . . . . . . . . . . . 0.060-0.062

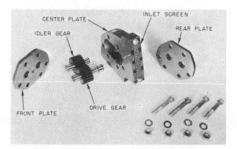

*Fig. 54–Disassembled view of Hydra-Power drive oil pump. Dowels are not shown.*

Rear thrust washer
  thickness . . . . . . . . . . . . . 0.060-0.062
Countershaft support
  shaft O.D. . . . . . . . . . 1.3745-1.3750
Countershaft O.D.
  clutch end . . . . . . . . . 2.8429-2.8434
  Max. allowable taper . . . . . . . 0.0002
Clutch gear I.D. . . . . . . . 3.4988-3.4998
  Max. allowable I.D. taper . . . 0.0003
Front and rear clutch gear bearing:
  Bearing surface O.D.   3.4968-3.4978

To reassemble countershaft, proceed as follows: Install needle bearings in bore of countershaft only far enough to allow installation of snap rings. Place rear clutch gear bearing (69) on front of countershaft with bearing surface toward front. Install the over-running clutch (71) on front of countershaft so drag strips point toward the previously installed rear bearing (69). Install clutch gear (70) over the over-running clutch and rotate gear clockwise to assist in installing gear. After installation, clutch should over-run (turn freely) when clutch gear is turned clockwise, but should lock up when counterclockwise gear rotation is attempted. Install front clutch gear bearing (69), then install countershaft output gear (75) on rear of countershaft.

158. **REASSEMBLY PROCEDURE.** Install oil filler tube, control valve spool lever shaft or input shaft support studs if removed.

Use heavy grease on the countershaft thrust washers and position washers in housing with tangs in the slots provided. Place assembled countershaft in case with output gear to rear and allow the assembly to rest in bottom of case. Take care not to dislodge the thrust washers. Delay installation of countershaft support shaft until input shaft assembly has been installed.

Use a new bearing retainer "O" rings, and start output shaft into rear of housing. As shaft is moved forward, install output gear with hub to rear and clutch assembly with oil collector to rear. Be sure small "O" ring is in place in counterbore in front face of bearing retainer and aligned with drain hole in housing. Install lock washers and nuts on bearing retaining

studs and tighten the nuts securely. Install clutch spider retaining ring on front of output shaft.

159. Align the internal spline clutch plates in clutch assembly, place new "O" ring on input shaft support and install input shaft assembly. Rotate input shaft as it is being installed to be sure it correctly engages clutch plates. All five internal clutch plates should rotate when input shaft is properly installed. Be sure notch in support mates with notch in housing so control spool will have operating room.

160. Install the countershaft support shaft as follows: Install a new "O" ring in shaft bore in front of housing. Install new "O" ring on rear of shaft, and if necessary, install oil line elbow in rear of shaft. Lubricate both "O" rings, make certain that both thrust washers are in position, then lift countershaft assembly to align bores and install countershaft support shaft from rear. Install shaft retainer with cap screw.

161. To install housing cover assembly, install new oil collector stem "O" ring and place new cover gasket on housing. Lift stem of oil collector and start stem into bore of cover, then carefully install cover to housing by rotating cover slightly as stem enters cover bore. Tighten cover retaining cap screws securely.

Fill unit with 6 quarts (with filter) of Type A Automatic Transmission Fluid.

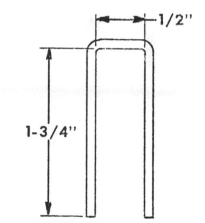

*Fig. 55–Service on clutch spider piston can be facilitated by using U-pins made from ⅛-inch brazing rod formed to dimensions shown. Refer to text.*

# HYDRAUL SHIFT

The Hydraul Shift is a three-speed auxiliary unit interposed between engine clutch and transmission and provides on-the-go shifting in over, under or direct drive. This auxiliary unit is a self-contained unit and includes its own reservoir, lubrication and pressure systems and control valving. Oil operating temperature is controlled by a cooler located in front of the radiator. Gears are in constant mesh and are helical cut. Drive selection is accomplished by engaging or disengaging multiple disc clutches and the operation of a planetary gear set and an over-running (Sprag) clutch. The pressurized oil used to operate clutches, as well as oil for lubrication is furnished by a gear type pump mounted on inner side of housing cover.

In overdrive position, oil pressure engages overdrive (rear) clutch and power flow goes through countershaft which drives the planet gear carrier. Overdrive clutch holds sun gear shaft stationary and the speed increase is obtained by the planet gears driving output shaft through the ring gear.

In underdrive position, both direct and overdrive clutches are disengaged and power flow is through the counter shaft which drives the planet gear carrier. The over-running clutch locks sun gear shaft and output shaft together causing them to rotate as a unit. This makes the planet drive inactive and results in a speed reduction.

In direct drive, the direct (front) clutch is engaged. This locks input shaft and output shaft together and power flow is straight through the unit. Over-running clutch over-runs and permits free rotation of sun gear shaft planet gears and countershaft.

## TROUBLE SHOOTING
**All Models So Equipped**

162. The following symptoms and their causes will be helpful in trouble shooting the Hydraul Shift unit.

1. OPERATING TEMPERATURE TOO HIGH.
   a. Coolant hoses, oil cooler or oil lines obstructed.
   b. Low pump pressure or pump intake clogged.
   c. One or both clutches slipping.
   d. Engine cooling system overheated.
   e. Defective or improperly adjusted bearings.
   f. Reservoir oil level low.

2. INOPERATIVE OR HESITATES IN OVERDRIVE.
   a. Overdrive clutch slipping or defective.
   b. Low pump pressure or pump intake clogged.
   c. Oil passage obstructed.

3. INOPERATIVE OR HESITATES IN UNDERDRIVE.
   a. Over-running clutch worn or defective.
   b. Damaged or sticking underdrive and/or overdrive clutch.

4. INOPERATIVE OR HESITATES IN DIRECT DRIVE
   a. Low pump pressure.
   b. Direct drive clutch faulty or slipping.
   c. Oil passage obstructed.

5. LUBRICATION CIRCUIT PRESSURE ABNORMAL (TOO HIGH OR TOO LOW)
   a. Defective by-pass valve spring.
   b. By-pass valve plunger sticking or not seating.
   c. Pump defective or pump intake clogged.
   d. Oil passage obstructed.

6. LOW CLUTCH OPERATING PRESSURE.
   a. Pump defective or pump intake clogged.
   b. Reservoir oil level low.
   c. Defective regulator valve spring.
   d. Regulator valve spool stuck open.

7. HIGH CLUTCH OPERATING PRESSURE.
   a. Defective regulator valve spring.
   b. Regulator valve spool stuck closed or sticking in bore.
   c. Oil too heavy or too cold.

8. ERRATIC CLUTCH OPERATING PRESSURE.

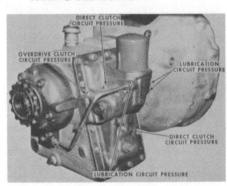

Fig. 56–View showing ports available for checking pressures of Hydraul Shift unit. Refer to text for specifications.

Fig. 57–Exploded view of Hydraul Shift unit cover, pump, valves and filter. Refer also to Fig. 59.

1. Knob
2. Lever
3. Pin
4. Shaft
5. Rod
7. Bellcrank
8. Pin
10. Rod
12. Washer
13. Lever
14. Snap ring
15. Pin
16. Bushing
17. Filter element
18. Adapter
19. Stud
21. "O" ring
22. Spool
23. Insert
24. Spring
25. Plug
26. Ball
27. Seat
28. Plug
29. Cover
32. Gasket
33. "O" ring
34. Seat
35. By-pass valve plunger
36. Spring
37. Plug
38. Cap
39. Washer

a. Regulator valve spool incorrectly installed.
b. Sticky regulator valve spool.

## SYSTEM PRESSURE CHECKS
**All Models So Equipped**

163. The Hydraul Shift unit has two oil pressure circuits; a lubrication circuit and a clutch operating pressure circuit. When making any system pressure check, be sure reservoir oil is at correct level, oil is at operating temperature and engine is operating at rated rpm.

To check lubrication circuit, refer to Fig. 56 and proceed as follows: Install a pressure gage either in place of the cover-to-countershaft oil line, or in place of heat exchanger (oil cooler) line, at the housing cover. Location will depend on space available as either position will suffice.

**NOTE: If desired a hole can be cut in the top of an oil filter and a gage welded in place. This assembly can be installed in place of the regular oil filter and the lubrication circuit pressure checked.**

Start engine, run at rated rpm and observe gage. Gage should register 20-60 psi with control valve in any drive position. If lubrication pressure is not as stated, refer to item 5 in TROUBLE SHOOTING section. Also, be sure hose on pressure gage has at least ½-inch I.D. as a smaller hose might restrict oil flow and give a false reading.

164. To check clutch circuit, refer to Fig. 56 and proceed as follows: Install pressure gage at port below filter element boss which will give the pump output pressure available for either clutch; or install gage at front port

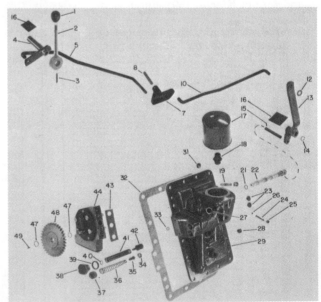

40. Shim
41. Spring
42. Regulator valve spool
43. Gasket
44. Pump
47. Snap ring
48. Pump gear
49. Key

located behind filter element boss which will give pressure available for direct drive clutch; or at rear port located behind filter element boss which will give pressure available for overdrive clutch.

Start engine, run at rated rpm and observe gage while moving control spool to direct drive and overdrive positions. Gage should register 140-160 psi in either position. If clutch pressure is not as stated, refer to items 6 and 7 in TROUBLE SHOOTING section. Shims (40—Fig. 57) behind the pressure regulator spring (41) can be added or subtracted to adjust pressure. Shims are 0.0359 thick and each shim will vary pressure about 5 psi.

### REMOVE AND REINSTALL
### All Models So Equipped

165. To remove the Hydraul Shift drive unit it is necessary to remove the engine and Hydraul Shift unit as outlined in paragraph 55. With engine removed, support it on wood blocks so oil pan will not be damaged, then remove lower dust shield and starter. Unbolt and remove clutch housing and Hydraul Shift unit from engine. Remove clutch release bearing. Remove cap screw and key from release bearing fork and pull clutch shaft from fork and clutch housing. Unbolt and remove the clutch housing from Hydraul Shift unit.

Reinstall by reversing removal procedure.

### OVERHAUL
### All Models So Equipped

166. **DISASSEMBLE HYDRAUL SHIFT UNIT.** With Hydraul Shift unit removed as outlined in paragraph 165, drain unit and remove dipstick and filler cap assembly. Remove valve spool control lever. Straighten tabs of lock washer (6—Fig. 58), then use a spanner wrench (Oliver No. ST-171, or equivalent) and remove nut (5). Remove sprocket (7) and "O" ring (8) from output shaft. Unscrew filter element (17—Fig. 57) from adapter (18), remove the filter to countershaft line, then remove cover (29) and pump (44) assembly from unit housing (45—Fig. 58). Pull input shaft support (93) and input shaft (104) from housing and direct drive clutch. Remove cap screws which retain overdrive clutch housing (12—Fig. 58), then rotate housing slightly to allow removal of countershaft support (44). Support the countershaft (75—Fig. 58) to prevent binding the support, bump on front end of support until sealing plug (42) is removed, then remove support (44) from front of housing. Allow counter-

shaft to rest on bottom of housing. Remove output shaft (77), direct drive clutch and internal ring gear (91) assembly from housing. Remove planet gear carrier (66), drive gear (62) and bearing (61) if not removed with output shaft, clutch and internal ring gear assembly. Remove thrust bushing (60) from housing. Remove overdrive clutch housing (12), clutch and sun gear shaft (37) from housing by bumping lightly on front end of sun gear shaft. Remove support (39) from housing if it was not removed with the overdrive clutch and housing assembly. Remove countershaft and countershaft thrust washers (73) from housing. Remove filler neck (59), studs, dowels or control lever shaft (47) as necessary.

With Hydraul Shift unit disassembled, discard all "O" rings, seals and gaskets and refer to the appropriate following paragraphs to service subassemblies. Refer to paragraph 173 for

reassembly procedure.

167. **COVER AND VALVES.** NOTE: If only pressure regulator valve, by-pass valve or oil filter are to be serviced, it can be accomplished without removing cover and with Hydraul Shift unit in tractor.

Remove plug (38—Fig. 57), washer (39) spring (41) and pressure regulator spool (42) from bore in cover. DO NOT lose any shims (40) which may be used. A 5/16-inch cap screw can be used to remove spool if it is stuck. Remove countersunk plug (37), spring (36) and by-pass plunger (35). Seat (34) is renewable, if necessary. Remove plug (25), spring (24) and detent ball (26), then pull control spool (22) from bore and discard "O" ring (21). DO NOT attempt to remove spool until detent assembly is removed or damage to spool could result. Remove oil filter adapter (18) so oil passages can be inspected and cleaned.

Clean and inspect all parts. Control

*Fig. 58–Exploded view of the Hydraul Shift unit showing component parts and their relative positions.*

| | | |
|---|---|---|
| 1. Rear sprocket | 30. Retainer | 54. Gasket |
| 2. Coupling chain | 31. Separator plates | 55. Cap |
| 3. Master link | 32. Clutch plates | 56. Baffle |
| 5. Spanner nut | 33. Thrust washer | 57. Dipstick |
| 6. Lock washer | 34. Over-running | 58. "O" ring |
| 7. Front sprocket | clutch | 59. Filler tube |
| 8. "O" ring | 35. Spacer | 60. Bushing |
| 9. Oil seal | 36. Bearing | 61. Bearing |
| 10. Snap ring | 37. Sun gear shaft | 62. Drive gear |
| 11. Bearing | 38. Seal rings | 63. Planet gear pin |
| 12. Clutch housing | 39. Support | 64. Bearing |
| 15. Cap screw | 40. Seal ring | 65. Planet gear |
| 16. Lock washer | 41. "O" ring | 66. Planet carrier |
| 19. "O" ring | 42. Plug | 69. Snap ring |
| 20. "O" ring | 43. "O" ring | 70. Seal rings |
| 21. Seat | 44. Support | 71. Thrust washer |
| 22. Oil line | 45. Housing | 72. Bushing |
| 23. Seal ring | 46. Magnetic plug | 73. Thrust washer |
| 24. "O" ring | 47. Lever shaft | 74. Bearing |
| 25. Piston | 48. Dowel | 75. Countershaft |
| 26. "O" ring | 50. Cap | 76. Retaining ring |
| 27. Seal ring | 51. Spacer | 77. Output shaft |
| 28. Spring plate | 52. Filter element | 78. Seal ring |
| 29. Return spring | 53. Cup | 79. "O" ring |

| | |
|---|---|
| 80. Piston | |
| 81. "O" ring | |
| 82. Seal ring | |
| 83. Thrust washer | |
| 84. Spring guide | |
| 85. Return spring | |
| 86. Retainer | |
| 87. Snap ring | |
| 88. Pressure plate | |
| 89. Clutch plates | |
| 90. Separator plates | |
| 91. Internal ring gear | |
| 92. Back plate | |
| 93. Support | |
| 96. Tube | |
| 97. "O" ring | |
| 98. Oil seal | |
| 99. Snap ring | |
| 100. Bearing | |
| 101. Bearing retainer | |
| 104. Input shaft | |
| 105. Oil seal | |
| 106. Bearing | |

spool (22) and pressure regulator spool (42) should be free of nicks or scoring and should slide freely in their bores when lubricated. By-pass plungers (35) and seat (34) should be free of nicks, burrs or grooves. Springs (24, 36 and 41) should be free of rust, bends or fractures.

Refer to the following data as a guide for parts renewal.

Pressure Regulator Valve Spring
    Free length-in. . . . . . . . . . . . . . . . .3⅜
    Spring test lbs. @ in. . . .55-57 @ 2¾
By-Pass Valve Spring
    Free length-in. . . . . . . . . . . . . . . . .3½
    Spring test lbs. @ in. 5¼-6½ @ 2 7/16
Detent Ball Spring
    Free length-in. . . . . . . . . . . . . . 1 9/32
    Spring test lbs. @ in.   15.3-18.7 @ ⅞
Pressure Regulator Spool
    Diameter . . . . . . . . . . 0.6865-0.6869
Control Spool Land
    Diameter . . . . . . . . . . 0.6865-0.6869
Pressure Regulator Spool
    Bore-I.D. . . . . . . . . . 0.6875-0.6883
Control Spool Bore I.D.   0.6875-0.6883

Reassemble by reversing the disassembly procedure.

**168. PUMP.** Unbolt pump from housing cover and discard gasket. Remove drive gear outer retainer snap ring, remove gear and key, then remove gear inner retainer snap ring.

Remove through bolts and separate pump. Dowels can be driven out and screen removed, if necessary. See Fig. 59.

NOTE: If same pump gears and shafts are to be reinstalled, mark idler gear so it will be reinstalled in original position. Do not remove gears from shafts as they cannot be serviced separately.

Pump dimensional data is as follows:
    Gear width . . . . . . . . 0.9975-0.9980
    Pump body width . . . 0.9995-1.0000
    Gear shaft diameter . 0.6247-0.6250
    Shaft bore diameter . 0.6265-0.6275
    Shaft operating
      clearance . . . . . . . . . 0.0015-0.0028

Reassemble by reversing the disassembly procedure and use new gasket when mounting pump on housing cover. Heads of bolts should be on drive gear side of pump and dowels should be installed prior to tightening through bolts.

**169. INPUT SHAFT AND SUPPORT.** Remove "O" ring (97—Fig. 58) from support (93), then remove bearing (106) and seal (105) from bore in rear of input shaft. Unbolt bearing retainer (101) and press input shaft (104) and bearing (100) from support. Oil seal (98) can now be removed from support, and if renewal is required, clutch release bearing tube (96) can

also be removed at this time. Do not remove bearing (100) from input shaft unless renewal of bearing or bearing retainer (101) is indicated. If either part is to be renewed, remove snap ring (99) and press input shaft from bearing.

To reassemble input shaft and support, place bearing retainer on input shaft, then press bearing on shaft and install snap ring. Press bearing tube, (if renewed) into support until tube protrudes 3⅜-inches from front flange of support and be sure tube is parallel with bore. Install oil seal (98) in support with lip of seal toward rear. Press input shaft and bearing into support and secure with bearing retainer. Install oil seal (105) in rear bore of input shaft with lip facing toward rear, then install bearing (106). Install "O" ring (97) in groove of support.

Refer to paragraph 173 for installation procedure.

**170. OUTPUT SHAFT, INTERNAL RING GEAR AND DIRECT DRIVE CLUTCH.** To disassemble the output shaft, internal ring gear and direct drive clutch, proceed as follows: If planet gear carrier remained with output shaft assembly during disassembly, separate it from output shaft, then remove snap ring (76—Fig. 58) and output shaft (77) from internal ring gear (91). With a press, compress spring (85), remove snap ring (87), retainer (86), return spring (85), guide (84) and thrust washer (83) from output shaft. Remove clutch piston (80) from output shaft (gear), either by bumping end of output shaft on a block of wood, or by directing compressed air into oil passage of output shaft, then remove seal rings (78 and 82) and "O" rings (79 and 81) from I.D. and O.D. of piston. Remove seal rings (70) and thrust washer (71) from rear of output shaft. NOTE: Bushing (72) is catalogued as a separate part, however, if renewal is not indicated, do not attempt to remove it from output shaft. Remove pressure plate (88)

clutch plates (89) and separator plates (90) from internal ring gear.

Clean and inspect all parts. Sprag (over-running) clutch surface of output shaft must be smooth and show no signs of wear. Clutch plates must have no broken lugs, splines, or warpage. Refer to the following data as a guide for parts renewal.

Return Spring
    Free length . . . . . . . . . . . . . . . . . 2.270
    Spring test lbs. @ in. . . 230 @ 1.750
Clutch Plate Thickness . . . 0.076-0.084
Separator Plate Thickness 0.087-0.093
Pressure Plate Thickness . 0.248-0.252
Piston Ring Groove-Outside
    Width . . . . . . . . . . . . . . . .0.156-0.162
    Depth . . . . . . . . . . . . . . . .0.193-0.194
Piston Ring Groove-Inside
    Width . . . . . . . . . . . . . . . .0.128-0.132
    Depth . . . . . . . . . . . . . . . .0.127-0.128

Reassemble output shaft, internal ring gear and direct drive clutch assembly as follows: Start with a clutch plate (89) and alternately install the six clutch plates and five separator plates (90) in internal ring gear (91), then install pressure plate (88). Install "O" rings (79 and 81) and seals (78 and 82) in O.D. and I.D. grooves of piston (80), grease seals and press piston in front side of output shaft gear. Use caution when installing piston not to damage seals. Install washer (83), guide (84), return spring (85) and retainer (86) on front end of output shaft, compress spring and install snap ring (87). Install output shaft assembly in internal ring gear with clutch piston next to clutch plates and install snap ring (76). Install thrust washer (71) over splined end of output shaft, then install the two seal rings (70) in grooves of output shaft.

**171. PLANET GEAR CARRIER.** To disassemble the planet gear carrier assembly, first remove drive gear (62 —Fig. 58) from carrier (66) by removing the four retaining cap screws. Remove the planet gear pin retaining rings (69) and drive planet gear pins (63) from carrier. Remove planet gears

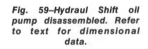
Fig. 59—Hydraul Shift oil pump disassembled. Refer to text for dimensional data.

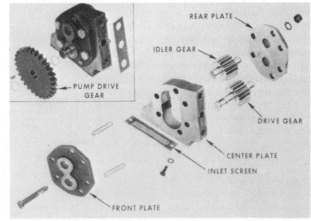

(65) from carrier and bearings (64) from planet gears.

Inspect planet gears, bearings and pins for wear, scoring or other damage and renew as necessary. Inspect drive gear teeth and inside diameter. Inside diameter of gear should not be unduly worn or scored. Check condition of bearing (61) and with bearing in gear, check fit of gear and bearing on support (39). Gear and bearing should have a free fit on support without excessive clearance. At this time, also inspect bushing (60) for wear or scoring.

To reassemble planet carrier, place planet gear pin retainer in groove at front of pin bore. Place bearing in planet gear, position gear in carrier, then insert planet gear pin from rear of carrier until pin is against retainer and step on rear of shaft is flush with lower surface of carrier. Repeat operation for the three remaining planet gears. Place drive gear on rear of carrier, align cap screw holes, then install retaining cap screws and tighten them to 48-52 ft.-lbs. torque. Bushing (60) and bearing (61) are placed on support (39) when unit is being assembled.

**172. OVERDRIVE CLUTCH AND SUN GEAR.** If removed with overdrive clutch assembly, remove sun gear shaft (37—Fig. 58) from overdrive clutch. Remove over-running clutch (34), spacer (35), bearing (36) and seal rings (38) from sun gear shaft. Bump support (39) from housing, if necessary, and remove seal ring (40). Remove clutch plates (32), separator plates (31), retainer (30), return springs (29) and return spring plate (28) from housing (12). Remove clutch piston (25—Fig. 58) from housing by bumping housing on a wood block. Remove seal rings (23 and 27) and "O" rings (24 and 26) from O.D. and I.D. of piston. Remove oil seal (9), snap ring (10) and bearing (11) from bore of housing.

With overdrive clutch disassembled, discard oil seal (9), seals (23 and 27) and "O" rings (24 and 26), then clean and inspect all parts. Clutch plates and separator plates must not be warped, scored or unduly worn. Clutch splines of sun gear shaft must not have deep grooves worn by clutch plates and bore for over-runnning clutch must be smooth and show no wear. Refer to the following data as a guide for parts renewal:

Return springs
    Free length ................ 1.320
    Spring test lbs. @ in. ... 10 @ 1.159
Clutch Plate Thickness ...0.073-0.078
Separator Plate Thickness 0.077-0.082

Piston Ring Groove—Outside
    Width ................. 0.128-0.132
    Depth ................. 0.195-0.197
Piston Ring Groove—Inside
    Width ................. 0.128-0.132
    Depth ................. 0.151-0.152

Partially assemble overdrive assembly as follows: Install bearing (11) in housing (12) and secure with snap ring (10). Install oil seal (9) with lip toward bearing. Install "O" rings in outer and inner grooves of piston (25), install seals over both "O" rings, then grease seals and install piston in housing (12). Install seals (38) in grooves of sun gear shaft (37). Place bearing (36) and spacer (35) in bore of sun gear shaft, then install the overrunning clutch so the drag strips (bronze) are toward bearing (36). Delay any further assembly of the overdrive clutch until the Hydraul Shift unit is assembled.

**173. REASSEMBLE HYDRAUL SHIFT UNIT.** When reassembling the Hydraul Shift unit, proceed as follows: Install seal ring (40—Fig. 58) on hub of support and the small "O" ring in oil hole counterbore on front side of support flange, then install support in rear bore of housing so oil hole in support flange (with "O" ring) is aligned with the oil hole in housing which is on right side of lower cap screw hole. Temporarily install one of the short cover retaining cap screws finger tight to keep support from shifting when planet carrier and output shaft assembly are installed.

Coat countershaft thrust washers (73) with grease and stick them to inner sides of countershaft support bores and be sure tangs of thrust washers are in slots. With bearings (74) installed, carefully place countershaft in housing, with double gear rearward, so thrust washers are not dislodged, then pull countershaft toward cover opening until gear teeth clear front bore of housing. Place bushing (60) on support with cupped side rearward, then install bearing (61) on support and use fingers to compress seal ring, if necessary. Install planet carrier and drive gear (gear rearward) in housing with gear (62) over bearing (61). Be sure bushing (72), thrust washer (71) and seal rings (70) are installed on splined (long) end of output shaft, start output shaft through planet carrier and support and install internal ring gear over the planet gears.

Place new "O" ring (41) on front end of countershaft support (44). Move countershaft into position, align thrust washers, lift countershaft so support will not bind, then slide support rearward until support is ready to enter rear bore. At this time, position tapped hole in rear end of support at the nine o'clock position and push shaft rearward until front of support is flush with housing.

NOTE: When overdrive clutch housing (12) is finally installed, holes in housing, plug (42) and support (44) must be aligned to allow installation of the ¼ x 1½-inch support retaining cap screw, as well as to place the oil hole in support on top side. Support is slotted at forward end if subsequent positioning is required.

With countershaft support positioned, place new "O" ring (41) on O.D. of plug (42), then use grease to stick the small "O" ring (43) in counterbore of cap screw hole in front of plug. Insert cap screw (15) through plug and small "O" ring (43) and thread a few turns into support, then bump plug into housing bore. Remove cap screw after plug is installed.

Obtain a small ½-inch thick wood block about 3x3 inches square, place block against forward end of output shaft, then tip assembly forward so output shaft is supported by the wood block. With small end forward, place sun gear shaft assembly over output shaft and push sun gear shaft on output shaft until it bottoms.

NOTE: Installation of over-running clutch can be checked at this time. Sun gear shaft will turn (clutch overrun) clockwise but will lock up when attempt is made to turn shaft counterclockwise.

Remove cap screw that was installed earlier in flange of support (39). To ease assembly of overdrive clutch, install two guide studs in cap screw holes of housing (45). Install retainer (30) over guide studs with chamfer in I.D. of retainer on top side. Start with a separator plate (31) and alternately install the six separator plates and the five clutch plates (32). Place thrust washer (33) over end of output shaft with flat side toward sun gear shaft and be sure it does not hang up on shoulder of output shaft. Place return springs (29) in blind holes of retainer (30), then place plate (28) over return springs. Place new "O" ring (19) over flange of support (39) and small "O" rings (20) in counterbores of front face of housing (12), then carefully install housing (12) over guide studs. Remove guide studs and secure clutch housing with cap screws. Install countershaft support retaining cap screw (15). Place new "O" ring (8), sprocket (7), lock washer (6) and nut (5) on end of output shaft and using a spanner wrench (Oliver No. ST-171 or equivalent), tighten nut (5). Secure nut by bending a tab of lock washer into slot in nut.

Set unit in an upright position, align splines and center direct drive clutch plates as nearly as possible. Install new "O" ring (97) on pilot of front support (93) then insert input shaft hub into direct drive clutch and rotate input shaft as necessary to engage the clutch plates. NOTE: Two cap screws can be temporarily installed to help in overcoming resistance of "O" ring; however BE SURE input shaft can be rotated while drawing support into housing to prevent damage to inner clutch plates. Cut-out in O.D. of support flange must mate with a similar cut-out at right side of housing to permit operation of control spool.

Place the wood block used earlier under front of housing to tip it slightly rearward, then remove the two temporarily installed cap screws from support. Place engine clutch housing on support and secure clutch housing and support to Hydraul Shift unit housing. Install clutch throw-out bearing, throw-out bearing fork and clutch shaft. Use new gasket and install cover and pump assembly. Install cover-to-countershaft oil line and the control spool lever. When installing new oil filter, tighten filter not more than ½-turn after gasket contacts base. Install drain plug if necessary and fill unit with 3½ quarts (with filter) of Automatic Transmission Fluid, Type "A".

on plug which is threaded into bottom of differential and final drive compartment just to rear of the transmission compartment drain plug. See Fig. 61.

Recommended transmission, differential and final drive lubricant is SAE 80 multi-purpose gear lubricant. SAE 90 Multi-purpose gear lubricant may be used in temperatures above 32° F. All gear lubricant must conform to Military Specification MIL-L-2105. Sump capacity is 43 quarts. Lubricant should be changed yearly or after each 1000 hours of operation. Oil filter should be renewed after each 250 hours of operation.

176. **PRESSURE LUBRICATION SYSTEM.** The transmission and differential are lubricated by a pressure oiling system. Pressure is provided by a G-rotor type pump which is chain driven from hydraulic pump driven gear. Oil from pump is directed from pump to an externally mounted oil filter. From the oil filter, oil is directed to the differential shaft, transmission input shaft, bevel pinion shaft, pto and hydraulic pump driven gear. See Fig. 62. A by-pass (pressure relief) valve in the transmission circuit limits system pressure to 8-14 psi. Oil by-passed by this valve is returned to the transmission compartment via a hole in the tractor main frame. Refer to Fig. 63.

# TRANSMISSION

Series 1755, 1855 and 1955 tractors are equipped with a constant mesh, helical gear type transmission having six forward and two reverse speeds. When equipped with an auxiliary Hydra-Power drive, or a Hydraul Shift drive, two or three different speed ratios are available for each transmission gear. Optional transmission gears are available for some models which provide different transmission speed ratios.

On tractors with no auxiliary drive or four-wheel drive, it is not absolutely necessary to split tractor for transmission service. However, due to space limitations and the shaft adjustments required, it is generally advisable, and most mechanics prefer, to make the tractor split. Consequently, the following information is based on the tractor being split. See paragraph 187.

### LUBRICATION

175. **LUBRICANT.** Transmission and differential are lubricated from a common oil supply. Oil level should be maintained at check plug opening located as shown in Fig. 60. Filler cap is located at upper left of pto housing as shown in Fig. 60. Two drain plugs are used; front plug is located under front center of main frame and drains transmission compartment and rear plug is located at rear end of main frame and drains the differential and final drive compartment. Also, pressure lubrication pump suction screen should be removed and cleaned whenever lubricating oil is drained. Screen is located

Fig. 63–Transmission lubrication circuit by-pass valve removed from bore in rear main frame. Valve plunger and spring are retained in main frame by an expansion plug (not shown).

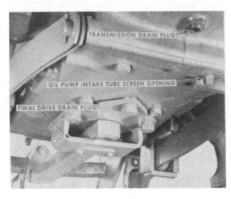

Fig. 61–Drain plugs and lubrication pump screen plug are located as shown.

*Fig. 60–Transmission and differential compartment test plug and filler plug are located as shown.*

*Fig. 62–Schematic view showing the transmission lubrication circuit. Note location of by-pass valve.*

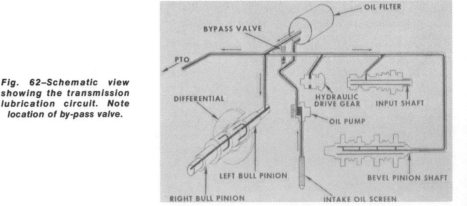

With the lubricating oil at operating temperature and engine running at 2400 rpm, system pressure should be 8-14 psi with the transmission shift lever in either neutral position. Pressure can be checked by installing a gage in port located directly below oil filter element. If pressure is below 8 psi, drain transmission and differential compartments, renew the transmission oil filter and refill to proper level with specified lubricant. Refer to paragraph 175. Faulty oil pump, broken oil tube or faulty by-pass valve could also cause low oil pressure. Refer to paragraph 186 for pump servicing information. If pressure is higher than 14 psi, check the by-pass valve, (Fig. 63). If valve is OK, check for obstruction in oil tubes and passages.

## OVERHAUL TRANSMISSION

To perform complete service on transmission will require removal of the hydraulic pump drive shaft or pto

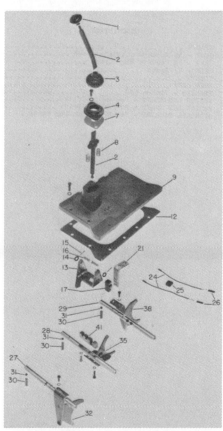

*Fig. 64–Exploded view of transmission top cover assembly. Shift rails operate in bores of transmission housing (rear main frame).*

| | |
|---|---|
| 1. Knob | 24. Switch wire |
| 2. Shift lever | 25. Wire seal |
| 3. Boot | 26. Connector |
| 4. Spring cover | 27. Right hand shift |
| 7. Gasket | rail |
| 8. Spring | 28. Center shift rail |
| 9. Top cover | 29. Left shift rail |
| 12. Gasket | 30. Detent spring |
| 13. Interlock | 31. Detent ball |
| 14. Washer | 32. Right fork |
| 15. Cotter pin | 35. Center fork |
| 16. Pin | 38. Left fork |
| 17. Safety switch | 41. Shifter lug |
| 21. Support | |

drive shaft, transmission top cover, hydraulic lift system housing, transmission lubrication pump, hydraulic pump, hydraulic pump drive bevel gears, bull gears, bull pinion shafts and differential.

Refer to the following appropriate paragraphs.

**177. R&R HYDRAULIC PUMP AND DRIVEN GEAR SUPPORT.** Remove pump shield, then remove plug from bottom of filter and drain reservoir. Disconnect pump drain line and remove pump outlet line. Loosen hose clamps and disconnect inlet hose. Loosen inlet line locknut, move tube out of the way, then remove pump retaining cap screws and remove pump and splined coupling from pump drive gear support.

Remove pump support retaining cap screws and remove support and shim pack from rear main frame.

178. To reinstall units, place shims on support, align cap screw holes, then place Permatex No. 3, or equivalent, on inside diameter of shims. Position support with lube holes on top side and install support to rear main frame.

NOTE: If backlash of hydraulic pump drive gears requires checking and/or adjusting, the Oliver Corporation recommends that the hydraulic lift system housing be removed and the backlash of gears checked with a dial indicator. Recommended gear backlash is 0.006-0.012 and is obtained by varying the shims between support and rear main frame. Add shims to increase, or remove shims to decrease backlash.

When reinstalling hydraulic pump, cement new gasket to pump with Permatex No. 3, or equivalent, and coat main frame side of gasket with grease. Fill pump with hydraulic fluid and install pump by reversing removal procedure. Fill hydraulic reservoir, bleed filter and operate system. Recheck fluid level and add as necessary.

**179. OVERHAUL HYDRAULIC PUMP AND SUPPORT ASSEMBLY.** To overhaul the hydraulic pump, refer to paragraph 257 in HYDRAULIC LIFT SYSTEM section of this manual.

To overhaul the driven gear and support, refer to Fig. 66 and proceed as follows: Straighten washer (2) and remove outer nut (1) and washer (2). Remove adjusting (inner) nut (1) and remove driven shaft (9) and bearings from support. Bearing cups can be removed from support, if necessary.

Clean and inspect all parts and renew as necessary. If driven gear (9) is renewed, also renew drive gear (10).

To reassemble, press new bearing cups in support with tapers facing out. Place inner bearing on driven gear shaft with taper facing away from gear. Insert shaft and bearing in support, install outer bearing with taper toward inside and install adjusting nut. Install a ⅜-inch cap screw in end of shaft so bearing preload can be checked, then tighten adjusting nut until shaft has a rolling torque of 3-5 inch pounds. Install lock washer and lock nut, recheck rolling torque of shaft, and if satisfactory, bend lock washer to secure nuts.

**180. R&R PTO OR HYDRAULIC PUMP DRIVE SHAFT.** To remove pto drive shaft, remove cover, or pto output shaft if so equipped, from top opening of pto housing. Extract the internal snap ring which retains plug in pto pinion, then pull plug using a ½-inch cap screw threaded into plug. Remove "O" ring from pinion bore, then use a ½-inch cap screw at least 7 inches long threaded into pto drive shaft to remove shaft.

Reinstall by reversing removal procedure.

181. To remove hydraulic pump drive shaft on tractors having no pto, remove cap screws which retain drive shaft bearing housing to main frame rear cover. Turn bearing housing until there is room enough behind housing for fingers or wedge bar and pull bearing housing and drive shaft as an assembly from rear main frame.

Reinstall by reversing removal procedure and in some cases it may be necessary, after disabling the engine, to momentarily engage starter to align shaft splines and drive hub splines.

**182. R&R HYDRAULIC PUMP DRIVE GEAR.** To remove hydraulic pump drive gear, remove hydraulic lift housing as outlined in paragraph 262, the pto drive shaft as in paragraph 180, or the hydraulic pump drive shaft as in paragraph 181. Disconnect

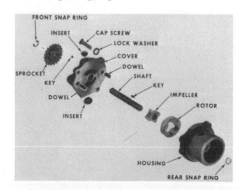

*Fig. 65–Exploded view of transmission lubrication pump. Pump is chain driven by hydraulic pump drive gear and shaft. Pump retaining lock and spring are not shown.*

master link and remove transmission lubrication pump drive chain. Remove hydraulic pump and support as in paragraph 177. Use a spanner wrench (Oliver No. ST-202 or equivalent) to hold spanner nut (14—Fig. 66) and unscrew sprocket (15) from shaft (10). Use Oliver tool No. STS-152 or STS-186 adapter to prevent shaft from turning and remove spanner nut. Remove gear and bearings. Bearing cups, spacer (13) and snap ring (12) can now be removed from rear main frame if necessary.

Clean and inspect all parts and renew as necessary. If drive gear (10) is renewed, also renew driven gear (9).

To reinstall pump drive gear, install snap ring (12) spacer (13) and bearing cups in rear main frame with tapers of cups facing away from each other. Install front bearing on shaft with taper facing rearward, place shaft and bearing in position and install rear bearing with taper facing forward. Install spanner nut and tighten until shaft has a rolling torque of 3-5 inch pounds. Hold nut in this position and install sprocket (15). Recheck the shaft rolling torque after sprocket has been tightened.

Reinstall remaining items by reversing the removal procedure.

183. **R&R TRANSMISSION TOP COVER.** To remove transmission top cover, remove gear shift lever (2—Fig. 64), spring cover (4), gasket (7) and the two lever springs (8) from top cover (9). Remove platform section and insulation. Disconnect the wiring and hydraulic lines which are routed over top cover. Disconnect the starter safety switch wires, then unbolt and remove top cover and gasket.

Interlock (13), safety switch (17), switch bracket, wiring and wiring seal (25) can now be removed if necessary and procedure for doing so is obvious.

Before reinstalling top cover, check adjustment of safety switch and if necessary, readjust position of switch until it opens when roller is about one-half way up side of interlock ramp. Switch has an audible click when it opens and closes.

Use new gasket and reinstall top cover by reversing removal procedure.

184. **SHIFTER RAILS AND FORKS.** Removal of shifter rails on models equipped with auxiliary drive will require removal of engine or splitting tractor.

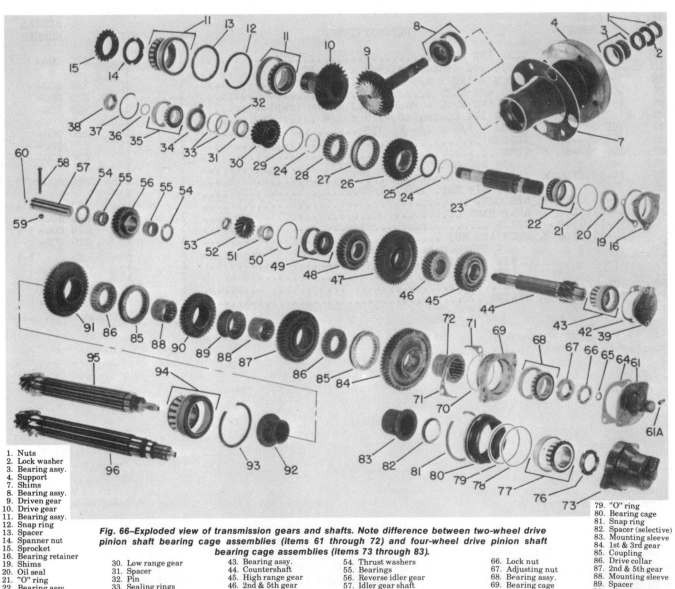

*Fig. 66—Exploded view of transmission gears and shafts. Note difference between two-wheel drive pinion shaft bearing cage assemblies (items 61 through 72) and four-wheel drive pinion shaft bearing cage assemblies (items 73 through 83).*

| | | | |
|---|---|---|---|
| 1. Nuts | 30. Low range gear | 43. Bearing assy. | 54. Thrust washers |
| 2. Lock washer | 31. Spacer | 44. Countershaft | 55. Bearings |
| 3. Bearing assy. | 32. Pin | 45. High range gear | 56. Reverse idler gear |
| 4. Support | 33. Sealing rings | 46. 2nd & 5th gear | 57. Idler gear shaft |
| 7. Shims | 34. Oil collector | 47. Low range gear | 58. Cap screw |
| 8. Bearing assy. | 35. Bearing assy. | 48. 4th & 6th gear | 59. Self-locking nut |
| 9. Driven gear | 36. "O" ring | 49. Bearing assy. | 60. Plug |
| 10. Drive gear | 37. Snap ring | 50. Snap ring | 61. Speedometer drive housing |
| 11. Bearing assy. | 38. Nut | 51. Locating sleeve | 61A. Plug |
| 12. Snap ring | 39. Bearing retainer | 52. Reverse gear | 64. Gasket |
| 13. Spacer | 42. Shims | 53. Nut | 65. Oil seal |
| 14. Spanner nut | | | |
| 15. Sprocket | | | 66. Lock nut |
| 16. Bearing retainer | | | 67. Adjusting nut |
| 19. Shims | | | 68. Bearing assy. |
| 20. Oil seal | | | 69. Bearing cage |
| 21. "O" ring | | | 70. "O" ring |
| 22. Bearing assy. | | | 71. Shims |
| 23. Input shaft | | | 72. Mounting sleeve |
| 24. Snap ring | | | 73. Covers |
| 25. Thrust washer | | | 76. Nut |
| 26. High range gear | | | 77. Bearing assy. |
| 27. Coupling | | | 78. "O" ring |
| 28. Drive collar | | | 79. "O" ring |
| 29. Sealing ring | | | 80. Bearing cage |

| | |
|---|---|
| 81. Snap ring | 89. Mounting sleeve |
| 82. Spacer (selective) | 90. 4th & 6th gear |
| 83. Mounting sleeve | 91. Reverse gear |
| 84. 1st & 3rd gear | 92. Mounting sleeve |
| 85. Coupling | 93. Snap ring |
| 86. Drive collar | 94. Bearing assy. |
| 87. 2nd & 5th gear | 95. Pinion shaft (2 WD) |
| 88. Mounting sleeve | 96. Pinion shaft (4 WD) |

Remove transmission top cover, as outlined in paragraph 183. Remove hydraulic lift housing as outlined in paragraph 262. Remove pinion shaft external oil line. Remove cap screws which retain shift forks to shifter rails. Push the two outer rails forward and push out the expansion plugs located at forward ends of shifter rails. Be careful not to lose poppet balls and springs as rails clear rear bores. Prior to removing the center shifter rail, remove shifter lug (41—Fig. 64). Lift out shifter forks after shift rails are removed.

Reinstall shift rails and forks as follows: Place center fork in case, place poppet spring in center poppet hole and place poppet ball on top of spring. Insert center shift rail part way into rear frame. Turn shifter rail so flat on inner end is downward, then use a small punch to depress poppet ball and spring and push shifter rail rearward until rail goes over poppet ball. Turn shifter rail until poppet ball seats in detent, then install shifter lug and fork. Install right and left shifter rails and forks in the same manner. Use sealing compound and install new expansion plugs in shifter rail front bores.

Complete balance of reassembly by reversing disassembly procedure.

**185. LUBRICATION BY-PASS VALVE.** The by-pass valve for the transmission lubrication circuit is located in a bore on left side of rear main frame directly above oil filter. Valve is accessible after removing the hydraulic lift system housing. See Fig. 63.

Valve spring and plunger can be removed after removing expansion plug. Inspect plunger for damage or wear. Valve spring has a free length of 2 15/32-inches and should test 1.7-2.3 pounds when compressed to a length of 1 25/32-inches. Valve will maintain a lubrication pressure of 8-14 psi at an engine rpm of 2400.

Lubrication pressure can be checked at a port located directly below oil filter element. Check pressure in both neutral positions of transmission. Refer to paragraph 176 if pressure is not within specifications.

**186. LUBRICATION PUMP.** To remove the lubrication pump, first remove the hydraulic lift housing. Disconnect master link and remove drive chain. Remove the pump to filter line and the pump suction tube. Use a small punch or rod to depress lock and withdraw pump from bore in rear main frame.

With pump removed, remove front snap ring and remove sprocket and

Woodruff key. Remove cap screws and separate cover from housing. Remove rear snap ring and remove shaft, impeller and rotor from housing. If necessary, remove dowels and inserts from cover.

Clean and inspect all parts. Impeller should be a free fit on shaft, and if necessary, polish shaft with crocus cloth or file drive key. Renew any parts which show undue wear or any other damage.

When reassembling pump, coat all parts with oil and place rotor in housing. Install key and impeller on shaft, insert assembly into housing and install rear snap ring. If necessary, install dowels and inserts in cover, slide cover over shaft and secure to housing. Install sprocket key, sprocket and front snap ring. Place spring and lock in bore of housing and install pump in bore of rear main frame. Complete reassembly by reversing the disassembly procedure.

**187. TRACTOR SPLIT.** To split tractor for complete service on transmission it is recommended that the fuel tank and instrument panel assembly be removed rather than the engine as this will provide a better working condition for the transmission service.

Make tractor split as follows: Drain transmission and final drive compartments and remove pto shaft, or the hydraulic pump drive shaft if tractor has no pto. Remove hood, hood side panels and fuel tank. Disconnect wiring from engine, starter safety switch and tail lights. Disconnect engine controls, clutch rod and auxiliary drive unit cable. Disconnect the necessary hydraulic lines, then unbolt and remove instrument panel.

Disconnect hydraulic lines attached to hydraulic lift unit and remove sprocket coupling chain. Remove any front frame weights which may be installed, then support front frame so it will not tilt. Attach hoist to rear frame, then remove the front to rear frame cap screws and separate front and rear frames.

When reassembling, tighten the front to rear frame cap screws to 280-300 ft.-lbs. torque.

**188. R&R INPUT SHAFT.** To remove input shaft, split tractor as outlined in paragraph 187. Drain rear main frame, then refer to the appropriate sections and remove pto drive shaft or the hydraulic pump drive shaft, transmission top cover and hydraulic lift system housing. Remove the hydraulic pump drive gear and input shaft oil collector oil lines. In some cases, it may be desirable to

remove pinion shaft oil lines and fittings. Remove the hydraulic pump, pump support and pump drive gear.

Remove input shaft bearing retainer (16—Fig. 66), then remove oil seal (20) and "O" ring (21) from retainer. Save shims (19) for reinstallation. Unstake nut (38) from rear of input shaft and remove nut and rear bearing cup snap ring (37). Slide coupling (27) forward onto gear (26), then lift front snap ring (24) from its groove and slide it forward over input shaft. Use a shaft driver (Oliver No. ST-138, or equivalent) and bump input shaft rearward until gear (30) and drive collar (28) can be separated far enough to expose Teflon ring (29) and snap ring (24). Cut the Teflon ring, lift rear snap ring from its groove and slide it rearward on input shaft. Place a spacer between front input shaft gear and housing so gears cannot move forward, then using shaft driver, bump input shaft forward until front bearing cup is free of housing and input shaft is free of rear bearing. Pull shaft from front of housing and lift shaft components from top of housing as shaft is being removed.

NOTE: The spacer used to block the gears can be made from a piece of 3½-inch I.D. pipe, 1½-inches long with a longitudinal section cut out to permit it to slip over the input shaft.

189. Reinstall input shaft assembly as follows: Press front bearing cone on input shaft with smaller diameter forward. Place front snap ring (24) on shaft about ½-inch forward of its groove, then place thrust washer (25) next to the snap ring. Place seal rings (33) in grooves of spacer (31), oil seal rings freely, then install spacer and rings in oil collector (34) from flat side of oil collector using fingers to compress sealing rings. Push center shift rail to its forward position. Place shift coupling (27) in its hub of high range gear (26), start rear of input shaft in front bearing bore and install gear and collar on shaft with collar rearward and mated with its shifter fork. Install drive collar (28) with counterbore rearward. Start rear snap ring (24) on shaft spline with sharp ends forward but do not slide it into its groove. Place Teflon seal ring (29) on shaft at rear of snap ring (24). Install low range gear (30) on shaft with helical teeth toward rear. Place pin (32) in spacer, align it with keyway of input shaft and install the oil collector and spacer assembly on shaft with counterbored side of oil collector forward. Install the rear snap ring (24) in its groove and be sure Teflon sealing ring (39) is on splined portion of shaft. In-

stall front bearing cup in its bore, install front bearing retainer (without seal and "O" ring) and shims and tighten retaining cap screws finger tight. Lubricate "O" ring (36) and install in groove next to threads on rear end of shaft. Start rear bearing on shaft with smaller diameter rearward and bump bearing on shaft until nut (38) can be started. Be sure large diameter flange of nut is against bearing, tighten nut until bearing is tight against spacer (31), then complete tightening of nut to a torque of 90-100 ft.-lbs. Stake nut into keyway of shaft.

NOTE: While tightening nut be sure sealing ring (29) stays on splined portion of input shaft or damage to sealing ring will occur.

Install rear bearing cup and snap ring (37), then bump on forward end of shaft to seat cup against snap ring. Slide drive collar (28), high range input gear (26) and thrust washer (25) rearward and install front snap ring (24) in its groove.

Further tighten the front bearing retainer cap screws, if necessary, until input shaft has only a slight end play. Demesh input shaft gears from countershaft gears, then using an inch pound torque wrench and adapter, check and record the rotational resistance of the input shaft. Completely tighten the bearing retainer cap screws and recheck the shaft rotational drag which should be not more than 5 inch pounds greater than the previously recorded value. Vary shims (19) as required to obtain this adjustment. Shims are available in thicknesses of 0.004, 0.007, 0.0149. Note: If shims are varied be sure shaft is bumped slightly forward after each check to insure that bearing cup contacts flange of bearing retainer.

With input shaft adjustment made, remove front bearing retainer and install oil seal (20) with lip rearward and "O" ring (12) in retainer. Lubricate oil seal and "O" ring and install retainer. Install input shaft oil line between oil collector and tee and be sure oil line does not cause oil collector to bind. Install any other oil lines that were removed. Shift transmission to neutral.

**190. R&R COUNTERSHAFT.** To remove the countershaft, first remove input shaft as outlined in paragraph 188.

Remove bearing retainer (39—Fig. 66) and shims (42). On four-wheel drive models, remove bevel pinion shaft oil line and the transfer case breather and neck. Unstake and remove nut (53). Use two ⅜ x 1 ⅛-inch cap screws, fitted with nuts, between

gear (45) and front wall of transmission housing to prevent loading of gears during shaft removal.

Thread screw of puller (Owatonna No. 927, or equivalent) into tapped hole in front of countershaft, then making sure gears are not binding, pull countershaft from rear bearing and front bearing cup from main frame. Complete removal of countershaft and lift shaft components from top of main frame as shaft is withdrawn. Remove front bearing from countershaft and rear bearing cup and snap ring from housing, if necessary.

Reinstall by reversing removal procedure and tighten nut (new) to a torque of 150 ft.-lbs. Vary shims (42) to provide countershaft with 0.001-0.003 end play. Shims are available in thicknesses of 0.004, 0.007 and 0.0149.

**191. R&R BEVEL PINION SHAFT.** Bevel pinion shaft and bearing cage assembly used in two-wheel drive tractors differ somewhat from those used in four-wheel drive tractors and can be noted in Fig. 66. However, in some cases where conversion from two-wheel to four-wheel drive is being considered, two-wheel drive tractors may be equipped with the four-wheel drive bevel pinion shaft.

Refer to the appropriate following paragraphs.

**192. TWO-WHEEL DRIVE.** To remove bevel pinion shaft, remove input shaft, countershaft and the differential as outlined in paragraph 195.

NOTE: While it is not mandatory that shifter rails and forks be removed, their removal will simplify procedure.

Remove bevel pinion shaft oil line, then remove speedometer drive housing (61—Fig. 66) and gasket (64). Remove oil seal (65) from speedometer housing. Wrap speedometer drive gear and seal surface on end of bevel pinion shaft with tape to prevent damage during disassembly.

Use spanner wrenches (Oliver ST-127 and a modified ST-142) as shown in Fig. 67 and remove lock nut (66—Fig. 66) and adjusting nut (67). Install two ⅜-16 cap screws in tapped holes of bevel pinion shaft front bearing cage (69) and pull bearing cage and front bearing assembly (68). Save shims (71) for subsequent reinstallation.

NOTE: Do not hammer on front end of bevel pinion shaft as shaft will be damaged.

Remove "O" ring (70), and if necessary, the front bearing cup from bearing cage. Slide bevel pinion shaft rearward and remove shaft components from top of main frame as shaft

is withdrawn. Remove rear bearing from bevel pinion shaft and rear bearing cup and snap ring from main frame.

**193.** Reinstall bevel pinion shaft assembly by reversing removal procedure. However, before installing speedometer drive housing, adjust shaft bearing preload and gear end play as follows. Use short cap screws to temporarily secure front bearing cage to housing. Block rear end of bevel pinion shaft and drive front bearing into position. Install adjusting nut, tighten nut until a slight amount of end play remains, then bump shaft both ways to seat bearings and bearing cups. Slide front gear coupling rearward on to gear (87) and place rear coupling (85) in neutral position. Tighten adjusting nut until bevel pinion shaft has a rolling torque of 4-6 lbs. measured with a spring scale. Spring scale is attached to center output gear (87—Fig. 66) with a length of cord to make shaft rolling torque adjustment. If adjusting nut tightens before correct preload is obtained, add an equal amount of shims (71) behind flanges of bearing retainer (69) until the correct bearing preload can be obtained. Shims are 0.0149 thick.

With bearing preload adjusted, mount a dial indicator so a reading can be taken from spacer (89), then move gears back and forth by prying behind gears (84 and 91) and check indicator which should show an end play of 0.001-0.017. If sleeve end play is incorrect, vary shims (71) behind bearing retainer as required and reset the shaft bearing preload.

With the preceding adjustments made, loosen adjusting lock nut about ¼-turn, install lock nut (66) and while holding adjusting nut, tighten lock nut to a minimum of 150 ft.-lbs. Recheck the bearing preload, and if necessary, readjust nut (67). Stake lock nut to bevel pinion shaft when adjustment is complete.

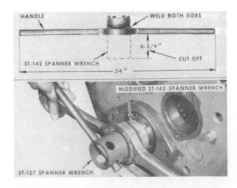

*Fig. 67—Modify an ST-142 spanner wrench as shown for use in adjusting bevel pinion shaft.*

193A. FOUR-WHEEL DRIVE. Four-wheel drive tractors use a bevel pinion shaft which is similar, except for front bearing cage assembly (items 73 through 83), to the bevel pinion shaft used in two-wheel drive tractors. The four wheel drive pinion shaft assembly may be installed in some two-wheel drive tractors.

To remove bevel pinion shaft, remove transfer drive assembly as in paragraph 32, input shaft, countershaft and the differential as in paragraph 195. Tape seal contact surface of bevel pinion shaft to prevent damage to shaft.

NOTE. While it is not mandatory that the shifter rails and forks be removed, their removal will simplify procedure.

Unstake nut (76—Fig. 66), then using a driving sleeve over oil tube on front of shaft, bump shaft (96) rearward and remove front bearing,

bearing cage (80) and spacer (82). Pull shaft rearward and lift gears and spacers from housing as shaft is withdrawn. Snap rings (81 and 93) and rear bearing cup can be removed from housing, if necessary.

Reinstall by reversing the removal procedure, however prior to complete assembly the bevel pinion shaft bearing preload and shaft gear end play must be adjusted as follows: Use the original spacer (82) and tighten nut (76) until only a slight end play remains, then bump shaft both ways to seat bearings and cups. Slide front shift collar (85) rearward on to gear (57) and place rear shift collar (85) in neutral position. Wrap a cord around center output gear (87) and attach a spring scale to cord. Tighten nut (76) until a spring scale reading of 4-6 lbs. is required to keep shaft in motion. If nut tightens before correct bearing preload is reached, remove nut, front bearing and spacer (82) and install

next smaller spacer. Spacers are available in thicknesses of 0.288, 0.314 and 0.340.

With bevel pinion bearing preload adjusted, mount a dial indicator so reading can be taken from spacer (89), then move gears back and forth by prying behind gears (84 and 91) and check indicator which should show an end play of 0.001-0.027. If end play of gears on shaft is not as stated, change spacer (89) as required and readjust bevel pinion shaft bearing preload. Stake nut into slot of bevel pinion shaft when preceding adjustments are correct.

194. R&R REVERSE IDLER GEAR. To remove reverse idler gear, first remove countershaft and bevel pinion shaft. Remove nut and cap screw (58—Fig. 66) which retains idler gear shaft in housing, then remove shaft, gear (56), bearings (55) and thrust washers (54).

# DIFFERENTIAL, BEVEL GEARS, FINAL DRIVE & REAR AXLE

### DIFFERENTIAL
### Series 1755-1855-1955

195. REMOVE AND REINSTALL. To remove the differential, first drain transmission and final drive compartments. Remove the pto drive shaft, or the hydraulic pump drive shaft if tractor has no pto. Remove the hydraulic lift assembly. Remove drawbar support and pto control valve linkage, then remove pto assembly from main frame. Disconnect wiring and remove both fenders, then support rear frame and remove both wheel and tire units. On models with internal draft sensing hydraulics, remove draft control spring, support and spacers. On either side of tractor, remove snap ring fron inner end of axle and the cap screws retaining axle carrier to main frame: then using a pusher, push axle from bull gear and remove axle and carrier. Repeat operation on opposite side of tractor using a piece of bar stock to support pusher. Lift bull gears from rear frame.

Disconnect hand brake linkage at brake housing cover and remove hydraulic line at fitting. Remove brake housing cover and insulating disc. Pull adjusting plate, disc, separator, second disc and four springs from housing. NOTE: Series 1955 tractors use three discs and two separators.

Remove cap screws from brake housings and wedge them out far enough to remove the two sets of shim packs from behind each brake housing. Identify shims as to their locations and tie them together to prevent intermixing. Install a lifting eye in the tapped hole in one of the differential pinion pins, attach a hoist to lifting eye and take slack from hoist. Carefully pull either brake housing and bull pinion out until differential side gear can be grasped, then hold gear and bull pinion in place and pull entire assembly from rear frame. Remove opposite brake housing and bull pinion in the same manner. Remove thrust bearings and thrust washers from differential shafts and identify them as

to their location. On Series 1955 tractors, unbolt brake housing supports (23—Fig. 69) from rear frame and save shims (26) located under supports. Remove differential (and supports 1955) from rear frame. If necessary remove the thrust bearing assembly (items 11 through 16—Fig. 68) used in Series 1955 tractors.

196. If any of the loose roller bearings drop out of bull pinion inner bearing, refer to paragraph 204 for reassembly. Reinstall differential by reversing the removal procedure and adjust bearings and backlash as outlined in paragraph 197.

**NOTE: When installing differential in Series 1955, be sure the shims (26—Fig. 69) located under brake housing supports**

Fig. 68–View showing component parts of typical differential assembly. Items 11 through 16 are used on Series 1955 only.

1. Side gear
2. Thrust bearing
3. Thrust washer
4. Bolt
5. Bevel gear
6. Pinion gear
7. Pinion pin
8. Retaining ring
9. Pinion pin
10. Conelock nut
11. Nut
12. Washer
13. Clip (lock)
14. Support
15. Bearing
16. Eccentric shaft

are the same thickness as the values stamped on each top rear side of rear main frame. Values may not be the same for both sides. Shims are available in thicknesses of 0.004, 0.007 and 0.0149.

197. BEARINGS ADJUST. Differential end play and gear backlash are controlled by the two shim packs (21—Fig. 69) interposed between the two brake housings and rear main frame. When adjusting bearings, do not depend on shim count or prior assembly. Use a micrometer to measure shim pack thickness to be sure that both shim packs installed between brake housing and rear frame are identical in thickness. Shims are available in thicknesses of 0.004, 0.007 and 0.0149.

Mount a dial indicator and check the differential end play which should be 0.001-0.003. Add or subtract shims (21) as necessary to obtain the specified end play, however BE SURE to maintain some gear backlash during this operation.

With differential end play adjusted, use the dial indicator to check backlash between bevel pinion and bevel gear at four points around drive gear; any large differences in backlash readings will indicate runout of drive gear due to improper tightening of gear mounting bolts or foreign material between gear and spider. Average the four readings, if no large differences exist, to determine gear backlash. Transfer shims, in equal amounts, from one brake housing to the other until the average of the four dial indicator readings agrees with the backlash specifications etched on the bevel drive (ring) gear.

Mesh position of bevel gears is fixed and nonadjustable.

After bearing and backlash adjustments have been made on Series 1955 tractors, use dial indicator on back side of bevel drive gear and locate high spot on gear. Mark high spot, and turn gear until high spot aligns with thrust bearing (15—Fig. 68), then adjust the eccentric bearing shaft (16) until bearing just contacts bevel gear (5). Tighten lock nut (11) and secure with clip (13).

198. OVERHAUL. Remove differential as outlined in paragraph 195. Remove pinion pin retaining rings (8 —Fig. 68), then with puller Oliver No. STS-100 or equivalent, threaded into tapped hole of pinion pin, pull pins and remove pinions (6). Be sure to note location of pinion pin (7) so it can be reinstalled in same location. Differential spider shaft is drilled to accept the pilot on end of pinion pin (7).

When reassembling, install pinions with taper facing inward and tapped

Fig. 69–Exploded view of brake housings, brakes and bull pinion shafts. Item (23) is used in Series 1855 and 1955. Series 1955 are equipped with three discs (12) and two separators (13) in each brake housing.

1. Cap
2. Bleed screw
3. Cover
4. Cap screw
5. Lock washer
7. "O" ring
8. "O" ring
9. Piston
10. Insulator
11. Adjuster
12. Disc
13. Separator
14. Pin
15. Spring
16. Pin
17. Seal
18. Brake housing
19. Cap screw
20. Lock washer
21. Shim
22. Dowel
23. Support
24. Washer
25. Cap screw
26. Shim
27. "O" ring
28. Bearing assy.

29. Teflon ring
30. "O" ring
31. Bull pinion shaft
32. Expansion plug
33. Roller
34. Bearing race

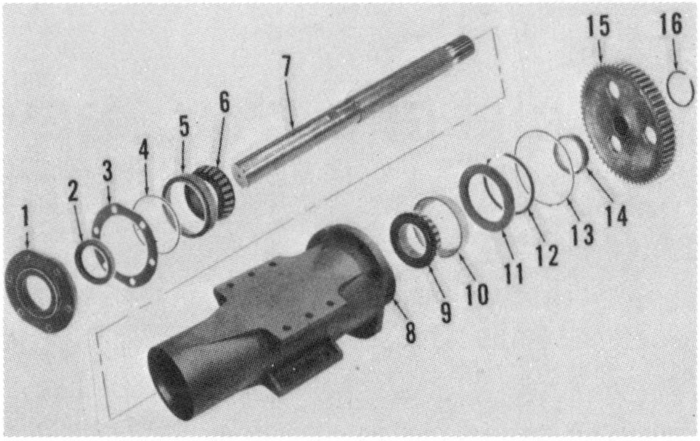

Fig. 70–Exploded view of rear axle, carrier and bull gear assembly. Axle bearing preload is adjusted by shims (3).

1. Retainer
2. Oil seal
3. Shim
4. "O" ring
5. Bearing cup
6. Bearing cone
7. Axle shaft
8. Carrier
9. Bearing cone
10. Bearing cup
11. Retainer
12. Snap ring
13. "O" ring
14. Spacer
15. Bull gear
16. Snap ring

hole of pinion pins facing outward. When installing bevel drive gear on spider be sure there is no foreign material between gear and spider and install bolts (4) with heads next to bevel drive gear. Tighten nuts alternately and evenly to a torque of 77-82 ft.-lbs. (oiled).

### BEVEL GEARS
### Series 1755-1855-1955

199. Data for removal and overhaul of bevel pinion shaft is given in paragraph 191 and for differential in paragraph 198. Adjust differential backlash as outlined in paragraph 197.

### FINAL DRIVE AND REAR AXLES
### Series 1755-1855-1955

200. For the purposes of this section, the final drive will include those components of the drive line which follow the differential. That is, the bull pinion shafts, bull gears and the wheel axle shafts. Due to their inter-relationship, they will be treated in the order of disassembly.

201. **WHEEL AXLE SHAFTS.** To remove an axle shaft, first drain final drive compartment, remove pto drive shaft or hydraulic pump drive shaft and the hydraulic lift assembly. Remove the drawbar support and pto control valve linkage, then remove pto assembly from rear main frame. Remove fenders, then support tractor rear frame and remove the wheel and tire assemblies. On models with internal draft control, remove draft control spring, support and spacers. Remove cap screws from axle outer bearing cap (1—Fig. 70), then remove snap ring (16) from inner end of axle. Use a suitable pusher and push axle from bull gear (15). Remove wheel hub, bearing retainer, shims (13) and bearing assembly from outer end of axle. Remove seals (2) from bearing retainer. Remove bearing and spacer (14) from inner end of axle.

NOTE: If desired, the axle and axle carrier can be removed as a unit by unbolting the axle carrier from the rear main frame and leaving the outer bearing retainer attached to axle carrier. Because of the weight involved, use a hoist for this operation.

If necessary to check and/or adjust the axle bearing preload, proceed as follows: Remove axle carrier from tractor, if not already done, then install axle in carrier and bull gear on inner end of axle. Strike both ends of axle with a soft faced hammer to seat bearings. Wrap a length of strong cord around bull gear outer diameter and attach a spring scale to cord. Pull on scale and note reading on scale required to keep axle rotating. This reading should be 10-12 ft.-lbs. for Series 1755 with regular tread and Series 1955 with a 45 3/16 inch axle; 13-17 ft.-lbs. for Series 1755 with 60-105 inch tread and Series 1855; and 7-9 ft.-lbs. for Series 1955 with 53 inch axle.

**202. AXLE CARRIERS.** To R&R and overhaul the axle carriers, first remove wheel axle shafts as outlined in paragraph 201, then unbolt and remove axle carrier.

Remove "O" ring (13—Fig. 70) from pilot of axle carrier and the snap ring (12) and retainer (11) which retains axle inner bearing cup.

NOTE: Retainer used in Series 1955 tractors is secured with cap screws and snap ring is not used.

Use a suitable puller and remove inner bearing cup.

When reassembling, adjust axle bearing preload as outlined in paragraph 201, and the draft control spring as outlined in paragraph 273, if tractor has internal draft control.

**203. BULL GEARS.** To remove bull gears, first drain hydraulic lift system and tractor rear frame. Remove axle shafts as outlined in paragraph 201. Bull gears can now be lifted from rear main frame.

**204. BULL PINIONS.** To remove bull pinions, proceed as outlined in paragraph 195 except that differential can rest in bottom of housing.

To disassemble bull pinion, remove differential side gear from inner end of

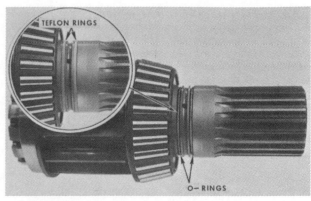

Fig. 72—Left hand bull pinion shaft is fitted with two "O" rings and two Teflon seal rings. Refer to text.

bull pinion (31—Fig. 69) and remove the 57 loose bearing rollers (33). Equally space three ¾×3 inch rods around the bull pinion gear as shown in Fig. 71, with rods between outer bearing cone and inner bearing outer race. Be sure housing is supported so that inner bearing race will clear, then press on spline (outer) end of bull pinion shaft to remove shaft, outer bearing cone and inner bearing race from the brake housing. With bull pinion removed, remove outer bearing cone from the bull pinion shaft and outer bearing cup from brake housing if necessary. Remove plug (32) from outer end of bull pinion shaft only if there is evidence of leakage at this point.

NOTE: The inner bearing race for the 57 loose roller bearings is pressed on the bull pinion shaft, then ground during manufacture, and must not be removed from shaft. If renewal of inner bearing race is required, install new bull pinion shaft and inner bearing race assembly.

Bull pinion shafts have two seal ring grooves. However, only the left bull pinion shaft has "O" rings and Teflon rings installed as it acts as an oil collector for the differential lubri-

cation circuit. When installing seals on left bull pinion shaft, install "O" rings in both grooves, then install the Teflon rings over the "O" rings. See Fig. 72. Stretch Teflon rings only enough to allow them to slide over shaft and allow them to return to normal size before attempting to install bull pinion shaft in brake housing. DO NOT install "O" rings or Teflon rings in grooves of bull pinion shaft to be installed on right hand side of tractor.

Install new bearing cup in brake housing with taper facing inward, and install new oil seal with lip inward. Install new bearing cone on bull pinion shaft. Install new plug (32) if plug was removed during disassembly. Be sure bearing cup and cone are seated, then lubricate lip of oil seal, and the sealing rings on left bull pinion shaft, and carefully install shaft taking care not to damage seals. Drive inner bearing outer race into brake housing until seated against shoulder in housing, then insert the 57 loose roller bearings and place differential side gear on bull pinion shaft. Refer to paragraph 197 for reinstallation and adjustment.

# BRAKES

The brakes, which are mounted on outer ends of bull pinions, are a hydraulically actuated disc type. Series 1755 and 1855 tractors have brakes using two brake discs, whereas the Series 1955 tractors use three brake discs plus an additional separator. Service brakes are self-adjusting, however adjustments for pedal free travel and the hand brake can be made.

tory and should remain constant. However, should it become necessary to adjust pedal free travel, proceed as follows: Remove hood side panels, hood and fuel tank. Refer to Fig. 73 and disconnect rods at spool connection. Loosen rod clevis jam nut and turn rods as necessary to obtain a pedal free travel of 1¼ inches.

**206. HAND BRAKE.** To adjust hand brake, refer to Fig. 74 and remove cap from brake housing cover. Loosen adjusting screw lock nuts and back-out adjusting screws slightly to relieve tension on linkage. Position equalizing bar parallel to the centerline of cross-shaft, then hand tighten

### ADJUSTMENT
### Series 1755-1855-1955

**205. PEDAL FREE TRAVEL.** Brake pedal free travel is set at fac-

Fig. 71—Removing bull pinion shaft from brake housing. Refer to text.

the adjusting screws. Operate hand brake lever several times, recheck adjusting screws and retighten if necessary. Be sure equalizing bar is parallel with centerline of cross-shaft when retightening adjusting screws. Hand brake adjustment is correct when lever travel is restricted to 3 or 4 notches on latch and equalizing bar is parallel with cross-shaft.

Tighten adjusting screw lock nuts and reinstall cover caps.

### OVERHAUL
### Series 1755-1855-1955

207. **SERVICE BRAKES.** To remove service brake assembly, disconnect hand brake linkage at brake housing cover and hydraulic line at fitting. Refer to Fig. 75 and remove housing cover (3) and insulator (10). Remove adjusting plate (11), outer disc (12), separator (13) inner disc (12) and the four springs (15) from brake housing.

NOTE: Series 1955 tractors are fitted with three discs (12) and two separators (13).

If necessary, remove spring clips from adjusting plate and remove the self-adjuster screws from adjusting plate. Remove cap from housing cover, loosen adjusting screw lock nut and turn adjusting screw in as far as possible. Actuate hand brake lever and force brake piston (9) from cover. Remove "O" rings (7 and 8) from cover.

To reassemble brakes, screw self-adjusting screws into adjuster plate, install spring clips, then back-out adjuster screws one or two notches. Install "O" rings in brake housing cover, lubricate piston and push piston into cover. Place return springs over locating pins, install inner disc, then install separator with taper inward and notches aligned on dowels. Install outer disc and adjusting plate with notches over dowels and self-adjuster screw tips between springs. Place insulator on piston, then install brake

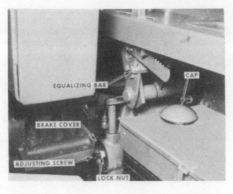

Fig. 74–To adjust hand brake, remove brake housing cap as shown. Refer to text for procedure.

housing cover over dowels and BE SURE insulator does not slip out of place. Secure cover and connect hydraulic line and hand brake linkage.

Start engine and hold brake pedals down to fill brake assemblies with oil. Continue to hold brake pedals down, then open bleed screw until air free oil flows from port, then close bleed screw.

208. **HAND BRAKE CROSS-SHAFT.** Hand brake cross shaft can be withdrawn after driving groove pin from hub on left end of shaft. If neces-

sary, pry out shaft seals and install new seals with lips facing toward inside.

209. **BRAKE VALVE.** To remove brake valve, remove hood side panels, hood and fuel tank. See Fig. 73. Disconnect rods at valve spools, then remove linkage to allow access to valve. Disconnect inlet line, return line and pressure (output) lines, then remove the two mounting cap screws and lift out valve.

210. With valve removed, remove the cap screws, cap (1—Fig. 76) and gasket (3). Identify spool assemblies as to their bores and remove spool assemblies from valve body. Remove large pistons (13), piston springs (14) and return springs (11) from valve body, then remove "O" ring (12) from pistons. Use a punch in hole at linkage end of spool to hold spool, then remove screw (4) at opposite end of spool. (Note: Screw is installed with Loctite). Remove retainer (5), reaction spring (6), small piston (7) and spring seat (10). Remove "O" ring (not shown) from bore of small piston (7). Remove fitting (21), ball (19), spring (18) and orifice plate (17) from inlet port. Remove "O" ring (20) from fitting and

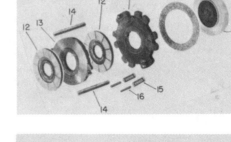

Fig. 75–View showing Series 1755 and 1855 service brake. Series 1955 uses three discs (12) and two separators (13).

1. Cap
2. Bleed screw
3. Cover
7. "O" ring
8. "O" ring
9. Piston
10. Insulator
11. Adjusting plate
12. Disc
13. Separator
14. Dowel (pin)
15. Spring
16. Spring pin

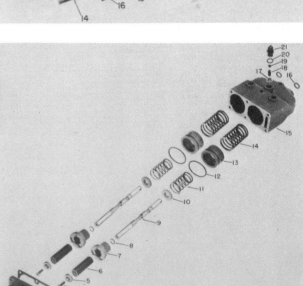

Fig. 76–Exploded view of hydraulic brake valve. Refer to Fig. 73 for mounting location.

1. Cap
2. Cap screw
3. Gasket
4. Screw
5. Retainer
6. Reaction spring
7. Piston, small
8. Snap ring
9. Spool
10. Spring seat
11. Spring
12. "O" ring
13. Piston, large
14. Spring
15. Body
16. "O" ring
17. Orifice plate
18. Spring
19. Ball (5/16)
20. "O" ring
21. Fitting

Fig. 73–Brake valve and control linkage are accessible after removing fuel tank.

"O" rings (16) from spool bores.

211. Discard all "O" rings, clean and inspect all parts and renew as necessary. If using used parts during reassembly, lap the large and small pistons to each other as follows: Place a small amount of lapping compound on lip of small piston, install large piston over small piston, then with assembly held between palms of hands, use a twisting motion of the hands to lap pistons. Thoroughly clean parts after lapping.

NOTE: New pistons will not require lapping compound however pistons should be mated by using the twisting motion already described.

212. Reassemble brake valve as follows: Install small snap rings (8) if removed. Install spring seat (10), small piston (7) with new "O" ring in bore, reaction spring (6) and retainer (5) in spool, then using Loctite (grade AV) on screw (4), install screw and tighten to 5-8 ft.-lbs. of torque.

Slide large piston (13) over linkage end of spool and be sure lapped surface mates with lip of small piston. Install return spring and piston spring. Install new spool "O" ring (16) in body, then oil spool assembly and insert in body (15). Use new gasket (3), install cap (1) and tighten retaining cap screws to 5-8 ft.-lbs. of torque.

Place new "O" ring (20) on fitting (21), then install orifice plate (17), spring (18) and ball (19) in inlet port and secure with fitting.

# POWER TAKE-OFF

Single speed 540 rpm, single speed 1000 rpm, or dual speed 540 or 1000 rpm pto units are available for all Series 1755, 1855 and 1955 tractors. In addition, an engine speed pto may be used with any unit. However, if tractor is equipped with a single speed 540 rpm pto, a 1000 rpm output shaft and bearing assembly must be obtained to perform engine speed pto operations.

NOTE: Use of the 540 rpm pto must be limited to not more than 60 horsepower.

Pto clutch and brake are hydraulically actuated and are controlled by a single spool, closed center valve mounted on pto housing.

## Series 1755-1855-1955

215. **R&R PTO DRIVE SHAFT.** To remove the pto drive shaft, first remove cover (21—Fig. 83), or the engine rpm pto output shaft, if so equipped, from upper bearing bore of pto housing. Remove snap ring (32) from inside diameter of pinion (35). Install puller assembly (Oliver No. STS-100), or a ½-13 inch cap screw, and pull plug (33). Oil trapped in the lubrication circuit will flow from

pinion when plug is removed. Remove "O" ring (34), then use a ½-13 inch cap screw at least 7 inches long, threaded into end of pto drive shaft and pull the pto drive shaft (1—Fig. 84).

Reinstall by reversing removal procedure, however be sure "O" ring (34—Fig. 83) is in front groove in I.D. of pinion (35) and that projection on rear of plug (33) fits into spline of pinion.

216. **R&R PTO UNIT.** To remove the pto unit, first drain the transmission and final drive compartments. Remove pto drive shaft as outlined in paragraph 215. Remove linkage and oil lines from pto control valve. Remove drawbar support, then unbolt and remove pto unit from tractor rear main frame.

When reinstalling pto unit, secure unit to rear main frame with the four cone head cap screws and tighten them to 135-150 ft.-lbs. of torque, then install remaining cap screws and lock washers. Complete installation by reversing the removal procedure, fill transmission and final drive compartments, then start engine and check pto operation.

217. **OVERHAUL PTO UNIT.** Remove pto unit from tractor as out-

lined in paragraph 216 and remove safety shield, bearing retainer and the output (stub) shaft and bearing assembly. Remove cap screws and pull control valve from housing and be careful not to lose the detent ball. Remove detent spring and the three "O" rings.

NOTE: Disassembly of pto unit will require the use of a special tool which can be fabricated by using the dimensions shown in Fig. 80.

With pto unit removed, remove snap ring (2—Fig. 84) and pull hub (3) from driver (4). Attach special tool as shown

**Fig. 81—Attach special tool as shown to compress pto clutch spring.**

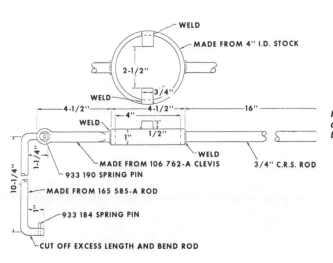

**Fig. 80—Special tool used to disassemble pto unit can be fabricated using the dimension shown.**

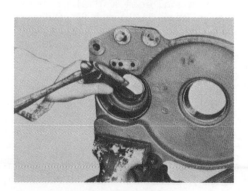

**Fig. 82—Drive clutch and brake pistons from bearing support as shown.**

in Fig. 81 and compress clutch spring, then remove snap ring (9—Fig. 84) and remove driver, spring (12) and spring guide (13). Snap ring (11) and bearing (10) can now be removed from driver. Remove retaining ring (5), then remove washer (6), clutch plates (7 and 8), collar (15) and spring washer (14) from clutch housing (17). Remove snap ring (16) from pinion shaft and remove clutch housing (17), actuator (18), friction plate (19), reaction plate (25) and the thrust bearing assembly (20 and 21). Unbolt and remove inner support (30) from pto housing and discard the two "O" rings (28—Fig. 83) which are located between inner support and housing. Clamp inner support in a vise as shown in Fig. 82, and bump out clutch piston (23—Fig. 84), brake piston (26) and remove "O" rings and quad rings (27) from pistons. Remove pinion seal (29) and if necessary, the bearing cups from inner support.

Remove shifter rod lock screw (dual speed units). Remove output gear, or gears, and output gear shaft (54—Fig. 83) and on dual speed units, remove shifter rail and fork. Remove drive pinion (35), then remove bearing cones from pinion and bearing cups and seals from pto housing. Remove bearing cones, output gear, or gears, and drive coupling (53) from output gear shaft (54).

On 540 rpm single speed units, remove snap ring (52) from output gear shaft drive collar and cup plug from I.D. of output gear shaft. Remove cup plug (45) from shift rod bore in housing.

On 1000 rpm single speed units, remove snap ring (52) from output gear shaft drive collar, then remove snap ring (62) and remove shifter lock mechanism, seal and alignment washer (items 56 through 61). Remove cup plug (45) from shift rod bore in pto housing.

On dual speed units, remove snap ring (62) from front bore of output gear shaft and remove shifter lock mechanism, seal and alignment washer (items 56 through 61). Remove shifter rod and seal from pto housing.

Refer to the following as a guide in renewing parts:

### CLUTCH
Drive plate thickness . . . . . 0.060-0.065
Driven plate thickness . 0.0665-0.0710
Spring free length . . . . . . . . . . . . 2.629

### BRAKE
Plate thickness . . . . . . . . . 0.183-0.187
Friction disc thickness . . . . 0.122-0.127

218. To reassemble pto unit, install bearing cups in pto housing and support with tapers facing toward each

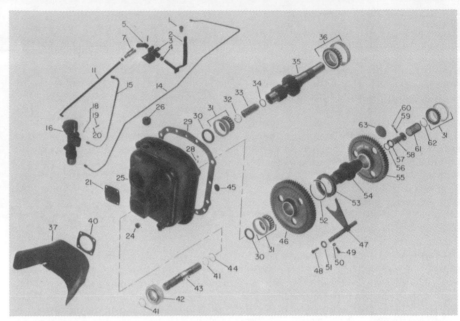

**Fig. 83–Exploded view showing component parts and their relative position of the dual speed pto. The single speed (540 or 1000) pto does not include all parts shown, however, basic construction remains similar.**

| | | |
|---|---|---|
| 1. Knob | 25. Housing | 42. Bearing |
| 2. Lever | 26. Filler plug | 43. Output shaft |
| 3. Support | 28. "O" ring | 44. Snap ring |
| 4. Bushing | 29. Gasket | 45. Plug (single speed) |
| 5. Bellcrank | 30. Oil seal | 46. Output gear (540) |
| 7. Yoke | 31. Bearing assy. | 47. Shift rail & fork |
| 11. Rod | 32. Retaining ring | (dual speed) |
| 14. Oil line | 33. Plug | 48. Cap screw |
| 15. Oil line | 34. "Nut" | 49. Cap screw |
| 16. Control valve | 35. Pinion shaft | 50. Nut |
| 18. "O" ring | 36. Bearing assy. | 51. Oil seal (dual |
| 19. Detent spring | 37. Safety shield | speed) |
| 20. Detent ball | 40. Bearing retainer | 52. Snap ring (single |
| 21. Cover | 41. Snap ring | speed) |
| 24. Plug | | 53. Drive coupling |

| |
|---|
| 54. Output gear shaft |
| 55. Output gear (1000) |
| 56. Locating washer (dual speed) |
| 57. Oil seal (dual speed) |
| 58. Locking cam (dual speed) |
| 59. Pin (dual speed) |
| 60. Spring (dual speed) |
| 61. Spring (dual speed) |
| 62. Snap ring (dual speed) |
| 63. Plug (single speed) |

other. Install oil seals in pto housing with lips facing forward.

On 540 rpm, single speed units, install cupped plug in shifter rod bore

with cupped side forward. If new cup plug is installed in output gear shaft, install with cupped side forward in front bore of shaft. Install snap (lock)

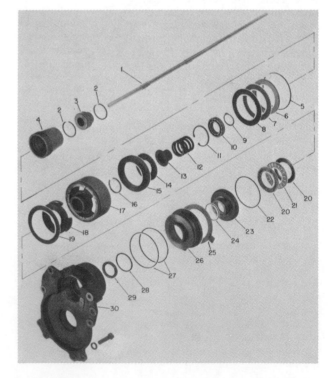

**Fig. 84–Exploded view of pto hydraulic clutch and brake assembly and inner support.**

1. Drive shaft
2. Snap ring
3. Hub
4. Driver
5. Retaining ring
6. Washer
7. Separator plate
8. Clutch plate
9. Snap ring
10. Bearing
11. Snap ring
12. Return spring
13. Spring guide
14. Plate
15. Collar
16. Snap ring
17. Clutch housing
18. Actuator
19. Friction plate
20. Bearing race
21. Thrust bearing
22. "O" ring
23. Clutch piston
24. "O" ring
25. Reaction plate
26. Brake piston
27. Quad rings
28. "O" ring
29. Oil seal
30. Support

ring (52—Fig. 83) on output gear shaft drive collar, then place drive coupling over rear end of shaft drive collar and against snap ring. Install 540 rpm output gear (46) on rear of output gear shaft and mesh teeth on gear with teeth of coupling. Install bearing cones on output gear shaft.

On 1000 rpm single speed units, install cup plug in shifter rod bore in pto housing with cupped side forward. Install alignment washer (56) and seal (57) in front bore of output gear shaft, then install lock pin spring (60) and lock pin (59). Depress lock pin and install locking cam (58) and spring (61). Secure parts in shaft bore with snap ring (62). Install snap (lock) ring (52) on drive collar of output gear drive shaft, then place drive coupling over front end of shaft drive collar and against snap ring. Install 1000 rpm output gear (55) on front of output gear shaft and mesh teeth on gear with teeth of coupling. Install bearing cones on output gear shaft.

On dual speed units, install oil seal in shifter rod bore in pto housing with lip facing forward. Install alignment washer (56) and seal (57) in front bore of output gear shaft, then install lock pin spring (60) and lock pin (59). Depress lock pin and install locking cam (58) and spring (61). Secure parts in bore with snap ring (62). Place 1000 rpm gear (55) in front of output gear shaft with drive coupling teeth rearward and secure with bearing cone. Slide drive coupling (53) on output gear shaft drive collar and mesh teeth of coupling with teeth of 1000 rpm gear. Place 540 rpm gear (46) on rear of output gear shaft with drive coupling teeth forward and secure with bearing cone.

On all units, position output gear and shaft in pto housing, and on dual speed units, mate shift fork with groove in drive coupling and install shifter rod and fork along with output gear and shaft assembly. Use caution not to damage shifter rod seal. Install bearing cones on pinion shaft (35) with tapers facing away from each other, then with teeth of pinion and output gears meshed, position pinion in pto housing. Install new pinion seal (29— Fig. 84) in support with lip facing forward. Install new "O" rings and quad rings on brake and clutch pistons and lubricate with Lubriplate No. 210, or equivalent. Push clutch piston (23) into brake piston (26), then push assembly into support (30). Place new "O" rings (28—Fig. 83) in pto housing, then install inner support on pto housing and secure with the six cap screws.

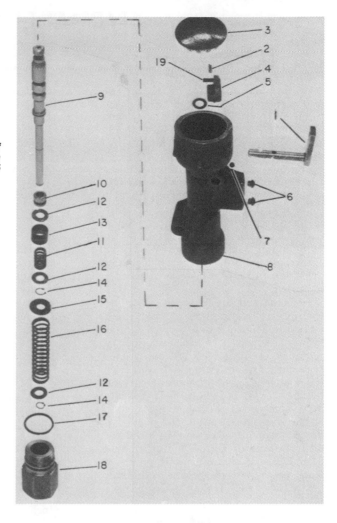

Fig. 85—Exploded view of pto control valve used on all 1755, 1855 and 1955 tractors.

1. Arm assembly
2. Pin
3. Cap
4. Arm with pin
5. "O" ring
6. Seat
7. Ball (3/16)
8. Body
9. Spool
10. Pressure control sleeve
11. Pressure control spring
12. Washer
13. Stop sleeve
14. Snap ring
15. Washer
16. Return spring
17. "O" ring
18. Plug
19. Pin

Install thrust bearing (20 and 21— Fig. 84). Install reaction plate (25) with flat side up and lug in recess in support. Install friction plate (19), actuator (18) and clutch housing and secure clutch housing with snap ring (16). Install collar (15), spring plate (14), spring guide (13) and return spring (12). Install bearing (10) in driver (4) and secure with snap ring (11). Place driver over return spring, use special tool (Fig. 81) to compress spring and install snap ring (9) with flat side of snap ring facing forward. Start with an external lug clutch (separator) plate and alternately install the external lug and internal lug clutch plates (7 and 8—Fig. 84). Install washer (6) and secure assembly with retaining ring (5). Place hub (3) in driver (4) and secure with snap ring (2).

**219. R&R AND OVERHAUL CONTROL VALVE.** To remove control valve, disconnect linkage and oil lines, then remove the two cap screws and lift control valve from pto

housing. Be careful not to drop the detent ball. Remove detent spring and the three "O" rings.

With valve removed, disassemble as follows: Remove cap (3—Fig. 85), drive out spring pin, then remove arm assembly (1) from housing and lift out arm and pin assembly (4). Remove plug (18) and discard "O" ring (17). Pull spool assembly from bottom of valve body, then remove and discard "O" ring (5). Compress return spring (16) and remove snap ring (14), washer (12), return spring (16) and washer (15). Compress pressure control spring (11) and remove snap ring (14), washer (12), pressure control spring (11), stop sleeve (13), washer (12) and pressure control sleeve (10).

Clean and inspect all parts and renew any that show signs of undue wear or other damage. If body is worn or damaged, renew complete valve assembly.

Reassemble by reversing disassembly procedure and install pressure control sleeve (10) with flattened sides toward top of spool. Lubricate "O" ring (5) before installing spool assembly.

# HYDRAULIC LIFT SYSTEM

**Series 1755-1855-1955**

The closed center hydraulic lift system is designed for draft control and position control of a three-point hitch and system will operate in conjunction with externally mounted control valves which provide remote hydraulic operations.

When tractor is equipped with a Category II hitch, the hydraulic lift system uses a single internal lift cylinder and draft sensing mechanism is located inside the tractor rear main frame. Tractors equipped with the 6000 pound Category III hitch use two externally mounted lift cylinders and draft sensing mechanism is externally mounted.

Pressurized oil for operating the hydraulic lift system is provided by a variable displacement, axial piston type pump which is externally mounted on lower left side of tractor rear main frame. Pump is driven from a set of bevel gears which are driven by the tractor pto shaft, or a separate drive shaft if tractor has no pto.

The externally mounted pressure detent control valve can be used for either single or double acting remote cylinders and the externally mounted float valve provides raise and lower positions as well as float position.

## TROUBLE SHOOTING

225. The following will be of assistance should trouble develop in the operation of the hydraulic lift system.

Also refer to the Quick Check section which follows:

226. **SYSTEM DOES NOT DEVELOP FULL PRESSURE OR DELIVER OIL.** Could be caused by:
1. Fluid level low.
2. Restriction in outlet port.
3. Filter plugged.
4. Suction hose collapsed or line damaged.
5. Air leak in inlet lines.
6. Hydraulic pump shaft broken.
7. Hydraulic pump pistons or valve plate worn, scored or damaged.
8. Compensator valve sticking, worn or scored.
9. System pressure relief valve set too low or leaking.
10. Priority valve pilot spool spring broken.
11. Low pressure relief valve cap unscrewed.
12. Unloader valve faulty.

227. **HYDRAULIC SYSTEM NOISY.** Could be caused by:
1. Fluid level low.
2. Outlet port restricted.

3. Hydraulic reservoir breather plugged.
4. Suction hose collapsed or line damaged.
5. Filter plugged.
6. Bleed line leaking air.
7. Cooler pump inlet line loose or pump has lost prime.
8. System pressure relief valve set too low.
9. Hydraulic pump bearings defective.
10. Hydraulic pump shaft misaligned.
11. Hydraulic pump bevel gear set incorrectly adjusted.
12. Compensator valve set too high or valve stuck in pumping condition.

228. **HITCH WILL NOT RAISE.** Could be caused by:
1. Servo valve restrictor closed.
2. Servo valve spool stuck.
3. Lift cylinder seals leaking.
4. Control linkage bent or broken.
5. Cylinder safety valve leaking.
6. Internal oil line or fitting failure (Cat. II hitch).

229. **HITCH WILL NOT LOWER.** Could be caused by:
1. Servo valve spool stuck.
2. Servo valve check ball stuck.
3. Servo valve interlock plunger stuck.

230. **DRAFT ACTION TOO SLOW.** Could be caused by:
1. Servo valve restrictor turned in too far.
2. Servo valve spool sticking.
3. Draft spring preload set too high.
4. Linkage worn or not properly adjusted.

231. **DRAFT ACTION TOO FAST.** Could be caused by:
1. Servo valve restrictor turned out too far.
2. Linkage not properly adjusted.

232. **REMOTE CYLINDER WILL NOT RAISE.** Could be caused by:
1. Control valve spool stuck.
2. Control valve restrictor closed.
3. Control valve restrictor ball incorrectly installed.
4. Breakaway coupler not installed properly or coupler damaged.
5. Faulty remote cylinder.

233. **REMOTE CYLINDER WILL NOT LOWER.** Could be caused by:
1. Control valve spool stuck.
2. Control valve interlock pin broken.
3. Control valve interlock plunger stuck.

234. **REMOTE CONTROL VALVE LEVERS WILL NOT RETURN TO NEUTRAL.** Could be caused by:
1. Control levers binding.
2. Incorrect torque on control valve retaining cap screws.
3. Control valve detent piston adjusted too high.
4. Incorrect number of shims between control valve detent piston and spring.

235. **HYDRAULIC SYSTEM OVERHEATS.** Could be caused by:
1. Oil cooler dirty or clogged.
2. Oil cooler pump defective.
3. Oil cooling circuit restricted.
4. Hydraulic pump defective.
5. Excessive leakage of one or more hydraulic control valves.
6. Incorrect application of remote control valves or remote hydraulic motor.

236. **TRANSFER OF HYDRAULIC OIL TO TRANSMISSION.** Could be caused by:
1. Hydraulic pump seal faulty.
2. Faulty pto seals.
3. Broken pto clutch hub.

237. **QUICK CHECK.** A quick check of the hydraulic lift system can be made with a minimum of disassembly and an analysis of this quick check will, when used with the TROUBLE SHOOTING Section, determine to a large extent what service is required on the hydraulic lift systems.

238. To perform the quick check, proceed as follows: Obtain a protractor and modify it as shown in Fig. 90. Mount the protractor on hydraulic pump yoke pintle as shown in Fig. 91 and fabricate a pointer so movement of protractor can be checked. Install a loop line to breakaway couplings as shown in Fig. 92. Install a 3000 psi gage at the pressure control valve.

Note the number of degrees at pointer with engine off. This locates yoke angle in full pumping stroke as system is not pressurized. Start engine and run at 2400 rpm. Note pressure

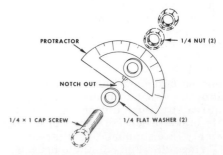

*Fig. 90–Modify a protractor as shown when using to make hydraulic system check. Also see Fig. 91.*

gage which should show system pressure of 2175-2225 psi, and protractor should have moved 15-16 degrees from full stroke position.

NOTE: It requires 1-2 degrees of yoke angle from minimum (zero) stroke to make up normal system leak-off.

If pressure is not as stated, refer to TROUBLE SHOOTING Section. If protractor has moved less than the 15-16 degrees, one or more of the hydraulic system components is leaking excessively.

Open restrictor on remote control valve manifold, move control lever rearward and allow oil to circulate through remote lift circuit and loop line. With engine operating at 2400 rpm, yoke angle should move to full pumping stroke position and pressure on gage should be reduced to priority pressure (approximately 1600 psi). If pressure is not reduced, lift circuit may be restricted causing pump to compensate. Check remote control valve, breakaway couplers and loop line.

NOTE: Air in system will cause hydraulic pump to destroke and yoke angle may not reach full pumping stroke position. Yoke (protractor) and pressure gage will oscillate rapidly, making hydraulic system noisy. Check pump, suction lines, cooling circuit and reservoir oil level for possible entrance of air.

238A. With items outlined in paragraph 237 completed, open remote control valve and allow oil to circulate through the loop line. Reduce engine speed to 800 rpm and check pressure gage which should show a priority valve pressure of 1500 psi. Adjust priority valve if necessary.

## LUBRICATION AND MAINTENANCE

239. When performing service or maintenance of any kind on the hydraulic lift system, the practice of cleanliness is of the utmost importance. Pumps, valves and cylinders are manufactured to very close tolerances and are mirror finished. The induction of dirt and foreign material is most detrimental and should be avoided at all times.

Capacity of hydraulic reservoir is 27 quarts. Add ¾-quart for each 3 inch cylinder, 1½ quarts for each 4 inch cylinder and ½-pint for each 5 feet of hose.

Recommended fluid is Oliver Type 55 hydraulic oil.

240. **DRAIN AND REFILL.** Drain and refill hydraulic lift system each year, or 1000 hours, as follows: Place lift arms in down position and remove plug from bottom of filter. Reinstall plug when fluid flow has stopped.

Disconnect and drain all hoses. Work piston rods in and out of cylinders to clear cylinders, then reconnect all hoses and fill reservoir. Open bleed port on filter and when air free oil flows from bleed port, close bleeder. Start engine and cycle hydraulic equipment several times to work any trapped air from system. Recheck fluid level and add if necessary to bring fluid level to "FULL" mark on dipstick.

241. OIL FILTER. Hydraulic fluid filter is renewed after the first 50 hours, and whenever hydraulic fluid is changed thereafter.

To renew filter, turn bottom of filter counterclockwise to remove center bolt. Separate housing from head and discard element and gaskets. Wash all parts, allow to dry, then coat gaskets and seals with grease. Place upper seal on filter head. Place lower seal housing, spring, washer, element and gaskets on center post, then install assembly on filter head and tighten center post to a maximum of 20 ft.-lbs. torque. Run engine and check for leaks.

NOTE: A by-pass valve is incorporated in head of filter assembly. Should service ever be required, remove inlet hose, connector and outlet line, then remove nut, valve, guide, spring, spacer, cap and screw from filter head.

242. **BREATHER.** A breather is located on top rear side of hydraulic lift cover and under normal operating conditions the breather element should be renewed each time hydraulic fluid is renewed. When operating under extremely dusty conditions, renew element more frequently. The elastic stop nut should be tightened until breather cap is snug against spacer.

## SYSTEM ADJUSTMENTS

The adjustments in this section will include those external adjustments which can be accomplished with no disassembly. For those internal adjustments which require disassembly, refer to the appropriate section.

243. **LIFT ARM ADJUSTMENT.** Remove any attached implement, then start engine and run at about 1000 rpm. Move hitch control lever to up position and allow lift arms to raise to upper limit. Grasp lift links and manually raise lower links and check the free play at ends of lift arms. This free play should be ½-1 inch. If free play is not as stated, refer to Fig. 93 and loosen lock nut on control rod. Shorten control rod to increase, or lengthen control rod to decrease, the lift arm free travel. Tighten control rod lock nut, recycle hitch, then recheck. Readjust control rod if necessary.

244. **POSITION CONTROL.** To convert system from draft control to position control, disconnect control rod from balance link, move end to top hole in balance link and connect it directly to position arm. See Fig. 93. Check lift arm free travel as outlined in paragraph 243.

245. **DRAFT SPRING PRELOAD.** Draft control springs for the 6000

*Fig. 91–Mount protractor as shown when making hydraulic system quick check. Refer to text for procedure.*

*Fig. 92–Use a loop line installed as shown when making hydraulic system quick check. Refer to text.*

pound Category III hitch are externally mounted and must be preloaded. Left hand adjusting screw is located in left hand axle flange and right hand adjusting screw is located in right hand actuating hub arm.

Adjust draft control springs as follows: Cut safety wire, loosen lock nuts and turn both adjusting screws in until all tension on draft springs is released. Now back-out left hand adjusting screw until left draft spring just begins to preload. Continue to back-out the left hand screw until top of draft signal arm has moved forward 9/16-inch. Back-out right hand adjusting screw until right hand draft springs are just beginning to preload, then back screw out an additional ¼-turn. Tighten lock nuts and secure cap screws with safety wire. Check lift arm free travel.

NOTE: As field conditions vary, it may be preferable to lengthen or shorten the 9/16-inch adjustment outlined above as required.

**246. RESTRICTOR VALVE.** A restrictor valve which controls lift speed of hitch is located in hydraulic lift manifold under left side of seat. See Fig. 94. To increase hitch lift speed, turn the adjusting screw counterclockwise. To decrease hitch lift speed, turn adjusting screw clockwise.

**247. EXTERNAL VALVE RESTRICTOR VALVE.** A restrictor valve is located on rear of each pressure detent (remote) control valve to

control fluid flow returning from remote cylinder. See Fig. 95. Implement may bounce when being lowered if fluid flow is not sufficiently restricted.

If implement bouncing occurs, turn the adjusting screw clockwise to increase restriction. If implement lowers too slowly, turn the adjusting screw counterclockwise. Implement should lower as rapidly as possible without bouncing.

NOTE: Remove load on cylinders before adjusting restrictors.

**248. EXTERNAL VALVE MANIFOLD RESTRICTOR.** A restrictor valve is provided on the rear of the external valve manifold as shown in Fig. 95. This restrictor controls fluid flow to the external control valves. To change remote cylinder speed, turn the adjusting screw clockwise to decrease, or counterclockwise to increase, cylinder speed.

NOTE: This adjustment will not decrease implement lowering speed when a float type external (remote) valve is used.

## SYSTEM OPERATING PRESSURE

The high pressure relief valve is located in the pressure control valve body which is mounted on left side of transmission housing directly below the hydraulic pump. The pressure control valve body contains the system

high pressure relief valve, the priority valve and the pressure reducing and low pressure valve. The high pressure valve is the valve located furtherest from tractor center line (i.e. the outermost valve).

As the compensator valve, mounted on outer end of hydraulic pump, controls the pump output, the high pressure relief valve functions to protect system from high pressure surges and to provide pressure relief in case the compensator valve malfunctions.

**249.** To check and/or adjust high pressure relief valve, refer to Fig. 96 and proceed as follows: Drain oil from reservoir, remove the ⅛-inch pipe plug below steering valve support and install a 3000 psi test gage. Remove reservoir line from unloader valve, then plug line and cap valve. Disconnect unloader valve pressure line at tee fitting on steering port, then connect a test unloader valve to steering port using ⅜-inch steel tubing and ¾-16 inch "O" ring straight pipe and ⅜-inch 90 degree elbow. Connect a ⅝-inch steel tube to unloader valve, attach a ⅝x½-inch male connector, then attach a ½-inch high pressure hose to male connector. Install a female breakaway coupler to hose.

Attach a hose to lower rear reservoir port with a ½-inch pipe and a male breakaway coupler. Attach a flow-rator to breakaway couplers, fill reservoir with fluid, then run engine at 2400 rpm and pump oil through flow-rator until oil temperature reaches 130 degrees.

Disconnect flow-rator, break seal on compensator valve in until pressure gage reads 2450 psi. At this time, not more than 100 ml/min. should escape from unloader valve pressure line. Use a graduate container and a watch when making this check.

NOTE: Unloader valve pressure line is open to back side of relief valve. Oil escaping past relief valve is allowed to pass out of pressure control valve

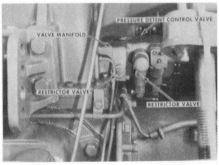
Fig. 95–External (remote) valve and valve manifold restrictors are located as shown. Refer to text for adjustment.

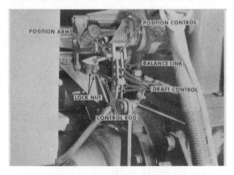
Fig. 93–Shorten or lengthen control rod to adjust lift arm free travel to ½-1 inch.

Fig. 94–Restrictor valve to control hitch lift speed is located as shown.

Fig. 96–View showing high pressure relief valve ready to be tested. Refer to text.

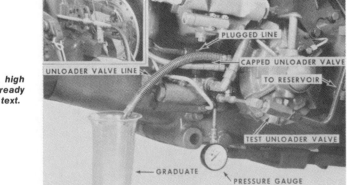

through unloader valve and pressure line.

If escaping oil exceeds 100 ml/min., check high pressure relief valve for the following: Damaged "O" ring on relief valve seat. Scored or damaged relief valve seat. Chipped or broken relief valve ball. Damaged relief valve body and cap. Renew parts as necessary.

Adjust compensator valve until the high pressure relief valve opens, however do not exceed 2600 psi gage pressure. Relief valve opening can be noted by an increase in the amount of oil escaping and an increase in noise. Gage reading should be 2470—2530 psi. If necessary, adjust valve by varying the number of shims located between spring and head of plunger. Adding shims will increase pressure, removing shims will decrease pressure. Shims are available in thickness of 0.004 and 0.0149.

With high pressure relief valve adjusted, readjust the compensator valve to 2225—2275 psi and reseal.

## PRESSURE CONTROL VALVE

Pressure control valve is mounted on left side of transmission housing below hydraulic pump. Valve body contains the system high pressure relief valve, priority valve, pressure reducing valve and low pressure relief valve.

If desired, pressure control valve can be disassembled without removing valve body from tractor. All parts of valve are available for service and while the flared seats in body can be renewed, a new valve body has the flared seats installed.

250. **REMOVE AND REIN-STALL.** To remove pressure control valve, either drain the hydraulic lift reservoir, or plug lines as they are disconnected. With all lines disconnected, remove the valve retaining cap screws, pull valve from housing, then remove unloader valve from pressure control valve.

When reinstalling, mount unloader valve on pressure control valve, then install pressure control valve by reversing removal procedure.

251. **OVERHAUL.** Remove cap (22—Fig. 97) with "O" ring (15), then remove spring (21), shims (14) and plunger (20) from cap. Remove ball (19) from bore. Use a stiff wire with a ¼-inch hook on one end and pull seat (18) and "O" ring (17) from bore.

Remove priority valve cap (16) and "O" ring (15), then remove shims (14) and spring (13) from cap. Remove pilot spool (12), thread a cap screw into plug (11), then remove plug and "O" rings (9 and 10). Remove spring (4) and spool (3) from bore. Remove plug (1) and "O" ring (2), if necessary.

Remove plug (34) and "O" ring (33). Remove spring (32), then slide pressure reducing spool (25) and relief valve assembly (items 26 through 31) out of valve body. Use caution not to mar the pressure reducing spool and remove cap (31), then remove washer (26) from spool and ball (27), plunger (28), shims (29) and spring (30) from cap (31).

252. Remove and discard all "O" rings. Clean and inspect all parts for nicks, burrs, scoring or any undue wear. Check all springs for distortion or fractures. Check bores in pressure control valve body. Renew any defective parts, then lubricate all parts and reassemble by reversing the disassembly procedure.

## UNLOADER VALVE

The unloader valve is mounted on the pressure control valve. Function of unloader valve is to eliminate build-up of hydraulic pressure during engine cranking. Hydraulic pressure will be relieved by the orifice in the unloader valve plunger until pump output reaches about 4 gpm at which time the unloader valve closes and hydraulic pressure raises to normal. Valve remains closed until engine is stopped and pressure is reduced in system.

253. **R&R AND OVERHAUL.** To remove unloader valve, drain hydraulic reservoir, then disconnect steering line, relief valve return line and low pressure return line. Remove line between unloader valve and tee-fitting at steering port of pressure control valve. Loosen "O" ring locknut on unloader valve elbow, turn unloader valve so elbow will clear brake drums, then remove elbow from unloader valve and "O" ring from elbow. Remove unloader valve by turning reducer bushing out of pressure control valve. Remove pipe nipple from unloader valve.

254. With unloader valve removed, remove snap ring (1—Fig. 98), then remove piston (2) and spring (3) from body (4).

Clean all parts and inspect piston and seat for nicks, burrs, scoring or undue wear. Check spring for distortion or fractures. Renew parts as necessary and reassemble by reversing disassembly procedure.

255. Reinstall unloader valve as follows: Coat pipe threads with Permatex No. 3, or equivalent, and install pipe nipple and reducing bushing. Install unloader valve assembly in pressure control body and turn assembly far enough past final position to allow installation of elbow. Place new "O" ring on elbow, lubricate "O" ring and turn elbow into unloader valve. Back the reducing bushing out of pressure control valve far enough to allow connection of relief valve return line. Connect relief valve return line and low pressure return line. Install line connecting tee-fitting at steering port of pressure control valve to unloader valve elbow. Tighten locknut on unloader valve elbow against unloader valve and connect steering line.

Refill hydraulic reservoir.

## HYDRAULIC PUMP

256. **REMOVE AND REIN-STALL.** To remove hydraulic pump, first remove shield under left platform, then remove drain plug from filter and drain hydraulic reservoir. Disconnect pump drain line and disconnect both ends of pump outlet line.

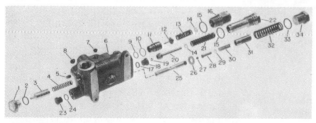

*Fig. 97—Exploded view of pressure control valve assembly showing high pressure relief valve, priority valve and the low pressure and pressure reducing valve.*

| | |
|---|---|
| 1. Plug | 12. Pilot spool |
| 2. "O" ring | 13. Spring |
| 3. Spool (priority valve) | 14. Shim (.004, .0149) |
| 4. Spring | 15. "O" ring |
| 5. Pipe plug | 16. Plug |
| 6. Body | 17. "O" ring |
| 7. Insert | 18. Seat |
| 8. Seat | 19. Ball (7/16) |
| 9. "O" ring | 20. Plunger (high press.) |
| 10. "O" ring | 21. Spring |
| 11. Plug | 22. Cap |
| | 23. Plug |

| | |
|---|---|
| 24. "O" ring | |
| 25. Spool (press. reducing) | |
| 26. Washer | |
| 27. Ball (½) | |
| 28. Plunger (low press.) | |
| 29. Shim (.004, .0149) | |
| 30. Spring | |
| 31. Cap | |
| 32. Spring | |
| 33. "O" ring | |
| 34. Cap | |

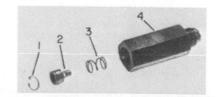

*Fig. 98—Exploded view of the hydraulic lift system unloader valve.*

| | |
|---|---|
| 1. Snap ring | 3. Spring |
| 2. Piston | 4. Body |

Disconnect hose clamps at inlet port of pump and slide coupling hose off inlet fitting and onto filter tube. Remove pump retaining cap screws, pull pump from pump drive support, then remove splined drive coupling.

When reinstalling pump, use Permatex No. 3, or equivalent on pump side of mounting gasket, grease transmission side of gasket, then install pump by reversing the removal procedure.

NOTE: For information on the hydraulic pump drive gears, refer to TRANSMISSION Section.

257. **OVERHAUL.** To disassemble hydraulic pump, proceed as follows: Remove compensator housing (6—Fig. 99) to gain access to valve cover cap screws, then remove valve cover (14). Remove piston (16) from tube (12) and remove tube (rod) from valve plate. Use a bearing puller if bearing (15) is to be removed. Tilt housing, turn cylinder block (21) slightly to free parts from swash plate and remove cylinder assembly. Lift shoe plate (25) and piston (26) assembly from cylinder block and be careful not to damage pistons during removal or subsequent handling. If limiter (18), spring (19) and washer (20) are to be removed, place a 1-inch washer on a ⅜ x 3½ inch cap screw and insert cap screw in

shaft hole of cylinder block at front end. Place another 1-inch washer on threaded end of cap screw and install nut. Tighten nut until tension against snap ring (17) is removed, then remove snap ring. Remove nut slowly and remove parts from cylinder block.

Remove swash plate (27) from yoke (29). NOTE: Swash plate may be oil locked in yoke and removal may be somewhat difficult until oil lock is broken. Remove snap ring (42), then use a soft-faced hammer to bump drive shaft (39) and bearing (40) from housing. Remove snap ring (41) and if necessary, use a puller to remove bearing. Removal of bearing (40) often destroys it so be sure a replacement bearing is available prior to attempting removal. Spacer (38) and oil seal (37) can now be removed. Remove cap screws (29) from yoke, then thread ¼-1-inch cap screws fitted with lock nuts into pintles and remove pintles. Remove yoke, spring seat (30) and the inner and outer springs (31).

258. Thoroughly clean all parts prior to inspection and after any lapping operations.

Inspect surface of valve plate which mates with cylinder block. Minor defects can be removed by lapping, however not more than 0.0004 of material

should be removed as this surface is hardened and excessive lapping will remove this hardening. Renew valve plate if wear or scoring is extensive.

Inspect piston bores and valve plate mating surface of cylinder block. Minor defects on valve plate mating surface can be removed by lapping, however if damage is extensive, or if piston bores are scored or worn, renew cylinder block.

Inspect piston and shoe assemblies. Worn or damaged pistons will require renewal. Variations between shoes must not be more than 0.001 to insure that all shoes ride properly on swash plate. Shoes can be lapped if necessary, however when pump is being overhauled, it is recommended that all nine pistons, shoe sub-assemblies and cylinder block be renewed to provide maximum overhaul life.

Swash plate can be lapped if necessary, however if more than 0.0004 material must be removed, renew the swash plate.

Inspect and renew pump shaft bearings as necessary. Inspect pump shaft oil seal area and splines. Renew pump shaft if it shows signs of wear or scoring, or if shaft is bent.

259. Reassemble hydraulic pump as follows: Use new seals, "O" rings and gaskets and coat all parts with hydraulic oil when reassembling.

Place springs (31—Fig. 99), seat (30) and yoke (28) in housing (33). Place new "O" rings (35) on pintles (36) and lubricate the "O" rings. Push pintles through housing into yoke, align grooves in pintles with cap screw holes in yoke, then install cap screws (29) and tighten to 20- 25 ft.-lbs. torque. Remove the previously installed cap screws from pintles. Install new oil seal (37) in housing with lip toward yoke and place spacer (38) on top of oil seal. Be sure bearing (40) is about half filled with high temperature grease, install bearing on shaft and secure with snap ring (41). Install shaft and bearing in housing, secure with snap ring (42) and be sure curved side of snap ring is away from bearing.

Place swash plate (27) in yoke with chamfered edge toward shaft bearing and be sure swash plate can be turned freely with fingers. If spring, washer and limiter (18) were removed, compress them with the ⅜ x 3½ inch cap screw (with washers and nut) and secure them with snap ring (17) installed with curved side out. Set cylinder block on valve plate end and insert the three pins (22), then grease washers (23 and 24) and place them on pins so curved side of washer (24) faces shoe plate (25). Assemble pistons and

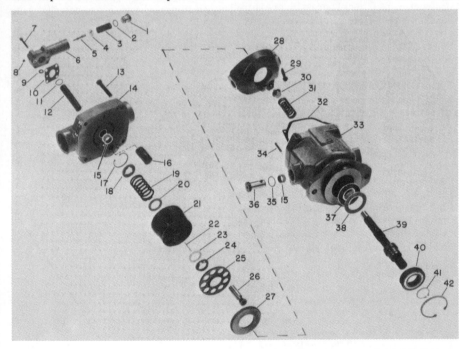

Fig. 99–View of the variable displacement piston type hydraulic pump used on all models. Kits are available for the rotating group (items 18 through 27) and for "O" rings and gaskets.

| | | |
|---|---|---|
| 1. Plug | 12. Rod (tube) | 23. Washer |
| 2. "O" ring | 14. Valve plate | 24. Washer |
| 3. Spring | 15. Bearing | 25. Shoe plate |
| 4. Seat | 16. Piston | 26. Piston (9 used) |
| 5. Spool | 17. Snap ring | 27. Swash plate |
| 6. Body | 18. Limiter | 28. Yoke |
| 8. Plug | 19. Spring | 29. Cap screw |
| 9. "O" ring | 20. Washer | 30. Seat |
| 10. Gasket | 21. Cylinder block | 31. Springs (inner & outer) |
| 11. Snap ring | 22. Dowel | 32. Gasket |

| | |
|---|---|
| 33. Housing | |
| 34. Dowel | |
| 35. "O" ring | |
| 36. Pintle | |
| 37. Seal | |
| 38. Spacer | |
| 39. Pump shaft | |
| 40. Bearing | |
| 41. Snap ring | |
| 42. Snap ring | |

shoes (26) in shoe plate (25), then while holding shoe plate over cylinder block, insert pistons in their bores. Pistons should slide easily in the lubricated bores. Hold the assembled cylinder block assembly together to prevent parts from shifting, then install assembly on the pump drive shaft. Rock drive shaft back and forth slightly if necessary to help align splines of shaft, spherical washer (24) and cylinder block.

Be sure snap ring (11) is on tube (12), and tube is in valve plate (14), then place piston (16) on tube. Use new gasket (32), install valve plate and tighten retaining cap screws to 42 —45 ft.-lbs. torque. Place new "O" ring (9) in compensator valve body (6), use new gasket (10) and install compensator valve. Be sure plug (8) is installed.

Fill pump through vent line with Oliver Type 55 Hydraulic Oil.

## COMPENSATOR VALVE

260. Compensator valve, shown in exploded view in Fig. 99, is mounted on lower side of hydraulic pump valve plate. Function of the compensator valve is to vary stroke (output) of the hydraulic pump to meet the demands of the hydraulic system.

261. **R&R AND OVERHAUL.** Compensator valve can be removed from hydraulic pump valve plate if necessary by removing the four screws (7—Fig. 99).

To disassemble valve, break seal and remove adjusting plug (1) with "O" ring (2). Remove spring (3), spool seat (4) and spool (5).

Check spool and spool bore for nicks, burrs, scoring or wear. Check seat for wear or other damage and be sure orifice is open. Check spring for distortion or fractures. Renew parts as necessary.

To reassemble, slide spool into bore small diameter end first, then install spool seat and spring. Install new "O" ring on adjusting plug, lubricate "O" ring, then install adjusting plug so slotted end is flush with valve body. To check and adjust compensator valve, refer to paragraph 249. Reseal the compensator valve after adjusting.

## HYDRAULIC RESERVOIR

262. **REMOVE AND REINSTALL.** To remove hydraulic reservoir, lower hitch to expel oil from cylinder or cylinders. Remove drain plug from hydraulic system filter and drain reservoir.

Remove upper link, seat and heat shield and control lever console. Disconnect pto control rod from control lever. Disconnect balance link from draft signal arm. On models with internal lift cylinder, disconnect lift links from lift arms. On models with external lift cylinders, disconnect lift cylinders from lift arms. Separate remote cylinder coupling holders from mounting bracket and if tractor has external lift cylinders, disconnect cylinder hoses from fitting at manifold plate. Cap and plug openings.

Remove gear shift lever and center platform. Disconnect brake return line, steering return line, cooler pump inlet line, relief valve return line, main pump inlet hose, pump vent line and pressure control vent line from bottom of reservoir. Disconnect lift circuit line from tee-fitting on left side of manifold plate. Disconnect pto clutch return line from rear of reservoir and cooler return line from remote valve manifold. Plug and cap openings as lines are disconnected. Remove clip retaining cooler return line to transmission cover and move line out of the way.

Remove right hand lift arm, then while noting location of cone head capscrews, remove capscrews retaining hydraulic reservoir to rear main frame. Install lifting eyes in right front and left rear seat mounting capscrew holes. Attach a hoist and remove hydraulic reservoir from tractor.

Reinstall by reversing the removal procedure and tighten the cone head reservoir retaining cap screws to a torque of 135-150 ft.-lbs.

## ROCKSHAFT

263. Rockshaft can be removed without removing hydraulic reservoir, if desired. However, if rocker arm is to be removed, it will be necessary to remove reservoir cover, servo valve and servo valve manifold.

To remove rockshaft without removing hydraulic reservoir, proceed as follows: Remove left fender and drain reservoir. On models with internal lift cylinder, disconnect lift links from lift arms. On models with external lift cylinders, disconnect lift cylinders from lift arms. Remove hitch position arm from left end of rockshaft, then remove lift arm retaining snap rings and pull lift arms from rockshaft. Drive rockshaft out left side of housing which will also force oil seal and bushing out of same side. The opposite seal and bushing can be removed in a similar manner, if necessary.

NOTE: This operation will damage oil seal and will require that a new seal be used. Bushing can be salvaged if caution is used.

Remove rockshaft from left side of housing.

264. When reinstalling rockshaft, insert rockshaft from left side, align blind splines of rockshaft and rocker arm and slide rockshaft into position. Use Oliver driver ST-144, or equivalent, and install bushing so outer end is flush with bottom of oil seal counterbore. Lubricate lip of oil seal and coat outer diameter with sealant. Lubricate rockshaft splines, then with lip of seal toward bushing, carefully slide seal over rockshaft and drive into counterbore.

Complete reassembly by reversing the disassembly procedure.

## HITCH CONTROL LEVER

265. The hitch control lever and friction disc can be removed without removing hydraulic reservoir cover by removing lever console, retaining nut, washer, spring, lever and friction disc. Stationary friction surface is removed by removing shaft key and the retaining cap screws.

If control shaft is to be removed, remove the hydraulic reservoir cover, drive out roll pin securing servo valve control arm to control shaft and pull shaft from manifold.

After reassembly, tighten lever retaining nut until it requires 18 pounds of force to move control lever through the last one inch of downward travel.

## DRAFT INPUT SHAFT

266. The "O" rings on draft input shaft and shaft support can be renewed without removing hydraulic reservoir. Proceed as follows: Disconnect draft control rod from draft input arm. Remove snap ring from outer end of draft input shaft, drive roll pin from the secondary draft input arm and pull input arms from input shaft. Remove cap screws from draft input shaft support and pull support out of reservoir housing and off shaft. "O" rings on shaft and support can now be removed.

If draft input shaft is to be removed from reservoir, remove reservoir cover and drive out roll pin securing servo valve input arm to shaft. Disconnect draft control rod from draft input arm, then pull shaft from support. If desired, support can be unbolted from reservoir housing and complete assembly removed.

Reassemble by reversing the disassembly procedure.

## SERVO VALVE AND MANIFOLD

The servo valve is mounted on a manifold plate which is positioned between the hydraulic reservoir and the reservoir cover. The servo valve can be removed without removing manifold plate, or if desired, the servo valve and manifold plate can be removed together.

**267. REMOVE AND REINSTALL.** Remove drain plug from hydraulic filter and drain hydraulic reservoir. Remove seat assembly and heat shield, and if manifold plate is to be removed, remove control lever console. Remove line running from tee-fitting on left side of manifold plate and remote control valves manifold. If manifold plate is being removed, also disconnect front oil line from tee-fitting. Also, if tractor is equipped with external lift cylinders and manifold plate is to be removed, disconnect lines leading to external cylinders.

Remove capscrews retaining cover to hydraulic reservoir and remove cover by sliding it toward the tee-fitting (left) and lifting cover.

Unscrew return line packing nut (Fig. 100) and pull return line from servo valve body. Remove snap ring from pin in servo valve actuating arm, then remove servo valve retaining capscrews and remove servo valve. Discard the "O" rings located between servo valve and manifold plate.

If manifold plate is to be removed, disconnect line leading to internal cylinder (if so equipped), then remove snap ring from pin extending through slot in connecting link and lift off manifold plate.

**268. OVERHAUL SERVO VALVE.** Remove spool (15—Fig. 101) from plug end. Disconnect spring (3) from plug (2), then drive out pin (16) and remove spring from spool. Remove cap (14), washer (5), spring guide (13), spring (12) and check ball (11). Remove restrictor (6) and "O" ring (8). Remove plug (4) and washer (5), then use a brass drift and drive piston (9) and check ball seat (10) from bore.

Clean and inspect all parts for scor-

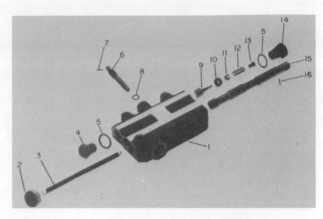

*Fig. 101–Exploded view of servo valve. Pin (7) in restrictor (6) mates with seat of knob located in reservoir cover.*

1. Body
2. Plug
3. Spring
4. Plug
5. Washer
6. Restrictor
7. Pin
8. "O" ring
9. Piston
10. Seat
11. Check ball
12. Spring
13. Spring guide
14. Cap
15. Spool
16. Pin

ing, nicks, burrs or other damage. If spool (15), piston (9) or body (1) require renewal, it will be necessary to renew the complete valve assembly.

When reassembling, install piston (9) with sealing (flat) side toward restrictor (6), then drive check ball seat into position using a proper size brass drift. Complete reassembly by reversing the disassembly procedure.

## HYDRAULIC LIFT CYLINDERS

The single hydraulic lift cylinder for tractors equipped with a Category II hitch is internally mounted in bottom side of hydraulic reservoir housing as shown in Fig. 102. Service on this assembly requires the removal of hydraulic reservoir.

The two hydraulic lift cylinders for tractors equipped with a Category III hitch are externally mounted and can be removed for service by disconnecting hoses and removing attaching pins.

**269. R&R AND OVERHAUL INTERNAL CYLINDER.** To remove internal lift cylinder, first remove hydraulic reservoir as outlined in paragraph 262. With hydraulic reservoir removed, stand unit on end, remove the shouldered dowels which retain oil pan to housing and remove oil pan. See Fig. 102. Remove hydraulic pump inlet line, then disconnect lift cylinder line. Remove the cylinder retaining

capscrews, pull cylinder from dowels and lift cylinder from housing.

Bump cylinder (9—Fig. 103) against a wood block to remove piston. Teflon ring (17) and "O" ring (16) can now be removed from groove in piston.

When renewing piston rings, install "O" ring first, then install Teflon ring on top of "O" ring. Installation of Teflon ring will be facilitated if ring is warmed, however allow Teflon ring to shrink to normal size before installing piston.

Reassemble by reversing the disassembly procedure.

**270. CYLINDER SAFETY RELIEF VALVE.** The internal lift cylinder is fitted with a safety relief valve (1—Fig. 103) which is designed to prevent damage to system due to overload or pressure increase due to a raise in temperature. Valve can be removed from cylinder after hydraulic reservoir is removed and oil pan is off. See Fig. 102.

To disassemble valve, refer to Fig. 104 and break loose the three stakes and remove seat from body. Remove steel ball, plunger and shims and spring.

Clean and inspect all parts and renew as necessary. Shims are available in thickness of 0.010.

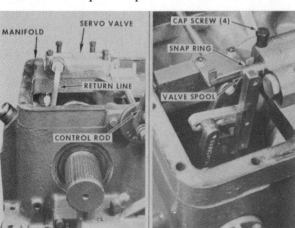

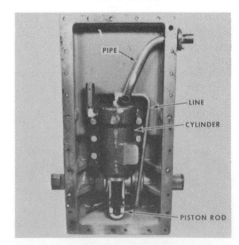

*Fig. 100–Hydraulic system servo valve is located as shown. Valve can be removed separately or with manifold plate.*

*Fig. 102–Internal lift cylinder for Category II hitch is internally mounted as shown.*

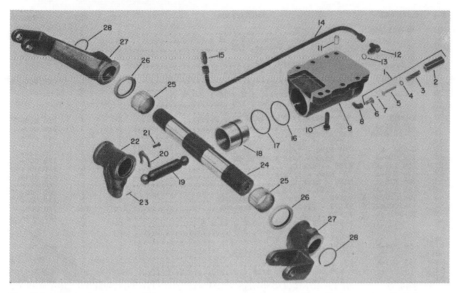

Fig. 103–Exploded view showing component parts of internal lift cylinder and rockshaft assemblies. See Fig. 102 for location of lift cylinder. Note safety relief valve (1).

| | | |
|---|---|---|
| 1. Safety relief valve | 11. Dowel | 19. Connecting rod |
| 2. Body | 12. Elbow | 20. Retainer |
| 3. Spring | 13. "O" ring | 22. Rocker arm |
| 4. Shim | 14. Oil line | 23. Pin |
| 5. Plunger | 15. Adapter | 24. Rockshaft |
| 6. Seat | 16. "O" ring | 25. Bushing |
| 7. Ball (¼) | 17. Teflon ring | 26. Oil seal |
| 8. Elbow | 18. Piston | 27. Lift arm |
| 9. Cylinder | | 28. Snap ring |

Reassemble by reversing disassembly procedure and tighten seat to 75-90 ft.-lbs. torque. However before staking seat in place, check and/or adjust the valve cracking pressure as follows: Attach valve to a hydraulic hand pump, or a diesel injector tester and check the valve cracking pressure which should be 3500-5000 psi. If adjustment is required, vary shims as required. If valve shows any signs of leakage below the minimum limit, valve must be repaired or renewed.

Stake valve seat in place after valve adjustment is complete.

**271. R&R AND OVERHAUL EXTERNAL CYLINDERS.** Cylinders can be removed by disconnecting hoses and removing the attaching pins.

With cylinder removed, remove elbow (2—Fig. 105), then pull piston rod out until snap ring on bottom of piston rod is located directly under port as shown in Fig. 106. Insert a screwdriver in port and move snap

ring from lower groove to the upper (deepest) groove, then pull piston rod from barrel. Remove and discard wiper and seal from barrel.

Clean and inspect piston rod and barrel for scoring, nicks, burrs or undue wear and renew parts as necessary.

When reassembling, install new seal and wiper in barrel and install snap ring in upper (deepest) groove of piston rod. Lubricate all parts and push piston rod into barrel until snap ring is located directly below port, then use screwdriver to move snap ring from upper groove to lower groove. Check to see that snap ring is completely engaged in groove and that piston rod cannot be pulled from barrel.

Reinstall by reversing the removal procedure.

NOTE: If bushings (13—Fig. 105) require renewal, renew them prior to installing lift cylinders.

## DRAFT CONTROL ACTUATING ASSEMBLY (CATEGORY II HITCH)

The draft control actuating assembly used with tractors having an internal hydraulic lift cylinder (Category II hitch) is located in the bottom rear of the rear main frame. The component parts which comprise this assembly are shown in Fig. 107.

**272. R&R AND OVERHAUL.** Drain oil from rear main frame and

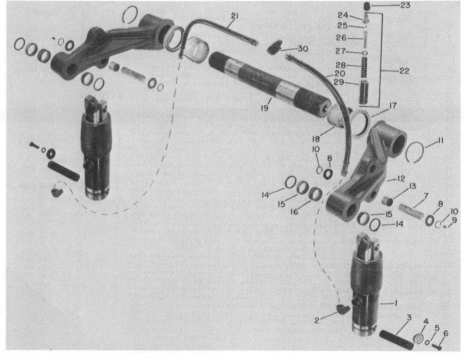

Fig. 105–Exploded view of external lift cylinders, lift arms and rockshaft. Safety relief valve (22) is same as valve used with internal lift cylinder systems.

| | | | |
|---|---|---|---|
| 1. Lift cylinder | 8. Washer | 16. Spacer | 24. Seat |
| 2. Elbow | 9. Grease fitting | 17. Seal | 25. Ball (¼) |
| 3. Pin | 10. Snap ring | 18. Bushing | 26. Plunger |
| 4. Washer | 11. Snap ring | 19. Rockshaft | 27. Shims |
| 5. Lock washer | 12. Lift arm | 20. Hose, right | 28. Spring |
| 6. Capscrew | 13. Bushing | 21. Hose, left | 29. Body |
| 7. Pin | 14. Seal | 22. Safety relief valve | 30. Elbow |
| | 15. Bearing | 23. Reducer bushing | |

Fig. 104–Exploded view of safety relief valve used with both internal and external lift cylinder systems. Shims are 0.010 thick.

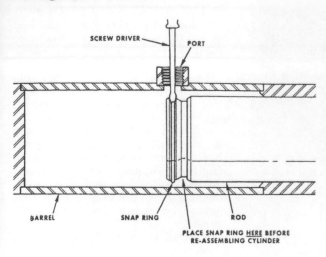

Fig. 106–Use screwdriver as shown to move snap ring when assembling or disassembling cylinder. Refer to text.

remove the pto assembly, or rear cover if tractor has no pto. Remove hitch lower links from supports (37 and 42—Fig. 107). Disconnect clevis link (31) from draft signal arm (34), then remove cap screws and keys (46) from actuating lever (43). Pull lower links (37 and 42) from rear main frame. Unbolt spring support (51) from bottom of rear main frame and remove spacers (50), then remove actuating arm (43), support (51) and draft spring (54) as an assembly. Remove lock nut, adjusting nut, retainer and spring from support. Disconnect rod (49) from actuating arm. Remove oil seals (39) from bores in rear frame and discard. Remove snap rings (40) and bump bearings (41) out of rear frame.

When reassembling, apply Permatex No. 3, or equivalent, to outer diameter of oil seals (39) and using a suitable driver, carefully drive seals into counterbore with lips of seals facing inward. If necessary, dress keyways or inner ends of supports (37 and 42) and wrap with tape. Lubricate lips of oil seals and carefully slide supports through seals and bearings. Adjust draft spring (54) after assembly as outlined in paragraph 273.

273. ADJUST DRAFT SPRING. Prior to installing pto, or rear frame rear cover, adjust draft control spring as follows: Turn adjusting nut (56) on rod (49) until it contacts retainer (55) and all slack is removed from spring assembly. Measure from top of spring support (51) to top of spring retainer, then turn adjusting nut until the spring is compressed ¼-inch. Hold adjusting nut in this position and tighten lock nut (57).

## DRAFT CONTROL ACTUATING ASSEMBLY (CATEGORY III HITCH)

The draft control actuating assembly used with tractors having ex-

ternal lift cylinders (Category III hitch) has a one-piece shaft running through lower rear of tractor rear main frame and the draft control springs are mounted externally. The component parts which comprise this assembly are shown in Fig. 108.

274. R&R AND OVERHAUL. To disassemble draft control mechanism, first drain the rear main frame. On some tractors, it may be necessary to remove left wheel, or slide it to outer end of axle, to provide room to remove shaft (37) from rear main frame.

Disconnect clevis link (31) from draft signal arm (34), then remove draft signal arm from hub on left end of shaft (37). Remove lower lift links, turn adjusting screws (38 and 50) in until all tension is removed from draft springs (40), then remove pins (41) from tractor rear frame and shaft hubs and remove draft springs. Remove nut (49), washer (48), hub (47) and thrust washers (43) from right hand end of shaft (37), then pull shaft from left side of tractor main frame. Pry out oil seals (44), then remove snap rings (45) and bearings (46). Discard oil seals.

When reassembling, apply Permatex No. 3, or equivalent, to outer diameter of oil seals (44), and using a suitable driver, carefully drive seals into counterbores with lips of seals facing inward. Place a thrust washer on left end of shaft (37), tape splines on right end of shaft, then lubricate lips of oil seals and install shaft through rear main frame. Install thrust washers (usually three) on right end of shaft, install hub (47) and be sure hub is aligned with hub on opposite end of shaft. Install washer (48) and nut (49), tighten nut and check to be sure there is a small amount of end play in shaft. If end

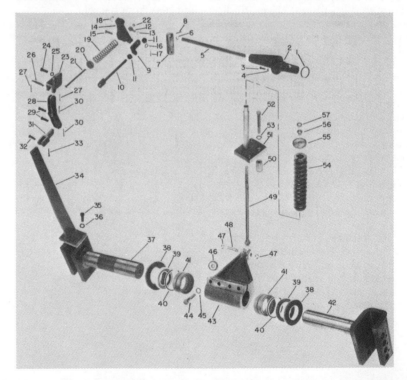

Fig. 107–View showing component parts which comprise the draft control actuating assembly used with tractors equipped with a Category II hitch.

| | | |
|---|---|---|
| 1. "O" ring | 14. Arm | 31. Clevis link | 46. Key |
| 2. Support | 15. Pin | 32. Pin | 47. Retainer ring |
| 5. Shaft | 18. Snap ring | 34. Arm | 48. Pin |
| 6. "O" ring | 19. Spring | 37. Support, left | 49. Rod |
| 7. Arm | 20. Retainer | 38. Washer | 50. Spacer |
| 8. Pin | 21. Capscrew | 39. Oil seal | 51. Support |
| 9. Link | 22. Nut | 40. Snap ring | 54. Draft spring |
| 10. Rod | 23. Arm | 41. Bearing | 55. Retainer |
| 12. Swivel | 26. Pin | 42. Support, right | 56. Adjusting nut |
| 13. Retainer ring | 28. Balance link | 43. Lever | 57. Lock nut |

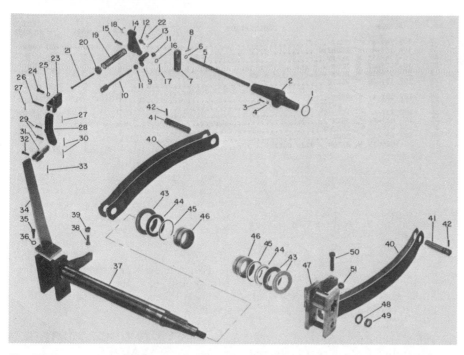

*Fig. 108–View showing component parts which comprise the draft control actuating assembly used with tractors equipped with a Category III hitch.*

| | | | |
|---|---|---|---|
| 1. "O" ring | 15. Pin | 31. Clevis link | 43. Washer |
| 2. Support | 18. Snap ring | 32. Pin | 44. Oil seal |
| 5. Shaft | 19. Spring | 34. Arm | 45. Snap ring |
| 6. "O" ring | 20. Retainer | 37. Shaft | 46. Bearing |
| 7. Arm | 21. Capscrew | 38. Adjusting screw | 47. Hub |
| 9. Link | 22. Nut | 39. Lock nut | 48. Washer |
| 10. Rod | 23. Arm | 40. Draft springs | 49. Nut |
| 12. Swivel | 26. Pin | 41. Pin | 50. Adjusting screw |
| 13. Retainer ring | 28. Balance link | 42. Spring pin | 51. Lock nut |
| 14. Arm | | | |

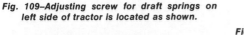

*Fig. 109–Adjusting screw for draft springs on left side of tractor is located as shown.*

*Fig. 110–Draft control springs are preloaded until draft signal arm moves forward 9/16-inch.*

*Fig. 111–Adjusting screw for draft springs on right side of tractor is located as shown.*

*Fig. 112–View showing the basic parts which make up the hydraulic oil cooling system.*

3. Return line, rear
4. Return line, front
5. Inlet line, rear
7. Hose
8. Inlet line
9. Connector
10. Elbow
11. Outlet line
12. Pump
15. Spacer, diesel
16. Support, diesel
21. Pulley
25. Bracket, gas
27. Support, gas
31. Hose
32. Line
33. Line
35. Hose
36. Drive belt
37. Cooler
38. Mounting strip
41. Clamp
45. Hose

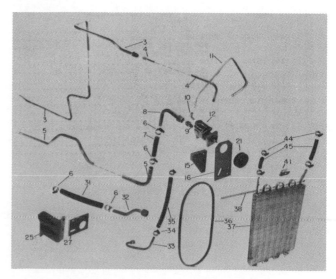

play of shaft requires adjustment, vary the thrust washers on right end of shaft. Thrust washers are available in thicknesses of 7/64 and ¼-inch.

Complete assembly by reversing the disassembly procedure and adjust the preload on draft control springs as outlined in paragraph 275.

275. ADJUST DRAFT CONTROL SPRINGS. Draft spring adjustment can be initiated on either side of tractor, however the left side is generally used because of the measurement involved with the draft signal arm. Proceed as follows:

Refer to Fig. 109 and back the adjusting screw out of the left hub until screw contacts the flange on axle carrier and left draft springs are just beginning to preload. Continue to back out adjusting screw until top of draft signal arm has moved forward 9/16-inch as shown in Fig. 110. Move to right side of tractor, refer to Fig. 111, and back out adjusting screw until right draft springs begin to preload, then back adjusting screw out an additional ¼-turn. Tighten lock nuts on both adjusting screws and secure screws with safety wire.

Check the lift arm free travel as outlined in paragraph 243.

## HYDRAULIC OIL COOLING SYSTEM

The basic components of the hydraulic oil cooling system are a gear type pump which circulates oil from reservoir, and a cooler which is mounted in front of the tractor radiator. Pressure in the system is controlled by a pressure relief valve incorporated within the circulating pump. Although the same pump (and cooler) is used on all models, the pump mounting bracket and the connecting lines and hoses differ between diesel and non-diesel tractors. See Fig. 112.

**276. SYSTEM PRESSURE.** To check system (pump) pressure, disconnect outlet line from pump and connect a flow-rator between pump outlet port and the disconnected pump outlet line. Start engine and run until oil reaches operating temperature, then close restrictor of flow-rator until pump relief valve pops. Note pressure gage reading which should be 100-200 psi. If pressure is not as stated, remove plug (1—Fig. 113) and vary shims (3) as required. If varying shims will not correct pressure, overhaul pump as outlined in paragraph 277.

**277. R&R AND OVERHAUL PUMP.** To remove pump, loosen support capscrews and disengage drive belt from pulley. Disconnect inlet and outlet lines from pump, then unbolt support bracket and remove support and pump. Remove pump pulley, then separate pump from support.

278. With pump removed, remove pulley key from shaft, if necessary, then place a scribe line across back plate (8—Fig. 113) and front plate (19). Remove plug (1) with "O" ring (2), shims (3), spring (4) and poppet (6). DO NOT attempt to remove poppet seat (8) from back plate. Remove capscrews (9) and separate pump. If necessary, bump pump shaft against a wood block. Remove backup gasket (16), protector gasket (17), diaphragm seal (18), check ball springs (14) and check balls (15) from front plate (19). Remove shaft seal (20).

Clean and inspect all parts. If drive gear shaft, or idler gear shaft is less than 0.4360 at seal areas, renew shafts and gears. If gear width is less than 0.566 renew shafts and gears. If gear pocket in back plate exceeds 1.695 in diameter, or more than 0.001 wear at bottom, renew back plate. Renew back plate or front plate if bushing inside diameter is more than 0.4375. Renew relief valve spring if distorted or fractured. Back plate and relief valve seat (8) are available as an assembly only; renew back plate if seat is worn or damaged. Relief valve components are available in kit form or are available separately.

Reassemble pump as follows: Using a dull pointed tool, tuck diaphragm seal (18) into grooves of front plate with open part of "V" sections down. Drop the steel balls (15) into their bores and insert springs (14) on top of balls. Install protector gasket (17) and backup gasket (16) into diaphragm seal and install diaphragm (13) over backup gasket with bronze side up and grooves at shaft holes toward low pressure side of pump. Install gear and shaft assemblies (10 and 11) in bores

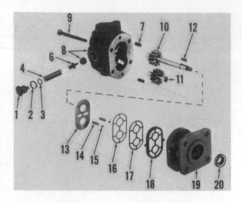

**Fig. 113–Exploded view of hydraulic oil cooler circulating pump. Relief valve seat or back plate are not available separately.**

| | |
|---|---|
| 1. Plug | |
| 2. "O" ring | 12. Key |
| 3. Shims | 13. Diaphragm |
| 4. Spring | 14. Spring |
| 6. Poppet | 15. Check ball |
| 7. Dowel | 16. Backup gasket |
| 8. Back plate | 17. Protector gasket |
| 9. Capscrew | 18. Seal |
| 10. Drive gear | 19. Front plate |
| 11. Idler gear | 20. Oil seal |

of front plate. Apply a thin coat of grease to milled faces of rear plate and front plate and install rear plate to front plate with the previously affixed scribe lines aligned. Install retaining cap screws and tighten to a torque of 7-10 ft.-lbs. Lubricate lip of shaft seal (20), work seal over drive shaft with

lip of seal inward and drive seal into counterbore of front plate. Install poppet (6) with largest diameter end toward spring, then install spring (4), shims (3) and plug (1) with "O" ring (2). Install support to pump front plate, then install pulley key, pulley and pulley retaining nut. Install support to support bracket and install assembly on tractor. Connect lines and adjust drive belt to allow ½-inch deflection between pulleys.

**279. OIL COOLER.** The hydraulic oil cooler is located at left front of tractor radiator and is secured to mounting strips which are attached to radiator supports. See Fig. 112.

With hood and front side panels off, oil cooler can be removed by disconnecting hoses and removing cooler from mounting strips.

Renew oil cooler if damage is extensive.

## REMOTE CYLINDER CONTROL VALVES

The remote cylinder valves and the valve manifold are mounted on the right side of the hydraulic reservoir and valves are mounted outboard of the valve manifold. See Fig. 114. Normally, tractors are equipped with

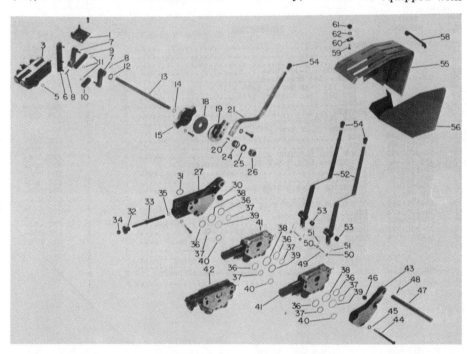

**Fig. 114–Exploded view of control valves, manifold and control levers. Two control valves are standard, however one or two additional valves can be added if desired.**

| | | | |
|---|---|---|---|
| 1. Stop | 19. Plate | 36. "O" ring | 49. Pin |
| 3. Servo valve | 20. Key | 37. Backup ring | 50. Washer |
| 5. "O" ring | 21. Lever | 38. "O" ring | 51. Cotter pin |
| 6. Arm | 24. Spring | 39. Backup ring | 52. Lever |
| 7. Arm | 25. Washer | 40. "O" ring | 53. Bushing |
| 8. Spring pin | 26. Nut | 41. Control valve | 54. Knob |
| 9. Link (3¾") | 27. Manifold | 42. Float valve | 55. Console |
| 10. Link (2") | 30. Plug | 43. End plate | 56. Plate |
| 11. Retainer ring | 31. "O" ring | 44. Cap screw | 58. Strip |
| 12. Washer | 32. Plug | 45. Plate | 59. Bolt |
| 13. Shaft | 33. Restrictor | 46. Plug | 60. Stop |
| 14. "O" ring | 34. Jam nut | 47. Lever shaft | 61. Knob |
| 15. Support | 35. "O" ring | 48. Pin | 62. Washer |
| 18. Friction disc | | | |

two external valves, however an additional one or two valves can be installed if desired.

Valves may be of two types; a pressure detent type which is detented in the raise and lower position and is automatically returned to neutral when the preset pressure (2100-2200 psi) is reached; or a float valve which is spring centered from raise and lower positions and detented in float position. Valve must be held in raise and lower positions and manually returned from float position.

Refer to paragraph 247 to adjust the restrictor in pressure detent valve and to paragraph 248 to adjust restrictor in valve manifold.

**280. R&R CONTROL VALVES.** To remove control valves, drain hydraulic reservoir, then remove control lever console and support. Clean area around control valves prior to removal. Disconnect remote coupling holders from mounting bracket. Disconnect control levers from control valve spools, then unscrew valve retaining capscrews from manifold but do not pull them from valves. Pull the valves and end cap assembly off the control shaft lever, then remove capscrews and separate control valves. Discard all "O" rings and backup rings.

To reinstall, use all new "O" rings and backup rings and reverse removal procedure.

NOTE: Bushings in control levers are renewable and renewal procedure is obvious.

**281. OVERHAUL PRESSURE DETENT VALVE.** Remove valve as outlined in paragraph 280, and proceed as follows: Unscrew detent cap (1—Fig. 115) and remove complete detent assembly and valve spool (20) from valve body (34). Clamp flat end of valve spool in a vise, pull detent cap from spool assembly and catch the three detent balls (3) as cap is removed. Unscrew detent retainer (8) with "O" ring (9) by using a 1-inch spanner wrench. Remove washers (10 and 12) and centering spring (11) from detent retainer. Remove snap ring (4) from I.D. of detent retainer and remove washer (5), spring (6), shim (45), release plunger (7) and piston (15) with "O" ring (16). Slide bushing (13) and "O" ring (14) from spool. Use Allen wrench and piston (15) and turn guide (17) out of spool and remove spring (18) and poppet (19). Remove "O" ring (21) from spool bore.

Remove plug (36) with backup ring (27) and "O" ring (28), then remove spring (29), poppet (30), seat (35) with "O" ring (32) and interlock plunger (33).

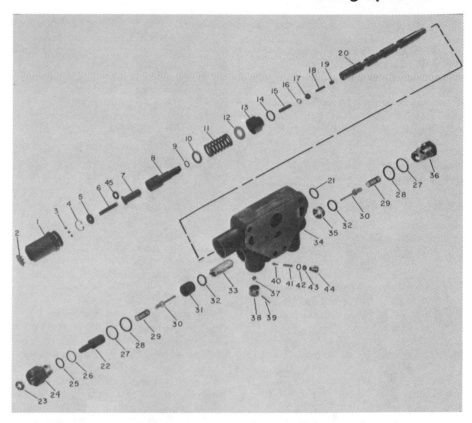

**Fig. 115–Exploded view of pressure detent control valve. Restrictor valve must be individually adjusted for each valve. Adjusting screw is (22).**

| | | | |
|---|---|---|---|
| 1. Cap | 13. Bushing | 24. Plug | 35. Seat |
| 2. Button plug | 14. "O" ring | 25. Backup washer | 36. Plug |
| 3. Detent balls | 15. Piston | 26. "O" ring | 37. Check ball |
| 4. Snap ring | 16. "O" ring | 27. Backup washer | 38. Retainer |
| 5. Washer | 17. Guide | 28. "O" ring | 39. Pin (dowel) |
| 6. Spring | 18. Spring | 29. Spring | 40. Poppet |
| 7. Plunger | 19. Poppet | 30. Poppet | 41. Spring |
| 8. Retainer | 20. Spool | 31. Seat | 42. "O" ring |
| 9. "O" ring | 21. "O" ring | 32. "O" ring | 43. Shim |
| 10. Washer | 22. Adjusting screw | 33. Plunger | 44. Plug |
| 11. Centering spring | 23. Jam nut | 34. Body | 45. Shim |
| 12. Washer | | | |

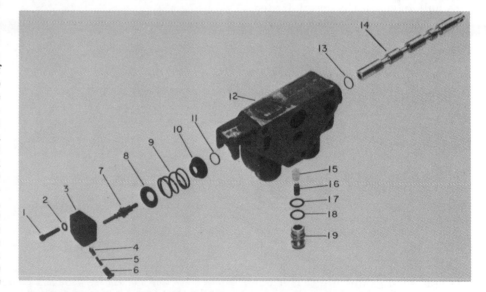

**Fig. 116–Exploded view of float valve which can be used in conjunction with pressure detent control valve.**

| | | | |
|---|---|---|---|
| 1. Allen screw | 6. Plug | 10. Washer | 15. Check valve |
| 2. Lock washer | 7. Detent screw | 11. "O" ring | 16. Spring |
| 3. Pawl block | 8. Washer | 12. Body | 17. "O" ring |
| 4. Pawl | 9. Centering spring | 13. "O" ring | 18. Backup washer |
| 5. Spring | | 14. Spool | 19. Body (plug) |

Remove restrictor plug (24) with backup ring (27) and "O" ring (28), then remove spring (29), poppet (30), and seat (31) with "O" ring (32). Remove jam nut (23) and remove restrictor screw (22) from plug. Remove "O" ring (26) and backup washer (27) from I.D. of plug (24). Remove check ball retainer (38) and check ball (37). Remove pin (39) from retainer, if necessary.

Remove both thermal relief valves (items 40 through 44) and note location of shims (43).

NOTE: The thermal relief valve unit shown in Fig. 115 (items 40 through 44) was used on early valve units and can be identified by the slotted plugs (44). Later (current) valve units use a cartridge type thermal relief valve and can be identified by the plug being hex headed. The two types of thermal relief valves ARE NOT interchangeable.

282. Thoroughly clean and dry all parts. Discard all "O" rings, backup rings and lockout poppets. If valve body or valve spool are damaged, the entire valve assembly must be renewed as spool is a select fit in body and parts are not available separately.

Inspect poppet (19) and seat in valve spool for excessive wear or damage.

Check detent grooves inside detent cap for excessive wear. Mechanical over-ride force required to manually return detented spool to neutral is 100-250 lbs. and can be adjusted by varying the shims located between detent plunger spring and detent plunger.

Inspect sensing port and small orifice hole in spool, the small drilled holes in thermal relief cavities and small hole in rear of lockout seat to be sure they are open and clean. Check large hole in side of rear lockout seat for wear or damage from check ball. Inspect all springs for distortion, excessive weakness or fractures. Spool should fit bore in body with a slight hand pressure with no appreciable side clearance.

283. Reassemble valve as follows: Use new "O" rings, backup washers and lockout poppets and lubricate parts prior to installation.

Install the thermal relief valve assemblies and if necessary, the early type (with slotted head plugs) can be adjusted by varying the shims located between spring and plug. Normal pressure relief is 3500-5000 psi.

Install backup washer (25) then "O" ring (26) in I.D. of plug (24) and install adjusting screw (22) in plug. Install jam nut (23) on adjusting screw. Install "O" ring (32) on seat (31) and install seat in body so that large hole is aligned with cylinder port and will allow check ball (37) to seat. Install poppet (30) and spring (29). Install backup washer (27) and "O" ring (28) on adjusting screw and install screw and plug assembly in body. Place check ball (37) in rear port, and with pin (39) in retainer (38), install retainer.

Install plunger (33) in front bore of body, install "O" ring (32) on seat (35) and install seat. Install poppet (30), spring (29) and plug (36) with "O" ring (28) and backup washer (27) installed.

Place poppet (19) and spring (18) in open end of spool (20), be sure poppet is properly seated, then using release piston (15) and an Allen wrench, turn guide (17) into spool until it bottoms. Install "O" ring (16) on release piston (15) and install piston in retainer (8) with tang end of piston toward guide (17). Place release plunger (7) in retainer (8) and be careful not to push piston out the other end of retainer. Place shims (45), if used, spring (6) and washer (5) inside retainer and secure with snap ring (4). Place washer (10), centering spring (11) and washer (12) on outside of retainer.

Clamp flat end of spool (20) in a vise and place "O" ring (14) and bushing (13) on spool. Install the detent retainer assembly in spool and with the pin spanner wrench, tighten retainer to 5-8 ft.-lbs. Use Allen wrench and engage tang on piston (15) with slot in guide (17), then back guide out of spool until it is against retainer.

Use heavy grease to hold detent balls (3) in holes of retainer (8), then carefully place detent cap (1) over retainer. Use a plastic hammer and drive cap over retainer until detent balls engage second detent groove in cap. Install button plug (2).

Install "O" ring (21) in spool bore, then install spool and detent assembly and tighten cap securely.

NOTE: Arrow on flat end of spool must point toward cylinder ports when assembly is complete.

284. **OVERHAUL FLOAT VALVE.** If necessary, disconnect hoses from valve body. Pull plug (19—Fig. 116) with backup washer (18) and "O" ring (17) and remove spring (16) and check valve (15).

Remove Allen screws (1) and pull detent block (3) from detent screw (7). Remove plugs (6), springs (7) and pawls (4) from detent block. Clamp flat end of spool (14) in a vise and remove detent screw (7) from spool. Remove spool (14), washers (8 and 10) and centering spring (9) from body (12). Remove the two "O" rings (11 and 13) from spool bore.

285. Discard all "O" rings and backup rings, then clean and dry all parts. Inspect spool and spool bore in body for scoring, excessive wear, or other damage. If either part requires renewal, it will be necessary to renew the complete valve as spool is a selec-

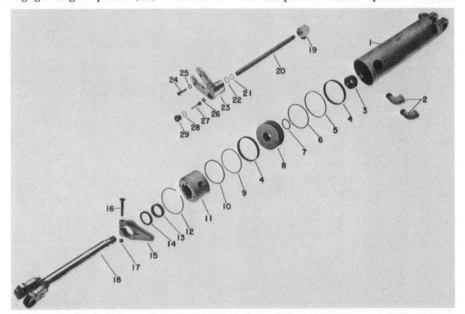

Fig. 117—Exploded view of remote cylinder. Stop arm (15) can be adjusted to vary length of piston stroke. When stop arm contacts valve (27), pressure build up in remote control (pressure detent) valve opens poppet in valve spool releasing detent and allowing valve to return to neutral.

| | | | |
|---|---|---|---|
| 1. Cylinder tube | | | |
| 2. Elbows | 8. Piston | 14. Wiper seal | 22. "O" ring |
| 3. Nut | 9. "O" ring | 15. Stop arm | 23. Stop valve housing |
| 4. Wear rings | 10. Backup washer | 18. Piston rod | 26. Seal |
| 5. Teflon ring | 11. Bearing | 19. Elbow | 27. Stop valve |
| 6. "O" ring | 12. Snap ring | 20. Tube | 28. "O" ring |
| 7. "O" ring | 13. Seal | 21. Backup washer | 29. Guide |

tive fit in body and these parts are not available separately. Spool should fit bore with a slight hand pressure and no appreciable side play.

Inspect detent pawls and groove in detent screw for wear or damage. Inspect check valve for wear or damage. Inspect springs for distortion, excessive weakness or fractures. Renew parts as necessary.

286. Use new "O" rings and backup washers, lubricate all parts and reassemble valve as follows: Install new "O" rings (11 and 13) in spool bore and be sure they are not twisted. Install centering spring (9) and washers (8 and 10) in body with the flattest washer (8) on detent block side. Push spool (14) into bore and clamp flat end of spool in a vise. Apply Loctite to threads of detent screw (7), install detent screw in end of spool and tighten detent screw to 5-8 ft.-lbs. Install pawls (4), springs (5) and plugs (6) in detent block (3), then slide detent block over detent screw and start Allen screws. Be sure detent screw is centered in detent block, then tighten

Allen screws to 10-13 ft.-lbs. torque.

Install backup washer (18) and "O" ring (17) on plug (19). Install check valve (15) and spring (16) in body, then install plug (19) so that groove in plug aligns with retaining cap screw holes in body.

Connect hoses to valve body, if necessary.

287. **REMOTE VALVE MANIFOLD.** To remove control valve manifold, remove remote control valves as outlined in paragraph 280. Disconnect oil cooler return line and lift circuit pressure line, then unbolt and remove manifold and "O" ring from hydraulic reservoir. Remove restrictor plug (32—Fig. 114), with restrictor (33), "O" ring and washer. Remove jam nut (34), turn restrictor out of plug and remove restrictor "O" ring. Control lever shaft (47) can be removed if necessary, by driving out roll pin (48).

Use new "O" rings and reassemble by reversing disassembly procedure. Coat reservoir side of manifold with hydraulic oil and tighten the retaining cap screws to 26-28 ft.-lbs. torque.

Refer to paragraph 248 to adjust

restrictor.

### REMOTE CYLINDER

288. **OVERHAUL CYLINDER.** Refer to exploded view of cylinder in Fig. 117 and proceed as follows:

Loosen nut (17) and slide stop arm (15) to yoke end of piston rod (18). Remove cap screws (24), retaining stop valve housing (23) to bearing (11) and withdraw valve housing from end of tube (20). Unscrew nut (29) and remove stop valve (27).

Push bearing (11) into cylinder tube (1) far enough to allow removal of snap ring (12), then bump bearing from cylinder tube with piston (8) by working piston rod in and out.

Remove nut (3), piston (8) and bearing (11) from piston rod. Need and procedure for further disassembly is evident on inspection of unit.

Install new piston rod seals (13 and 14) in bearing (11) and reassemble unit using all new "O" rings, wear rings and backup rings. Reverse disassembly procedure to reassemble the cylinder.

# NOTES

# OLIVER
## (Includes some Minneapolis-Moline Models)

**Oliver Model** ■ 2255
**Minneapolis-Moline Models** ■ G955 ■ G1355

Previously contained in I&T Shop Manual No. O-26

I&T

SHOP MANUALS

# SHOP MANUAL

# OLIVER

## MODEL 2255

## Also Covers MINNEAPOLIS-MOLINE
## Models G955-G1355

**Tractor ser. no. plate is located on rear side of instrument panel support.**

**Engine serial number plates for Oliver tractors are located on left side of cylinder block and on top side of left valve cover.**

**Engine serial number plate for Minneapolis-Moline tractors is located on right side of engine crankcase.**

## INDEX (By Starting Paragraph)

# INDEX CONT.

# CONDENSED SERVICE DATA

| | Model G955 | Model G1355 | Model 2255 |
|---|---|---|---|
| **GENERAL** | | | |
| Engine Make | Own | Own | Caterpillar |
| Number of Cylinders | 6 | 6 | 8 |
| Bore & Stroke—Inches (Non-Diesel) | 4¼x5 | 4-5/8x5 | ..... |
| Bore & Stroke—Inches (Diesel) | 4-3/8x5 | 4¾x5½ | *4½x5 |
| Displacement—Cu. In. (Non-Diesel) | 425 | 504 | ..... |
| Displacement—Cu. In. (Diesel) | 451 | 585 | *636 |
| Cylinders Sleeved | No | No | No |
| Main Bearings, Number of | 4 | 4 | 5 |
| **TUNE-UP** | | | |
| Firing Order | 1-5-3-6-2-4 | 1-5-3-6-2-4 | 1-2-7-3-4-5-6-8 |
| Valve Tappet Gap—Inlet (Cold) | | | |
|   Diesel | 0.015 | 0.010 | 0.015 |
|   Non-Diesel | 0.012 | 0.010 | ..... |
| Valve Tappet Gap—Exhaust (Cold) | | | |
|   Diesel | 0.022 | 0.020 | 0.025 |
|   Non-Diesel | 0.028 | 0.028 | ..... |
| Compression At Cranking Speed (PSI) | | | |
|   LPG | 185-215 | 185-215 | ..... |
|   Diesel | 440-460 | 440-460 | 640 |
| Ignition Distributor Make | Delco-Remy | Delco-Remy | ..... |
| Breaker Contact Gap | 0.021 | 0.021 | ..... |
| Ignition Timing (LP-Gas) | 16° at 1200 | 17° at 1500 | ..... |
| Injection Timing | 2° BTDC | 4° ATDC | 16° BTDC Static |
| Timing Mark Location | Flywheel | Flywheel | Par. 138 |
| Spark Plug Electrode Gap | | | |
|   LP-Gas | 0.015 | 0.015 | ..... |
| Carburetor Make (LP-Gas) | Century | Century | ..... |
| Engine Low Idle RPM (Non-Diesel) | 600 | 600 | ..... |
| Engine Low Idle RPM (Diesel) | 600 | 700 | 800 |
| Engine High Idle RPM (Non-Diesel) | 2040 | 2400 | ..... |
| Engine High Idle RPM (Diesel) | 1950 | 2335 | 2820 |
| Engine Loaded RPM (Non-Diesel) | 1800 | 2200 | ..... |
| Engine Loaded RPM (Diesel) | 1800 | 2200 | 2625 |
| **SIZES—CAPACITIES—CLEARANCES** | | | |
| **(CLearances in Thousandths)** | | | |
| Crankshaft Journal Diameter | | | |
|   (Non-Diesel) | 2.9110-2.9120 | 2.9110-2.9120 | ..... |
| Crankshaft Journal Diameter (Diesel) | 2.9110-2.9120 | 3.4980-3.4990 | Par. 91 |
| Crankpin Diameter (Non-Diesel) | 2.9115-2.9120 | 2.9115-2.9120 | ..... |
| Crankpin Diameter (Diesel) | 2.9115-2.9120 | 3.4980-3.4990 | Par. 91 |
| Camshaft Journal Diameter | | | |
|   No. 1 (Front) | 3.3380-3.3390 | 3.3380-3.3390 | Par. 85 |
|   No. 2 | 3.3075-3.3085 | 3.3075-3.3085 | ..... |
|   No. 3 | 3.2750-3.2760 | 3.2750-3.2760 | ..... |
|   No. 4 | 1.9960-1.9970 | 1.9960-1.9970 | ..... |
| Main Bearings, Diameter Clearance | | | |
|   Non-Diesel | 1.4-4.4 | 1.4-4.4 | ..... |
|   Diesel | 1.4-4.4 | 3.3-5.3 | 1.5-4.5 |
| Rod Bearing, Diameter Clearance | | | |
|   Non-Diesel | 1.4-3.9 | 1.4-3.9 | ..... |
|   Diesel | 1.4-3.9 | 2.3-5.3 | 1.5-4.5 |
| Cooling System—Gallons | | | |
|   Non-Diesel | 7.5 | 7.5 | ..... |
|   Diesel | 8.25 | 8.25 | 10.0 |
| Crankcase Oil—Quarts (Including Filter) | | | |
|   Non-Diesel | 15 | 15 | ..... |
|   Diesel | 15 | 15 | 12 |
| Transmission and Differential—Gals. | 10.75 | 10.75 | 10.75 |

*For 3208 engine; 4½x4½ bore and stroke, 573 C.I.D. for 3150 engine.

# FRONT AXLE
# (Two Wheel Drive)

The adjustable front axle used on Models G955, G1355 and 2255 tractors are similar except for supports as shown in Fig. 1 and Fig. 2.

## AXLE MAIN MEMBER AND PIVOT PIN

### Models G955-G1355-2255

1. To remove front axle and wheels as an assembly, support front of tractor and unbolt stay rod support (2) from side rails or front main frame. Disconnect inner ends of tie rods from center steering arm. Remove pivot pin (3) from front support (4) and axle, raise front of tractor and roll front axle and wheels assembly from under tractor. Remove rear pivot pin (16) from rear support and remove support from stay rod.

Bushings (13) and (17) are renewable and should be reamed after installation (if necessary) to provide 0.001 to 0.002 diametral clearance on pivot pins.

## FRONT AXLE CARRIER

### Models G955-G1355-2255

2. Front axle carriers (bolsters) are shown in Figs. 1 and 2.

To remove front carrier remove front axle as outlined in paragraph 1.

On Model G1355, the front support (4) can now be removed from steering shaft housing.

On Model 2255, remove the power steering cylinder as outlined in paragraph 39, then unbolt front support from front main frame. Refer to paragraph 45 for overhaul information.

## TIE RODS AND TOE-IN

### Models G955-G1355-2255

3. Tie rod ends are non-adjustable and are serviced either by renewing

*Fig. 2—Exploded view of adjustable front axle used on Model 2255.*

2. Bracket (support)
3. Pivot pin
5. U-bolt
6. Steering arm
7. Washer
8. Bushing
9. Spindle
10. Clamp
11. Front axle
13. Bushing
14. Extension
15. Thrust bearing
16. Reach pin
17. Bushing
18. Center steering arm
19. Lock pin
20. Bearing
21. Seal
22. Retainer
23. Lock nut
24. Shaft
25. Seal
26. Bearing
27. Wheel hub
28. Bearing cup
29. Bearing cone
30. Adjusting nut
31. Gasket

32. Hub cap
33. Seal

34. Bearing cone
35. Bearing cup

the end assembly or the complete tie rod assembly.

Toe-in for all models is 3/16-inch. Obtain correct toe-in by adjusting each tie rod an equal amount.

## STEERING KNUCKLES AND BUSHINGS

### Models G955-G1355-2255

4. To remove steering knuckles and renew bushings (8—Figs. 1 and 2), support front end of tractor and remove tire and wheel assembly. Remove bolts from steering arms (6), remove arms and washers (7) from knuckles, then withdraw knuckles from axle extensions (14). Remove thrust bearings (15) from knuckles.

Drive bushings (8) from axle extensions and install new bushings flush with outer ends of bore. Clearance of knuckle in bushings should be 0.002-0.006. Bushings are presized and should not require reaming if installed with a suitable piloted arbor.

Thrust bearing (15) must be installed with lettering toward top end of knuckle.

## WHEEL BEARINGS AND SEAL

### Models G955-G1355-2255

5. Refer to Fig. 2 for an exploded view of front wheel hub, bearings and seal.

When installing new bearing cups (28 and 35), press them into hub until they bottom. Seal (33) is installed with lip toward inside of hub. Pack wheel bearings with ½-pound of No. 1 grade multi-purpose lithium base grease.

To adjust wheel bearings, tighten nut while rotating wheel until a preload is felt on the wheel bearings, then back-off nut until nearest slot in nut aligns with hole in spindle and install cotter pin. Bend one leg of cotter pin against spindle and the other leg against nut. Use new gasket (31) when installing hub cap (32).

# FRONT SYSTEM
# (Four-Wheel Drive)

Model 2255 tractors are available with a front drive axle which is driven from the transmission bevel pinion shaft via a transfer case and a drive shaft. A shifting mechanism in the transfer case allows connecting or disconnecting power to the front axle as required.

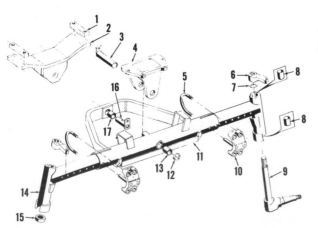

*Fig. 1—Exploded view of adjustable front axle used on Models G955 and G1355.*

1. Strip (spacer)
2. Trunnion (support)
3. Pivot (clevis) pin
4. Clevis (support)
5. U-bolt
6. Steering arm
7. Washer
8. Bushing
9. Spindle
10. Clamp
11. Front axle
12. Spacer
13. Bushing
14. Axle extension
15. Thrust bearing
16. Reach pin
17. Bushing

Front wheel toe-in on tractors equipped with the front drive axle is fixed and non-adjustable. See Fig. 3 for view of front drive axle, carrier, bolster and support.

## FRONT AXLE AND CARRIER

### Model 2255

6. **R&R AXLE ASSEMBLY.** The complete front axle assembly can be removed from tractor as follows: Disconnect drive shaft from companion flange of differential pinion shaft. Disconnect both power steering cylinders from axle and spindle supports and secure cylinders to front frame. Remove bolts retaining axle to axle carrier, then raise tractor and roll complete axle and wheels assembly forward away from tractor.

NOTE: A rolling floor jack can be placed under differential pinion shaft to keep axle from rotating as tractor is lifted from axle.

If necessary, wheels and tie-rod can now be removed and procedure for doing so is obvious.

Reinstall axle by reversing the removal procedure and be sure piston rod ends of steering cylinders are attached to steering spindle supports. Tighten the cylinder attaching bolt lock nuts until they just contact the mounting flanges. Further tightening may distort mounting flanges and cause cylinder to bind.

7. **R&R AXLE CARRIER, BOLSTER AND SUPPORT.** To remove axle carrier, first remove front axle as outlined in paragraph 6. Place a rolling floor jack under axle carrier and take weight of carrier. Remove pivot pin retaining cap screws, slide pivot pin from bolster, then slide carrier rearward until pilot at rear of carrier clears rear support and lower axle carrier from tractor. If necessary, support (9—Fig. 3) can be removed from tractor frame.

Bushings in axle carrier and rear support can now be renewed. Bushings are pre-sized and should require no final sizing if carefully installed.

8. **OVERHAUL FRONT AXLE.** Overhaul of the front axle drive axle assembly will be discussed as four operations; the plant spider assembly, the hub assembly, the spindle support assembly and the differential and carrier assembly. All operations except the differential and carrier overhaul can be accomplished without removing the front drive axle from tractor. Both outer ends of axle are identical, hence, only one outer end will be discussed.

9. **PLANET SPIDER.** To overhaul the planet spider assembly, support outer end of axle and remove the tire

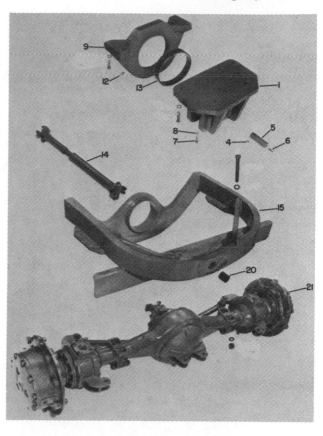

Fig. 3—View showing the basic components which comprise the four wheel drive front axle.

1. Bolster
4. Grease fitting
5. Pivot pin
6. Grease fitting
9. Rear pivot assy.
12. Grease fitting
13. Bushing
14. Drive shaft
15. Carrier
20. Bushing
21. Axle assembly

and rim. Remove relief valve from center of planet spider, remove plug from wheel hub and drain planet spider. Remove cap screws which retain planet spider to wheel hub and the two puller hole cap screws. Use two of the removed retaining cap screws in the puller holes and pull planet spider assembly from wheel hub.

With unit removed, remove the three pinion shaft lock pins by driving them toward center of unit. Remove pinion shafts and expansion plugs by driving pinion shafts toward outside of planet spider. Remove planet pinions, rollers (34 each pinion) and thrust washers. Discard the expansion plugs. See Fig. 4.

Clean and inspect all parts and renew as necessary. Pay particular attention to the pinion rollers and thrust washers.

10. When reassembling, use heavy grease to hold rollers in inner bore of pinions. Be sure tangs of thrust washers are in the slots provided for them and that holes in pinion shaft and mounting boss are aligned before pinion shafts are finally positioned. Coat mating surfaces of planet spider and wheel hub with Permatex No. 2 or equivalent sealer and install planet spider in wheel hub. Tighten retaining cap screws to a torque of 52-57 ft.-lbs.

11. **HUB ASSEMBLY.** To overhaul the wheel hub assembly, first remove planet spider assembly as outlined in

paragraph 9. With planet spider removed, remove snap ring and sun gear (46—Fig. 5 or 6) from outer end of axle shaft. Straighten tabs of spindle nut lock washer, then use OTC tool JD-4, or its equivalent, and remove spindle outer nut and the lock washer. Now loosen, but DO NOT remove, the spindle inner nut. Unbolt spindle from spindle support and remove wheel hub assembly from spindle support as shown in Fig. 7.

Place hub assembly on bench with spindle nut on top side and block up assembly so spindle will be free to drop several inches. Remove the spindle inner nut, then place a wood block over end of spindle and bump spindle from the internal gear hub. Lift internal gear hub and bearing from wheel hub and be careful not to allow bearing to

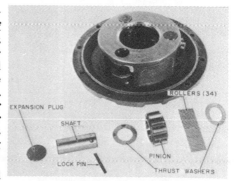

Fig. 4—View showing one pinion assembly removed from planet spider.

5

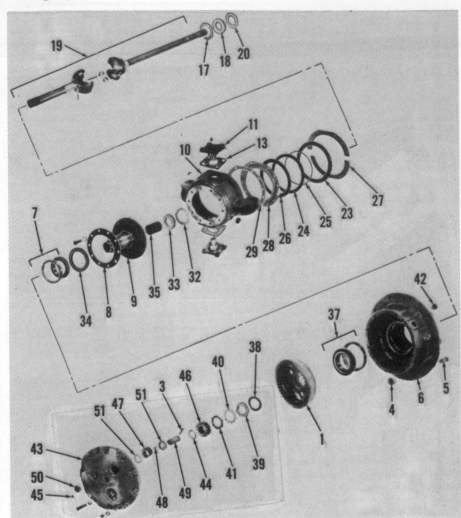

*Fig. 5—Exploded view showing component parts of support and hub assemblies of front drive axle. Trunnion bearings are not shown.*

| | |
|---|---|
| 1. Internal gear | |
| 3. Lock pin | 29. Gasket |
| 4. Nut | 32. Thrust washer |
| 5. Wheel bolt | 33. Oil seal |
| 6. Wheel hub | 34. Oil seal |
| 7. Inner bearing | 35. Bushing |
| 8. Dirt shield | 37. Outer bearing |
| 9. Spindle | 38. Thrust washer |
| 10. Support | 39. Inner nut |
| 11. Upper trunnion | 40. Lock washer |
| 13. Shims | 41. Outer nut |
| 17. Thrust bearing | 42. Filler plug |
| 18. Oil seal | 43. Planet spider |
| 19. Axle shaft & U-joint | 44. Snap ring |
| 20. Washer | 45. Relief valve |
| 23. Dust seal retainer | 46. Sun gear |
| 24. Dust seal | 47. Pinion |
| 25. Dust seal spring | 48. Pinion rollers |
| 26. Felt seal | 49. Pinion shaft |
| 27. Retainer ring | 50. Expansion plug |
| 28. Seal retainer | 51. Thrust washers |

12. Use Fig. 5 or 6 as a reference and reassemble wheel hub assembly as follows: Install bearing cups (7 and 37) in hub with smallest diameters toward inside of hub. Place inner bearing in inner bearing cup, then install oil seal (34) with lip facing bearing. Bump seal into bore until it bottoms. Place dirt shield (8) on hub so flat side is toward flange of spindle, then using caution not to damage seal, install spindle in wheel hub. Hold spindle in position and turn unit over so threaded end of spindle shaft is on top side. Place outer bearing over end of spindle and push down into bearing cup. Start bearing hub of internal gear hub (1) into outer

drop from hub of internal gear hub. Complete removal of spindle from wheel hub. All bearings and seals, thrust washers and dirt shield can now be removed and renewed, if necessary, and procedure for doing so is obvious. Bushing (35—Fig. 5 or 6) and oil seal (33) in inner bore of spindle can also be renewed at this time.

*Fig. 6—Cross sectional view showing components of front drive axle outer end.*

| | |
|---|---|
| 1. Internal gear hub | |
| 2. Thrust washer | 27. Retaining ring |
| 3. Pinion shaft lock pin | 28. Seal retainer |
| 6. Wheel hub | 29. Gasket |
| 7. Hub inner bearing assy. | 30. Shims |
| 8. Dirt shield | 31. Lower trunnion |
| 9. Spindle | 32. Thrust washer |
| 10. Spindle support | 33. Oil seal |
| 11. Upper trunnion | 34. Oil seal |
| 12. Grease fitting | 35. Bushing |
| 13. Shims | 36. Axle shaft (outer) |
| 14. Universal joint | 37. Hub outer bearing assy. |
| 15. Upper trunnion bearing assy. | 38. Thrust washer |
| 16. Grease fitting | 39. Inner nut |
| 17. Thrust washer | 40. Lock washer |
| 18. Oil seal | 41. Outer nut |
| 19. Axle shaft (inner) | 42. Filler plug hole |
| 20. Washer | 43. Planet spider |
| 21. Axle housing | 44. Snap ring |
| 22. Lower trunnion bearing | 45. Relief valve |
| 23. Dust retainer | 46. Sun gear |
| 24. Dust seal | 47. Pinion |
| 25. Spring | 48. Pinion rollers |
| 26. Seal (left) | 50. Expansion plug |
| | 51. Thrust washer |
| | 52. Internal gear |

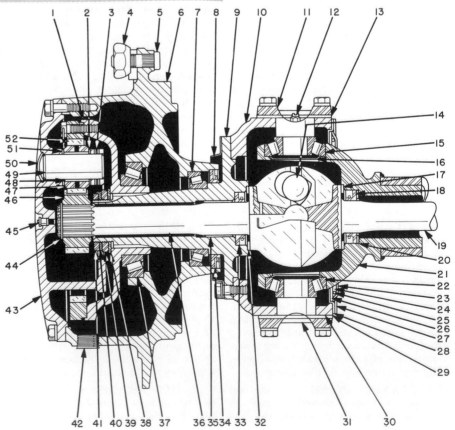

bearing cone, and if necessary, tap gear lightly with a soft faced hammer to position. Install thrust washer (38) and spindle inner nut (39) and tighten nut only finger tight. Coat mating surfaces of spindle and spindle support with Permatex No. 2 or equivalent sealer and install dirt shield and spindle on spindle support. Tighten retaining cap screws to a torque of 80-88 ft.-lbs.

Adjust inner nut as required until a pull of 33-38 pounds on a spring scale attached to a wheel stud is required to keep hub in motion. See Fig. 8. Install lock washer (40—Fig. 5 or 6) and outer nut (41). Tighten outer nut and recheck the hub rolling torque. When adjustment is correct, bend tabs on lock washer to secure both nuts. Install sun gear (46) and snap ring (44) on outer end of axle shaft. Coat mating surfaces of wheel hub and planet spider with Permatex No. 2 or its equivalent sealer and install planet spider. Tighten retaining cap screws to a torque of 52-57 ft.-lbs. Install the puller hole cap screws and the tire and rim.

13. **SPINDLE SUPPORT.** The spindle support can be serviced after planet spider and wheel hub assemblies are removed as outlined in paragraphs 9 and 11. However, if service is required only on the spindle support, the planet spider, wheel hub and axle shaft can be removed as an assembly as follows:

Raise outer end of axle and remove tire and rim. Attach a hoist to wheel stud, then unbolt spindle from spindle support and pull complete hub assembly and axle shaft from outer end of axle. See Fig. 9. Do not allow weight of the hub assembly to be supported by axle shaft or damage to oil seal in axle housing outer end will result.

With complete hub assembly and axle shaft removed, disconnect tie-rod and power steering cylinder from spindle support. Remove the cap screws from the two-piece retainer ring on inner side of spindle support and separate retainers, seals and gasket from spindle support as shown in Fig. 10.

NOTE: At this time it is desirable to remove the grease from the cavity formed by the spindle support and outer end of axle housing.

Remove upper trunnion, pull top of spindle support outward and remove spindle support from outer end of axle housing as shown in Fig. 11. Keep shims present under top trunnion tied to the trunnion for use during reassembly. Remove upper trunnion bearing from axle housing. Remove lower trunnion, shims and bearing from spindle support. Both trunnion bearing cups and upper trunnion bearing grease retainer can now be removed, if necessary. If necessary to remove axle shaft thrust washer, oil seal and oil seal

washer from axle outer end, a slide hammer and puller can be used. Seals (Fig. 10) on outer end of axle housing can also be removed at this time.

Clean and inspect all parts and renew as necessary. It is recommended that new seals be used during assembly.

14. To reassemble spindle support, proceed as follows: Install axle shaft seal washer and oil seal with lip toward inside and be sure oil seal is bottomed. Install axle shaft thrust washer. Install seal components over outer end of axle housing in the following order: Inner seal retainer with step toward inside of tractor. Dust seal spring, rubber dust seal and felt grease seal and be sure that bevel on inside diameter of both seals is toward the bell of axle outer end. Outer seal retainer with step toward outside of tractor and gasket. Install grease retainer, cup side down, and upper bearing cup, smallest I. D. down, in the upper trunnion bearing bore. Bolt the lower trunnion to spindle support using the original shims and tighten the cap screws to a torque of 80-88 ft.-lbs. Place the lower trunnion bearing over lower trunnion. Install lower trunnion bearing cup in outer end of axle with smallest I. D. up. Place upper trunnion bearing in the upper trunnion bearing cup, then while tipping upper side of spindle support slightly outward, position spindle support over outer end of axle and install the upper trunnion and original shim pack. Tighten trunnion retaining cap screws to a torque of 80-88 ft.-lbs.

Before attaching tie-rod, power steering cylinder or seal assembly to spindle support check adjustment of trunnion bearings as follows: Connect a spring scale to tie-rod hole of spindle support and check the effort required to rotate the spindle support as shown in Fig. 12. This effort should be 12-18 pounds on the spring scale. If adjustment is required, vary the number of shims located under trunnions. Keep the number of shims equal on upper and

Fig. 7—Removing wheel hub assembly front spindle support.

Fig. 9—Planet spider, wheel hub and axle shaft assembly can be removed as shown.

Fig. 8—Use method shown to check wheel bearing preload which should be 33-38 lbs. pull on spring scale.

Fig. 10—Axle outer end seals and retainers are removed from spindle support as shown.

Fig. 11—Support and upper trunnion bearing are removed from axle outer end as shown.

lower trunnions. Shims are available in thicknesses of 0.003 (green), 0.005 (blue) and 0.010 (brown).

Use grease to hold seal retainer gasket in place, then install the seal assembly on spindle support and be sure none of the split ends of outer seal assembly are aligned. Attach tie-rod and tighten tie-rod stud nut to a torque of 165 ft.-lbs. Attach power steering cylinder and tighten attaching bolt lock nut until it just contacts mounting flange.

Place approximately four pounds of grease in the cavity of spindle support and pack universal joint of axle shaft. Coat mating surfaces of spindle and spindle support with Permatex No. 2 or equivalent sealer and install the planet spider, wheel hub and axle assembly on the spindle support. Tighten the attaching cap screws to a torque of 80-88 ft.-lbs.

Complete assembly by installing tire and rim.

## DIFFERENTIAL AND CARRIER

### Model 2255

The Four Wheel Drive front axle can be equipped with either a conventional differential assembly or a "No-Spin" differential assembly. Removal procedure will be the same for both types. For service information, refer to paragraphs 16 and 20.

15. **REMOVE AND REINSTALL.** To remove the differential and carrier assembly, raise front of tractor until tires and rims can be removed and support in this position. Block axle carrier to prevent axle assembly from rocking, then remove front tires and rims. Drain differential housing. Disconnect power steering cylinders and lay them on top side of axle carrier. Disconnect drive shaft from companion flange of differential pinion shaft.

Attach a hoist to one of the wheel studs and take up slack of hoist. Unbolt spindle from spindle support and pull complete assembly from outer end of axle. See Fig. 9. Repeat operation for opposite side.

Fig. 12—Use a spring scale in the tie-rod stud hole to check trunnion bearing adjustment.

**NOTE: Do not allow weight of hub assembly to be supported by axle shaft or damage to oil seal in axle housing outer end will result.**

Place a rolling floor jack under front axle, unbolt axle from axle carrier and lower axle away from tractor. Position axle on supports with differential pinion shaft up and secure in this position with blocks. Disconnect one end of tie-rod and swing it out of the way. Remove cap screws which retain differential carrier to axle housing and remove assembly from housing.

Reinstall by reversing the removal procedure and coat carrier retaining capscrews and mating surfaces of carrier and axle housing with Permatex No. 2 or equivalent sealer. Tighten cap screws to a torque of 37-41 ft.-lbs. Tighten the tie-rod stud nut to a torque of 165 ft.-lbs. When joining spindle to spindle support, coat the mating surfaces with Permatex No. 2 or equivalent sealer and tighten the retaining cap screws to a torque of 80-88 ft.-lbs. Piston rod ends of power steering cylinders are attached to spindle supports. Tighten cylinder attaching bolt lock nuts until they just contact the mounting flanges.

16. **OVERHAUL (CONVENTIONAL).** With differential and carrier removed as outlined in paragraph 15, unit is disassembled as follows: If used, straighten tab of lock washer (90—Fig. 14) and remove lock nut, lock washer and thrust screw (91). Punch mark the carrier bearing caps (89) so they can be reinstalled in their original positions, then remove cotter pins and the adjusting nut lock pins (95). Cut lock wires and remove the carrier bearing caps. Lift differential from carrier and keep bearing cups identified with their bearing cones (94). Bearing cones can now be removed from differential case, if necessary. Unbolt and remove bevel ring gear from differential case if necessary. Drive pinion pin (shaft) lock pin (101) out of differential case and remove pinion pin (100), pinions (107), side gears (105) and thrust washers (104 and 106) from differential case.

Remove cotter pin and nut (76) from pinion shaft (87), then using a puller, remove the companion flange (78) and dust shield. Remove pinion shaft bearing retainer (81) and press shaft and bearings from carrier. Use a split bearing puller to support pinion bearing cup on edge nearest pinion shaft gear and press pinion shaft from rear bearing (85) and bearing cup. Remove spacer (86) and any shims which may be present from pinion shaft. Remove front bearing (85) and inner (pilot)

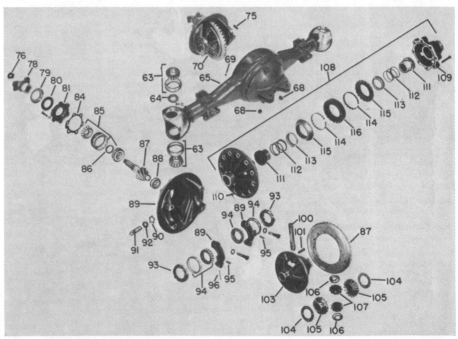

Fig. 14—Exploded view showing the conventional and no-spin differential assemblies which are optionally available for front drive axle.

| | | | |
|---|---|---|---|
| 63. Trunnion bearing | 80. Seal | 92. Jam nut | 108. No-spin differential assy. |
| 64. Grease retainer | 81. Retainer | 93. Adjusting nut | 109. Case half |
| 65. Housing | 84. Gasket | 94. Carrier bearing | 110. Case half |
| 68. Plug | 85. Bearing assy. | 95. Lock pin | 111. Side gear |
| 69. Breather | 86. Spacer & shim kit | 96. Cotter pin | 112. Spring |
| 70. Differential and carrier | 87. Ring gear and pinion | 100. Pinion pin | 113. Spring retainer |
| 75. Tapered dowel | 88. Bearing | 101. Lock pin | 114. Hold-out ring |
| 76. Nut | 89. Carrier & caps | 104. Thrust washer | 115. Clutch |
| 78. Flange | 90. Lockwasher | 105. Side gear | 116. Central driver & center cam |
| 79. Shield | 91. Thrust screw | 106. Thrust washer | |
| | | 107. Pinion gears | |

bearing (88) in a similar manner.

Clean and inspect all parts. Pay particular attention to bearings, bearing cups and thrust washers. If any of the differential side gears or pinions are damaged or excessively worn, renew all gears and thrust washers. Pinion shaft and bevel ring gear are available only as a matched set.

17. The differential and carrier unit is assembled as follows: Place the inner (pilot) bearing on inner end of pinion shaft and stake in four places. Use a piece of pipe the size of inner race of pinion shaft bearing and press the forward pinion shaft bearing (85) on shaft with taper facing threaded end of pinion shaft. Place bearing spacer, and any shims which were present during disassembly over pinion shaft, then position the bearing cup over forward bearing. Press the rear pinion shaft bearing (85) on shaft with taper away from threaded end of pinion shaft. Check, and renew if necessary, the oil seal (80) in pinion shaft bearing retainer (81). Seal is installed with lip toward inside. Place gasket and bearing retainer over pinion shaft. Check, and renew if necessary, the dust shield (79) on companion flange. Position companion flange so it will not obstruct cotter pin hole in end of pinion shaft, slide bearing retainer oil seal on its land on companion flange, then press companion flange on pinion shaft. Install retaining nut, clamp companion flange in a vise and tighten nut to a torque of 300 ft.-lbs.

**NOTE: Pressure of oil seal will generally hold bearing retainer away from bearing. If it does not, tie retainer to companion flange.**

18. With pinion shaft assembled as outlined above, clamp the bearing cup in a soft jawed vise only tight enough to prevent rotation, then using an inch-pound torque wrench attached to companion flange retaining nut, check the rolling torque (bearing preload) of the pinion shaft. This rolling torque should be 13-23 in.-lbs.

If rolling torque of shaft is not as specified, it will be necessary to disassemble the shaft assembly and vary the spacer and/or shims as required. A spacer and shim kit is available under Oliver part number 155 342-A.

With pinion shaft assembled and the rolling torque of shaft determined, press pinion shaft assembly into carrier, install bearing retainer and tighten cap screws to a torque of 25 ft.-lbs. Install cotter pin to lock nut in place.

19. Reassemble differential case assembly as follows: Place side gears, pinions and their thrust washers in differential case and install pinion pin

(shaft). Secure pinion pin in place with lock pin and if lock pin is straight type, stake pin in position. If lock pin is a spring type, staking is not necessary. If bevel ring gear was removed, reinstall with bolt heads next to bevel ring gear and tighten nuts to a torque of 78-86 ft.-lbs. Press bearings on differential case with tapers facing away from case. Place bearing cups over differential bearings and place differential assembly in carrier. Position bearing adjusting nuts in carrier and install the carrier bearing caps. Tighten the bearing cap retaining cap screws until caps are snug but BE SURE threads of caps and adjusting nuts are in register. Maintain some clearance between gear teeth and tighten adjusting nuts until bearing cups are seated and all end play of differential is eliminated. Mount a dial indicator and shift differential assembly as required to obtain a backlash of 0.008-0.011 between bevel pinion shaft gear and bevel ring gear. Differential is shifted by loosening one adjusting nut and tightening the opposite an equal amount.

**NOTE: Fore and aft position of the bevel pinion shaft is not adjustable.**

With gear backlash adjusted, tighten the bearing cap retaining cap screws to a torque of 65 ft.-lbs. and secure with lock wire. Install adjusting nut lock pins and cotter pins.

**NOTE: If lock pins will not enter slots of adjusting nuts after backlash adjustment has been made, tighten rather than loosen the adjusting nut, or nuts. Recheck gear backlash.**

If used, install thrust screw and turn screw in until it contacts bevel ring gear, then back it out ¼-½ turn. Apply sealant to threads of thrust screw at surface of carrier, then install lock washer and lock nut. Secure lock nut by bending one tang of lock washer over nut and one tang over boss of carrier.

20. **OVERHAUL (NO-SPIN).** With differential and carrier removed as outlined in paragraph 15, unit is disassembled as follows: If used, straighten tab of lock washer (90—Fig. 14) and remove lock nut, lock washer and thrust screw (91). Punch mark the carrier bearing caps (89) so they can be reinstalled in their original positions, then remove cotter pins and the adjusting nut lock pins (95). Cut lock wires and remove the carrier bearing caps. Lift differential from carrier and keep bearing cups identified with their bearing cones (94). Bearing cones can now be removed from differential case, if necessary.

Remove cotter pin and nut (76) from pinion shaft (87), then using a puller, remove the companion flange (78) and dust shield. Remove pinion shaft bearing retainer (81) and press shaft and bearings from carrier. Use a split bearing puller to support pinion bearing cup on edge nearest pinion shaft gear and press pinion shaft from rear bearing (85) and bearing cup. Remove spacer (86) and any shims which may be present from pinion shaft. Remove front bearing (85) and inner (pilot) bearing (88) in a similar manner.

Unbolt and remove bevel ring gear from differential, if necessary. Remove differential case bolts and hold case together as last bolt is removed to keep assembly from flying apart due to the internal spring pressure. Hold out rings (114) can be removed with snap ring spreaders.

21. Clean and inspect all parts. Check splines on side gears and clutch members and remove any burrs or chipped edges with a stone or burr grinder. Renew any parts which have sections of splines broken away. Inspect springs (112) for fractures, or other damage, and renew springs which do not have a free height of 2¼-2½ inches. Center cam in central driver (116) must be free to rotate within the limits of keys in central driver. Check the weld between driven clutch (115) and cam ring on clutch by tapping lightly on cams of cam ring. If cam ring rotates in driven clutch, weld is defective (failed). Inspect teeth on central driver and driven clutches. Small defects can be dressed with a stone. If central driver or driven clutch is renewed, also renew the part it mates with. A smooth wear pattern up to 50 percent of face width is acceptable for the cams on center cam and driven clutch.

22. The differential and carrier unit is assembled as follows: Place the inner (pilot) bearing on inner end of pinion shaft and stake in four places. Use a piece of pipe the size of inner race of pinion shaft bearing and press the forward pinion shaft bearing (85—Fig. 14) on shaft with taper facing threaded end of pinion shaft. Place bearing spacer, and any shims which were present during disassembly over pinion shaft, then position the bearing cup over forward bearing. Press the rear pinion shaft bearing (85) on shaft with taper away from threaded end of pinion shaft. Check, and renew if necessary, the oil seal (80) in pinion shaft bearing retainer (81). Seal is installed with lip toward inside. Place gasket and bearing retainer over pinion shaft. Check, and renew if necessary, the dust shield (79) on companion flange. Position

companion flange so it will not obstruct cotter pin hole in end of pinion shaft, slide bearing retainer oil seal on its land on companion flange, then press companion flange on pinion shaft. Install retaining nut, clamp companion flange in a vise and tighten nut to a torque of 300 ft.-lbs.

**NOTE: Pressure of oil seal will generally hold bearing retainer away from bearing. If it does not, tie retainer to companion flange.**

23. With pinion shaft assembled as outlined above, clamp the bearing cup in a soft jawed vise only tight enough to prevent rotation, then using an inch-pound torque wrench attached to companion flange retaining nut, check the rolling torque (bearing preload) of the pinion shaft. This rolling torque should be 13-23 in.-lbs.

If rolling torque is not as specified, it will be necessary to disassemble the shaft assembly and vary the spacer and/or shims as required. A spacer and shim kit is available under Oliver part number 155 342-A.

With pinion shaft assembled and the rolling torque of shaft determined, press pinion shaft assembly into carrier, install bearing retainer and tighten cap screws to a torque of 25 ft.-lbs. Install cotter pin to lock nut (1) in place.

24. Lubricate all parts and reassemble no-spin differential by reversing disassembly procedure. Be sure to position spring retainers so that spring seats inside cupped section of retainer. Be sure key in central driver is aligned with slot in holdout ring. Tighten the case bolts to 36 ft.-lbs. torque and use a single wire through holes in bolt heads to lock bolts. Place an axle shaft into each side of differential and turn axle back and forth. It should be possible to feel backlash between clutch teeth. Backlash should be about 5/32-inch. If no backlash can be felt and side gear is locked solid, check differential for incorrect assembly.

If bevel ring gear was removed, reinstall with bolt heads next to bevel ring gear and tighten nuts to a torque of 78-86 ft.-lbs. Press bearings on differential case with tapers facing away from case. Place bearing cups over differential bearings and place differential assembly in carrier. Position bearing adjusting nuts in carrier and install the carrier bearing caps. Tighten the bearing cap retaining cap screws until caps are snug but BE SURE threads of caps and adjusting nuts are in register. Maintain some clearance between gear teeth and tighten adjusting nuts until bearing cups are seated and all end play of differential is eliminated. Mount a dial indicator and shift differential assembly as required to obtain a backlash of 0.008-0.011 between bevel pinion shaft gear and bevel ring gear. Differential is shifted by loosening one adjusting nut and tightening the opposite an equal amount.

**NOTE: Fore and aft position of the bevel pinion shaft is not adjustable.**

With gear backlash adjusted, tighten the bearing cap retaining cap screws to a torque of 65 ft.-lbs. and secure with lock wire. Install adjusting nut lock pins and cotter pins.

**NOTE: If lock pins will not enter slots of adjusting nuts after backlash adjustment has been made, tighten rather than loosen the adjusting nut, or nuts. Recheck gear backlash.**

Install thrust screw if used and turn screw in until it contacts bevel ring gear, then back it out ¼-½ turn. Apply sealant to threads of thrust screw at surface of carrier, then install lock washer and lock nut. Secure lock nut by bending one tang of lock washer over nut and one tang over boss of carrier.

## TRANSFER DRIVE

### Model 2255

25. **R&R AND OVERHAUL.** To remove the transfer drive assembly, it will be necessary to split the tractor front main frame from the rear main frame. Tractor engine must be removed prior to making split in order to gain access to the cap screws which retain front frame to rear frame.

26. To split tractor, first remove engine and auxiliary drive unit as outlined in paragraph 76, then proceed as follows: Disconnect speedometer drive cable from adapter. Disconnect shifting cable from transfer drive shifter arm and bracket and pull it from front main frame. Disconnect drive shaft from yoke of transfer drive output shaft, unclip and disconnect power steering oil lines. Disconnect safety starting switch wires. Disconnect safety starting switch wires. Disconnect rear light wires and remove clips. Support tractor front frame in a manner which will prevent tipping, then support rear frame with a rolling floor jack. Remove front frame to rear frame retaining cap screws and separate tractor.

27. With tractor split as outlined in paragraph 26, drain transmission and transfer drive housings, then refer to Fig. 15 and proceed as follows: Place transmission and transfer drive in gear to unscrew nut (54), then remove yoke (53) and spacer (52) from output shaft (21). Remove bevel pinion shaft oil line, then unbolt transfer drive cover (39) from case (6) and remove cover along with idler gear (18), output gear (23) and output shaft (21). Remove shifter coupling (22) from fork if necessary.

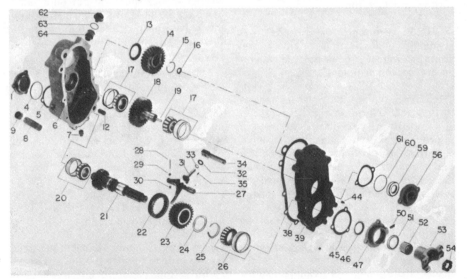

*Fig. 15—Exploded view of the four-wheel drive transfer drive assembly. Breather (62) element should be renewed yearly or after each 1000 hours of operation.*

| | | |
|---|---|---|
| 1. Bearing retainer | 18. Idler gear | 30. Plug | 50. Grease fitting |
| 4. "O" ring | 19. Shaft | 31. Actuator | 51. Seal |
| 5. Shim | 20. Bearing | 32. Washer | 52. Spacer |
| 6. Case | 21. Output shaft | 33. Seal | 53. Yoke |
| 7. Drain plug | 22. Coupling | 34. Arm | 54. Nut |
| 8. Pipe nipple | 23. Gear (27 teeth) | 35. Woodruff key | 56. Retainer |
| 9. Pipe cap | 24. Washer | 38. Gasket | 59. Seal |
| 12. Dowel | 25. Snap ring | 39. Cover | 60. "O" ring |
| 13. Oil seal | 26. Bearing | 44. Plug | 61. Shim |
| 14. Gear (28 teeth) | 27. Shifter fork | 45. Gasket | 62. Breather |
| 15. Snap ring | 28. Spring | 46. Seal | 63. "O" ring |
| 16. Oil seal | 29. Steel ball (7/16) | 47. Retainer | 64. Neck |
| 17. Bearing | | | |

Remove snap ring (15) and gear (14) from forward end of transmission bevel pinion shaft and discard snap ring. Unbolt and remove case (6) from tractor rear frame. Shifter fork (27) and actuator arm (31) can be removed after shifter arm (34) and woodruff key (35) are removed. Any further disassembly will be evident. Save shims (5 and 61) located under bearing retainers (1 and 56) for use during reassembly.

28. Clean and inspect all parts and renew as necessary. Use new seals, "O" rings and gaskets during assembly.

The unit should be partially reassembled and the end play of idler and output shafts checked and adjusted before unit is installed on tractor rear main frame.

Install idler shaft rear bearing cup in transfer drive case and install output shaft rear bearing cup in shaft rear bearing retainer. Be sure both bearing cups are firmly seated and install output shaft rear bearing retainer (1) along with original shim pack (5). Install output shaft front bearing retainer (47) and idler shaft front bearing retainer (56) with original shim pack. Be sure both bearing cups are against the bearing retainers. Install dowels (12) in transfer case if removed. Place output shaft and idler shaft assemblies in the transfer drive case cover, then using a new gasket (38), secure cover to case. Use dial indicator to check end play of both shafts; end play should be 0.001-0.003 and is adjusted by varying number of shims under the bearing retainers. If shims are added, be sure bearing cups are seated against retainers before rechecking shaft end play.

With shaft bearings (end play) adjusted, remove cover, shafts and bearing retainers. Renew the "O" rings (4 and 60) on bearing retainers and install new oil seals as follows: Bevel pinion shaft oil seal (13) in case with lip rearward. Bevel pinion shaft seal (16) in cover with lip forward. Shift fork actuator seal (33) with lip inward. Install wide seal (46) at rear with lip facing rearward and the narrow oil seal (51) at front with lip facing forward in the output shaft front bearing retainer.

Coat all seals and "O" rings with grease prior to installation and fill cavity between the two seals in output shaft front bearing retainer with grease.

Reassemble unit by reversing the disassembly procedure. Renew the "O" ring on bevel pinion shaft front bearing retainer and install case with new gasket. Use a new snap ring (15) when installing drive gear (14) on transmission bevel pinion shaft. Use heavy grease on sliding coupling (22) to hold

it in position in shifter fork. When installing companion yoke (53), be sure not to obstruct cotter pin hole.

## DRIVE SHAFT

### Model 2255

29. The front axle drive shaft is of conventional design and removal and overhaul procedure is evident after an examination of the unit. Spider and bearing assemblies are available as units only. All other parts are available separately.

When installing drive shaft, yoke end is installed to rear.

# POWER STEERING

### Models G955-G1355-2255

30. Models G955, G1355 and 2255 are equipped with power steering that

utilizes a gerotor type control unit which furnishes pressurized fluid in metered amounts to actuate the front wheel steering cylinder, or cylinders. Control units for two-wheel and four-wheel drive tractors are the same except for the width (displacement) of the metering section (gerotor).

Hydraulic system fluid is used as power steering fluid and is furnished by hydraulic system pump via a priority valve located in the pressure control valve.

## LUBRICATION AND BLEEDING

### Models G955-G1355-2255

31. Steering system can be considered self-bleeding and any air trapped in lines, cylinder(s) or control unit can be eliminated by turning steering wheel to move steering cylinder or cylinders, through full range of travel several times in each direction.

**NOTE: Prior to bleeding the power steering system, be sure the hydraulic system is bled as outlined in paragraph 237.**

## SYSTEM PRESSURE TEST

### Models G955-G1355-2255

32. Steering system operating pressure should be 1775-1825 psi when

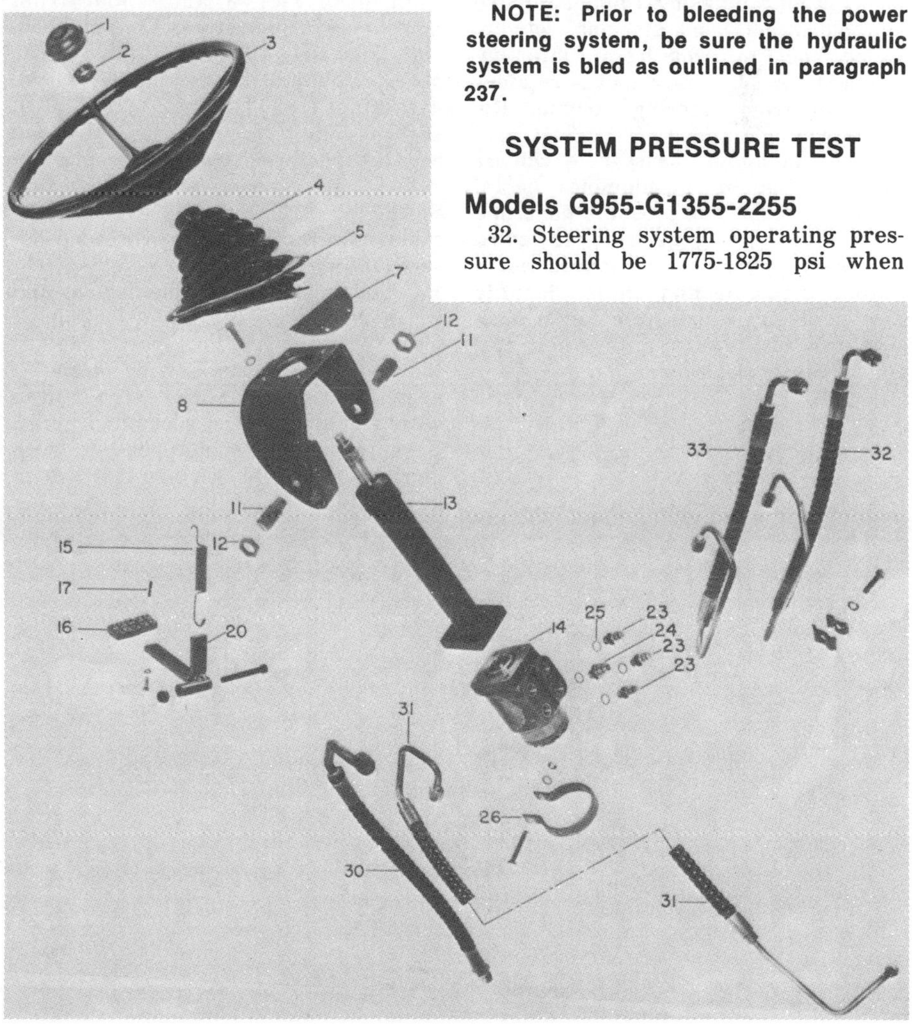

Fig. 16—View showing component parts of the power steering system. Refer to Fig. 17 for an exploded view of control unit (14). Model 2255 is shown. Inlet and return hoses vary somewhat for Models G955 and G1355.

| | | |
|---|---|---|
| 1. Cap | 8. Carrier | 16. Pad |
| 2. Nut | 11. Pivot pin | 17. Roll pins (2) |
| 3. Steering wheel | 12. Nut | 20. Lever |
| 4. Boot | 13. Steering column | 23. Connector (3/8) |
| 5. Retainer | 14. Control unit | 24. Connector (1/2) |
| 7. Stop | 15. Spring | |

| | |
|---|---|
| 25. "O" rings | |
| 26. Clamp | |
| 30. Return hose | |
| 31. Inlet hose | |
| 32. Left turn hose | |
| 33. Right turn hose | |

checked with engine running at approximately 2000 rpm.

To check steering system pressure (priority valve), refer to paragraph 233 of HYDRAULIC LIFT section of this manual.

## PUMP

### Models G955-G1355-2255

33. The hydraulic lift system pump also furnishes the pressurized oil for operation of the power steering system.

For information on the hydraulic lift system pump, refer to paragraph 253.

## CONTROL UNIT

### Models G955-G1355-2255

34. **R&R CONTROL UNIT.** Disconnect battery ground strap. Remove cap from steering wheel, unscrew wheel retaining nut and using a puller remove steering wheel from steering column. Unbolt boot retainer from center instrument panel and remove retainer and boot from steering column. Remove dash side panels.

**NOTE: In some cases there will be enough slack in speedometer cable, tachometer cable, oil pressure gage line and temperature gage wire to allow dash side panels to be moved aside without disconnecting units. Identify and disconnect wiring from center dash panel gages, lights and fuses and remove center dash panel.**

Tag the four control unit hoses for ease in reconnecting, then disconnect hoses from unit and immediately cap all openings. Loosen clamp and pivot pin lock nuts, unscrew pivot pins until they clear control unit carrier, then lift control unit, carrier and steering column from opening in instrument panel support.

Reverse removal procedure to install control unit assembly and bleed unit as outlined in paragraph 31.

35. **OVERHAUL CONTROL UNIT.** With steering unit removed as outlined in paragraph 34, proceed as follows: Remove the cap screws which retain steering column and control unit to carrier, remove carrier, then pull control unit from steering column.

With control unit removed, remove cap screws (22—Fig. 17), then remove end cap (21), gerotor set (20), spacer (18) and plate (19). Pull drive (17) from sleeve and spool assembly (14 and 16), then push sleeve and spool from housing (9). Remove pin (13), pull spool (14), from sleeve (16), then remove the six springs (15) from spool. Remove retaining ring (1), wiper seal (2), seal bushing (3), "O" ring (4), quad ring (5) and thrust bearing assembly (6 and 7) from upper end of housing. Spring pin (10) and ball (11) can be removed after removing threaded insert (12).

Clean parts in a suitable solvent and carefully inspect all parts for undue wear or damage. Pay particular attention to the thrust bearing (6 and 7) assembly. Check splines for undue wear and all parts for nicks and burrs. Be sure pin (13) is not bent or worn and that centering springs (15) are in good condition. Clearance between

teeth of rotor and inside diameter of stator should be 0.001-0.005. Plate (19) should be within 0.0002 of being flat, and if carefully done, plate can be lapped providing not more than 0.005 of material is removed.

36. Use new "O" rings and seals when reassembling. A seal kit containing items 2, 4 and 5 is available. Assemble springs (15) in two parts of three each and install in spool (14) with arches facing each other, then install spool and spring assembly in sleeve (16) and install pin (13). Slide spool and sleeve assembly into body, then install thrust bearing (6 and 7), quad ring (5), "O" ring (4), wiper seal and seal bushing (2 and 3) and secure with retaining ring (1). Invert unit, install drive (17) in spool and note position of pin slot. Place plate (19) over drive, then install rotor on drive so pin slot in drive is aligned with valley between rotor teeth. Install stator over rotor, place end cap (21) on stator and install cap screws (22) only finger tight at this time. Carefully align parts, then complete tightening of cap screws (22—Fig. 17).

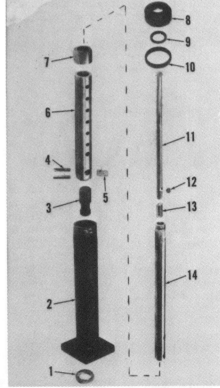

Fig. 18—Exploded view of steering column assembly. Refer also to Fig. 16.

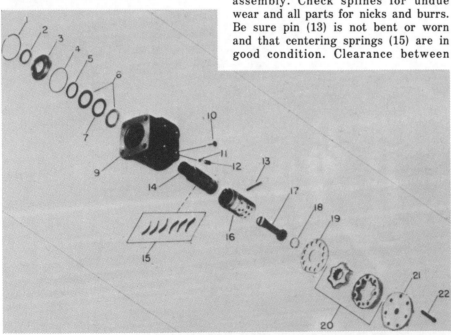

Fig. 17—Exploded view of power steering control unit.

| | | | |
|---|---|---|---|
| 1. Retaining ring | | 13. Pin | 18. Spacer |
| 2. Wiper seal | 7. Thrust bearing | 14. Spool | 19. Plate |
| 3. Seal bushing | 9. Housing | 15. Centering springs | 20. Gear set (gerotor) |
| 4. "O" ring | 10. Spring pin | 16. Sleeve | 21. End cap |
| 5. Quad ring | 11. Steel ball (¼) | 17. Drive | 22. Cap screw |
| 6. Bearing races | 12. Threaded insert | | |

| | |
|---|---|
| 1. Thrust washer | 8. End cap |
| 2. Tube and flange | 9. Wiper seal |
| 3. Splined shaft | 10. Lock ring |
| 4. Dowel pins | 11. Lock rod |
| 5. Guide pin | 12. Steel ball (3/8) |
| 6. Carrier | 13. Spring |
| 7. Guide bushing | 14. Steering shaft |

## STEERING COLUMN

**37. R&R AND OVERHAUL.** To remove the steering column assembly, refer to paragraph 34.

To disassemble steering column assembly, remove end cap (8—Fig. 18), lock ring (10) and thrust washer (1). Pull steering shaft assembly from tube and flange. Disassembly can be completed by removing dowel pins (4), wiper seal (9) and bushing (7) and separating steering shaft assembly.

Clean and inspect parts. All parts are available separately. Reassemble by reversing disassembly procedure.

## STEERING CYLINDERS

### Models G955-G1355

**38. R&R AND OVERHAUL.** To remove power steering cylinder or cylinders, identify and disconnect hydraulic lines and cap lines. Unpin cylinder from side rails and vertical steering shaft and remove cylinder.

With cylinder removed, move piston rod in and out several times to clear fluid from cylinder. Clamp flat end of cylinder tube (16—Fig. 19) in a vise. Remove retaining ring (6) and spacer (7). Bump cylinder head (11) inward and remove retaining ring (8). Pull outward on piston rod (2) and withdraw rod, cylinder head and piston assembly from cylinder tube (16). Remove piston retaining nut (not shown), piston (13), cylinder head, retaining rings and spacer from piston rod (2).

Clean and inspect all parts for excessive wear, scoring or other damage and renew parts as necessary. Renew all "O" rings and back-up rings and reassemble by reversing the disassembly procedure. Reinstall cylinder and bleed system as outlined in paragraph 31.

### Model 2255 Two-Wheel Drive

**39. R&R AND OVERHAUL.** With hood and side panels removed, disconnect oil coolers from their supports and remove grille from front frame. Remove radiator as outlined in paragraph 147. Loosen steering cylinder line clamps, disconnect lines from power steering cylinder and immediately cap all openings. Refer to Fig. 21; remove plug (4) from cap (1) and remove cap from cylinder (14). Pry plug (7) from pinion (11). Pull pinion from splines of vertical shaft (8—Fig. 24) in front carrier (bolster) by turning puller bolt (9—Fig. 21) counterclockwise. Remove nut and washer from torque link and lift steering cylinder from front frame.

Reverse removal procedure to reinstall steering cylinder. The index mark on pinion (11) must be aligned with centering mark on piston (21). Tighten

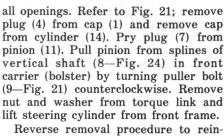

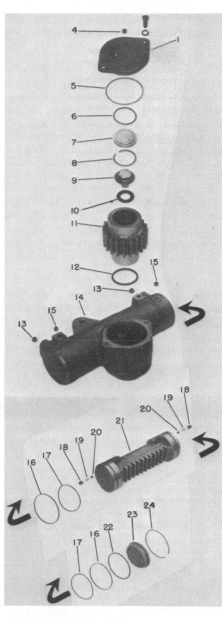

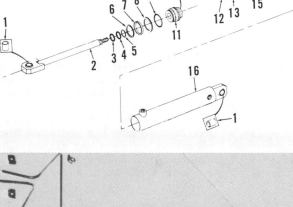

Fig. 19—Exploded view of power steering cylinder. Two cylinders are used on Models G955 and G1355 as shown in Fig. 20.

1. Bushing
2. Piston rod
3. Wiper ring
4. Back-up ring
5. "O" ring
6. Retaining ring
7. Spacer
8. Retaining ring
9. "O" ring
10. Bushing
11. Cylinder head
12. "O" ring
13. Piston
14. Back-up rings
15. "O" ring
16. Cylinder tube

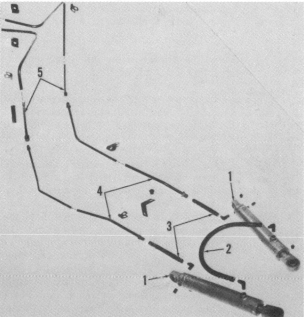

Fig. 20—Models G955 and G1355 use two power steering cylinders as shown.

Fig. 21—Exploded view of power steering cylinder used on Model 2255 two-wheel drive tractors.

| | |
|---|---|
| 1. Cap | 14. Cylinder |
| 4. Plug | 15. Plugs |
| 5. "O" ring | 16. Teflon rings |
| 6. "O" ring | 17. "O" rings |
| 7. Plug | 18. Spring pins |
| 8. Snap ring | 19. Springs |
| 9. Puller bolt | 20. Check valves |
| 10. Washer | 21. Piston |
| 11. Pinion | 22. "O" ring |
| 12. Quad ring | 23. End plug |
| 13. Connectors | 24. Snap ring |

the puller bolt (9) to a torque of 200 ft.-lbs., and tighten nut on torque link securely. Use new "O" ring (6) when installing plug (7). Fill cavity around pinion with power steering fluid before installing cap (1) with new "O" ring (5).

40. OVERHAUL POWER STEERING CYLINDER. After removing the cylinder as outlined in paragraph 39, proceed as follows: If necessary, remove pinion from cylinder then remove snap ring (8), puller bolt (9) and washer (10) from pinion. To remove end plug retaining snap ring (24) insert pin or punch in hole in end of housing (14) and push snap ring from groove in cylinder. Grasp one of the lugs on end plug with vise-grip pliers and, with a twisting motion, pull end plug from cylinder. Remove plug sealing "O" ring (22) from end of cylinder and remove the piston (21). Remove the Teflon seals (16) and "O" rings (17) from groove at each end of piston and remove pinion sealing quad ring (12) from bottom of cylinder. If necessary to remove the check valve assembly (18, 19 and 20) from either end of cylinder, insert punch in hole from inner side of piston head and drive valve ball, spring and spring pin from piston. If the hydraulic line connector seats (13) are damaged, they may be removed by threading inside diameter of seats and using puller bolts. Carefully clean and inspect all parts and renew any that are excessively worn, deeply scored or damaged beyond further use.

Reassemble cylinder, using all new sealing rings as follows: Insert check ball (20) into bore in end of piston (21), insert spring (19) with small end next to ball and drive the retaining spring pin (18) in flush with end of piston. Install "O" ring (17) in bottom of

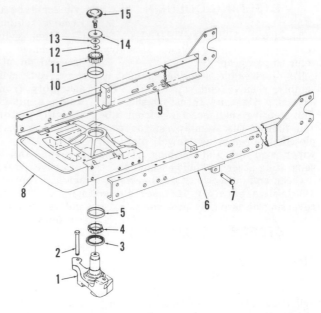

Fig. 23—Exploded view of front support, vertical shaft and side rails used on Models G955 and G1355.

1. Vertical shaft
2. Pin
3. Oil seal
4. Lower bearing
5. Bearing cup
6. Side race L.H.
7. Pin
8. Front support
9. Side race R.H.
10. Bearing cup
11. Upper bearing
12. Shim (0.005)
13. Shim (0.010)
14. Washer
15. Cover

groove in end of piston, then install Teflon ring (16) on top of the "O" ring. Lubricate piston and install in bore of cylinder. Install "O" ring (22) in groove in open end of cylinder, lubricate and install end plug (23) and install plug retaining snap ring (24). Install quad ring (12) in bottom of pinion bore in cylinder casting and lubricate ring and bore. Install pinion (11), and washer (10), puller bolt (9) and retaining snap ring in top of pinion. Install pinion in cylinder with timing marks on pinion and piston aligned and reinstall cylinder as outlined in paragraph 39.

## Model 2255 Four-Wheel Drive

41. R&R AND OVERHAUL. To remove either power steering cylinder, disconnect the hoses, cap all openings, then remove the bolts from cylinder

end assemblies and remove the cylinder from tractor. Loosen the lock nut on rod end of assembly and the clamp on cylinder end, then remove the end assemblies. Bushings in either end assembly may be renewed if worn or damaged.

To disassemble cylinder, refer to Fig. 22 and proceed as follows: Remove elbow from cylinder if not already removed. Remove nut (1), push bearing (4) into barrel to expose snap ring (2) and remove snap ring. Pull the piston rod (9), piston (12) and bearing (4) from barrel as an assembly. Clamp flat of piston rod in a vise, remove self-locking nut (14), then remove washer (13) and piston (12). Remove bearing assembly (4) from piston end of piston rod. Remove all rings and seals from piston and bearing. Clean and carefully inspect all parts and renew any that are excessively worn, scored or damaged.

Reassemble using all new seals as follows: Install back-up washer (5) and "O" ring (6) in bore of bearing with back-up washer at outer (threaded) side of bearing. Install wiper seal (3) with lip outward, lubricate bore of bearing and install bearing over piston end of rod with threaded end of bearing toward flat of piston rod. Install back-up washer (7) and "O" ring (8) on outer diameter of bearing with back-up washer toward threaded end. Clamp flat of piston rod in vise, install piston, washer (13) and self-locking nut (14) and tighten nut securely. Install "O" rings (11) on piston and install rings (10) over "O" rings. Lubricate barrel bore, piston and bearing and carefully insert piston and bearing into barrel but avoid damaging seals on snap ring groove. Use a dull tool to compress

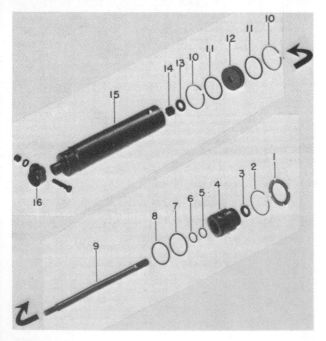

Fig. 22—Exploded view of power steering cylinders used on Model 2255 four-wheel drive tractors.

1. Nut
2. Snap ring
3. Seal
4. Bearing
5. Back-up washer
6. "O" ring
7. Back-up ring
8. "O" ring
9. Piston rod
10. Piston ring
11. "O" ring
12. Piston
13. Washer
14. Nut
15. Barrel
16. Clamp

sealing rings as they pass hole for elbow in barrel. Push bearing far enough in barrel to install snap ring (2), then bump bearing with piston until it is snug against snap ring. Install elbow in bearing; install nut (1) and tighten securely.

Before reinstalling cylinder, install the end assemblies and, with cylinder rod fully retracted, adjust position of end assemblies so that center-to-center measurement of the mounting holes in 18½ inches. Then, tighten the lock nut and clamp to secure end assemblies in this position and install cylinder on tractor.

## FRONT SUPPORT AND VERTICAL SHAFT

The front support (bolster) contains the vertical steering shaft and provides the means for attaching front axle to tractor as well as accommodating the power steering cylinders. See Figs. 23 and 24.

## Models G955-G1355

42. R&R AND OVERHAUL. Refer to Fig. 23. To remove either the front support (8) or the vertical steering shaft (1) it will be necessary to first remove hood, grille, oil coolers and radiator.

43. To remove front support (8), remove steering cylinders, support front of tractor and remove front axle assembly. Attach a hoist to front support, unbolt side rails and slide front support out of side rails.

NOTE: In some cases it may be necessary to loosen the front engine mounting bolts and wedge the side rails

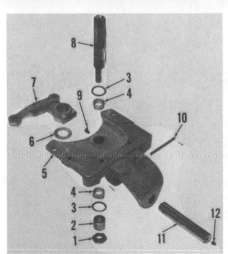

Fig. 24—Exploded view of front support (bolster) and vertical steering shaft used on Model 2255 two-wheel drive tractors.

1. Lock nut
2. Retainer
3. Seal
4. Bearing
5. Front support
6. Washer
7. Steering arm
8. Vertical shaft
9. Grease fitting
10. Retaining pin
11. Pivot pin
12. Grease fitting

outward slightly to facilitate sliding front support from side rails.

44. To remove the vertical steering shaft, refer to paragraph 42 and Fig. 23 and proceed as follows: Disconnect tie rods and steering cylinders from vertical shaft (1). Remove cover (15), cap screw, washer (14) and shims (12 and 13). Bump shaft downward out of bearing cone (11) and remove shaft from below. Remove bearing cone (4) and seal (3) from shaft. Bearing cups (5 and 10) can now be removed from front support.

Clean and inspect all parts and renew as necessary. Note shims (12 and 13). Shim (12) is 0.005 thick and shim (13) is 0.010 thick.

When reassembling, install bearing cups in housing and pack bearings with chassis lubricant. Place bearing cone (4) in bearing cup (5), then install new oil seal (3) in front support. Carefully insert vertical shaft through oil seal and lower bearing (4), then install upper bearing (11). Install the previously removed shims (12 and 13), washer (14) and cap screw. Tighten cap screw while rotating vertical shaft. If bearings bind before cap screw is tightened securely, add shims as necessary. If vertical shaft has excessive end play with cap screw securely tightened, remove shims as necessary. Shaft should have zero end play and yet

rotate without binding. Install cover (15) and complete reassembly by reversing disassembly procedure.

## Model 2255 Two-Wheel Drive

45. R&R AND OVERHAUL. To remove the front support (5—Fig. 24) remove front axle as outlined in paragraph 1 and the power steering cylinder as outlined in paragraph 39. Front support can now be removed from front frame.

To remove vertical shaft (8), remove lock nut (1) and retainer (2), pull vertical shaft from support and remove center steering arm (7) and washer (6). Remove seals (3) and bearings (4) from support and note location of bearings prior to removal.

Clean and inspect parts and renew any that show undue wear or damage. Pay particular attention to the caged bearings (4) and retainer (2).

When installing new bearings, drive a press only on lettered end of bearing cage and be sure all holes in bearing cages are aligned with grease fittings when bearings are installed. Seals (3) are installed with lips toward ends of steering shaft (outward). Be sure washer (6) is located on bottom side of center steering arm (7) and that vertical shaft turns without binding when lock nut (1) is tightened.

Complete assembly of tractor by reversing disassembly procedure.

# ENGINE AND COMPONENTS

## Models G955-G1355

The six cylinder engines used in Models G955 and G1355 conform to the same general design, consisting of a one-piece cast crankcase, with cylinder blocks and cylinder heads cast in pairs. Engines are available either in LP-Gas or diesel versions.

Model G955 may be equipped with an LP-Gas engine having a bore and stroke of 4¼ x 5 inches with a displacement of 425 cubic inches, or a diesel engine having a bore and stroke of 4-3/8 x 5 inches with a displacement of 451 cubic inches.

Model G1355 may be equipped with an LP-Gas engine having a bore and stroke of 4-5/8 x 5 inches with a displacement of 504 cubic inches, or a diesel engine having a bore and stroke of 4¾ x 5½ inches with a displacement of 585 cubic inches.

## R&R ENGINE AND CLUTCH

## Models G955-G1355

50. To remove engine and clutch,

first drain radiator and remove hood. Disconnect battery cables. Disconnect wiring harness, tachometer cable, fuel lines and throttle linkage from engine. Disconnect and remove hydraulic lines and power steering lines as necessary. Remove pto shaft as in paragraph 180 or hydraulic pump drive shaft as in paragraph 181. Block up fuel tank, then unbolt and remove the fuel tank front support.

Install split stands, if available, or block up under clutch housing (center frame) and support engine with a chain hoist. Unbolt frame side rails and engine from clutch housing and separate the tractor halves.

CAUTION: Before removing engine from front frame, block up under front support or front weights to prevent radiator and front frame assembly from tipping forward as engine is removed.

Disconnect radiator hoses, then unbolt engine from front of side rails and lift engine from frame assembly.

## CYLINDER HEAD

### Models G955-G1355

51. To remove any cylinder head, first remove hood and drain cooling system. Remove upper radiator hose, water manifold and air cleaner. On non-diesels, disconnect throttle control rod and fuel line from carburetor, and remove exhaust and inlet manifolds. On diesels, disconnect nozzle lines and remove the manifolds. On all models, remove rocker arm cover, oil lines and rocker arms and shaft assembly. Remove cylinder head retaining nuts and lift cylinder head from engine.

When reinstalling cylinder head, tighten inlet and exhaust manifold nuts finger tight to align mounting faces, then tighten cylinder head stud nuts. After cylinder head nuts are tightened to the correct torque, tighten the manifold nuts.

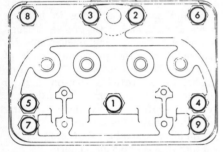

Fig. 30—Cylinder head nut tightening sequence for Models G955 and G1355 LP-Gas engines.

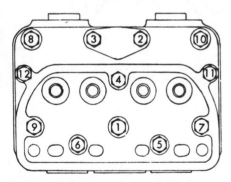

Fig. 31—Cylinder head nut tightening sequence for Model G955 diesel engine.

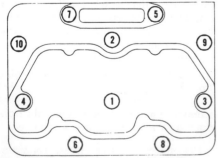

Fig. 32—Cylinder head nut tightening sequence for Model G1355 diesel engine.

On G955 and G1355 models equipped with LP-Gas engines tighten all cylinder head nuts to a torque of 170-175 ft.-lbs. using the sequence shown in Fig. 30.

On G955 models equipped with diesel engines, tighten all 9/16-inch stud nuts to a torque of 130-135 ft.-lbs., and all 5/8-inch stud nuts to a torque of 170-175 ft.-lbs. using the sequence shown in Fig. 31.

On G1355 models equipped with diesel engines, tighten all cylinder head nuts to a torque of 170-175 ft.-lbs. using the sequence shown in Fig. 32.

On all models, adjust valve tappet gap (cold) as outlined in pagagraph 52. Retighten cylinder head stud nuts using the indicated tightening sequence and torque, after engine is at operating temperature. Then, allow engine to cool and readjust valve tappet (cold) as outlined in paragraph 52.

**NOTE: Cylinder heads are widely interchangeable and compression ratio can be varied on spark ignition engines by installing heads of a different configuration. Cylinder heads are identified by the part number cast into the head. Make certain that identical heads are used when engine is assembled.**

### VALVES AND SEATS

#### Models G955-G1355

52. On all model G955 LP-Gas and diesel engines, cylinder head is fitted with renewable seat inserts for the exhaust valves, while intake valves seat directly in the cylinder head. All G1355 engines are equipped with renewable seat inserts on both intake

and exhaust valves. Intake and exhaust valves are not interchangeable. Valves have a face and seat angle of 45 degrees and a desired seat width of 3/32-inch. Valve stem diameter is 0.4335-0.4345.

When installing valves in a diesel cylinder head, make certain that intake valve heads are at least 0.015 below the surface of the cylinder head and are at the same depth within 0.015. The depth of the exhaust valve heads also must not vary more than 0.015.

**NOTE: When servicing the G955 diesel engine valves, be sure to inspect valve stem caps. Install caps prior to installing rocker arm and shaft assembly.**

Stem seals are used on intake valves on all engines.

To adjust valve tappet gap on all models except G1355 diesel, crank engine until No. 1 piston in T.D.C. of compression stroke. At this time, adjust tappet gap of the six valves indicated in Fig. 33. Crank engine one revolution to position No. 6 piston at T.D.C. of compression stroke and adjust the tappet gap of the remaining six valves indicated in Fig. 34.

On Model G1355 diesel engines, adjust valve tappet gap as follows: Crank engine until No. 1 piston is at T.D.C. of compression stroke. At this time, adjust tappet gap of the six valves indicated in Fig. 35. Crank engine one revolution to position No. 6 piston at T.D.C. of compression stroke and adjust tappet gap of the remaining six valves indicated in Fig. 36.

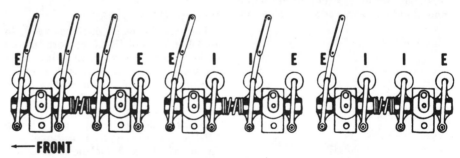

Fig. 33—On all models except G1355 diesel, position No. 1 piston at TDC (compression) and adjust six valves as indicated. Refer to text for correct tappet gap.

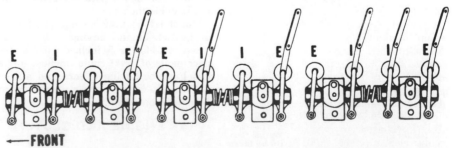

Fig. 34—On all models except G1355 diesel, position No. 6 piston at TDC (compression) and adjust six valves as indicated. Refer to text for correct tappet gap.

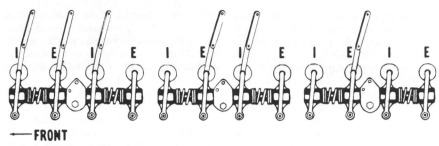

*Fig. 35—On Model G1355 diesel, position No. 1 piston at TDC (compression) and adjust six valves as indicated.*

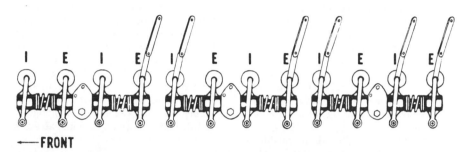

*Fig. 36—On Model G1355 diesel, position No. 6 piston at TDC (compression) and adjust six valves as indicated. Refer to text for correct tappet gap.*

For greatest accuracy, adjust valve tappet gap when engine is cold and not operating. Valve tappet gap (cold) is as follows:

### Model G955
Non-diesel (static),
Intake . . . . . . . . . . . . . . . . . . . . . .0.012
Exhaust. . . . . . . . . . . . . . . . . . . . .0.028
Diesel (static),
Intake . . . . . . . . . . . . . . . . . . . . . .0.015
Exhaust. . . . . . . . . . . . . . . . . . . . .0.022

### Model G1355
Non-diesel (static),
Intake . . . . . . . . . . . . . . . . . . . . . .0.010
Exhaust . . . . . . . . . . . . . . . . . . . . .0.028
Diesel (static),
Intake . . . . . . . . . . . . . . . . . . . . . .0.010
Exhaust. . . . . . . . . . . . . . . . . . . . .0.020

NOTE: Valve arrangement on G1355 diesel engine is intake—exhaust—intake—exhaust, starting at front of each cylinder head. On all G955 and G1355 LP-Gas engines, valve arrangement is exhaust—intake—intake—exhaust on each cylinder head.

## VALVE GUIDES AND SPRINGS

### Models G955-G1355

53. Valve guides for all engines except Model G1355 diesel are available either finished or semi-finished. Valve guides for Model G1355 diesel are available only as semi-finished. Valve guides for LP-Gas engines are interchangeable while valve guides for diesel engines are not. Ream guides after installation to provide a stem to guide clearance of 0.0015-0.0035 for LP-Gas engines, or 0.002-0.004 for diesel engines.

Valve springs are interchangeable for all LP-Gas engines and should test 61-69 lbs. at 2-15/16 inches. Valve springs for the G955 diesel engine should test 46-54 lbs. at 2-7/8 inches. Valve springs for the G1355 diesel engine should test 76-84 lbs. at 2¼ inches.

## VALVE TAPPETS
### (Cam Followers)

### Models G955-G1355

54. The barrel type tappets are supplied in standard size only and should have a clearance of 0.0018-0.0033 in the case bores. Any tappet can be removed after removing either the camshaft or the cylinder block.

## ROCKER ARMS AND SHAFT

### Models G955-G1355

55. Rocker arms and shaft assembly can be removed after removing hood, valve cover and rocker arm shaft support retaining nuts. On all engines except G1355 diesel, desired clearance between rocker arm bushing and the 0.9645-0.9670 diameter shaft is 0.001-0.0045. On G1355 diesel, clearance between rocker arm bushing and the 0.996-0.997 diameter shaft should be 0.002-0.004. Rocker arm bushings are not supplied separately for service; therefore, if clearance between bushing and shaft exceeds 0.008, it will be necessary to renew rocker arms and/or shaft. On all engines except G1355 diesel, be sure plugs are in place in ends of rocker shaft.

On G955 diesel engine, the end rocker arms for each pair of cylinders are offset on their bushings. When reassembling, make certain that long part of bushings is toward rocker arm shaft supports. On G1355 diesel engines, intake valve rocker arms are offset and offsets will face toward exhaust valve rocker arm.

## VALVE TIMING

### Models G955 - G1355

56. Valves are properly timed when mark "1" on camshaft is in mesh with mark "1" on crankshaft gear.

## TIMING GEAR COVER

### Models G955-G1355

57. To remove timing gear cover, remove hood and drain cooling system. Disconnect radiator hoses, then remove grille, oil coolers and radiator. Disconnect power steering cylinders from vertical steering shaft and front axle rear support from side rails. Support front of tractor, unbolt side rails from front support and slide front support forward. Remove fan and water pump assemblies.

Use a suitable puller and remove crankshaft pulley. Remove the timing gear cover retaining cap screws, then remove timing gear cover.

Camshaft end play is controlled by a thrust screw located in front face of timing gear cover. To adjust, turn screw in until it contacts the camshaft gear retaining cap screw, then backout screw ¼-turn and lock in place with lock nut.

## TIMING GEARS

### Models G955-G1355

58. The timing gear train consists of the camshaft gear, crankshaft gear and water pump and distributor or injection pump drive gear. Gears are available in standard size only and if backlash is excessive, renew gear or gears.

To remove the timing gears, proceed as outlined in the appropriate following paragraphs.

59. CAMSHAFT GEAR. To remove the camshaft gear, first remove timing gear cover as outlined in paragraph 57. Remove cap screw and washer from front of camshaft, then use a suitable puller to remove the camshaft gear.

When installing gear, heat gear to approximately 300°F. and use a longer cap screw if necessary, to push gear against camshaft shoulder.

NOTE: If camshaft is pushed to rear it may cause damage to the welch plug in crankcase and cause an oil leak.

Make sure timing marks are aligned as shown in Fig. 37 when gear is installed.

60. **CRANKSHAFT GEAR.** To remove the crankshaft gear, first remove the camshaft gear as outlined in paragraph 59. If not already off, remove distributor or injection pump and drive housing unit. Remove alternator.

Remove cap screws retaining timing gear housing to crankcase. Remove the four cap screws securing oil pan to timing gear housing and loosen the remaining oil pan cap screws; then carefully separate oil pan gasket from timing gear housing. Remove timing gear housing.

Remove the crankshaft gear using a suitable puller. Assemble by reversing the disassembly procedure, making sure timing marks are aligned as shown in Fig. 37 when gears are installed.

61. **DISTRIBUTOR DRIVE GEAR & HOUSING.** Refer to Fig. 38 for exploded view of distributor drive gear and housing. The water pump is driven by a slot in front face of drive gear (13). If only the gear, drive housing and associated parts are to be overhauled, the unit can be removed from the rear without major disassembly of the tractor; proceed as follows.

Note (mark) position of distributor body and rotor and remove distributor. Remove oil line (5), then remove retaining cap screws and withdraw housing and gear from timing gear housing. To reinstall, align water pump tang (pin) with slot in front of gear (13) and insert gear in timing gear housing.

Complete assembly by reversing disassembly and if necessary retime engine as outlined in paragraph 153.

Bushing (4) is renewable and can be renewed as follows: With drive assembly removed, remove the expansion plug from rear of housing (3), remove

snap ring (14) and withdraw drive gear (13), then remove bushing. Install new bushing so oil hole in bushing aligns with oil hole in housing. Use a new expansion plug during reassembly.

62. **INJECTION PUMP DRIVE.** Refer to Fig. 39. The injection pump drive gear (11) meshes with the camshaft gear, and the entire unit can be removed from rear without removing timing gear cover.

To remove the unit, first remove injection pump, adapter (8) and gear (10) as outlined in paragraph 134, being careful not to lose thrust plunger and spring which are inserted in bore of injection pump shaft.

Disconnect the lubrication lines, remove retaining cap screws and lift off housing (12) and gear (11). The gear can be withdrawn from housing bore after removing the external snap ring.

Gears, bushings and housing are pressure lubricated. If shaft bushings must be renewed, make certain that lube hole in bushing aligns with passage hole in housing when bushing is installed. Ream the bushings after installation to provide the recommended 0.0007-0.0019 shaft diametral clearance. Also check to be sure the cup plug in bore of gear shaft (11) is in place and to the rear of the two holes in bearing area of shaft.

To install the injection pump drive unit, align the water pump drive tang with drive slot in front of gear (11), and install gear and snap ring; then install housing (12) by reversing the removal procedure. Install and time injection pump as outlined in paragraph 134.

## CAMSHAFT

### Models G955-G1355

63. To remove the camshaft, first re-

move the timing gear cover as outlined in paragraph 57. Remove rocker arms and shaft assemblies, push rods, governor, oil pan and oil pump. Then, on all models, block up cam followers and withdraw camshaft with camshaft gear through front of timing gear case.

On G1355 LP-Gas and all G955 engines, camshaft bushings are used on the front and front intermediate journals. On G1355 diesel engines, the camshaft rides in five renewable bushings. Bushings are presized and require no reaming if carefully installed. When installing new bushings, align oil hole in bushings with holes in crankcase and make certain the two rear intermediate bushings are installed with split to the top. Recommended camshaft bearing clearance is 0.003-0.006 on all models except G1355 diesel. Recommended clearance on G1355 diesel is 0.002-0.007.

On all models, camshaft journal sizes

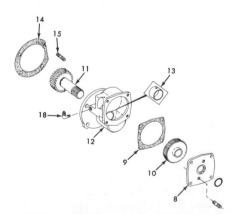

*Fig. 38—Exploded view of distributor drive gear and housing.*

| | |
|---|---|
| 1. Distributor | 7. Elbow |
| 2. Drive gear | 10. Gasket |
| 3. Housing | 11. Clamp |
| 4. Bushing | 12. Gasket |
| 5. Oil line | 13. Drive gear |
| 6. Connector | 14. Snap ring |

*Fig. 37—Non-diesel engine timing gear train. Mesh the No. 1 marks on camshaft gear and crankshaft gear as shown at (T). Timing marks are similar on diesel engines.*

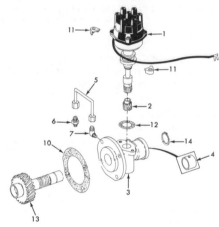

*Fig. 39—Exploded view of injection pump drive gears, adapter housing and associated parts.*

| | |
|---|---|
| 8. Mounting plate | |
| 9. Gasket | 13. Bushing |
| 10. Pump gear | 14. Gasket |
| 11. Drive gear | 15. Stud |
| 12. Adapter housing | 18. Elbow |

are as follows: Front, 3.338-3.339; Second, 3.3075-3.3085; Third (and fourth G1355 diesel), 3.275-3.276; Rear, 1.996-1.997.

When reinstalling camshaft and gear assembly, mesh gears as shown in Fig. 37. After timing gear cover is installed, adjust camshaft end play by turning the adjusting screw (located in case cover) IN until screw contacts cap screw in end of camshaft; then retract the screw ¼-turn and lock in place with the locking nut.

## ROD AND PISTON UNITS

### Models G955-G1355

64. To remove piston and connecting rod units, first remove cylinder heads as outlined in paragraph 51. Drain the oil and remove oil pan and connecting rod caps. Remove water manifold and lift off each cylinder block and the two rod and piston units as an assembly. Withdraw pistons from below after cylinder block has been removed from crankcase.

Piston and connecting rod assemblies are installed with the rod correlation marks facing the camshaft on all models except G1355 diesel. On G1355 diesel, rod correlation marks face side opposite from camshaft. Tighten the commecting rod bolts to 70-75 ft.-lbs. torque on all G955 models, 70-75 ft.-lbs on G1355 non-diesel models, and 55-60 ft.-lbs. on G1355 diesel.

Mark new rods and caps on the camshaft side (all models except G1355 diesel; opposite camshaft side on G1355 diesel), corresponding to the cylinder number.

## PISTONS AND RINGS

### Models G955-G1355

65. Pistons and rings are available in standard size and oversizes of 0.020, 0.040 and 0.060. Check pistons, cylinders and rings against the values which follow:

#### G955

Standard Cylinder Bore Diameter
Diesel . . . . . . . . . . . . . . . . 4.376-4.3770
Non-Diesel . . . . . . . . . . . 4.251-4.2520
Recommended Piston Skirt Clearance
Diesel . . . . . . . . . . . . . . . . 0.005-0.007
Non-Diesel . . . . . . . . . . 0.0045-0.0065
Maximum Cylinder Wear
Bottom of Cylinder . . . . . . . . . . 0.003
Top of Cylinder . . . . . . . . . . . . . . 0.010
Top Ring Side Clearance
Diesel . . . . . . . . . . . . . . . 0.0025-0.0045
Non-Diesel . . . . . . . . . . . 0.0035-0.005
Second Ring Side Clearance
Diesel . . . . . . . . . . . . . . . 0.0025-0.0045
Non-Diesel . . . . . . . . . . . 0.0035-0.005

Third Ring Side Clearance
Diesel . . . . . . . . . . . . . . . . 0.002-0.004
Non-Diesel . . . . . . . . . . . 0.003-0.0045
Fourth Ring Side Clearance
Diesel . . . . . . . . . . . . . . . 0.0015-0.0035
Non-Diesel . . . . . . . . . . . 0.002-0.0035
Compression Ring End Gap
Diesel . . . . . . . . . . . . . . . . 0.015-0.025
Non-Diesel . . . . . . . . . . . 0.020-0.030
Oil Ring End Gap
Diesel . . . . . . . . . . . . . . . . 0.010-0.020
Non-Diesel . . . . . . . . . . . 0.015-0.025

#### G1355

Standard Cylinder Bore Diameter
Diesel . . . . . . . . . . . . . . . . 4.752-4.753
Non-Diesel . . . . . . . . . . . 4.626-4.627
Recommended Piston Skirt Clearance
Diesel . . . . . . . . . . . . . . . . 0.007-0.009
Non-Diesel . . . . . . . . . . . 0.006-0.008
Maximum Cylinder Wear
Bottom of Cylinder . . . . . . . . . . 0.003
Top of Cylinder . . . . . . . . . . . . . . 0.010
Top Ring Side Clearance
Diesel . . . . . . . . . . . . . . . 0.0025-0.0045
Non-Diesel . . . . . . . . . . . 0.0025-0.0045
Second Ring Side Clearance
Diesel . . . . . . . . . . . . . . . 0.0025-0.0045
Non-Diesel . . . . . . . . . . . 0.0025-0.0045
Third Ring Side Clearance
Diesel . . . . . . . . . . . . . . . 0.0020-0.0040
Non-Diesel . . . . . . . . . . . 0.0020-0.0040
Fourth Ring Side Clearance
Diesel . . . . . . . . . . . . . . . 0.0015-0.0035
Non-Diesel . . . . . . . . . . . 0.0010-0.0030
Compression Ring End Gap
Diesel
Top Ring . . . . . . . . . . . . . . 0.017-0.027
2nd & 3rd Rings . . . . . . 0.017-0.032
Non-Diesel . . . . . . . . . . . 0.016-0.026
Oil Ring End Gap
Diesel . . . . . . . . . . . . . . . . 0.015-0.025
Non-Diesel . . . . . . . . . . . 0.013-0.023

## CYLINDER BLOCK

### Models G955-G1355

66. To remove any cylinder block, remove hood, rocker arms and shaft assembly, push rods and cylinder head. Drain the oil and remove oil pan. Remove connecting rod bearing caps and lift off cylinder block and two rod and piston units as an assembly. Withdraw pistons from below after block has been removed from crankcase.

On model G955, cylinder blocks should be renewed or rebored if wear exceeds 0.003 at bottom of cylinder or 0.010 at top of cylinder. On all G1355 models, cylinder blocks should be renewed or rebored if wear or taper exceeds 0.008. Reboring should be done from bottom of block to duplicate factory alignment. Rebore to standard oversizes listed in paragraph 65, or check parts stock for available oversize. Install cylinder heads as outlined in

paragraph 51 and adjust valves as outlined in paragraph 52.

## PISTON PINS

### Models G955-G1355

67. Piston pins are of the full floating type which are retained in the piston pin bosses by snap rings. The 1.2495-1.2496 diameter pins for non-diesels, 1.8745-1.8746 diameter pins for G1355 diesel are available in standard size and oversizes of 0.005 and 0.010.

Install piston pin bushing in rod so that oil hole in bushing registers with oil hole in top end of connecting rod. The bushings must be honed after installation to obtain the correct piston pin clearance of 0.008-0.0012.

## ROD BEARINGS

### Models G955-G1355

68. Connecting rod bearings are of precesion type, renewable from below after removing oil pan. When installing new bearing shells, be sure that bearing shell projection engages milled slot in rod and cap, and that rod and cap correlation marks face toward camshaft of engine on all models except G1355 diesel and opposite camshaft side on G1355 diesel. Bearings are available in 0.002, 0.010, 0.020, 0.030 and 0.040 undersizes as well as standard.

Standard crankpin diameter is 3.498-3.499 on G1355 diesel and 2.9115-2.9120 on all other models. Recommended connecting od bearing diametral clearance is 0.0023-0.0053 for G1355 diesel, 0.0014-0.0039 for G1355 non-diesel and all G955 engines. Connecting rod side clearance should be 0.008-0.013. Tighten connecting rod cap nuts or cap screws to a torque of 70-75 ft.-lbs. on all G955 models, 70-75 ft.-lbs. on G1355 non-diesel models, 55-60 ft.-lbs on G1355 diesel models.

## CRANKSHAFT AND BEARINGS

### Models G955-G1355

69. **MAIN BEARINGS.** Crankshaft main bearings are precision, trimetal type on all models, which may be renewed from below. Bearing caps are marked "FRONT" and should be installed with markings toward front of engine.

**CAUTION: Make certain that main bearing insert with oil hole is installed in the crankcase. Bearings are available in 0.002, 0.010, 0.020, 0.030 and 0.040 undersizes as well as standard.**

70. **CRANKSHAFT.** Crankshaft end play is controlled by rear main bearing flange. Recommended end play is 0.008-

0.012 for G1355 diesel and 0.006-0.010 for all other models. To remove crankshaft, it is necessary to remove engine, oil pan, timing gear cover, timing gear case, oil pump, flywheel, connecting rod bearing caps and main bearing caps.

Check crankshaft, crankpins and main bearing journals for wear, scoring and out-of-round condition against values listed below.

Crankpin diameter . . . See paragraph 68
Maximum allowable taper or
  out-of-round . . . . . . . . . . . . . . . . . . 0.003
Main journal diameter,
  G1355 Diesel . . . . . . . . . . . 3.498-3.499
  All other models . . . . . . . . 2.911-2.912
Main bearing oil clearance,
  G1355 Diesel . . . . . . . . 0.0033-0.0053
  All other models . . . . . . . 0.0014-0.0044
Main bearing bolt torque (ft.-lbs.)
  with lubricated threads.
G955
  Front (¾-inch cap screws) . . . . . . 240
  Intermediates and rear
    (5/8-inch cap screws) . . . . . . . . . 140
G1355 NON-DIESEL
  Front (¾-inch cap screws) . . . . . . 240
  Intermediates and rear
    (5/8-inch cap screws) . . . . . . . . . 140
G1355 DIESEL
  Front (¾-inch cap screws) . . . 195-205
  Intermediates and rear
    (5/8-inch cap screws) . . . . . . 108-112

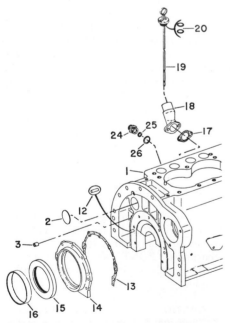

Fig. 40—The one-piece rear seal is pressed into adapter (14). Wear sleeve (16) is pressed on rear of crankshaft.

2. Expansion plug
3. Plug
12. Dowel
13. Gasket
14. Adapter
15. Rear seal
16. Wear sleeve
17. Gasket
18. Oil filler
19. Dipstick
20. "O" rings
24. Adapter (tach.)
25. Seal
26. Annular gasket

## CRANKSHAFT OIL SEALS

### Models G955-G1355

71. **FRONT SEAL.** The felt-type seal is contained in a metal retainer which is pressed into timing gear case. Procedure for renewing the seal is evident after sliding the radiator, pedestal and front wheels assembly forward and removing the crankshaft pulley. Install seal so that felt insert is toward outside of timing gear case.

72. **REAR SEAL.** The one-piece rear oil seal (15—Fig. 40) is pressed into adapter (14) and wear sleeve (16) is pressed on rear flange of crankshaft.

To renew oil seal, split engine from clutch housing as outlined in paragraph 50 and remove clutch and flywheel. Adapter and oil seal can now be removed from crankcase and wear sleeve can be removed from crankshaft.

When renewing wear sleeve, use suitable installation tool to insure that sleeve is parallel with flange on crankshaft. Install oil seal in adapter with lip toward front of engine. Be sure dowels (12) are in place, lubricate lip of seal, then using a new gasket (13) install adapter and seal assembly to crankcase.

## FLYWHEEL

### Models G955-G1355

73. The flywheel can be removed after separating engine from clutch housing (center frame) as outlined in paragraph 50. The flywheel is retained by six equally spaced bolts and nuts or cap screws, and can be installed in any of six positions; only one position is right as follows. When installing flywheel, turn crankshaft until No. 1 piston is at TDC; then install flywheel with DC timing mark at 9 o'clock position as viewed from clutch side of flywheel.

Fig. 41—Exploded view of oil pump typical of those used on Models G955 and G1355. Shim (19) not used on LP-Gas models.

1. Drive pinion
2. Tachometer drive sleeve
3. Drive shaft
4. Pump body
5. Regulator valve
6. Shim
7. Regulator spring
8. Spring retainer
9. Snap ring
10. Intake tube
11. Return tube
12. Discharge tube
13. Pump drive gear
14. Idler gear
15. Pump cover
16. By-pass valve
17. By-pass spring
18. Idler gear shaft
19. Shim (diesel)

To install a new flywheel ring gear, heat same to approximately 500 degrees F. and install on flywheel with beveled edge of teeth facing front of engine.

## OIL PAN

### Models G955-G1355

74. To remove the oil pan, first unbolt and remove the oil filter cover, element and filter can from oil pan. Unbolt and remove the front axle rear support.

Support oil pan with a floor jack, remove cap screws retaining oil pan to crankcase and remove oil pan.

## OIL PUMP

### Models G955-G1355

75. The gear type pump, shown in Fig. 41, can be removed after removing oil pan.

To disassemble the oil pump, remove cotter pin and withdraw intake tube (10). Unbolt and remove cover (15) and idler gear (14). Place the pump in a press in an inverted position, apply pressure to the shaft (3) and press it out of pump gear (13). Withdraw shaft from pump body (4). Remove snap ring (9) from side of pump body and remove the pressure regulator spring (7) and valve (5). On diesel models, note number of shims (6) removed with regulator valve. To remove the filter by-pass vlave, pull the groove pin out far enough to release the valve assembly, and remove piston (16), shim (19) (diesel) and spring (17) from bottom of pump.

Check all parts and renew as necessary. The pressure regulator spring (7) free length is 2-61/64-inches for all models except G1355 diesel which is 2-59/64-inches. The by-pass valve

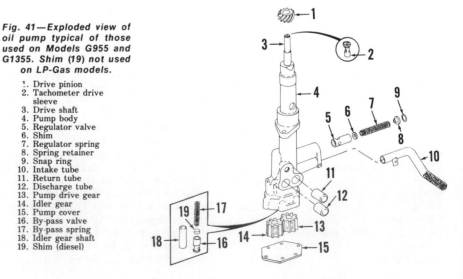

spring (16) free length is 2-7/16-inches on all models.

When reassembling, press pump gear (13) on shaft until shaft is flush to 0.010 below face of gear. With all other parts installed, press drive pinion (1) on shaft until pinion is 0.010-0.030 from top of pump body.

**NOTE: Normal oil pressure is 40-45 psi for all G955 models and G1355 LP-Gas models, or 50-60 psi for G1355 diesel models.**

## OIL COOLER
### Model G1355
75A. Engines used in model G1355

tractors are fitted with a tubular type engine oil cooler which is shown in exploded view 41A. All engine oil entering the engine oil gallery must pass through the oil cooler. Engine coolant water is passed through oil cooler via lines attached to water pump and water inlet manifold.

Oil cooler is mounted on lower left rear of crankcase and can be removed after draining cooling system and oil cooler (LPG engines). Except for cleaning, service of oil cooler consists of renewing faulty units. BE SURE adapter gaskets are satisfactory and renew if any doubt exists.

# ENGINE AND COMPONENTS

The Model 2255 tractors are equipped with Caterpillar diesel engines. Early tractors were equipped with a Model 3150 engine whereas the later tractors were equipped with a Model 3208 engine.

The Model 3150 engine has a bore and stroke of 4½x4½ inches with a displacement of 573 cubic inches. The Model 3208 engine has a bore and stroke of 4½x5 inches with a displacement of 636 cubic inches.

Both engines are the same general design and have a high degree of parts interchangeability.

## R & R ENGINE

### Model 2255
76. To remove engine, first disconnect battery ground strap, then drain cooling system and if engine is to be disassembled, drain oil pan. On models with auxiliary drive unit, drain drive

unit if it is to be disassembled. Remove hood and hood side panels.

Remove the following hydraulic lines and hoses: Power steering control unit to steering cylinder; hydraulic lift unit to circulating pump; circulating pump to cooler and auxiliary transmission to

cooler. Disconnect radiator hoses and remove grille, oil coolers and radiator. Disconnect fuel supply line, leak-off line and wiring from fuel tank and remove fuel tank. Disconnect tachourmeter cable, wiring, battery cables and injection pump controls from engine. Disconnect clutch rod, and if so equipped, controls for auxiliary transmission.

Remove pto shaft as outlined in paragraph 180, or the hydraulic pump drive shaft as outlined in paragraph 181 if tractor has no pto.

Attach hoist to lifting brackets on engine and unbolt engine front support from front main frame. Lift engine slightly, then move engine forward until output shaft housing clears trunnion mount on front of transmission housing and lift engine from tractor.

After engine is removed, remove the auxiliary transmission or the straight through drive assembly from rear of engine.

Reinstall engine by reversing the removal procedure. Be sure to renew

"O" ring on output shaft housing and lubricate it thoroughly prior to reinstallation.

## CYLINDER HEAD

### Model 2255
77. Cylinder heads can be used on either right or left cylinder bank by installing a plug in the unused water outlet hole in end of cylinder head. Inlet manifold is cast into cylinder head.

To remove cylinder head, drain cooling system and remove exhaust manifold shield (guard) and exhaust manifold. On right cylinder head, disconnect hose from ventilator valve. Remove cap screws and lift off valve (tappet) cover. Remove fuel return manifold retaining cap screws, remove fuel return manifold, and rocker arm and shaft assembly. Disconnect and remove injection pump to injector lines as an assembly and remove injectors.

**NOTE: Cap all openings in injection pump, injectors and lines.**

Remove clamp from water sleeve located between cylinder head and timing gear cover and use special tool (Cat. No. 856692) to slide water sleeve into front cover. Remove air inlet manifold (tube) then remove cylinder head retaining bolts and lift off cylinder head.

**NOTE: Cylinder head bolts are various lengths. Note location of bolts as they are removed.**

Prior to installing cylinder head, clean and inspect water sleeve bores and lubricate "O" rings prior to installing water sleeve.

Use new cylinder head gasket (dry) and install cylinder head. Use an anti-sieze compound on cylinder head bolts and tighten bolts in the sequence shown in Fig. 42 as follows: Tighten bolts 1 thru 18 in the sequence shown to 50-70 ft.-lbs. of torque, then retighten to 90-100 ft.-lbs. of torque. Again, recheck bolts 1 through 18 to insure the 90-100 ft.-lbs. of torque. Tighten

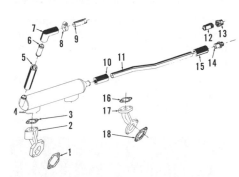

Fig. 41A—Exploded view of engine oil cooler used on Model G1355 tractors. Unit is connected between water pump and water inlet manifold.

1. Gasket
2. Adapter
3. Gasket
4. Oil cooler
5. Hose
6. Tube
7. Hose (elbow)
8. Clamp
9. Nipple
11. Tube
12. Nipple
13. Bushing (LPG)
14. Connector (diesel)
15. Hose
16. Gasket
17. Adapter
18. Gasket

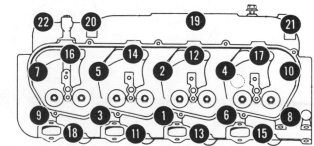

Fig. 42—When tightening cylinder head bolts follow the numbered sequence shown. Refer to text for torque values.

bolts 19 through 22 to 27-37 ft.-lbs. of torque.

Complete reassembly by reversing the disassembly procedure. Tighten rocker arm shaft support bolts to 18 ft.-lbs. of torque. Valve tappet gap cold is 0.015 for intake valves and 0.025 for exhaust valves. Tighten the air inlet manifold (tube) bolts in the sequence shown in Fig. 43. Bolts (3 and 4) are tightened to 15 ft.-lbs. of torque and the remaining bolts to 32 ft.-lbs. of torque. Tighten exhaust manifold bolts to 32 ft.-lbs. of torque.

## VALVES, GUIDES AND SEATS

### Model 2255

78. Intake valves have a face angle of $30 \pm \frac{1}{4}$ degrees and exhaust valves have a face angle of $45 \pm \frac{1}{4}$ degrees. Valve stem diameter for intake valves is 0.3720-0.3730 with a minimum allowable diameter of 0.3710. Valve stem for exhaust valve is tapered and measures 0.3700-0.3710 at head end and 0.3710-0.3720 at lock end of stem. Minimum allowable diameter of valve stem is 0.3690.

Exhaust valve seat inserts are standard equipment. The outside diameter of valve insert is 1.9115-1.9125 and bore in head for insert has a diameter of 1.9085-1.9095 and a depth of 0.440-0.444. Maximum allowable seat width is 0.105. Valve seat angle is $45\frac{1}{4} \pm \frac{1}{2}$ degrees.

Intake valve seat inserts are available for service and have an outside diameter of 2.1495-2.1505. Bore in head for insert should have a diameter of 2.1465-2.1475 and a depth of 0.440-0.444. Maximum allowable seat width is 0.120. Valve seat angle is $30\frac{1}{4} \pm \frac{1}{2}$ degrees.

Valve seats can be regulated by using 15 and 60 degree stones and when service is completed, exhaust valves should be 0.050-0.080 and intake valves 0.036-0.068, below surface of cylinder head.

Valve guides are integral with cylinder head with a valve stem bore diameter of 0.3740-0.3750. Maximum allowable valve stem bore for all valves is 0.3760, measured ¾-inch deep in

bore from both ends. Valve stem bores that are worn can be brought to specification by knurling the bore and then reaming to size. Normal operating clearance of intake valve stem in bore is 0.0010-0.0025. Normal operating clearance of exhaust valve stem in bore is 0.0030-0.0050 (head end) and 0.0020-0.0040 (lock end).

## VALVE SPRINGS

### Model 2255

79. All valves are fitted with two valve springs which are interchangeable between intake and exhaust valves.

Inner valve spring has a free length of 1.83-1.87 inches and should test 16.1-16.9 lbs. when compressed to a length of 1.715 inches. Outer valve spring has a free length of 1.855 inches and should test 30-40 lbs. when compressed to a length of 1.715 inches. Renew any spring which does not meet specifications, does not have square ends, is rusted or shows any signs of fracture.

## VALVE LIFTERS
## (Cam Followers)

### Model 2255

80. Valve lifters are cylindrical type and operate directly in machined bores in the cylinder block. Lifters are supplied only in standard size of 1.1585-1.1593 with a minimum allowable diameter of 1.1575. Bore in cylinder block for lifters is 1.1618-1.1630 with a maximum allowable diameter of 1.1650. Normal operating clearance of lifters in their bores is 0.0025-0.0045. If clearance of new lifter in bore of cylinder block exceeds 0.006, renew cylinder block.

Valve lifters can be removed with a magnet after cylinder head is removed.

**NOTE: It is recommended that lifters be identified with their bores so they can be reinstalled in the same position.**

## VALVE ROCKER
## ARMS AND SHAFT

### Model 2255

81. The rocker arms and shaft assembly can be removed as follows: Remove hood side panels. Remove exhaust manifold shield (guard) if desired. On right side, disconnect ventilator valve hose, then unbolt and remove valve tappet cover. Remove capscrews retaining fuel return manifold and rocker arm shaft assembly, remove fuel return manifold and lift off rocker arm shaft assembly. If push rods are removed identify them so they

can be reinstalled in their original position.

Disassemble rocker arm assembly by removing cap screws and sliding parts off shaft. Rocker arm shaft diameter is 0.8580-0.8588 for all engines. Minimum allowable diameter for rocker arm shaft is 0.8570. Inside diameter of rocker arm bushings is 0.8595-0.8611 with a maximum allowable inside diameter of 0.8630. Normal operating clearance of rocker arm on shaft is 0.0007-0.0031 for all engines. Renew rocker arms and/or shaft if clearance exceeds 0.005. Rocker arm bushings are not available separately.

## VALVE TIMING

### Model 2255

82. Valves are correctly timed when "V" mark on camshaft tooth is aligned with an identical mark on the crankshaft gear as shown in Fig. 44.

## TIMING GEAR COVER

### Model 2255

83. To remove timing gear cover, first remove engine as outlined in paragraph 76. Remove fan, all drive belts and alternator. Disconnect fuel filter bracket from front cover. Remove coolant temperature sending unit and tachourmeter drive housing (adapter). Remove clamps from water sleeves and using sleeve tool (Cat. No. 8S6692), slide sleeves into front cover. Disconnect oil cooler water outlet line elbow from rear side of front cover. Remove crankshaft pulley, engine oil pan and oil pump suction tube. Remove engine front support, then remove cap screws retaining front cover to cylinder block. Identify capscrews with holes as they vary in length. Pry front cover from dowels and remove from cylinder block.

**NOTE: Front cover weighs approximately 85-90 pounds and is recommended that a hoist be used during removal and installation.**

Front oil seal can be renewed with front cover either on or off the cylinder block. It is recommended that seal be

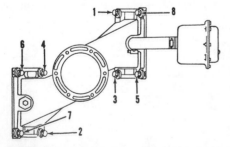

*Fig. 43—Tighten air inlet manifold bolts in the numbered sequence shown. Refer to text for torque values.*

*Fig. 44—Valves are correctly timed when timing marks are aligned as shown.*

installed with seal installer (Cat. No. 8S2276) after front cover is installed to insure correct position of seal. See paragraph 92.

If seal is renewed while front cover is removed, carefully measure depth of seal in bore prior to removal and install new seal, lip to rear, to the same dimension. However, if the crankshaft pulley has a wear track, the seal can be installed an additional 1/32-inch deeper. Lubricate lip of seal and mating surface of crankshaft pulley prior to installing pulley.

## TIMING GEARS

### Model 2255

84. Timing gear train consists of the crankshaft gear, camshaft gear and timing (weight) gear mounted in front of the camshaft gear. Gears are accessible for service after removing the timing gear cover as outlined in paragraph 83.

To remove camshaft gear, first remove injection pump drive gear as follows: Turn crankshaft in direction of normal rotation until timing marks on crankshaft gear and camshaft gear are aligned, then remove plug from timing pin hole in injection pump housing and install timing pin.

On early model 3150 engines having injection pump drive gear is retained hub, remove the tachourmeter drive adapter shaft, then use puller group (Cat. No. 9S8520, or equivalent) and thread bolt 9S8528 into end of injection pump camshaft. DO NOT force bolt as it should thread easily into shaft. Thread bolt 9S8527 into injection pump drive gear, then tighten bolt until gear pops loose from shaft. Remove injection pump drive gear and puller bolts.

On later Model 3150 engines, the injection pump drive gear is retained on injection pump camshaft by three capscrews and can be removed after removing tachourmeter drive adapter shaft and the three capscrews.

On model 3208 engines fitted with the vee block injection pump, the injection pump drive gear has two tapped puller bolt holes and gear can be removed with puller (Cat. No. 5P2371, or equivalent) after removing tachourmeter adapter drive shaft and washer.

With injection pump drive gear removed, remove retaining screw and the washer and pin from front of outer timing (weight) gear and pull gear from camshaft. Use sleeve (Cat. No. 9S9155) over camshaft extension, place step-plate (Cat. No. 8S5579) over sleeve, then use puller to remove camshaft gear.

Use a step-plate with puller and remove crankshaft gear from crankshaft.

To install crankshaft gear, heat gear in oil to approximately 300 degrees F. and install gear on crankshaft with timing mark on front side.

To install camshaft, heat gear in oil to approximately 400 degrees F., align slot in gear with pin in camshaft and install gear on camshaft so timing mark on camshaft gear is aligned with timing mark on crankshaft gear. Be sure the camshaft gear is seated against shoulder of camshaft.

NOTE: DO NOT drive camshaft gear on camshaft as damage to camshaft or camshaft thrust pin could occur.

Align holes in timing (weight) gear with dowels in camshaft gear and install timing gear. Align pin in washer with hole in camshaft (extension shaft) and install pin and washer and retaining screw. Tighten retaining screw to 6-12 ft.-lbs. (72-144 in.-lbs.) of torque and stake retaining screw in two places.

NOTE: Use caution when staking the retaining screw. Too much force could drive camshaft extension into camshaft and eliminate the required gear end clearance. After retaining screw is staked, outer gear should have 0.003-0.027 end clearance.

Place injection pump drive gear on injection pump camshaft.

On early Model 3150 engines, install tachourmeter drive adapter shaft and tighten to 30-34 ft.-lbs. of torque.

On later Model 3150 engines, tighten the three gear retaining capscrews to 27-37 ft.-lbs. of torque and the tachourmeter drive adapter shaft to 40-60 ft.-lbs. of torque.

On Model 3208 engines, tighten the tachourmeter drive adapter shaft to 75-85 ft.-lbs. of torque.

Complete balance of assembly by reversing disassembly procedure.

## CAMSHAFT

### Model 2255

85. To remove camshaft, first remove

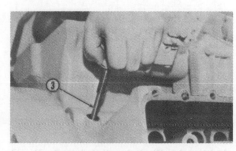

Fig. 45—Camshaft thrust pin (3) is located at right rear of cylinder block.

cylinder heads as outlined in paragraph 77, then remove tappets (cam followers) with a magnet. Keep tappets identified with their bores so they can be reinstalled in their original position. Remove timing gear cover as outlined in paragraph 83. Turn crankshaft in direction of normal rotation until timing marks on crankshaft and camshaft are aligned, then remove plug from timing pin hole in injection pump housing and install timing pin. Remove injection pump drive gear as outlined in paragraph 84.

Remove camshaft thrust pin from right rear of cylinder block. (See Fig. 45), then carefully pull camshaft from cylinder block. Timing (weight) gear and camshaft gear can now be removed from camshaft. See paragraph 84.

Clean and inspect camshaft and camshaft bushings. Specifications are as follows:

| | |
|---|---|
| Camshaft journal dia. | 2.4495-2.5005 |
| Camshaft end play | 0.004-0.010 |
| Max. allowable | 0.020 |
| Camshaft bushing I.D. | 2.5020-2.5040 |
| Camshaft operating clearance | 0.0015-0.0045 |
| Max. allowable | 0.007 |
| Lobe lift, exhaust | 0.3071 |
| Min. allowable | 0.2821 |
| Lobe lift, intake | 0.3077 |
| Min. allowable | 0.2827 |

Camshaft bushings are presized and should not require final sizing if carefully installed. Install bushings from chamfered side of bearing bores and be sure to align oil holes in bushings with oil holes in cylinder block.

Inspect thrust pin for wear or damage prior to installation and tighten pin to 30-40 ft.-lbs. of torque.

## ROD AND PISTON UNITS

### Model 2255

86. Connecting rod and piston assemblies are removed from above after removing cylinder head(s), oil pan and oil pump suction tube.

Prior to removal of rod and piston units, remove ridge from top of cylinder bore and check to see that rod and cap are identified and that numbers are on same side.

When installing units, lubricate crankshaft journal, rod bearings and rod bolts and nuts. Tighten rod nuts to 27-33 ft.-lbs. of torque, place match marks on rod cap and rod nut, then tighten rod nut an additional 55-65 degrees.

## PISTONS, PISTON RINGS AND CYLINDERS

### Model 2255

87. Pistons operate in unsleeved bores in cylinder block. Standard cylinder bore is 4.5000-4.5010 and cylinder block can be rebored to 0.020 and 0.040 oversizes. Maximum allowable bore diameter is 4.5490 which will require renewal of the cylinder block.

Pistons are fitted with one compression ring and one oil control ring. Pistons are available in standard size as well as ovesizes of 0.020 and 0.040. When cylinder wear, measured in area of piston ring travel, reaches a diameter of 4.5060 it is recommended that cylinder be rebored to the 0.020 oversize diameter of 4.5200-4.5210 and a 0.020 oversize piston installed. It is considered mandatory to rebore to the 0.020 oversize when cylinder diameter reaches 4.5090. When cylinder wear, measured in area of ring travel, reaches a diameter of 4.5260, it is recommended that cylinder be rebored to a diameter of 4.5400-4.5410 and a 0.040 oversize piston installed. It is considered mandatory to rebore to the 0.040 oversize when cylinder diameter reaches 4.5290.

88. PISTON RINGS. Two piston rings are used on each piston and are availble in standard size as well as 0.020 and 0.040 oversizes.

Side clearance for compression (top) ring is 0.0045-0.0070 with a maximum allowable side clearance of 0.014. End clearance for compression ring is 0.0150-0.0300 with a maximum allowable end clearance of 0.0550.

Side clearance for oil (lower) ring is 0.0015-0.0035 with a maximum allowable side clearance of 0.008. End clearance for oil ring is 0.0100-0.0250 with a maximum allowable end clearance of 0.0450.

An oil ring spring is used under oil ring and oil ring should be installed with gap 180 degrees from oil ring spring joint. Install compression ring with side marked "TOP" toward top of piston and gap approximately 120 degrees from oil ring gap.

## PISTON PINS AND BUSHINGS

89. The 1.4998-1.5000 diameter full floating type piston pin is retained in piston bosses by snap rings and is available in standard size only.

Piston pin end of connecting rod is fitted with a renewable bushing having an inside diameter of 1.5007-1.5013 which provides operating clearance of 0.0007-0.0015 for the piston pin. Maximum allowable clearance of piston pin

in bushing is 0.0030.

Piston pin bore in piston bosses is 1.5006-1.5009 which provides an operating clearance for piston pin of 0.0006-0.0011. Maximum allowable clearance of piston pin in piston is 0.0030.

Two types of connecting rods have been used; one having a boss on the piston pin end, or one which does not have the boss. When assembling connecting rod and piston, install the connecting rod having boss so the boss is on the same side as the crater in crown of piston. When installing connecting rods that do not have boss, install connecting rod so that cylinder identification marks on rod and cap are opposite to crater in crown of piston.

## CONNECTING RODS AND BEARINGS

### Model 2255

90. Connecting rod bearings are the non-adjustable insert type and are available in standard size as well as 0.010 and 0.020 undersize. Bearings can be removed after removing oil pan and connecting rod caps. When installing new bearings make certain the bearing insert projections fit into slot in connecting rod and cap and that cylinder numbers are in register and face away from center line of engine.

NOTE: When installing new connecting rod, affix correct cylinder number and refer to paragraph 89.

Lubricate rod bolts and nuts, tighten nuts to 27-33 ft.-lbs. of torque, place match marks on rod cap and rod nut, then tighten rod nut an additional 55-65 degrees.

Check the crankshaft crankpin and the bearing inserts against the values which follow:

```
Crankpin diameter . . . . . . . 2.7496-2.7504
Rod bearing running
   clearance . . . . . . . . . . . . . 0.0015-0.0045
Max. running clearance . . . . . . . . . 0.007
```

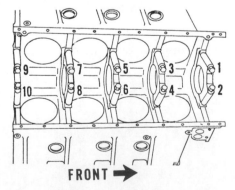

Fig. 46—Tighten main bearing cap bolts in the sequence shown. Refer to text for procedure and torque values.

## CRANKSHAFT AND BEARINGS

### Model 2255

91. Crankshaft is supported in five non-adjustable insert bearings which are available in standard size as well as undersizes of 0.010 and 0.020. Bearings can be removed without removing crankshaft. Normal crankshaft end play is 0.003-0.009 and is controlled by the flanged number four main bearing. Maximum allowable crankshaft end play is 0.012 and is corrected by renewing the number four main bearing.

NOTE: Removal of front main bearing cap with timing gear cover and oil pump installed will require use of an adapter wrench (Cat. No. 8S5131, or equivalent) as left front bearing bolt is partially hidden by oil pump.

To remove crankshaft, remove engine as outlined in paragraph 76, then remove clutch housing (with auxiliary drive or straight through drive), clutch, flywheel, flywheel housing, oil pan and timing gear cover (with oil pump). Remove main bearing caps and lower bearing halves, then attach a hoist and sling to crankshaft and lift shaft from cylinder block. Remove upper bearing halves from cylinder block and note that upper bearing half has an oil hole.

Check crankshaft and main bearings against the values which follow:

```
Crankshaft end play . . . . . . . . 0.003-0.009
   Max. allowable . . . . . . . . . . . . . . 0.012
Crankpin diameter, std. . . 2.7496-2.7504
   Min. allowable . . . . . . . . . . . . . 2.7486
Crankpin diameter,
   .010 under . . . . . . . . . . . . 2.7396-2.7404
   Min. allowable . . . . . . . . . . . . . 2.7386
Crankpin diameter,
   .020 under . . . . . . . . . . . . 2.7296-2.7304
   Min. allowable . . . . . . . . . . . . . 2.7286
Main journal dia., std. . . . . 3.4995-3.5005
   Min. allowable . . . . . . . . . . . . . 3.4985
Main journal dia.,
   .010 under . . . . . . . . . . . . 3.4895-3.4905
   Min. allowable . . . . . . . . . . . . . 3.4885
Main journal dia.,
   .020 under . . . . . . . . . . . . 3.4795-3.4805
   Min. allowable . . . . . . . . . . . . . 3.4785
Main bearing
   operating clearance . . . . 0.0015-0.0045
   Max. allowable . . . . . . . . . . . . . . 0.006
```

When installing crankshaft, be sure bearing halves with oil hole are installed in cylinder block and that identification marks on main bearing caps mate with number on cylinder block. Lubricate bearings and main bearings bolts, refer to Fig. 46 for tightening sequence and tighten main bearing bolts as follows:

On series 3150 engines prior to serial number 97M197M2403, use sequence shown and tighten (first time) main bearing bolts to 55-65 ft.-lbs. of torque, then using same sequence, tighten (second time) bolts to 165-185 ft.-lbs. of torque.

On series 3150 engines serial number 97M197M2403 and up, and series 3208 engines, use sequence shown and tighten (first time) main bearing bolts to 27-33 ft.-lbs of torque, then place match marks on bolt and cap and tighten (second time) bolts an additional 115-125 degrees from previously affixed marks using the same tightening sequence.

## CRANKSHAFT OIL SEALS

### Model 2255

92. **FRONT SEAL.** The front oil seal can be renewed with the timing gear cover either on or off the engine. However, it is recommended that front oil seal be installed with timing gear cover on engine and using seal installer (Cat. No. 8S2276, or equivalent) to insure seal is installed to correct depth.

NOTE: Beginning with serial number 97M375, the series 3150 engines and series 3208 engines were fitted with a timing gear cover that will permit installing front oil seal to two depths should a new seal wear track be required on the crankshaft pulley. To obtain the deeper oil seal position use spacer (Cat. No. 9S6012, or equivalent) between seal and shoulder in installer.

To remove front oil seal, drain cooling system, then remove hood, grille, oil coolers and radiator. Remove the crankshaft pulley and use a step plate with puller to preclude damage to crankshaft. Remove front oil seal using puller (Cat. No. 1P3075, or equivalent).

To install front oil seal, place seal over short end of installer with lip toward engine and use sealant on outside diameter of seal. Place installer over end of crankshaft and use the crankshaft pulley retaining bolt and washer to push seal into position. Seal is correctly positioned when installer bottoms against crankshaft gear. Lubricate lip of oil seal and mating surface of crankshaft pulley and install pulley.

NOTE: DO NOT use pulley retaining bolt and washer to push pulley onto crankshaft. Use puller bolt, hardened nut, thrust bearing and flat washers.

Crankshaft pulley is correctly installed when hub of pulley contacts crankshaft gear. Remove pulley and install pulley retaining bolt and washer. Tighten bolt to 230-295 ft.-lbs. of torque.

93. **REAR SEAL.** To remove rear oil seal, remove engine as outlined in paragraph 76, then remove clutch and flywheel. Seal can now be removed from bore in flywheel housing.

NOTE: It is recommended that a new wear sleeve be installed each time the rear oil seal is renewed. To remove wear sleeve from flange of crankshaft, use puller (Cat. No. 8S7164, or equivalent) to remove wear sleeve.

To install new wear sleeve, place wear sleeve between end of crankshaft and installing pilot (Cat. No. 8S5134, or equivalent) with lead chamfer for rear seal toward flywheel end of crankshaft. Install pilot retaining bolts and tighten them evenly about one half turn at a time until pilot butts against end of crankshaft. At this time, wear sleeve will be flush with end of crankshaft.

To install new rear oil seal, lubricate lip of seal, outside diameter of pilot and wear sleeve, then carefully install seal on pilot with lip toward front of engine. Apply sealant to outside diameter of rear seal. Place cap (Cat. No. 8S2287, or equivalent) over stud of pilot, install forcing nut, then tighten nut until cap butts against flywheel housing.

When installing flywheel, align paint marks (early) or stamped match marks, coat retaining cap screw threads with sealant, and tighten cap screws to 50-70 ft.-lbs. of torque.

## FLYWHEEL AND HOUSING

### Model 2255

94. Flywheel housing can be removed after removing flywheel as follows: Note paint marks or stamped marks which locate flywheel on crankshaft, then unbolt and remove flywheel. Straighten tabs of locks located on the two lower flywheel housing retaining cap screws, then unbolt and remove flywheel housing.

To install flywheel housing, first clean gasket surface of cylinder block and flywheel housing. Use new locks on the two bottom flywheel housing retaining cap screws and using a new gasket, install flywheel housing. If necessary, trim gasket so it will be flush with oil pan mounting surface of cylinder block.

NOTE: Gasket must not protrude beyond oil pan mounting surface or leakage will occur.

Radial run-out and clutch housing mounting surface run-out must not exceed 0.008.

To install flywheel, position on crankshaft with locating marks on crankshaft and flywheel aligned, then coat threads of flywheel retaining cap screws with sealant and install cap screws to a torque of 50-70 ft.-lbs. Check flywheel, after installation, with the following: Radial run-out of inside diameter and axial run-out of clutch cover mounting surface must not exceed 0.006. Radial run-out of pilot bore must not exceed 0.005.

## OIL PAN

### Model 2255

95. To remove oil pan, it will be necessary to remove engine as outlined in paragraph 76. With engine removed, oil pan can be unbolted and removed from cylinder block. When installing oil pan, tighten bolts to 14-17 ft.-lbs of torque.

## OIL PUMP

### Model 2255

96. Oil pump is contained in the timing gear (front) cover and can be serviced after removing front cover as outlined in paragraph 83.

With front cover removed, remove suction pipe and bell, if necessary, then refer to Fig. 47 and measure end play of pump rotors prior to removing pump cover. End play should be 0.002-0.006. Remove pump cover and check rotor tip clearance as shown in Fig. 48. Rotor tip clearance should be 0.002-0.004 with a maximum allowable of 0.009. Remove both rotors and check inside diameter of bearing in front cover which should

*Fig. 47—Measure end play (0.002-0.006) of oil pump gears prior to removing pump cover.*

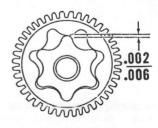

*Fig. 48—Oil pump rotor tip clearance of 0.002-0.004 is measured as shown.*

be 2.802-2.806. See Fig. 49. If necessary, bearing can be renewed using driver (Cat. No. 2285, or equivalent).

Clean all parts, lubricate throughly and reassemble by reversing disassembly procedure. Use new locks on cover retaining cap screws, tighten cap screws to 13-23 ft.-lbs. of torque and secure with locks.

NOTE: Capacity of oil pump in series 3208 engines has been increased by widening pump gears.

### OIL PRESSURE RELIEF VALVE

#### Model 2255

97. The oil pressure relief valve is located in the oil pump cover and can be removed after oil pan is off. To remove relief valve assembly, loosen set screw and pull guide, spring and valve from pump cover. See Fig. 50.

Relief valve spring should have an approximate free length of 3.50 inches and should test 32-36 pounds when compressed to a length of 2.66 inches.

To assemble relief valve, place spring in valve and slide spring and valve into guide. Push guide into pump cover so lip is flush with oil pump face and flat on guide is aligned with set screw then tighten set screw.

### OIL COOLER

#### Model 2255

98. The engine oil cooler is located at

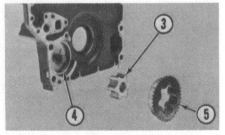

Fig. 49—Oil pump rotor bearing inside diameter is 2.802-2.806.

   3. Inner rotor
   4. Bearing
   5. Outer rotor

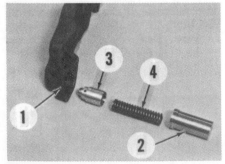

Fig. 50—Exploded view of oil pump relief valve assembly. Refer to text for spring (4) specifications.

1. Cover         3. Valve
2. Guide        4. Spring

lower left front of cylinder block. To remove oil cooler, drain cooling system and remove left exhaust manifold shield (guard). Disconnect water lines from cooler, then unbolt and remove cooler from cylinder block.

NOTE: If desired, oil filters can be removed prior to removing cooler assembly. Unbolt cover from cooler base and remove core. By-pass spring and valve can be removed after removing cap and fitting.

If oil cooler core is faulty, it is recommended that core be renewed rather than repaired.

By-pass spring has a free length of approximately 2.28 inches and should test 3.5 pounds when compressed to a length of 1.70 inches. Oil cooler by-pass valve will open at 14-22 psi.

When assembling core cover to base, be sure index point (tab) of gasket (Fig. 51) is on top side and to front side of cooler. Tighten cover to base retaining nuts to a torque of 14-20 ft.-lbs. Install by-pass valve and spring, tighten fitting to 15-23 ft.-lbs. and cap to 11-19 ft.-lbs. of torque. Install oil cooler to cylinder block, tighten retaining cap screws to 27-37 ft.-lbs. torque. Install water lines.

NOTE: It is recommended that water line hoses be renewed every year, or 1500 hours, whichever occurs first.

If new filters are being installed, be sure filter and filter base are clean and apply engine oil to filter gasket. Turn filter on base until gasket contacts base, then by hand, tighten filter an additional 1/2-3/4 turn. Each new filter will require an additional one quart of oil.

# LP-GAS SYSTEM

Models G955 and G1355 tractors are available with factory installed LP-Gas systems using Century equipment. These systems are designed to operate with the fuel tank not more than 80% filled.

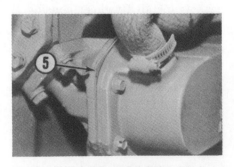

Fig. 51—Oil cooler gasket index tab is positioned as shown.

Model G955 tractors use Century Model 3C706UD carburetors and Century Model H regulators.

Model G1355 tractors use Century Model 3C707UD carburetors and Century Model M5 regulators.

It is important when starting LP-Gas tractors to open the vapor valve on the supply tank SLOWLY; if opened too fast, the fuel supply to the regulator will be shut off. Too rapid opening of vapor or liquid valves may cause freezing.

### CARBURETOR ADJUSTMENTS

#### Models G955-G1355

99. To make the preliminary adjustments, close power adjusting screw (3—Fig. 55) completely, then back the screw out 6 turns on G955, and 5½ turns on G1355.

On all models, place the speed control lever in slow idle position and adjust idle speed stop screw (2) to obtain an idle speed of 600 rpm. Loosen lock nuts on drag link (4) and rotate drag link barrel to the longest position possible to obtain a smooth idle.

The carburetor is not equipped with a choke valve. Depressing the primer button (1) manually opens the regulator low pressure valve and allows fuel to enter carburetor mixing chamber at 5-9 psi pressure. A hiss should be heard as this primer button is depressed.

Final adjustments must be made with engine at operating temperature and liquid withdrawal valve open to assure adequate fuel supply. Fuel for part-throttle operation is controlled by the metering valve and balanced to the throttle valve by adjusting the length of drag link (4).

To adjust the idle mixture after engine is at operating temperature,

Fig. 55—Location of adjustment points on LP-Gas carburetor.

1. Primer button
2. Idle speed stop screw
3. Power adjusting screw
4. Idle drag link

loosen locknuts on drag link (4) and turn knurled center section to the point of smoothest idle. Adjust idle speed stop screw (2) to provide a slow idle speed of 600 rpm, then recheck drag link adjustment. Leave drag link as long as possible to still produce a smooth idle.

Turn the power adjusting screw (3) to produce a reading of 12.5-13.5 on gasoline scale of exhaust analyzer, with throttle wide open and engine under full load. If an exhaust gas analyzer is not used, adjust the power mixture screw out until maximum rpm is reached with throttle wide open and engine under load. Turn power adjusting screw in until engine speed starts to drop from too lean a mixture. Then back screw out until power is restored and tighten lock nut.

Recheck idle settings after power mixture has been adjusted and correct as necessary.

## TROUBLE SHOOTING

### Model G955

100. Trouble in this fuel system is normally one of the following conditions: No fuel, too much fuel or freezing.

If engine will not start, due to lack of fuel, close both withdrawal valves on tank and check fuel supply in tank. Remove drain plug from fuel filter. Slowly and momentarily open the liquid withdrawal valve. Fuel should emerge from drain plug opening as a white mist.

Fig. 56—On Century Model H regulator, primary pressure should be 4-6 psi with inlet pressure of 130-180 psi. Pressure is checked at secondary valve orifice.

**CAUTION: The escaping fuel can cause severe freezing and damage to unprotected skin. Keep clear of escaping LP-Gas fuel.**

Repeat the test using the vapor withdrawal valve. If no fuel emerges, trouble is in withdrawal valves or lines. Renew parts as necessary. If fuel emerges with drain plug removed, reinstall drain plug and disconnect fuel inlet line from regulator. Slowly open vapor withdrawal valve and check for fuel at disconnected line. If no fuel is present, close the valve, then clean or renew fuel filter element. If fuel is present, remove and overhaul regulator as outlined in paragraph 103.

If too much fuel (poor idling and poor engine performance) is the trouble, close both withdrawal valves and operate engine until fuel is exhausted from carburetor and regulator. Thoroughly clean exterior of regulator assembly. Remove primer and front cover assembly, then carefully remove the secondary diaphragm. Remove retaining screws, then remove the secondary valve lever assembly. Attach a compressed air hose, with regulated pressure of 130-180 psi, to fuel inlet fitting on regulator. Place a 0-15 psi test gage over secondary valve orifice. See Fig. 56.

**NOTE: Use Century test gage No. M-508 or equivalent gage fitted with a conical rubber end to seat on secondary valve orifice.**

With inlet air pressure at 130-180 psi, test gage pressure should be 4-6 psi. If pressure reading is too low, primary valve orifice is clogged or primary valve spring is defective. If pressure reading is too high, primary valve seat is defective or dirty, diaphragm is ruptured or valve lever is distorted. If pressure reading is within the 4-6 psi range, the secondary valve seat could be dirty or defective, valve level could be incorrectly set or valve spring could be defective. Renew faulty parts and reassemble regulator as outlined in paragraph 103.

If regulator freezing (with engine operating) is the trouble, check for restricted regulator water lines, low radiator coolant level, defective thermostats or defective water pump.

**NOTE: The regulator back cover gasket seals coolant passages from expansion areas in back cover. In case of freezing coolant, this gasket is forced into the expansion area which absorbs expansion of the frozen coolant. This prevents damage to vital parts of the regulator.**

If regulator "freezes up," the assembly should be removed, disassembled and a new back cover gasket installed. Repeated freezing can cause distortion of rear cover which could result in leakage of fuel and coolant. Refer to paragraph 103 for overhaul procedure for regulator.

### Model G1355

101. To locate trouble in the fuel system, shut off the liquid and vapor withdrawal valves and run engine until gas is exhausted from carburetor and regulator. Connect a low reading pressure gage (0-15 psi) in 1/8-inch pipe plug opening as shown in Fig. 57.

With gage installed, open the vapor withdrawal valve and note gage reading which should hold steady within the range of 6-8 psi. Very slightly depress primer button for a few seconds, then release. Gage pressure should drop slightly and then return to original point when primer button is released. If pressure is not as indicated, overhaul regulator as outlined in paragraph 104.

If pressure continues to rise after withdrawal valve is opened; or primer button fully depressed, then released, a leaky primary valve is indicated. Overhaul the regulator as outlined in paragraph 104.

If the regulator becomes noticeably cool to the touch, or frosts over, after standing with engine stopped and withdrawal valve open, secondary valve is leaking or there are external leaks in regulator. Remove and overhaul regulator as outlined in paragraph 104.

If regulator freezing (with engine operating) is the trouble, check for restricted regulator water lines, low

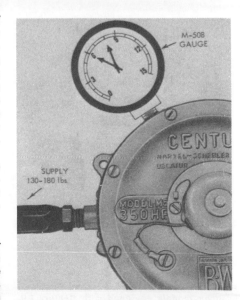

Fig. 57—Century Model M5 regulator showing test gage installed. Primary pressure should be 6-8 psi with inlet pressure of 130-180 psi. Refer to text.

radiator coolant level, defective thermostats or defective water pump.

NOTE: The regulator back cover gasket seals coolant passages from expansion areas in back cover. In case of freezing coolant, this gasket is forced into the expansion area which absorbs expansion of the frozen coolant. This prevents damage to vital parts of the regulator. If regulator "freezes up," the assembly should be removed, disassembled and a new back cover gasket installed. Repeated freezing can cause distortion of rear cover which could result in leakage of fuel and coolant. Refer to paragraph 104 for overhaul procedure for regulator.

If engine will not start and appears to not be getting fuel, close both withdrawal valves and remove drain plug (1—Fig. 58) from fuel filter. Slowly and momentarily open the liquid withdrawal valve. Fuel should emerge from drain plug opening as a white mist and purge the filter trap of any foreign material.

CAUTION: The escaping fuel can cause severe freezing and damage to unprotected skin. Keep clear of escaping LP-Gas fuel.

If no fuel emerges, the trouble is in withdrawal valve or lines, or tank is empty. Repeat the test using the vapor withdrawal valve. Repair or renew the parts found to be malfunctioning. If fuel emerges with drain plug removed, disassemble the filter and clean or renew filter element (5).

If filter is clean and lines open, tractor should fire when turned with

starter while intermittently depressing primer button on regulator.

## FUEL TANK, FUEL FILTER AND LINES

102. The pressure tank is fitted with a fuel filler, vapor return, pressure relief valve, 80% bleeder valve, and liquid and vapor withdrawal valves which can only be serviced as complete assemblies. Before renewal is attempted on any of these units, drive the tractor to an open area and allow engine to run until the fuel is exhausted; then, open bleeder valve and allow any remaining pressure to escape. Fuel gage can only be renewed if the fuel tank is completely empty. The safety relief valve is set to open at 312 psi pressure to protect the tank against excessive pressures. U-L regulations in most states prohibit any welding or repair on LP-Gas containers and the tank must be renewed or returned to the tank manufacturer for repair in the event of damage. Fuel lines can be safely renewed at any time without emptying tank if liquid and vapor withdrawal valves are closed and the engine run until all fuel is exhausted from the lines.

The fuel filter contains a renewable type element (5—Fig. 58) which may be removed and cleaned in a suitable solvent if in good condition. Thoroughly air-dry the element before reinstalling in the filter body.

## VAPORIZER-REGULATOR AND CARBURETOR

### Model G955

103. MODEL H VAPORIZER—REGULATOR OVERHAUL. To remove the regulator, close withdrawal valves and operate engine until fuel is exhausted. Drain cooling system, then disconnect fuel and coolant lines from regulator. Unbolt and remove regulator assembly.

To disassemble the regulator, refer to Fig. 59 and remove primer assembly (1 thru 4A). Remove front cover (5), secondary diaphragm assembly (6 thru 9) and gasket (10). Remove two screws which retain lever pivot pin (12) and lift out valve lever (11), secondary

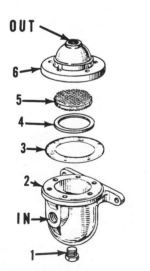

Fig. 58—Exploded view of typical fuel filter used on models equipped with Century LP-Gas system.

| | |
|---|---|
| 1. Drain plug | 4. Retainer |
| 2. Body | 5. Filter |
| 3. Gasket | 6. Cover |

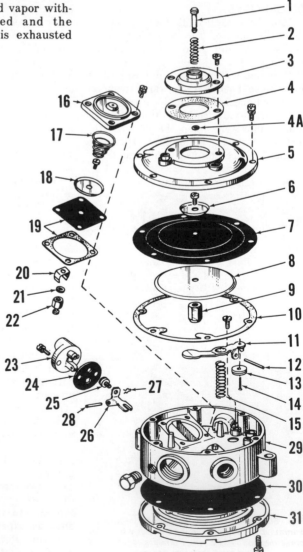

Fig. 59—Exploded view of Century Model H regulator.

1. Primer plunger
2. Spring
3. Primer cover
4. Gasket
4A. "O" ring
5. Front cover
6. Diaphragm plate (small)
7. Secondary diaphragm
8. Diaphragm plate (large)
9. Diaphragm button
10. Gasket
11. Secondary valve lever
12. Pivot pin
13. Secondary valve
14. Pin
15. Spring
16. Cover
17. Spring
18. Diaphragm plate
19. Primary diaphragm & gasket assy.
20. Damper spring
21. Washer
22. Diaphragm link
23. Primary valve cover
24. Gasket
25. Primary valve
26. Primary valve lever
27. Retainer
28. Pivot pin
29. Body
30. Gasket
31. Back cover

valve (13) and spring (15). Remove inlet and primary valve assembly (23 thru 28), then remove cover (16), spring (17) and primary diaphragm assembly (18 thru 22). Remove retaining screws, then separate back cover (31) and gasket (30) from body (29). Further disassembly of the primary and secondary valves and diaphragms is obvious after examination of the units and reference to Fig. 59.

Wash metal parts in solvent and dry with compressed air. Renew all gaskets when reassembling. Renew diaphragms (7 and 19) if their condition is questionable and valves (13 and 25) if their sealing surface is damaged. Check back cover and back surface of regulator, using a straight edge, to make certain they are flat. If slightly warped condition exists, lap castings on a surface plate until they are flat. Severely warped parts must be renewed.

To reassemble, lay regulator body (29) with front face down and install new back gasket (30) and back cover (31).

**NOTE: Do not use sealing compound on any gasket or diaphragm. Tighten back cover screws securely.**

When installing new primary diaphragm, assemble diaphragm plate (18) with flanged outer edge toward screw head, diaphragm and gasket assembly (19), damper spring (20), washer (21) and diaphragm link (22). Make certain that legs of damper spring (20) are parallel to any two flat sides of diaphragm link (22) and two square sides of diaphragm. Place diaphragm assembly in position on primary cavity so that legs of damper spring contact sides of slotted opening. To insure proper installation and prevent damage to diaphragm, four aligning pins (Century Part No. M-501) should be used to position diaphragm and cover during assembly. With diaphragm and gasket assembly (19) in position, insert the four aligning pins in screw holes.

Place spring (17) on diaphragm and cover (16) over the aligning pins. Press down on cover, remove one aligning pin and install retaining screw before removing the next aligning pin. Tighten all four screws evenly and securely.

Assemble and install inlet and primary valve assembly (23 thru 28), using new gasket (24) and new primary valve (25) if necessary. Make certain that fork on primary lever (26) straddles the diaphragm link (22). Connect a compressed air hose, with regulated pressure of 130-180 psi, to fuel inlet fitting on regulator. Place a 0-15 psi test gage (Century No. M-508) over secondary orifice as shown in Fig. 56. Pressure gage should read 4-6 psi. If pressure creeps upward beyond 6 psi, primary valve is leaking and must be reworked or renewed.

To install a new secondary valve (13—Fig. 59) on secondary valve lever (11), insert retaining pin (14) through valve and lever. Press face of valve lightly against a clean flat surface and while holding lever firmly, bend upper end of pin sharply over top of lever. Clip bent portion of pin to a length of about 1/8-inch after bending. The valve is self-aligning and should be secured to lever with a minimum of end play, but should not bind. Do not hammer or rivet the retaining pin. Place spring (15) in position, insert pivot pin (12) in lever (11), install the assembly and secure pivot pin with two screws. Open valve and allow it to snap closed several times to align seat with orifice. Using special gage (Century No. 2V-01) or straight edge and steel rule, measure lever height as shown in Fig. 60. Bend secondary lever as necessary to obtain the 5/16-inch distance between lever and machined surface of regulator body. See Fig. 61. Reconnect air supply to fuel inlet on regulator. Plug one water fitting opening and apply soap bubble to other water opening. Any

continuous growth of bubble will indicate internal leakage of back cover gasket. Immerse entire unit in water or check all joints, including secondary valve, with soap solution and correct any leakage noted before proceeding further.

Assemble secondary diaphragm as follows: Place small diaphragm plate (6—Fig. 59) on screw with flanged outer edge of plate toward screw head, secondary diaphragm (7) with dished side away from screw head, large diaphragm plate (8) with flanged outer edge away from diaphragm and screw diaphragm button (9) on exposed screw threads. Tighten securely. Insert three alignment pins (M-501) in alternate holes in front face of regulator body. Place gasket (10) down over the pins so that the two ears on gasket are over the primary cover screws. Install secondary diaphragm over pins so that diaphragm button is contacting valve lever. Place cover (5) on the alignment pins and install three cover retaining screws. Remove pins and install the other three screws. Refer to Fig. 63 and while pulling upward on diaphragm screw with pliers, tighten cover retaining screws securely.

Use new gasket (4—Fig. 59) and install primer plunger and cover assembly (1, 2, 3 & 4A). Apply air pressure at fuel inlet and check for leakage at vapor outlet port with soap bubble. Any growth in bubble would indicate secondary valve leakage. Depress

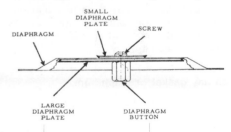

*Fig. 62—View showing correct assembly of secondary diaphragm components.*

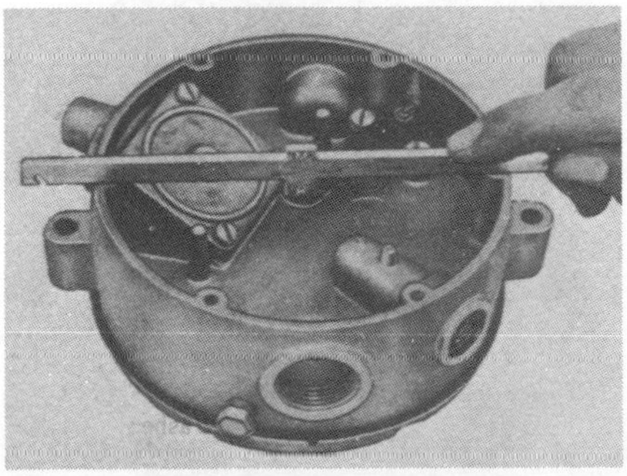

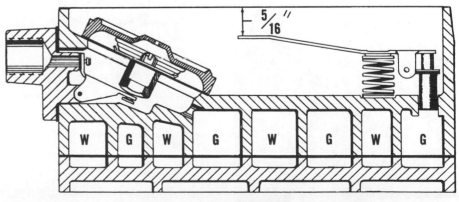

*Fig. 60—Checking secondary valve lever height with Century No. 2V-01 gage. Alternate method is to use straight edge and steel rule. Refer to Fig. 61.*

*Fig. 61—Cross sectional drawing of regulator body, primary regulator and secondary regulator valve assembly. Distance between machined face of regulator body and secondary valve lever should be 5/16-inch as shown. Hot engine coolant in passages (W) vaporize LP-Gas in passages (G).*

primer and check for air flow from vapor outlet port. Volume need not be great, but escaping air should be audible.

Reinstall regulator assembly by reversing removal procedure. Bleed any trapped air from regulator by loosening water hose at upper connection on regulator.

### Model G1355

**104. MODEL M5 VAPORIZER-REGULATOR OVERHAUL.** To remove the regulator, first close withdrawal valves and allow engine to run until fuel is exhausted. Drain the cooling system and disconnect fuel and coolant lines. Then, unbolt and remove the assembly.

Disassemble the regulator as follows: Remove primer cover (23—Fig. 64) and primer button (25) as a unit, then remove secondary regulator cover (20) and diaphragm assembly. Remove the two screws (S) which retain secondary valve lever shaft (10) and lift out the lever (9), valve (11) and spring (6). Compound lever (8) can be removed in the same manner if desired. Remove inlet cover (31—Fig. 65); then remove primary diaphragm cover (40), spring (39) and diaphragm (36). Separate the secondary valve body (5—Fig. 64), primary valve body (3) and back cover (1) only if leaks are suspected or complete overhaul is indicated.

Wash metal parts in solvent and dry with compressed air. Renew all gaskets when reassembling. Renew diaphragms (17—Fig. 64 or 36—Fig. 65) if their condition is questionable, and valves (11—Fig. 64 or 29—Fig. 65) if sealing surface is damaged.

**NOTE: Do not use sealing compound on any gasket or diaphragm.**

If high pressure diaphragm must be disassembled, make sure that legs of damper spring (34) align with flats of

link (32) and any flat side of diaphragm (36). The damper spring legs apply pressure to post in primary cover and correct positioning is essential. Proper installation is also important. Four aligning pins (Century Part No. M-501) should be used to position diaphragm and cover during assembly. Proceed as follows: Install aligning pins in screw holes in primary valve body (3—Fig. 64). Install a new gasket (35—Fig. 65) over aligning pins. Then, install the assembled diaphragm, making sure the legs of damper spring are in position to contact the round posts in primary cavity. Insert a screwdriver or similar flat tool through opening for inlet cover (31) and hold up on diaphragm link (32); then install diaphragm spring (39) and cover (40). Hold diaphragm (36) with the flat tool, press down on cover (40) and remove aligning pins one at a

time, installing retaining screw before removing the next aligning pin. Release the pressure on diaphragm link (32) and cover (40) only after all screws are tightened.

To install a new low pressure valve (11—Fig. 64) on secondary valve lever (9), proceed as follows: Insert the retaining pin (12) through valve (11) and lever (9). Press face of valve lightly against a clean, flat surface and while firmly supporting lever, bend upper end of retaining pin sharply over top edge of lever. Clip bent portion of pin to a length of approximately 1/8-inch after bending. The valve is self-aligned and should be secured to lever with a minimum of end play, but should not bind. Do not hammer or rivet the retaining pin. Install the assembled secondary lever (9), spring (6) and compound lever (8), if removed, by reversing the disassembly procedure.

After levers are installed, the free end of compound lever (8) should be 1/8 to 5/32-inch below gasket surface of secondary valve body (5) as shown at (A—Fig. 66). If adjustment is required, bend secondary lever (9) at (B). DO NOT bend compound lever (8) to obtain the measurement. The setting can be measured with Century gage 2V-01 as shown in Fig. 67, or with a straight edge and rule.

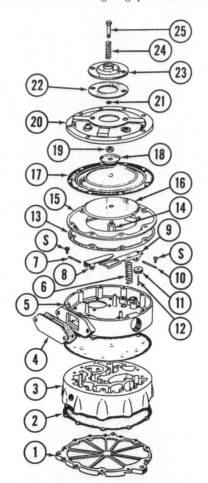

Fig. 64—Exploded view of Century Model M5 regulator showing secondary regulator and associated parts.

| | |
|---|---|
| 1. Back cover | |
| 2. Gasket | 14. Diaphragm button |
| 3. Primary body | 15. Backing plate |
| 4. Cover | 16. Diaphragm plate |
| 5. Secondary body | 17. Diaphragm |
| 6. Spring | 18. Diaphragm plate |
| 7. Pivot pin | 19. Nut |
| 8. Compound lever | 20. Secondary cover |
| 9. Secondary lever | 21. Ring |
| 10. Pivot pin | 22. Gasket |
| 11. Secondary valve | 23. Primer cover |
| 12. Pin | 24. Primer spring |
| 13. Gasket | 25. Primer button |

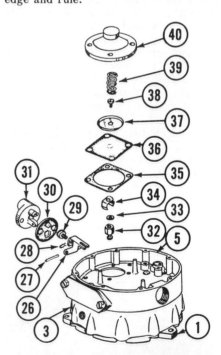

Fig. 65—Century Model M5 regulator primary body, primary valve and associated parts.

| | |
|---|---|
| 1. Back cover | 32. Diaphragm link |
| 3. Primary body | 33. Washer |
| 5. Secondary body | 34. Damper spring |
| 26. Primary lever | 35. Gasket |
| 27. Pivot pin | 36. Primary diaphragm |
| 28. Hairpin retainer | 37. Diaphragm plate |
| 29. Primary valve | 38. Screw |
| 30. Gasket | 39. Spring |
| 31. Primary cover | 40. Diaphragm cover |

Fig. 63—Pull upward on diaphragm screw with pliers while tightening cover retaining screws.

105. **CARBURETOR.** Refer to Fig. 68. The carburetor is simply constructed and overhaul procedures are self-evident. Use a thread sealant when installing reducer bushing in metering valve housing (9) and do not overtighten. Overtightening may distort the housing and cause valve to bind. If metering valve (10) does not move freely, remove lever (7) and plug (12) and push metering valve from housing. Clean the valve and housing bore with an oil base solvent. Coat valve with Lubriplate and reinsert in housing bore. If valve is not now free, chuck lever end of valve in a slow-speed drill and reseat the valve, using tallow or penetrating oil as a lubricant. DO NOT use a lapping compound, or allow valve and housing to become excessively warm. If condition is questionable, renew the parts.

# DIESEL SYSTEM

The diesel fuel system consists of three basic units: the fuel filter, injection pump and injection nozzles. When servicing any unit associated with the fuel system, the maintenance of absolute cleanliness is of utmost importance. Of equal importance is the avoidance of nicks or burrs on any of the working parts.

Probably the most importance precaution that service personnel can impart to owners of diesel powered tractors, is to urge them to use an approved fuel that is absolutely clean and free from foreign material. Because of the extreme pressures involved in the operation of pump and injectors, the working parts must be hand fitted with utmost care. While the filtering system will easily remove the larger particles of

foreign material, the greater danger exists in the presence of water or fine dust particles which might pass through an overloaded filter system. Proper care in fuel handling will pay big dividends in better service and performance.

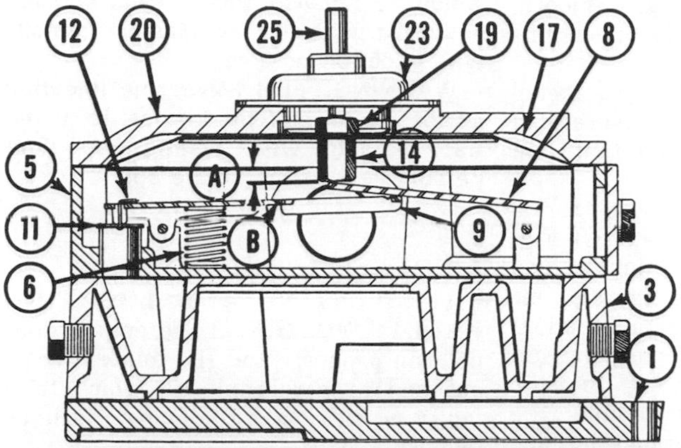

## TROUBLE SHOOTING

### All Models

106. **QUICK CHECKS-UNITS ON TRACTOR.** If the diesel engine does not start or does not run properly, and the fuel system is suspected as the source of trouble, refer to the following list of troubles and their possible causes. Many of the troubles are self-explanatory; however, if the difficulty points to the fuel filters, injection pump and/or injection nozzles, refer to the appropriate following paragraphs.

1. Sudden Stopping of Engine
   a. Lack of fuel
   b. Clogged fuel filters and/or lines
   c. Faulty injection pump
2. Lack of Power
   a. Inferior fuel
   b. Clogged fuel filters and/or lines
   c. Improper injection pump timing
   d. Faulty injection pump
3. Engine Hard to Start
   a. Inferior fuel
   b. Clogged fuel filters and/or lines
   c. Improper injection pump timing
   d. Faulty injection pump
4. Irregular Engine Operation
   a. Inferior fuel
   b. Clogged fuel filters and/or lines
   c. Faulty nozzle
   d. Improper injection pump timing
   e. Faulty injection pump
5. Engine Knocks
   a. Inferior fuel
   b. Improper injection pump timing
   c. Faulty nozzle
6. Excessive Smoking
   a. Inferior fuel
   b. Improper injection pump timing
   c. Faulty nozzle
   d. Improperly adjusted smoke stop
7. Excessive Fuel Consumption
   a. Inferior fuel
   b. Improper injection pump timing
   c. Faulty nozzle

## FUEL FILTERS AND BLEEDING

### Models G955-G1355

107. **FILTERS.** Models G955 and G1355 are equipped with a fuel filter such as that shown in Fig. 69. Filter element should be renewed after every 250 hours of operation.

To renew filter element, close fuel supply valve and clean filter and

Fig. 66—Cross sectional view of Century Model M5 regualtor showing details of secondary valve lever adjustment. Distance (A) between diaphragm surface of body (5) and end of compound lever (8) should be 1/8 to 5/32-inch. Adjust by bending secondary valve lever (9) at point shown at (B). Refer to Fig. 64 for legend.

Fig. 67—Using Century gage No. 2V-01 to adjust secondary valve levers. Refer to text.

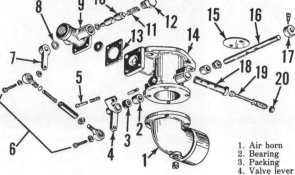

Fig. 68—Exploded view of typical Century carburetor and associated parts.

5. Idle speed screw
6. Drag link
7. Valve lever
8. Thrust washer
9. Metering valve body
10. Metering valve
11. Spring
12. Valve plug
13. Gasket
14. Carburetor body
15. Throttle valve
16. Throttle shaft
17. Retainer
18. Spray bar
19. Power adjusting screw
20. Jam nut

1. Air horn
2. Bearing
3. Packing
4. Valve lever

surrounding area. Refer to Fig. 70 and depress extended tab with heel of hand, then lift the slotted tab to release clamp from locking slot in filter base.

NOTE: A screwdriver may be used if additional leverage is required to remove clamp, however, do not attempt to remove clamp without depressing the extended tab.

Clean the three sealing spots on filter base and be sure no foreign material enters locating pin at top of filter base. Install new element by carefully locating fuel outlet passage in top of element over locating pin at top of filter base and installing clamp. Be sure clamp is fully engaged, then open fuel supply valve and bleed filter as outlined in paragraph 109.

108. Model 2255 tractors fitted with the series 3208 engine are equipped with a fuel filter as shown in Fig. 72. Fuel filter on earlier model tractors fitted with series 3150 engine is similar but does not include the hand primer pump. Fuel filter should be renewed after every 250 hours of operation.

To renew element, close fuel supply valve and clean filter and surrounding area. Turn filter element counterclockwise to remove it from base. Clean filter base as necessary. Fill new filter element with clean diesel fuel, lubricate element gasket, then thread element clockwise onto base and tighten it not

*Fig. 69—View showing type of fuel filter used on models G955 and G1355. Note location of filter bleed screw.*

more than 1/2-turn after gasket first contacts filter base. Open fuel supply valve, check for leaks, then bleed fuel system as outlined in paragraph 110.

109. **BLEEDING.** To bleed fuel system on models G955 and G1355, refer to Fig. 69 and proceed as follows: Be sure fuel tank has adequate fuel and that fuel supply valve is open, then loosen air bleed plug about two turns. Close (tighten) air bleed plug after filter has filled and while fuel is still running.

With filter filled, attempt to start engine. If engine does not start, open throttle and turn engine with starting motor while loosening the injector lines at injector. Tighten lines after air-free fuel flows from loosened connection.

If engine starts but runs unevenly, let engine run at idle speed and loosen each injector line in turn until lines are purged of air and engine runs smoothly. Be sure to tighten all connections when bleeding is complete.

110. To bleed fuel system on model 2255, proceed as follows: Be sure fuel tank has adequate fuel and that fuel supply valve is open.

On tractors equipped with series 3150 engine, loosen vent cap (C—Fig. 71) and crank engine until bubble free fuel flows from filter, then tighten vent cap. Attempt to start engine and if engine does not start, or runs unevenly, loosen injector lines at injector ends, one at a time, until bubble free fuel emerges from connection. Be sure to tighten connections when bleeding is completed.

On tractors equipped with series 3208 engines, open bleed valve located on top side of injection pump, then loosen plunger of hand primer pump and operate primer pump until air-free fuel emerges from bleed valve. Close bleed valve after injection pump is purged of air.

Attempt to start engine and if engine does not start, loosen each line at injector in turn and operate primer pump until air-free fuel emerges from

loosened connection. Tighten fitting after line has been purged of air.

NOTE: Either the injection pump bleed screw or an injection line fitting must be open when operating the hand primer pump or damage to the fuel filter will result.

If engine starts but runs unevenly, run engine at idle speed and loosen each line at injector in turn until air-free fuel flows from loosened connection. Tighten connection after line is purged of air.

## INJECTION NOZZLES

Model 955 diesel engines are of the indirect injection type and are equipped with Bosch injection nozzles used in conjunction with energy cells.

Model G1355 and Model 2255 diesel engines are of the direct injection type which do not use energy cells and are equipped with Roosa Master "pencil" type injection nozzles.

WARNING: Fuel leaves the injection nozzles with sufficient force to penetrate the skin. When testing, keep your person clear of the nozzle spray.

### All Diesel Models

111. **TESTING AND LOCATING FAULTY NOZZLE.** If the engine does not run properly and the quick checks, outlined in paragraph 106 point to a faulty nozzle, locate the faulty nozzle as follows:

If one engine cylinder is misfiring, it is reasonable to suspect a faulty nozzle. Generally, a faulty nozzle can be located by loosening the high pressure line fitting on each nozzle holder in turn, thereby allowing fuel to escape at the union rather than enter the

*Fig. 72—Hand primer pump can be actuated after pump plunger is loosened (unlatched). Refer to text for fuel system bleeding procedure.*

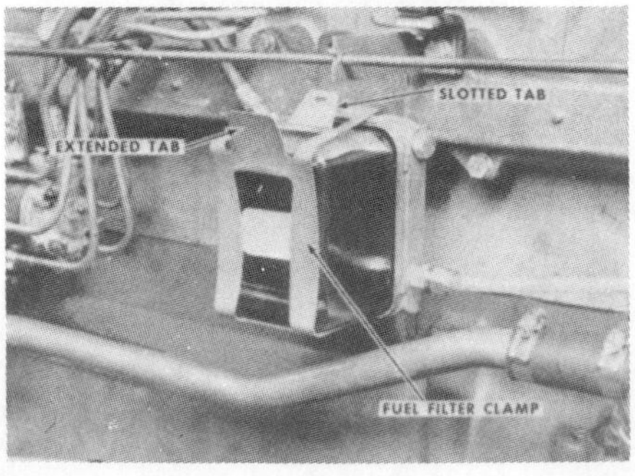

*Fig. 70—Depress extended tab of fuel filter clamp before attempting to remove clamp. Refer to text.*

*Fig. 71—Loosen vent cap (C) when bleeding fuel system of series 3150 engine. Refer to text.*

cylinder. As in checking spark plugs in a spark ignition engine, the faulty nozzle is the one which, when its line is loosened, least affects the running of the engine.

If a faulty nozzle is found and considerable time has elapsed since the injectors have been serviced, it is recommended that all injectors be removed and new or reconditioned units be installed, or the nozzles be serviced as outlined in the following paragraphs.

**Model G955**

112. **REMOVE AND REINSTALL NOZZLES.** Before loosening any lines, wash the nozzle holder and connections with clean diesel fuel or solvent. After disconnecting the high pressure and leak-off lines, cover open ends of connections with caps to prevent the entrance of dirt or other foreign material. Remove the nozzle holder stud nuts and carefully withdraw the nozzle from cylinder head, being careful not to strike the tip end of the nozzle against any hard surface.

Remove "O" ring (1—Fig. 73) and thoroughly clean the nozzle recess in the cylinder head. It is important that the seating surface of recess be free of even the smallest particle of carbon which could cause the nozzle to be cocked and result in blow-by of hot gases. No hard or sharp tools should be used for cleaning. A piece of wood dowel or brass stock properly shaped is

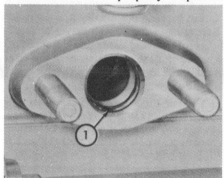

*Fig. 73—Renew "O" ring (1) before installing injection nozzle on indirect injection models.*

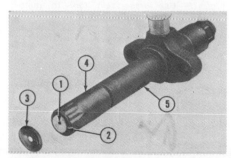

*Fig. 74—Typical Bosch injection nozzle used on G955 models.*

1. Pintle
2. Valve body
3. Heat shield washer
4. Holder nut
5. Nozzle holder assy.

very effective. Do not reuse heat shield washer (3—Fig. 74).

Renew "O" ring seal (1—Fig. 73) and install nozzle assembly. Tighten nozzle holder stud nuts evenly and to a torque of 15 ft.-lbs.

113. **NOZZLE TEST.** A complete job of testing and adjusting the injector requires the use of a special tester such as that shown in Fig. 75. Only clean approved testing oil should be used in tester tank.

The injector should be tested for spray pattern, seat leakage, back leakage and opening pressure as follows:

114. **SPRAY PATTERN.** Operate tester handle until oil flows from injector assembly. Close the valve to tester gage and operate tester handle a few quick strokes to purge air from injector and tester pump, and to make sure injector is not plugged or inoperative.

If a straight, solid core of oil flows from nozzle tip without undue pressure on tester handle, open valve to tester gage and remove cap-nut (1—Fig. 76). Slowly depress the tester handle and observe the pressure at which core emerges. If opening pressure is not within the recommended range of 1800-1850 psi, loosen locknut (3) and turn adjusting screw (5) in or out until opening pressure is within the recommended range. Tighten locknut.

When opening pressure has been set, again close valve to tester gage and operate tester handle at approximately 100 strokes per minute while examining spray core. Fuel should emerge from nozzle opening in one solid core,

*Fig. 75—Adjusting opening pressure on Bosch injection nozzle using nozzle tester.*

in a straight line with injector body, with no branches, splits or atomization.

NOTE: The tester pump cannot duplicate the injection velocity necessary to obtain the operating spray pattern of the delay type nozzles. Also absent will be the familiar popping sound associated with nozzle opening of multi-hole nozzles. Under operating velocities, the observed solid core will cross the combustion chamber and enter the energy cell. In addition, a fine conical mist surrounding the core will ignite in the combustion chamber area above the piston. The solid core cannot vary more than 7½ degrees in any direction and still enter the energy cell. While the core, is the only spray characteristic which can be observed on the tester, absence of core deviation is of utmost importance.

115. SEAT LEAKAGE. The nozzle valve should not leak at pressures less than 1700 psi. To check for seat leakage, open the valve to tester gage and actuate tester handle slowly until gage pressure approaches 1700 psi. Maintain this pressure for at least 10 seconds, then observe the flat surface of valve body and the pintle tip for drops or undue wetness. If drops or wetness appear, the injector must be

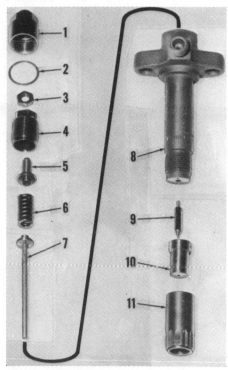

*Fig. 76—Exploded view of typical Bosch injection nozzle assembly.*

1. Cap nut
2. Copper gasket
3. Lock nut
4. Spring retainer nut
5. Adjusting screw
6. Pressure spring
7. Spindle
8. Nozzle body
9. Valve pintle
10. Valve body
11. Holder nut

disassembled and overhauled as outlined in paragraph 118.

116. BACK LEAKAGE. A back leak test will indicate the condition of the internal sealing surfaces of the nozzle assembly. Before checking the back leakage, first check for seat leakage as outlined in paragraph 115, then proceed as follows:

Turn the adjusting screw (5—Fig. 76) inward until nozzle opening pressure is set at 2350 psi. Release the tester handle and observe the length of time required for gage needle to drop from 2200 psi to 1500 psi. The time should be not less than 6 seconds. A faster drop would indicate wear or scoring between piston surface of valve pintle (9) and body (10), or improper sealing of pressure face surfaces (A, B and J—Fig. 77).

NOTE: Leakage at tester connections or tester check valve will show up as fast leak back in this test. If all injectors tested fail to pass this test, the tester, rather than the injector, should be suspected.

117. OPENING PRESSURE. To assure peak engine performance, it is recommended that all of the injectors installed in any engine be adjusted as nearly as possible, to equal opening pressures. The recommended opening pressure range is 1800-1850 psi. When a new spring (6—Fig. 76) is removed from parts stock and installed in an injector assembly, the injection pressure will drop quickly as the spring becomes

seated under the constant compression. This rate of pressure drop is approximately 10 per cent. It is recommended that injectors containing new springs be initially set at 2000 psi opening pressure, and injectors with used springs at 1850 psi. After the opening pressure has been adjusted, tighten locknut (3) and reinstall cap nut (1), then recheck opening pressure to make sure adjusting screw has not moved.

118. OVERHAUL. The maintenance of absolute cleanliness in the overhaul of injector assemblies is of utmost importance. Of equal importance is the avoidance of nicks or scratches on any of the lapped surfaces. To avoid any damage to any of the highly machined parts, only the recommended cleaning kits and oil base carbon solvents should be used in the injector repair sections of the shop. The valve pintle (9—Fig. 76) and body (10) are individually fit and hand lapped, and these two parts should always be kept together as mated parts.

Before disassembling a set of injectors, cap the pressure and leak-off line connections with a line nut with the hole soldered shut, or with a special metal cap, and immerse the units in a clean carbon solvent. While the injectors are soaking, clean the work area and remove any accumulation of discarded parts from previous service jobs. Remove the injectors one at a time from the solvent and thoroughly clean the outer surfaces with a brass wire brush. Be extremely careful not to damage the pintle end of the nozzle valve extending out of nozzle body. Rinse the injector in clean diesel fuel and test the injector as outlined in paragraphs 114 through 117. Never disassemble an injector which can be adjusted and returned to service without disassembly.

If the injector unit must be disassembled, clamp the injector body in a soft jawed vise, tightening only enough to keep injector from slipping, or use a holding fixture. Remove cap nut (1—Fig. 76), loosen locknut (3) and back off

the adjusting screw (5) until all tension is removed from the spring, then remove the nozzle holder nut (11). Withdraw the valve pintle (9) from valve body (10). If valve is stuck, use a special extractor or soak in solvent. NEVER loosen valve by tapping exposed pintle end of valve on a hard surface.

Examine the lapped pressure faces (A, B and J—Fig. 77) of nozzle body and holder for nicks or scratches, and the piston (larger) portion of valve pintle for scratches or scoring. Clean the fuel gallery (F) with the special hooked scraper as shown in Fig. 78, by applying side pressure while the body is rotated. Clean the valve seat with the brass seat tool as shown in Fig. 79. Polish the seat with the pointed wooden polishing stick and a small amount of tallow as shown in Fig. 80. Clean the pintle orifice from the inside, using the proper size probe. Polish the nozzle valve seat and pintle with a piece of felt and some tallow, loosening any particle of hardened carbon with a pointed piece of brass stock. Never use a hard or sharp object such as a knife blade as any scratches will cause distortion of the injection core.

As the parts are cleaned, immerse in clean diesel fuel in a compartmented pan. Insert the valve pintle into body underneath the fuel level and assemble valve body to nozzle holder while wet. Do not attempt to dry the parts with towels or compressed air because of the danger of dust particles remaining on the pressure faces of nozzle holder

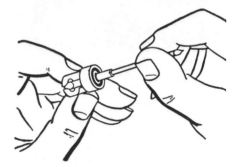

Fig. 79—Use brass seat tool to remove carbon from valve seat.

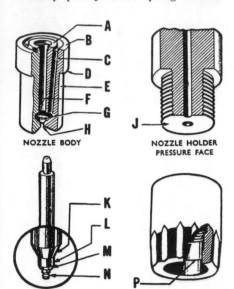

Fig. 77—Inspect disassembled Bosch injector at the points shown in above views.

A. Nozzle body
   pressure face
B. Nozzle body
   pressure face
C. Fuel feed hole
D. Shoulder
E. Nozzle trunk
F. Fuel gallery
G. Valve seat

H. Pintle orifice
J. Holder pressure face
K. Valve cone
L. Stem
M. Valve seat
N. Pintle
P. Nozzle retaining
   shoulder

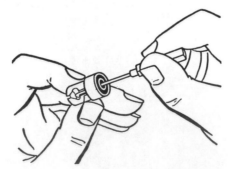

Fig. 78—Use special scraper to remove carbon from fuel gallery.

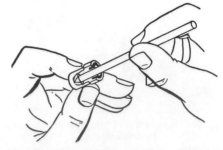

Fig. 80—Use pointed wooden stick and small amount of tallow to polish valve seat.

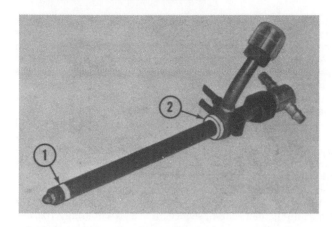

*Fig. 81—Typical Roosa Master "pencil" type injection nozzle used on G1355 Diesel engine.*

1. Teflon carbon dam seal ring
2. Teflon compression seal ring

and valve body. Use a centering sleeve or shim stock when reassembling. To use the sleeve, tighten the holder nut with the fingers while rotating centering sleeve in the opposite direction. When the nut is finger tight remove the sleeve and tighten the holder nut to a torque of approximately 55-60 ft.-lbs. Retest the injector as previously outlined. If injector fails to meet the test and no leaks because of dust were found upon disassembly, renew the valve body and pintle assembly and any other parts suspected of being faulty.

## Model G1355

119. **REMOVE AND REINSTALL NOZZLES.** Before loosening any lines, wash the nozzle holder and connections with clean diesel fuel or solvent. After disconnecting high pressure lines and leak-off hoses, plug or cap all openings to prevent entrance of dirt or other foreign material. Remove nozzle clamp cap screw, clamp and spacer, then withdraw nozzle assembly.

Before reinstalling the injection nozzle, clean nozzle bore in cylinder head and blow out with compressed air. Renew carbon dam seal (1—Fig. 81) and compression seal ring (2) on nozzle assembly. Do not lubricate seal rings.

Using a gentle rotating motion, insert injection nozzle into cylinder head. Install spacer, clamp and cap screw engaging the locating plate on nozzle. Connect high pressure line and tighten connection finger tight. Tighten clamp cap screw to a torque of 20 ft.-lbs. Connect leak-off hoses. Crank engine with starter until fuel emerges from the loosened high pressure line connection, then tighten the connection.

120. **NOZZLE TEST.** A complete job of testing and adjusting an injection nozzle requires the use of special test equipment. Only clean approved testing oil should be used in tester tank. Injection nozzle should be tested for opening pressure, seat leakage, back leakage and spray pattern. When tested, the nozzle should open with a

sharp popping or buzzing sound and cut off quickly at end of injection with a minimum of seat leakage and a controlled amount of back leakage. Check injection nozzle as outlined in the following paragraphs.

121. **OPENING PRESSURE.** Before conducting test, operate tester lever until fuel flows from tester output line, then attach injection nozzle as shown in Fig. 82. Unscrew nut (2—Fig. 83) and remove leak-off tee (1). Close the valve to tester gage and pump the tester lever a few quick strokes to be sure nozzle valve is not stuck and that the spray hole is open.

Open valve to tester gage and operate tester lever slowly while observing gage reading. Opening pressure should

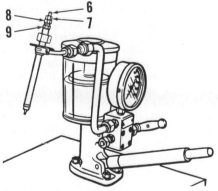

*Fig. 82—Roosa Master injection nozzle attached to nozzle tester.*

6. Valve lift adjusting screw
7. Locknut
8. Pressure adjusting screw
9. Locknut

be 2600 psi, if it is not, adjust opening pressure and valve lift as follows:

Loosen locknut (7—Fig. 83) and back out lift adjusting screw (6) at least two turns to assure against bottoming. Loosen locknut (9) and turn pressure adjusting screw (8) in to increase or out to decrease pressure until specified opening pressure is obtained. Tighten locknut (9). Turn lift adjusting screw (6) in until it bottoms, then back out one turn and tighten locknut (7). Recheck opening pressure.

**NOTE: When adjusting a new injection nozzle or an overhauled nozzle with a new pressure spring, set opening pressure at 2800 psi to allow for initial pressure loss as the spring takes a set.**

122. SPRAY PATTERN. The injection nozzle tip has a single spray hole which is located 35 degrees from centerline of injection nozzle. The conical shaped spray should be finely atomized and distributed at the 35 degree angle from nozzle tip. If a solid type irregular spray pattern is observed, check for partially clogged or damaged spray hole or an improperly seating nozzle valve. If spray pattern is not satisfactory, disassemble and overhaul injection nozzle as outlined in paragraph 125.

123. VALVE SEAT LEAKAGE. Operate tester slowly to maintain a gage pressure of 2400 psi while examining nozzle tip for fuel accumulation. If nozzle is in good condition, there should be no noticeable accumulation for a period of 10 seconds. However, slight dampness is permissible on a used injection nozzle. If a drop forms or undue wetness appears on nozzle tip, disassemble and overhaul injection nozzle as in paragraph 125.

124. BACK LEAKAGE. Reposition injection nozzle on tester so that nozzle tip is slightly higher than adjusting screw end of nozzle. Operate tester slowly to maintain a gage pressure of 1500 psi. Observe fuel leakage at leak-off area (flat side of valve lift adjusting screw). After the first drop falls, leakage should be at the rate of 3 to 10

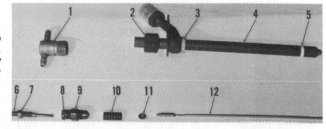

*Fig. 83—Exploded view of Roosa Master injection nozzle assembly. An "O" ring seal is used in nut (2).*

1. Leak-off tee
2. Nut
3. Teflon compression seal ring
4. Nozzle body
5. Teflon carbon dam seal ring
6. Valve lift adjusting screw
7. Locknut
8. Pressure adjusting screw
9. Locknut
10. Pressure spring
11. Spring seat
12. Valve

drops in 30 seconds. If back leakage is less than specified rate, disassemble and overhaul injection nozzle as in paragraph 125.

125. **OVERHAUL INJECTION NOZZLE.** First clean outside of injection nozzle thoroughly. Place nozzle in a holding fixture and clamp the fixture in a vise. NEVER tighten vise jaws on nozzle body without the fixture. Refer to Fig. 83, unscrew nut (2) and remove leak-off tee (1). Loosen locknut (9) and back out pressure adjusting screw (8) with valve lift adjusting screw (6). Slip nozzle body from fixture, invert the body and remove pressure spring (10), spring seat (11) and valve (12) from body. If valve will not slide from body, use a special retractor (Roosa Master No. 16481) to remove the valve.

Nozzle body and valve are a matched set and should never be intermixed. Keep parts for each injection nozzle separate and immerse in clean diesel fuel in compartmented pan as nozzle is disassembled. Remove nut (2) with its "O" ring seal, carbon dam seal ring (5) and compression seal ring (3) from nozzle body.

Clean all parts thoroughly in clean diesel fuel using a brass wire brush and lint-free wiping towels. Hard carbon or varnish can be loosened with a non-corrosive solvent. Use a cleaning wire to remove carbon from the 0.028 diameter spray hole.

Refer to Fig. 84 and use a sac hole drill to clean carbon from the 0.042 diameter sac hole. Clean the valve seat with a special brass seat scraper.

Piston area of valve can be lightly polished by hand if necessary, using Roosa Master No. 16489 lapping compound. Use the special valve retractor to turn valve. Move valve in and out of body slightly while rotating valve but do not apply down pressure while valve is in contact with seat.

Valve and seat are ground to a slight interference angle. Seating areas may be cleaned if necessary, using a small amount of No. 16489 lapping compound, very light pressure and no more than 3 to 5 turns of valve on seat. Thoroughly flush all compound from valve body after polishing.

Before reassembling, loosen locknut (7—Fig. 83) and back out lift adjusting screw (6) two turns. Reassemble by reversing disassembly procedure using Fig. 83 as a guide. Adjust opening pressure and valve lift as outlined in paragraph 121 after valve is assembled.

## Model 2255

126. **REMOVE AND REINSTALL NOZZLES.** Before loosening or removing any fuel lines, wash lines, connections and surrounding cylinder head area with clean diesel fuel.

Remove valve cover, fuel return manifold and the rocker arm assembly. Remove the injection nozzle clamp, then disconnect nozzle inlet line and pressure line from adapter. Remove adapter from cylinder head then using a slight twisting motion pull injection nozzle straight out of cylinder head.

Before reinstalling injection nozzle, clean nozzle bore in cylinder head and renew carbon dam and compression seal ring on nozzle assembly. Do not lubricate seal rings. Install new "O" ring on injection nozzle (inlet line), then with a slight twisting motion, install injection nozzle in bore of cylinder head. Install new "O" ring on adapter, install adapter in cylinder head, then connect injection nozzle to adapter and tighten nozzle to adapter nut to 25-35 ft.-lbs. of torque. Install the injection nozzle pressure line but do not tighten at this time. Install injection nozzle clamp, then install rocker arm assembly and fuel manifold. Tighten rocker arm assembly retaining cap screws to 15-23 ft.-lbs. of torque and secure with locks. Adjust intake valves to 0.015 and exhaust valves to 0.025, install valve cover and tighten retaining cap screws to 8-12 ft.-lbs. of torque.

Bleed lines by cranking engine with starting motor or with hand primer pump as outlined in paragraph 110.

127. **NOZZLE TEST.** A complete job of testing and adjusting an injection nozzle requires the use of special test equipment. Only clean approved testing oil should be used in tester tank. Injection nozzle should be tested for opening pressure, seat leakage, back leakage and spray pattern. When tested, the nozzle should open with a sharp popping or buzzing sound and cut off quickly at end of injection with a minimum of seat leakage and a controlled amount of back leakage. Check injection nozzle as outlined in the following paragraph.

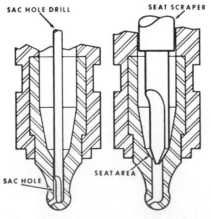

Fig. 84—Use special drill to clean sac hole and brass scraper to clean seat area.

128. **OPENING PRESSURE.** Before conducting test, operate tester lever until fuel flows from tester outlet line, then attach injection nozzle. Close the valve to tester gage and pump the tester lever a few quick strokes to be sure nozzle valve is not stuck and that spray holes (orifices) are open.

Open valve to tester gage and operate tester lever slowly while observing gage reading. Opening pressure should be 2400-2900 (2650) psi. If pressure is not as stated, adjust opening pressure and valve lift as follows:

Remove injection nozzle from tester and secure it in a holding fixture. Loosen lift adjusting screw lock nut and back out lift adjusting screw two or three turns to assure against bottoming during reassembly. Loosen pressure screw and remove injector from holding fixture. Tilt the injector slightly and remove pressure screw and shims. Add shims as required to correct pressure.

**NOTE: Shims are 0.005 thick and each shim will increase pressure about 250 pounds. Do not use more than two shims to increase the opening pressure.**

When reassembling, make sure the thickest shim is next to the pressure screw. Place injector in holding fixture and tighten pressure screw to 75-80 in.-lbs. of torque. Turn lift adjusting screw in until it bottoms. Then back out 5/8-7/8-turn.

Hold lift adjusting screw in this position and tighten lock nut to 35-40 in.-lbs.

**NOTE: When adjusting a new injection nozzle or an overhauled nozzle with a new pressure spring, set opening pressure at 2800 psi to allow for initial pressure loss as the spring takes a set.**

129. SPRAY PATTERN. The injection nozzle has four equally spaced spray holes (orifices) at nozzle tip. The spray must be equal and uniform through all four orifices. Any deviation either vertically or horizontally indicates an unsatisfactory nozzle. If spray pattern is not satisfactory, disassemble and overhaul injection nozzle as outlined in paragraph 132.

130. VALVE SEAT LEAKAGE. Operate tester slowly to maintain a gage pressure of 2400 psi while examining tip of nozzle for fuel accumulation. If more than three drops appear on tip in 15 seconds, disassemble and overhaul injection nozzle as outlined in paragraph 132.

131. BACK LEAKAGE. Reposition injection nozzle on tester so that nozzle tip is slightly higher than adjusting screw end of nozzle. Operate tester slowly to maintain a gage pressure of

1500 psi. Observe fuel leakage at leak-off area. After the first two drops fall, leakage should be 1 to 10 drops in 15 seconds. If back leakage is less than stated, disassemble and overhaul nozzle as outlined in paragraph 132. If back leakage is more than stated, worn or faulty parts are indicated and the injection nozzle should be renewed.

132. **OVERHAUL INJECTION NOZZLE.** First clean outside of injection nozzle thoroughly. Place nozzle in a holding fixture and clamp the fixture in a vise. NEVER tighten vise jaws on nozzle body without the fixture. Refer to Fig. 85, loosen lift adjusting screw lock nut (2) and back out lift adjusting screw (1) two or three turns, then loosen the pressure screw (3). Remove injection nozzle from fixture, invert the body and remove pressure screw, spring, spring seat and shims. If valve (9) does not slide out of body, use retractor to remove valve.

Nozzle body and valve are a matched set and should never be intermixed. Keep parts for each injection nozzle separate and immerse in clean diesel fuel in compartment pan as nozzle is disassembled. Remove carbon dam seal (13) and compression seal ring (8) from nozzle body.

Clean all parts thoroughly in clean diesel fuel using a brass wire brush and lint-free wiping towels. Hard carbon or varnish can be loosened with a non-corrosive solvent. Use cleaning wire (0.011) in pin vise to clean orifices.

NOTE: Use a rotating motion when cleaning orifices.

Use seat scraper to remove any deposits from seat area and sac hole drill to remove any deposits from sac hole area. Repeat orifice cleaning procedure after cleaning sac hole.

Piston area of valve can be lightly polished by hand if necessary, using lapping compound. Use the special valve retractor to turn valve. Move valve in and out of body slightly while rotating valve but do not apply down pressure while valve is in contact with seat.

Valve and seat are ground to slight interference angle. Seating areas may be cleaned if necessary, using a small amount of lapping compound, very light pressure and no more than 3 to 5 turns of valve on seat. Thoroughly flush all compound from valve body after polishing.

Before reassembling, be sure valve lift adjusting screw has been backed out two or three turns from its original location in the pressure screw, then reassemble the injection nozzle by reversing the disassembly procedure. Adjust opening pressure and valve lift as outlined in paragraph 128 after valve is assembled.

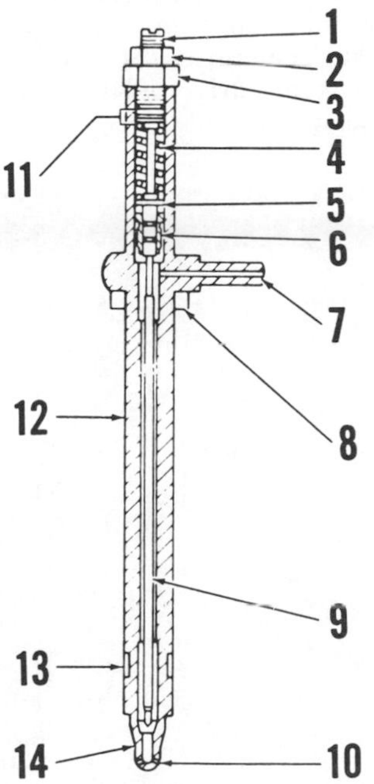

## INJECTION PUMP

### Models G955-G1355

Models G955 and G1355 tractors are

*Fig. 86—Typical Roosa Master injection pump installation showing pump timing mark.*

equipped with Roosa-Master injection pumps.

**The following paragraphs will outline ONLY the injection pump service work which can be accomplished without the use of special pump testing equipment. If additional service work is required, the pump should be turned over to a properly equipped diesel service station for overhaul. Inexperienced service personnel should never attempt to overhaul a diesel injection pump.**

133. **TIMING TO ENGINE.** The correct timing for Model G955 tractors is 2 degrees BTDC, or 4 degrees ATDC for Model G1355 tractors.

To time the pump to the engine, first locate and mark the correct timing position on flywheel. Crank engine until No. 1 piston is coming up on the compression stroke, and continue cranking until the indicated timing mark is centered in timing window on left side of flywheel housing. Shut off the fuel and remove timing window cover on side of injection pump body. Pump is correctly timed when scribe lines on governor cage and cam ring are aligned as shown at (1—Fig. 86). If marks are visible but not perfectly aligned, loosen the two stud nuts attaching pump to mounting plate and shift pump body slightly until marks are perfectly aligned.

Reinstall pump timing window and turn on fuel. Injection pump is self-bleeding. Injection lines or filter should not require bleeding if pump connections are not broken.

134. **REMOVE AND REINSTALL.** Before attempting to remove the injection pump, thoroughly wash the pump and connections with clean diesel fuel. Disconnect fuel lines and control rods, then cap all exposed fuel line connections to prevent entrance of dirt. Pressure lines must be disconnected at both the pump and injectors. Loosen fuel line clamps, if necessary, and shift the lines rearward as far as possible.

Remove the four cap screws securing mounting plate (8-Fig. 87) to adapter housing (3) and withdraw the pump, mounting plate and drive gear (7—Fig.

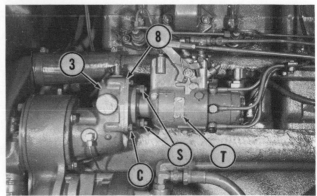

*Fig. 85—Cross-sectional view of the "pencil" type injection nozzle used in Model 2255 Diesel engines.*

| | |
|---|---|
| 1. Lift adjusting screw | 8. Compression seal |
| 2. Locknut | 9. Valve |
| 3. Pressure screw | 10. Orifices |
| 4. Spring | 11. Shims |
| 5. Spring seat | 12. Body |
| 6. Valve guide | 13. Carbon dam |
| 7. Fuel inlet | 14. Nozzle tip |

*Fig. 87—Left side view of an early diesel engine equipped with Roosa Master injection pump. Later models are similar.*

C. Cap screw
S. Stud nuts
T. Timing window cover
3. Adapter housing
8. Mounting plate

88) straight to rear out of adapter housing. Be careful not to lose thrust button (5) and spring (6) as pump is removed.

The pump adapter shaft and seals are a part of injection pump. The mounting plate (8—Fig. 89), drive gear (7) and "O" rings (9) are not a part of the pump and must be transferred if a new pump is installed. DO NOT withdraw adapter shaft and drive gear (7) from pump unless required for service. If adapter shaft must be withdrawn, make sure assembly marks on shaft and rotor are aligned, and that lip of cuptype seal is not turned back during installation.

When installing the pump, make sure that No. 1 piston is on compression stroke and that the correct flywheel timing mark is aligned with timing pointer as outlined in paragraph 133. Remove pump timing window cover (T—Fig. 87) and turn drive shaft until timing marks on cam ring and governor cage are aligned as shown in Fig. 86. Turn drive gear backward (clockwise as viewed from front) until governor cage timing marks move downward approximately 1/16-inch, then install the pump. With mounting plate cap screws installed and flywheel timing marks aligned, pump timing marks

*Fig. 88—Roosa Master injection pump removed from tractor. Shaft (A), spring (6) and thrust button (5) are serviced as part of pump (P) but gear (7) and mounting plate (8) are not.*

should be within 3/32-inch of alignment. If they are not, remove pump, turn drive gear 1 tooth in proper direction and reinstall. If marks are less than on gear tooth out of alignment, tighten the mounting plate retaining cap screws, loosen the two stud nuts (S—Fig. 87) and shift pump housing slightly until marks are aligned.

135. **GOVERNOR.** Adjustments on the injection pump are sealed and warranty on engine and fuel injection equipment will be voided if seals are broken by other than authorized diesel service personnel.

136. IDLE SPEED. Engine low idle speed is adjusted as follows: Run engine until it reaches operating temperature, then move throttle lever to idle position and disconnect governor rod from injection pump lever. Loosen locknut on idle speed adjusting screw, hold injection pump lever against idle speed stop and turn adjusting screw until an idle speed of 700 rpm is registered on tachourmeter. Tighten idle adjusting screw locknut.

Loosen lock nut on governor rod clevis and adjust clevis, if necessary, until clevis pin will enter clevis and injection pump lever while throttle lever and injection pump lever are both in idle position, then tighten locknut. The governor rod bellcrank located under fuel tank support and the injection pump lever must reach their stops simultaneously.

The engine high idle no-load rpm can now be adjusted.

137. HIGH IDLE NO-LOAD SPEED. With engine idle speed adjusted as outlined in paragraph 136, adjust the engine high idle no-load rpm as follows: Loosen locknut on bellcrank stop screw located on fuel tank support and turn stop screw (out) to its minimum position. Disconnect governor rod from bellcrank. Loosen the injection pump high idle adjusting screw locknut, hold injection pump lever against high idle stop and turn high idle adjusting screw until a high idle no-load speed of 2335 rpm is registered on tachourmeter. Tighten adjusting screw locknut.

Adjust (move) throttle lever, if necessary, so that governor rod will freely enter hole in bellcrank when throttle lever and injection pump lever are both in high idle position, then turn stop screw until it contacts bellcrank and tighten locknut. The bellcrank and the injection pump lever must reach their stops simultaneously.

NOTE: Do not alter the length of governor rod as this will change the engine idle speed previously adjusted.

## Model 2255

138. **TIMING TO ENGINE.** To check and/or adjust injection pump timing, proceed as follows: Remove timing slot plug from rear of injection pump and insert timing pin, then turn engine in direction of normal rotation until timing pin drops into slot in the injection pump camshaft. Remove timing hole plug from front of timing gear cover, then using one of the cover retaining capscrews, insert the capscrew through timing hole and thread it into tapped hole in camshaft gear. See Fig. 89A. If timing pin is in slot of injection pump camshaft and capscrew is centered in timing hole after being threaded into camshaft gear, the injection pump is correctly timed to engine.

139. If capscrew will not screw into engine camshaft as outlined in paragraph 138, adjust injection pump timing as follows: Be sure timing pin is in slot of injection pump camshaft. Disconnect tachourmeter cable from tachourmeter drive adapter and remove adapter.

*Fig. 89A—Timing hole in timing gear cover is located as shown.*

C. Oil cooler
H. Thermostat housing
T. Timing hole

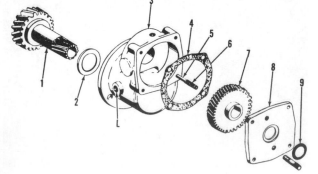

*Fig. 89—Exploded view of typical injection pump drive gears, adapter housing and associated parts used on models equipped with Roosa Master injection pump.*

L. Lube passage
1. Drive gear
2. Thrust washer
3. Adapter housing
4. Gasket
5. Thrust button
6. Spring
7. Pump gear
8. Mounting plate
9. "O" ring

On early model 3150 engines, remove the tachourmeter adapter drive shaft using a 5/8-inch deep wall socket. Use puller group (Cat. No. 9S8520) and thread bolt (9S8528) into injection pump camshaft.

**NOTE: DO NOT force bolt as it should turn easily by hand.**

Thread bolt (Cat. No. 9S8527) into timing gear adapter, then tighten bolt until gear loosens. Now turn engine clockwise until capscrew inserted through timing hole will thread into tapped hole in engine camshaft and is centered in timing hole. Injection pump is now timed to engine.

Remove puller group, install tachourmeter adapter drive shaft and tighten to 30-31 ft.-lbs. of torque.

On late Model 3150 engines, loosen the tachourmeter adapter drive shaft about two turns and remove the three gear retaining capscrews. Turn engine clockwise until bolt inserted through timing hole will thread into tapped hole in engine camshaft and is centered in timing hole. Injection pump is now timed to engine.

Install gear retaining capscrews and tighten to 27-37 ft.-lbs. of torque. Tighten the tachourmeter drive adapter shaft to 40-60 ft.-lbs. of torque.

On Model 3208 engines, remove the tachourmeter adapter drive shaft and washer. Install puller (Cat. No. 5P2371, or equivalent) and tighten puller bolts until gear loosens, then remove puller. Turn engine clockwise until capscrew inserted through timing hole will thread into tapped hole in engine camshaft and is centered in timing hole. Injection pump is now timed to engine.

Install washer and tachourmeter adapter drive shaft and tighten shaft to 75-85 ft.-lbs of torque.

Install tachourmeter drive adapter and connect cable.

**140. REMOVE AND REINSTALL.** Before attempting to remove injection pump, thoroughly clean pump, connections and surrounding area with clean diesel fuel. Remove timing plug from injection pump and insert timing pin, then turn engine in direction of normal rotation until timing pin drops into slot in injection pump camshaft. Remove timing hold plug from front of timing gear cover, then remove one of the cover retaining capscrews. Insert capscrew in timing hole and thread it into tapped hole in engine camshaft. With capscrew centered in timing hole, injection pump is timed to engine. (Engine is at TDC.)

Remove air cleaner. Remove injection lines as an assembly, disconnect fuel supply line and return line, then disconnect fuel shut-off solenoid wire.

Cap all fuel line openings. Disconnect control linkage. Remove air inlet (tube) and cover intake manifold openings in cylinder heads.

Disconnect cable from tachourmeter drive adapter and remove adapter. Loosen injection pump drive gear from injection pump camshaft as outlined in paragraph 139.

Attach a sling to injection pump, remove the pump retaining capscrews and remove pump from engine.

Reinstall injection pump by reversing the removal procedure.

**141. GOVERNOR.** Adjustments on the injection pump are sealed and warranty on engine and fuel injection equipment will be voided if seals are broken by other than authorized diesel service personnel.

**142. IDLE SPEED.** Start engine and run until it reaches operating temperature. Loosen lock nut on idle speed adjusting screw and governor rod adjusting clevis, then disconnect governor rod from injection pump lever. Hold injection pump lever against idle speed stop and turn idle speed adjusting screw as required until an idle speed of 800 rpm registers on tachourmeter. Tighten lock nut on idle speed adjusting screw. With throttle lever and injection pump lever in idle position, adjust clevis, if necessary, so clevis pin will freely enter clevis and injection pump lever, then tighten clevis lock nut. Bellcrank located under fuel tank and injection pump lever must reach their stops simultaneously.

The engine high idle (control linkage) adjustment can now be made, if necessary.

**143. HIGH IDLE SPEED (CONTROL LINKAGE).** The engine high idle speed is controlled by a governor located on rear of injection pump. If engine high idle rpm is not correct after it has been determined that the control linkage adjustment is correct, the injection pump governor will require internal adjustment and the adjustment should be made by an authorized diesel service station.

To check and/or adjust the control linkage, proceed as follows: Loosen the lock nut on stop screw located in fuel tank support and back out stop screw (outward) to its minimum position. Move throttle control lever toward full throttle position until injection pump lever shaft stops turning, then continue to move throttle control lever until end of injection pump lever has 1/8-inch overtravel after lever shaft stops. Maintain this throttle control lever position, then turn the stop screw located in fuel tank support in until it contacts bellcrank and tighten locknut. Do not change length of governor rod as it will alter the idle speed adjustment.

Check the engine high idle speed which should be 2770-2870 rpm.

## ENERGY CELLS

### Model G955

**144. R&R AND CLEAN.** The necessity for cleaning the energy cells is usually indicated by excessive exhaust smoke or a drop in fuel economy. To remove the energy cells, unscrew the retainer plug and remove the cap holder. Grasp the energy cell cap (3—Fig. 90) with a pair of pliers and withdraw cap from cylinder head. Using a 15/16-inch NEF-2 puller screwed into energy cell (4), pull cell from head as shown in Fig. 91.

Clean all parts in a suitable carbon solvent and renew any which are cracked or damaged. Clean the small orifice in the cell chamber with a brass brush or a piece of hard wood. Check seating surface of cell body in cylinder and mating surfaces of energy cell and cell cap. If the surfaces are rough, they can be cleaned by lapping with a mixture of jeweler's rouge and diesel fuel.

When reinstalling the energy cell, make certain that it fits squarely into cylinder head. Tighten the retainer plug to a torque of 100 ft.-lbs.

Fig. 90—Exploded view of energy cell used in Model G955 diesel engincs.

1. Plug
2. Holder
3. Cell cap
4. Energy cell

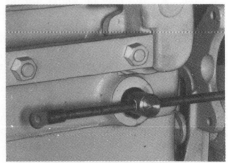

Fig. 91—Using special threaded puller to remove energy cell from Model G955 diesel engine. Refer to text.

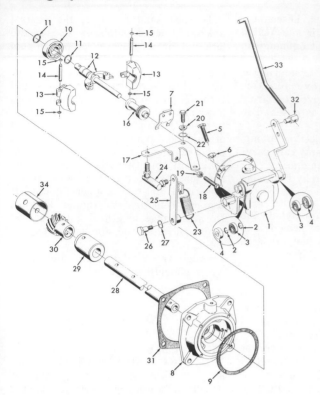

1. Housing & rocker shaft assy.
2. Retainers
3. Bearings
4. Oil seals
5. Idle position screw
6. Locknut
7. Fork
8. Driver base
9. Gasket
10. Bearing
11. Retainers
12. Shaft assembly
13. Flyweights
14. Weight pins
15. Retainers
16. Thrust sleeve & bearing
17. Speed change lever
18. High speed stop screw
19. Locknut
20. Bushing
21. Cap screw
22. Wave washer
23. Governor spring
24. Balljoint link
25. Variable speed lever
26. Shoulder cap screw
27. Wave washer
28. Extension shaft
29. Bushing
30. Drive gear
31. Gasket
32. Adjusting pin
33. Linkage rod
34. Bushing (in crankcase)

# NON-DIESEL GOVERNOR

## ADJUSTMENT

### Models G955-G1355

145. Before attempting to adjust the governor, warm up engine and adjust carburetor as outlined in paragraph 99. Also check and correct any binding in the throttle or governor linkage.

With engine not running, move hand throttle lever to high speed position. Disconnect governor to carburetor link (rod) and adjust link so that, with carburetor throttle arm against high speed stop, the adjusting pin on link will be a full hole below the hole in governor arm. It will be necessary to move the governor fork arm down the distance of a full hole in order to insert the adjusting pin in hole in governor arm.

On all models, start engine and adjust high speed stop screw (18—Fig. 92) to obtain the recommended high idle, no load engine rpm.

**G-955**

High idle, no load . . . . . . . . 2040 rpm
Fully loaded . . . . . . . . . . . . . 1800 rpm
Low idle . . . . . . . . . . . . . . . . . 600 rpm

**G-1355**

High idle, no load . . . . . . . 2400 rpm
Fully loaded . . . . . . . . . . . . . 2200 rpm
Low idle . . . . . . . . . . . . . . . . . 600 rpm

## R&R AND OVERHAUL

### Models G955-G1355

146. The flyweight governor is mounted on the right side of the engine crankcase and is driven by a gear on the engine camshaft. To remove the governor, disconnect linkage, remove the four retaining cap screws and withdraw governor assembly.

To disassemble the removed governor, unbolt and remove housing (1—Fig. 92) from the drive base (8). Most of the governor can be examined for wear and damage at this time. Check weights (13), weight pins (14) and driver assembly for binding or excessive wear. The balance of disassembly is obvious after an examination and reference to Fig. 92.

After assembly and installation are completed, check and adjust engine governed speeds as outlined in paragraph 145.

# COOLING SYSTEM

## RADIATOR

### Models G955-G1355-2255

147. To remove radiator, drain cooling system and remove hood and hood side panels. Disconnect radiator hoses from radiator. Disconnect oil coolers mounting strips and remove grille. Provide support for oil coolers, then unbolt radiator supports and remove supports and radiator.

## WATER PUMP

### Models G955-G1355

148. **R&R AND OVERHAUL.** To remove the water pump, drain cooling system and disconnect water lines and hoses from pump. Unbolt and remove pump cover (13—Fig. 93), then remove cap screw (12) and washer (11). Unbolt water pump from timing gear case and remove pump from tractor.

To disassemble the pump, remove the three cap screws and washers securing ball bearing (2) to pump body (6). Then, press shaft (3) with bearing from impeller (10) and pump body. Remove oil seal (4) and pump seal (8) from body and ceramic seal (9) from impeller. Clean and inspect all parts and renew any showing excessive wear or other damage.

Renew all seals and reassemble by reversing disassembly procedure. When reinstalling pump, use new seal ring (5) and make certain drive pin in shaft (3) is properly engaged in drive gear. Secure pump to timing gear case, then install cap screw (12) with washer (11). Use new gasket (7) and install cover (13).

Refer to the following torque values when reassembling and reinstalling water pump:

Bearing retaining cap screws . . . 9 ft-lbs.
Impeller to drive
  gear cap screw . . . . . . . . . . . . 26 ft.-lbs.
Pump mounting stud nuts . . . . 25 ft.-lbs.
Pump cover cap screws . . . . . . 14 ft.-lbs.

*Fig. 93—Exploded view of water pump used on Models G955 and G1355. Cap screw (12) threads into injection pump or distributor drive gear.*

1. "O" ring
2. Ball bearing
3. Pump shaft
4. Oil seal
5. Seal ring
6. Pump body
7. Gasket
8. Pump seal
9. Ceramic impeller seal
10. Impeller
11. Washer
12. Cap screw
13. Cover

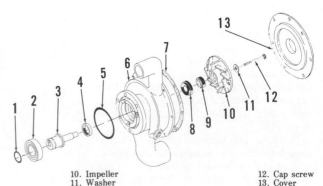

## Model 2255

NOTE: Water pumps for Model 2255 tractors are available as a unit only and individual parts are not furnished for service. However, rebuilt pumps may be available from various sources and if they are used, they must conform to Caterpillar specifications. See Fig. 94.

The water pumps used on Model 3208 engines have a larger capacity than those used on Model 3150 engines and while any service involved will be similar, specifications of the two pumps differ.

Specifications for the Model 3150 water pump are: Bearing end clearance, 0.001-0.025; clearance between impeller and front timing gear cover, 0.005-0.045; distance from rear face of impeller to rear face of housing, 1.370-1.400.

Specifications for the Model 3208 water pump are: Bearing end clearance, 0.001-0.025; clearance between impeller and front timing gear cover, 0.011-0.035; distance from rear face of impeller to rear face of housing, 1.265-1.275.

Clearance between impeller and front cover can be checked as follows: Place a ball of clay or wax on two impeller vanes directly opposite each other. Use a new gasket, install pump and tighten mounting capscrews, then push on front of pulley with about 15 pounds of force. Without turning pump, carefully remove pump and measure the clay or wax where it has been compressed between vanes and front cover. This thickness should be 0.005-0.045 (3150) or 0.011-0.035 (3208).

**149. REMOVE AND REINSTALL.** To remove water pump, proceed as follows: Remove belts, then remove fan and spacer from front of pulley and let

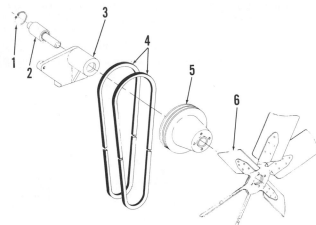

Fig. 95—Exploded view of fan assembly used on Models G955 and G1355.

1. Snap ring
2. Bearing & Shaft
3. Fan bracket
4. Fan belts
5. Pulley
6. Fan blade

fan rest in fan shroud. Unbolt water pump from front cover and remove from right side of tractor.

## FAN ASSEMBLY

### Models G955-G1355

150. Procedure for removal and/or overhaul of the fan assembly is evident after an examination of the unit and reference to Fig. 95.

The fan shaft and bearing is serviced only as an assembly. Fan belts should be adjusted until a deflection of 1/2-3/4-inch is obtained when a force of 10 pounds is applied midway between alternator and fan pulleys.

## THERMOSTAT

### Models G955-G1355-2255

151. All models use two thermostats. Thermostats for Models G955 and G1355 are located in housing bolted to front flange of the water outlet manifold on top of engine. Thermostats for Model 2255 tractors equipped with Model 3150 engines are located in front

(timing gear) cover at lower right side and are retained in position by the water inlet elbow. Thermostats for Model 2255 tractors equipped with Model 3208 engines are located at upper left of front (timing gear) cover and are retained in position by the water outlet elbow. See Fig. 95A.

Thermostats for Models G955 and G1355 should start to open at 170 degrees F. while thermostats for Model 2255 should start to open at 175 degrees F.

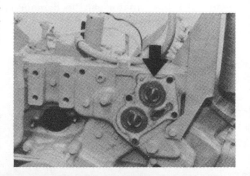

Fig. 95A—Thermostates in Model 3208 engines are located as shown.

# IGNITION AND ELECTRICAL SYSTEM

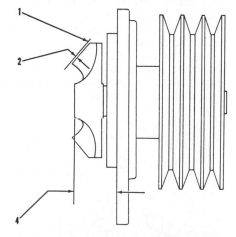

Fig. 94—View of water pump used on Model 2255. Bearing end play should be 0.001-0.025. Refer to text for other specifications.

1. Front cover
2. Impeller clearance
4. Impeller height

## DISTRIBUTOR

### Models G955-G1355

152. Delco-Remy distributors are used on all non-diesel (LP-Gas) models. Specification data follows:

**Delco-Remy 1112662**

| | |
|---|---|
| Breaker arm contact gap | 0.022 |
| Breaker arm spring tension (measured at center of contact) | 17-21 oz. |
| Cam angle | 31-34° |
| Start advance | 0-3.5 at 300 |
| Max. advance | 7-9 at 480 |

Advance data is in distributor degrees and distributor rpm.

## IGNITION TIMING

### Models G955-G1355

153. To time ignition, it is strongly recommended that a timing light be used. Timing procedure is as follows:

Adjust breaker contact gap to 0.022 and remove timing hole cover. Install timing light and start engine. Run engine at 1200 rpm, or above for Model G955, or 1500 rpm, or above for Model G1355, and hold timing light close to the flywheel. Timing pointer should register with 16 degree mark for Model G955, or with 17 degree mark for Model G1355. If the stated timing mark does not align with timing pointer, loosen distributor clamps and rotate distributor as required. Tighten distributor clamps when adjustment is complete.

Stop engine, remove timing light and install timing hole cover.

**NOTE: Final timing setting may need to be varied slightly from that stated to compensate for variations in fuel, altitude and mechanical condition of engine.**

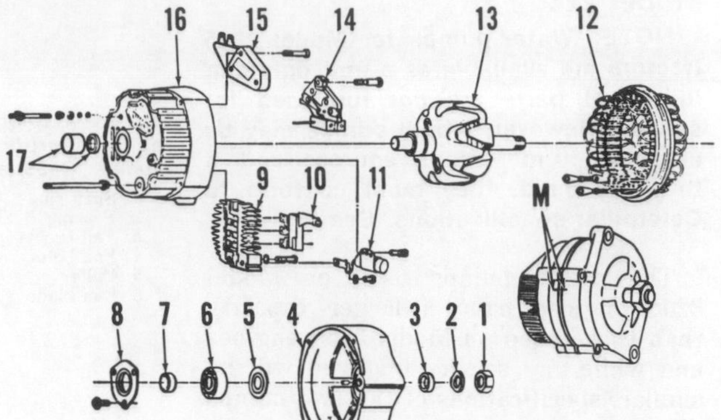

# ELECTRICAL SYSTEM

## ALTERNATOR AND REGULATOR

### All Models

Delco-Remy "DELCOTRON" alternators are used on all models. Alternators are equipped with a solid state regulator which is mounted internally and has no provision for adjustment. Rated output of alternators is 61 amperes.

**CAUTION: Because certain components of the alternator can be damaged by procedures that will not affect a D.C. generator, the following precautions MUST be observed.**

**a. When installing batteries or connecting a booster battery, the negative post of battery must be grounded.**

**b. Never short across any terminal of the alternator or regulator.**

**c. Do not attempt to polarize the alternator.**

**d. Disconnect all battery ground straps before removing or installing any electrical unit.**

**e. Do not operate alternator on an open circuit and be sure all leads are**

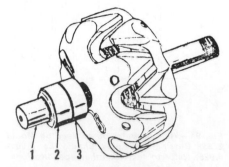

Fig. 95C—Exploded view of alternator showing component parts and their relative positions. Note match marks (M) on end frames.

1. Pulley nut
2. Washer
3. Spacer
4. Drive end frame
5. Grease slinger
6. Ball bearing
7. Spacer
8. Bearing retainer
9. Bridge rectifier
10. Diode trio
11. Capacitor
12. Stator
13. Rotor
14. Brush holder
15. Solid state regulator
16. Slip ring end frame
17. Bearing & seal assy.

properly connected before starting engine.

**154. ALTERNATOR TESTING AND OVERHAUL.** The only test which can be made without removal and disassembly of alternator is the regulator. If there is a problem with the battery not being charged, and the battery and cable connector have been checked and are good. Check the regulator as follows: Operate engine at moderate speed and turn all accessories on and check the ammeter. If ammeter reading is within 10 amperes of rated output as stamped on alternator frame, alternator is not defective. IF ampere output is not within 10 amperes of rated output, ground the field winding by inserting a screwdriver into test hole, Fig. 95B. If output is then within 10 amperes of rated output, replace the regulator.

**CAUTION: When inserting screwdriver in test hole the tab is within ¾-inch of casting surface. Do not force screwdriver deeper than one inch into end frame.**

If output is still not within 10 amperes of rated output, the alternator will have to be disassembled. Check the field windings, diode trio, rectifier bridge and stator as follows:

To disassemble the alternator, first scribe match marks (M—Fig. 95C) on the two frame halves (4 and 16), then remove the four through-bolts. Pry frame apart with a screwdriver between stator frame (12) and drive end frame (4). Stator assembly (12) must remain with slip ring end frame (16) when unit is separated.

**NOTE: When frames are separated, brushes will contact rotor shaft at bearing area. Brushes MUST be cleaned of lubricant with a soft dry cloth if they are to be reused.**

Clamp the iron rotor (13) in a protected vise, only tight enough to permit loosening of pulley nut (1). Rotor and end frame can be separated after pulley and fan are removed. Check bearing surface of rotor shaft for visible wear or scoring. Examine slip ring surface for scoring or wear and rotor winding for overheating or other damage. Check rotor for grounded, shorted or open circuits using an ohmmeter as follows:

Refer to Fig. 95D and touch the ohmmeter probes to points (1-2) and (1-3); a reading near zero will indicate a ground. Touch ohmmeter probes to the

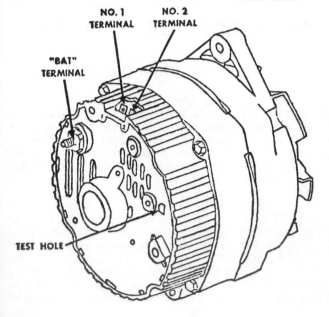

Fig. 95B—View showing the terminal and test hole locations. Refer to text.

Fig. 95D—Test points for rotor. Refer to text.

slip rings (2-3); reading should be 5.3-5.9 ohms. A higher reading will indicate an open circuit and a lower reading will indicate a short. If windings are satisfactory, mount rotor in a lathe and check runout at slip rings using a dial indicator. Runout should not exceed 0.002. Slip ring surfaces can be trued if runout is excessive or if surfaces are scored. Finish with 400 grit or finer polishing cloth until scratches or machine marks are removed.

Before removing stator, brushes of diode trio, refer to Fig. 95E and check for grounds between points A to C and B to C with an ohmmeter, using the lowest range scale. Then, reverse the lead connections. If both A to C readings or both B to C readings are the same, the brushes may be grounded because of defective insulating washer and sleeve at the two screws. If the

screw assembly is not damaged or grounded, the regulator is defective.

To test the diode trio, first remove the stator. Then, remove the diode trio, noting the insulator positions. Using an ohmmeter, refer to Fig. 95G and check between points A and D. Then, reverse the ohmmeter lead connections. If diode trio is good, it will give one high and one low reading. If both readings are the same, the diode trio is defective. Repeat this test at points B and D and at C and D.

The rectifier bridge (Fig. 95H) has a grounded heat sink (A) and an insulated heat sink (E) that is connected to the output terminal. Connect ohmmeter to the grounded heat sink (A) and to the flat metal strip (B). Then, reverse the ohmmeter lead connections. If both readings are the same, the rectifier bridge is defective. Repeat this test between points A and C, A and D, B and E, C and E, and D and E.

Test the stator (12—Fig. 95C) windings for grounded or open circuits as follows: Connect ohmmeter leads successively between each pair of leads. A

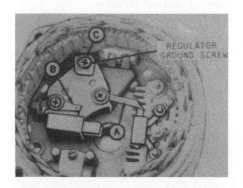

Fig. 95E—Test points for brush holder. Refer to text.

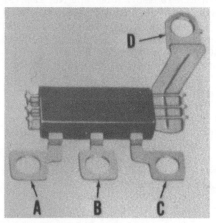

Fig. 95G—Test points for diode trio. Refer to text.

Fig. 95H—Test points for bridge rectifier. Refer to text.

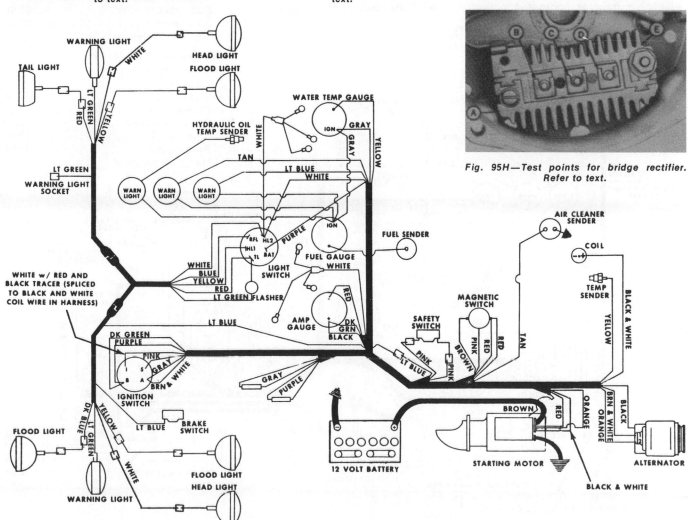

Fig. 95F—Wiring diagram for LP-Gas tractors.

high reading would indicate an open circuit.

NOTE: The three stator leads have a common connection in the center of the windings. Connect ohmmeter leads between each stator lead and stator frame. A very low reading would indicate a grounded circuit. A short circuit within the stator windings cannot be readily determined by test because of the low resistance of the windings.

Brushes and springs are available only as an assembly which includes brush holder (14—Fig. 95C). If brushes are reused, make sure all grease is removed from surface of brushes before unit is reassembled. When reassembling, install regulator and then brush holder, springs and brushes. Push brushes up against spring pressure and insert a short piece of straight wire through the hole and through end frame to outside. Be sure that the two screws at Points A and B (Fig. 95E) have insulating washers and sleeves.

NOTE: A ground at these points will cause no output or controlled output.

Withdraw the wire only after alternator is assembled.

Capacitor (11—Fig. 95C) connects to the rectifier bridge and is grounded to the end frame. Capacitor protects the diodes from voltage surges.

Remove and inspect ball bearing (6—Fig. 95C). If bearing is in satisfactory condition, fill bearing ¼-full with Delco-Remy lubricant No. 1948791 and reinstall. Inspect needle bearing (17) in slip ring end frame. This bearing should be renewed if its lubricant supply is exhausted; no attempt should be made to relubricate and reuse the bearing. Press old bearing out towards inside and press new bearing in from outside until bearing is flush with outside of end frame. Saturate felt seal with SAE 20 oil and install seal.

Reassemble alternator by reversing the disassembly procedure. Tighten pulley nut to a torque of 50 ft.-lbs.

## STARTING MOTORS

### All Models

155. Delco-Remy starting motors are used on all models and specification data follows:

**Starting Motor 1107575**
Brush spring tension (min.) . . . . . . 35 oz.
No load test:
    Volts . . . . . . . . . . . . . . . . . . . . . . .9.0
    Amperes (min.) . . . . . . . . . . . . . . . . .50*
    Amperes (max.) . . . . . . . . . . . . . . . . .80*
    Rpm (min.) . . . . . . . . . . . . . . . .5500
    Rpm (max.) . . . . . . . . . . . . . . . .9000
    *Includes solenoid

**Starting Motor 1113698**
Brush spring tension . . . . . . . . . . .80 oz.
No load test:
    Volts . . . . . . . . . . . . . . . . . . . . . . .9.0
    Amperes (min.) . . . . . . . . . . . . . . . .130*
    Amperes (max.) . . . . . . . . . . . . . . . .160*
    Rpm (min.) . . . . . . . . . . . . . . . .5000
    Rpm (max.) . . . . . . . . . . . . . . . .7000
    *Includes solenoid

**Starting Motor 1114110**
Brush spring tension . . . . . . . . . . .80 oz.
No load test:
    Volts . . . . . . . . . . . . . . . . . . . . . .11.0
    Amperes (min.) . . . . . . . . . . . . . . . .115*
    Amperes (max.) . . . . . . . . . . . . . . . .170*
    Rpm (min.) . . . . . . . . . . . . . . . .6300
    Rpm (max.) . . . . . . . . . . . . . . . .9500
    *Includes solenoid

**Starting Motor 1114156**
Brush spring tension . . . . . . . . . . .80 oz.
No load test:
    Volts . . . . . . . . . . . . . . . . . . . . . . .9.0
    Amperes (min.) . . . . . . . . . . . . . . . .140*
    Amperes (max.) . . . . . . . . . . . . . . . .190*
    Rpm (min.) . . . . . . . . . . . . . . . .4000
    Rpm (max.) . . . . . . . . . . . . . . . .7000
    *Includes solenoid

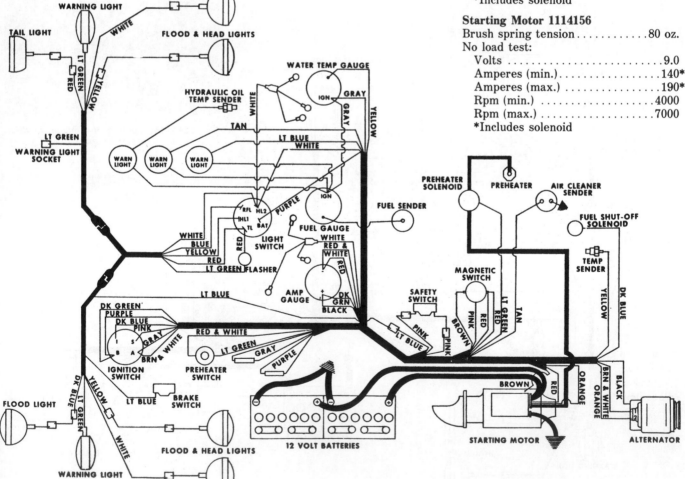

*Fig. 95J—Wiring diagram for G1355 diesel tractors.*

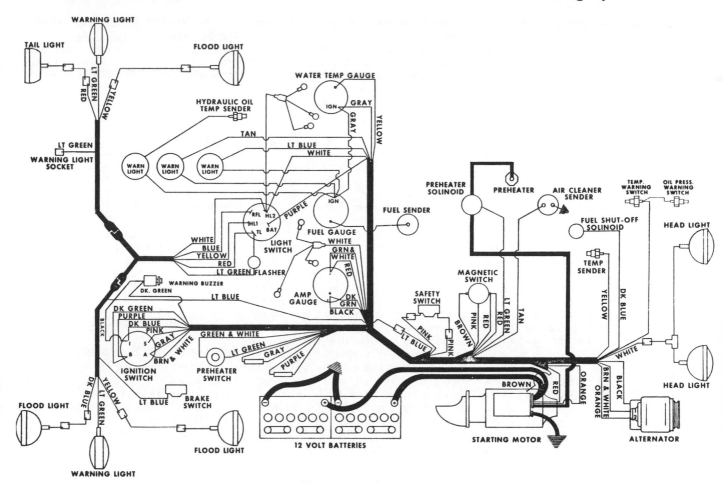

*Fig. 95K—Wiring diagram for 2255 diesel tractors.*

# ENGINE CLUTCH

Engine clutch for all models is a 14 inch dry, spring loaded type utilizing a button type driven plate having six buttons.

## ADJUSTMENT

### All Models

156. Clutch linkage should be adjusted when pedal free travel has decreased to ¾-inch.

To adjust clutch linkage, refer to Fig. 96 and proceed as follows: Loosen clevis jam nut (N) at clutch lever end of the clutch control rod, remove cotter pin and clevis pin and rotate clevis as required to obtain a clutch pedal free travel of 1½ inches. Clutch pedal free travel is measured vertically at the clutch pedal pad.

## REMOVE AND REINSTALL

### Models G955-G1355-2255

157. To remove clutch on Models G955 and G1355, split engine from center frame as outlined in paragraph 50. To remove clutch on Model 2255,

remove engine as outlined in paragraph 76, then remove clutch housing along with auxiliary transmission or straight through drive housing.

Punch mark clutch cover and flywheel so clutch can be reinstalled in the same position and use three 3/8-16x2½ inch capscrews to unload (compress) clutch cover assembly. Unbolt and remove clutch cover and driven plate from flywheel.

Reinstall by reversing removal procedure. Install driven plate so widest hub faces rearward. Use spare clutch shaft or a pilot to align the clutch driven plate. Remove the three unloading capscrews, when used.

## OVERHAUL

### Models G955-G1355-2255

158. With clutch removed as outlined in paragraph 157, place clutch on a press with a block under pressure plate so cover will be free to move downward. Place a bar across top of clutch cover and compress assembly slightly. Remove the three unloading capscrews.

Remove nuts and lock washers from pivot lever blocks. Slowly release the press pressure and lift clutch cover from clutch springs and pressure plate. Lift off clutch springs. Remove release lever pins and release levers from pressure plate.

Inspect all parts and renew as necessary. Fifteen clutch springs are used

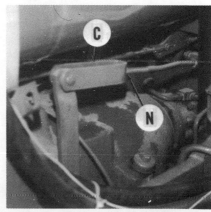

*Fig. 96—To adjust clutch free travel, loosen lock nut (N) and turn clevis (C) as required. Refer to text.*

and are color coded yellow. Springs have a solid height of 1½ inches and should test 176-186 pounds when compressed to a length of 1-11/16 inches.

Reassemble by reversing the disassembly procedure. Tighten the pivot lever block nuts to 45 ft.-lbs. torque and install the three unloading cap screws before releasing the press pressure. Refer to paragraph 159 for information on clutch release lever adjustment.

159. ADJUST RELEASE LEVERS. To adjust release levers, place clutch cover assembly on a flat surface, then adjust the three unloading cap screws until bottom edge of cover measures 0.724 from the flat surface. Check this measurement at several places around cover to insure accuracy. With cover height adjusted, turn each release lever adjusting screw as required until distance between contacting surface of adjusting screw and the flat surface is 1.897-1.927 inches. Be sure all levers are adjusted equally.

## CLUTCH RELEASE SHAFT, FORK AND BEARINGS

### Models G955-G1355-2255

160. Renewal of clutch throwout bearing, or other related parts, will require separating engine from clutch housing on Models G955 and G1355, or removal of tractor engine and unbolting clutch housing from engine on Model 2255.

Release bearing and sleeve can be pulled from fork and support tube. Press bearing from sleeve if renewal is required. Remove cap screw and key from release fork and pull release shaft from clutch housing and fork.

Release bearing support tube is renewable and can be removed by unbolting clutch housing from auxiliary

drive unit or straight drive unit. Press tube from support and be careful not to cock new tube when installing.

Press new release bearing on sleeve if necessary and fill grease cavity of sleeve prior to installation on support tube. Release bearing is a sealed type and need not be lubricated.

## PILOT BEARING
### Models G955-G1355-2255

161. The sealed clutch shaft pilot bearing is secured in the pto drive hub by a snap ring. With clutch removed, the pto drive hub can be unbolted from the flywheel and the pilot bearing pressed from pto drive hub.

# HYDRAUL SHIFT

The Hydraul Shift is a three-speed auxiliary unit interposed between engine clutch and transmission and provides on-the-go shifting in over, under or direct drive. This auxiliary unit is a self-contained unit and includes its own reservoir, lubrication and pressure systems and control valving. Oil operating temperature is controlled by a cooler located in front of the radiator. Gears are in constant mesh and are helical cut. Drive selection is accomplished by engaging or disengaging multiple disc clutches and the operation of a planetary gear set and an over-running (Sprag) clutch. The pressurized oil used to operate clutches, as well as oil for lubrication is furnished by a gear type pump mounted on inner side of housing cover.

In overdrive position, oil pressure engages overdrive (rear) clutch and power flow goes through countershaft which drives the planet gear carrier. Overdrive clutch holds sun gear shaft stationary and the speed increase is obtained by the planet gears driving output shaft through the ring gear.

In underdrive position, both direct and overdrive clutches are disengaged and power flow is through the counter shaft which drives the planet gear carrier. The over-running clutch locks sun gear shaft and output shaft together causing them to rotate as a unit. This makes the planet drive

inactive and results in a speed reduction.

In direct drive, the direct (front) clutch is engaged. This locks input shaft and output shaft together and power flow is straight through the unit. Over-running clutch over-runs and permits free rotation of sun gear shaft planet gears and countershaft.

## TROUBLE SHOOTING

### All Models So Equipped

162. The following symptoms and their causes will be helpful in trouble shooting the Hydraul Shift unit.
1. OPERATING TEMPERATURE TOO HIGH.
   a. Coolant hoses, oil cooler or oil lines obstructed.
   b. Low pump pressure or pump intake clogged.
   c. One or both clutches slipping.
   d. Engine cooling system overheated.
   e. Defective or improperly adjusted bearings.
   f. Reservoir oil level low.
2. INOPERATIVE OR HESITATES IN OVERDRIVE.
   a. Overdrive clutch slipping or defective.
   b. Low pump pressure or pump intake clogged.
   c. Oil passage obstructed.
3. INOPERATIVE OR HESITATES IN UNDERDRIVE.
   a. Over-running clutch worn or defective.
   b. Damaged or sticking underdrive and/or overdrive clutch.
4. INOPERATIVE OR HESITATES IN DIRECT DRIVE.
   a. Low pump pressure.
   b. Direct drive clutch faulty or slipping.
   c. Oil passage obstructed.
5. LUBRICATION CIRCUIT PRESSURE ABNORMAL (TOO HIGH OR TOO LOW)
   a. Defective by-pass valve spring.
   b. By-pass valve plunger sticking or not seating.
   c. Pump defective or pump intake clogged.
   d. Oil passage obstructed.

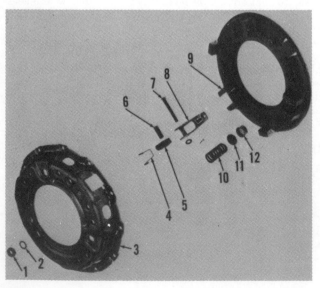

**Fig. 97—Exploded view of clutch assembly used on all models.**

1. Nut
2. Lock washer
3. Cover
4. Spring
5. Lever block
6. Block pin
7. Lever pin
8. Lever with screw
9. Pressure plate
10. Clutch spring
11. Insulating washer
12. Retainer

6. LOW CLUTCH OPERATING PRESSURE.
   a. Pump defective or pump intake clogged.
   b. Reservoir oil level low.
   c. Defective regulator valve spring.
   d. Regulator valve spool stuck open.

7. HIGH CLUTCH OPERATING PRESSURE.
   a. Defective regulator valve spring.
   b. Regulator valve spool stuck closed or sticking in bore.
   c. Oil too heavy or too cold.

8. ERRATIC CLUTCH OPERATING PRESSURE.
   a. Regulator valve spool incorrectly installed.
   b. Sticky regulator valve spool.

## SYSTEM PRESSURE CHECKS

### All Models So Equipped

163. The Hydraul Shift unit has two oil pressure circuits; a lubrication circuit and a clutch operating pressure circuit. When making any system pressure check, be sure reservoir oil is at correct level, oil is at operating temperature and engine is operating at rated rpm.

To check lubrication circuit, refer to Fig. 98 and proceed as follows: Install a pressure gage either in place of the cover-to-countershaft oil line, or in place of heat exchanger (oil cooler) line, at the housing cover. Location will depend on space available as either position will suffice.

NOTE: If desired a hole can be cut in the top of an oil filter and a gage welded in place. This assembly can be installed in place of the regular oil filter and the lubrication circuit pressure checked.

Start engine, run at rated rpm and observe gage. Gage should register 27-33 psi with control valve in any

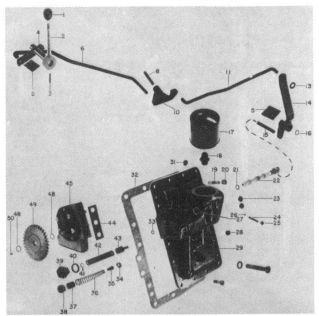

Fig. 99—Exploded view of Hydraul Shift unit cover, pump, valves and linkage. Refer also to Fig. 101.

1. Knob
2. Control lever
3. Roll pin
4. Lever shaft
5. Bushing
6. Rod
8. Pin
10. Bellcrank
11. Rod (lower)
13. Washer
14. Control lever
15. Lever pin
16. Snap ring
17. Oil filter
18. Adapter
19. Stud
20. Nut
21. "O" ring
22. Spool
23. Seat
24. Detent spring
25. Plug
26. Detent ball
27. Seat
28. Plug
29. Cover
31. Plug
32. Gasket
33. "O" ring
34. Seat
35. By-pass valve plunger
36. By-pass valve spring
37. Spacer
38. Plug
39. Plug
40. Crush washer
41. Shim (.035)
42. Pressure regulator spring
43. Pressure regulator spool
44. Gasket
45. Pump
48. Snap ring
49. Pump drive gear
50. Woodruff key

drive position. If lubrication pressure is not as stated, refer to item 5 in TROUBLE SHOOTING section. Also, be sure hose on pressure gage has at least ½-inch I.D. as a smaller hose might restrict oil flow and give a false reading.

164. To check clutch circuit, refer to Fig. 98 and proceed as follows: Install pressure gage at port below filter element boss which will give the pump output pressure available for either clutch; or install gage at front port located behind filter element boss which will give pressure available for direct drive clutch; or at rear port located behind filter element boss which will give pressure available for overdrive clutch.

Start engine, run at rated rpm and observe gage while moving control spool to direct drive and overdrive positions. Gage should register 140-160

psi in either position. If clutch pressure is not as stated, refer to items 6 and 7 in TROUBLE SHOOTING section. Shims (99) behind the pressure regulator spring (41) can be added or subtracted to adjust pressure. Shims are 0.035 thick and each shim will vary pressure about 5 psi.

### REMOVE AND REINSTALL

#### All Models So Equipped

165. To remove Hydraul Shift drive unit on Models G955 and G1355, first remove engine as outlined in paragraph 50, then disconnect clutch housing (center frame) from rear frame, and remove the clutch housing and Hydraul Shift drive unit.

Remove clutch release bearing assembly. Remove capscrew and key from release bearing fork and pull clutch lever shaft from fork and clutch housing. Unbolt and remove the Hydraul Shift drive unit from clutch housing.

Reinstall by reversing removal procedure.

166. To remove the Hydraul Shift drive unit on Model 2255, it is necessary to remove the engine and Hydraul Shift unit as outlined in paragraph 76. With engine removed, support on wood blocks so oil pan will not be damaged. Unbolt and remove clutch housing and Hydraul Shift unit from engine. Remove clutch release bearing. Remove capscrew and key from release bearing fork and pull clutch shaft from fork and clutch

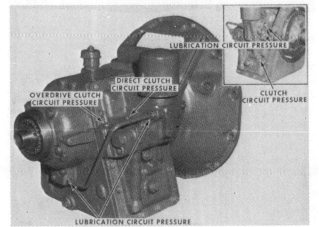

Fig. 98—Oil pressure check points of Hydraul Shift unit. Refer to text for checking procedure.

housing. Unbolt and remove the clutch housing from Hydraul Shift unit.

Reinstall by reversing removal procedure.

## OVERHAUL

### All Models So Equipped

167. **DISASSEMBLE HYDRAUL SHIFT UNIT.** With Hydraul Shift unit removed as outlined in paragraph 165 or 166, drain unit and remove dipstick and filler cap assembly. Remove valve spool control lever. Straighten tabs of lock washer (5—Fig. 100), then use a spanner wrench (Oliver No. ST-171, or equivalent) and remove nut (4). Remove sprocket (6) and "O" ring (7) from output shaft. Unscrew filter element (17—Fig. 99) from adapter (18), remove the filter to countershaft line, then remove cover (29) and pump (45) assembly from unit housing (45—Fig. 100). Pull input shaft support (95)

and input shaft (87) from housing and direct drive clutch. Remove cap screws which retain overdrive clutch housing (12—Fig. 100), then rotate housing slightly to allow removal of countershaft support (44). Support the countershaft (67—Fig. 100) to prevent binding the support, bump on front end of support until sealing plug (42) is removed, then remove support (44) from front of housing. Allow countershaft to rest on bottom of housing. Remove output shaft direct drive clutch and internal ring gear (68) assembly from housing. Remove planet gear carrier (58), drive gear (54) and bearing (53) if not removed with output shaft, clutch and internal ring gear assembly. Remove thrust bushing (52) from housing. Remove overdrive clutch housing (12), clutch and sun gear shaft (37) from housing by bumping lightly on front end of sun gear shaft. Remove support (39) from housing if it was not

removed with the overdrive clutch and housing assembly. Remove countershaft and countershaft thrust washers (65) from housing. Remove filler neck (50), studs, dowels or control lever shaft (47) as necessary.

With Hydraul Shift unit disassembled, discard all "O" rings, seals and gaskets and refer to appropriate following paragraphs to service subassemblies. Refer to paragraph 174 for reassembly procedure.

168. **COVER AND VALVES.**

**NOTE: If only pressure regulator valve, by-pass valve or oil filter are to be serviced, it can be accomplished without removing cover and with Hydraul Shift unit in tractor.**

Remove plug (39—Fig. 99), washer (40), spring (42) and pressure regulator spool (43) from bore in cover. DO NOT lose any shims (41) which may be used. A 5/16-inch cap screw can be used to remove spool if it is stuck. Remove countersunk plug (38), spacer (37), spring (36) and by-pass plunger (35). Seat (34) is renewable, if necessary. Remove plug (25), spring (24) and detent ball (26), then pull control spool (22) from bore and discard "O" ring (21). DO NOT attempt to remove spool until detent assembly is removed or damage to spool could result. Remove oil filter adapter (18) so oil passages can be inspected and cleaned.

Clean and inspect all parts. Control spool (22) and pressure regulator spool (43) should be free of nicks or scoring and should slide freely in their bores when lubricated. By-pass plunger (35) and seat (34) should be free of nicks, burrs or grooves. Springs (24, 36 and 42) should be free of rust, bends or fractures.

Refer to the following data as a guide for parts renewal.

Pressure Regulator Valve Spring
    Free length—in................3-5/8
    Spring test lbs. at in.....55-57 at 2¾
By-Pass Valve Spring
    Free length—in.............2-15/32
    Spring test lbs.
    at in.............1.7-2.3 at 1-25/32
Detent Ball Spring
    Free length—in.............1-9/32
    Spring test lbs.
    at in.............15.3-18.7 at 7/8
Pressure Regulator Spool
    Diameter.............0.6865-0.6869
Control Spool Land
    Diameter.............0.6865-0.6869
Pressure Regulator Spool
    Bore-I.D.............0.6875-0.6883
Control Spool Bore I.D....0.6875-0.6883

Reassemble by reversing the disassembly procedure.

Fig. 100—Exploded view of Hydraul Shift unit showing component parts and their relative positions.

| | | |
|---|---|---|
| 1. Sprocket, rear | 26. "O" ring | 49. "O" ring | 72. "O" ring |
| 2. Snap ring | 27. Seal ring | 50. Filler tube | 73. Seal ring |
| 3. Coupling sleeve | 28. Spring plate | 51. Dipstick and | 74. Thrust washer |
| 4. Nut | 29. Return spring |    breather | 75. Spring guide |
| 5. Lock washer | 30. Retainer | 52. Bushing | 77. Return spring |
| 6. Sprocket, front | 31. Separator plate | 53. Bearing | 78. Spring retainer |
| 7. "O" ring | 32. Clutch plate | 54. Drive gear | 79. Snap ring |
| 8. "O" ring | 33. Thrust washer | 55. Planet gear pin | 80. Pressure plate |
| 9. Oil seal | 34. Over-running clutch | 56. Bearing | 81. Clutch plate |
| 10. Snap ring | 35. Spacer | 57. Planet gear | 82. Separator plate |
| 11. Bearing | 36. Bearing | 58. Planet carrier | 83. Back plate |
| 12. Clutch housing | 37. Sun gear shaft | 61. Snap ring | 84. Retaining ring |
| 16. Capscrew | 38. Seal rings | 62. Seal rings | 85. Bearing |
| 17. Lock washer | 39. Support | 63. Thrust washer | 86. Oil seal |
| 18. Copper washer | 40. Seal ring | 64. Bushing | 87. Input shaft |
| 19. "O" ring | 41. "O" ring | 65. Thrust washer | 88. Bearing retainer |
| 20. "O" ring | 42. Plug | 66. Bearing | 91. Bearing |
| 21. Seat | 43. "O" ring | 67. Countershaft | 92. Snap ring |
| 22. Oil line | 44. Support shaft | 68. Output shaft | 93. Oil seal |
| 23. Seal ring | 45. Housing | 69. Seal ring | 94. "O" ring |
| 24. "O" ring | 46. Magnetic plug | 70. "O" ring | 95. Support |
| 25. Piston | 47. Lever shaft | 71. Piston | 98. Tube |
| | 48. Dowel | | |

169. PUMP. Unbolt pump from housing cover and discard gasket. Remove drive gear outer retainer snap ring, remove gear and key, then remove gear inner retainer snap ring.

Remove through bolts and separate pump. Dowels can be driven out and screen removed, if necessary. See Fig. 101.

**NOTE: If same pump gears and shafts are to be reinstalled, mark idler gear so it will be reinstalled in original position. Do not remove gears from shafts as they cannot be serviced separately.**

Pump dimensional data is as follows:

Gear width . . . . . . . . . . . 0.9975-0.9980
Pump body width . . . . . . 0.9995-1.0000
Gear shaft diameter . . . . 0.6247-0.6250
Shaft bore diameter . . . . 0.6265-0.6275
Shaft operating
clearance . . . . . . . . . . . . . 0.0015-0.0028

Reassemble by reversing the disassembly procedure and use new gasket when mounting pump on housing cover. Heads of bolts should be on drive gear side of pump and dowels should be installed prior to tightening through bolts.

170. INPUT SHAFT AND SUPPORT. Remove "O" ring (94—Fig. 100) from support (95), then remove bearing (85) and seal (86) from bore in rear of input shaft. Unbolt bearing retainer (88) and press input shaft (87) and bearing (91) from support. Oil seal (93) can now be removed from support, and if renewal is required, clutch release bearing tube (98) can also be removed at this time. Do not remove bearing (91) from input shaft unless renewal of bearing or bearing retainer (88) is indicated. If either part is to be renewed, remove snap ring (92) and press input shaft from bearing.

To reassemble input shaft and support, place bearing retainer on input shaft, then press bearing on shaft and install snap ring. Press bearing tube, (if renewed) into support until tube pro-

trudes 3-5/8 inches from front flange of support and be sure tube is parallel with bore. Install oil seal with lip of seal toward rear. Press input shaft and bearing into support and secure with bearing retainer. Install oil seal (86) in rear bore of input shaft with lip facing toward rear, then install bearing (85). Install "O" ring (94) on support.

Refer to paragraph 174 for installation procedure.

171. OUTPUT SHAFT, INTERNAL RING GEAR AND DIRECT DRIVE CLUTCH. To disassemble the output shaft, internal ring gear and direct drive clutch, proceed as follows: If planet gear carrier remained with output shaft assembly during disassembly, separate it from output shaft, then remove snap ring (84—Fig. 100) and remove back plate (83), clutch plates (81 and 82) and pressure plate (80) from internal ring gear of output shaft assembly (68). With a press, compress spring (77), remove snap ring (79), retainer (78), return spring (77), guide (75) and thrust washer (74) from output shaft. Remove clutch piston (71) from output shaft (gear), either by bumping end of output shaft on a block of wood, or by directing compressed air into oil passage of output shaft, then remove seal rings (69 and 73) and "O" rings (70 and 72) from I.D. and O.D. of piston. Remove seal rings (62) and thrust (63) from rear of output shaft.

**NOTE: Bushing (64) is catalogued as a separate part, however, if renewal is not indicated, do not attempt to remove it from output shaft.**

Clean and inspect all parts. Sprag (over-running) clutch surface of output shaft must be smooth and show no signs of wear. Clutch plates must have no broken lugs, splines, or warpage. Refer to the following data as a guide for parts renewal.

Return Spring
    Free length . . . . . . . . . . . . . . . . . 2.270
    Spring test lbs. at in. . . . . 230 at 1.750
Clutch Plate Thickness . . . . . 0.076-0.084
Separator Plate Thickness . . 0.065-0.073

Pressure Plate Thickness . . . 0.205-0.209
Piston Ring Groove—Outside
    Width . . . . . . . . . . . . . . . . 0.156-0.162
    Depth . . . . . . . . . . . . . . . . 0.193-0.194
Piston Ring Groove—Inside
    Width . . . . . . . . . . . . . . . . 0.128-0.132
    Depth . . . . . . . . . . . . . . . . 0.128-0.137

Reassemble output shaft and direct drive clutch assembly as follows: Install "O" rings (70 and 72) and seals (69 and 73) in O.D. and I.D. grooves of piston (71), grease seals and press piston in its bore. Use caution when installing piston not to damage seals. Install thrust washer (74), guide (75), return spring (77) and retainer (78) on front end of output shaft, compress spring and install snap ring (79). Install pressure plate (80), then starting with an internal splined clutch plate (81) alternately install the clutch plates (81 and 82). Six internally splined and five externally splined clutch plates are used. Install back plate (83) and snap (retaining) ring (84). Install thrust washer (63) on rear end of output shaft, then install sealing rings (62) in grooves of output shaft.

172. PLANET GEAR CARRIER. To disassemble the planet gear carrier assembly, first remove drive gear (54—Fig. 100) from carrier by removing the four retaining capscrews. Remove the planet gear pin retaining rings (61) and drive planet pins (55) from carrier. Remove planet gears (57) from carrier and bearings (56) from planet gears.

Inspect planet gears, bearings and pins for wear, scoring or other damage and renew as necessary. Inspect drive gear teeth and inside diameter. Inside diameter of gear should not be unduly worn or scored. Check condition of bearing (53) and with bearing in gear, check fit of gear and bearing on support (39). Gear and bearing should have a free fit on support without excessive clearance. At this time, also inspect bushing (52) for wear or scoring.

To reassemble planet carrier, place planet gear pin retainer in groove at front of pin bore. Place bearing in planet gear, position gear in carrier, then insert planet gear pin from rear of carrier until pin is against retainer and step on rear of shaft is flush with lower surface of carrier. Repeat operation for the three remaining planet gears. Place drive gear on rear of carrier, align cap screw holes, then install retaining cap screws and tighten them to 48-52 ft.-lbs. torque. Bushing (52) and bearing (53) are placed on support (39) when unit is being assembled.

173. OVERDRIVE CLUTCH AND SUN GEAR. If removed with overdrive clutch assembly, remove sun gear shaft (37—Fig. 100) from overdrive clutch.

**Fig. 101—Exploded view of Hydraul Shift oil pump. Refer to text for dimensional data.**

1. Front plate
5. Dowels
6. Screen
9. Center plate
10. Drive gear
11. Idler gear
12. Rear plate

Remove over-running clutch (34), spacer (35), bearing (36) and seal rings (38) from sun gear shaft. Bump support (39) from housing, if necessary, and remove seal ring (40). Remove clutch plates (32), separator plates (31), retainer (30), return springs (29) and return spring plate (28) from housing (12). Remove clutch piston (25—Fig. 100) from housing by bumping housing on a wood block. Remove seal rings (23 and 27) and "O" rings (24 and 26) from O.D. and I.D. of piston. Remove oil seal (9), snap ring (10) and bearing (11) from bore of housing.

With overdrive clutch disassembled, discard oil seal (9), seals (23 and 27) and "O" rings (24 and 26), then clean and inspect all parts. Clutch plates and separator plates must not be warped, scored or unduly worn. Clutch splines of sun gear shaft must not have deep grooves worn by clutch plates and bore for over-running clutch must be smooth and show no wear. Refer to the following data as a guide for parts renewal:

Return Springs
    Free length . . . . . . . . . . . . . . . . . .1.320
    Spring test lbs. at in. . . . . .10 at 1.159
Clutch Plate Thickness . . . . .0.073-0.078
Separator Plate Thickness . .0.077-0.082
Piston Ring Groove—Outside
    Width . . . . . . . . . . . . . . . . . .0.128-0.132
    Depth . . . . . . . . . . . . . . . . . .0.195-0.197
Piston Ring Groove—Inside
    Width . . . . . . . . . . . . . . . . . .0.128-0.132
    Depth . . . . . . . . . . . . . . . . . .0.151-0.152

Partially assemble overdrive assembly as follows: Install bearing (11) in housing (12) and secure with snap ring (10). Install oil seal (9) with lip toward bearing. Install "O" rings in outer and inner grooves of piston (25), install seals over both "O" rings, then grease seals and install piston in housing (12). Install seals (38) in grooves of sun gear shaft (37). Place bearing (36) and spacer (35) in bore of sun gear shaft, then install the over-running clutch so the drag strips (bronze) are toward bearing (36). Place new "O" ring (8) in groove on rear of housing (12). Delay any further assembly of the overdrive clutch until the Hydraul Shift unit is assembled.

**174. REASSEMBLE HYDRAUL SHIFT UNIT.** When reassembling the Hydraul Shift unit, proceed as follows: Install seal ring (40—Fig. 100) on hub of support and the small "O" ring in oil hole counterbore on front side of support flange, then install support in rear bore of housing so oil hole in support flange (with "O" ring) is aligned with the oil hole in housing which is on right side of lower cap

screw hole. Temporarily install one of the short cover retaining cap screws finger tight to keep support from shifting when planet carrier and output shaft assembly are installed.

Coat counter shaft thrust washers (65) with grease and stick them to inner sides of countershaft support bores and be sure tangs of thrust washers are in slots. With bearings (66) installed, carefully place countershaft in housing, with double gear rearward, so thrust washers are not dislodged, then pull countershaft toward cover opening until gear teeth clear front bore of housing. Place bushing (52) on support with cupped side rearward, then install bearing (53) on support and use fingers to compress seal ring, if necessary. Install planet carrier and drive gear (gear rearward) in housing with gear (54) over bearing (53). Be sure bushing (64), thrust washer (63) and seal rings (62) are installed on splined (long) end of output shaft, start output shaft through planet carrier and support and install internal ring gear over the planet gears.

Place new "O" ring (41) on front end of countershaft support (44). Move countershaft into position, align thrust washers, lift countershaft so support will not bind, then slide support rearward until support is ready to enter rear bore. At this time, position tapped hole in rear end of support at the nine o'clock position and push shaft rearward until front of support is flush with housing.

**NOTE: When overdrive clutch housing (12) is finally installed, holes in housing, plug (42) and support (44) must be aligned to allow installation of the ¼ x 1½-inch support retaining cap screw, as well as to place the oil hole in support on top side. Support is slotted at forward end if subsequent positioning is required.**

With countershaft support positioned, place new "O" ring (41) on O.D. of plug (42), then use grease to stick the small "O" ring (43) in counterbore of cap screw (16) through plug and small "O" ring (43) and thread a few turns into support, then bump plug into housing bore. Remove cap screw after plug is installed.

Obtain a small ½-inch thick wood block about 3x3 inches square, place block against forward end of output shaft, then tip assembly forward so output shaft is supported by the wood block. With small end forward, place sun gear shaft assembly over output shaft and push sun gear shaft on output shaft until it bottoms.

**NOTE: Installation of over-running**

clutch can be checked at this time. Sun gear shaft will turn (clutch over-run) clockwise but will lock up when attempt is made to turn shaft counterclockwise.

Remove cap screw that was installed earlier in flange of support (39). To ease assembly of overdrive clutch, install two guide studs in cap screw holes of housing (45). Install retainer (30) over guide studs with chamfer in I.D. of retainer on top side. Start with a separator plate (31) and alternately install the six separator plates and the five clutch plates (32). Place thrust washer (33) over end of output shaft with flat side toward sun gear shaft and be sure it does not hang up on shoulder of output shaft. Place return springs (29) in blind holes of retainer (30), then place plate (28) over return springs. Place new "O" ring (20) over flange of support (39) and small "O" rings (19) in counterbores of front face of housing (12), then carefully install housing (12) over guide studs. Remove guide studs and secure clutch housing with cap screws. Install countershaft support retaining cap screw (16). Place new "O" ring (7), sprocket (6), lock washer (5) and nut (4) on end of output shaft and using a spanner wrench (Oliver No. ST-171 or equivalent), tighten nut (4). Secure nut by bending a tab of lock washer into slot in nut.

Set unit in an upright position, align splines and center direct drive clutch plates as nearly as possible. Install new "O" ring (94) on pilot of front support (95) then insert input shaft hub into direct drive clutch and rotate input shaft as necessary to engage the clutch plates.

**NOTE: On earlier units, two cap screws can be temporarily installed to help in overcoming resistance of "O" ring; however BE SURE input shaft can be rotated while drawing support into housing to prevent damage to inner clutch plates. Cut-out in O.D. of support flange must mate with a similar cut-out at right side of housing to permit operation of control spool.**

Place the wood block used earlier under front of housing to tip it slightly rearward, then remove the two temporarily installed cap screws from support. Place engine clutch housing on support and secure clutch housing and support to Hydraul Shift unit housing. Install clutch throw-out bearing, throw-out bearing fork and clutch shaft. Use new gasket and install cover and pump assembly. Install cover-to-countershaft oil line and the control spool lever. When installing new oil filter, tighten filter not more than ½-turn after gasket contacts base. Install drain plug if necessary and fill unit with 3½ quarts (with filter) of Automatic Transmission Fluid, Type "A'

# TRANSMISSION

Models G955, G1355 and 2255 tractors are equipped with a constant mesh, helical gear type transmission having six forward and two reverse speeds. When equipped with a Hydraul Shift drive, three different speed ratios are available for each transmission gear.

## LUBRICATION

175. **LUBRICANT.** Transmission and differential are lubricated from a common oil supply. Oil level should be maintained at check plug opening located as shown in Fig. 102. Filler cap is located at upper left of pto housing as shown in Fig. 102. Two drain plugs are used; front plug is located under front center of main frame and drains transmission compartment and rear plug is located at rear end of main frame and drains the differential and final drive compartment. Also, pressure lubrication pump suction screen should be removed and cleaned whenever lubricating oil is drained. Screen is located on plug which is threaded into bottom of differential and final drive compartment just to rear of the transmission compartment drain plug. See Fig. 103.

Recommended transmission, differential and final drive lubricant is SAE 80 multi-purpose gear lubricant. SAE 90 multi-purpose gear lubricant may be used in temperatures above 32° F. All gear lubricant must conform to Military Specification MIL-L-2105 or API GL-5 specification. Sump capacity is 43 quarts. Lubricant should be changed yearly or after each 1000 hours of operation. Oil filter should be renewed after each 250 hours of operation.

176. **PRESSURE LUBRICATION SYSTEM.** The transmission and differential are lubricated by a pressure oiling system. Pressure is provided by a G-rotor type pump which is chain driven from hydraulic pump driven

*Fig. 102—Transmission and differential compartment test plug and filler plug are located as shown.*

gear. Oil from pump is directed from pump to an externally mounted oil filter. From the oil filter, oil is directed to the differential shaft, transmission input shaft, bevel pinion shaft, pto and hydraulic pump driven gear. See Fig. 104. A by-pass (pressure relief) valve in the transmission circuit limits system pressure to 8-14 psi. Oil by-passed by this valve is returned to the transmission compartment via a hole in the tractor main frame. Refer to Fig. 105.

With the lubricating oil at operating temperature and engine running at 2400 rpm, system pressure should be 8-14 psi with the transmission shift lever in either neutral position. Pressure can be checked by installing a gage in port located directly below oil filter element. If pressure is below 8 psi, drain transmission and differential compartments, renew the transmission oil filter and refill to proper level with specified lubricant. Refer to paragraph 175. Faulty oil pump, broken oil tube or faulty by-pass valve could also cause low oil pressure. Refer to paragraph 186 for pump servicing information. If pressure is higher than 14 psi, check the by-pass valve, (Fig. 105). If valve is OK, check for obstruction in oil tubes and passages.

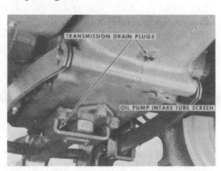

*Fig. 103—Drain plugs and lubrication pump screen plug are located as shown.*

## OVERHAUL TRANSMISSION

To perform complete service on transmission will require removal of the hydraulic pump drive shaft or pto drive shaft, splitting tractor, and removing transmission top cover, hydraulic lift system housing, transmission lubrication pump, hydraulic pump, hydraulic pump drive bevel gears, bull gears, bull pinion shafts and differential.

Refer to the following appropriate paragraphs.

177. **R&R HYDRAULIC PUMP AND DRIVEN GEAR SUPPORT.**

NOTE: Beginning with tractor serial number 248 722-, the hydraulic oil cooler pump has been incorporated into the hydraulic pump driven gear support. This assembly, shown in exploded view Fig. 108A, replaces the earlier assembly (items 1 through 15—Fig. 108). Removal and reinstallation procedure will be basically similar for both assemblies.

Remove plug from bottom of filter and drain hydraulic reservoir. Disconnect pump drain line and remove pump outlet line. Remove bleed line. Loosen

*Fig. 105— Transmission lubrication circuit by-pass valve removed from bore in rear main frame. Valve plunger and spring are retained in main frame by an expansion plug [not shown].*

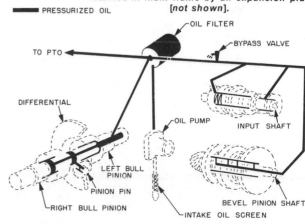

*Fig. 104—Schematic view of the transmission lubrication circuit. Note location of by-pass valve.*

PRESSURIZED OIL

OIL FILTER
BYPASS VALVE
TO PTO
DIFFERENTIAL
OIL PUMP
INPUT SHAFT
LEFT BULL PINION
PINION PIN
RIGHT BULL PINION
INTAKE OIL SCREEN
BEVEL PINION SHAFT

hose clamps and disconnect inlet hose. Loosen inlet line lock nut, move tube out of the way, then remove pump retaining capscrews and remove pump and splined coupling from pump driven gear support.

Remove pump support retaining capscrews and remove support (and pump) and shim pack from rear main frame.

178. To reinstall units, place shims on support, align capscrew holes, then place Permatex No. 3, or equivalent, on inside diameter of shims. Position support with lube holes on top sides and install support to rear main frame.

NOTE: If backlash of hydraulic pump drive gears requires checking and/or adjusting, the Oliver (White Farm Equipment) Corporation recommends that the hydraulic lift system housing be removed and the backlash of gears checked with a dial indicator. Recommended gear backlash is 0.006-0.012

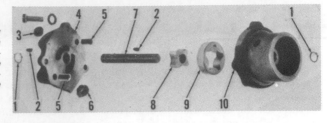

Fig. 107—Exploded view of transmission lubrication pump. Pump is chain driven by hydraulic pump drive gear and shaft. Pump retaining lock and spring are not shown.

1. Snap ring
2. Woodruff key
3. Insert
4. Cover
5. Dowels
6. Insert
7. Shaft
8. Impeller
9. Rotor
10. Housing

and is obtained by varying the shims between support and rear main frame. Shims are available in thicknesses of 0.004, 0.007 and 0.0149. Add shims to increase, or remove shims to decrease backlash.

When reinstalling hydraulic pump, cement new gasket to pump with Permatex No. 3, or equivalent, and coat main frame side of gasket with grease. Fill pump with hydraulic fluid and install pump by reversing removal procedure. Fill hydraulic reservoir, bleed filter and operate system. Recheck fluid level and add as necessary.

179. **OVERHAUL HYDRAULIC PUMP, COOLING PUMP AND SUPPORT ASSEMBLY.** To overhaul the hydraulic pump, refer to paragraph 254 in HYDRAULIC LIFT SECTION of this manual.

To overhaul the driven gear support on models prior to serial number 248 722-, refer to Fig. 108 and proceed as follows: Straighten washer (2) and remove outer nut (1) and washer (2). Remove adjusting (inner) nut (1) and remove driven shaft (9) and bearings from support. Bearing cups can be removed from support, if necessary.

Clean and inspect all parts and renew as necessary. If driven gear (9) is renewed, also renew drive gear (10).

To reassemble, press new bearing cups in support with tapers facing out. Place inner bearing on driven gear shaft with taper facing away from gear. Insert shaft and bearing in support, install outer bearing with taper toward inside and install adjusting nut. Install a 3/8-inch capscrew in end of shaft so bearing preload can be checked, then tighten adjusting nut until shaft has a rolling torque of 3-5 inch pounds. Install lock washer and lock nut, recheck rolling torque of shaft, and if satisfactory, bend lock washer to secure nuts.

To overhaul the driven gear, hydraulic oil cooling pump and support on models with serial number 248 722- and up, refer to Fig. 108A and proceed as follows: Remove the 12-point capscrews which retain inner support (26) to outer support (5), then remove inner support and gear, pump housing (1) and gerotor

assembly (2) from outer support. Remove and discard "O" rings (3 and 25). Remove gerotor assembly from driven shaft (31) and remove woodruff key (30) and pump housing (1). Relief valve plunger (7) and spring (8) can be removed from outer support after removing plug (9). Relief valve seat (6) is not available separately and if renewal is required, renew the complete outer support (5).

Straighten washer (22) and remove nuts (21) and washer. Remove driven shaft (31) and bearings from inner support (26). Bearing cups (24 and 28) and oil seal (27) can be removed from support, if necessary.

Clean and inspect all parts and renew as necessary. If driven gear (31) is renewed, also renew drive gear (32). Check pump gerotor assembly for wear or damage. Clearance between lobes of inner and outer rotors should be 0.002-0.006.

To reassemble, install new oil seal (27) in inner support (26) with lip toward inside. Press bearing cups in support with tapers facing out. Place inner bearing on driven gear shaft with taper facing away from gear. Insert shaft and bearing in support, install outer bearing with taper facing toward inside and install adjusting nut. Install a 3/8-inch capscrew in end of shaft so bearing preload can be checked, then tighten adjusting nut until shaft has a rolling torque of 3-5 inch pounds. Install lock washer and lock nut, recheck rolling torque of shaft and if satisfactory, bend lock washer to secure nuts.

Place new "O" ring (25) on face of pump housing (1) and position housing over end of driven shaft. Install woodruff key (30), then install gerotor assembly on shaft. Use new "O" ring (3), then align capscrew holes in inner support, pump housing and outer support and install the four 12-point capscrews. Tighten capscrews securely.

Install relief valve plunger (7), spring (8) and plug (9).

180. **R&R PTO OR HYDRAULIC PUMP DRIVE SHAFT.** To remove pto drive shaft, remove cover, or pto output shaft if so equipped, from top opening of pto housing. Extract the

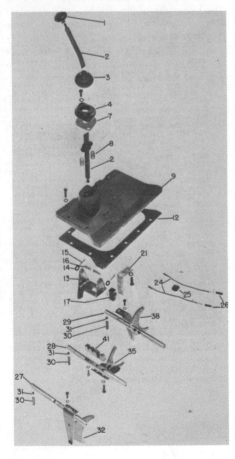

Fig. 106—Exploded view of transmission top cover assembly. Shift rails operate in bores of transmission housing (rear main frame).

1. Knob
2. Shift lever
3. Boot
4. Spring cover
7. Gasket
8. Spring
9. Top cover
12. Gasket
13. Interlock
14. Washer
15. Cotter pin
16. Pin
17. Safety switch
21. Support
24. Switch wire
25. Wire seal
26. Connector
27. Right shift rail
28. Center shift rail
29. Left shift rail
30. Detent spring
31. Detent ball
32. Right fork
35. Center fork
38. Left fork
41. Shifter lug

internal snap ring which retains plug in pto pinion, then pull plug using a ½-inch cap screw threaded into plug. Remove "O" ring from pinion bore, then use a ½-inch cap screw at least 7 inches long threaded into pto drive shaft to remove shaft.

Reinstall by reversing removal procedure.

181. To remove hydraulic pump drive shaft on tractors having no pto, remove cap screws which retain drive shaft bearing housing to main frame rear cover. Turn bearing housing until there is room enough behind housing for fingers or wedge bar and pull bearing housing and drive shaft as an assembly from rear main frame.

Reinstall by reversing removal procedure and in some cases it may be necessary, after disabling the engine, to momentarily engage starter to align shaft splines and drive hub splines.

182. **R&R HYDRAULIC PUMP DRIVE GEAR.** To remove hydraulic pump drive gear, remove hydraulic lift housing as outlined in paragraph 259,

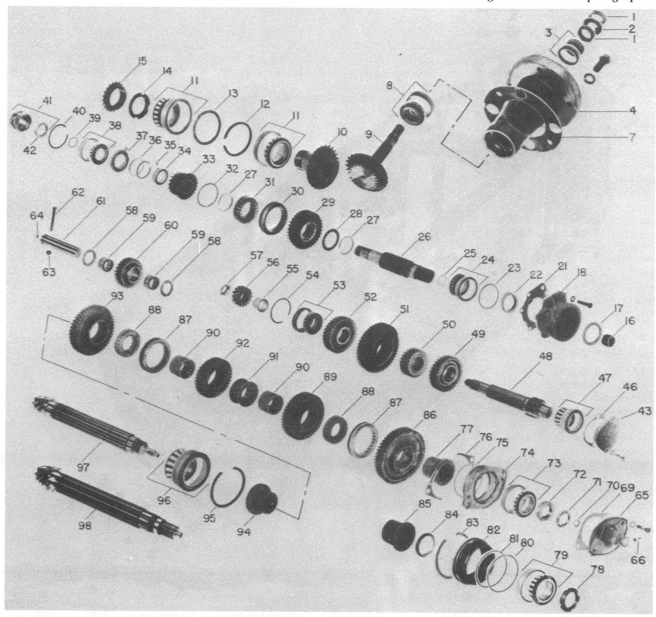

*Fig. 108—Exploded view of transmission gears and shafts. Note difference between two-wheel pinion shaft bearing cage assemblies (items 65 through 77) and four-wheel drive pinion shaft bearing cage assemblies (items 78 through 85). Refer to Fig. 108A for a view of the hydraulic pump driven shaft support used on tractors serial number 248 722- and up.*

| | | | |
|---|---|---|---|
| 1. Nut | 18. Support | 36. Sealing rings | 54. Snap ring | 71. Lock nut | 85. Mounting sleeve |
| 2. Lock washers | 21. Shims | 37. Oil collector | 55. Locating sleeve | 72. Adjusting nut | (4WD) |
| 3. Bearing assy. | 22. Seal | 38. Bearing assy. | 56. Reverse gear | 73. Bearing assy. | 86. 1st & 3rd gear |
| 4. Support | 23. "O" ring | 39. "O" ring | 57. Lock nut | 74. Bearing cage | 87. Coupling |
| 7. Shims | 24. Bearing assy. | 40. Snap ring | 58. Thrust washers | 75. "O" ring | 88. Drive collar |
| 8. Bearing assy. | 25. "O" ring | 41. Nut and seal | 59. Bearings | 76. Shims | 89. 2nd & 5th gear |
| 9. Driven gear | 26. Input shaft | 42. Seal | 60. Reverse idler gear | 77. Mounting sleeve | 90. Mounting sleeve |
| 10. Drive gear | 27. Snap ring | 43. Cover | 61. Idler gear shaft | (4WD) | 91. Spacer |
| 11. Bearing assy. | 28. Thrust washer | 46. Shims | 62. Cap screw | 78. Nut (4WD) | 92. 4th & 6th gear |
| 12. Snap ring | 29. High range gear | 47. Bearing assy. | 63. Self-locking nut | 79. Bearing assy. | 93. Reverse gear |
| 13. Spacer | 30. Coupling | 48. Countershaft | 64. Plug | (4WD) | 94. Mounting sleeve |
| 14. Spanner nut | 31. Drive collar | 49. High range gear | 65. Speedometer drive | 80. "O" ring (4WD) | 95. Snap ring |
| 15. Sprocket | 32. Sealing ring | 50. 2nd & 5th gear | housing | 81. "O" ring (4WD) | 96. Bearing assy. |
| 16. Plug (no pto or | 33. Low range gear | 51. Low range gear | 66. Plug | 82. Bearing cage (4WD) | 97. Pinion shaft (2WD) |
| hyd.) | 34. Spacer | 52. 4th & 6th gear | 69. Gasket | 83. Snap ring (4WD) | 98. Pinion shaft (4WD) |
| 17. Seal | 35. Pin | 53. Bearing assy. | 70. "O" ring | 84. Spacer (selective) | |
| | | | | (4WD) | |

the pto drive shaft as in paragraph 180, or the hydraulic pump drive shaft as in paragraph 181. Disconnect master link and remove transmission lubrication pump drive chain. Remove hydraulic pump and support as in paragraph 177. Use a spanner wrench (Oliver No. ST-202 or equivalent) to hold spanner nut (14—Fig. 108 or 39—Fig. 108A) and unscrew sprocket from shaft. Use Oliver tool No. STS-152 or STS-186 adapter to prevent shaft from turning and remove spanner nut. Remove gear and bearings. Bearing cups, spacer and snap ring can now be removed from rear main frame if necessary.

Clean and inspect all parts and renew as necessary. If drive gear (10 or 32) is renewed, also renew driven gear (9 or 31).

To reinstall pump drive gear, install snap ring (12 or 35) spacer (13 or 36) and bearing cups in rear main frame with tapers of cups facing away from each other. Install front bearing on shaft with taper facing rearward, place shaft and bearing in position and install rear bearing with taper facing forward. Install spanner nut and tighten until shaft has a rolling torque of 3-5 inch pounds. Hold nut in this position and install sprocket (15 or 40). Recheck the shaft rolling torque after sprocket has been tightened.

Reinstall remaining items by reversing the removal procedure.

**183. R&R TRANSMISSION TOP COVER.** To remove transmission top cover, remove gear shift lever (2—Fig. 106), spring cover (4), gasket (7) and the two lever springs (8) from top cover (9). Remove platform section and insulation. Disconnect the wiring and hydraulic lines which are routed over top cover. Disconnect the starter safety switch wires, then unbolt and remove top cover and gasket.

Interlock (13), safety switch (17), switch bracket, wiring and wiring seal (25) can now be removed if necessary and procedure for doing so is obvious.

Before reinstalling top cover, check adjustment of safety switch and if necessary, readjust position of switch until it opens when roller is about one-half way up side of interlock ramp. Switch has an audible click when it opens and closes.

Use new gasket and reinstall top cover by reversing removal procedure.

**184. SHIFTER RAILS AND FORKS.** Removal of shifter rails on models equipped with auxiliary drive will require removal of engine or splitting tractor.

Remove transmission top cover, as outlined in paragraph 183. Remove hydraulic lift housing as outlined in paragraph 259. Remove pinion shaft

external oil line. Remove cap screws which retain shift forks to shifter rails. Push the two outer rails forward and push out the expansion plugs located at forward ends of shifter rails. Be careful not to lose poppet balls and springs as rails clear rear bores. Prior to removing the center shifter rail, remove shifter lug (41—Fig. 106). Lift out shifter forks after shift rails are removed.

Reinstall shift rails and forks as follows: Place center fork in case, place poppet spring in center poppet hole and place poppet ball on top of spring. Insert center shift rail part way into rear frame. Turn shifter rail so flat on inner end is downward, then use a small punch to depress poppet ball and

spring and push shifter rail rearward until rail goes over poppet ball. Turn shifter rail until poppet ball seats in detent, then install shifter lug and fork. Install right and left shifter rails and forks in the same manner. Use sealing compound and install new expansion plugs in shifter rail front bores.

Complete balance of reassembly by reversing disassembly procedure.

**185. LUBRICATION BY-PASS VALVE.** The by-pass valve for the transmission lubrication circuit is located in a bore on left side of rear main frame directly above oil filter. Valve is

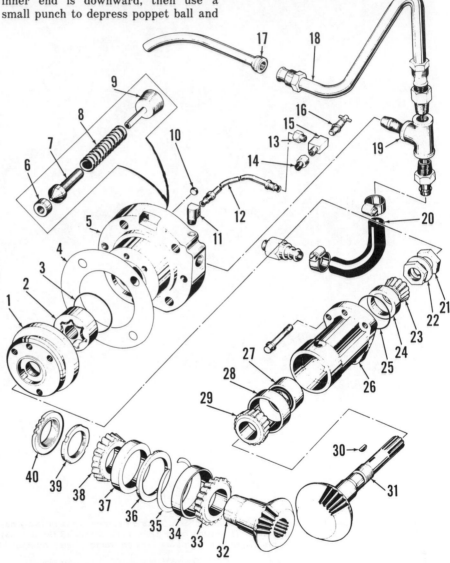

**Fig. 108A—Tractors beginning with serial number 248 722- have the hydraulic oil cooling pump incorporated into the hydraulic pump driven gear support. Refer to Fig. 108 for a view of the support used on earlier tractors.**

| | | |
|---|---|---|
| 1. Pump housing | 11. Elbow | 21. Nut | 31. Driven shaft |
| 2. Gerotor assembly | 12. Bleed tube | 22. Lock washer | 32. Drive shaft |
| 3. "O" ring | 13. Connector | 23. Bearing cone | 33. Bearing cone |
| 4. Shim | 14. Adapter | 24. Bearing cup | 34. Bearing cup |
| 5. Support (outer) | 15. Tee | 25. "O" ring | 35. Snap ring |
| 6. Relief valve seat | 16. Bleeder valve | 26. Support (inner) | 36. Spacer |
| 7. Relief valve plunger | 17. Sump tube | 27. Oil seal | 37. Bearing cup |
| 8. Relief valve spring | 18. Inlet tube | 28. Bearing cup | 38. Bearing cone |
| 9. Plug assembly | 19. Tee | 29. Bearing cone | 39. Nut |
| 10. Plug | 20. Hose | 30. Woodruff key | 40. Sprocket |

accessible after removing the hydraulic lift system housing. See Fig. 105.

Valve spring and plunger can be removed after removing expansion plug. Inspect plunger for damage or wear. Valve spring has a free length of 2-15/32-inches and should test 1.7-2.3 pounds when compressed to a length of 1-25/32-inches. Valve will maintain a lubrication pressure of 8-14 psi at an engine rpm of 2400.

Lubrication pressure can be checked at a port located directly below oil filter element. Check pressure in both neutral positions of transmission. Refer to paragraph 176 if pressure is not within specifications.

186. **LUBRICATION PUMP.** To remove the lubrication pump, first remove the hydraulic lift housing. Disconnect master link and remove drive chain. Remove the pump to filter line and the pump suction tube. Use a small punch or rod to depress lock and withdraw pump from bore in rear main frame.

NOTE: In some cases it is necessary to remove plug in bottom of rear frame before suction tube can be removed.

With pump removed, remove front snap ring (1—Fig. 107) and remove sprocket and Woodruff key (2). Remove cap screws and separate cover (4) from housing (10). Remove rear snap ring and remove shaft (7), impeller (8) and rotor (9) from housing. If necessary, remove dowels and inserts from cover.

Clean and inspect all parts. Impeller should be a free fit on shaft, and if necessary, polish shaft with crocus cloth or file drive key. Renew any parts which show undue wear or any other damage.

When reassembling pump, coat all parts with oil and place rotor in housing. Install key and impeller on shaft, insert assembly into housing and install rear snap ring. If necessary, install dowels and inserts in cover, slide cover over shaft and secure to housing. Install sprocket key, sprocket and front snap ring. Place spring and lock in bore of housing and install pump in bore of rear main frame. Complete reassembly by reversing the disassembly procedure.

187. **TRACTOR SPLIT.** To split tractor for complete service on transmission it is recommended that the fuel tank and instrument panel assembly be removed rather than the engine.

Make tractor split as follows: Drain transmission and final drive compartments and remove pto shaft, or the hydraulic pump drive shaft if tractor has no pto. Remove hood, hood side panels and fuel tank. Remove platforms, then disconnect wiring from engine, starter safety switch and tail lights. Disconnect engine controls, clutch rod and auxiliary drive unit cable. Disconnect the necessary hydraulic lines, then unbolt and remove instrument panel.

Disconnect hydraulic lines attached to hydraulic lift unit. Remove any front frame weights which may be installed, then support front frame so it will not tilt. On Model 2255, also support rear of engine. Attach hoist to rear frame, then remove the front to rear frame cap screws and separate front and rear frames.

When reassembling, tighten the front to rear frame cap screws to 280-300 ft.-lbs. torque. Bleed power steering and diesel fuel systems.

188. **R&R INPUT SHAFT.** To remove input shaft, split tractor as outlined in paragraph 187. Drain rear main frame, then refer to the appropriate sections and remove pto drive shaft or the hydraulic pump drive shaft, transmission top cover and hydraulic lift system housing. Remove the hydraulic pump drive gear and input shaft oil collector oil lines. In some cases, it may be desirable to remove pinion shaft oil lines and fittings. Remove the hydraulic pump, pump support and pump drive gear.

Remove support (18—Fig. 108), then remove oil seal (22) and "O" ring (23) from retainer. Save shims (21) for reinstallation. Unstake nut (41) from rear of input shaft and remove nut, seal (42) and rear bearing cup snap ring (40). Slide coupling (30) forward onto gear (29), then lift front snap ring (27) from its groove and slide it forward over input shaft. Use a shaft driver (Oliver No. ST-138, or equivalent) and bump input shaft rearward until gear (33) and drive collar (31) can be separated far enough to expose Teflon ring (32) and snap ring (27). Cut the Teflon ring, lift rear snap ring from its groove and slide it rearward on input shaft. Place a spacer between front input shaft gear and housing so gears cannot move forward, then using shaft driver, bump input shaft forward until front bearing cup is free of housing and input shaft is free of rear bearing.

NOTE: Beginning with tractor serial number 248 925-, a spacer is used with bearing (38) which is not shown in Fig. 108.

Pull shaft from front of housing and lift shaft components from top of housing as shaft is being removed.

NOTE: The spacer used to block the gears can be made from a piece of 3½-inch I.D. pipe, 1½-inches long with a longitudinal section cut out to permit it to slip over the input shaft.

189. Reinstall input shaft assembly as follows: Press front bearing cone on input shaft with smaller diameter forward. Place front snap ring (27) on shaft about ½-inch forward of its groove, then place thrust washer (28) next to the snap ring. Place seal rings (36) in grooves of spacer (34), oil seal rings freely, then install spacer and rings in oil collector (37) from flat side of oil collector using fingers to compress sealing rings. Push center shift rail to its forward position. Place shift coupling (30) in its hub of high range gear (29), start rear of input shaft in front bearing bore and install gear and collar on shaft with collar rearward and mated with its shifter fork. Install drive collar (31) with counterbore rearward. Start rear snap ring (27) on shaft spline with sharp ends forward but do not slide it into its groove. Place Teflon seal ring (32) on shaft at rear of snap ring (27). Install low range gear (33) on shaft with helical teeth toward rear. Place pin (35) in spacer, align it with keyway of input shaft and install the oil collector and spacer assembly on shaft with counterbored side of oil collector forward. Install the rear snap ring (27) in its groove and be sure Teflon sealing ring (32) is on splined portion of shaft. Install front bearing cup in its bore, install support (18) (without seal and "O" ring) and shims and tighten retaining cap screws finger tight. Lubricate "O" ring (39) and install in groove next to threads on rear end of shaft and on later models, place spacer on shaft. Start rear bearing on shaft with smaller diameter rearward and bump bearing on shaft until nut seal (42) and nut (41) can be started. Be sure large diameter flange of nut is toward bearing, tighten nut until bearing is tight against spacer (34), then complete tightening of nut to a torque of 90-100 ft.-lbs. Stake nut into keyway of shaft.

NOTE: While tightening nut be sure sealing ring (32) stays on splined portion of input shaft or damage to sealing ring will occur.

Install rear bearing cup and snap ring (40), then bump on forward end of shaft to seat cup against snap ring. Slide drive collar (31), high range input gear (29) and thrust washer (28) rearward and install front snap ring (27) in its groove.

Further tighten the support retainer cap screws, if necessary, until input shaft has only a slight end play. Demesh input shaft gears from countershaft gears, then using an inch pound torque wrench and adapter, check and record the rotational resistance of the input shaft. Completely tighten the bearing retainer cap screws and re-check the shaft rotational drag which should be not more than 5 inch pounds greater than the previously recorded value. Vary shims (21) as required to obtain this adjustment. Shims are available in thicknesses of 0.004, 0.007, 0.0149.

**NOTE: If shims are varied be sure shaft is bumped slightly forward after each check to insure that bearing cup contacts flange of bearing retainer.**

With input shaft adjustment made, remove support and install oil seal (22) with lip rearward and "O" ring (23). Lubricate oil seal and "O" ring and install support. Install input shaft oil line between oil collector and tee and be sure oil line does not cause oil collector to bind. Install any other oil lines that were removed. Shift transmission to neutral.

190. **R&R COUNTERSHAFT.** To remove the countershaft, first remove input shaft as outlined in paragraph 188.

Remove bearing retainer (43—Fig. 108) and shims (46). On four-wheel drive models, remove bevel pinion shaft oil line and the transfer case breather and neck. Unstake and remove nut (57). Use two 3/8 x 1-1/8-inch cap screws, fitted with nuts, between gear (49) and front wall of transmission housing to prevent loading of gears during shaft removal.

Thread screw of puller (Owatonna No. 927, or equivalent) into tapped hole in front of countershaft, then making sure gears are not binding, pull countershaft from rear bearing and front bearing cup from main frame. Complete removal of countershaft and lift shaft components from top of main frame as shaft is withdrawn. Remove front bearing from countershaft and rear bearing cup and snap ring from housing, if necessary.

Reinstall by reversing removal procedure and tighten nut (new) to a torque of 150 ft.-lbs. Vary shims (46) to provide countershaft with 0.001-0.003 end play. Shims are available in thicknesses of 0.004, 0.007 and 0.0149.

191. **R&R BEVEL PINION SHAFT.** Bevel pinion shaft and bearing cage assembly used in two-wheel drive tractors differ somewhat from those used in four-wheel drive tractors and can be noted in Fig. 108.

Refer to the appropriate following paragraphs.

192. **TWO-WHEEL DRIVE.** To remove bevel pinion shaft, remove input shaft, countershaft and the differential as outlined in paragraph 195.

**NOTE: While it is not mandatory that shifter rails and forks be removed, their removal will simplify procedure.**

Remove bevel pinion shaft oil line, then remove speedometer drive housing (65—Fig. 108) and gasket (69). Remove oil seal (70) from speedometer housing. Wrap speedometer drive gear and seal surface on end of bevel pinion shaft with tape to prevent damage during disassembly.

Use spanner wrenches (Oliver ST-127 and a modified ST-142) as shown in Fig. 109 and remove lock nut (71—Fig. 108) and adjusting nut (72). Install two 3/8-16 cap screws in tapped holes of bevel pinion shaft front bearing cage (74) and pull bearing cage and front bearing assembly (73). Save shims (76) for subsequent reinstallation.

**NOTE: Do not hammer on front end of bevel pinion shaft as shaft will be damaged.**

Remove "O" ring (75), and if necessary, the front bearing cup from bearing cage. Slide bevel pinion shaft rearward and remove shaft components from top of main frame as shaft is withdrawn. Remove rear bearing from bevel pinion shaft and rear bearing cup and snap ring from main frame.

193. Reinstall bevel pinion shaft assembly by reversing removal procedure. However, before installing speedometer drive housing, adjust shaft bearing preload and gear end play as follows. Use short cap screws to temporarily secure front bearing cage to housing. Block rear end of bevel pinion shaft and drive front bearing into position. Install adjusting nut, tighten nut until a slight amount of end play remains, then bump shaft both ways to seat bearings and bearing cups. Slide front gear coupling rearward on to gear (89) and place rear coupling (87) in neutral position. Tighten adjusting nut until bevel pinion shaft has a rolling torque of 4-6 lbs. measured with a spring scale. Spring scale is attached to center output gear (89—Fig. 108) with a length of cord to make shaft rolling torque adjustment. If adjusting nut tightens before correct

preload is obtained, add an equal amount of shims (76) behind flanges of bearing retainer (74) until the correct bearing preload can be obtained. Shims are 0.0149 thick.

With bearing preload adjusted, mount a dial indicator so a reading can be taken from spacer (91), then move gears back and forth by prying behind gears (86 and 93) and check indicator which should show an end play of 0.001-0.017. If sleeve end play is incorrect, vary shims (76) behind bearing retainer as required and reset the shaft bearing preload.

With the preceding adjustments made, loosen adjusting lock nut about ¼-turn, install lock nut (71) and while holding adjusting nut, tighten lock nut to a minimum of 150 ft.-lbs. Recheck the bearing preload, and if necessary, readjust nut (72). Stake lock nut to bevel pinion shaft when adjustment is complete.

193A. **FOUR-WHEEL DRIVE.** Four-wheel drive tractors use a bevel pinion shaft which is similar, except for front bearing cage assembly (items 78 through 85), to the bevel pinion shaft used in two-wheel drive tractors.

To remove bevel pinion shaft, remove transfer drive assembly as in paragraph 25), input shaft, countershaft and the differential as in paragraph 195. Tape seal contact surface of bevel pinion shaft to prevent damage to shaft.

**NOTE: While it is not mandatory that the shifter rails and forks be removed, their removal will simplify procedure.**

Unstake nut (78—Fig. 108), then using a driving sleeve over oil tube on front of shaft, bump shaft (98) rearward and remove front bearing, bearing cage (82) and spacer (84). Pull shaft rearward and lift gears and spacers from housing as shaft is withdrawn. Snap rings (83 and 95) and rear bearing cup can be removed from housing, if necessary.

Reinstall by reversing the removal procedure, however prior to complete assembly the bevel pinion shaft bearing preload and shaft gear end play must be adjusted as follows: Use the original spacer (84) and tighten nut (78) until only a slight end play remains, then bump shaft both ways to seat bearings and cups. Slide front shift collar (87) rearward on to gear (89) and place rear shift collar (87) in neutral position. Wrap a cord around center output gear (89) and attach a spring scale to cord. Tighten nut (78) until a spring scale reading of 4-6 lbs. is required to keep shaft in motion. If nut tightens before correct bearing preload is reached, remove nut, front bearing and spacer

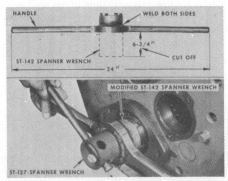

*Fig. 109—Modify an ST-142 spanner wrench as shown for use in adjusting bevel pinion shaft.*

(84) and install next smaller spacer. Spacers are available in thicknesses of 0.288, 0.314 and 0.340.

With bevel pinion bearing preload adjusted, mount a dial indicator so reading can be taken from spacer (91), then move gears back and forth by prying behind gears (86 and 93) and check indicator which should show an end play of 0.001-0.027. If end play of gears on shaft is not as stated, change spacer (84) as required and readjust bevel pinion shaft bearing preload. Stake nut into slot of vebel pinion shaft when preceding adjustments are correct.

**194. R&R REVERSE IDLER GEAR.** To remove reverse idler gear, first remove countershaft and bevel pinion shaft. Remove nut and cap screw (62—Fig. 108) which retains idler gear shaft in housing, then remove shaft, gear (60), bearings (59) and thrust washers (58).

# DIFFERENTIAL, BEVEL GEARS, FINAL DRIVE & REAR AXLE

## DIFFERENTIAL

### Models G955-G1355-2255

**195. REMOVE AND REINSTALL.** To remove the differential, first drain transmission and final drive compartments. Remove the pto drive shaft, or the hydraulic pump drive shaft if tractor has no pto. Remove the hydraulic lift assembly. Remove drawbar support and pto control valve linkage, then remove pto assembly from main frame. Disconnect wiring and remove both fenders, then support rear frame and remove both wheel and tire units. On either side of tractor, remove snap ring from inner end of axle and the cap screws retaining axle carrier to main frame: then using a pusher, push axle from bull gear and remove axle and carrier. Repeat operation on opposite side of tractor using a piece of bar stock to support pusher. Lift bull gears from rear frame.

Disconnect hand brake linkage at brake housing cover and remove hydraulic line at fitting. Remove brake housing cover and insulating disc. Pull adjusting disc, brake discs, separators and the four springs from housing.

Remove cap screws from brake housings and wedge them out far enough to remove the two sets of shim packs from behind each brake housing. Identify shims as to their locations and tie them together to prevent intermixing. Install a lifting eye in the tapped hole in one of the differential pinion pins, attach a hoist to lifting eye and take slack from hoist. Remove brake housings and bull pinions. Remove cap screws which retain inner supports

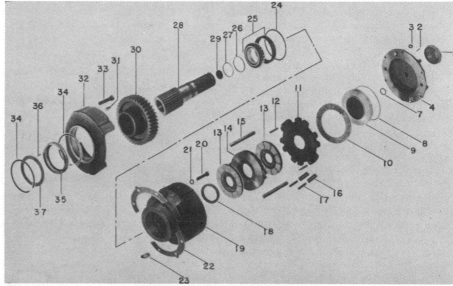
Fig. 111—Exploded view of brakes, brake housings and bull pinions. Supports (32) are doweled to inner sides of rear main frame.

| | | | |
|---|---|---|---|
| 1. Cap | 11. Adjuster disc | 19. Brake housing | 29. Cup plug |
| 2. Bleed screw | 12. Roll pin | 22. Shims | 30. Pinion gear |
| 3. Seal | 13. Brake disc | 23. Dowel | 31. Dowel |
| 4. Brake cover | 14. Plate | 24. "O" ring | 32. Support |
| 7. "O" ring | 15. Torque pin | 25. Bearing assy. | 34. Snap ring |
| 8. "O" ring (piston) | 16. Release spring | 26. Teflon ring | 35. Bearing race |
| 9. Piston | 17. Pin | 27. "O" ring | 36. Rollers |
| 10. Insulator | 18. Oil seal | 28. Pinion shaft | 37. Retainers |

(32—Fig. 111), pry left support off dowels (31) and turn support 90 degrees. Pry right support off dowels and lift differential and supports from rear frame. Hold bull pinion gears (30) against supports and remove supports and gears from differential shafts. Remove thrust bearings (2—Fig. 110) and thrust washers (3). Identify all parts as to right and left.

**196.** If any of the loose roller bearings drop out of inner support

Fig. 110—View component parts of differential assembly. Pinion pin (7) has a pilot on inner end which seats in a bore in spider shaft.
1. Side gear
2. Thrust bearing
3. Thrust washer
4. Bevel gear
5. Pinion gear
6. Spider and shaft
7. Pinion pin
8. Retaining ring
9. Pinion pin

refer to paragraph 207. Reinstall differential by reversing removal procedure and adjust differential bearings and bevel gear backlash as outlined in paragraph 197.

**197. BEARINGS ADJUST.** Differential end play and gear backlash are controlled by the two shim packs (22—Fig. 111) interposed between the two brake housings and rear main frame. When adjusting bearings, do not depend on shim count or prior assembly. Use a micrometer to measure shim pack thickness to be sure that both shim packs installed between brake housing and rear frame are identical in thickness. Shims are available in thicknesses of 0.004, 0.007 and 0.0149.

Mount a dial indicator and check the differential end play which should be 0.001-0.003. Add or subtract shims (22) as necessary to obtain the specified end play, however, BE SURE to maintain some gear backlash during this operation.

With differential end play adjusted, use the dial indicator to check backlash between bevel pinion and bevel gear at four points around drive gear; any large differences in backlash readings will indicate runout of drive gear due to improper tightening of gear mounting bolts or foreign material between gear and spider. Average the four readings, if no large differences exist, to determine gear backlash. Transfer shims, in equal amounts, from one brake housing to the other until the average of the four dial indicator readings agrees with the backlash specifications etched on the bevel drive (ring) gear.

Mesh position of bevel gears is fixed and nonadjustable.

198. **OVERHAUL.** Remove differential as outlined in paragraph 195. Remove pinion pin retaining rings (8—Fig. 110), then with puller Oliver No. STS-100 or equivalent, threaded into tapped hole of pinion pin, pull pins and remove pinions (5). Be sure to note location of pinion pin (7) so it can be reinstalled in same location. Differential spider shaft is drilled to accept the pilot on end of pinion pin (7).

When reassembling, install pinions with taper facing inward and tapped hole of pinion pins facing outward. When installing bevel drive gear on spider be sure there is no foreign material between gear and spider and install bolts with heads next to bevel drive gear. Tighten nuts alternately and evenly to a torque of 77-82 ft.-lbs. (oiled).

### BEVEL GEARS

#### Models G955-G1355-2255

199. Data for removal and overhaul of bevel pinion shaft is given in paragraph 191 and for differential in paragraph 198. Adjust differential backlash as outlined in paragraph 197.

### FINAL DRIVE AND REAR AXLES

#### Models G955-G1355-2255

200. For the purposes of this section, the final drive will include those components of the drive line which follow the differential. That is, the bull pinion shafts, bull gears and the wheel axle shafts. Due to their inter-relationship, they will be treated in the order of disassembly.

201. **WHEEL AXLE SHAFTS.** To remove wheel axle, proceed as follows: Drain final drive compartment and remove pto drive shaft, or the hydraulic pump drive shaft. Remove pto unit and snap ring from inner end of axle. Drain planetary unit, remove

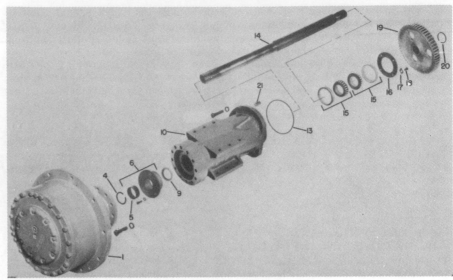

Fig. 112—Exploded view showing component parts of rear axle assembly. Refer to Fig. 113 for exploded view of planetary unit (1).

| | | | |
|---|---|---|---|
| 1. Planetary unit | | | |
| 4. Snap ring | 9. Oil seal | 14. Rear axle | 17. Lock |
| 5. Bushing | 10. Axle carrier | 15. Bearing assy. | 19. Bull gear |
| 6. Bushing support | 13. "O" ring | 16. Retainer | 20. Snap ring |

fender and tire and wheel and hub. Remove cover (30—Fig. 113) from planet carrier (16), then remove snap ring (21) and gear (22) from outer end of axle. Support planetary unit with a hoist, unbolt spindle (1) and pull planetary unit from carrier. Attach hoist to axle carrier, remove snap ring from inner end of rear axle, (if not done), unbolt carrier from rear frame, then using pusher, Oliver STS-147, or equivalent, push axle from bull gear and remove axle and carrier from tractor.

**NOTE: If left axle carrier is being removed, remove draft control linkage.**

To remove axle from carrier, remove snap ring (4—Fig. 112) from outer end of axle, then remove inner bearing retainer (16). Push axle out toward

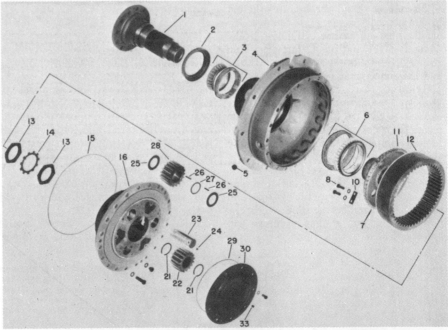

Fig. 113—Exploded view of the planetary unit. Spindle (1) is bolted to outer end of axle carrier.

| | | |
|---|---|---|
| 1. Spindle | 10. Plate | 16. Planet carrier | 26. Roller |
| 2. Oil seal | 11. Ring gear hub | 21. Snap ring | 27. Spacer |
| 3. Bearing assy. | 12. Ring gear | 22. Drive gear | 28. Planet pinion gear |
| 4. Hub | 13. Nut | 23. Pinion shaft | 29. "O" ring |
| 5. Plug | 14. Lock washer | 24. Retainer ball | 30. Cover |
| 6. Bearing assy. | 15. "O" ring | 25. Washer | 33. Breather |
| 8. Cap screw | | | |

inside of carrier which will also remove the inner bearing cup. Remove remaining inner bearing cup and the bearing support (6), bushing (5) and seal (9) assembly. Remove and discard seal (9). Do not remove bushing (5) from support unless renewal is required. Remove bearing cones from axle, if necessary.

When reassembling, install oil seal (9) in support (6) with lip of seal facing toward inner end of carrier. Inner bearings are installed on axle with tapers facing away from each other. Complete reassembly by reversing the disassembly procedure.

**202. AXLE CARRIERS.** To remove axle carriers, first drain final drive compartment and remove pto drive shaft, or the hydraulic pump drive shaft. Remove pto unit and hydraulic lift unit. Remove pto oil supply line. Remove fenders, support tractor and remove tires and wheels and hubs. Remove draft control signal arm. Remove pto actuator levers and shaft.

**203.** Refer to paragraph 201 to remove axle and carrier. Lift bull gear from rear frame, if necessary.

**204. OVERHAUL PLANETARY DRIVE.** Remove and overhaul planetary drive as follows: Remove tire and wheel and drain unit. Remove cover (30—Fig. 113) and discard "O" ring (29). Remove outer snap ring (21) from end of axle shaft and pull drive gear (22) from axle. Support planetary drive with a hoist, unbolt spindle (1) from axle carrier and pull planetary drive unit from carrier. Unbolt planet carrier (16) from hub (4) and discard "O" ring (15). Remove nuts (13) and lock (tang) washer and separate spindle and ring gear assembly from hub (4). Remove seal (2), inner bearing assembly (3) and outer bearing cup from hub. Cut lockwire and remove cap screws (8) and plates (10) and separate ring gear hub (11) from ring gear (12). Remove bearing from ring gear hub. Drive pinion shafts (23) out toward outside of planet carrier (16) and remove planet gears (28), rollers (26), spacer (27) and washers (25). Do not lose retaining balls (24).

Clean and inspect all parts and renew as necessary. Refer to paragraph 205 for assembly information.

**205.** Assemble planetary drive unit as follows: Place a washer (25) on a flat surface with raised portion of washer on bottom side. Lubricate bore of a planet gear (28) and place gear on top of washer. Insert pinion shaft (23) in bore of gear and install a row of rollers (26). Place spacer (27) on top of rollers and install second row of rollers. Place second washer (25) on top of second row of rollers with raised portion of

washer on top side. Align raised portions of washers, carefully withdraw pinion shaft, then install planet gear and roller assembly in carrier (16) with raised portions of washers (25) aligned with grooves in carrier. Place retaining ball (24) in hole in pinion shaft. Start plain end of pinion shaft into carrier and planet gear, align retaining ball with slot in carrier, then press pinion shaft into carrier until retaining ball seats. Stake face of carrier to secure pinion pin in place. Install remaining two pinions in same manner.

Place bearing cups in hub (4) with tapers facing away from each other. Place inner bearing cone in bearing cup and install new seal (2) with lip facing toward bearing. Press outer bearing cone on ring gear hub (11) with taper facing away from hub. Place hub in ring gear (12) and secure with plates (10) and cap screws (8). Secure cap screws in pairs with lockwire. Stand spindle (1) on flange end, place hub (4) over spindle, then install ring gear and hub assembly. Install adjusting nut (13) and tighten nut until it requires 9-17 lbs. (21-28 lbs. with new bearings) pull on a spring scale (See Fig. 114) to keep hub in motion. Install lock (tang) washer and remaining nut (13). Tighten nut and secure adjusting nut and lock nut with tangs of lock washer.

Place new "O" ring (15) around carrier (16) and install carrier in hub (4). Mount planetary drive unit on axle carrier, install drive gear (22) and outer snap ring (21) on end of axle, then use new "O" ring (29) and install cover (30).

Use SAE-80 Multi-Purpose gear lubricant to fill planetary drive unit. Model 2255 units hold 8 quarts. Model G955 or G1355 units hold 6 quarts.

**206. BULL GEARS.** To remove bull gears, first drain hydraulic lift system and tractor rear frame. Remove axle shafts as outlined in paragraph 201. Bull gears can now be lifted from rear main frame.

**207. BULL PINIONS.** To remove the bull pinions, proceed as outlined in paragraph 195.

With brake housing, bull pinion and inner supports removed, disassemble bull pinions and inner supports as follows:

Remove bearing from bull pinion shaft. Remove cup plug (29—Fig. 111) if it shows any signs of leakage. Remove Teflon rings (26) and "O" rings (27) from left bull pinion. Remove oil seal (18), bearing cup and "O" ring (24) from brake housing. Discard "O" ring.

Place gear and support assembly gear side down on a flat surface. Remove upper snap ring (34) and retainer (37), then lift support (32) off gear (30). Be sure not to lose any rollers (36) as they will fall out when support and gear are separated. Remove bottom snap ring (34) and press bearing race (35) from support.

To reassemble bull pinion and inner support assemblies, proceed as follows: Install lower (inner) snap ring (34) in support, then with lip of race (35) toward snap ring, press bearing race into support snug against snap ring. Set bull pinion gear on a flat surface, hub end up, and place support over gear. Install rollers (36) and retainer (37) and secure retainer with upper snap ring (34). Be sure to keep gear against support during any subsequent handling to prevent rollers from being displaced. Press outer bearing on bull pinion shaft with taper toward outside and install new cup plug (29), if necessary. On left bull pinion only, install new "O" rings in grooves and then install Teflon rings on top of the "O" rings. Press bearing cup in brake housing with taper toward inside. Install oil seal in housing with lip toward inside.

Install new "O" ring (24) on brake housing and reinstall bull pinion assemblies by reversing removal procedure.

# BRAKES

The brakes, which are mounted on outer ends of bull pinions, are a hydraulically actuated disc type and utilize three brake discs and two sepa-

*Fig. 114—Attach cord to wheel stud to check bearing preload.*

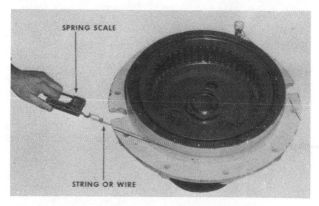

rators. Service brakes are self-adjusting, however, adjustments for pedal free travel and the hand brake can be made.

## ADJUSTMENT

### Models G955-G1355-2255

208. **PEDAL FREE TRAVEL.** Brake pedal free travel is set at factory and should remain constant. However, should it become necessary to adjust pedal free travel, proceed as follows: Remove hood side panels, hood and fuel tank. Refer to Fig. 115 and disconnect rods at spool connection. Loosen rod clevis jam nut and turn rods as necessary to obtain a pedal free travel of 1¼ inches.

209. **HAND BRAKE.** To adjust hand brake, refer to Fig. 116 and remove cap from brake housing cover. Loosen adjusting screw lock nuts and back-out adjusting screws slightly to relieve tension on linkage. Position equalizing bar parallel to the centerline of cross-shaft, then hand tighten the adjusting screws. Operate hand brake lever several times, recheck adjusting screws and retighten if necessary. Be sure equalizing bar is parallel with centerline of cross-shaft when retightening adjusting screws. Hand brake adjustment is correct when lever travel is restricted to 3 or 4 notches on latch and equalizing bar is parallel with cross-shaft.

Fig. 115—Brake valve and control linkage are accessible after removing fuel tank.

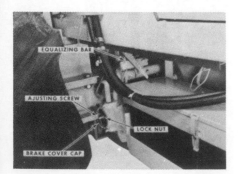

Fig. 116—To adjust hand brake, remove brake housing cap as shown. Refer to text for procedure.

Fig. 117—View showing brake components. Brakes are fitted with three discs (12) and two separators (13) instead of the number shown.

1. Cap
2. Bleed screw
3. Cover
6. Seal
7. "O" ring
8. "O" ring
9. Piston
10. Insulator
11. Adjusting plate
11A. Roll pin
12. Disc
13. Separator
14. Pin (dowel)
15. Spring
16. Spring pin

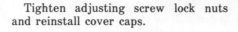

Tighten adjusting screw lock nuts and reinstall cover caps.

## OVERHAUL

### Models G955-G1355-2255

210. **SERVICE BRAKES.** To remove service brake assembly, disconnect hand brake linkage at brake housing cover and hydraulic line at fitting. Refer to Fig. 117 and remove housing cover (3) and insulator (10). Remove adjusting plate (11), brake discs (12), separators (13) and the four springs (15) from brake housing.

If necessary, remove spring clips from adjusting plate and remove the self-adjuster screws from adjusting plate. Remove cap from housing cover, loosen adjusting screw lock nut and turn adjusting screw in as far as possible. Actuate hand brake lever and force brake piston (9) from cover. Remove "O" rings (7 and 8) from cover.

To reassemble brakes, screw self-adjusting screws into adjuster plate, install spring clips, then back-out adjuster screws one or two notches. Install "O" rings in brake housing cover, lubricate piston and push piston into

cover. Place return springs over locating pins. Start with brake disc and alternately install brake disc and separators. Separators are installed with taper inward and notches aligned on dowels. Install adjusting plate with notches over dowels and self-adjuster screw tips between springs. Place insulator on piston, then install brake housing cover over dowels and BE SURE insulator does not slip out of place. Secure cover and connect hydraulic line and hand brake linkage.

Start engine and hold brake pedals down to fill brake assemblies with oil. Continue to hold brake pedals down, then open bleed screw until air free oil flows from port, then close bleed screw.

211. **HAND BRAKE CROSS-SHAFT.** Hand brake cross-shaft can be withdrawn after driving groove pin from hub on left end of shaft. If necessary, pry out shaft seals and install new seals with lips facing toward inside.

212. **BRAKE VALVE.** To remove brake valve, remove hood side panels, hood and fuel tank. See Fig. 115. Disconnect rods at valve spools, then remove linkage to allow access to valve. Disconnect inlet line, return line and

Fig. 118—Exploded view of hydraulic brake valve. Refer to Fig. 115 for mounting location.

1. Cap
2. Cap screw
3. Gasket
4. Screw
5. Retainer
6. Reaction spring
7. Piston, small
8. Snap ring
9. Spool
10. Spring seat
11. Spring
12. "O" ring
13. Piston, large
14. Spring
15. Body
16. "O" ring
17. Orifice plate
18. Spring
19. Ball (5/16)
20. "O" ring
21. Fitting

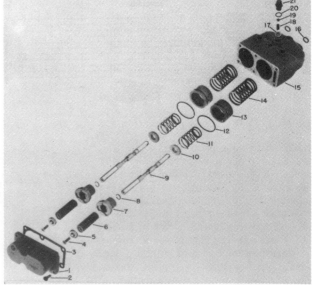

pressure (output) lines, then remove the two mounting cap screws and lift out valve.

213. With valve removed, remove the cap screws, cap (1—Fig. 118) and gasket (3). Identify spool assemblies as to their bores and remove spool assemblies from valve body. Remove large pistons (13), piston springs (14) and return springs (11) from valve body, then remove "O" ring (12) from pistons. Use a punch in hole at linkage end of spool to hold spool, then remove screw (4) at opposite end of spool.

**NOTE: Screw is installed with Loctite.**

Remove retainer (5), reaction spring (6), small piston (7) and spring seat (10). Remove "O" ring (not shown) from bore of small piston (7). Remove

fitting (21), ball (19), spring (18) and orifice plate (17) from inlet port. Remove "O" ring (20) from fitting and "O" rings (16) from spool bores.

214. Discard all "O" rings, clean and inspect all parts and renew as necessary. If using used parts during reassembly, lap the large and small pistons to each other as follows: Place a small amount of lapping compound on lip of small piston, install large piston over small piston, then with assembly held between palms of hands, use a twisting motion of the hands to lap pistons. Thoroughly clean parts after lapping.

**NOTE: New pistons will not require lapping compound, however, pistons should be mated by using the twisting motion already described.**

215. Reassemble brake valve as fol-

lows: Install small snap rings (8) if removed. Install spring seat (10), small piston (7) with new "O" ring in bore, reaction spring (6) and retainer (5) in spool, then using Loctite (grade AV) on screw (4), install screw and tighten to 5-8 ft.-lbs. of torque.

Slide large piston (13) over linkage end of spool and be sure lapped surface mates with lip of small piston. Install return spring and piston spring. Install new spool "O" ring (16) in body, then oil spool assembly and insert in body (15). Use new gasket (3), install cap (1) and tighten retaining cap screws to 5-8 ft.-lbs. of torque.

Place new "O" ring (20) on fitting (21), then install orifice plate (17), spring (18) and ball (19) in inlet port and secure with fitting.

# POWER TAKE-OFF

Single speed 1000 rpm, or dual speed 540 or 1000 rpm pto units are available for all Models G955-G1355-2255 tractors. In addition, an engine speed pto may be used with any unit.

**NOTE: Use of the 540 rpm pto must be limited to not more than 60 horsepower.**

Pto clutch and brake are hydraulically actuated and are controlled by a single spool, closed center valve mounted on pto housing.

**Models G955-G1355-2255**

216. **R&R PTO DRIVE SHAFT.** To remove the pto drive shaft, first remove cover (18—Fig. 119), or the engine rpm pto output shaft, if so equipped, from upper bearing bore of pto housing. Remove snap ring (30) from inside diameter of pinion (33). Install puller assembly (Oliver No. STS-100), or a ½-13 inch cap screw, and pull plug (31). Oil trapped in the lubrication circuit will flow from pinion when plug is removed. Remove "O" ring (32), then use a ½-13 inch cap screw at least 7 inches long, threaded into end of pto drive shaft and pull the pto drive shaft (1—Fig. 120).

Reinstall by reversing removal procedure, however, be sure "O" ring (32—Fig. 119) is in front groove in I.D. of pinion (33) and that projection on rear of plug (31) fits into spline of pinion.

217. **R&R PTO UNIT.** To remove the pto unit, first drain the transmission and final drive compartments. Remove pto drive shaft as outlined in paragraph 216. Remove linkage and oil lines from pto control valve. Remove drawbar

support, then unbolt and remove pto unit from tractor rear main frame.

When reinstalling pto unit, secure unit to rear main frame with the four cone head cap screws and tighten them to 135-150 ft.-lbs. of torque, then install

remaining cap screws and lock washers. Complete installation by reversing the removal procedure, fill transmission and final drive compartments, then start engine and check pto operation.

218. **OVERHAUL PTO UNIT.** Re-

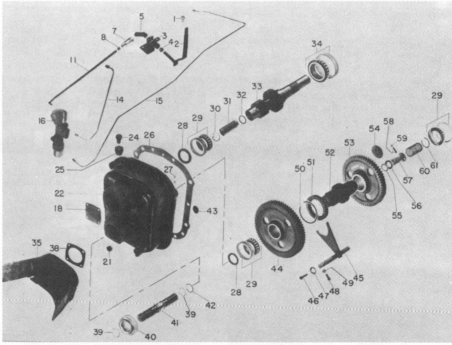

**Fig. 119—Exploded view showing component parts and their relative positions of the dual speed pto. The single speed (1000 rpm) pto does not include all parts shown, however, basic construction remains the same.**

| | | | |
|---|---|---|---|
| 1. Knob | 24. Breather | 39. Snap ring | speed) |
| 2. Lever | 25. Reducer | 40. Bearing | 51. Drive coupling |
| 3. Support | 26. Gasket | 41. Output shaft | 52. Output gear shaft |
| 4. Bushing | 27. "O" rings | 42. Snap ring | 53. Output gear (1000) |
| 5. Bellcrank | 28. Oil seal | 43. Plug (single speed) | 54. Plug |
| 7. Clevis (end) | 29. Bearing assy. | 44. Output gear (540) | 55. Washer |
| 11. Rod | 30. Snap ring | 45. Shift race & fork | 56. Oil seal |
| 14. Oil line | 31. Plug | (dual speed) | 57. Locking cam (dual |
| 15. Oil line | 32. "O" ring | 46. Cap screw | speed) |
| 16. Control valve | 33. Pinion shaft | 47. Oil seal | 58. Pin |
| 18. Cover | 34. Bearing assy. | 48. Cap screw | 59. Spring |
| 21. Plug (ctsk.) | 35. Safety shield | 49. Nut | 60. Spring |
| 22. Housing | 38. Bearing retainer | 50. Snap ring (single | 61. Snap ring |

move pto unit from tractor as outlined in paragraph 217 and remove safety shield, bearing retainer and the output (stub) shaft and bearing assembly. Remove cap screws and pull control valve from housing. Discard "O" rings.

NOTE: Disassembly of pto unit will require the use of a special tool which can be fabricated by using the dimensions shown in Fig. 121.

With pto unit removed, remove snap ring (2—Fig. 120) and pull hub (3) from driver (4). Attach special tool as shown in Fig. 122 and compress clutch spring, then remove snap ring (9—Fig. 120) and remove driver, spring (12) and spring guide (13). Snap ring (11) and bearing (10) can now be removed from driver. Remove retaining ring (5), then remove washer (6), clutch plates (7 and 8), collar (15) and spring washer (14) from clutch housing (17). Remove snap ring (16) from pinion shaft and remove clutch housing (17), actuator (18), friction plate (19), reaction plate (25) and the thrust bearing assembly (20 and 21). Unbolt and remove inner support (30) from pto housing and discard the two "O" rings (27—Fig. 119) which are located between inner support and housing. Clamp inner support in a vise as shown in Fig. 123, and bump out clutch piston (23—Fig. 120), brake piston (26) and remove "O" rings and quad rings (27) from pistons. Remove pinion seal (29) and if necessary, bearing cups from inner support.

Remove shifter rod lock screw (dual speed units). Remove output gear, or gears, and output gear shaft (52—Fig. 119) and on dual speed units, remove

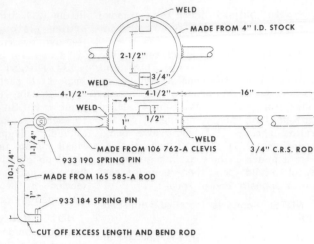

*Fig. 121—Special tool used to disassemble pto unit can be fabricated using the dimensions shown.*

shifter rail and fork. Remove drive pinion (33), then remove bearing cones from pinion and bearing cups and seals from pto housing. Remove bearing cones, output gear, or gears, and drive coupling (51) from output gear shaft (52).

On 1000 rpm single speed units, remove snap ring (50) from output gear shaft drive collar, then remove snap ring (61) and remove shifter lock mechanism, seal and alignment washer (items 55 through 60). Remove cup plug (43) from shift rod bore in pto housing.

On dual speed units, remove snap ring (61) from front bore of output gear shaft and remove shifter lock mechanism, seal and alignment washer (items 55 through 60). Remove shifter rod and seal from pto housing.

Refer to the following as a guide in renewing parts:

**CLUTCH**

| | |
|---|---|
| Drive plate thickness | 0.060-0.065 |
| Driven plate thickness | 0.0665-0.0710 |
| Spring free length | 2.629 |

**BRAKE**

| | |
|---|---|
| Plate thickness | 0.183-0.187 |
| Friction disc thickness | 0.122-0.127 |

219. To reassemble pto unit, install bearing cups in pto housing and support with tapers facing toward each other. Install oil seals in pto housing with lips facing forward.

On 1000 rpm single speed units, install cup plug in shifter rod bore in pto housing with cupped side forward. Install alignment washer (55) and seal (56) in front bore of output gear shaft,

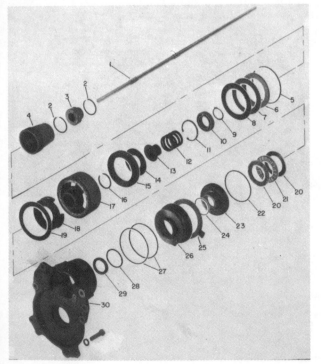

*Fig. 120—Exploded view of pto hydraulic clutch and brake assembly and inner support.*

1. Drive shaft
2. Snap ring
3. Hub
4. Driver
5. Retaining ring
6. Washer
7. Separator plate
8. Clutch plate
9. Snap ring
10. Bearing
11. Snap ring
12. Return spring
13. Spring guide
14. Plate
15. Collar
16. Snap ring
17. Clutch housing
18. Actuator
19. Friction plate
20. Bearing plate
21. Thrust bearing
22. "O" ring
23. Clutch piston
24. "O" ring
25. Reaction plate
26. Brake piston
27. Quad rings
28. "O" ring
29. Oil seal
30. Support

*Fig. 122—Attach special tool as shown to compress pto clutch spring.*

*Fig. 123—Drive clutch and brake pistons from bearing support as shown.*

then install lock pin spring (59) and lock pin (58). Depress lock pin and install locking cam (57) and spring (60). Secure parts in shaft bore with snap ring (61). Install snap (lock) ring (50) on drive collar of output gear drive shaft, then place drive coupling over front end of shaft drive collar and against snap ring. Install 1000 rpm output gear (53) on front of output gear shaft and mesh teeth on gear with teeth of coupling. Install bearing cones on output gear shaft.

On dual speed units, install oil seal in shifter rod bore in pto housing with lip facing forward. Install alignment washer (55) and seal (56) in front bore of output gear shaft, then install lock pin spring (59) and lock pin (58). Depress lock pin and install locking cam (57) and spring (60). Secure parts in bore with snap ring (61). Place 1000 rpm gear (53) on front of output gear shaft with drive coupling teeth rearward and secure with bearing cone. Slide drive coupling (51) on output gear shaft drive collar and mesh teeth of coupling with teeth of 1000 rpm gear. Place 540 rpm gear (44) on rear of output gear shaft with drive coupling teeth forward and secure with bearing cone.

On all units, position output gear and shaft in pto housing, and on dual speed units, mate shift fork with groove in drive coupling and install shifter rod and fork along with output gear and shaft assembly. Use caution not to damage shifter rod seal. Install bearing cones on pinion shaft (33) with tapers facing away from each other, then with teeth of pinion and output gears meshed, position pinion in pto housing. Install new pinion seal (29—Fig. 120) in support with lip facing forward. Install new "O" rings and quad rings on brake and clutch pistons and lubricate with Lubriplate No. 210, or equivalent. Push clutch piston (23) into brake piston (26), then push assembly into support (30). Place new "O" rings (27—Fig. 119) in pto housing, then install inner support on pto housing and secure with the six cap screws.

Install thrust bearing (20 and 21—Fig. 120). Install reaction plate (25) with flat side up and lug in recess in support. Install friction plate (19), actuator (18) and clutch housing and secure clutch housing with snap ring (16). Install collar (15), spring plate (14), spring guide (13) and return spring (12). Install bearing (10) in driver (4) and secure with snap ring (11). Place driver over return spring, use special tool (Fig. 122) to compress spring and install snap ring (9) with flat side of snap ring facing forward. Start with an external lug clutch (separator) plate and alternately install the external lug

and internal lug clutch plates (7 and 8—Fig. 120). Install washer (6) and secure assembly with retaining ring (5). Place hub (3) in driver (4) and secure with snap ring (2).

**220. R&R AND OVERHAUL CONTROL VALVE.** To remove control valve, disconnect linkage and oil lines, then remove the two cap screws and lift control valve from pto housing. Remove the three "O" rings.

With valve removed, disassemble as follows: Remove cap (1—Fig. 124), drive out spring pin, then remove arm assembly (6) from housing and lift out arm and pin assembly (2). Remove plug (29) and discard "O" ring (27), then remove feel disc (26) and feel adjusting sleeve (25). Pull spool assembly from

bottom of valve body, then remove and discard "O" ring (4). Compress feel spring (23) and remove snap ring (21), washer (24), spring (23) and washer (22). Compress springs (17 and 19) and remove snap ring (21), washer (20), springs (17 and 19), stop sleeve (18), washer (16) and pressure control sleeve (15).

Clean and inspect all parts and renew any that show signs of undue wear or other damage. If body is worn or damaged, renew complete valve assembly.

Reassemble by reversing disassembly procedure and install pressure control sleeve (15) with flattened sides toward top of spool. Lubricate "O" ring (4) before installing spool assembly.

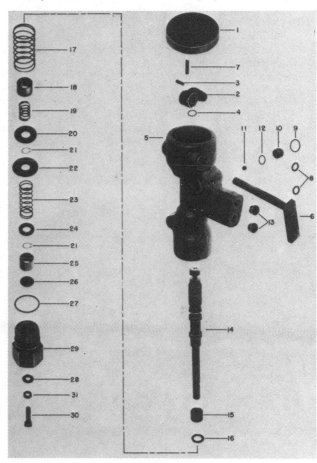

**Fig. 124—Exploded view of pto control valve.**

1. Cap
2. Arm
3. Pin
4. "O" ring
5. Body
6. Arm assembly
7. Pin
8. "O" rings
9. "O" rings
10. Detent piston
11. Detent ball
12. "O" ring
13. Seats
14. Spool
15. Pressure control sleeve
16. Washer
17. Spring (outer)
18. Stop sleeve
19. Spring (inner)
20. Retaining washer
21. Snap ring
22. Stop washer
23. Feel spring
24. Retaining washer
25. Feel adjusting sleeve
26. Feel disc
27. "O" ring
28. Seal
29. End cap
30. Cap (adjusting) screw
31. Nut

# HYDRAULIC LIFT SYSTEM

## Models G955-G1355-2255

The closed center hydraulic lift system is designed for draft control and position control of a three-point hitch and system will operate in conjunction with externally mounted control valves which provide remote hydraulic operations.

When tractor (G955) is equipped with a Category II hitch, the hydraulic lift system uses a single internal lift

cylinder and draft sensing mechanism is located inside the tractor rear main frame. Tractors equipped with the 6000 pound Category III hitch (G1355 and 2255) use two externally mounted lift cylinders and draft sensing mechanism is externally mounted.

Pressurized oil for operating the hydraulic lift system is provided by a variable displacement, axial piston type pump which is externally mounted on lower left side of tractor rear main frame. Pump is driven from a set of bevel gears which are driven by the tractor pto shaft, or a separate drive shaft if tractor has no pto.

The externally mounted pressure detent control valve can be used for either single or double acting remote cylinders and the externally mounted float valve provides raise and lower positions as well as float position.

## TROUBLE SHOOTING

221. The following will be of assistance should trouble develop in the operation of the hydraulic lift system.

Also refer to the Quick Check section which follows:

222. SYSTEM DOES NOT DEVELOP FULL PRESSURE OR DELIVER OIL. Could be caused by:
1. Fluid level low.
2. Restriction in outlet port.
3. Filter plugged.
4. Suction hose collapsed or line damaged.
5. Air leak in inlet lines.
6. Hydraulic pump shaft broken.
7. Hydraulic pump pistons or valve plate worn, scored or damaged.
8. Compensator valve sticking, worn or scored.
9. System pressure relief valve set too low or leaking.
10. Priority valve pilot spool spring broken.
11. Low pressure relief valve cap unscrewed.
12. Unloader valve faulty.

223. HYDRAULIC SYSTEM NOISY. Could be caused by:
1. Fluid level low.
2. Outlet port restricted.
3. Hydraulic reservoir breather plugged.
4. Suction hose collapsed or line damaged.
5. Filter plugged.
6. Bleed line leaking air.
7. Cooler pump inlet line loose or pump has lost prime.
8. System pressure relief valve set too low.

9. Hydraulic pump bearings defective.
10. Hydraulic pump shaft misaligned.
11. Hydraulic pump bevel gear set incorrectly adjusted.
12. Compensator valve set too high or valve stuck in pumping condition.

224. HITCH WILL NOT RAISE. Could be caused by:
1. Servo valve restrictor closed.
2. Servo valve spool stuck.
3. Lift cylinder seals leaking.
4. Control linkage bent or broken.
5. Cylinder safety valve leaking.
6. Internal oil line or fitting failure (Cat. II hitch).

225. HITCH WILL NOT LOWER. Could be caused by:
1. Servo valve spool stuck.
2. Servo vlave check ball stuck.
3. Servo valve interlock plunger stuck.

226. DRAFT ACTION TOO SLOW. Could be caused by:
1. Servo valve restrictor turned in too far.
2. Servo valve spool sticking.
3. Draft spring preload set too high.
4. Linkage worn or not properly adjusted.

227. DRAFT ACTION TOO FAST. Could be caused by:
1. Servo valve restrictor turned out too far.
2. Linkage not properly adjusted.

228. REMOTE CYLINDER WILL NOT RAISE. Could be caused by:
1. Control valve spool stuck.
2. Control valve restrictor closed.
3. Control valve restrictor ball incorrectly installed.
4. Breakaway coupler not installed properly or coupler damaged.
5. Faulty remote cylinder.

229. REMOTE CYLINDER WILL NOT LOWER. Could be caused by:
1. Control valve spool stuck.
2. Control valve interlock pin broken.
3. Control valve interlock plunger stuck.

230. REMOTE CONTROL VALVE LEVERS WILL NOT RETURN TO NEUTRAL. Could be caused by:
1. Control levers binding.
2. Incorrect troque on control valve retaining cap screws.
3. Control valve detent piston adjusted too high.
4. Incorrect number of shims between control valve detent piston and spring.

231. HYDRAULIC SYSTEM OVERHEATS. Could be caused by:
1. Oil cooler dirty or clogged.
2. Oil cooler pump defective.
3. Oil cooling circuit restricted.

4. Hydraulic pump defective.
5. Excessive leakage of one or more more hydraulic control valves.
6. Incorrect application of remote control valves or remote hydraulic motor.

232. TRANSFER OF HYDRAULIC OIL TO TRANSMISSION. Could be caused by:
1. Hydraulic pump seal faulty.
2. Faulty pto seals.
3. Broken pto clutch hub.

233. QUICK CHECK. A quick check of the hydraulic lift system can be made with a minimum of disassembly and an analysis of this quick check will, when used with the TROUBLE SHOOTING Section, determine to a large extent what service is required on the hydraulic lift systems.

234. To perform the quick check, proceed as follows: Obtain a protractor and modify it as shown in Fig. 125. Mount the protractor on hydraulic pump yoke pintle as shown in Fig. 126 and fabricate a pointer so movement of protractor can be checked.

NOTE: On later models which use cup plug to seal pintle bore, remove plug and install protractor using special tool ST214.

Install a loop line to breakaway couplings as shown in Fig. 127. Install a 3000 psi gage at the pressure control valve.

Fig. 126—Mount protractor as shown when making hydraulic system quick check. Refer to text for procedure.

Fig. 127—Use a loop line as shown when making hydraulic system quick check. Refer to text.

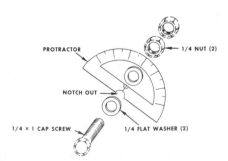

Fig. 125—Modify a protractor as shown when using to make hydraulic system check. Also see Fig. 126.

Note the number of degrees at pointer with engine off. This locates yoke angle in full pumping stroke as system is not pressurized. Start engine and run at 2400 rpm. Note pressure gage which should show system pressure of 2225-2275 psi, and protractor should have moved 15-16 degrees from full stroke position.

**NOTE: It requires 1-2 degress of yoke angle from minimum (zero) stroke to make up normal system leak-off.**

If pressure is not as stated, refer to TROUBLE SHOOTING Section. If protractor has moved less than the 15-16 degrees, one or more of the hydraulic system components is leaking excessively.

Open restrictor on remote control valve manifold, move control lever rearward and allow oil to circulate through remote lift circuit and loop line. With engine operating at 2400 rpm, yoke angle should move to full pumping stroke position and pressure on gage should be reduced to priority pressure (approximately 2000 psi). If pressure is not reduced, lift circuit may be restricted causing pump to compensate. Check remote control valve, breakaway couplers and loop line.

**NOTE: Air in system will cause hydraulic pump to destroke and yoke angle may not reach full pumping stroke position. Yoke (protractor) and pressure gage will oscillate rapidly, making hydraulic system noisy. Check pump, suction lines, cooling circuit and reservoir oil level for possible entrance of air.**

235. With items outlined in paragraph 234 completed, open remote control valve and allow oil to circulate through the loop line. Reduce engine speed to 800 rpm and check pressure gage which sould show a priority valve pressure of 1900 psi. Adjust priority valve if necessary.

## LUBRICATION AND MAINTENANCE

236. When performing service or maintenance of any kind on the hydraulic lift system, the practice of cleanliness is of the utmost importance. Pumps, valves and cylinders are manufactured to very close tolerances and are mirror finished. The induction of dirt and foreign material is most detrimental and should be avoided at all times.

Capacity of hydraulic reservoir is 28 quarts for Model G955 with Category II hitch; 33 quarts for Models G1355 and 2255 with Category III three point hitch, or 36 quarts for Models G1355

and 2255 with remote only hydraulics.

Recommended fluid is Oliver Type 55 hydraulic oil.

237. **DRAIN AND REFILL.** Drain and refill hydraulic lift system each year, or 500 hours, as follows: Place lift arms in down position and remove plug from bottom of filter. Reinstall plug when fluid flow has stopped.

Disconnect and drain all hoses. Work piston rods in and out of cylinders to clear cylinders, then reconnect all hoses and fill reservoir. Open bleed port on filter and when air free oil flows from bleed port, close bleeder. Start engine and cycle hydraulic equipment several times to work any trapped air from system. Recheck fluid level and add if necessary to bring fluid level to "FULL" mark on dipstick.

238. **OIL FILTER.** Hydraulic fluid filter is renewed after first 50 hours, and whenever hydraulic fluid is changed thereafter.

To renew filter, turn bottom of filter counterclockwise to remove center bolt. Separate housing from head and discard element and gaskets. Wash all parts, allow to dry, then coat gaskets and seals with grease. Place upper seal on filter head. Place lower seal, housing, spring, washer, element and gaskets on center post, then install assembly on filter head and tighten center post to a maximum of 20 ft.-lbs. torque. Run engine and check for leaks.

**NOTE: A by-pass valve is incorporated in head of filter assembly. Should service ever be required, remove inlet hose, connector and outlet line, then remove nut, valve, guide, spring, spacer, cap and screw from filter head.**

239. **BREATHER.** A breather is located on top rear side of hydraulic lift cover and under normal operating conditions the breather element should be renewed each time hydraulic fluid is renewed. When operating under extremely dusty conditions, renew element more frequently. The elastic stop nut should be tightened until breather cap is snug against spacer.

## SYSTEM ADJUSTMENTS

The adjustments in this section will include those external adjustments which can be accomplished with no disassembly. For those internal adjustments which require disassembly, refer to the appropriate section.

240. **LIFT ARM ADJUSTMENT.** Remove any attached implement, then start engine and run at about 1000 rpm. Move hitch control lever to up position and allow lift arms to raise to upper limit. Grasp lift links and manually raise lower links and check the free play at ends of lift arms. This

free play should be ½-inch. If free play is not as stated, refer to Fig. 128 and loosen lock nut on control rod. Shorten control rod to increase, or lengthen control rod to decrease, the lift arm free travel. Tighten control rod lock nut, recycle hitch, then recheck. Readjust control rod if necessary.

241. **POSITION CONTROL.** To convert system from draft control to position control, disconnect control rod from balance link, move end to top hole in balance link and connect it directly to position arm. See Fig. 129. Check lift arm free travel as outlined in paragraph 240.

242. **DRAFT SPRING PRELOAD.** Draft control springs for the 6000 pound Category III hitch are externally mounted and must be preloaded. Left hand adjusting screw is located in left hand axle flange and right hand adjusting screw is located in right hand actuating hub arm.

Adjust draft control springs as follows: Cut safety wire, loosen lock nuts and turn both adjusting screws in until all tension on draft springs is released. Now back-out left hand adjusting screw until left draft spring just begins to preload. Continue to back-out the left hand screw until top of draft signal arm has moved forward 9/16-inch. Back-out right hand adjusting screw until right hand draft springs are just beginning to preload, then back screw out an additional ¼-turn. Tighten lock nuts and secure cap

*Fig. 128—Shorten or lengthen control rod to adjust lift arm free travel to ½-inch.*

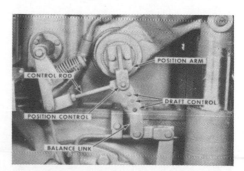

*Fig. 129—To convert from draft control to position control, connect control rod directly to position arm.*

screws with safety wire. Check lift arm free travel.

NOTE: As field conditions vary, it may be preferable to lengthen or shorten the 9/16-inch adjustment outlined above as required.

243. RESTRICTOR VALVE. A restrictor valve which controls lift speed of hitch is located in hydraulic lift manifold under left side of seat. See Fig. 130. To increase hitch lift speed, turn the adjusting screw counterclockwise. To decrease hitch lift speed, turn adjusting screw clockwise.

244. EXTERNAL VALVE RESTRICTOR VALVE. A restrictor valve is located on rear of each pressure detent (remote) control valve to control fluid flow returning from remote cylinder. See Fig. 131. Implement may bounce when being lowered if fluid flow is not sufficiently restricted.

If implement bouncing occurs, turn the adjusting screw clockwise to increase restriction. If implement lowers too slowly, turn the adjusting screw counterclockwise. Implement should lower as rapidly as possible without bouncing.

NOTE: Remove load on cylinders before adjusting restrictors.

245. EXTERNAL VALVE MANIFOLD RESTRICTOR. A restrictor valve is provided on the rear of the external valve manifold as shown in Fig. 131. This restrictor controls fluid flow to the external control valves. To change remote cylinder speed, turn the adjusting screw clockwise to decrease, or counterclockwise to increase, cylinder speed.

NOTE: This adjustment will not decrease implement lowering speed when a float type external (remote) valve is used.

## SYSTEM OPERATING PRESSURE

The high pressure relief valve is located in the pressure control valve body which is mounted on left side of transmission housing directly below the hydraulic pump. The pressure control valve body contains the system high pressure relief valve, the priority valve and the pressure reducing and low pressure valve. The high pressure valve is the valve located furtherest from tractor center line (i.e. the outermost valve).

As the compensator valve, mounted on outer end of hydraulic pump, controls the pump output, the high pressure relief valve functions to protect system from high pressure surges and to provide pressure relief in case the compensator valve malfunctions.

246. To check and/or adjust high pressure relief valve, refer to Fig. 132 and proceed as follows: Drain oil from reservoir, remove the 1/8-inch pipe plug below steering valve support and install a 3000 psi test gage. Remove reservoir line from unloader valve, then plug line and cap valve. Disconnect unloader valve pressure line at tee fitting on steering port, then connect a test unloader valve to steering port using 3/8-inch steel tubing and ¾-16 inch "O" ring straight pipe and 3/8-inch 90 degree elbow. Connect a 5/8-inch steel tube to unloader valve, attach a 5/8x½-inch male connector, then attach a ½-inch high pressure hose to male connector. Install a female breakaway coupler to hose.

Attach a hose to lower rear reservoir port with a ½-inch pipe and a male breakaway coupler. Attach a flowrator to breakaway couplers, fill reservoir

Fig. 131—External (remote) valve and valve manifold restrictors are located as shown. Refer to text for adjustment.

with fluid, then run engine at 2400 rpm and pump oil through flowrator until oil temperature reaches 130 degrees.

Disconnect flow-rator, break seal on compensator valve and turn compensator valve in until pressure gage reads 2450 psi. At this time, not more than 100 ml/min. should escape from unloader valve pressure line. Use a graduate container and a watch when making this check.

NOTE: Unloader valve pressure line is open to back side of relief valve. Oil escaping past relief valve is allowed to pass out of pressure control valve through unloader valve and pressure line.

If escaping oil exceeds 100 ml/min., check high pressure relief valve for the following: Damaged "O" ring on relief valve seat. Scored or damaged relief valve seat. Chipped or broken relief valve ball. Damaged relief valve body and cap. Renew parts as necessary.

Adjust compensator valve until the high pressure relief valve opens, however, do not exceed 2600 psi gage pressure. Relief valve opening can be noted by an increase in the amount of oil escaping and an increase in noise. Gage reading should be 2470-2530 psi. If necessary, adjust valve by varying the number of shims located between spring and head of plunger. Adding shims will increase pressure, removing shims will decrease pressure. Shims are available in thickness of 0.004 and 0.0149.

With high pressure relief valve adjusted, readjust the compensator valve to 2250 psi and reseal. Redrill a 1/16-inch hole in compensator valve case if necessary for wire and lead seal installation.

## PRESSURE CONTROL VALVE

Pressure control valve is mounted on left side of transmission housing below hydraulic pump. Valve body contains the system high pressure relief valve,

Fig. 132—View showing high pressure relief valve ready to be tested. Refer to text.

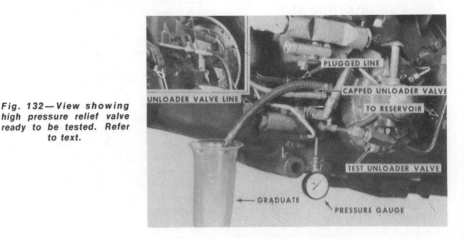

Fig. 130—Restrictor valve to control hitch lift speed is located as shown.

priority valve, pressure reducing valve and low pressure relief valve.

If desired, pressure control valve can be disassembled without removing valve body from tractor. All parts of valve are available for service and while the flared seats in body can be renewed, a new valve body has the flared seats installed.

247. REMOVE AND REINSTALL. To remove pressure control valve, either drain the hydraulic lift reservoir, or plug lines as they are disconnected. With all lines disconnected, remove the valve retaining cap screws, pull valve from housing, then remove unloader valve from pressure control valve.

When reinstalling, mount unloader valve on pressure control valve, then install pressure control valve by reversing removal procedure.

248. OVERHAUL. Remove cap (22—Fig. 133) with "O" ring (15), then remove spring (21), shims (14) and plunger (20) from cap. Remove ball (19) from bore. Use a stiff wire with a ¼-inch hook on one end and pull seat (18) and "O" ring (17) from bore.

Remove priority valve cap (16) and "O" ring (15), then remove shims (14) and spring (13) from cap. Remove pilot spool (12), thread a cap screw into plug (11), then remove plug and "O" rings (9 and 10). Remove spring (4) and spool (3) from bore. Remove plug (1) and "O" ring (2), if necessary.

Remove plug (34) and "O" ring (33). Remove spring (32), then slide pressure reducing spool (25) and relief valve assembly (items 26 through 31) out of valve body. Use caution not to mar the pressure reducing spool and remove cap (31), then remove washer (26) from spool and ball (27), plunger (28), shims (29) and spring (30) from cap (31).

249. Remove and discard all "O" rings. Clean and inspect all parts for nicks, burrs, scoring or any undue wear. Check all springs for distortion or fractures. Check bores in pressure control valve body. Renew any defective parts, then lubricate all parts

and reassemble by reversing the disassembly procedure.

## UNLOADER VALVE

The unloader valve is mounted on the pressure control valve. Function of unloader valve is to eliminate build-up of hydraulic pressure during engine cranking. Hydraulic pressure will be relieved by the orifice in the unloader valve plunger until pump output reaches about 4 gpm at which time the unloader valve closes and hydraulic pressure raises to normal. Valve remains closed until engine is stopped and pressure is reduced in system.

250. R&R AND OVERHAUL. To remove unloader valve, drain hydraulic reservoir, then disconnect steering line, relief valve return line and low pressure return line. Remove line between unloader valve and tee-fitting at steering port of pressure control valve. Loosen "O" ring locknut on unloader valve elbow, turn unloader valve so elbow will clear brake drums, then remove elbow from unloader valve and "O" ring from elbow. Remove unloader valve by turning reducer bushing out of pressure control valve. Remove pipe nipple from unloader valve.

251. With unloader valve removed, remove snap ring (1—Fig. 134), then remove piston (2) and spring (3) from body (4).

Clean all parts and inspect piston and seat for nicks, burrs, scoring or undue wear. Check spring for distortion or fractures. Renew parts as necessary and reassemble by reversing disassembly procedure.

252. Reinstall unloader valve as follows: Coat pipe threads with Permatex No. 3, or equivalent, and install pipe nipple and reducing bushing. Install unloader valve assembly in pressure control body and turn assembly far enough past final position to allow installation of elbow. Place new "O" ring on elbow, lubricate "O" ring and turn elbow into unloader valve. Back the reducing bushing out of

pressure control valve far enough to allow connection of relief valve return line. Connect relief valve return line and low pressure return line. Install line connecting tee-fitting at steering port of pressure control valve to unloader valve elbow. Tighten locknut on unloader valve elbow against unloader valve and connect steering line.

Refill hydraulic reservoir.

## HYDRAULIC PUMP

253. REMOVE AND REINSTALL. To remove hydraulic pump, remove drain plug from filter and drain hydraulic reservoir. Disconnect pump drain line and disconnect both ends of pump outlet line. Disconnect hose clamps at inlet port of pump and slide coupling hose off inlet fitting and onto filter tube. Remove pump retaining cap screws, pull pump from pump drive support, then remove splined drive coupling.

When reinstalling pump, use Permatex No. 3, or equivalent on pump side of mounting gasket, grease transmission side of gasket, then install pump by reversing the removal procedure.

NOTE: For information on the hydraulic pump drive gears, refer to TRANSMISSION Section.

254. OVERHAUL. To disassemble hydraulic pump, proceed as follows: Remove compensator housing (6—Fig. 135) to gain access to valve cover cap screws, then remove valve cover (14). Remove piston (16) from tube (12) and remove tube (rod) from valve plate. Remove check valve assembly (items 17, 18, 19 and 20). Use a bearing puller if bearing (15) is to be removed. Tilt housing, turn cylinder block (25) slightly to free parts from swash plate and remove cylinder assembly. Lift shoe plate (29) and piston (30) assembly from cylinder block and be careful not to damage pistons during removal or subsequent handling. If limiter (22), spring (23) and washer (24) are to be removed, place a 1-inch washer on a 3/8 x 3½ inch cap screw and insert cap screw in shaft hole of cylinder block at front end. Place another 1-inch washer on threaded end of cap screw and install nut. Tighten nut until tension

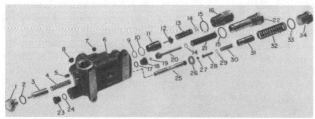

1. Plug
2. "O" ring
3. Spool (priority valve)
4. Spring
5. Pipe plug
6. Body
7. Insert
8. Seat
9. "O" ring
10. "O" ring
11. Plug

12. Pilot spool
13. Spring
14. Shim (.004, .0149)
15. "O" ring
16. Plug
17. "O" ring
18. Seat
19. Ball (7/16)
20. Plunger (high press.)
21. Spring
22. Cap

23. Plug
24. "O" ring
25. Spool (press. reducing)
26. Washer
27. Ball (½)
28. Plunger (low press.)
29. Shim (.004, .0149)
30. Spring
31. Cap
32. Spring
33. "O" ring
34. Cap

Fig. 133—Exploded view of pressure control valve assembly showing high pressure control valve, priority valve and low pressure and pressure reducing valve.

Fig. 134—Exploded view of the hydraulic lift system unloader valve.

1. Snap ring
2. Piston
3. Spring
4. Body

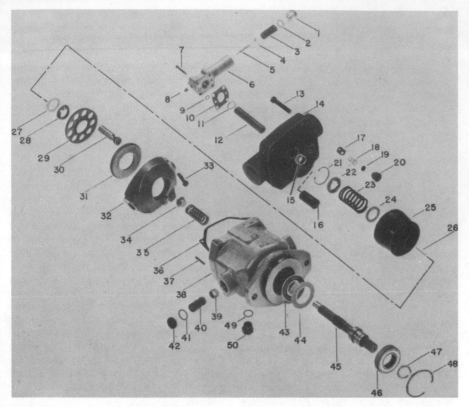

*Fig. 135—View of the variable displacement piston type hydraulic pump. Kits are available for valve plate and check valve assembly, rotating group and seals ("O" rings and gaskets).*

| | | | |
|---|---|---|---|
| 1. Plug | 15. Bearing | 27. Washer | 39. Bearing |
| 2. "O" ring | 16. Piston | 28. Spherical washer | 40. Pintle |
| 3. Spring | 17. Cup | 29. Shoe plate | 41. "O" ring |
| 4. Seat | 18. Spring | 30. Pistons | 42. Plug |
| 5. Spool | 19. Check vlave | 31. Swash plate | 43. Seal |
| 6. Body | 20. Seat | 32. Yoke | 44. Spacer |
| 8. Plug | 21. Snap ring | 33. Cap screw | 45. Pump shaft |
| 9. "O" ring | 22. Limiter | 34. Spring seat | 46. Bearing |
| 10. Gasket | 23. Spring | 35. Spring (outer) | 47. Snap ring |
| 11. Snap ring | 24. Thrust washer | 36. Gasket | 48. Snap ring |
| 12. Piston rod | 25. Cylinder block | 37. Dowel | 49. "O" ring |
| 14. Valve plate | 26. Dowel | 38. Housing | 50. Plug (refill) |

against snap ring (21) is removed, then remove snap ring. Remove nut slowly and remove parts from cylinder block.

Remove swash plate (31) from yoke (32).

**NOTE: Swash plate may be oil locked in yoke and removal may be somewhat difficult until oil lock is broken.**

Remove snap ring (48), then use a soft-faced hammer to bump drive shaft (45) and bearing (46) from housing. Remove snap ring (47) and if necessary, use a puller to remove bearing. Removal of bearing (46) often destroys it so be sure a replacement bearing is available prior to attempting removal. Spacer (44) and oil seal (43) can now be removed. Remove cap screws (33) from yoke, then remove plugs (42), "O" rings (41) and pintles (40). Remove yoke, spring seat (34) and the inner and outer springs (35).

255. Thoroughly clean all parts prior to inspection and after any lapping operations.

Inspect surface of valve plate which mates with cylinder block. Minor defects can be removed by lapping, however, not more than 0.0004 of material should be removed as this surface is hardened and excessive lapping will remove this hardening. Renew valve plate if wear or scoring is extensive.

Inspect piston bores and valve plate mating surface of cylinder block. Minor defects on valve plate mating surface can be removed by lapping, however, if damage is extensive, or if piston bores are scored or worn, renew cylinder block.

Inspect piston and shoe assemblies. Worn or damaged pistons will require renewal. Variations between shoes must not be more than 0.001 to insure that all shoes ride properly on swash plate. Shoes can be lapped if necessary, however, when pump is being overhauled, it is recommended that all nine pistons, shoe sub-assemblies and cylinder block be renewed to provide maximum overhaul life.

Swash plate can be lapped if necessary, however, if more than 0.0004 material must be removed, renew the swash plate.

Inspect and renew pump shaft bearings as necessary. Inspect pump shaft oil seal area and splines. Renew pump shaft if it shows signs of wear or scoring, or if shaft is bent.

256. Reassemble hydraulic pump as follows: Use new seals, "O" rings and gaskets and coat all parts with hydraulic oil when reassembling.

Place springs (35—Fig. 135), seat (34) and yoke (32) in housing (38). Push pintles through housing into yoke, align grooves in pintles with cap screw holes in yoke, then install cap screws (33) and tighten to 20-25 ft.-lbs. torque. Install "O" rings (41) and plugs (42). Install new oil seal (43) in housing with lip toward yoke and place spacer (44) on top of oil seal. Be sure bearing (46) is about half filled with high temperature grease, install bearing on shaft and secure with snap ring (47). Install shaft and bearing in housing, secure with snap ring (49) and be sure curved side of snap ring is away from bearing.

Place swash plate (31) in yoke with chamfered edge toward shaft bearing and be sure swash plate can be turned freely with fingers. If spring, washer and limiter (22) were removed, compress them with the 3/8 x 3½ inch cap screw (with washers and nut) and secure them with snap ring (21) installed with curved side out. Set cylinder block on valve plate end and insert the three pins, if necessary, then grease washers (27 and 28) and place them on pins so curved side of washer (28) faces shoe plate (29). Assemble pistons and shoes (30) in shoe plate (29), then while holding shoe plate over cylinder block, insert pistons in their bores. Pistons should slide easily in the lubricated bores. Hold the assembled cylinder block assembly together to prevent parts from shifting, then install assembly on the pump drive shaft. Rock drive shaft back and forth slightly if necessary to help align splines of shaft, spherical washer (28) and cylinder block.

Be sure snap ring (11) is on tube (12), and tube is in valve plate (14), then place piston (16) on tube. Install check valve assembly (items 17, 18, 19 and 20). Use new gasket (36), install valve plate and tighten retaining cap screws to 42-45 ft.-lbs torque. Place new "O" ring (9) in compensator valve body (6), use new gasket (10) and install compensator valve. Be sure plug (8) is installed.

Fill pump through vent line with Oliver Type 55 Hydraulic Oil.

## COMPENSATOR VALVE

257. Compensator valve, shown in exploded view in Fig. 135, is mounted on lower side of hydraulic pump valve plate. Function of the compensator valve is to vary stroke (output) of the hydraulic pump to meet the demands of the hydraulic system.

258. **R&R AND OVERHAUL.** Compensator valve can be removed from hydraulic pump valve plate if necessary by removing the four screws (7—Fig. 135).

To disassemble valve, break seal and remove adjusting plug (1) with "O" ring (2). Remove spring (3), spool seat (4) and spool (5).

Check spool and spool bore for nicks, burrs, scoring or wear. Check seat for wear or other damage and be sure orifice is open. Check spring for distortion or fractures. Renew parts as necessary.

To reassemble, slide spool into bore small diameter end first, then install spool seat and spring. Install new "O" ring on adjusting plug, lubricate "O" ring, then install adjusting plug so slotted end is flush with valve body.

To check and adjust compensator valve, refer to paragraph 246. Reseal the compensator valve after adjusting. Redrill a 1/16-inch hole in compensator valve case if necessary for wire and lead seal installation.

## HYDRAULIC RESERVOIR

259. **REMOVE AND REINSTALL.** To remove hydraulic reservoir, lower hitch to expel oil from cylinder or cylinders. Remove drain plug from hydraulic system filter and drain reservoir.

Remove upper link, seat and heat shield and control lever console. Disconnect pto control rod from control lever. Disconnect balance link from draft signal arm. On models with internal lift cylinder, disconnect lift links from lift arms. On models with external lift cylinders, disconnect lift cylinders from lift arms. Separate remote cylinder coupling holders from mounting bracket and if tractor has external lift cylinders, disconnect cylinder hoses from fitting at manifold plate. Cap and plug openings.

Remove gear shift lever and center platform. Disconnect brake return line, steering return line, cooler pump inlet line, relief valve return line, main pump inlet hose, pump vent line and pressure control vent line from bottom of reservoir. Disconnect lift circuit line from tee-fitting on left side of manifold plate. Disconnect pto clutch return line from rear of reservoir and cooler return line from remote valve manifold. Plug and cap openings as lines are disconnected. Remove clip retaining cooler

return line to transmission cover and move line out of the way.

Remove right hand lift arm, then while noting location of cone head capscrews, remove capscrews retaining hydraulic reservoir to rear main frame. Install lifting eye in right front and left rear seat mounting capscrew holes. Attach a hoist and remove hydraulic reservoir from tractor.

Reinstall by reversing the removal procedure and tighten the cone head reservoir retaining cap screws to a torque of 135-150 ft.-lbs.

## ROCKSHAFT

260. Rockshaft can be removed without removing hydraulic reservoir, if desired. However, if rocker arm is to be removed, it will be necessary to remove reservoir cover, servo valve and servo valve manifold.

To remove rockshaft, without removing hydraulic reservoir, proceed as follows: Remove left fender and drain reservoir. On models with internal lift cylinder, disconnect lift links from lift arms. On models with external lift cylinders, disconnect lift cylinders from lift arms. Remove hitch position arm from left end of rockshaft, then remove lift arm retaining snap rings and pull lift arms from rockshaft. Drive rockshaft out left side of housing which will also force oil seal and bushing out of same side. The opposite seal and bushing can be removed in a similar manner, if necessary.

**NOTE: This operation will damage oil seal and will require that a new seal be used. Bushing can be salvaged if caution is used.**

Remove rockshaft from left side of housing.

261. When reinstalling rockshaft, insert rockshaft from left side, align blind splines of rockshaft and rocker arm and slide rockshaft into position. Use Oliver driver ST-144, or equivalent, and install bushing so outer end is flush with bottom of oil seal counterbore. Lubricate lip of oil seal and coat outer diameter with sealant. Lubricate rockshaft splines, then with lip of seal toward bushing, carefully slide seal over rockshaft and drive into counterbore.

Complete reassembly by reversing the disassembly procedure.

## HITCH CONTROL LEVER

262. The hitch control lever and friction disc can be removed without removing hydraulic reservoir cover by removing lever console, retaining nut, washer, spring, lever and friction disc. Stationary friction surface is removed

by removing shaft key and the retaining cap screws.

If control shaft is to be removed, remove the hydraulic reservoir cover, drive out roll pin securing servo valve control arm to control shaft and pull shaft from manifold.

After reassembly, tighten lever retaining nut until it requires 18 pounds of force to move control lever through the last one inch of downward travel.

## DRAFT INPUT SHAFT

263. The "O" rings on draft input shaft and shaft support can be renewed without removing hydraulic reservoir. Proceed as follows: Disconnect draft control rod from draft input arm. Remove snap ring from outer end of draft input shaft, drive roll pin from the secondary draft input arm and pull input arms from input shaft. Remove cap screws from draft input shaft support and pull support out of reservoir housing and off shaft. "O" rings on shaft and support can now be removed.

If draft input shaft is to be removed from reservoir, remove reservoir cover and drive out roll pin securing servo valve input arm to shaft. Disconnect draft control rod from draft input arm, then pull shaft from shaft support. If desired, support can be unbolted from reservoir housing and complete assembly removed.

Reassemble by reversing the disassembly procedure.

## SERVO VALVE AND MANIFOLD

The servo valve is mounted on a manifold plate which is positioned between the hydraulic reservoir and the reservoir cover. The servo valve can be removed without removing manifold plate, or if desired, the servo valve and manifold plate can be removed together.

264. **REMOVE AND REINSTALL.** Remove drain plug from hydraulic filter and drain hydraulic reservoir. Remove seat assembly and heat shield, and if manifold plate is to be removed, remove control lever console. Remove line running from tee-fitting on left side of manifold plate and remote control valves manifold. If manifold plate is being removed, also disconnect front oil line from tee-fitting. Also, if tractor is equipped with external lift cylinders and manifold plate is to be removed, disconnect lines leading to external cylinders.

Remove capscrews retaining cover to hydraulic reservoir and remove cover by sliding it toward the tee-fitting (left) and lifting cover.

Unscrew return line packing nut

(Fig. 136) and pull return line from servo valve body. Remove snap ring from pin in servo valve actuating arm, then remove servo valve retaining capscrews and remove servo valve. Discard the "O" rings located between servo valve an manifold plate.

If manifold plate is to be removed, disconnect line leading to internal cylinder, then remove snap ring from pin extending through slot in connecting link and lift off manifold plate.

### 265. OVERHAUL SERVO VALVE.
Remove spool (15—Fig. 137) from plug end. Disconnect spring (3) from plug (2), then drive out pin (16) and remove spring from spool. Remove cap (14), washer (5), spring guide (13), spring (12) and check ball (11). Remove restrictor (6) and "O" ring (8). Remove plug (4) and washer (5), then use a brass drift and drive piston (9) and check ball seat (10) from bore.

Clean and inspect all parts for scoring, nicks, burrs or other damage. If spool (15), piston (9) or body (1) require renewal, it will be necessary to renew the complete valve assembly.

When reassembling, install piston (9) with sealing (flat) side toward restrictor (6), then drive check ball seat into position using a proper size brass drift. Complete reassembly by reversing the disassembly procedure.

## HYDRAULIC LIFT CYLINDERS
The single hydraulic lift cylinder for tractors equipped with a Category II hitch is internally mounted in bottom side of hydraulic reservoir housing. Service on this assembly requires the removal of hydraulic reservoir.

The two hydraulic lift cylinders for tractors equipped with a Category III hitch are externally mounted and can be removed for service by disconnecting hoses and removing attaching pins.

### 266. R&R AND OVERHAUL INTERNAL CYLINDER.
To remove internal lift cylinder, first remove hydraulic reservoir as outlined in paragraph 259. With hydraulic reservoir removed, stand unit on end, remove the shouldered dowels which retain oil pan to housing and remove oil pan. Remove hydraulic pump inlet line, then disconnect lift cylinder line. Remove the cylinder retaining capscrews, pull cylinder from dowels and lift cylinder from housing.

Remove snap ring (18A—Fig. 138), then bump cylinder (9) against a wood block to remove piston. Teflon ring (18) and "O" ring (17) can now be removed from groove in piston.

When renewing piston rings, install "O" ring first, then install Teflon ring on top of "O" ring. Installation of Teflon ring will be facilitated if ring is warmed, however, allow Teflon ring to

shrink to normal size before installing piston.

Reassemble by reversing the disassembly procedure.

### 267. CYLINDER SAFETY RELIEF VALVE.
The internal lift cylinder is fitted with a safety relief valve (1—Fig.

138) which is designed to prevent damage to system due to overload or pressure increase due to a raise in temperature. Valve can be removed from cylinder after hydraulic reservoir is removed and oil pan is off.

To disassemble valve, refer to Fig.

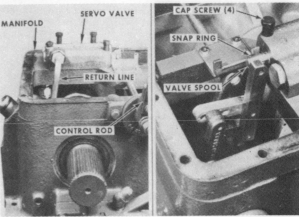
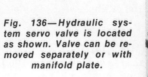

Fig. 136—Hydraulic system servo valve is located as shown. Valve can be removed separately or with manifold plate.

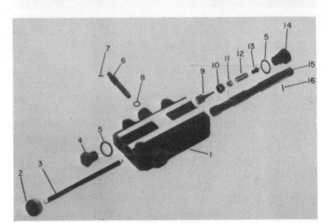

Fig. 137—Exploded view of servo valve. Pin (7) in restrictor (6) mates with seat of knob located in reservoir cover.

1. Body
2. Plug
3. Spring
4. Plug
5. Washer
6. Restrictor
7. Pin
8. "O" ring
9. Piston
10. Seat
11. Check ball
12. Spring
13. Spring guide
14. Cap
15. Spool
16. Pin

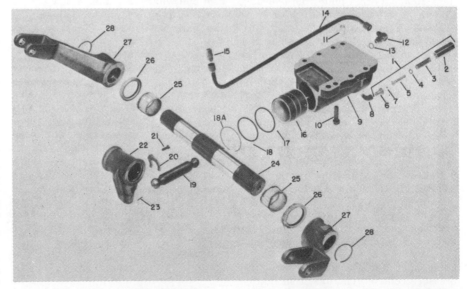

Fig. 138—Exploded view showing component parts of internal lift cylinder and rockshaft assemblies. Note safety relief valve (1).

| | | |
|---|---|---|
| 1. Safety relief valve | 8. Elbow | 16. Piston |
| 2. Body | 9. Cylinder | 17. "O" ring |
| 3. Spring | 11. Dowel | 18. Teflon ring |
| 4. Shim | 12. Elbow | 18A. Snap ring |
| 5. Plunger | 13. "O" ring | 19. Connecting rod |
| 6. Seat | 14. Oil line | 20. Retainer |
| 7. Ball (¼) | 15. Adapter | 22. Rocker arm |
| | | 23. Pin |
| | | 24. Rockshaft |
| | | 25. Bushing |
| | | 26. Oil seal |
| | | 27. Lift arm |
| | | 28. Snap ring |

139 and break loose the three stakes and remove seat from body. Remove steel ball, plunger and shims and spring.

Clean and inspect all parts and renew as necessary. Shims are available in thickness of 0.010.

Reassemble by reversing disassembly procedure and tighten seat to 75-90 ft.-lbs. torque. However, before staking seat in place, check and/or adjust the valve cracking pressure as follows: Attach valve to a hydraulic hand pump, or a diesel injector tester and check the valve cracking pressure which should be 2540-2660 psi. If adjustment is required, vary shims as required. If valve shows any signs of leakage below the minimum limit, valve must be repaired or renewed.

Stake valve seat in place after valve adjustment is complete.

**268. R&R AND OVERHAUL EXTERNAL CYLINDERS.** Cylinders can be removed by disconnecting hoses and removing the attaching pins.

With cylinder removed, remove elbow (2—Fig. 140), then pull piston rod out until snap ring on bottom of piston rod is located directly under port as shown in Fig. 141. Insert a screwdriver in port and move snap ring from lower groove to the upper (deepest) groove, then pull piston rod from barrel. Remove and discard wiper and seal from barrel.

Clean and inspect piston rod and barrel for scoring, nicks, burrs or undue wear and renew parts as necessary.

When reassembling, install new seal and wiper in barrel and install snap ring in upper (deepest) groove of piston rod. Lubricate all parts and push piston rod into barrel until snap ring is located directly below port, then use screwdriver to move snap ring from upper groove to lower groove. Check to see that snap ring is completely engaged in groove and that piston rod cannot be pulled from barrel.

Reinstall by reversing the removal procedure.

**NOTE: If bushings (13—Fig. 140) require renewal, renew them prior to installing lift cylinders.**

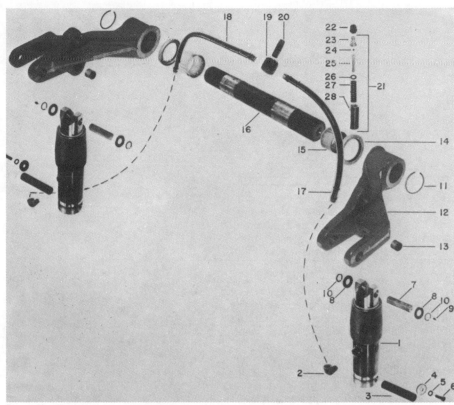

Fig. 140—Exploded view of lift cylinders, lift arms and rockshaft. Safety relief valve (21) is internally mounted on servo valve manifold.

| | | |
|---|---|---|
| 1. Cylinder | 8. Washer | 15. Bushing | 22. Reducer bushing |
| 2. Elbow | 9. Grease fitting | 16. Rockshaft | 23. Seat |
| 3. Pin | 10. Snap ring | 17. Hose, right | 24. Ball (¼) |
| 4. Washer | 11. Snap ring | 18. Hose, left | 25. Plunger |
| 5. Lock washer | 12. Lift arm | 19. Elbow | 26. Shims |
| 6. Capscrew | 13. Bushing | 20. Nipple | 27. Spring |
| 7. Pin | 14. Oil seal | 21. Safety relief valve | 28. Body |

## DRAFT CONTROL ACTUATING ASSEMBLY (CATEGORY II HITCH)

The draft control actuating assembly used with tractors having an internal hydraulic lift cylinder (Category II hitch) is located in the bottom rear of the rear main frame. The component parts which comprise this assembly are shown in Fig. 142.

**269. R&R AND OVERHAUL.** Drain oil from rear main frame and remove

the pto assembly, or rear cover if tractor has no pto. Remove hitch lower links from support (37 and 42—Fig. 142). Disconnect clevis link (31) from draft signal arm (34), then remove cap screws and keys (46) from actuating lever (43). Pull lower links (37 and 42) from rear main frame. Unbolt spring support (51) from bottom of rear main frame and remove spacers (50), then remove actuating arm (43), support (51) and draft spring (54) as an assembly.

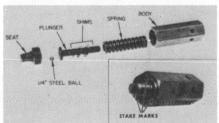

Fig. 139—Exploded view of safety relief valve used with both internal and external lift cylinder systems. Shims are 0.010 thick.

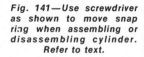

Fig. 141—Use screwdriver as shown to move snap ring when assembling or disassembling cylinder. Refer to text.

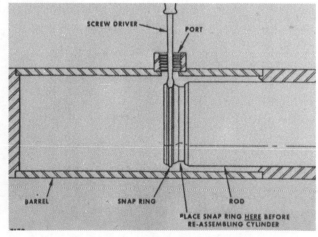

1. "O" ring
2. Support
5. Shaft
6. "O" ring
7. Arm
8. Pin
9. Link
10. Rod
12. Swivel
13. Retainer ring
14. Arm
15. Pin
18. Snap ring
19. Spring
20. Retainer
21. Capscrew
22. Nut
23. Arm
26. Pin
28. Balance link

31. Clevis link
32. Pin
34. Arm
37. Support, left
38. Washer
39. Oil seal
40. Snap ring
41. Bearing
42. Support, right
43. Lever
46. Key
47. Retainer ring
48. Pin
49. Rod
50. Spacer
51. Support
54. Draft spring
55. Retainer
56. Adjusting nut
57. Lock nut

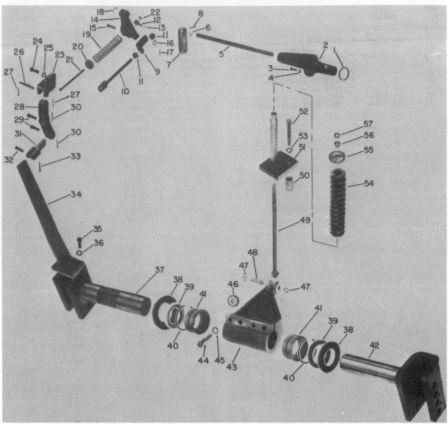

Fig. 142—View showing component parts which comprise the draft control actuating assembly used with tractors equipped with a Category II hitch.

Remove lock nut, adjusting nut, retainer and spring from support. Disconnect rod (49) from actuating arm. Remove oil seals (39) from bores in rear frame and discard. Remove snap rings (40) and bump bearings (41) out of rear frame.

When reassembling, apply Permatex No. 3, or equivalent, to outer diameter of oil seals (39) and using a suitable driver, carefully drive seals into counterbore with lips of seals facing inward. If necessary, dress keyways or inner ends of supports (38 and 42) and wrap with tape. Lubricate lips of oil seals and carefully slide supports through seals and bearings. Adjust draft spring (54) after assembly as outlined in paragraph 270.

270. ADJUST DRAFT SPRING. Prior to installing pto, or rear frame rear cover, adjust draft control spring as follows: Turn adjusting nut (56) on rod (49) until it contacts retainer (55) and all slack is removed from spring assembly. Measure from top of spring support (51) to top of spring retainer, then turn adjusting nut until the spring is compressed ¼-inch. Hold adjusting nut in this position and tighten lock nut (57).

## DRAFT CONTROL ACTUATING ASSEMBLY (CATEGORY III HITCH)

The draft control actuating assembly used with tractors having external lift cylinders (Category III hitch) has a one-piece shaft running through lower rear of tractor rear main frame and the draft control springs are mounted externally. The component parts which comprise this assembly are shown in Fig. 143.

271. R&R AND OVERHAUL. To disassemble draft control mechanism, first drain the rear main frame. On some tractors, it may be necessary to remove left wheel, or slide it to outer end of axle, to provide room to remove shaft (37) from rear main frame.

Disconnect clevis link (31) from draft signal arm (34), then remove draft signal arm from hub on left end of

1. "O" ring
2. Support
5. Shaft
6. "O" ring
7. Arm
9. Link
10. Rod
12. Swivel
13. Retainer ring
14. Arm
15. Pin
18. Snap ring
19. Spring
20. Retainer
21. Capscrew
22. Nut
23. Arm
26. Pin
28. Balance Link

31. Clevis link
32. Pin
34. Arm
37. Shaft
38. Adjusting screw
39. Lock nut
40. Draft springs
41. Pin
42. Spring pin
43. Washer
44. Oil seal
45. Snap ring
46. Bearing
47. Hub
48. Washer
49. Nut
50. Adjusting screw
51. Lock nut

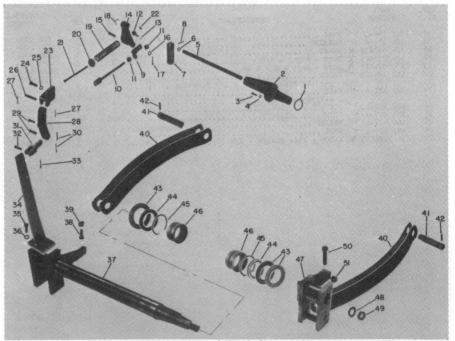

Fig. 143—View showing component parts of the draft control actuating assembly used with tractor equipped with a Category III hitch.

shaft (37). Remove lower lift links, turn adjusting screws (38 and 50) in until all tension is removed from draft springs (40), then remove pins (41) from tractor rear frame and shaft hubs and remove draft springs. Remove nut (49), washer (48), hub (47) and thrust washers (43) from right hand end of shaft (37), then pull shaft from left side of tractor main frame. Pry out oil seals (44), then remove snap rings (45) and bearings (46). Discard oil seals.

When reassembling, apply Permatex No. 3, or equivalent, to outer diameter of oil seals (44), and using a suitable driver, carefully drive seals into counterbores with lips of seals facing inward. Place a thrust washer on left end of shaft (37), tape splines on right end of shaft, then lubricate lips of oil seals and install shaft through rear main frame. Install thrust washers (usually three) on right end of shaft, install hub (47) and be sure hub is aligned with hub on opposite end of shaft. Install washer (48) and nut (49), tighten nut and check to be sure there is a small amount of end play in shaft. If end play of shaft requires adjustment, vary the thrust washers on right end of shaft. Thrust washers are available in thicknesses of 7/64 and 1/4-inch.

Complete assembly by reversing the disassembly procedure and adjust the preload on draft control springs as outlined in paragraph 272.

272. ADJUST DRAFT CONTROL SPRINGS. Draft spring adjustment can be initiated on either side of tractor, however, the left side is generally used because of the measurement involved with the draft signal arm. Proceed as follows:

Refer to Fig. 144 and back the adjusting screw out of left hub until screw contacts the flange on axle carrier and left draft springs are just beginning to preload. Continue to back out adjusting screw until top of draft signal arm has moved forward 9/16-inch as shown in Fig. 145. Move to right side of tractor, refer to Fig. 146, and back out adjusting screw until right draft springs begin to preload, then back adjusting screw out an additional 1/4-turn. Tighten lock nuts on both adjusting screws and secure screws with safety wire.

Check the lift arm free travel as outlined in paragraph 240.

## HYDRAULIC OIL COOLING SYSTEM

The basic components of the hydraulic oil cooling system for tractors prior to serial number 248 721- are a gear type pump which circulates oil from reservoir, and a cooler which is mounted in front of the tractor radiator. Pressure in the system is controlled by a pressure relief valve incorporated within the circulating pump. See Figs. 147 and 148.

Tractors with serial number 248 721- and up, differ in that the circulating pump is a gerotor type and is incorporated into the hydraulic pump driven shaft support. See Fig. 108A for an exploded view and to paragraph 179 for service information.

273. SYSTEM PRESSURE (PRIOR SER. NO. 248 721-). To check system (pump) pressure, disconnect outlet line from pump and connect a flow-rator between pump outlet port and the disconnected pump outlet line. Start engine and run until oil reaches operating temperature, then close restrictor of flow-rator until pump relief valve pops. Note pressure gage reading which should be 100-200 psi.

If pressure is not as stated on pumps used on Models G955 and G1355, refer to Fig. 149 and proceed as follows: Loosen jam nut (11) and turn adjusting screw (9) as required.

If pressure is not as stated on pumps used on Models 2255, refer to Fig. 150 and proceed as follows: Remove plug (1) and vary shims (3) as required.

If regulating adjusting screw (9—Fig. 149) or varying shims (3—Fig. 150) will not correct pressure, overhaul pump as outlined in paragraph 274 or 275.

### Models G955-G1355

274. R&R AND OVERHAUL PUMP. To remove pump, disconnect lines from pump, then unbolt adapter (16—Fig. 147) from engine. Remove pump drive

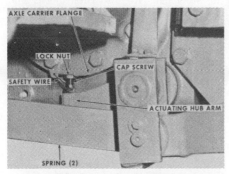

*Fig. 144—Adjusting screw for draft springs on left side of tractor is located as shown.*

*Fig. 145—Draft control springs are preloaded until draft signal arm moves forward 9/16-inch.*

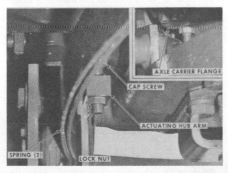

*Fig. 146—Adjusting screw for draft springs on right side of tractor is located as shown.*

*Fig. 147—Basic parts of hydraulic oil cooling system used on G955 and G1355 models prior to serial number 248 721-.*

1. Cooler
2. Mounting strip
5. Mounting clamp
7. Hose
8. Hose clamp
9. Pump to cooler line
10. Flare adapter
11. Pump drive gear
12. Woodruff key
13. Washer
14. Nut
15. Gasket
16. Adapter
19. Gasket
20. Pump
23. Nipple
24. Tee
25. Reducing bushing
26. Connector
27. Reservoir to pump line
31. Bleed line
32. Elbow
44. Connecting hose
45. Cooler to reservoir line
46. Clamp
47. Spacer
48. Clip
51. Cooler to reservoir line

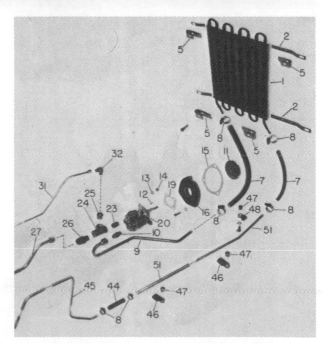

gear (11) and adapter (16) from pump (20). Discard gaskets (15 and 19).

With pump removed, remove plug (8—Fig. 149), "O" ring (7), spring seat (6), relief valve spring (5) and relief valve poppet (4). Remove Woodruff key (15) from pump drive shaft (14), then remove plate and seal (2) from front of pump body (3). Place a scribe line across pump halves, then remove through bolts and separate pump. Anti-extrusion block (13), wear plate and seal (12), pump gears and shafts (14 and 16), gasket insert (17), bridging insert (18) and "O" ring (20) can now be removed. Dowels (19) need not be removed unless renewal is indicated.

Clean and inspect all parts. Check fit of pump shafts in front plate (3) and cover (21) and if clearance is excessive, renew pump halves and/or pump shafts and gears. Renew relief valve spring (5) if distorted or shows signs of fracture. Seat for relief valve poppet (4) is not available separately.

Reassemble by reversing the disassembly and lubricate parts during assembly. Tighten through bolts to 11-15 ft.-lbs. torque.

## Model 2255

**275. R&R AND OVERHAUL PUMP.** To remove pump, loosen support (bracket) capscrews and disengage drive belt from pulley. Disconnect inlet and outlet lines from pump, then unbolt support bracket and remove support and pump. Remove pump pulley, then separate pump from support.

With pump removed, remove pulley

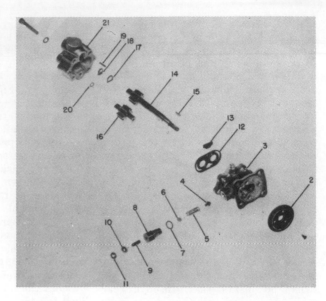

Fig. 149—Exploded view of hydraulic oil cooler pump used on models G955 and G1355 prior to serial number 248 721-. Pressure adjusting screw is (9).

2. Plate and seal
3. Front body
4. Relief valve
5. Relief valve spring
6. Spring seat disc
7. "O" ring
8. Plug
9. Relief valve screw
10. Seal
11. Jam nut
12. Wear plate and seal
13. Anti-extrusion block
14. Drive shaft and gear
15. Woodruff key
16. Idler shaft and gear
17. Gasket insert
18. Bridging insert
19. Dowel pins
20. "O" ring
21. Rear cover

key from shaft, if necessary, then place a scribe line across back plate (8—Fig. 150) and front plate (19). Remove plug (1) with "O" ring (2), shims (3), spring (4) and poppet (6). DO NOT attempt to remove poppet seat (8) from back plate. Remove capscrews (9) and separate pump. If necessary, bump pump shaft against a wood block. Remove backup gasket (16), protector gasket (17), diaphragm seal (18), check ball spring (14) and check ball (15) from front plate (19). Remove shaft seal (20).

Clean and inspect all parts. If drive gear shaft, or idler gear shaft is less than 0.4360 at seal areas, renew shafts and gears. If gear width is less than 0.566 renew shafts and gears. If gear

pocket in back plate exceeds 1.695 in diameter, or more than 0.001 wear at bottom, renew back plate. Renew back plate or front plate if bushing inside diameter is more than 0.4375. Renew relief valve spring if distorted or fractured. Back plate and relief valve seat (8) are available as an assembly only; renew back plate if seat is worn or damaged. Relief valve components are available in kit form or are available separately.

Reassemble pump as follows: Using a dull pointed tool, tuck diaphragm seal (18) into grooves of front plate with open part of "V" sections down. Drop the steel ball (15) into their bores and insert spring (14) on top of ball. Install protector gasket (17) and backup gasket (16) into diaphragm seal and install diaphragm (13) over backup gasket with bronze side up and grooves

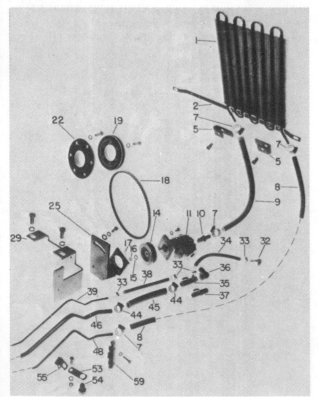

Fig. 148—Basic parts of hydraulic oil cooling system used on 2255 models prior to serial number 248 721-.

1. Cooler
2. Mounting strip
5. Mounting clamp
7. Hose clamp
8. Hose
9. Hose
10. Fitting
11. Pump
14. Pulley
15. Washer
16. Nut
17. Cotter pin
18. Drive belt
19. Pulley
22. Pulley adapter
25. Bracket
29. Bracket
32. Elbow (2WD)
33. Hose clamp
34. Hose (2WD)
35. Adapter (2WD)
36. Elbow
37. Adapter (4WD)
38. Hose
39. Bleed tube
44. Hose clamp
45. Inlet hose
46. Reservoir to pump line
48. Cooler to reservoir line
53. Strap
55. Clip
59. Line clamp

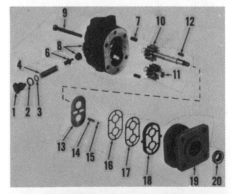

Fig. 150—Exploded view of hydraulic oil cooler circulating pump used on Model 2255 prior to serial number 248 721-. Relief valve seat or back plate are not available separately.

| | |
|---|---|
| 1. Plug | 12. Key |
| 2. "O" ring | 13. Diaphragm |
| 3. Shims | 14. Spring |
| 4. Spring | 15. Check ball |
| 6. Poppet | 16. Backup gasket |
| 7. Dowel | 17. Protector gasket |
| 8. Back plate | 18. Seal |
| 9. Capscrew | 19. Front plate |
| 10. Drive gear | 20. Oil seal |
| 11. Idler gear | |

at shaft holes toward low pressure side of pump. Install gear and shaft assemblies (10 and 11) in bores of front plate. Apply a thin coat of grease to milled faces of rear plate and front plate and install rear plate to front plate with the previously affixed scribe lines aligned. Install retaining cap screws and tighten to a torque of 7-10 ft.-lbs. Lubricate lip of shaft seal (20), work seal over drive shaft with lip of seal inward and drive seal into counterbore of front plate. Install poppet (6) with largest diameter end toward spring, then install spring (4), shims (3) and plug (1) with "O" ring (2). Install support to pump front plate, then install pulley key, pulley and pulley retaining nut. Tighten pulley retaining nut to 18-22 ft.-lbs. torque. Install support to support bracket and install assembly on tractor. Connect lines and adjust drive belt to allow ½-inch deflection between pulleys using eight pounds pull.

**276. OIL COOLER.** The hydraulic oil cooler is located at left front of tractor radiator and is secured to mounting strips which are attached to radiator supports. See Figs. 147 and 148.

With hood and front side panels off, oil cooler can be removed by disconnecting hoses and removing cooler from mounting strips.

Renew oil cooler if damage is extensive.

## REMOTE CYLINDER CONTROL VALVES

The remote cylinder valves and the valve manifold are mounted on the right side of the hydraulic reservoir and valves are mounted outboard of the valve manifold. See Fig. 151. Normally, tractors are equipped with two external valves, however, an additional one or two valves can be installed if desired.

Valves may be of two types; a pressure detent type which is detented in the raise and lower position and is automatically returned to neutral when the preset pressure (2100-2200 psi) is reached; or a float valve which is spring centered from raise and lower positions and detented in float position. Valve must be held in raise and lower positions and manually returned from float position.

Refer to paragraph 244 to adjust the restrictor in pressure detent valve and to paragraph 245 to adjust restrictor in valve manifold.

**277. R&R CONTROL VALVES.** To remove control valves, drain hydraulic reservoir, then remove control lever console and support. Clean area around control valves prior to removal. Disconnect remote coupling holders from mounting bracket. Disconnect control

levers from control valve spools, then unscrew valve retaining capscrews from manifold but do not pull them from valves. Pull the valves and end cap assembly off the control shaft lever, then remove capscrews and separate control valves. Discard all "O" rings and backup rings.

To reinstall, use all new "O" rings and backup rings and reverse removal procedure.

**NOTE: Bushings in control levers are renewable and renewal procedure is obvious.**

**278. OVERHAUL PRESSURE DETENT VALVE.** Remove valve as outlined in paragraph 277, and proceed as follows: Unscrew detent cap (1—Fig. 152) and remove complete detent assembly and valve spool (21) from valve body (35). Clamp flat end of valve spool in a vise, pull detent cap from spool assembly and catch the three detent balls (3) as cap is removed.

Unscrew detent retainer (9) with "O" ring (10) by using a 1-inch spanner wrench. Remove washers (11 and 13) and centering spring (12) from detent retainer. Remove snap ring (4) from I.D. of detent retainer and remove washer (5), spring (6), shim (7), release plunger (8) and piston (16) with "O" ring (17). Slide bushing (14) and "O" ring (15) from spool. Use Allen wrench and piston (16) and turn guide (18) out of spool and remove spring (19) and poppet (20). Remove "O" ring (22) from spool bore.

Remove plug (37) with backup ring (28) and "O" ring (29), then remove spring (30), poppet (31), seat (36) with "O" ring (33) and interlock plunger (34).

Remove restrictor support (25) with backup ring (28) and "O" ring (29), then remove spring (30), poppet (31), and seat (32) with "O" ring (33). Remove jam nut (24) and remove restrictor screw (23) from plug. Remove "O" ring

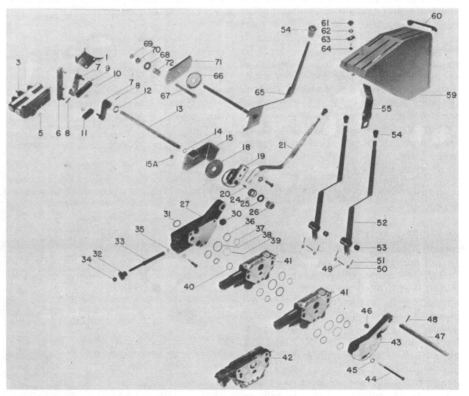

*Fig. 151—Exploded view showing control valves, valve manifold and control levers. Two control valves are standard, however, one or two additional valves can be used if desired. Note differences between early servo control lever and support (21 and 15) and late control lever and support (65 and 71).*

| | | |
|---|---|---|
| 1. Stop | 20. Key | 39. Backup ring | 55. Brace |
| 3. Servo valve | 21. Control lever | 40. "O" ring | 59. Console |
| 5. "O" ring | (early) | 41. Control valve | 60. Trim |
| 6. Arm | 24. Spring | 42. Float valve | 61. Knob |
| 7. Arm (2) | 25. Washer | 43. End plate | 62. Washer |
| 8. Pin | 26. Elastic nut | 44. Cap screw (4) | 63. Stop |
| 9. Link (3/4") | 27. Manifold | 45. Lock washer | 64. Bolt |
| 10. Retaining rings | 30. Plug | 46. Plug | 65. Lever & shaft |
| 11. Link (2") | 31. "O" ring | 47. Lever shaft | (late) |
| 12. Washer | 32. Restrictor support | 48. Pin | 66. Friction disc |
| 13. Shaft (early) | 33. Restrictor | 49. Pin | 67. Bolt |
| 14. "O" ring | 34. Jam nut | 50. Washer | 68. Washer |
| 15. Support (early) | 35. "O" ring | 51. Cotter pin | 69. Nut |
| 15A. Jam nut | 36. "O" ring | 52. Lever | 70. Elastic nut |
| 18. Friction disc | 37. Backup ring | 53. Nylon bushing | 71. Support (late) |
| 19. Plate (early) | 38. "O" ring | 54. Knob | 72. Spring |

(27) and backup washer (28) from I.D. of plug (25). Remove check ball retainer (39) and check ball (38). Remove pin (40) from retainer, if necessary.

Remove both thermal relief valve (43) and "O" rings (41 and 42).

279. Thoroughly clean and dry all parts. Discard all "O" rings, backup rings and lockout poppets. If valve body or valve spool are damaged, the entire valve assembly must be renewed as spool is a select fit in body and parts are not available separately.

Inspect poppet (20) and seat in valve spool for excessive wear or damage.

Check detent grooves inside detent cap for excessive wear. Mechanical over-ride force required to manually return detented spool to neutral is 100-250 lbs. and can be adjusted by varying the shims located between detent plunger spring and detent plunger.

Inspect sensing port and small orifice hole in spool, the small drilled holes in thermal relief cavities and small hole in rear of lockout seat to be sure they are open and clean. Check large hole in side of rear lockout seat for wear or damage from check ball. Inspect all springs for distortion, excessive weakness or fractures. Spool should fit bore in body with a slight hand pressure with no appreciable side clearance.

280. Reassemble valve as follows: Use new "O" rings, backup washers and lockout poppets and lubricate parts prior to installation.

Install the thermal relief valve assemblies. Normal pressure relief is 3500-5000 psi.

Install backup washer (26) then "O" ring (27) in I.D. of support (25) and install adjusting screw (23) in support. Install jam nut (24) on adjusting screw. Install "O" ring (33) on seat (32) and install seat in body so that large hole is aligned with cylinder port and will allow check ball (38) to seat. Install poppet (31) and spring (30). Install backup washer (28) and "O" ring (29) on adjusting screw and install screw and support assembly in body. Place check ball (38) in rear port, and with pin (40) in retainer (39), install retainer.

Install plunger (34) in front bore of body, install "O" ring (33) on seat (36) and install seat. Install poppet (31), spring (30) and plug (37) with "O" ring (29) and backup washer (28) installed.

Place poppet (20) and spring (19) in open end of spool (21), be sure poppet is properly seated, then using release piston (16) and an Allen wrench, turn guide (18) into spool until it bottoms. Install "O" ring (17) on release piston (16) and install piston in retainer (9) with tang end of piston toward guide

(18). Place release plunger (8) in retainer (9) and be careful not to push piston out the other end of retainer. Place shims (7), if used, spring (6) and washer (5) inside retainer and secure with snap ring (4). Place washer (11), centering spring (12) and washer (13) on outside of retainer.

Clamp flat end of spool (21) in a vice and place "O" ring (15) and bushing (14) on spool. Install the detent retainer assembly in spool and with the pin spanner wrench, tighten retainer to 5-8 ft.-lbs. Use Allen wrench and engage tang on piston (16) with slot in guide (18), then back guide out of spool until it is against retainer.

Use heavy grease to hold detent balls (3) in holes of retainer (9), then carefully place detent cap (1) over retainer. Use a plastic hammer and drive cap over retainer until detent balls engage second detent groove in cap. Install button plug (2).

Install "O" ring (22) in spool bore, then install spool and detent assembly and tighten cap securely.

**NOTE: Arrow on flat end of spool must point toward cylinder ports when assembly is complete.**

281. **OVERHAUL FLOAT VALVE.** If necessary, disconnect hoses from valve body. Pull plug (19—Fig. 153) with backup washer (18) and "O" ring (17) and remove spring (16) and check valve (15).

Remove Allen screws (1) and pull detent block (3) from detent screw (7). Remove plugs (6), springs (5) and pawls (4) from detent block. Clamp flat end of spool (14) in a vise and remove detent screw (7) from spool. Remove spool (14), washers (8 and 10) and centering spring (9) from body (12). Remove the two "O" rings (11 and 13) from spool bore.

282. Discard all "O" rings and backup rings, then clean and dry all parts. Inspect spool and spool bore in body for scoring, excessive wear, or other damage. If either part requires renewal, it will be necessary to renew the complete valve as spool is a selective fit in body and these parts are not available separately. Spool should fit bore with a slight hand pressure and no appreciable side play.

Inspect detent pawls and groove in detent screw for wear or damage. Inspect check valve for wear or damage. Inspect springs for distortion, excessive

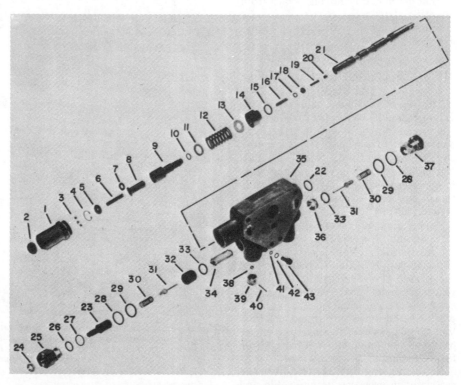

Fig. 152—Exploded view of remote (pressure detent) control valve. Restrictor valve must be adjusted individually for each valve. Restrictor valve adjusting screw is (23).

| | | |
|---|---|---|
| 1. Cap | 12. Centering spring | 23. Adjusting screw |
| 2. Plug | 13. Washer | 24. Jam nut |
| 3. Detent balls | 14. Bushing | 25. Plug |
| 4. Snap ring | 15. "O" ring | 26. Backup washer |
| 5. Washer | 16. Piston | 27. "O" ring |
| 6. Spring | 17. "O" ring | 28. Backup washer |
| 7. Shims | 18. Guide | 29. "O" ring |
| 8. Plunger | 19. Spring | 30. Spring |
| 9. Retainer | 20. Poppet | 31. Poppet |
| 10. "O" ring | 21. Spool | 32. Seat |
| 11. Washer | 22. "O" ring | 33. "O" ring |

| |
|---|
| 34. Plunger |
| 35. Body |
| 36. Seat |
| 37. Plug |
| 38. Check ball |
| 39. Retainer |
| 40. Pin (dowel) |
| 41. "O" ring |
| 42. "O" ring |
| 43. Thermal relief valve |

weakness or fractures. Renew parts as necessary.

283. Use new "O" rings and backup washers, lubricate all parts and reassemble valve as follows: Install new "O" rings (11 and 13) in spool bore and be sure they are not twisted. Install centering spring (9) and washers (8 and 10) in body with the flattest washer (8) on detent block side. Push spool (14) into bore and clamp flat end of spool in a vise. Apply Loctite to threads of detent screw (7), install detent screw in end of spool and tighten detent screw to 5-8 ft.-lbs. Install pawls (4), springs (5) and plugs (6) in detent block (3), then slide detent block over detent screw and start Allen screws. Be sure detent screw is centered in detent block, then tighten Allen screws to 10-13 ft.-lbs. torque.

Install backup washer (18) and "O" ring (17) on plug (19). Install check valve (15) and spring (16) in body, then install plug (19) so that groove in plug aligns with retaining cap screw holes in body.

Connect hoses to valve body, if necessary.

### 284. REMOTE VALVE MANIFOLD.

To remove control valve manifold, remove remote control valves as outlined in paragraph 277. Disconnect oil cooler return line and lift circuit pressure line, then unbolt and remove manifold and "O" ring from hydraulic reservoir. Remove restrictor support (32—Fig. 151), with restrictor (33), "O" ring and washer. Remove jam nut (34), turn restrictor out of support and remove restrictor "O" ring. Control lever shaft (47) can be removed if necessary, by driving out roll pin (48).

Use new "O" rings and reassemble by reversing disassembly procedure. Coat reservoir side of manifold with hydraulic oil and tighten the retaining cap screws to 26-28 ft.-lbs. torque.

Refer to paragraph 245 to adjust restrictor.

### REMOTE CYLINDERS

Two remote (depth stop) cylinders are available, both having an 8-inch stroke. Refer to Fig. 154 for an exploded view of the 3½-inch cylinder and to Fig. 155 for an exploded view of the 4-inch cylinder. Overhaul information for the 3½-inch cylinder is given in paragraph 285, and for the 4-inch cylinder in paragraph 286.

285. **OVERHAUL 3½-INCH CYLINDER.** Refer to exploded view of 3½-inch cylinder in Fig. 154 and proceed as follows:

Loosen nut (17) and slide stop arm (15) to yoke end of piston rod (18). Remove cap screws (24), retaining stop

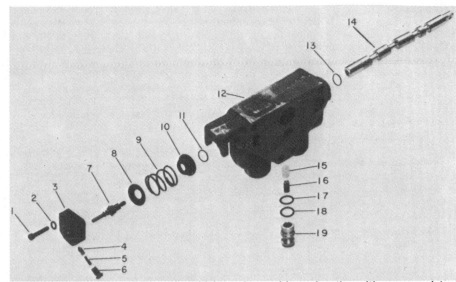

Fig. 153—Exploded view of float valve which can be used in conjunction with pressure detent control valve.

| | | | |
|---|---|---|---|
| 1. Allen screw | 6. Plug | 11. "O" ring | 16. Spring |
| 2. Lock washer | 7. Detent screw | 12. Body | 17. "O" ring |
| 3. Pawl block | 8. Washer | 13. "O" ring | 18. Backup washer |
| 4. Pawl | 9. Centering spring | 14. Spool | 19. Body (plug) |
| 5. Spring | 10. Washer | 15. Check valve | |

valve housing (23) to bearing (11) and withdraw valve housing from end of tube (20). Unscrew nut (29) and remove stop valve (27).

Push bearing (11) into cylinder tube (1) far enough to allow removal of snap ring (12), then bump bearing from cylinder tube with piston (8) by working piston rod in and out.

Remove nut (3), piston (8) and bearing (11) from piston rod. Need and procedure for further disassembly is evident on inspection of unit.

Install new piston rod seals (13 and 14) in bearing (11) and reassemble unit using all new "O" rings, wear rings and backup rings. Reverse disassembly procedure to reassemble the cylinder.

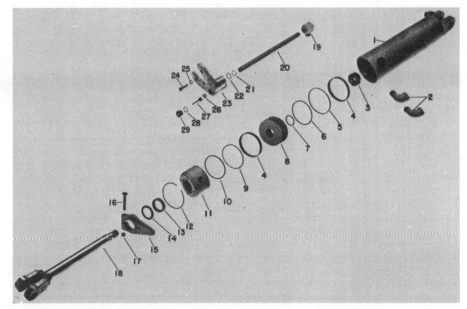

Fig. 154—Exploded view of 3½-inch remote cylinder. Stop arm (15) can be adjusted to vary length of piston stroke. When stop arm contacts valve (27), pressure build up in remote control (pressure detent) valve open poppet in valve spool releasing detent and allowing valve to return to neutral.

| | | | |
|---|---|---|---|
| 1. Cylinder tube | 8. Piston | 14. Wiper seal | 22. "O" ring |
| 2. Elbows | 9. "O" ring | 15. Stop arm | 23. Stop valve housing |
| 3. Nuts | 10. Backup washer | 18. Piston rod | 26. Seal |
| 4. Wear rings | 11. Bearing | 19. Elbow | 27. Stop valve |
| 5. Teflon rings | 12. Snap ring | 20. Tube | 28. "O" ring |
| 6. "O" ring | 13. Seal | 21. Backup washer | 29. Guide |
| 7. "O" ring | | | |

**286. OVERHAUL 4-INCH CYLIN-DER.** Refer to exploded view of 4-inch cylinder in Fig. 155 and proceed as follows:

Loosen nut (33) and slide stop clamp (32) to yoke end of piston rod (10). Remove end plate (31) from bearing (3), then remove lock ring (4) and bump bearing (3) from barrel (2) with piston by working piston rod in and out. Remove nut (11) and piston (12) from piston rod and pull bearing from piston

rod. "O" rings, backup rings, wear rings and seals can now be removed from piston and I.D. of bearing.

Stop valve assembly (items 17 through 28) removed after removing plug (17). Push rod (29) and spring (30) can be removed after driving roll pin that is nearest stop valve end out of push rod.

Clean and inspect all parts. Renew all "O" rings, wear rings and seals and reassemble by reversing disassembly procedure.

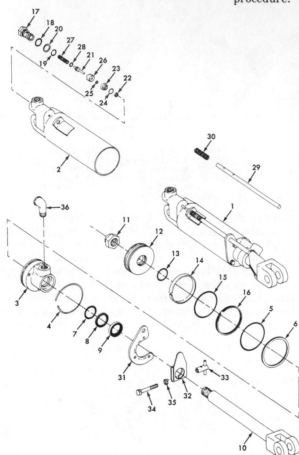

*Fig. 155—Exploded view of remote cylinder. Stop arm (15) can be adjusted to vary length of piston stroke. When stop arm contacts valve (27), pressure build up in remote control (pressure detent) valve opens poppet in valve spool releasing detent and allowing valve to return to neutral.*

1. Cylinder assy.
2. Barrel
3. Bearing
4. Lock ring
5. "O" ring
6. Backup washer
7. "O" ring
8. Backup washer
9. Wiper seal
10. Piston rod
11. Lock nut
12. Piston
13. "O" ring
14. Wear ring
15. "O" ring
16. Teflon ring
17. Plug
18. "O" ring
19. "O" ring
20. Backup washer
21. Plunger
22. Wiper seal
23. Seal retainer
24. "O" ring
25. "O" ring
26. Spacer
27. Spring
28. "O" ring
29. Push rod
30. Spring
31. End plate
32. Stop clamp
33. Adjusting nut
34. Cap screw
35. Lock washer
36. Elbow